AF577270

Contemporary Mathematicians

Gian-Carlo Rota
Editor

Nathan Jacobson, 1981

Nathan Jacobson
Collected Mathematical Papers
Volume 3
(1965–1988)

Birkhäuser
Boston · Basel · Berlin
1989

Nathan Jacobson
Department of Mathematics
Yale University
New Haven, CT 06520

Library of Congress Cataloging-in-Publication Data
Jacobson, Nathan, 1910–
[Selections. 1989]
Collected mathematical papers / Nathan Jacobson.
p. cm.—(Contemporary mathematicians)
Includes bibliographies.
ISBN 0-8176-3446-0 (v. 3)
1. Mathematics. I. Title. II. Series.
QA3.J3325 1989
510—dc20 89-14889
CIP

Printed and bound by Edwards Brothers Inc., Ann Arbor, Michigan.
Printed in the United States of America.

ISBN 0-8176-3446-0
ISBN 3-7643-3446-0

Bibliography of Nathan Jacobson's Books and Papers

Note: **Boldface** numbers at the end of each entry denote the volume in which the entry appears.

Books

The Theory of rings: *Mathematical Surveys, No. II*, Amer. Math. Soc., 1943 (Russian Translation, 1947).

Lectures in Abstract Algebra: *Vol. 1, Basic Concepts*, D. Van Nostrand Co. Inc., 1951 (Springer-Verlag reprint, 1975; Chinese translation, 1966).

Lectures in Abstract Algebra: *Vol. 2, Linear Algebra*, D. Van Nostrand Co. Inc., 1953 (Springer-Verlag reprint, 1975; Chinese translation, 1960).

Lectures in Abstract Algebra: *Vol. 3, Theory of Fields and Galois Theory*, D. Van Nostrand Co. Inc., 1964 (Springer-Verlag reprint, 1975).

Structure of Rings, Amer. Math. Soc. Colloquium Publications, Vol. 37, 1956, 1964 (Russian translation, 1961).

Lie Algebras, Interscience Publishers (John Wiley and Sons), 1962 Interscience Tracts in Pure and Applied Mathematics, No. 10 (Dover reprint, 1979; Russian translation, 1964; Chinese translation, 1964).

Structure and Representations of Jordan Algebras, Amer. Math. Soc. Colloquium Publications, Vol. 39, 1968.

Lectures on Quadratic Jordan Algebras, Tata Institute of Fundamental Research, Bombay, 1969.

Exceptional Lie Algebras, Lecture Notes in Pure and Applied Mathematics, Marcel Dekker Inc., New York, 1971.

Basic Algebra I, W. H. Freeman and Co., New York, 1974; second edition, 1985.

Pi-Algebras: *An Introduction*, Springer Verlag, 1975.

Basic Algebra II, W. H. Freeman and Co., New York, 1980; second edition, 1989.

Structure Theory of Jordan Algebras, University of Arkansas Lecture Notes in Mathematics, 1981.

Finite Dimensional Division Algebras (with David Saltman), Springer-Verlag Grundlehre Series, in press.

Papers

[1] "Non-commutative polynomials and cyclic algebras", (Princeton University dissertation) *Annals of Math.* **35** (1934) 197–208. **(1)**

[2] "A note on non-commutative polynomials", *Annals of Math.* **35** (1934) 209–210. **(1)**

[3] "Locally compact rings" (with O. Taussky), *Proc. Nat. Acad. Sci.* **21** (1935) 106–108. **(1)**

[4] "Rational methods in the theory of Lie algebras", *Annals of Math.* **36** (1935) 875–881. **(1)**

[5] "On pseudo-linear transformations", *Proc. Nat. Acad. Sci.* **21** (1935) 667–670. **(1)**

[6] "Totally disconnected locally compact rings", *Amer. J. Math.* **58** (1936) 433–449. **(1)**

[7] "Simple Lie algebras of type *A*", *Proc. Nat. Acad. Sci.* **23** (1937) 240–242. **(1)**

[8] "Pseudo-linear transformations", *Annals of Math.* **38** (1937) 484–507. **(1)**

[9] "A class of normal simple Lie algebras of characteristic zero", *Annals of Math.* **38** (1937) 508–517. **(1)**

[10] "A note on non-associative algebras", *Duke Math. J.* **3** (1937) 544–548. **(1)**
[11] "Abstract derivation and Lie algebras", *Trans. Amer. Math. Soc.* **42** (1937) 206–224. **(1)**
[12] "*p*-Algebras of exponent *p*", *Bull. Amer. Math. Soc.* **43** (1937) 667–670. **(1)**
[13] "A note on topological fields", *Amer. J. Math.* **59** (1937) 889–894. **(1)**
[14] "Simple Lie algebras of type *A*", *Annals of Math.* **39** (1938) 181–188. **(1)**
[15] "Simple Lie algebras over a field of characteristic zero", *Duke Math. J.* **4** (1938) 534–551. **(1)**
[16] "Normal semi-linear transformations", *Amer. J. Math.* **61** (1939) 45–58. **(1)**
[17] "An application of E. H. Moore's determinant of a Hermitian matrix", *Bull. Amer. Math. Soc.* **45** (1939) 745–848. **(1)**
[18] "Structure and automorphisms of semi-simple Lie groups in the large", *Annals of Math.* **40** (1939) 755–763. **(1)**
[19] "Cayley numbers and normal simple Lie algebras of type *G*", *Duke Math. J.* **5** (1939) 775–783. **(1)**
[20] "The fundamental theorem of the Galois theory for quasi-fields", *Annals of Math.* **41** (1940) 1–7. **(1)**.
[21] "A note on hermitian forms", *Bull. Amer. Math. Soc.* **46** (1940) 264–268. **(1)**
[22] "Restricted Lie algebras of characteristic *p*", *Trans. Amer. Math. Soc.* **50** (1941) 15–25. **(1)**
[23] "Classes of restricted Lie algebras of characteristic *p* I", *Amer. J. Math.* **63** (1941) 481–515. **(1)**
[24] "Classes of restricted Lie algebras of characteristic *p* II", *Duke Math. J.* **10** (1943) 107–121. **(1)**
[25] "An extension of Galois theory to non-normal and non-separable fields", *Amer. J. Math.* **66** (1944) 1–29. **(1)**
[26] "Schur's theorems on commutative matrices", *Bull. Amer. Math. Soc.* **50** (1944) 431–436. **(1)**
[27] "Relations between the composites of a field and those of a subfield", *Amer. J. Math.* **66** (1944) 636–644. **(1)**
[28] "Galois theory of purely inseparable fields of exponent one", *Amer. J. Math.* **66** (1944) 645–648. **(1)**
[29] "Construction of central simple associative algebras", *Annals of Math.* **45** (1944) 658–666. **(1)**
[30] "The equation $x' = xd - dx = b$", *Bull. Amer. Math. Soc.* **50** (1944) 902–905. **(1)**
[31] "Structure theory of simple rings without finiteness assumptions", *Trans. Amer. Math. Soc.* **57** (1945) 228–245. **(1)**
[32] "The radical and semi-simplicity for arbitrary rings", *Amer. J. Math.* **67** (1945) 300–320. **(1)**
[33] "Structure theory for algebraic algebras of bounded degree", *Annals of Math.* **46** (1945) 695–707. **(1)**
[34] "A topology for the set of primitive ideals in an arbitrary ring", *Proc. Nat. Acad. Sci.* **31** (1945) 333–338. **(1)**
[35] "On the theory of primitive rings", *Annals of Math.* **48** (1947) 8–21. **(1)**
[36] "A note on division rings", *Amer. J. Math.* **69** (1947) 27–36. **(1)**
[37] "Isomorphisms of Jordan rings", *Amer. J. Math.* **70** (1948) 317–326. **(1)**
[38] "The center of a Jordan ring", *Bull. Amer. Math. Soc.* **54** (1948) 316–322. **(1)**
[39] "Lie and Jordan triple systems", *Amer. J. Math.* **71** (1949) 149–170. **(2)**
[40] "Classification and representation of semi-simple Jordan algebras" (with F. D. Jacobson), *Trans. Amer. Math. Soc.* **65** (1949) 141–169. **(2)**
[41] "Derivation algebras and multiplication algebras of semi-simple Jordan algebras", *Annals of Math.* **50** (1949) 866–874. **(2)**
[42] "Enveloping algebras of semi-simple Lie algebras", *Can. J. Math.* **2** (1950) 257–266. **(2)**
[43] "Some remarks on one-sided inverses", *Proc. Amer. Math. Soc.* **1** (1950) 352–355. **(2)**
[44] "Jordan homomorphisms of rings" (with C. E. Rickart), *Trans. Amer. Math. Soc.* **69** (1950) 479–502. **(2)**

[45] "Completely reducible Lie algebras of linear transformations", *Proc. Amer. Math. Soc.* **2** (1951) 105–113. **(2)**

[46] "General representation theory of Jordan algebras", *Trans. Amer. Math. Soc.* **70** (1951) 509–530. **(2)**

[47] "Une généralization du théorème d'Engel", *C. R. Acad. Sci.* **234** (1952) 679–681. **(2)**

[48] "Homomorphisms of Jordan rings of self-adjoint elements" (with C. E. Rickart), *Trans. Amer. Math. Soc.* **72** (1952) 310–322. **(2)**

[49] "A note on Lie algebras of characteristic p", *Amer. J. Math.* **74** (1952) 357–359. **(2)**

[50] "Operator commutativity in Jordan algebras", *Proc. Amer. Math. Soc.* **3** (1952) 973–976. **(2)**

[51] "A Kronecker factorization theorem for Cayley algebras and the exceptional simple Jordan algebra", *Amer. J. Math.* **76** (1954) 447–452. **(2)**

[52] "Structure of alternative and Jordan bimodules", *Osaka Math. J.* **6** (1954) 1–71. **(2)**

[53] "A note on automorphisms and derivations of Lie algebras", *Proc. Amer. Math. Soc.* **6** (1955) 281–283. **(2)**

[54] "Commutative restricted Lie algebras", *Proc. Amer. Math. Soc.* **6** (1955) 476–481. **(2)**

[55] "A note on two dimensional division ring extensions", *Amer. J. Math.* **77** (1955) 593–599. **(2)**

[56] "A theorem on the structure of Jordan algebras", *Proc. Nat. Acad. Sci.* **42** (1956) 140–147. **(2)**

[57] "Generation of separable and central simple algebras", *J. Math. Pures Appl.* **36** (1957) 217–227. **(2)**

[58] "On reduced exceptional simple Jordan algebras" (with A. A. Albert), *Annals of Math.* **66** (1957) 400–417. **(2)**

[59] "On Jordan algebras with two generators" (with L. J. Paige), *J. Rat. Mech. Anal.* **6** (1957) 895–906. **(2)**

[60] "Composition algebras and their automorphisms", *Rend. Circ. Math. Palermo* **7 (Series II)** (1958) 1–26. **(2)**

[61] "Nilpotent elements in semi-simple Jordan algebras", *Math. Ann.* **136** (1958) 375–386. **(2)**

[62] "A note on three dimensional simple Lie algebras", *J. Math. Mech.* **7** (1958) 823–832. **(2)**

[63] "Some groups of transformations defined by Jordan algebras I", *J. Reine Angew. Math.* **201** (1959) 178–195. **(2)**

[64] "Some groups of transformations defined by Jordan algebras II", *J. Reine Angew. Math.* **204** (1960) 74–98. **(2)**

[65] "Some groups of transformations defined by Jordan algebras III", *J. Reine Angew. Math.* **207** (1961) 61–85. **(2)**

[66] "Macdonald's theorem on Jordan algebras", *Archiv Math.* **13** (1962) 241–250. **(2)**

[67] "A coordinatization theorem for Jordan algebras", *Proc. Nat. Acad. Sci.* **48** (1962) 1154–1160. **(2)**

[68] "A note on automorphisms of Lie algebras", *Pac. J. Math.* **12** (1962) 303–315. **(2)**

[69] "Generic norm of an algebra", *Osaka Math. J.* **15** (1963) 25–50. **(2)**

[70] "Clifford algebras for algebras with involution of type D", *J. Algebra* **1** (1964) 288–300. **(2)**

[71] "Triality and Lie algebras of type D_4", *Rend. Circ. Math. Palermo* **13 (Series II)** (1964) 1–25. **(2)**

[72] "Cartan subalgebras of Jordan algebras", *Nagoya Math. J.* **27** (1966) 591–609. **(3)**

[73] "Structure theory for a class of Jordan algebras", *Proc. Nat. Acad. Sci.* **55** (1966) 243–251. **(3)**

[74] "Quadratic Jordan algebras of quadratic forms with base points" (with K. McCrimmon), *Indian Math. Soc.* **35** (1971) 1–45. **(3)**

[75] "Generally algebraic quadratic Jordan algebras" (with J. Katz), *Scri. Math.* **29** (1971) 215–227. **(3)**

[76] "Structure groups and Lie algebras of Jordan algebras of symmetric elements of associative algebras with involution", *Adv. Math.* **20** (1976) 106–150. **(3)**

[77] "Localization of Jordan algebras" (with K. McCrimmon and M. Parvathi), *Commun. Algebra* **6 (9)** (1978) 911–958. **(3)**

[78] "Bimodule structure of certain Jordan algebras relative to subalgebras with one generator", *Hokkaido Math. J.* **10** (1981) 333–342. **(3)**

[79] "Some application of Jordan norms to associative algebras", *Adv. Math.* **48** (1983) 149–165. **(3)**

[80] "Forms of the generic norm of a separable algebra", *J. Algebra* **86** (1984) 76–84. **(3)**

[81] "Some projective varieties defined by Jordan algebras", *J. Algebra* **97** (1985) 565–598. **(3)**

[82] "Jordan algebras of real symmetric matrices", *Algebras, Groups and Geometries*, **4** (1987) 291–304. **(3)**

Expository Papers

[83] "*The Classical Groups* by Hermann Weyl" (A book review), *Bull. Amer. Math. Soc.* **46** (1940) 592–595. **(3)**

[84] "Representation theory for Jordan rings", *Proc. Intl. Congr. Math.* (1950) **2** 37–43. **(3)**

[85] "Le problème de Kurosch", *Séminaire Bourbaki* **64** (1951–1952) 295–303. **(3)**

[86] "Some aspects of the theory of representations of Jordan algebras", *Proc. Intl. Congr. Math.* (1954) **3** 28–33. **(3)**

[87] "Jordan algebras", Report of a Conference on Linear Algebras, *Nat. Acad. Sci.–Nat. Research Council* **502** (1957) 12–19. **(3)**

[88] "Representation theory of Jordan algebras, Some Aspects of Ring Theory", *Cen. Int. Mat. Est.* 1966. **(3)**

[89] "Forms of algebras; Some Recent Advances in Basic Sciences", *Academic Press* (1966) 41–71. **(3)**

[90] "Associative algebras with involution and Jordan algebras", *Proc. K. Akad. van Wetenschappen* **69** (1966) 202–212. **(3)**

[91] "Connections between associative and Jordan rings", *Ist. Naz. Alta Math. Symp. Math.* **9** (1972) 261–268. **(3)**

[92] "Abraham Adrian Albert" (an obituary), *Bull. Amer. Math. Soc.* **80** (1974) 1076–1093. **(3)**

[93] "PI-Algebras, Ring Theory", Proceedings of Oklahoma Conference, edited by B. R. MacDonald, Marcel Dekker (1974) 1–30. **(3)**

[94] "Some recent developments in the theory of PI-algebras", Proceedings Winter School, Reinhardsbrunn, German Democratic Republic (1976) 17–21. **(3)**

[95] "Some recent developments in the theory of algebras with polynomial identities, I. Razmyslov's central polynomial, II. The Artin-Procesi theorem, III. On Shirshov's local finiteness theorems", *Proceedings* 18*th SRI* Springer-Verlag Canberra, 1978, 8–46. **(3)**

[96] "Survey of Jordan structure theory", *Southeast Asian Bull. Math.* **5** (1981) 27–38. **(3)**

[97] *Emmy Noether Collected Papers* (Introduction), Springer-Verlag, 1983, 16–26. **(3)**

[98] "Brauer factor sets, Noether factor sets and crossed products, *Emmy Noether in Bryn Mawr*, edited by Srinivasan and Sally, Springer-Verlag, 1983, 1–20. **(3)**

Papers Listed by Topic

(Ore) Skew Polynomial Domains: 1, 2, 5, 8, 16.

Topological Algebra: 3, 6, 13.

Lie Algebras and General Non-associative Algebras: 4, 7, 9, 10, 11, 14, 15, 19, 22, 23, 24, 39, 42, 45, 47, 49, 53, 54, 62, 68, 69, 71.

Galois Theory: 20, 25, 27, 28, 29, 36, 55.

General Structure Theory of Rings: 31, 32, 33, 34, 35, 43, 55, 93, 94, 95.

Jordan Rings: 37, 38, 39, 40, 41, 44, 46, 48, 50, 51, 61, 63, 64, 65, 66, 67, 69, 72, 73, 74, 75, 76, 77, 78, 79, 80, 81, 82, 84, 86, 87, 88, 90, 91, 96.

Associative Algebras: 1, 12, 29, 57, 70, 79, 85, 98.

Miscellaneous: 17, 18, 21, 26, 30, 60, 92, 97.

Doctoral Students of Nathan Jacobson

University of North Carolina

Charles L. Carroll, Jr. Normal simple Lie algebras of type D and order 28 over a field of characteristic zero (1945).

Yale University

Eugene Schenkman A theory of subinvariant Lie algebras (1950).
Charles W. Curtis Additive ideal theory in general rings (1951).
William G. Lister A structure theory of Lie triple systems (1951).
Henry G. Jacob A theorem on Kronecker products (1953).
George B. Seligman Lie algebras of prime characteristic (1954).
Morris Weisfeld Derivations in division rings (1954).
Bruno Harris Galois theory of Jordan algebras (1956).
Earl J. Taft Invariant Wedderburn factors (1956).
Dallas W. Sasser On Jordan matrix algebras (1957).
Maria J. Woneburger On the group of similitudes and its projective group (1957).
Tae-Il Suh On isomorphisms of little projective groups of Cayley planes (1961).
Herbert F. Kreimer, Jr. Differential, difference, and related operational rings (1962).
Charles M. Glennie Identities in Jordan algebras (1963).
David A. Smith On Chevalley's method in the theory of Lie algebras and linear groups of prime characteristic (1963).
Dominic C. Soda Groups of type D_4 defined by Jordan algebras (1964).
Harry P. Allen Jordan algebras and Lie algebras of type D_4 (1965).
Eugene A. Klotz Isomorphisms of simple Lie rings (1965).
Kevin M. McCrimmon Norms and noncommutative Jordan algebras (1965).
Joseph C. Ferrar On Lie algebras of type E_6 (1966).
Daya-Nand Verma Structure of certain induced representations of complex semi-simple Lie algebras (1966).
Lynn Barnes Small Mapping theorems in simple rings with involution (1967).
John R. Faulkner Octonion planes defined by quadratic Jordan algebras (1969).
Samuel R. Gordon On the automorphism group of a semi-simple Jordan algebra of characteristic zero (1969).
Michel Racine The Arithmetics of quadratic Jordan algebras (1971).
Jerome M. Katz Automorphisms of the lattice of inner ideals of certain quadratic Jordan algebras (1972).
Ronald Infante Strongly normal difference extensions (1973).
Louis H. Rowen On algebras with polynomial identity (1973).
Georgia M. Benkart Inner ideals and the structure of Lie algebras (1974).
David J. Saltman Azumaya algebras over rings of characteristic p (1976).
Robert A. Bix Separable Jordan algebras over commutative rings (1977).
Leslie Hogben Radical classes of Jordan algebras (1978).
Craig L. Huneke Determinantal ideals and questions related to factoriality (1978).

Chronology

September 8, 1910	Born, Warsaw, Poland (U.S. Citizen)
1930	A.B. University of Alabama
1934	Ph.D. Princeton University
1934–1935	Assistant, Institute for Advanced Study
1935–1936	Lecturer, Bryn Mawr College
1936–1937	National Research Council Fellow, University of Chicago
1937–1938	Instructor, University of North Carolina
1938–1940	Assistant Professor, University of North Carolina
1940–1941	Visiting Associate Professor, Johns Hopkins University
1941–1942	Associate Professor, University of North Carolina
1942–1943	Associate Ground School Instructor, University of North Carolina
1943–1947	Associate Professor, Johns Hopkins University
1947–1949	Associate Professor, Yale University
1949–1961	Professor, Yale University
1961–1963	James E. English Professor, Yale University
1963–1981	Henry Ford II Professor, Yale University
1981–	Henry Ford II Professor Emeritus, Yale University
Summer, 1947	Visiting Professor, University of Chicago
1951–1952	Fulbright Scholar, University of Paris
1957–1958	Visiting Professor, University of Paris
Oct. 1964–Jan. 1965	Visiting Professor, University of Chicago
Spring, 1965	Lecturer, Mathematical Society of Japan
Spring, 1969	Visiting Professor, Tata Institute of Fundamental Research
Oct. 1981–Jan. 1982	Visiting Professor, ETH, Zurich
Sept.–Nov. 1983	Visiting Professor, Nanjing University, People's Republic of China
Nov.–Dec. 1983	Visiting Professor, Taiwan National University, Republic of China
Feb.–March 1984	Visitor, Center for Advanced Studies, University of Virginia
Sept.–Oct. 1985	Visiting Professor, Pennsylvania State University
April 1988	John Hasbrouck Van Vleck Distinguished Visiting Professor, Wesleyan University
Visiting Lecturer	Australia, Israel, Italy, People's Republic of China, University of Texas
1951–1952	Guggenheim Fellow
1954	Elected Member of National Academy of Sciences

1960	Elected Member of American Academy of Arts and Sciences
1971–1973	President of American Mathematical Society
1972	Honorary D.Sc., University of Chicago
1972	Honorary Member of London Mathematical Society
1972–1974	Vice President of International Mathematical Union
1981	Sesquicentennial Honorary Professor, University of Alabama

Preface

This collection contains all my published papers, both research and expository, that were published from 1934 to 1988. The research papers arranged in chronological order appear in Volume I and II and in the first part of Volume III. The expository papers, which are mainly reports presented at conferences, appear in chronological order in the last part of Volume III. Volume I covers the period 1910 to 1947, the year I moved to Yale, Volume II covers the period 1947 to 1965 when I became Chairman of the Department at Yale and Volume III covers the period from 1965 to 1989, which goes beyond my assumption of an emeritus status in 1981. I have divided the time interval covered in each volume into subintervals preceded by an account of my personal history during this period, and a commentary on the research papers published in the period. I have omitted commentaries on the expository papers and have sorted out the commentaries on the research papers according to the principal fields of my research.

The personal history has been based on my recollections, checked against written documentation in my file of letters as well as diaries. One of these was a diary I kept of my trip to the USSR in 1961; the others were diaries Florie (Florence) kept during other major visits abroad. I have also consulted Professor A. W. Tucker on historical details on Princeton during the 1930's.

The material on personal history and commentary has had critical readings by Florie, by George Seligman and by Kevin McCrimmon. I take this opportunity to express my sincere thanks to them for numerous suggestions which have resulted in an improved manuscript.

Nathan Jacobson
Hamden, Connecticut
September 25, 1989

Table of Contents

Personal History and Commentary

1965–1970

In January of 1965, when I was still at Chicago, I received a letter from Charles Rickart, the Chairman of our department, detailing some plans that he and George Seligman had made to facilitate a new policy of rotating chairmen for the department. The new element they proposed was the introduction of a position of Assistant or Vice Chairman with well-defined duties and with authority to make decisions on matters assigned to him. This would further decentralize the department that already had a delegation of authority to two other officers, the Director of Graduate Studies and the Director of Undergraduate Studies. It should also be mentioned that during his six years as Chairman, Rickart had increased and improved the secretarial staff and had succeeded in turning over many administrative tasks to our main secretary, Phyllis Stevens, who was superb in her position. In the last paragraph of his letter, Charles wrote: "I hope that these plans are such that you will feel that your work will not suffer too much by accepting the appointment of Chairman".

I did accept such an appointment for a term of three years beginning 1965, when this was offered to me by President Kingman Brewster. I wrote him that, though I had been reluctant to accept, I had decided to do so for two reasons: "In the first place, Professor Rickart has done a marvelous job in organizing the department and he has proposed further steps which will permit the Chairman to share the administrative burden with his colleagues.... My second reason for accepting the position is that I feel that our department now has the backing which it needs from you and the administration as a whole to achieve a top position in mathematics."

George Seligman was Assistant Chairman during my first year as Chairman, and Frank Hahn served in this capacity during the next year until his tragic death of cancer in 1968 at the age of 38.

For a number of years, the department was aware of the advantages of having special instructorships like the Moore Instructorships at M.I.T. and the Benjamin Peirce Instructorships at Harvard. The time appeared to be ripe to convince the administration of this fact. Accordingly, a committee consisting of Hedlund, Szczarba, and Seligman (as chairman) was charged with this task and with the task of suggesting the conditions for such special instructorships at Yale. They consulted the department chairmen at M.I.T., Harvard, Michigan, and Princeton, all of whom had special named instructorships. All responded that these gave them access to the cream of the new crop of Ph.D.'s each year. Princeton was planning to expand its program. The name suggested by the Seligman committee for our special instructorship was that of Josiah Willard Gibbs, who had been a professor at Yale and was perhaps the leading mathematical physicist produced in this country. Also, the committee suggested that the initial appointment be for a term of two years, with a possibility of renewal for one additional year. The teaching load suggested was 6–3, with assurance of an opportunity of giving a one-term graduate course during the two years in the area of the instructor's field. The salaries should be competitive with the existing special instructorships.

The administration accepted essentially all of these recommendations, and we established the Josiah Willard Gibbs Instructorships. The term for these was set at two years, without renewal, to begin with three appointments for the academic year 1967–1968 and thereafter to have three additional appointments for the following years to bring the total to six for the years after 1968. The funding for these, which entailed a substantial increase in the departmental budget, would come entirely from Yale University funds. This gave us an assurance of permanence for our new instructorships. Moreover, we had been assured by the Air Force of summer support amounting to two-ninths of academic salaries for the Gibbs Instructors we would appoint. These conditions made our instructorships competitive and in some respects more attractive than the existing ones. The program was an immediate success and substantially strengthened the department in the junior ranks. At the same time, while the Gibbs Instructorships were not "tenure-track," the additional training at a strong research center as well as the prestige associated with a "Gibbs" made them highly attractive to fresh Ph.D.'s.

In 1965, mathematical logic at Yale was located in the Department of Philosophy, with two tenured professors, F. B. Fitch and Alan Anderson. Anderson resigned his position to accept a professorship at the University of Pittsburgh. The philosophy department was not anxious to make another appointment in logic. An ad hoc committee consisting of Church, Kleene, Wang, and Montague recommended the transfer of the position in logic to our department. In this way our department, which had never had anyone in logic before, acquired an incremental tenure slot in this area. I was delighted with this development. In 1964, I had written Rickart that I thought it was highly desirable that we develop this area in our department.

We quickly decided on Abraham Robinson, who was at U.C.L.A., as our choice for a logician. Everyone knew of his creation of nonstandard analysis and of his application of this tool to obtain an important result in operator theory. We were also aware of his versatility, but perhaps not with all aspects of his scholarship, in particular, not with his earlier work on fluid dynamics. This work was much appreciated by Professor Wegener of the Department of Engineering and Applied Science.

In November, I wrote Robinson at St. Catherine's College in Oxford, where he was visiting, on leave from his home institution. I gave him the background of our new position in logic and wrote that "we feel very strongly that you would be the ideal person to help us develop in this new direction for our Department.... Accordingly, I am writing to ask if you might be interested in an offer of a Professorship in Mathematics at Yale." In his reply, dated December 10, he stated that he would give the most serious consideration to an offer from us, and he offered to stop off in New Haven with Mrs. Robinson on February 3 (1966) on their way back to California to discuss details and to have a brief look at the housing situation. The Robinsons did visit Yale on February 3. By this time, a formal offer had been approved by the provost, Charles Taylor. I had suggested a top salary of $25,000. Taylor did not quite go along with this and set the salary at $24,000 for the first year with an advance to $25,000 the second year. The department gave a luncheon for Robinson to which a number of colleagues from other departments were invited. Simultaneously, Florie hosted a luncheon for Mrs. Robinson and the department wives. At this luncheon, Mrs. Robinson made it clear that she was reluctant to leave the Los Angeles area, especially since they had recently purchased a beautiful house in a lovely part of Los Angeles.

Our formal offer was sent to Robinson on February 4. On February 19, I received his reply that "with a heavy heart" he had decided to decline our offer. This was a serious blow, but, because of Mrs. Robinson's attitude, it was not totally unexpected.

Our second choice of a logician was Dana Scott. He also came for a visit with his wife and baby, and his reaction to a possible offer from Yale was quite positive. While our negotiations with Scott were going on, I received a phone call from Robinson from New York to say that he wished to come to New Haven to talk to me alone about the offer he had declined. I met him that same evening at the train, and we had dinner together at a downtown restaurant. He told me that he would now accept the offer if this could be reinstated. I mentioned the complication of our negotiations with Scott that had gone quite far and, though I was confident that our offer could be reinstated, I needed to have some time to discuss this with the professors in the department and with the provost. All were enthusiastic about this turn of events and Provost Taylor agreed to making the offer again to Robinson, while at the same time continuing our effort to persuade Scott to come to Yale. In the end, Scott declined our offer and also an offer we made for him to try out Yale on a visiting basis. In order not to make too abrupt a break with UCLA, Robinson requested that his position at Yale begin in 1967. He began his tenure at Yale in that year.

James Kelsey Whittemore was a member of the Department of Mathematics from 1916 to 1943, when he retired as associate professor. He died in 1948, and in 1950 his widow Elizabeth L. Whittemore established the Professor James K. Whittemore Memorial Fund in his honor. The initial principal of the fund was $500. Over the years, she added to it until her death in 1966, when the fund amounted to $15,743.34. The terms of this gift was that the income was to be used "for the benefit of the Department of Mathematics for such purpose as the Chairman of the Department shall determine." The income of the Whittemore fund amounted to $708.45 in 1966.

After consulting my colleagues I decided that a good use of this amount of income would be to fund an annual lecture series. Accordingly, we established the James K. Whittemore Lectures. These were to consist of 3 to 5 lectures given over a period of a week by an outstanding mathematician on a topic of his own choosing. A committee was appointed to select the Whittemore Lecturer. Their choice for the first of these was Robert P. Langlands, whose lectures were on Euler Products. These were given during the week of April 3, 1967, and were subsequently published by the Yale University Press as the first monograph in a new Yale Mathematical Monograph Series. The series of Whittemore Lectures has been of a uniformly high quality and they have constituted a fine memorial to Professor Whittemore.

Robert Langlands received his doctorate at Yale in 1960. After this, he went of Princeton as Fine Instructor for two years. Then he became a tenured member of their faculty. At Princeton, he changed his field from functional analysis to automorphic forms, and he was recognized by the experts both here and abroad as a top man in this field. We approached him a number of times, seeking to persuade him to return to Yale. In particular, in 1965, we presented his name with strong supporting letters to the Science Advisory Committee. Since he indicated he was not interested in leaving Princeton, no action was taken.

A dramatic change occurred in March of 1967. On Friday, March 10, I received a telephone call from Burton Randol, who was visiting the Institute at Princeton at the time, to say that Langlands had resigned from Princeton as a result of a dispute he had had with the Chairman of the department. On Saturday we held a meeting of our department and voted to offer Langlands a professorship. I telephoned the provost about our action. He approved and, since Langlands had been considered favorably by the Science Advisory Committee in 1965, he authorized us to make the offer immediately. The formal offer was sent to Langlands on March 15, just five days after we had voted this action. This may have set a record for speed in making an appointment at Yale. Of course, this would be impossible at the present time.

Langlands also had an offer from Johns Hopkins when he was considering ours, and he was expecting an offer from M.I.T. On April 16, I received a letter from him accepting our offer. Since he had previously accepted a visiting position at Ankara for 1967–1968, he requested a leave of absence without pay for that year. This was granted by the administration, and he became an active member of the department in the fall of 1968. He remained at Yale until the summer of 1972, when he accepted a professorship at the Institute for Advanced Study at Princeton.

During my tenure as chairman (1965–1968), we had strong support from the administration. This should be clear from my account of the negotiations with Robinson and Langlands and the establishment of the Gibbs Instructorships. Also, President Brewster and Provost Taylor participated actively in an effort we made to persuade Phillip Griffiths to come to Yale. We were frustrated in this by an offer he received from Princeton. The favorable treatment accorded us by the administration was part of an effort they were making to improve the sciences at Yale, with the objective of catching up with our rivals, Chicago, Harvard, and Princeton. Also, this was the period shortly after the Soviet's orbiting of Sputnik led to a strong national effort to improve education in the sciences and mathematics in the United States.

In 1967, we had our first Oberwolfach conference on Jordan algebras. The idea of having such a conference originated with Max Koecher. He was the leader of the German school of Jordan algebras. He and Hel Braun had collaborated on a book *Jordan Algebren* that had been published in 1965 in the Springer Grundlehren series. They had been attracted to the subject by applications to analysis, differential geometry, and automorphic functions. Koecher asked me to serve with him as one of the organizers of the conference. A bit later, Lowell Paige also became a member of the Organizing Committee. Our first task was to draw up a list of possible participants and to sound them out on their interest in such a conference. Their response was very favorable, so we went ahead with our plans. Koecher was able to secure the dates August 17–26 for the conference on *Jordan Algebras and Non-associative Algebras* at Oberwolfach.

There were 42 participants in the conference, 30 of whom presented papers. A few of the talks were an hour in length, the others were for half an hour. I gave two half-hour lectures on Cartan Subalgebras and Some Remarks on Exceptional Jordan Algebras. A wide range of topics that fitted under the title of the conference was covered: Jordan structure theory (K. H. Helwig, U. Hirzebruch, N. Jacobson, K. McCrimmon, J. M. Osborn, E. J. Taft, A. Thedy, J. Tits, and C. Tsai); connections between Jordan algebras and Lie algebras and algebraic groups (R. B. Brown, M. Koecher, K Meyberg, and T. A. Springer); connections with geometry (H. Braun, A. Gray, and A. A. Sagle); and connections with analysis (H. L. Resnikoff and F. D. Veldkamp). The other speakers were H. P. Allen, E. Bönecke, H. Braun, A. H. Boers, W. Eichhorn, C. M. Glennie, P. Jonker, J. Knopfmacher, W. S. Martindale, L. J. Paige, R. D. Schafer, and F. Schweiger. The other participants were E. Behrens, R. Carlsson, E. Christensen, J. H. v. Doogeweert, G. Hoildan, G. Janssen, L. J. Kalmijn, J. Makkelsen, H. Rühaak, T. Smits, K. Scheele, and H. J. Weinert.

There were lectures both morning and afternoon, about five hours of lectures each day, except at the midpoint of the conference when no afternoon lectures were scheduled. At that point, most of the participants went on hikes in the beautiful hills of the Black Forest around Oberwolfach. Florie and I were not tempted to undertake anything as strenuous as this, and anyhow a pleasant alternative was offered us by the Koechers: a trip to Strasbourg. We accepted and spent the afternoon sight-seeing with them in this beautiful city. After that, we dined at a fine restaurant located on one of the canals of the city. We drove back to Oberwolfach after dinner, arriving quite late.

In the evenings, some of the participants who were scheduled to lecture the following day were busy putting the finishing touches on their talks. Others enjoyed socializing in the common room of the Lorenzenhof, the old mansion that served for the lectures, conferences, living quarters, and social activities. Unfortunately, its facilities were not adequate, and it had to be torn down a few years later to make room for some ultramodern buildings that were built with funds donated by the Volkswagon Foundation. On the evening of August 24, two days before the end of the conference, we had a grand gala that included refreshments, beer and wine, and dancing to the piano music of Jerry (W. S.) Martindale.

After the conference, I received a letter from Koecher suggesting another one for 1970. I thought it was a fine idea, but I was unable to serve on the organizing committee since I had planned to visit India for several months in 1969. The second Oberwolfach conference on Jordan algebras took place in August of 1970, and these conferences have been repeated every three years since then.

During the academic year 1967–68, the Mathematics Department of the University of Massachusetts at Amherst invited a number of mathematicians to be "Smolt" ("Senior mathematicians on lecture tour") lecturers. I was invited to be one of these, and I gave three lectures on Associative Rings with Involution and Jordan Rings in February of 1968.

We had a very pleasant summer in 1968, lecturing in Scotland and sight-seeing in Scotland and Ireland. I was invited to give a course of lectures at the St. Andrews Colloquium of the Edinburgh Mathematical Society. My invitation came from Professor E. M. Patterson of King's College, Old Aberdeen. He also invited me to give some lectures at his college prior to the St. Andrews Colloquium. I accepted both invitations. We spent a week at Aberdeen, where I gave two lectures. The colloquium ran from July 10 to July 20. There were four courses of eight lectures each, by F. Bonsall, N. G. deBruijn, P. Billingsley, and myself, a course of four lectures by H. Zassenhaus, and a single lecture by K. McCrimmon. The title of my course was Quadratic Jordan Algebras. Zassenhaus spoke on A 'Categorical' Treatment of J. H. M. Wedderburn's Structure Theorems on Semisimple Hypercomplex Systems and The Decomposition Theorem of J. H. Maclagan-Wedderburn. McCrimmon's lecture was on The Freudenthal–Springer–Tits Construction of Exceptional Jordan Algebras. The evenings were reserved for socializing and music.

Professor Patterson was our host at Aberdeen, and Professor J. B. Tatchell was in charge of the arrangements at St. Andrews.

After the colloquium we spent a few days in Edinburgh, where we were taken sight-seeing in the countryside by Professor Erdelyi. After Edinburgh, we rented a car and did some touring of the Scottish highlands on our own. From Scotland we flew to Dublin. We spent a week in Ireland and returned home on August 2.

In the sixties, Yale instituted a policy of triennial leaves for tenured faculty as an alternative to sabbaticals. The triennial leaves allowed one semester off out of every six with full pay. I was due to have one of these in the academic year of 1967–1968, my last year as chairman. I interpreted my term to include the full year, so I requested a postponement of the triennial leave until the following year with the understanding that I would be eligible for another such leave in 1970–1971. In November 1967, I received an invitation from Professor K. G. Ramanathan to be a Visiting Professor at the Tata Institute of Fundamental Research (TIFR) at Bombay for a period of three months between November 1968 and March 1969. They wanted me to give a three-hour course of my own choosing and offered to pay my tourist round trip air fare from New Haven to Bombay as well as an honorarium. I chose the topic Quadratic Jordan Algebras for my lectures. This was to be an expanded version of the lectures I gave the preceding summer at St. Andrews. I wrote up the notes for these lectures and published them in the TIFR monograph series. My lectures were scheduled for Tuesday, Wednesday, and Friday mornings, which left Friday afternoon to Monday free for visits to other parts of India for lectures and sight-seeing.

The foreign visitors who were at the Institute at approximately the same time as we were the Kervaires, the Mostows and Professor T. Schneider. We shared an apartment with the Mostows and Professor Schneider at Green Fields opposite the Oval. We had four servants, three of whom were men who lived in a single room off the kitchen. The fourth was a beautiful young woman of the lowest caste who came in daily to do the menial tasks. The landlord also had a room and bath in the apartment, and he kept four dogs in his quarters. He objected to our addressing the "bearer" (the number-one servant) by his name, rather than calling this 40-year-old man "boy" as he did. There were other problems in our living arrangements, but the location was an interesting one that afforded an excellent view of the colorful panorama of Indian life.

On the days I went to the Institute, I was picked up by the car assigned to the School of Mathematics. The driver was a Mr. Fernandes, who also had the responsibility of taking us to the airport and meeting us there when we took trips to other parts of India. He was a very pleasant person, and I became quite well acquainted with him on the numerous rides I took with him. The Institute is located on a point jutting out into the Arabian Sea. It has magnificent grounds and buildings isolated from the rest of the city. Its location made it easy to forget the poverty, the homelessness, and even the starvation that could be seen almost everywhere and especially around the Oval, where we lived.

The Institute gave a dinner for its foreign visitors on the evening of January 16. The next day we had planned to take a trip to the caves of Ellora and Ajanta with the Mostows. During the night Florie became quite ill, apparently from something she had eaten at the dinner. She developed a high fever so we had to cancel our trip. We were acquainted with a radiologist, Dr. Bajaj, who had been a resident at a hospital in New Haven. We called him to recommend an internist for Florie. He sent one who treated her with a variety of medicines including an antibiotic. Even so, it took a number of days for her to recover. Soon after, we were invited with the Mostows to the Bajajs' for dinner, and we saw quite a lot of them during our stay in Bombay and on a visit to Poona where Dr. Bajaj's sister lived.

Another Indian family who became our friends were the Ratnagars. Their daughter, Shernaz, was married to Jal Choksi, who visited Yale from time to time to work with Kakutani. Mr. Ratnagar was retired. One of his retirement activities was collecting bronzes. He introduced us to this great Indian art and helped us acquire a few pieces.

One of these, a beautiful Krishna from Gujurat, was a gift from him. Another, a fourteenth-century Vishnu from South India we could not have acquired without his help.

One of the junior members of the Institute was my former student, Daya-Nand Verma. He and the Public Relations Officer, Mr. R. G. Jalihal, had met us at the airport when we arrived, and Verma audited my course. We also went sight-seeing around Bombay with him. There were a number of music and dance concerts in Bombay when we were there. Verma got us tickets for these and accompanied us to them.

On February 9, a Sunday, we had planned to visit the Kanheri Caves with Verma. To reach these, we boarded a train at the Churchgate station. At the northern suburb of Dadar, we were told to pull down the metal shutters over the windows since we were approaching an area where there were riots. The rioters were from a political party called Shin Sena. They were demonstrating against Indians from the state of Madras who had come to Maharashtra state (which included Bombay). Because of their superior education, they had managed to fill positions that the Shin Sena believed rightfully belonged to the natives. Many South Indians lived in the Dadar area. Shortly after we passed the Dadar station, the train came to a halt, and we were told to leave the train and walk back along the tracks to the Dadar station where we could take a train back to Bombay. Along the way, we witnessed looting and burning of merchandise taken from stores belonging to South Indians. There were a number of policemen along the way. They were a reassuring sight for us, but they did nothing to stop the looting.

The next day a curfew was imposed in Bombay. The Institute closed and we were sent home. For several days, the city was deserted except for a few taxicabs and a few pedestrians, as well as the cows, who could now feed in the grassy areas of the streets undisturbed by the traffic. It was an eerie feeling to live in a paralyzed city. After several days, troops were called in, and on February 12, as suddenly as they had begun, the riots ended. More than 50 people had been killed, and the burning of buses and derailment of railroad cars severely damaged an already inadequate transportation system.

We never made it to Kanheri, but on February 14–16, we succeeded in visiting the caves of Ellora and Ajanta, which are among the greatest sights in India.

I was invited to lecture at Madurai University and at the Ramanujan Institute for Advanced Study in Mathematics at Madras during February 20–24. We had been to both places during our previous visit to India in 1965. We went first to Madurai, where we were met at the airport by M. Rajagopalan, N. S. Natarajan, and two other members of the department. Rajagopalan was a brother of the chairman, M. Venkataraman. He had been a student at Yale and more recently had been an associate professor at the University of Illinois. Natarajan was someone we had met at Chandigarh during our previous visit to India. Unfortunately, we missed seeing Venkatarman; he was visiting Pondicherry for a darshan (balcony appearance) of his spiritual leader. We had a very busy day and a half at Madurai. This included a second visit to the great Meenakshi temple, two lectures, and conferences with some advanced graduate students, who expected instant reactions to their research. The visit to the temple was fascinating because of its intense religious atmosphere. This was tempered somewhat by the commercialism of shops selling trinkets. We had the experience of mounting a temple elephant and of walking a long distance in the temple in our stocking feet.

At Madras, we were housed in the guest house of the university located on the Bay of Bengal at a distance from the center of the city. From our windows we had a marvelous view of an entire village, and we could observe the awakening of the village and the morning rituals of its inhabitants. Since we had a free day on Sunday, we decided to take a motor trip to Mahabalipuram and Kanchipuram. The former is the site of an enormous stone bas-relief variously called "The Descent of the Ganges," "The Penance of Bhaghirata," or "Arjuna's Penance." We had heard about this from Mr. Ratnagar. Kanchipuram is the site of two temples, one of which has 96 carved stone pillars, no two alike. We hired a car with a driver for the trip, and we invited two residents of the guest house, a Hungarian and a Czech, to accompany us. The Hungarian was a metallurgist, and the Czech was an Indian historian whose knowledge of the local dialects and customs came in handy on our trip. Our first stop was the village of Tirupporur, where we were held up for several hours by a religious festival celebrating the new moon. This was extremely interesting to watch, and the villagers were very friendly and did not at all mind our photographing the festival. After this stop, we drove on to Mahabalipuram. This was worth the trip, but the pleasure of viewing the sculptures was marred by the aggressive vendors and guides who preyed on the tourists. On the way to Kanchipuram, we made a brief stop at Thirukalikhunram to visit a temple located on a high hill. During the short ride back to Madras from Kanchipuram, we encountered a religious procession centered around an enormous covered wagon elaborately decorated with painted wood carvings. The wagon was drawn by a number of worshippers and was followed by many others. It was explained to us by our Czech friend that they were on their way to a temple, carrying an idol inside the wagon.

The next day, which was Monday, I gave my lecture at 11 A.M. at the Ramanujan Institute, and I talked to professor Balachandran and one of his students. We returned to Bombay in the afternoon.

On March 13–17, we had a wonderful trip to Udaipur. We stayed at the Lake Palace Hotel, a fabulous hotel built about 200 years ago as a palace for the Maharana of Udaipur. It was located on a man-made island on a man-made lake that had been created in the fourteenth century.

On March 29, we departed Bombay for New Delhi. We were driven to the airport by Fernandes, and we were accompained by R. C. Cowsik, who had assisted me in writing my lecture notes, and by Mr. Jalihal, the Public Relations Officer of the Institute. At the airport we were met by Dr. P. S. Rema of the Ramanujan Institute, who had been auditing my course. She had come with her brother to see us off. On the way to Delhi, we made a brief stop at Jaipur, a predominantly Muslim city in Rajasthan. We were there at the time of an important religious festival, and we experienced a rather hostile reception when we attempted to photograph the festivities. This was quite a different reaction from the one we had experienced earlier in the village of Tirupporur.

In the summer before this trip to India, I had received a letter from William R. Abel, Staff Scientist of the U.S. Agency for International Development (USAID). They had learned that I would be visiting the TIFR during the winter. As part of an ongoing program of aid in the development of research in mathematics, they proposed that I tour India for two or three weeks to lecture at some of the universities and consult with their mathematicians on their programs. In any case, he asked me to stop for a day or two in Delhi to report on my experiences at the TIFR. They offered to pay my travel expenses and a consulting fee. As a result of the correspondence that ensued, the

USAID paid our travel expenses to Madurai, Madras, and Delhi (which normally would have been paid by the universities), and Florie and I wrote reports on our impressions of the state of mathematics in India. Mine dealt with research and hers with education at the collegiate and secondary school levels.

After the stop at Jaipur, we flew to Delhi. We were met at the airport by Jerry Becker of USAID and his wife and by Drs. P. B. Bhattacharya and S. K. Jain of the Mathematics Department of the University of Delhi. We were driven to the guest house of the university, where we were guests for five nights. On Sunday, March 30, we had a day's exursion by plane to Khajuraho, the site of temples that date from the 10th and 11th centuries and are famous for their erotic sculptures. The next day we visited the NSF offices, where we conferred with some of their staff and wrote our reports. In my report, I mentioned the high quality of the faculty and students at the TIFR, but I was rather negative in my assessment of the state of mathematics in the rest of the country.

On April 1–5, there was a conference at the University of Delhi called A Seminar of Modern Algebra and Global Analysis. I attended the first day's session, which began at 10 A.M. and ended at 5 P.M., with a two-hour break for lunch. I gave a talk at the morning session on structure theory of Jordan algebras. The second morning of the conference was also devoted to algebra; I attended part of this session. We had lunch at the "Van Nostrand House," a residence that served as offices and home for the staff of the Indian branch of the Van Nostrand Publishers. They had published cheap editions of a number of the Van Nostrand books, including the three volumes of my *Lectures in Abstract Algebra*.

On April 3, we flew to Benares, where we were the guests of the Mathematics Department of the University of Benares. Verma had been a student at this university, and he had informed them of our visit to the TIFR. When we arrived in Benares, we were met at the airport by a party of six that included the Chairman of the Department, Professor R. S. Meshra, and two officials in uniform. We were taken to the university's guest house, where we stayed during our two days in Benares.

We have many vivid recollections of our stay at Benares:

A visit to Sarnath, which is one of the most important centers of Buddhism in India. Many Buddhist monks and temples.

A boat trip at dawn on the Ganges. Even at this early hour, there were many men and women on the banks of the river bathing and changing into fresh clothes in preparation for worshipping in one of the temples. From the river, we could see the smoke from a number of funeral pyres.

Morning tea with the Maharajah of Benares. Presumably this meeting was arranged because he was an important benefactor of the University. He sent a car to take us to his palace. We had a pleasant conversation with him—mainly on the subject of the student uprisings that were common in many parts of the world (but not in India) at that time.

A visit to the largest sari merchant in Benares. The city is famous for its saris. We were accompanied by a group from the Department. Florie bought an exquisite embroidered black sari.

At the insistence of Professor Meshra's young son, who was apparently very religious, we visited a new temple that had been donated to the city by the Birlas, perhaps the richest family in India. Because of its newness, the temple was quite garish. To us, it appeared superfluous in a city that had so many temples.

Several walks through crowded and noisy city lanes. On one of these, we were jostled by a funeral procession making its way to a crematorium with a body that was lightly wrapped in linen and was borne on bamboo poles.

The heat was oppressive during our stay in Benares. It got to me, especially during my lecture, when I almost fainted.

We decided to return to the United States by way of the Pacific, so we flew to Calcutta from Benares. We had a day's stop there as well as at Bangkok. After that, we stopped in Tokyo for three days. We arrived just before the end of a conference, The International Conference on Functional Analysis and Related Topics, held in honor of Professor K. Yosida on the occasion of his retirement from the University of Tokyo. I had written him of our intention to be in Tokyo from April 7–10, and he reserved a room for us in a Japanese inn, the Fukadaya, which was located near the Imperial Palace. He also invited us to the farewell dinner of the conference on April 8. We met many old friends there, including Yasua Akizuki, Michael Atiyah, Henri Cartan, Joe Doob, Peter Lax, Ralph Phillips, my colleague Shizuo Kakutani, and, of course, Yosida. The Fukadaya was a wonderful inn, an island of tranquility in a bustling and noisy city. After the conference, we had a sukiyaki dinner with Shizuo in a private room at the inn. Altogether we had a delightful second visit to Tokyo.

Form Tokyo we flew back to the United States by way of Honolulu. We stopped there briefly to visit with Leon Cohen and Paul Halmos, who were teaching at the University of Hawaii at the time.

For the academic year 1969–1970, Lowell Paige organized a year of concentration in algebra at UCLA. Among the visitors who participated were S. A. Amitsur, H. Bass, D. K. Harrison, I. Kaplansky, H. Zassenhaus, and myself. By stretching our spring recess, I was able to attend for a period of three weeks in April, and Florie came for two of these. During her stay, we had the use of a lovely house (with a cat) that belonged to Ed and Alice Bechenbach.

We attended the 1970 ICM at Nice. An innovation of this Congress was the elimination of oral presentations of contributed 10-minute papers. These were accepted as written communications that were available to the members. A somewhat unusual feature of this Congress was the emphasis placed on one particular field, finite groups, which it was recognized was in the midst of a spectacular development. One of the Fields medalists was John Thompson who was awarded a medal for his work on the problem of classifying the finite simple groups. One of the hour "Conferences Generales" was Walter Feit's address, The Current Situation in the Theory of Finite Groups. There was also a conference on finite groups with eight speakers.

Immediately preceding the Nice Congress we attended the meeting of the IMU Assembly at Menton. The welcoming speech by the mayor of Menton was a pleasant surprise. It was elegant in style and of a high level in content.

The papers I published during this period were [72] and [73], both on Jordan algebras. In [72], I introduced a concept of Cartan subalgebras for a finite dimensional Jordan algebra $\mathfrak{J}$ over a field of characteristic $\neq 2$, analogous to Cartan subalgebras of Lie algebras. In defining the Lie concept, one requires Engel's theorem and the Fitting decomposition of a vector space under the action of a nilpotent Lie algebra (see *Lie Algebras*, pp. 56–61). For Jordan algebras, the role of nilpotent algebras is taken by that of associator nilpotent algebras. If $a, b \in \mathfrak{J}$, we let R_a denote the multiplication $x \to x \cdot a$ and $R_{a,b} = R_a R_b - R_{a,b}$, so $xR_{a,b}$ is the associator $[x, a, b] = (x \cdot a) \cdot b - x \cdot (a \cdot b)$. Then $\mathfrak{J}$ is called associator nilpotent if there exists an N such that $R_{a_1,b_1} R_{a_2,b_2} \cdots R_{a_N,b_N} = 0$

for all a_i, $b_i \in \mathfrak{J}$. The element a is associator nilpotent if every R_{a^i,a^j} is nilpotent. I proved an analogue of Engel's theorem for Lie algebras: If $\mathfrak{J}$ has an identity element and is finite dimensional over an infinite field Φ, then $\mathfrak{J}$ is associator nilpotent if and only if every $a \in \mathfrak{J}$ is associator nilpontent. The assumption that Φ is infinite is used to show that the hypothesis that every $a \in \mathfrak{J}$ is associator nilpotent carries over to any $\mathfrak{J}_\Pi$, Π an extension field of Φ. Then, if Φ is algebraically closed, $\mathfrak{J}$ is a direct sum of ideals that are almost nil, that is, have the form $\Phi 1 + \mathfrak{N}$, where $\mathfrak{N}$ is a nil ideal. Any such algebra is readily seen to be associator nilpotent. To obtain a Fitting decomposition of a Jordan algebra $\mathfrak{J}$ relative to an associator nilpotent subalgebra $\mathfrak{K}$, one proves that the Lie algebra $\mathfrak{L}_\mathfrak{J}(\mathfrak{K})$ generated by the R_a, $a \in \mathfrak{K}$, (acting in $\mathfrak{J}$) is nilpotent. Then the Fitting decomposition of $\mathfrak{J}$ relative to $\mathfrak{K}$ is defined to be that of $\mathfrak{J}$ relative to $\mathfrak{L}_\mathfrak{J}(\mathfrak{K})$. The proof makes use of the following result of general interest. Let $\mathfrak{J}$ be a finite dimensional Jordan algebra, $\mathfrak{K}$ a solvable ($=$ nil) ideal in $\mathfrak{J}$. Then, the image of $\mathfrak{K}$ in the universal associative algebra $\mathfrak{U}(\mathfrak{J})$ is a nilpotent ideal in $\mathfrak{J}$. I also proved that if $\mathfrak{J}$ is a finite dimensional Jordan algebra with 1 over an algebraically closed of characteristic 0, then, if $\mathfrak{K}_1$ and $\mathfrak{K}_2$ are Cartan subalgebras of $\mathfrak{J}$, there exists an automorphism s in a certain subgroup (the inner automrphism group) such that $\mathfrak{K}_1^s = \mathfrak{K}_2$. Finally, I gave a proof in [72] of a trace criterion for separability of Jordan algebras that does not use the determination of the simple Jordan algebras over an algebraically closed field (cf. [69], Sec. 5). An improved proof of this is given in *Structure and Representations of Jordan Algebras*, pages 352–358.

In the paper "Existence and conjugacy of Cartan subalgebras of Jordan algebras" appearing in the *Proceedings of the American Mathematical Society* **50**, 1975, pp. 40–44, Ottmar Loos gave the following improvements of the results of [72]:

(i) the existence of Cartan subalgebras without the restriction that Φ is infinite,
(ii) conjugacy of Cartan subalgebras if the base field is algebraically closed of any characteristic ($\neq 2$).

In a monograph "Jordan algebras of self-adjoint operators" (*Memoir of the American Mathematical Society* **53**, 1965), D. M. Topping defined a quadratic ideal in a Jordan algebra $\mathfrak{J}$ of self-adjoint operators in a Hilbert space as a subspace $\mathfrak{B}$ of $\mathfrak{J}$ such that $bab \in \mathfrak{B}$ for all $a \in \mathfrak{J}$. This suggested the definition of a quadratic ideal (later called an inner ideal) of a Jordan algebra $\mathfrak{J}$ as a subspace $\mathfrak{B}$ such that $U_b a \in \mathfrak{B}$ for all $b \in \mathfrak{B}$, $a \in \mathfrak{J}$, where $U_b = 2R_a^2 - R_{a^2}(R_a\colon x \sim> x \cdot a)$. This is the starting point of [73], which gives a structure theory of (not necessarily finite dimensional) Jordan algebras that is analogous to the Wedderburn–Artin structure theory of semisimple Artinian associative rings.

If a is an element of a Jordan algebra $\mathfrak{J}$, then $\mathfrak{J}U_a$ is an inner ideal called the principal inner ideal determined by a. The element $z \in \mathfrak{J}$ is called an absolute zero divisor if $\mathfrak{J}U_z = 0$. In [73], I considered the structure of Jordan algebras (over fields of characteristic $\neq 2$) that satisfy the following conditions:

(i) $\mathfrak{J}$ has an identity element.
(ii) $\mathfrak{J}$ contains no absolute zero divisor $\neq 0$.
(iii) the minimum condition holds for principal quadratic ideals determined by idempotent elements, and every quadratic ideal determined by an idempotent contains a minimal quadratic ideal.

Condition (ii) is analogous to the requirement of semisimplicity in the associative theory, and holds if we have the descending chain condition on quadratic ideals.

A key result for the structure theory of Jordan algebras satisfying (i)–(iii) is the theorem that if $\mathfrak{J}$ is such an algebra, then the identity element 1 of $\mathfrak{J}$ has a decomposition as $1 = \sum_1^n e_i$, where the e_i are orthogonal idempotent elements such that every $\mathfrak{J}U_{e_i}$ is a division algebra, that is, the nonzero elements of $\mathfrak{J}U_{e_i}$ are regular in $\mathfrak{J}U_{e_i}$ (as defined in [56]). Later, Jordan algebras having this property were said to have a capacity, and the minimum n was called the capacity. I showed that any $\mathfrak{J}$ satisfying (i)–(iii) is direct sum of a finite number of simple $\mathfrak{J}$ satisfying (i)–(iii). In the case in which $\mathfrak{J}$ is simple and $n > 1$, any two e_i, e_j, $i \neq j$, are connected in the sense that there exists an a_{ij} such that $e_i \cdot a_{ij} = \frac{1}{2}a_{ij}$, which is regular in $\mathfrak{J}U_{e_i+e_j}$. Then, if $n \geq 3$, the coordinatization theorem of [67] implies that $\mathfrak{J}$ is isomorphic to an algebra $H(\mathfrak{D}_n, \gamma)$ as defined in [67]. The possibilities for $\mathfrak{D}$ can be sorted out using results of Kleinfeld and Osborn.

This analysis left open the determination of the Jordan algebras of capacity 1, that is, Jordan division algebras, and the simple Jordan algebras of capacity two. The second gap was filled by J. M. Osborn (who was visiting Yale for the academic year 1965–1966) in the paper "Jordan algebras of capacity two" in the *Proceedings of the National Academy of Sciences* **57**, 1967, pp. 582–588. The determination of the Jordan division algebras was given much later by Effim Zel'manov in "Jordan division algebras" (Russian) in *Algebra i Logika* **18**, 1979, pp. 286–310. An account of his proof is given in Chapter 8 of *StructureTheory of Jordan Algebras*. An improved proof of this result was given in a subsequent paper by Zel'manov: "On prime Jordan algebras," (*Siberian Mathematics Journal* **24**, 1983, pp. 89–104).

It is trivial to extend the foregoing theory to Jordan algebras over a commutative ring Φ in which $2 = 1 + 1$ is invertible. However, it is by no means trivial to remove this restriction on Φ. This was done by McCrimmon in "A general structure theory of Jordan rings," (*Proceedings of the National Academy of Sciences* **56**, 1966, pp. 1072–1079), by defining Jordan algebras over any commutative Φ by means a map $a \to U_a$ of $\mathfrak{J}$ into $\mathrm{End}_\Phi \mathfrak{J}$ that is quadratic in a and satisfies a simple set of axioms. These algebras were later called quadratic Jordan algebras and were the subjects of my lectures at St. Andrews and at the Tata Institute as mentioned before.

CARTAN SUBALGEBRAS OF JORDAN ALGEBRAS

N. JACOBSON[1]

Dedicated to the memory of TADASI NAKAYAMA

In this paper we shall give a definition of an analogue for Jordan algebras of the classical notion of a Cartan subalgebra of a Lie algebra. This is based on a notion of associator nilpotency of a Jordan algebra. A Jordan algebra $\mathfrak{J}$ is called associator nilpotent if there exists a positive (odd) integer M such that every associator of order M formed of elements of $\mathfrak{J}$ is 0 (§ 2). If $a, b \in \mathfrak{J}$ then we set $R_{a,b} = R_a R_b - R_{a \cdot b}$ where R_a is the mapping $x \to x \cdot a$ (product of x and a) in $\mathfrak{J}$. An element a is called associator nilpotent in $\mathfrak{J}$ if all the operators of the form $R_{a^{\cdot i}, a^{\cdot j}}$ are nilpotent. We prove an analogue of Engel's theorem on Lie algebras to the effect that a finite dimensional Jordan algebra $\mathfrak{J}$ (with 1) over an infinite field is associator nilpotent if and only if every element of $\mathfrak{J}$ is associator nilpotent (Theorem 4). If $\mathfrak{K}$ is an associator nilpotent subalgebra of the finite dimensional Jordan algebra $\mathfrak{J}$ then the Lie algebra $\mathfrak{L}_{\mathfrak{J}}(\mathfrak{K})$ of linear transformations in $\mathfrak{J}$ generated by the linear transformations in $\mathfrak{J}$ of the form $R_{b,c}$, $b, c \in \mathfrak{K}$ is nilpotent. We define a Cartan subalgebra of $\mathfrak{J}$ to be an associator nilpotent subalgebra $\mathfrak{K}$ of $\mathfrak{J}$ such that the Fitting null component $\mathfrak{J}_0$ of $\mathfrak{J}$ relative to the nilpotent Lie algebra of linear transformations $\mathfrak{L}_{\mathfrak{J}}(\mathfrak{K})$ coincides with $\mathfrak{K}$ (§ 5).

If $a \in \mathfrak{J}$ and $\dim \mathfrak{J} = n$ then we set $\mathfrak{Z}_a = \{z \in \mathfrak{J} | z(R_{a^{\cdot i}, a^{\cdot j}})^n = 0, i,j = 0, 1, 2, \ldots\}$. The element $a \in \mathfrak{J}$ is called associator regular in $\mathfrak{J}$ if $\dim \mathfrak{Z}_a$ is minimal. If the base field is infinite then $\mathfrak{Z}_a$ is a Cartan subalgebra for any associator regular element a (Theorem 6). If the base field is algebraically closed of characteristic 0 then any two Cartan subalgebras are conjugate in a strong sense (Theorem 7).

As applications of our results we obtain a formula for the generic trace

Receired July 7, 1965.

[1] This research has been supported by the Air Force Office of Scientific Research under Grant AF-AFOSR-402-64.

and we give a proof of the trace criterion for separability of a finite dimensional Jordan algebra, which is independent of the determination of the finite dimensional simple Jordan algebras over an algebraically closed field (§ 7). In § 1 we collect the results on Jordan algebras which are needed in the remainder of the paper.

1. Preliminaries. In this paper all algebras are algebras over fields of characteristic not two, all associative algebras are assumed to have identity elements, and the usual conventions for such algebras will be adopted: subalgebras will be assumed to contain 1, homomorphisms to map 1 into 1, etc. Beginning with § 2 we shall adopt these conventions also for Jordan algebras. However, we do not do this in the present section. We denote the product in a Jordan algebra by $a.b$, so, the defining identities are $a.b = b.a$, $(a^{\cdot 2}.b).a = a^{\cdot 2}.(b.a)$ where $a^{\cdot 2} = a.a$. These have the following consequences:

$$(1)\qquad \begin{aligned}&(x.a).(b.c) + (x.b).(a.c) + (x.c).(a.b)\\ &\quad = (x.(b.c)).a + (x.(a.c)).b + (x.(a.b)).c\end{aligned}$$

$$(2)\qquad \begin{aligned}&((x.a).b).c + ((x.c).b).a + x.((a.c).b)\\ &\quad = (x.a).(b.c) + (x.b).(a.c) + (x.c).(a.b).\end{aligned}$$[2]

If we denote the linear mapping $x \rightarrow x.a$ by R_a then $[R_a, R_{a^{\cdot 2}}] = 0$ where $[A, B] = AB - BA$, and (1) and (2) are equivalent to

$$(1')\qquad [R_a,\ R_{b.c}] + [R_b,\ R_{a.c}] + [R_c,\ R_{a.b}] = 0$$

$$(2')\qquad R_a R_b R_c + R_c R_b R_a + R_{(a.c).b} = R_a R_{b.c} + R_b R_{a.c} + R_c R_{a.b}.$$

Jordan algebras are power associative, that is, if we define $a^{\cdot k}$ by $a^{\cdot 1} = a$, $a^{\cdot k} = a^{\cdot k-1}.a$, $k = 2, 3, \ldots$, then $a^{\cdot k}.a^{\cdot l} = a^{\cdot k+l}$. We recall also that if $\mathfrak{J}$ is a Jordan algebra over a field Φ and Π is an extension field of Φ then the extension algebra $\mathfrak{J}_\Pi = \Pi \otimes_\Phi \mathfrak{J}$ is Jordan.

Let $\mathfrak{J}$ be a Jordan algebra with an identity element 1 and suppose $1 = \sum_1^n e_i$ where the e_i are orthogonal idempotent elements ($e_i^{\cdot 2} = e_i$, $e_i.e_j = 0$ if $i \neq j$). Let $\mathfrak{J}_{ii} = \{x_{ii} \mid x_{ii}.e_i = x_{ii}\}$ and $\mathfrak{J}_{ij} = \left\{x_{ij} \mid x_{ij}.e_i = \frac{1}{2}\, x_{ij} = x_{ij}.e_j\right\}$ if $i \neq j = 1, \ldots, n$. Then

[2] The results stated in this section without proof or reference can be found in Albert [2].

(3) $$\mathfrak{J} = \sum_{i \leq j} \oplus \mathfrak{J}_{ij}.$$

We call the $\mathfrak{J}_{ij} = \mathfrak{J}_{ji}$ the Peirce spaces of $\mathfrak{J}$ relative to the orthogonal idempotents e_i and (3) the Peirce decomposition of $\mathfrak{J}$ relative to these idempotents. We have the following multiplication table for the Peirce spaces:

(4) $$\begin{aligned} &\mathfrak{J}_{ii}^{\cdot 2} \subseteq \mathfrak{J}_{ii}, \quad \mathfrak{J}_{ii} \cdot \mathfrak{J}_{jj} = 0 \text{ if } i \neq j \\ &\mathfrak{J}_{ij} \cdot \mathfrak{J}_{ii} \subseteq \mathfrak{J}_{ij}, \quad \mathfrak{J}_{ij}^{\cdot 2} \subseteq \mathfrak{J}_{ii} + \mathfrak{J}_{jj} \text{ if } i \neq j \\ &\mathfrak{J}_{ij} \cdot \mathfrak{J}_{jk} \subseteq \mathfrak{J}_{ik}, \quad \mathfrak{J}_{ii} \cdot \mathfrak{J}_{jk} = 0, \quad \mathfrak{J}_{ij} \cdot \mathfrak{J}_{kl} = 0 \text{ if } i, j, k, l \text{ are unequal.} \end{aligned}$$

We shall call a Jordan (or associative) algebra $\mathfrak{J}$ over a field Φ *almost nil* if $\mathfrak{J}$ has an identity 1 and $\mathfrak{J} = \Phi 1 + \mathfrak{N}$ where $\mathfrak{N}$ is a nil ideal. If $\mathfrak{J}$ is a Jordan algebra with 1 such that every element of $\mathfrak{J}$ has the form $\alpha 1 + z$ where $\alpha \in \Phi$ and z is nilpotent, then $\mathfrak{J}$ is almost nil (Albert [3] p. 514, Jacobson [4], McCrimmon [1]). This implies that if $\mathfrak{J}$ is finite dimensional with 1 over an algebraically closed field and $1 = \sum_1^r e_i$ where the e_i are orthogonal idempotents $\neq 0$ and are primitive in the sense that we cannot write $e_i = e_i' + e_i''$ where $e_i' \neq 0$, $e_i'' \neq 0$ and $e_i' \cdot e_i'' = 0$, then the Peirce spaces $\mathfrak{J}_{ii}$ are almost nil Jordan algebras: $\mathfrak{J}_{ii} = \Phi e_i + \mathfrak{N}_i$, $\mathfrak{N}_i$ a nil ideal in $\mathfrak{J}_{ii}$. If $\mathfrak{J}$ is finite dimensional then the following three conditions on $\mathfrak{J}$ are equivalent: (1) $\mathfrak{J}$ is solvable (2) $\mathfrak{J}$ is a nil algebra (3) $\mathfrak{J}$ is nilpotent, in the sense that there exists an integer N such that every product of N elements (in any association) of $\mathfrak{J}$ is 0. A finite dimensional Jordan algebra contains a maximal solvable ideal $\mathfrak{S}$ called the radical of $\mathfrak{J}$. If $\mathfrak{J}$ is finite dimensional and $\mathfrak{R}$ is a non-zero solvable ideal then $\mathfrak{R}$ contains an ideal $\mathfrak{R}' \neq \mathfrak{R}$ of $\mathfrak{J}$ such that $\mathfrak{R}^{\cdot 2} \subseteq \mathfrak{R}'$ (Penico [1], p. 408).

A finite dimensional Jordan algebra is called semi simple if the radical $\mathfrak{S} = 0$. The algebra $\mathfrak{J}/\mathfrak{S}$ is semi-simple. If $\mathfrak{J}$ is semi-simple then $\mathfrak{J}$ has an identity element and is a direct sum of simple ideals (Albert [3]). If $\mathfrak{J}$ is a finite dimensional simple Jordan algebra over an algebraically closed field Φ then either $\mathfrak{J} = \Phi 1$ or $\mathfrak{J}$ contains $r > 1$ primitive orthogonal idempotent elements $e_i \neq 0$. Moreover, the Peirce spaces $\mathfrak{J}_{ij}$, $i \neq j$, determined by these elements are non-zero (Albert [3], Jacobson [4]).

Let $\mathfrak{J}$ be an arbitrary Jordan algebra and let $\mathfrak{T}(\mathfrak{J}) = \Phi \oplus \mathfrak{J} \oplus (\mathfrak{J} \otimes \mathfrak{J}) \oplus (\mathfrak{J} \otimes \mathfrak{J} \otimes \mathfrak{J}) \oplus \cdots$ be the tensor algebra based on the vector space $\mathfrak{J}$, $\mathfrak{R}$ the ideal in $\mathfrak{T}(\mathfrak{J})$ generated by all elements of the form

$$(5)\qquad \begin{gathered} a\otimes a^{\cdot 2}-a^{\cdot 2}\otimes a \\ a^{\cdot 3}.b+2\,a\otimes b.\otimes a-b\otimes a^{\cdot 2}-2\,a\otimes a.b, \end{gathered}$$

where $a,\ b\in\mathfrak{J}$. The associative algebra $\mathfrak{U}(\mathfrak{J})=\mathfrak{T}(\mathfrak{J})/\mathfrak{N}$ will be called the *universal multiplication envelope* of the Jordan algebra $\mathfrak{J}$.[3] If $a\in\mathfrak{J}$ we set $a^r=a+\mathfrak{N}$ in $\mathfrak{U}(\mathfrak{J})$. $\mathfrak{U}(\mathfrak{J})$ and the mapping $a\to a^r$ have the following universal mapping property: Let $a\to a^\rho$ be a linear mapping of $\mathfrak{J}$ into an associative algebra $\mathfrak{A}$ such that $[a^\rho,\ a^{\cdot 2\rho}]=0$ and $(a^{\cdot 2}.b)^\rho+2\,a^\rho b^\rho a^\rho=b^\rho a^{\cdot 2\rho}+2\,a^\rho(a.b)^\rho$ then there exists a unique homomorphism of $\mathfrak{U}(\mathfrak{J})$ into $\mathfrak{A}$ such that $a^r\to a^\rho$. It follows from this that if $\mathfrak{K}$ is an ideal in $\mathfrak{J}$ and $\mathfrak{B}$ is the ideal in $\mathfrak{U}(\mathfrak{J})$ generated by $\mathfrak{K}$ then we have an isomorphism of $\mathfrak{U}(\mathfrak{J}/\mathfrak{K})$ onto $\mathfrak{U}(\mathfrak{J})/\mathfrak{B}$ mapping $(a+\mathfrak{K})^r$, $a\in\mathfrak{J}$, into $a^r+\mathfrak{B}$ (cf. Jacobson, *Lie Algebras*, Th. 5.1 (4), p. 153). It is known that $\mathfrak{U}(\mathfrak{J})$ is finite dimensional if $\mathfrak{J}$ is finite dimensional (Jacobson [2], p. 519).

Let $\mathfrak{J}$ be a subalgebra of the Jordan algebra $\mathfrak{J}'$ and let R_a, $a\in\mathfrak{J}$, be the multiplication $x\to x.a$ in $\mathfrak{J}'$. Then $a\to R_a$ is a linear mapping of $\mathfrak{J}$ into $\mathrm{Hom}_\Phi(\mathfrak{J}',\ \mathfrak{J}')$ and we have $[R_a,\ R_{a^{\cdot 2}}]=0$ and $2\,R_aR_bR_a+R_{a^{\cdot 2}.b}=2\,R_aR_{a.b}+R_bR_{a^{\cdot 2}}$. Hence we have a homomorphism of $\mathfrak{U}(\mathfrak{J})$ onto the subalgebra $\mathfrak{E}$ of $\mathrm{Hom}_\Phi(\mathfrak{J}',\ \mathfrak{J}')$ generated by the R_a, $a\in\mathfrak{J}$. Let $\mathfrak{J}$ be finite dimensional and let $\mathfrak{K}$ be a solvable ideal in $\mathfrak{J}$. Then we wish to show that the elements R_b, $b\in\mathfrak{K}$ (acting in $\mathfrak{J}'$), generate a nilpotent ideal in $\mathfrak{E}$.[4] In view of the homomorphism we have just noted this will follow from the following

THEOREM 1. *Let $\mathfrak{J}$ be a finite dimensional Jordan algebra, $\mathfrak{K}$ a solvable ideal in $\mathfrak{J}$. Then the image $\mathfrak{K}^r$ of $\mathfrak{K}$ in $\mathfrak{U}(\mathfrak{J})$ generates a nilpotent ideal in $\mathfrak{U}(\mathfrak{J})$.*

We shall prove this result by induction on the dimensionality dim $\mathfrak{K}$ and we may assume $\mathfrak{K}\neq 0$. Let $\mathfrak{K}'$ be an ideal of $\mathfrak{J}$ properly contained in $\mathfrak{K}$ such that $\mathfrak{K}^{\cdot 2}\subseteq\mathfrak{K}'$. Then $\mathfrak{K}'^r$ generates a nilpotent ideal $\mathfrak{B}$ in $\mathfrak{U}(\mathfrak{J})$ and we have the isomorphism noted above of $\mathfrak{U}(\mathfrak{J}/\mathfrak{K}')$ onto $\mathfrak{U}(\mathfrak{J})/\mathfrak{B}$. In view of this it suffices to show that the image of $\mathfrak{K}/\mathfrak{K}'$ in $\mathfrak{U}(\mathfrak{J}/\mathfrak{K}')$ generates a nilpotent ideal. Accordingly, it is enough to prove the theorem for the case $\mathfrak{K}^{\cdot 2}=0$. In this case we base the proof on the following two lemmas.

---

[3] This is a slightly different definition from that of the universal associative algebra of representations of $\mathfrak{J}$ given in Jacobson [2] and [3]. The modification has been made to take care of the characteristic three case. The results of Jacobson [2] and [3] carry over without change to the present situation.

[4] For characteristic 0 this is proved in Jacobson [2], p. 522.

LEMMA 1. *Let $\mathfrak{J}$ be a finite dimensional Jordan algebra such that $\mathfrak{J}^{\cdot 2}=0$. Then the image $\mathfrak{J}^r$ of $\mathfrak{J}$ in $\mathfrak{U}(\mathfrak{J})$ generates a nilpotent ideal.*

Proof. Since any subspace of $\mathfrak{J}$ is an ideal the inductive argument we have just used gives a reduction to the case in which $\dim \mathfrak{J}=1$. Then $\mathfrak{J}=\Phi e$. The general relation

$$(a^{\cdot 2}.b)^r + 2\, a^r b^r a^r = b^r (a^{\cdot 2})^r + 2\, a^r (a.b)^r \tag{6}$$

in universal multiplication envelopes gives the relation $(e^r)^3=0$ in $\mathfrak{U}(\mathfrak{J})$. Since $\mathfrak{J}^r$ generates $\mathfrak{U}(\mathfrak{J})$, e^r generates $\mathfrak{U}(\mathfrak{J})$ in the present case. Hence it is clear that $\mathfrak{J}^r=(\Phi e)^r$ generates a nilpotent ideal in $\mathfrak{U}(\mathfrak{J})$.

LEMMA 2. *Let $\mathfrak{J}$ be a finite dimensional Jordan algebra, $\mathfrak{K}$ a non-zero ideal in $\mathfrak{J}$ such that $\mathfrak{K}^{\cdot 2}=0$. Let $(e_1, e_2, \ldots, e_n)$ be a basis for $\mathfrak{J}$ such that $(e_1, e_2, \ldots, e_m)$ is a basis for $\mathfrak{K}$, k a positive integer. Then any monomial in $\mathfrak{U}(\mathfrak{J})$ of the form $e_{j_1}^r e_{j_2}^r \cdots e_{j_l}^r$ in which $k+m$ of the j's are in the range $I=\{1, 2, \ldots, m\}$ is a linear combination of monomials of the form $e_{i_1}^r e_{i_2}^r \cdots e_{i_k}^r \cdots$ where $i_1, \ldots, i_k \in I$.*

Proof. This is clear if $l=k+m$; hence we may use induction on the formal degree l of the monomial. Now suppose the first h subscripts j in the given monomial $M=e_{j_1}^r e_{j_2}^r \cdots e_{j_l}^r$ are in I but the $h+1$-st subscript is not. Then the assertion holds if $h \geq k$, so for a given l we may use a downward induction on the integer h and we may assume $h<k$. Then M has the form

$$\cdots e^r \cdots e^r \cdots \tag{7}$$

where e is one of the e_i, $i \in I$, and the first displayed e^r in (7) occurs after the $h+1$-st place. Let $\mathfrak{B}$ denote the space of linear combinations of monomials of the form $e_{i_1}^r e_{i_2}^r \cdots e_{i_k}^r \cdots$ where $i_1, i_2, \ldots, i_k \in I$. Suppose first that there is just one e_j^r between the indicated e^r in (7). Then (6) for $a=e$, $b=e_j$ gives the relation $e^r e_j^r e^r = e^r (e.e_j)^r$. Since $e.e_j \in \mathfrak{K}$ and $e.e_j=0$ if $e_j \in \mathfrak{K}$ substitution of the relation just given in (7) and the induction hypothesis on l implies that $M \in \mathfrak{B}$. Next assume that the indicated e^r in (7) are consecutive. We have the following relation in universal multiplication envelopes which is a consequence of (6):

$$\begin{aligned} a^r b^r c^r = & -c^r b^r a^r - ((a.c).b)^r + a^r (b.c)^r \\ & + b^r (a.c)^r + c^r (a.b)^r. \end{aligned} \tag{8}$$

Setting $b=c=e$ and $a=e_j$ where e_j^r precedes the first e^r in (7) we obtain $e_j^r(e^r)^2 = -(e^r)^2e_j^r + 2e^r(e_j.e)^r$. A succession of replacements of this type shows that $M \equiv \pm M' \pmod{\mathfrak{B}}$ where M' has formal degree l, has $k+m$ factors of the form e_i^r, $i \in I$, and has $h+2$ such factors at the beginning of M'. The induction on h can be invoked to conclude that $M' \in \mathfrak{B}$. Hence $M \in \mathfrak{B}$. Finally, suppose we have at least two e_j^r between the displayed e^r in (7). Then we can use (8) with $c=e$ to show that $M \equiv -M' \pmod{\mathfrak{B}}$ where M' is obtained from M by moving the second e^r two places to the left. A succession of moves of this type shows that $M \equiv \pm N \pmod{\mathfrak{B}}$ where N is obtained from M by moving the second e^r either next to the first one or to the position two places after the first one. Then $N \in \mathfrak{B}$ by the first two cases. Hence $M \in \mathfrak{B}$ in all cases.

We can now complete the proof of Theorem 1 for the case $\mathfrak{R}^{\cdot 2}=0$. By Lemma 1 and the universal mapping property of $\mathfrak{U}(\mathfrak{R})$ there exists a positive integer N such that the product of any N elements of the form e^r, $e \in \mathfrak{R}$ is 0. By Lemma 2 it follows that the product of any elements a^r, $a \in \mathfrak{J}$, which includes $N+m$ ($m = \dim \mathfrak{R}$) elements of $\mathfrak{R}$ is 0. This implies that if $\mathfrak{B}$ is the ideal in $\mathfrak{U}(\mathfrak{J})$ generated by $\mathfrak{R}^r$ then $\mathfrak{B}^{N+m}=0$.

2. Associator nilpotent Jordan algebras. From now on we assume that all the Jordan algebras under consideration *are finite dimensional and contain* 1. The usual conventions for algebras with 1 are adopted. In particular, subalgebras necessarily contain 1.

Let $\mathfrak{J}$ be a Jordan algebra over the field Φ (finite dimensional with 1). If $a, b, c \in \mathfrak{J}$ we write $[a, b, c]$ for the associator $(a.b).c - a.(b.c)$. This defines a trilinear composition in $\mathfrak{J}$ which can be iterated to define n-linear compositions for any positive odd integer n. Let n be a positive odd integer, $(a_1, a_2, \ldots, a_n)$ an ordered set of elements $a_i \in \mathfrak{J}$. Then we define an *associator of order n* $A_n(a_1, \ldots, a_n)$ inductively by $A_1(a_1) = a_1$,

$$A_n(a_1, \ldots, a_n) = [A_{n_1}(a_1, \ldots, a_{n_1}), A_{n_2}(a_{n_1+1}, \ldots, a_{n_1+n_2}), A_{n_3}(a_{n_1+n_2+1}, \ldots, a_{n_1+n_2+n_3})] \tag{9}$$

where $n = n_1+n_2+n_3$, n_i a positive odd integer and $A_{n_i}(\cdots)$ is an associator of order n_i. We shall call $\mathfrak{J}$ *associator nilpotent* if there exists a positive odd integer M such that every associator of order M formed of elements of $\mathfrak{J}$ is 0.

The minimum M with this property is called the *index of associator nilpotency* of $\mathfrak{J}$. It is clear that subalgebras, homomorphic images and direct sums of associator nilpotent Jordan algebras are associator nilpotent. Also since the higher associator compositions are multilinear it is clear that $\mathfrak{J}$ is associator nilpotent if and only if $\mathfrak{J}_\Pi = \Pi \otimes_\Phi \mathfrak{J}$ is associator nilpotent for any extension field Π/Φ. This remark will permit us to reduce considerations on associator nilpotency to the case of an algebraically closed field. For these we have the following important criterion.

THEOREM 2. *A Jordan algebra over an algebraically closed field is associator nilpotent if and only if it is a direct sum of ideals which are almost nil algebras.*

Proof. To prove the sufficiency it is enough to show than any almost nil algebra is associator nilpotent. Assume $\mathfrak{J}$ is almost nil: $\mathfrak{J} = \Phi 1 + \mathfrak{N}$ where $\mathfrak{N}$ is a nil subalgebra. Since the higher associator compositions are multilinear and since any higher associator involving 1 is 0, to prove associator nilpotency it is enough to show that there is an odd integer M such that every associator of order M of elements $z_i \in \mathfrak{N}$ is 0. This is clear since $\mathfrak{N}$ nil implies that $\mathfrak{N}$ is nilpotent. Conversely, assume $\mathfrak{J}$ is associator nilpotent and let $\mathfrak{J} = \sum_{i \leq j} \mathfrak{J}_{ij}$ be a Peirce decomposition of $\mathfrak{J}$ relative to a set of orthogonal primitive idempotent elements $e_i \neq 0$, $i = 1, 2, \ldots, r$, such that $\sum e_i = 1$. Then every $\mathfrak{J}_{ii}$ is an almost nil algebra with identity element e_i. Let $a_{ij} \in \mathfrak{J}_{ij}$, $i \neq j$. Then $[a_{ij}, e_i, e_j] = (a_{ij}.e_i).e_j - a_{ij}.(e_i.e_j) = \frac{1}{4} a_{ij}$. Iteration of this gives

$$(10) \qquad [\cdots[a_{ij}, \overbrace{e_i, e_j], e_i, e_j], \ldots, e_i, e_j}^{k}] = \left(\frac{1}{4}\right)^k a_{ij}.$$

Since $\mathfrak{J}$ is associator nilpotent, k can be chosen so that the left-hand side is 0. Hence $a_{ij} = 0$ and so $\mathfrak{J}_{ij} = 0$ if $i \neq j$. Since $\mathfrak{J}_{ii} . \mathfrak{J}_{jj} = 0$ for $i = j$ it is now clear that the $\mathfrak{J}_{ii}$ are ideals and $\mathfrak{J} = \mathfrak{J}_{11} \oplus \mathfrak{J}_{22} \oplus \cdots \oplus \mathfrak{J}_{rr}$ is a direct sum of almost nil ideals.

We prove next the following necessary (but not sufficient) condition for associator nilpotency.

THEOREM 3. *If $\mathfrak{J}$ is associator nilpotent then $\mathfrak{J}/\mathfrak{S}$ is associative for $\mathfrak{S}$ the radical of $\mathfrak{J}$.*

Proof. Since $\mathfrak{J}/\mathfrak{S}$ is associator nilpotent it is enough to show that if $\mathfrak{J}$

is associator nilpotent and semi-simple then $\mathfrak{J}$ is associative. Since a semi-simple Jordan algebra is a direct sum of simple ideals we may assume $\mathfrak{J}$ simple. Then the center Γ of $\mathfrak{J}$ is a field and $\mathfrak{J}$ over Γ is central simple. Also it is clear that $\mathfrak{J}$ over Γ is associator nilpotent. Hence we may assume $\mathfrak{J}$ is central simple. Then $\mathfrak{J}_\Omega$ is simple for Ω the algebraic closure of the base field. We have noted in § 1 that if dim $\mathfrak{J}_\Omega > 1$, then $\mathfrak{J}_\Omega$ contains $r>1$ primitive orthogonal idempotent $e_i \neq 0$ such that $\sum e_i = 1$. Also the Peirce spaces $\mathfrak{J}_{ij}$, $i \neq j$, determined by the e_i are $\neq 0$. Then $[\cdots[a_{ij}, e_i, e_j], e_i, e_j]\cdots e_i, e_j] = \left(\frac{1}{4}\right)^k a_{ij} \neq 0$ for $a_{ij} \neq 0$ in $\mathfrak{J}_{ij}$. This contradicts the associator nilpotency of $\mathfrak{J}$. Hence dim $\mathfrak{J}_\Omega = 1$ so $\mathfrak{J}_\Omega$ is associative. Hence $\mathfrak{J}$ is associative.

The following example shows that the converse of Theorem 2 is false.

Example. Let $\mathfrak{J}$ be the Jordan algebra of triangular matrices $\begin{pmatrix} \alpha & \gamma \\ 0 & \beta \end{pmatrix}$ where $\alpha, \beta, \gamma \in \Phi$ and the multiplication is $a.b = \frac{1}{2}(ab+ba)$. The radical $\mathfrak{N} = \Phi e_{12}$ and $\mathfrak{J}/\mathfrak{N} \cong \Phi e_{11} \oplus \Phi e_{22}$ where e_{ij}, $i, j = 1, 2$ is the usual set of matrix units. Thus $\mathfrak{J}/\mathfrak{N}$ is associative. On the other hand, $\mathfrak{J}$ is not associator nilpotent since $]e_{12}, e_{11}, e_{22}] = \frac{1}{4} e_{12}$. This example shows also that the associator nilpotency of an ideal $\mathfrak{K}$ and of $\mathfrak{J}/\mathfrak{K}$ does not imply the associator nilpotency of $\mathfrak{J}$.

3. Analogue of Engel's theorem. If $a, b \in \mathfrak{J}$ we write $R_{a,b}$ for $R_a R_b - R_{a.b}$ where R_a is the linear mapping $x \to x.a$ in $\mathfrak{J}$. We have $xR_{a,b} = (x.a).b - x.(a.b) = [x, a, b]$. Hence if $\mathfrak{J}$ is associator nilpotent then there exists an integer M such that $R_{a_1,b_1} R_{a_2,b_2} \cdots R_{a_M,b_M} = 0$ for all $a_i, b_i \in \mathfrak{J}$. In particular, $R_{a,b}$ is nilpotent for all $a, b \in \mathfrak{J}$. We shall now call a an *associator nilpotent element relative to* $\mathfrak{J}$ if $a \in \mathfrak{J}$ and $R_{a^{\cdot i}, a^{\cdot j}}$ is nilpotent for all $i, j = 0, 1, 2, \ldots$. If dim $= n$ this will hold if and only if $R_{a^{\cdot i}, a^{\cdot j}}{}^n = 0$, $i, j = 0, 1, \ldots$. For any a we define the subspace

$$(11) \qquad \mathfrak{Z}_a = \{z \in J \mid z(R_{a^{\cdot i}, a^{\cdot j}})^n = 0,\ i, j = 0, 1, 2, \ldots\}.$$

Then a is an associator nilpotent element relative to $\mathfrak{J}$ if and only if $\mathfrak{Z}_a = \mathfrak{J}$. The following result is an analogue of Engel's theorem on Lie algebras.

THEOREM 4. *If Φ is infinite then $\mathfrak{J}$ over Φ is associator nilpotent if and only if every $a \in \mathfrak{J}$ is associator nilpotent relative to $\mathfrak{J}$.*

Proof. The hypothesis amounts to assuming the identities $(xR_{y^{\cdot i}, y^{\cdot j}})^n = 0$ in

$\mathfrak{J}$ where n is the dimensionality of $\mathfrak{J}$. We now observe that if the base field is infinite then any polynomial identity $p(x_1, x_2, \ldots, x_r) = 0$ which is valid in a finite dimensional algebra $\mathfrak{J}$ is valid also for any extension algebra $\mathfrak{J}_\Pi$. Here we let $(u_1, \ldots, u_n)$ be a basis for $\mathfrak{J}$, hence for $\mathfrak{J}_\Pi$, and we take $x_j = \sum \xi_{ji} u_i$ where ξ_{ji} are indeterminates. Thus the x_j are elements of $\mathfrak{J}_{\Phi(\xi_{ji})}$ and $p(x_1, \ldots, x_r) = \sum_i p_i(\xi_{jk}) u_i$. If $p(a_1, \ldots, a_r) = 0$ for all $a_j \in \mathfrak{J}$ then $p_i(\alpha_{jk}) = 0$ for all choices of the α_{jk} in Φ. It follows that $p_i(\xi_{jk}) = 0$ and consequently $p(a_1, \ldots, a_r) = 0$ for all $a_j \in \mathfrak{J}_\Pi$. In particular, we see that our hypothesis on $\mathfrak{J}$ carries over to $\mathfrak{J}_\Pi$ for any extension field Π/Φ. It therefore suffices to prove the theorem for algebraically closed Φ. Let $\Phi[a]$ be the subalgebra of $\mathfrak{J}$ generated by a and let $\mathfrak{L}_a$ be the space of Φ-linear combinations of the elements $R_{b,c}$, $b, c \in \Phi[a]$. Since every element of $\Phi[a]$ is a linear combination of the powers $a^{\cdot i}$, it is clear from the definition of $R_{b,c}$ that every $R_{b,c}$ with b, c in $\Phi[a]$ is a linear combination of the operators $R_{a^{\cdot i}, a^{\cdot j}}$. We recall also the operators $R_{a^{\cdot i}}$ commute. Hence the operators $R_{a^{\cdot i}, a^{\cdot j}}$ commute and consequently the hypothesis implies that every $R_{b,c}$ is nilpotent if $b, c \in \Phi[a]$. Now let $e_1, e_2, \ldots, e_r$ be a set of orthogonal primitive idempotents in $\mathfrak{J}$ such that $e_j \neq 0$ and $\sum e_i = 1$ and let $\mathfrak{J} = \sum_{i \leq j} \mathfrak{J}_{ij}$ be the corresponding Peirce decomposition of $\mathfrak{J}$. Take $a = \sum \alpha_i e_i$ where the α_i are distinct in Φ. Then the $e_i \in \Phi[a]$ so the foregoing result shows that R_{e_i, e_j} is nilpotent. As before, let $i \neq j$ and let $a_{ij} \in \mathfrak{J}_{ij}$. Then $a_{ij} R_{e_i, e_j} = [a_{ij}, e_i, e_j] = \frac{1}{4} a_{ij}$. Since R_{e_i, e_j} is nilpotent this implies that $a_{ij} = 0$. Hence every $\mathfrak{J}_{ij} = 0$ for $i \neq j$ and $\mathfrak{J} = \sum \mathfrak{J}_{ii}$ is a direct sum of almost nil ideals. Hence $\mathfrak{J}$ is associator nilpotent by Theorem 2.

4. Fitting decomposition relative to an associator nilpotent subalgebra. We recall that if $\mathfrak{L}$ is a nilpotent Lie algebra of linear transformations acting in an n ($< \infty$)dimensional vector space $\mathfrak{M}$ then we have the Fitting decomposition $\mathfrak{M} = \mathfrak{M}_0 \oplus \mathfrak{M}_1$ relative to $\mathfrak{L}$ where $\mathfrak{M}_0 = \{z \mid zA^n = 0,\ A \in \mathfrak{L}\}$ and $\mathfrak{M}_1 = \cap\, \mathfrak{M}\mathfrak{L}^{*i}$ where $\mathfrak{L}^*$ is the ideal generated by $\mathfrak{L}$ in the enveloping associative algebra $\mathfrak{E}(\mathfrak{L})$ of $\mathfrak{L}$. $\mathfrak{M}_0$ and $\mathfrak{M}_1$ are called respectively the Fitting null and one component of $\mathfrak{M}$ relative to $\mathfrak{L}$ (Jacobson, *Lie Algebras*, p. 39). These are invariant subspaces relative to $\mathfrak{L}$ (and $\mathfrak{L}^*$) and $\mathfrak{M}_0(\mathfrak{L}^*)^n = 0$. We remark that $\mathfrak{M}_1$ can be characterized as any complementary subspace of $\mathfrak{M}$, which is invariant under $\mathfrak{L}$. For, if $\mathfrak{N}$ is such a complement then $\mathfrak{M} = \mathfrak{M}_0 \oplus \mathfrak{N}$ and $\mathfrak{N}$ has the Fitting decomposition $\mathfrak{N}_0 \oplus \mathfrak{N}_1$ where $\mathfrak{N}_0$ is the Fitting null component of $\mathfrak{N}$

relative to $\mathfrak{L}$. Since $\mathfrak{M}_0$ is the Fitting null component of $\mathfrak{M}$ relative to $\mathfrak{L}$, $\mathfrak{N}_0 = 0$ so $\mathfrak{N} = \mathfrak{N}_1 \subseteq \mathfrak{M}_1$ by the definition of the 1-components. It follows from the direct decompositions that $\mathfrak{N} = \mathfrak{M}_1$. Let Π be an extension of the base field Φ of $\mathfrak{M}$ and $\mathfrak{L}$. Then $\mathfrak{L}_\Pi$ is a nilpotent Lie algebra of linear transformations in $\mathfrak{M}_\Pi$. Since $\mathfrak{M}_0(\mathfrak{L}^*)^n = 0$ and every element of $(\mathfrak{L}_\Pi)^*$ is a P-linear combination of the elements of $\mathfrak{L}^*$ it is clear that $\mathfrak{M}_{0\Pi}$ is the Fitting null component of $\mathfrak{M}_\Pi$ relative to $\mathfrak{L}_\Pi$. It is clear also that $\mathfrak{M}_{1\Pi}$ is the Fitting one component of $\mathfrak{M}_\Pi$ relative to $\mathfrak{L}_\Pi$.

Now let $\mathfrak{K}$ be a subalgebra of the Jordan algebra $\mathfrak{J}$. We let $\mathfrak{L}_{\mathfrak{J}}(\mathfrak{K})$ denote the Lie algebra of linear transformations in $\mathfrak{J}$ generated by the mappings $R_{b,c}$, $b, c \in \mathfrak{K}$. We have the following theorem.

THEOREM 5. *Let $\mathfrak{K}$ be an associator nilpotent subalgebra of a Jordan algebra $\mathfrak{J}$. Then the Lie algebra of linear transformations $\mathfrak{L}_{\mathfrak{J}}(\mathfrak{K})$ is nilpotent and if $\mathfrak{J} = \mathfrak{J}_0 \oplus \mathfrak{J}_1$ is the Fitting decomposition of $\mathfrak{J}$ relative to $\mathfrak{L}_{\mathfrak{J}}(\mathfrak{K})$ then $\mathfrak{J}_0$ is a subalgebra and $\mathfrak{J}_0 \cdot \mathfrak{J}_1 \subseteq \mathfrak{J}_1$. Moreover, if the base field is algebraically closed then $\mathfrak{J}_0 = \bigcap_{b\in\mathfrak{K}} \mathfrak{Z}_b$ where $\mathfrak{Z}_a$ is defined by* (11).

Proof. It is clear from our remarks that it is enough to prove the result assuming the base field Φ is algebraically closed. Then, by Theorem 2, $\mathfrak{K} = \sum \oplus \mathfrak{K}_i$ where $\mathfrak{K}_i = \Phi e_i + \mathfrak{N}_i$ is an ideal in $\mathfrak{K}$ and e_i is the identity of $\mathfrak{K}_i$. Then $\mathfrak{N} = \sum \mathfrak{N}_i$ is the radical of $\mathfrak{K}$. Hence R_z, $z \in \mathfrak{N}$, is in the radical of the enveloping associative algebra of the linear transformations R_b, $b \in \mathfrak{K}$, acting in $\mathfrak{J}$. It follows that $R_{z,b}$, $z \in \mathfrak{N}$, $b \in \mathfrak{K}$, is in this radical. Consequently, the mappings $R_{z,b}$, $z \in \mathfrak{N}$, $b \in \mathfrak{K}$, are in the radical $\mathfrak{S}$ of the enveloping associative algebra $\mathfrak{E}(\mathfrak{L}_{\mathfrak{J}}(\mathfrak{K}))$ of the Lie algebra $\mathfrak{L}_{\mathfrak{J}}(\mathfrak{K})$. Let $\mathfrak{J} = \sum_{i\leq j} \mathfrak{J}_{ij}$ be the Peirce decomposition of $\mathfrak{J}$ relative to the e_i $(1 = \sum e_i)$. Then $\mathfrak{K}_i \subseteq \mathfrak{J}_{ii}$ and every $\mathfrak{J}_{ij}$ is an invariant subspace of $\mathfrak{J}$ relative to the R_b, $b \in \mathfrak{K} = \sum \mathfrak{K}_i$. Hence to prove that $\mathfrak{L}_{\mathfrak{J}}(\mathfrak{K})$ is a nilpotent Lie algebra of linear transformations it suffices to show that for every i, j the restrictions $R^{ij}_{b,c}$ of $R_{b,c}$, b, c in $\mathfrak{K}$, generate a nilpotent Lie algebra of linear transformations acting in $\mathfrak{J}_{ij}$. Write $b = \sum_k (\beta_k e_k + z_k)$, $c = \sum_l (\gamma_k e_k + w_k)$, $\beta_k, \gamma_k \in \Phi$, $z_k, w_k \in \mathfrak{N}_k$. Then $R_{b,c} = \sum \beta_k \gamma_l R_{e_k, e_l} + S$ where $S \in \mathfrak{S}$. We have $R^{ij}_{b,c} = \sum \beta_k \gamma_l R^{ij}_{e_k, e_l} + S^{ij}$ where, in general, A^{ij} denotes the restriction of A to $\mathfrak{J}_{ij}$. One checks that $R^{ij}_{e_k, e_l}$ is a scalar. Hence $R^{ij}_{b,c} = \mu_{ij} 1^{ij} + S^{ij}$ where $\mu_{ij} \in \Phi$. Since the S_{ij} are contained in a nilpotent ideal of $\mathfrak{E}(\mathfrak{L}_{\mathfrak{J}}(\mathfrak{K}))^{ij}$ it is clear that the $R^{ij}_{b,c}$ generate a nilpotent Lie algebra of linear transformations acting in $\mathfrak{J}_{ij}$. This proves that $\mathfrak{L}_{\mathfrak{J}}(\mathfrak{K})$ is nilpotent. We now observe that $R^{ii}_{e_k, e_l}=0$ so $R^{ii}_{b,c}=S^{ii}\in\mathfrak{S}^{ii}$. Since S^{ii} is nilpotent it follows that $\mathfrak{J}_{ii}\subseteq\mathfrak{J}_0$ the Fitting null component of $\mathfrak{J}$ relative to $\mathfrak{L}_{\mathfrak{J}}(\mathfrak{K})$. Hence $\sum\mathfrak{J}_{ii}\subseteq\mathfrak{J}_0$. Since $\bigcap_{b\in\mathfrak{K}}\mathfrak{Z}_b\supseteq\mathfrak{J}_0$ we have $\bigcap_{b\in\mathfrak{K}}\mathfrak{Z}_b\supseteq\mathfrak{J}_0\supseteq\sum\mathfrak{J}_{ii}$. Next choose $b=\sum\beta_i e_i$ so that the β_i are distinct. Then $\mathfrak{J}=\sum_{i\leq j}\mathfrak{J}_{ij}$ is a decomposition of $\mathfrak{J}$ into $\mathfrak{L}_{\mathfrak{J}}(\Phi[b])$ invariant subspaces. Since $e_i\in\Phi[b]$ so that $R_{e_i,e_j}\in\mathfrak{L}_{\mathfrak{J}}(\Phi[b])$ and $\mathfrak{J}_{ij}R_{e_i,e_j}=\mathfrak{J}_{ij}$ if $i\neq j$, $\mathfrak{Z}_b\subseteq\sum\mathfrak{J}_{ii}$. Hence $\bigcap_{b\in\mathfrak{K}}\mathfrak{Z}_b\subseteq\sum\mathfrak{J}_{ii}$. This and the earlier inclusion imply that $\bigcap_{b\in\mathfrak{K}}\mathfrak{Z}_b=\mathfrak{J}_0=\sum\mathfrak{J}_{ii}$. Since $\sum_{i<j}\mathfrak{J}_{ij}$ is invariant under $\mathfrak{L}_{\mathfrak{J}}(\mathfrak{K})$ and $\mathfrak{J}=\mathfrak{J}_0\oplus\sum_{i<j}\mathfrak{J}_{ij}$, we see that $\sum_{i<j}\mathfrak{J}_{ij}$ is the Fitting one component of $\mathfrak{J}$ relative to $\mathfrak{L}_{\mathfrak{J}}(\mathfrak{K})$. Thus $\mathfrak{J}_0=\sum\mathfrak{J}_{ii}$ is a subalgebra and $\mathfrak{J}_0\cdot\mathfrak{J}_1\subseteq\mathfrak{J}_1$ follows from the properties of the Peirce decomposition.

5. Cartan subalgebras of Jordan algebras. We have noted (§2) that if $\mathfrak{J}$ is an associator nilpotent Jordan algebra then there exists an integer M such that $R_{a_1,b_1}R_{a_2,b_2}\cdots R_{a_M,b_M}=0$ for a_i, $b_i\in\mathfrak{J}$. It follows that $\mathfrak{L}_{\mathfrak{J}}(\mathfrak{J})$ is nilpotent (cf. Theorem 5) and $\mathfrak{J}$ coincides with the Fitting null component of $\mathfrak{J}$ relative to $\mathfrak{L}_{\mathfrak{J}}(\mathfrak{J})$. This implies that if $\mathfrak{K}$ is an associator nilpotent subalgebra of an arbitrary Jordan algebra $\mathfrak{J}$ then $\mathfrak{K}\subseteq\mathfrak{J}_0$ the Fitting null component of $\mathfrak{J}$ relative to the Lie algebra $\mathfrak{L}_{\mathfrak{J}}(\mathfrak{K})$. We now give the following

Definitions. A *Cartan subalgebra* of a Jordan algebra $\mathfrak{J}$ is an associator nilpotent subalgebra $\mathfrak{K}$ of $\mathfrak{J}$ such that $\mathfrak{K}=\mathfrak{J}_0$ the Fitting null component of $\mathfrak{J}$ relative to $\mathfrak{L}_{\mathfrak{J}}(\mathfrak{K})$. An element $a\in\mathfrak{J}$ is *associator regular* in $\mathfrak{J}$ if $\dim\mathfrak{Z}_a$ is minimal.

The proofs of Theorem 5 and Theorem 2 show that if the base field is algebraically closed then $\mathfrak{K}$ is a Cartan subalgebra of $\mathfrak{J}$ if and only if $\mathfrak{K}=\sum\mathfrak{J}_{ii}$ where $\mathfrak{J}=\sum_{i\leq j}\mathfrak{J}_{ij}$ is the Peirce decomposition of $\mathfrak{J}$ relative to a set of non-zero primitive orthogonal idempotents e_i such that $\sum e_i=1$. Hence the existence of Cartan subalgebras is clear in the algebraically closed case. We shall need a stronger result on the imbedding of associator regular elements in Cartan subalgebras.

Let $(u_1, u_2, \ldots, u_n)$ be a basis for $\mathfrak{J}/\Phi$, $\Xi=\Phi(\xi_1, \xi_2, \ldots, \xi_n)$ the field of rational expressions in n indeterminates ξ_i. The element $x=\sum\xi_i u_i$ of $\mathfrak{J}_\Xi$ is called a generic element of $\mathfrak{J}$. Let $M^{ij}(x)$ denote the $n\times n$ matrix of the linear transformation R_{x^i,x^j} in $\mathfrak{J}_\Xi$ relative to the basis $(u_1, u_2, \ldots, u_n)$. The entries of $M^{ij}(x)$ are polynomials in the ξ's and the same is true of the matrix $M(x)$ of n columns which is obtained by writing the $M^{ij}(x)$ in a single column of $n\times n$ matrices in some order. Let r be the rank of $M(x)$ and set $l=n-r$. Then l is the dimensionality of the subspace $\mathfrak{Z}_x$ in $\mathfrak{J}_\Xi$ (cf. (11)) and it is clear that $l\geq 1$ since $1\in\mathfrak{Z}_x$. Thus $r<n$. If $a\in\mathfrak{J}$, $a=\sum\alpha_i u_i$, $\alpha_i\in\Phi$, and the rank r' of the matrix $M(a)$ obtained by the specialization $\xi_i\to\alpha_i$ in $M(x)$ does not exceed r. Hence $\dim\mathfrak{Z}_a=n-r'\geq l$. If Φ is infinite we can choose the α's so that $r'=r$, $\dim\mathfrak{Z}_a=l$. Hence in this case a is associator regular if and only if $\dim\mathfrak{Z}_a=l$. Also we see that these elements form a Zariski open subset of $\mathfrak{J}$. We can now prove

THEOREM 6. *If a is an associator regular element of a Jordan algebra $\mathfrak{J}$ over an infinite field Φ then $\mathfrak{Z}_a$ is a Cartan subalgebra of $\mathfrak{J}$ (containing a).*

Proof. Let a be an associator regular element of $\mathfrak{J}$ and consider the associative, hence associator nilpotent sub-algebra $\Phi[a]$ of $\mathfrak{J}$. It is clear that the Fitting null component of $\mathfrak{J}$ relative to $\mathfrak{L}_{\mathfrak{J}}(\Phi[a])$ coincides with the subspace $\mathfrak{Z}_a$. Hence we have a Fitting decomposition $\mathfrak{J}=\mathfrak{Z}_a\oplus\mathfrak{J}_1$ where $\mathfrak{J}_1$ is the Fitting one component of $\mathfrak{J}$ relative to $\mathfrak{L}_{\mathfrak{J}}(\Phi[a])$. Suppose the basis $(u_1, u_2, \ldots, u_n)$ for $\mathfrak{J}$ is chosen so that $(u_1, u_2, \ldots, u_l)$ is a basis for $\mathfrak{Z}_a$ and $(u_{l+1}, \ldots, u_n)$ is a basis for $\mathfrak{J}_1$. Since $\mathfrak{Z}_a$ is a subalgebra of $\mathfrak{J}$ and $\mathfrak{J}_1.\mathfrak{Z}_a\subseteq\mathfrak{J}_1$ the matrices $M^{ij}(b)$, $b\in\mathfrak{Z}_a$, have the form

$$(12)\qquad \begin{pmatrix} N^{ij}(b) & 0 \\ 0 & P^{ij}(b)\end{pmatrix}$$

where $N^{ij}(b)$ is the matrix of the restriction of $R_{b\cdot i, b\cdot j}{}^{n}$ to $\mathfrak{Z}_a$ and $P^{ij}(b)$ is the matrix of the restriction of $R_{b\cdot i, b\cdot j}{}^{n}$ to $\mathfrak{J}_1$ relative to the indicated bases. We have $N^{ij}(a)=0$ and the rank of the matrix $P(a)$ which is a column of the matrices $P^{ij}(a)$ is $n-l$. We shall now show that every $N^{ij}(b)=0$, $b\in\mathfrak{Z}_a$. Suppose this is not the case and let b be an element of $\mathfrak{Z}_a$ such that one of the matrices $N^{ij}(b)$ has a non-zero entry ρ. Consider the elements $\xi a+\eta b$ in $\mathfrak{J}_{\Phi(\xi,\eta)}$, ξ, η indeterminates. Choose a non-zero minor of order $n-l$ in $P(a)$ and consider the same minor for $\xi a+\eta b$. Its value is a polynomial $f_1(\xi, \eta)$ such that $f_1(1, 0)\neq 0$. Also the entry in the column of matrices $N^{ij}(\xi a+\eta b)$ which is in the same position as ρ is a polynomial $f_2(\xi, \eta)$ such that $f_2(0, 1)=\rho\neq 0$. Hence $f_1(\xi, \eta)f_2(\xi, \eta)\neq 0$ and we can choose $\alpha, \beta\in\Phi$ such that $f_1(\alpha, \beta)f_2(\alpha, \beta)\neq 0$. Then it is clear that if $c=\alpha a+\beta b$ then the rank of $M(c)$

exceeds $n-l$, contrary to the regularity of a. Hence we have proved that $N^{ij}(b)=0$ for every $b\in \mathfrak{Z}_a$ and every $i, j=0, 1, 2, \ldots$. It follows from Theorem 4 that $\mathfrak{Z}_a$ is an associator nilpotent subalgebra of $\mathfrak{J}$. Since $a\in \mathfrak{Z}_a$ it is clear that the Fitting null component of $\mathfrak{J}$ relative to $\mathfrak{L}_{\mathfrak{J}}(\mathfrak{Z}_a)$ which is $\bigcap_{b\in \mathfrak{Z}_a} \mathfrak{Z}_b$ is contained in $\mathfrak{Z}_a$. Since the reverse inequality is general we see that $\mathfrak{Z}_a$ coincides with the Fitting null component relative to $\mathfrak{L}_{\mathfrak{J}}(\mathfrak{Z}_a)$. Hence $\mathfrak{Z}_a$ is a Cartan subalgebra.

6. Conjugacy of Cartan subalgebras. We recall that if a, b are in a Jordan algebra $\mathfrak{J}$ then $D_{a,b}=[R_aR_b]$ is a derivation (Jacobson [1], p. 867). Hence the linear mappings of the form $\sum D_{a_i,b_i}$ are derivations. We call such derivations inner. These constitute an ideal $\mathfrak{D}'$ in the derivation algebra $\mathfrak{D}(\mathfrak{J})$. We recall also that if the base field is of characteristic zero and $\mathfrak{L}$ is any Lie algebra of linear transformations in a finite dimensional vector space then the Lie algebra $\tilde{\mathfrak{L}}$ of the intersection G of all the algebraic groups of linear transformations whose Lie algebras contain $\mathfrak{L}$ satisfies $\tilde{\mathfrak{L}}\supseteq\mathfrak{L}$ (Chevalley [1] II, pp. 158-169, Hochschild [1]). Evidently G is an irreducible algebraic group. In particular, let I be the algebraic group determined in this way by the Lie algebra $\mathfrak{D}'$ of inner derivations of $\mathfrak{J}$. Since the group A of automorphisms of $\mathfrak{J}$ is an algebraic group whose Lie algebra is $\mathfrak{D}(\mathfrak{J})$ (Chevalley [1], II, p. 179) it is clear that $I\subseteq A$. Hence the elements of I are automorphisms. We shall call these *the inner automorphisms* of the Jordan algebra $\mathfrak{J}$ (for characteristic 0 only). We shall now prove the following analogue of classical conjugacy theorem for Cartan subalgebras of Lie algebras.

THEOREM 7. *Let $\mathfrak{J}$ be a finite dimensional Jordan algebra (with 1) over an algebraically closed field of characteristic 0. Then if $\mathfrak{R}_1$ and $\mathfrak{R}_2$ are Cartan subalgebras of $\mathfrak{J}$ there exists an inner automorphism s of $\mathfrak{J}$ such that $\mathfrak{R}_1^s=\mathfrak{R}_2$.*

Proof. Our proof will be patterned after Chevalley's proof of the Lie algebra result (Chevalley [1] III, pp. 215-219). We show first that the orbit Ω_1 of $\mathfrak{R}_1$ under I contains a Zariski open subset of $\mathfrak{J}$. We note first that since $\mathfrak{R}_1$ and I are irreducible, Ω_1 is épais, that is, it is irreducible and contains a non-vacuous open subset of its Zariski closure (Prop. 3, p. 193 of Chevalley [1], III). Hence our assertion will follow by showing that the dimensionality of the irreducible set Ω_1 is $n=\dim \mathfrak{J}$. We shall do this by showing that there exists an open subset of $\mathfrak{K}_1$ such that the tangent space $\mathfrak{T}$ to Ω_1 at any a in this set is $\mathfrak{J}$. It is known that the tangent space $\mathfrak{T}$ to Ω_1 at any $a \in \Omega_1$ contains $\mathfrak{K}_1 + a\mathfrak{D}'$ where $a\mathfrak{D}'$ is the set of images of the element a under the elements $D \in \mathfrak{D}'$ (Chevalley [1], III Prop. 2, p. 192). Now, we have a set of orthogonal idempotents $e_1, e_2, \ldots, e_r$ such that if $\mathfrak{J} = \sum_{i \leq j} \mathfrak{J}_{ij}$ is the corresponding Peirce decomposition, then $\mathfrak{K}_1 = \sum \mathfrak{J}_{ii}$. Also $\mathfrak{J}_{ii} = \Phi e_i + \mathfrak{N}_i$ where $\mathfrak{N}_i$ is a nil ideal in $\mathfrak{J}_{ii}$. Let 0_1 be the open subset of $\mathfrak{K}_1$ of elements $a = \sum_1^r \alpha_i e_i + z_i$, $\alpha_i \in \Phi$, $z_i \in \mathfrak{N}_i$, such that $\prod_{i \neq j} (\alpha_i - \alpha_j) \neq 0$. Let $a \in 0_1$ and consider $a[R_{e_k} R_{a_{kl}}]$ where $k \neq l$ and $a_{kl} \in \mathfrak{J}_{kl}$. We have

$$
\begin{aligned}
a[R_{e_k} R_{a_{kl}}] &= (\sum \alpha_i e_i + z_i)[R_{e_k} R_{a_{kl}}] \\
&= (\alpha_k e_k + z_k)[R_{e_k} R_{a_{kl}}] + (\alpha_l e_l + z_l)[R_{e_k} R_{a_{kl}}] \qquad (13) \\
&= \frac{1}{4}(\alpha_k - \alpha_l) a_{kl} + \frac{1}{2}(z_k - z_l) . a_{kl}.
\end{aligned}
$$

Hence the tangent space $\mathfrak{T}$ to Ω_1 at a contains $(\alpha_k - \alpha_l)a_{kl} + 2(z_k - z_l).a_{kl} = a_{kl}S_{kl}$ where S_{kl} is the linear operator $(\alpha_k - \alpha_l)1 + R_{2(z_k - z_l)}$. It is clear from the multiplication table (4) for Peirce spaces that S_{kl} maps $\mathfrak{J}_{kl}$ into itself. Also $z_k - z_l$ is nilpotent, hence $R_{2(z_k - z_l)}$ is nilpotent (Albert [2], p. 550). Since $\alpha_k - \alpha_l \neq 0$ it is clear that S_{kl} has an inverse. Hence S_{kl} maps $\mathfrak{J}_{kl}$ onto itself and consequently $\mathfrak{T}$ contains $\mathfrak{J}_{kl}$. It now follows that $\mathfrak{T} = \mathfrak{J}$. Since Ω_1 is the orbit of $\mathfrak{K}_1$ it is clear that $\mathfrak{K}_1$ contains a simple point of Ω_1. Since $\mathfrak{K}_1$ is irreducible the open subset 0_1 contains a simple point. Hence dim $\Omega_1 = n$ and Ω_1 contains a non-vacuous open subset of $\mathfrak{J}$. In the same manner we have that the orbit Ω_2 of the Cartan subalgebra $\mathfrak{K}_2$ contains a non-vacuous open set. Since the set of associator regular elements is Zariski dense we see that $\Omega_1 \cap \Omega_2$ contains an associator regular element b. Hence there exist inner automorphisms s_1, s_2 such that $b \in \mathfrak{K}_i^{s_i}$. Then $\mathfrak{Z}_b \supseteq \mathfrak{K}_i^{s_i}$ and since $\mathfrak{Z}_b$ and $\mathfrak{K}_i^{s_i}$ are Cartan subalgebras, $\mathfrak{Z}_b = \mathfrak{K}_1^{s_1} = \mathfrak{K}_2^{s_2}$. Hence $\mathfrak{K}_2 = \mathfrak{K}_1^{s}$ where $s = s_1 s_2^{-1} \in I$.

7. Applications to generic traces. Separability criterion. As in § 5, let $x = \sum_1^n \xi_i u_i$ be a generic element of $\mathfrak{J}$ over Φ where $(u_1, u_2, \ldots, u_n)$ is a basis and the ξ's are indeterminates. Then one knows that the minimum polynomial $m_x(\lambda)$ of x in $\mathfrak{J}_{\Xi}$, $\Xi = \Phi(\xi_1, \ldots, \xi_n)$, has the form $\lambda^m - \sigma_1(\xi)\lambda^{m-1} + \cdots + (-1)^m \sigma_m(\xi)$ where $\sigma_j(\xi) \equiv \sigma_j(\xi_1, \ldots, \xi_n)$ is a homogeneous polynomial of degree j in the ξ's. If $a = \sum \alpha_i u_i \in \mathfrak{J}$ where the $\alpha_i \in \Phi$ then the specialization $\xi_i \to \alpha_i$ gives a polynomial $m_a(\lambda) = \lambda^m - \sigma_1(\alpha)\lambda^{m-1} + \cdots$ such that $m_a(a) = 0$. The polynomial $m_a(\lambda)$ is called the generic minimum polynomial of a, $T(a) \equiv \sigma_1(\alpha)$ and $N(a) \equiv \sigma_m(\alpha)$ are the generic trace and norm respectively of a. The degree m of $m_x(\lambda)$ is called the degree of the algebra. The mapping $a \rightarrow T(a)$ is linear and the bilinear form $T(a, b) = T(a.b)$, which is called the *generic trace form*, is symmetric and associative in the sense that $T(a.b, c) = T(a, b.c)$ (Tits [1], p. 35). The generic minimum polynomial (hence the trace and norm) are unchanged under extension of the base field. If Φ is infinite then the set of elements a such that the generic minimum polynomial $m_a(\lambda)$ coincides with the minimum polynomial $\mu_a(\lambda)$ of a is a non-vacuous Zariski open subset of $\mathfrak{J}$. The same is true of the subset of associator regular elements. Hence Theorem 6 implies that for infinite base fields there exist Cartan sub-algebras satisfying the hypotheses of the following

THEOREM 8. *Let $\mathfrak{J}$ be a Jordan algebra over an algebraically closed field Φ and let $\mathfrak{K}$ be a Cartan subalgebra of $\mathfrak{J}$ which contains an element which is associator regular and has generic minimum polynomial equal to its minimum polynomial. Suppose $\mathfrak{K} = \sum \oplus \mathfrak{K}_i$ where $\mathfrak{K}_i$ is an ideal in $\mathfrak{K}$ of the form $\Phi e_i + \mathfrak{N}_i$ where e_i is the identity element of $\mathfrak{K}_i$ and $\mathfrak{N}_i$ is a nil ideal in $\mathfrak{K}_i$. Let $\mathfrak{J} = \mathfrak{K} \oplus \mathfrak{J}_1$ be the Fitting decomposition of $\mathfrak{J}$ relative to $\mathfrak{L}_{\mathfrak{J}}(\mathfrak{K})$. If $a \in \mathfrak{J}$ we write $a = \sum_1^r (\alpha_i e_i + z_i) + a_1$ where $\alpha_i \in \Phi$, $z_i \in \mathfrak{N}_i$ and $a_1 \in \mathfrak{J}_1$. Then the generic trace*

$$T(a) = \sum n_i \alpha_i \tag{14}$$

where n_i is the maximum index of nilpotency of the elements of $\mathfrak{N}_i$. Also the degree of $\mathfrak{J}$ is $\sum n_i$.

Proof. If $b \in \mathfrak{K}$ let $m_{b,\mathfrak{K}}(\lambda)$ be the generic minimum polynomial of b as element of $\mathfrak{K}$. Then $m_{b,\mathfrak{K}}(\lambda)$ is a factor of $m_b(\lambda)$, and since $\mathfrak{K}$ contains elements c such that $m_c(\lambda) = \mu_c(\lambda)$, it follows that $m_{b,\mathfrak{K}}(\lambda) = m_b(\lambda)$. If $b = \sum(\alpha_i e_i + z_i)$ and the minimum polynomial of z_i is λ^{m_i} then it is clear that $\mu_b(\lambda) =$ L.C.M. $(\lambda - \alpha_i)^{m_i}$. If we introduce a generic element for $\mathfrak{K}$ relative to an appropriate basis it follows easily that $m_{b,\mathfrak{K}}(\lambda) = \Pi(\lambda - \alpha_i)^{n_i}$ where n_i is the maximum index of nilpotency of the elements of $\mathfrak{N}_i$. Hence the generic trace of b relative to $\mathfrak{K}$, $T_{\mathfrak{K}}(b) = \sum n_i \alpha_i$. Since $m_{b,\mathfrak{K}}(\lambda) = m_b(\lambda)$ we have $T(b) = \sum n_i \alpha_i$. Next we recall that $\mathfrak{J}_1 = \sum_{i<j} \mathfrak{J}_{ij}$ where $\mathfrak{J} = \sum_{i \leq j} \mathfrak{J}_{ij}$ is the Peirce decomposition of $\mathfrak{J}$ relative to the e_i. Also we have seen that if $a_{ij} \in \mathfrak{J}_{ij}$, $i \neq j$, then a_{ij} is an associator. Hence $T(a_{ij}) = 0$ since $T(a, b) = T(a.b)$ is an associative form. It follows that

if $a=\sum_1^r(\alpha_i e_i+z_i)+a_1$, a_1 in $\mathfrak{J}_1$, then $T(a_1)=0$. Then $T(a)=T(b)$, $b=\sum(\alpha_i e_i+z_i)$ and so $T(a)=\sum n_i\alpha_i$.

If the base field is of characteristic 0 then the conjugacy theorem implies that every Cartan subalgebra $\mathfrak{K}$ of a Jordan over a field of characteristic 0 satisfies the hypothesis of Theorem 8. We conjecture that this holds also for Jordan algebras over infinite fields of characteristic $p \neq 0$.

We recall that a Jordan algebra $\mathfrak{J}$ is called separable if $\mathfrak{J}_{\Pi}$ is semi-simple for every extension field Π of the base field Φ of $\mathfrak{J}$. It is known that $\mathfrak{J}$ is separable if and only if the generic trace bilinear form $T(a, b)$ is non-degenerate (Jacobson [6], p. 41). The proof of the sufficiency of this condition is a general one. On the other hand, the known proof of the necessity is based on the classification of simple Jordan algebras over an algebraically closed field and consists of a case by case verification for these. We shall now give a general proof of this half of the theorem based on the formula of Theorem 8 and on a notion of a reduced Jordan algebra and reduced trace which we now define.

We shall call a Jordan algebra $\mathfrak{J}$ *reduced* if $\mathfrak{J}$ contains a Cartan subalgebra $\mathfrak{K}$ which is a direct sum of almost nil ideals. Then $\mathfrak{K}=\sum_1^r \oplus \mathfrak{K}_i$ where $\mathfrak{K}_i$ is an ideal of the form $\Phi e_i+\mathfrak{N}_i$, $e_i^2=e_i$, $\mathfrak{N}_i$ is a nil ideal in $\mathfrak{K}_i$. If $\mathfrak{J}=\mathfrak{K}\oplus\mathfrak{J}_1$ is the Fitting decomposition of $\mathfrak{J}$ relative to $\mathfrak{L}_{\mathfrak{J}}(\mathfrak{K})$ and $a=\sum_1^r(\alpha_i e_i+z_i)+a_1, z_i \in \mathfrak{N}_i$, $a_1\in\mathfrak{J}_1$, then we define the *reduced trace* $t(a)=\sum_1^r \alpha_i$. Clearly, $a\to t(a)$ is a linear function on $\mathfrak{J}$. We wish to show that t vanishes on all associators, or, equivalently, $t(a, b)=t(a.b)$ is an associative bilinear form on $\mathfrak{J}$. As in the algebraically closed case (proof of Theorem 5), $\mathfrak{J}_1=\sum_{i<j}\mathfrak{J}_{ij}$ where $\mathfrak{J}=\sum_{i\leq j}\mathfrak{J}_{ij}$ is the Peirce decomposition of $\mathfrak{J}$ relative to the e_i.

LEMMA. *If $a_{ij}\in\mathfrak{J}_{ij}$, $i\neq j$, then $a_{ij}^{.2}=\alpha(e_i+e_j)+z_i+z_j$ where $\alpha\in\Phi$, $z_k\in\mathfrak{N}_k$.*

Proof. We have $a_{ij}^{.2}\in\mathfrak{J}_{ii}+\mathfrak{J}_{jj}$ so $a_{ij}^{.2}=\alpha e_i+\beta e_j+z_i+z_j$, α, $\beta\in\Phi$, $z_k\in\mathfrak{N}_k$. Then $(a_{ij}^{.2}.e_i).a_{ij}=a_{ij}^{.2}.(e_i.a_{ij})$ gives

$$(\alpha e_i+z_i).a_{ij}=\frac{1}{2}(\alpha e_i+\beta e_j+z_i+z_j).a_{ij}.$$

Hence $(\alpha e_i-\beta e_j).a_{ij}=(z_j-z_i).a_{ij}$ so $2(z_j-z_i).a_{ij}=(\alpha-\beta)a_{ij}$. Since z_j-z_i is nilpotent, $R_{z_j-z_i}$ is nilpotent. Hence the last equation gives $(\alpha-\beta)^n a_{ij}=0$ for some positive integer n. Then $\alpha = \beta$ since this holds also if $a_{ij} = 0$. Thus $a_{ij}^{2} = \alpha(e_i + e_j) + z_i + z_j$.

We can now prove

THEOREM 9. *The reduced trace from $t(a, b) = t(a.b)$ is associative.*[5)]

Proof. We have to show that $t([a, b, c]) = 0$ for all $a, b, c \in \mathfrak{J}$. It is enough to prove this for a, b, c in the Peirce spaces $\mathfrak{J}_{ij}$ relative to the e_i. Now $t(a_{ij}) = 0$ if $a_{ij} \in \mathfrak{J}_{ij}$, $i \neq j$, and the product of three elements of Peirce spaces is 0 or is in a $\mathfrak{J}_{ij}$, $i \neq j$, except in the following cases: I. the three elements are in an a_{ii}, II. two of the elements are in a $\mathfrak{J}_{ij}$, $i \neq j$, and the third is in $\mathfrak{J}_{ii}$, III. one element is in $\mathfrak{J}_{ij}$, the second in $\mathfrak{J}_{jk}$ and the third in $\mathfrak{J}_{ik}$ where i, j, k are unequal. We note also that in any commutative algebra one has the "Jacobi identity": $[a, b, c] + [b, c, a] + [c, a, b] = 0$ for associators. Hence it suffices to verify that $t([a, b, c)] = 0$ in the following cases:

$$\begin{array}{ll} \text{I.} & t([a_{ii}, b_{ii}, c_{ii}]) = 0, \quad a_{ii}, b_{ii}, c_{ii} \in \mathfrak{J}_{ii} \\ \text{II}'. & t([a_{ij}, b_{ij}, c_{jj}]) = 0 \quad a_{ij}, b_{ij} \in \mathfrak{J}_{ij}, \quad c_{jj} \in \mathfrak{J}_{jj} \\ \text{II}''. & t([a_{ij}, c_{jj}, b_{ij}]) = 0, \quad a_{ij}, b_{ij} \in \mathfrak{J}_{ij}, \quad c_{jj} \in \mathfrak{J}_{jj} \\ \text{III.} & t([a_{ij}, b_{jk}, c_{ki}]) = 0, \quad a_{ij} \in \mathfrak{J}_{ij}, \ b_{jk} \in \mathfrak{J}_{jk}, \ c_{ik} \in \mathfrak{J}_{ik}, \quad i, j, k \neq . \end{array}$$

The first of these is clear since $[a_{ii}, b_{ii}, c_{ii}] \in \mathfrak{N}_i$ for $a_{ii} = \alpha e_i + z_i$, $b_{ii} = \beta e_i + z_i'$, $c_{ii} = \gamma e_i + z_i''$, $z_i, z_i', z_i'' \in \mathfrak{N}_i$. For II′ we note that we have the relation

$$[a_{ij}, b_{ij}, c_{jj}] = (e_j - e_i).((a_{ij}.c_{jj}).b_{ij}) \tag{15}$$

which is obtained from (2) by taking $x = c_{jj}$, $a = a_{ij}$, $b = b_{ij}$, $c = e_j$. By Lemma 1, $(a_{ij}.c_{jj}).b_{ij} = \alpha(e_i + e_j) + z_i + z_j$, $z_k \in \mathfrak{N}_{\mathfrak{K}}$. Multiplication by $e_j - e_i$ gives $\alpha(e_j - e_i) + z_j - z_i$. The reduced trace of this element is 0. Hence $t([a_{ij}, b_{ij}, c_{jj}]) = 0$. For II″ we use the relation

$$[a_{ij}, c_{jj}, b_{ij}].e_i = 0 \tag{16}$$

which is obtained from (15) by multiplying by e_i. By Lemma 1, $[a_{ij}, c_{jj}, b_{ij}] = \alpha(e_i + e_j) + z_i + z_j$, $z_k \in \mathfrak{N}_k$. This and the last equation give $\alpha = 0$. Hence $t([a_{ij}, c_{jj}, b_{ij}]) = 0$. For III we note that we have

$$[a_{ij}, b_{jk}, c_{ki}] = (e_k - e_j).((a_{ij}.c_{ki}).b_{jk}) \tag{17}$$

5) Cf. Albert [3], p. 522.

which is obtained by setting $x=a_{ij}$, $a=c_{ki}$, $b=b_{jk}$, $d=e_k$ in (2). By the lemma, $(a_{ij}.c_{ki}).b_{jk}=\alpha(e_j+e_k)+z_j+z_k$, $z_l\in\mathfrak{N}_l$. Then the result follows as for II′.

If $f(a,b)$ is any symmetric associative bilinear form on a Jordan algebra $\mathfrak{J}$ then it is clear that the radical $\mathfrak{J}^\perp$ of f is an ideal. In particular, this holds if $\mathfrak{J}$ is reduced and $f=t$ the reduced trace form. It is clear from the definition of t that $\sum\mathfrak{N}_i\subseteq\mathfrak{J}^\perp$ and that $t(e_i,e_i)=t(e_i)=1$ which shows that $t\neq 0$ and $\mathfrak{J}^\perp\neq\mathfrak{J}$. Now suppose $\mathfrak{J}$ is simple over an algebraically closed field. Let $\mathfrak{H}$ be a Cartan subalgebra of $\mathfrak{J}$ satisfying the hypotheses of Theorem 8. Then the formula (14) for the generic trace $T(a)$ is valid. Also $\mathfrak{H}$ is a direct sum of almost nil ideals so we have the reduced trace $t(a)$ defined by the Fitting decomposition relative to $\mathfrak{L}_{\mathfrak{J}}(\mathfrak{H})$. The relation $\mathfrak{J}^\perp\neq\mathfrak{J}$ for the radical $\mathfrak{J}^\perp$ relative to t and the simplicity of $\mathfrak{J}$ imply that $\mathfrak{J}^\perp=0$. Hence t is non-degenerate. Also $\sum\mathfrak{N}_i=0$ since $\sum\mathfrak{N}_i\subseteq\mathfrak{J}^\perp$. Hence (14) becomes $T(a)=\sum\alpha_i$ which coincides with the definition of $t(a)$. Hence we see that the generic trace form T of a simple Jordan algebra over an algebraically closed field is non-degenerate. We can now prove

THEOREM 10. *A Jordan algebra is separable if and only if the generic trace bilinear form T where $T(a,b)=T(a.b)$ is non-degenerate.*

Proof. We omit the proof of sufficiency since a general proof of this is given in Jacobson [6], p. 41. Also, the argument given in this reference shows that to prove the necessity it is enough to suppose $\mathfrak{J}$ is simple over an algebraically closed field. Hence the proof is complete by what we have just shown.

BIBLIOGRAPHY

A. A. Albert

[1] On Jordan algebras of linear transformations, Trans. Amer. Math. Soc., **59** (1946), 524-555.

[2] A structure theory for Jordan algebras, Ann. Math., **48** (1947), 546-567.

[3] A theory of power associative commutative algebras, Trans. Amer. Math. Soc., **69** (1950), 503-527.

C. Chevalley

[1] Théorie des groupes de Lie II (1951), III (1955), Act. Sci., Paris.

G. P. Hochschild

[1] On the algebraic hull of a Lie algebra, Proc. Amer. Math. Soc., **19** (1960), 195-200.

N. Jacobson

[1] Derivation algebras and multiplication algebras of semi simple Jordan algebras, Annals of Math., **50** (1949), 866-874.

[2] General representation theory of Jordan algebras, Trans. Amer. Math. Soc., **70** (1951), 509-530.

[3] Structure of alternative and Jordan bimodules, Osaka Math. Jour., **6** (1954), 1-71.

[4] A theorem on the structure of Jordan algebras, Proc. Nat. Acad. Sci., **42** (1956), 140-147.

[5] Some groups of transformations defined by Jordan algebras I, Jour. für reine und angew. Math., **201** (1959), 178-195.

[6] Generic norm of an algebra, Osaka Math. Jour., **15** (1963), 25-50.

K. McCrimmon

[1] Jordan algebras of degree 1, Bull. Amer. Math. Soc., **70** (1964), p. 702.

A. J. Penico

[1] The Wedderburn principal theorem for Jordan algebras. Trans. Amer. Math., Soc., **70** (1951), 404-420.

J. Tits

[1] A theorem on generic norms of strictly power associative algebras, Proc. Amer. Math. Soc., **15** (1964), 35-36.

Yale University

Reprinted from
Nagoya Mathematical Journal, July 1966.

Reprinted from the PROCEEDINGS OF THE NATIONAL ACADEMY OF SCIENCES
Vol. 55, No. 2, pp. 243–251. February, 1966.

STRUCTURE THEORY FOR A CLASS OF JORDAN ALGEBRAS

BY N. JACOBSON

DEPARTMENT OF MATHEMATICS, YALE UNIVERSITY

Communicated December 10, 1965

In this note, we shall give a structure theory for a class of Jordan algebras which corresponds rather closely to the class of semisimple Artinian rings in the associative theory. Throughout the paper, "algebra" will mean algebra over a field Φ of characteristic not two, which is not necessarily associative or of finite dimensionality. For Jordan algebras we write $a \cdot b$ for the product, so the defining identities are: $a \cdot b = b \cdot a$, $(a^{\cdot 2} \cdot b) \cdot a = a^{\cdot 2} \cdot (b \cdot a)$, where $a^{\cdot 2} = a \cdot a$. If $\mathfrak{A}$ is an arbitrary algebra with product ab, then we define $a \cdot b = {}^1/_2(ab + ba)$ and denote the algebra with this product as $\mathfrak{A}^+$. If $\mathfrak{A}$ is associative, then $\mathfrak{A}^+$ is Jordan. Such Jordan algebras and their subalgebras are called special. An important composition in Jordan algebras is the trilinear product $\{abc\} \equiv (a \cdot b) \cdot c + (b \cdot c) \cdot a - (a \cdot c) \cdot b$. We write U_a for the linear mapping $x \rightarrow \{axa\}$, so $U_a = 2R_a{}^2 - R_{a^{\cdot 2}}$, where R_a is $x \rightarrow x \cdot a$. We recall the following important identities:

$$\{aba\}^{\cdot 2} = \{a\{ba^{\cdot 2}b\}a\}. \tag{1}$$

$$U_b{}^n = U_{b^{\cdot n}}, \qquad n = 1,2,3,\ldots. \tag{2}$$

$$U_a U_b U_a = U_{bU_a}{}^{\cdot 1} \tag{3}$$

If $U_{a,b}$ denotes the mapping $x \rightarrow \{axb\} = \{bxa\}$, then $U_{a+b} = U_a + U_b + 2U_{a,b}$. If e is an idempotent in the Jordan algebra $\mathfrak{J}$, then it is easily seen that the range $\mathfrak{J}U_e$ of U_e coincides with the Peirce space $\mathfrak{J}_1(e) = \{x | x \cdot e = x\}$. We recall also that if $\mathfrak{J}$ has an identity element 1, then the element $a \in \mathfrak{J}$ is invertible with inverse b if $a \cdot b = 1$, $a^{\cdot 2} \cdot b = a$.[2] This relation is symmetric and b is unique. Also, a is invertible if and only if $1 \in \mathfrak{J}U_a$ and if a is invertible with inverse b, then $U_a U_b = 1 =$

U_bU_a. If a is invertible, so are all of its powers and the criterion for invertibility in terms of U_a shows that if $a^{\cdot n}$, n a positive integer, is invertible, then a is invertible. A Jordan algebra $\mathfrak{J}$ is called a division algebra if $\mathfrak{J}$ has a $1 \neq 0$ and every $a \neq 0$ in $\mathfrak{J}$ is invertible.

1. *Quadratic Ideals.*—We shall base our structure theory on a notion which has been introduced by Topping[3] for Jordan algebras of operators in Hilbert space and which is defined for arbitrary Jordan algebras as follows.

Definition: A subset $\mathfrak{B}$ of a Jordan algebra $\mathfrak{J}$ is called a *quadratic ideal* in $\mathfrak{J}$ if $\mathfrak{B}$ is a subspace and $\mathfrak{B} \supseteq \mathfrak{J}U_b$ for every $b \in \mathfrak{B}$.

It is clear that a subspace is a quadratic ideal if and only if $\{b_1ab_2\} \in \mathfrak{B}$ for all $b_i \in \mathfrak{B}$, $a \in \mathfrak{J}$. In particular, if $\mathfrak{B}$ is a quadratic ideal and $b_1, b_2, b_3 \in \mathfrak{B}$, then $\{b_1b_2b_3\} \in \mathfrak{B}$. Hence, $(b_1 \cdot b_2) \cdot b_3 = {}^1/_2\{b_1b_2b_3\} + {}^1/_2\{b_2b_1b_3\} \in \mathfrak{B}$. If $\mathfrak{J}$ has an identity element, then $\mathfrak{B}$ is a subalgebra of $\mathfrak{J}$. Clearly, any ideal in $\mathfrak{J}$ is a Jordan ideal. If $\mathfrak{J} = \mathfrak{A}^+$ where $\mathfrak{A}$ is associative, then any left or right ideal of $\mathfrak{A}$ is a quadratic ideal of $\mathfrak{A}^+$, since $aU_b = bab$. If $\mathfrak{B}$ is a quadratic ideal in $\mathfrak{J}$ and $a \in \mathfrak{J}$, then $\mathfrak{B}U_a$ is a quadratic ideal since if $b \in \mathfrak{B}$, then $\mathfrak{J}U_{bU_a} = \mathfrak{J}U_aU_bU_a \subseteq \mathfrak{J}U_bU_a \subseteq \mathfrak{B}U_a$. In particular, $\mathfrak{J}U_a$ is a quadratic ideal for any $a \in \mathfrak{J}$. We shall call this the *principal quadratic ideal determined by a*. If e is an idempotent, then $\mathfrak{J}U_e$ is the Peirce space $\mathfrak{J}_1(e)$ and this is a subalgebra of $\mathfrak{J}$ with e as identity element. Moreover, if $\mathfrak{B}$ is a quadratic ideal of $\mathfrak{J}_1(e)$, then it is a quadratic ideal of $\mathfrak{J}$ since for $b \in \mathfrak{B}$, $b = bU_e$, and $\mathfrak{J}U_b = \mathfrak{J}U_{bU_a} = ((\mathfrak{J}U_e)U_b)U_e \subseteq \mathfrak{B}U_e = \mathfrak{B}$. An element b of $\mathfrak{J}$ will be called an *absolute zero divisor* in $\mathfrak{J}$ if $\mathfrak{J}U_b = 0$ (that is, $U_b = 0$). The equations just noted show also that if $b \in \mathfrak{J}U_e$ is an absolute zero divisor in $\mathfrak{J}U_e$, then b is an absolute zero divisor in $\mathfrak{J}$.

For the idempotent e one has the Peirce decomposition $\mathfrak{J} = \mathfrak{J}_1(e) \oplus \mathfrak{J}_{1/2}(e) \oplus \mathfrak{J}_0(e)$, where $\mathfrak{J}_i(e) = \{x_i | x_i \cdot e = ix_i\}$. Then $\mathfrak{J}_0(e)$ is a subalgebra and it is well known that $\mathfrak{J}_0(e) \cdot \mathfrak{J}_1(e) = 0$, $\mathfrak{J}_0(e) \cdot \mathfrak{J}_{1/2}(e) \subseteq \mathfrak{J}_{1/2}(e)$ and if $x \in \mathfrak{J}_{1/2}(e)$, $b \in \mathfrak{J}_0(e)$, then $x \cdot b^{\cdot 2} = 2(x \cdot b) \cdot b$. Hence, $\mathfrak{J}_1(e)U_b = 0$ and $\mathfrak{J}_{1/2}(e)U_b = 0$ if $b \in \mathfrak{J}_0(e)$. Then $\mathfrak{J}U_b = \mathfrak{J}_0(e)U_b$. This implies that any quadratic ideal in $\mathfrak{J}_0(e)$ is a quadratic ideal of $\mathfrak{J}$. Also, if $b \in \mathfrak{J}_0(e)$ is an absolute zero divisor in $\mathfrak{J}_0(e)$, then it is an absolute zero divisor in $\mathfrak{J}$.

A quadratic ideal $\mathfrak{B}$ of $\mathfrak{J}$ is *minimal* if $\mathfrak{B} \neq 0$ and there exists no quadratic ideal $\mathfrak{C}$ such that $\mathfrak{B} \supset \mathfrak{C} \supset 0$.

THEOREM 1. *If $\mathfrak{B}$ is a minimal quadratic ideal in a Jordan algebra $\mathfrak{J}$, then one has one of the following possibilities:* I. $\mathfrak{B} = \Phi b$, *where b is an absolute zero divisor.* II. $\mathfrak{B} = \mathfrak{J}U_b$ *for every nonzero $b \in \mathfrak{B}$ and $b_1 \cdot b_2 \cdot b_3 = 0$ for all $b_i \in \mathfrak{B}$. Moreover, if either $\mathfrak{J}$ has an identity element or $\mathfrak{J}$ contains no absolute zero divisor $\neq 0$, then $\mathfrak{B}^{\cdot 2} = 0$.* III. $\mathfrak{B} = \mathfrak{J}U_e$ *for an idempotent e and $\mathfrak{B}$ is division subalgebra of $\mathfrak{J}$.*

Proof: Suppose $\mathfrak{B}$ contains an element $b \neq 0$ such that $U_b = 0$. Then Φb is a nonzero quadratic ideal contained in $\mathfrak{B}$ so $\mathfrak{B} = \Phi b$ and we have case I. Now assume that for every $b \in \mathfrak{B}$ we have $\mathfrak{J}U_b \neq 0$. Then $\mathfrak{J}U_b$ is a nonzero quadratic ideal contained in $\mathfrak{B}$, so $\mathfrak{J}U_b = \mathfrak{B}$ for every $b \neq 0$ in $\mathfrak{B}$. Also, $\mathfrak{B}U_b$ is a quadratic ideal contained in $\mathfrak{B}$, so either $\mathfrak{B}U_b = 0$ or $\mathfrak{B}U_b = \mathfrak{B}$. Suppose there exists a $b \neq 0$ in $\mathfrak{B}$ such that $\mathfrak{B}U_b = 0$, and let $b' \in \mathfrak{B}$. Since $\mathfrak{B} = \mathfrak{J}U_b$, $b' = aU_b$ for $a \in \mathfrak{J}$. Then $\mathfrak{B}U_{b'} = \mathfrak{B}U_{aU_b} = \mathfrak{B}U_bU_aU_b = 0$. Thus, $\mathfrak{B}U_b = 0$ for all $b \in \mathfrak{B}$ and $\mathfrak{B}U_{b_1,b_2} = 0$ for all $b_1, b_2 \in \mathfrak{B}$. This implies that $b_1 \cdot b_2 \cdot b_3 = 0$ for all $b_i \in \mathfrak{B}$ so we have case II. If $\mathfrak{J}$ has a 1, then $\mathfrak{B}$ is a subalgebra. Then in case II, $b^{\cdot 3} = 0$ for every $b \in \mathfrak{B}$, so if $b^{\cdot 2} \neq 0$, then $b^{\cdot 2}$ is a nonzero element of $\mathfrak{B}$. Hence, there exists an a in $\mathfrak{J}$ such that $b = \{b^{\cdot 2}ab^{\cdot 2}\}$. Then $b^{\cdot 2} = \{b^{\cdot 2}\{ab^{\cdot 4}a\}b^{\cdot 2}\} = 0$, by (1). This contradiction shows that $b^{\cdot 2} = 0$ for all $b \in \mathfrak{B}$. Since $\mathfrak{J}$ is a commutative algebra of characteristic $\neq 2$, this implies that $\mathfrak{B}^{\cdot 2} = 0$. Again suppose we have case II and $\mathfrak{J}$ contains no absolute zero divisor $\neq 0$. If $b \neq 0$ in $\mathfrak{B}$, then $\mathfrak{B} = \mathfrak{J}U_b \supseteq (\mathfrak{J}U_b)U_b = \mathfrak{J}U_{b^{\cdot 2}}$, by (2). If $\mathfrak{J}U_{b^{\cdot 2}} = 0$, we have $b^{\cdot 2} = 0$. Otherwise, $\mathfrak{J}U_{b^{\cdot 2}} = \mathfrak{B}$ by the minimality of $\mathfrak{B}$. Then $b = \{b^{\cdot 2}ab^{\cdot 2}\}$, $a \in \mathfrak{J}$, and we obtain $b^{\cdot 2} = 0$ contrary to $\mathfrak{J}U_{b^{\cdot 2}} = \mathfrak{B}$. Hence, again we have $\mathfrak{B}^{\cdot 2} = 0$. It remains to consider the remaining alternative for $\mathfrak{B}$: $\mathfrak{B} = \mathfrak{B}U_b$ for every $b \neq 0$ in $\mathfrak{B}$. Take $b \neq 0$ in $\mathfrak{B}$. Then there exists a $c \in \mathfrak{B}$ such that $b = cU_b$. This implies that $U_b = U_bU_cU_b$, so $E = U_bU_c$ and $F = U_cU_b$ are idempotent linear transformations in $\mathfrak{J}$ which map $\mathfrak{B}$ into itself and have surjective restrictions to $\mathfrak{B}$. Hence, these restrictions are the identity mapping $1_{\mathfrak{B}}$ on $\mathfrak{B}$, so $\bar{U}_b\bar{U}_c = 1_{\mathfrak{B}} = \bar{U}_c\bar{U}_b$, where $\bar{U}$ denotes the restriction of U to $\mathfrak{B}$. Consider $f = \{cb^{\cdot 2}c\} \in \mathfrak{B}$. We have $f^{\cdot 2} = \{cb^{\cdot 2}c\}^{\cdot 2} = \{c\{b^{\cdot 2}cb^{\cdot 2}\}c\} = \{c\{b\{bc^{\cdot 2}b\}b\}c\} = \{bc^{\cdot 2}b\}$, since $\{bc^{\cdot 2}b\} \in \mathfrak{B}$ and $\bar{U}_b\bar{U}_c = 1_{\mathfrak{B}}$. Similarly, $\{bc^{\cdot 2}b\}^{\cdot 2} = \{cb^{\cdot 2}c\}$. Hence, $f^{\cdot 4} = (f^{\cdot 2})^{\cdot 2} = \{bc^{\cdot 2}b\}^{\cdot 2} = \{cb^{\cdot 2}c\} = f$. Put $e = f^{\cdot 3}$. Then $e \in \mathfrak{B}$ and $e^{\cdot 2} = f^{\cdot 6} = f^{\cdot 3} = e$, so e is idempotent. Now $\bar{U}_f = \bar{U}_c\bar{U}_b{}^2\bar{U}_c = 1_{\mathfrak{B}}$ and $\bar{U}_e = \bar{U}_f{}^3 = 1_{\mathfrak{B}}$. Hence, $e \neq 0$. Then $\mathfrak{B} = \mathfrak{J}U_e = \mathfrak{J}_1(e)$ the Peirce 1-space of e. Hence, $\mathfrak{B}$ is a subalgebra of $\mathfrak{J}$. If $d \in \mathfrak{B}$ and $d \neq 0$, then $\mathfrak{B}U_d = \mathfrak{B}$ contains e. This implies that d is invertible in $\mathfrak{B}$ and so $\mathfrak{B}$ is a division algebra. Hence, we have case III.

We next obtain sufficient conditions that a Jordan algebra contain idempotents $\neq 0, 1$. For this we require

LEMMA 1. *Let $\mathfrak{J}$ be a Jordan algebra with identity element 1 and elements a,b such that $a^{\cdot 2} = 0 = b^{\cdot 2}$ and $2a\cdot b = 1$. Then $\mathfrak{J}$ contains an idempotent element $e \neq 0,1$.*

Proof: One sees directly that $1, a,b$ are linearly independent. Choose α,β in Φ, so that $\alpha\beta = 1$. Put $f = 1 + \alpha a + \beta b$. Then $f^{\cdot 2} = 2f$. Hence, $e = {}^1/_2 f$ is an idempotent $\neq 0,1$.

THEOREM 2. *Let $\mathfrak{J}$ be a Jordan algebra such that: (1) $\mathfrak{J}$ has an identity element 1 ($\neq 0$). (2) $\mathfrak{J}$ contains no absolute zero divisors $\neq 0$. (3) $\mathfrak{J}$ contains minimal quadratic ideals. Then either $\mathfrak{J}$ is a division algebra or $\mathfrak{J}$ contains an idempotent element $e \neq 0,1$.*

Proof: Let $\mathfrak{B}$ be a minimal quadratic ideal in $\mathfrak{J}$. Then $\mathfrak{B}$ is not of type I by condition (2). If $\mathfrak{B}$ is of type III, either $\mathfrak{J} = \mathfrak{B}$ is a division algebra or we have an idempotent $e \neq 0,1$. Hence, we may assume $\mathfrak{B}$ is of type II. Also, we have $\mathfrak{B}^{\cdot 2} = 0$. Now choose $b \neq 0$ in $\mathfrak{B}$ and $a \in \mathfrak{J}$ so that $b = \{bab\}$. Let $c = a\cdot b$. We have the identity

$$(a\cdot b)^{\cdot 2} = {}^1/_2 a\cdot\{bab\} + {}^1/_4\{ba^{\cdot 2}b\} + {}^1/_4\{ab^{\cdot 2}a\} \tag{4}$$

in any Jordan algebra, which is clear by Shirshov's theorem. In the present situation this gives $c^{\cdot 2} = {}^1/_2 c + {}^1/_4\{ba^{\cdot 2}b\}$. Since $\{ba^{\cdot 2}b\} \in \mathfrak{B}$, we have $(2c^{\cdot 2} - c)^{\cdot 2} = 0$. This implies that either $c^{\cdot 2} = 0$, or there exists an idempotent $e \neq 0$ of the form $\alpha c + \beta c^{\cdot 2} + \gamma c^{\cdot 3}$, $\alpha,\beta,\gamma \in \Phi$. In the first case we have $-2c = \{ba^{\cdot 2}b\}$, so $c \in \mathfrak{B}$. Then $b = aU_b = 2(a\cdot b)\cdot b - a\cdot b^{\cdot 2} = 2c\cdot b = 0$, since $\mathfrak{B}^{\cdot 2} = 0$. Hence, we have the second case: $e = \alpha c + \beta c^{\cdot 2} + \gamma c^{\cdot 3}$ is a nonzero idempotent. Suppose $e = 1$. Then c is a root of a cubic equation with nonzero constant term. Since $(2c^{\cdot 2} - c)^{\cdot 2} = 0$, this implies that $(2c - 1)^{\cdot 2} = 4c^{\cdot 2} - 4c + 1 = 0$. Then $\{ba^{\cdot 2}b\} = 4c^{\cdot 2} - 2c = 2a\cdot b - 1$. Since $b^{\cdot 2} = 0$, $xU_b = 2(x\cdot b)\cdot b$, so the foregoing equation gives $2(a^{\cdot 2}\cdot b)\cdot b = 2a\cdot b - 1$ or $2a'\cdot b = 1$ for $a' = a - a^{\cdot 2}\cdot b$. Then $2(a'\cdot b)\cdot b$

$= b$, so $a'U_b = b$ and replacing a by a' permits us to assume that $2a \cdot b = 1$. Then $c = {}^1/_2$ and $\{ba^{\cdot 2}b\} = 0$. Hence, $\{aba\}^{\cdot 2} = 0$ by (1). Also, $2\{aba\} \cdot b = 2a \cdot \{bab\} = 2a \cdot b = 1$ and $\{aba\}U_b = 2(\{aba\} \cdot b) \cdot b = b$. Hence, replacing a by $\{aba\}$, we have $a^{\cdot 2} = 0 = b^{\cdot 2}$ and $2a \cdot b = 1$. Then we obtain an idempotent $\neq 0, 1$ by Lemma 1.

2. *Axioms and Structure Theory.*—We shall now investigate the structure of Jordan algebras satisfying the following axioms:

(*i*) $\mathfrak{J}$ has an identity element 1.

(*ii*) $\mathfrak{J}$ contains no absolute zero divisors $\neq 0$.

(*iii*) The minimum condition holds for quadratic ideals determined by idempotent elements, and every quadratic ideal determined by an idempotent $\neq 0$ contains a minimal quadratic ideal.

We shall call a Jordan algebra $\mathfrak{J}$ *regular* if $a \in \mathfrak{J}U_a$ for every $a \in \mathfrak{J}$. Since $xU_a = axa$ in $\mathfrak{A}^+$, $\mathfrak{A}$ associative, it is clear that this is analogous to the well-known definition of regularity for associative rings due to von Neumann. Clearly, regularity implies (*ii*). If $\mathfrak{A} = \mathrm{Hom}_\Delta(\mathfrak{M}, \mathfrak{M})$, where $\mathfrak{M}$ is a finite dimensional vector space over the division algebra, then it is well known that $\mathfrak{A}$ is regular. Hence, $\mathfrak{A}^+$ is regular. It is easy to see that $\mathfrak{A}^+$ satisfies axioms (*i*)–(*iii*). It is easy to see also that if $\mathfrak{A} = \mathrm{Hom}_\Delta(\mathfrak{M}, \mathfrak{M})$ has an involution J (anti-automorphism of period two), then the subalgebra $\mathfrak{H}(\mathfrak{A}, J)$ of $\mathfrak{A}^+$ of J-symmetric elements satisfies the axioms.

THEOREM 3. *Let $\mathfrak{J}$ be a Jordan algebra satisfying axioms (i)–(iii). Then $1 = \sum_1^n e_i$ in $\mathfrak{J}$, where the e_i are orthogonal idempotent elements in $\mathfrak{J}$ such that $\mathfrak{J}U_{e_i}$ is a division algebra, $i = 1,2,\ldots,n$.*

Proof: It is clear from the remarks at the beginning of §1 that if e is a nonzero idempotent in $\mathfrak{J}$, then $\mathfrak{J}U_e$ satisfies the axioms. Let $\mathfrak{J}U_e$ be a minimal element in the collection of quadratic ideals determined by idempotents $e \neq 0$. We claim that $\mathfrak{J}U_e$ is a division algebra. Otherwise, Theorem 2 implies that $\mathfrak{J}U_e$ contains an idempotent $f \neq 0, e$. Then $\mathfrak{J}U_e \supset \mathfrak{J}U_f$ contrary to the minimality of $\mathfrak{J}U_e$. Therefore, we see that there exist idempotents e in $\mathfrak{J}$ such that $\mathfrak{J}U_e$ is a division algebra. Suppose $e_1, e_2, \ldots, e_r$ are orthogonal idempotents in $\mathfrak{J}$ such that $\mathfrak{J}U_{e_i}$ is a division algebra. Set $f_i = 1 - \sum_1^i e_j$. Then f_i is an idempotent and $\mathfrak{J}U_{f_1} \supset \mathfrak{J}U_{f_2} \supset \ldots \supset \mathfrak{J}U_{f_r}$. If $f_r = 0$, then $1 = \sum_1^r e_j$ and the theorem is proved. If $f_r \neq 0$, then $\mathfrak{J}U_{f_r}$ satisfies our conditions and so contains an idempotent e_{r+1} such that $\mathfrak{J}U_{e_{r+1}}$ is a division algebra. Then $e_1, e_2, \ldots, e_{r+1}$ is set of orthogonal idempotents and $\mathfrak{J}U_{f_1} \supset \ldots \supset \mathfrak{J}U_{f_r} \supset \mathfrak{J}U_{f_{r+1}}$ if $f_{r+1} = 1 - \sum_1^{r+1} e_j$. By (*iii*) this process must terminate with an $f_n = 0$ and this proves the theorem.

If $1 = \sum_1^n e_i$, where the e_i are orthogonal idempotents in $\mathfrak{J}$, then we have the Peirce decomposition $\mathfrak{J} = \sum_{i \leq j} \oplus \mathfrak{J}_{ij}$, where $\mathfrak{J}_{ii} = \mathfrak{J}U_{e_i}$, $\mathfrak{J}_{ij} = \mathfrak{J}U_{e_i, e_j} = \{x_{ij} | x_{ij} \cdot e_i = {}^1/_2 x_{ij} = x_{ij} \cdot e_j\}$ for $i \neq j$.

LEMMA 2. *Let $\mathfrak{J} = \mathfrak{J}_{11} \oplus \mathfrak{J}_{12} \oplus \mathfrak{J}_{22}$ be the Peirce decomposition of a Jordan algebra $\mathfrak{J}$ with 1 relative to the orthogonal idempotent elements e_1, e_2 such that $e_1 + e_2 = 1$*

and assume $\mathfrak{J}_{ii}$ is a division algebra for $i = 1,2$. Then any $a_{12} \in \mathfrak{J}_{12}$ is either invertible or $a_{12}^{\cdot 2} = 0$.

Proof: It is well known that $\mathfrak{J}_{ij}^{\cdot 2} \subseteq \mathfrak{J}_{ii} + \mathfrak{J}_{jj}$. Hence, $a_{12}^{\cdot 2} = a_1 + a_2$, $a_i \in \mathfrak{J}_{ii}$. Then $(a_{12}^{\cdot 2} \cdot e_1) \cdot a_{12} = a_{12}^{\cdot 2} \cdot (e_1 \cdot a_{12})$ gives $a_1 \cdot a_{12} = (a_1 + a_2) \cdot (^1/_2 a_{12})$, so $a_1 \cdot a_{12} = a_2 \cdot a_{12}$. Suppose $a_1 \cdot a_{12} = 0$. Then $a_2 \cdot a_{12} = 0$ and $a_{12}^{\cdot 3} = 0$. Then $a_1^{\cdot 2} + a_2^{\cdot 2} = a_{12}^{\cdot 4} = 0$. Since $\mathfrak{J}_{ii}$ is a division algebra, this implies that $a_i = 0$ so $a_{12}^{\cdot 2} = 0$. If $a_1 \cdot a_{12} \neq 0$, then $a_2 \cdot a_{12} \neq 0$ so the $a_i \neq 0$ and $a_{12}^{\cdot 2} = a_1 + a_2$ is invertible. This implies that a_{12} is invertible.

If $e_1, e_2, \ldots, e_n$ is a set of orthogonal idempotents such that $\sum e_i = 1$ and $\mathfrak{J} = \sum_{i \leq j} \mathfrak{J}_{ij}$ the corresponding Peirce decomposition, then we say that e_k and e_l are *connected* if there exists an element in $\mathfrak{J}_{kl}$ which is invertible in the subalgebra $\mathfrak{J}_{kk} + \mathfrak{J}_{kl} + \mathfrak{J}_{ll}$ of $\mathfrak{J}$, where we take $\mathfrak{J}_{kl} = \mathfrak{J}_{lk}$ if $k > l$.

Lemma 2 implies that, under the indicated hypotheses, either e_1 and e_2 are connected or $\mathfrak{J}_{12}$ is an ideal such that $\mathfrak{J}_{12}^{\cdot 2} = 0$.

LEMMA 3. *If $\mathfrak{J}$ is as in Lemma 2, then either $\mathfrak{J}$ is simple, $\mathfrak{J} = \mathfrak{J}_{11} \oplus \mathfrak{J}_{22}$, and the $\mathfrak{J}_{ii}$ are ideals, or $\mathfrak{J}$ contains an ideal $\mathfrak{N}$ such that $\mathfrak{N}^{\cdot 2} = 0$.*

Proof: If $\mathfrak{J}$ is simple or $\mathfrak{J}_{12}^{\cdot 2} = 0$, we have the desired result. Hence, assume $\mathfrak{J}$ not simple and $\mathfrak{J}_{12}^{\cdot 2} \neq 0$. Let $\mathfrak{N}$ be an ideal $\neq 0$, $\mathfrak{J}$. We have $\mathfrak{N} = \mathfrak{N}_{11} + \mathfrak{N}_{12} + \mathfrak{N}_{22}$, where $\mathfrak{N}_{ij} = \mathfrak{N} \cap \mathfrak{J}_{ij}$. By Lemma 2 and $\mathfrak{N} \neq \mathfrak{J}$ we have $\mathfrak{N}_{12}^{\cdot 2} = 0$. Also, since $\mathfrak{J}_{ii}$ is a division algebra and $\mathfrak{N}_{ii}$ is an ideal in $\mathfrak{J}_{ii}$, either $\mathfrak{N}_{ii} = 0$ or $\mathfrak{N}_{ii} = \mathfrak{J}_{ii}$. In the latter case $\mathfrak{N} \supseteq \mathfrak{J}_{12} = e_i \cdot \mathfrak{J}_{12}$. Then $\mathfrak{N}_{12} = \mathfrak{J}_{12} \neq 0$ contrary to $\mathfrak{N}_{12}^{\cdot 2} = 0$, $\mathfrak{J}_{12}^{\cdot 2} \neq 0$. Hence, $\mathfrak{N}_{ii} = 0$ and $\mathfrak{N} = \mathfrak{N}_{12}$ satisfies $\mathfrak{N}_{12}^{\cdot 2} = 0$.

If $\mathfrak{N}$ is an ideal in a Jordan algebra such that $\mathfrak{N}^{\cdot 2}$, then any $b \in \mathfrak{N}$ is an absolute zero divisor since $xU_b = 2(x \cdot b) \cdot b - x \cdot b^{\cdot 2} = 0$ since $x \cdot b \in \mathfrak{N}$. Hence, if the algebra $\mathfrak{J}$ of Lemma 2 has no absolute zero divisors $\neq 0$, then either $\mathfrak{J}$ is simple and $\mathfrak{J} = \mathfrak{J}_{11} \oplus \mathfrak{J}_{22}$.

We can now prove the

FIRST STRUCTURE THEOREM. *Let $\mathfrak{J}$ be a Jordan algebra satisfying the axioms (i)–(iii). Then $\mathfrak{J}$ is a direct sum of a finite number of ideals which are simple algebras satisfying (i)–(iii).*

Proof: Let $\mathfrak{J} = \sum_{i \leq j} \oplus \mathfrak{J}_{ij}$, where $\mathfrak{J}_{ii}$ is a division algebra as in Theorem 3. If $\mathfrak{J}$ is simple, there is nothing to prove, so we assume $\mathfrak{J}$ has an ideal $\mathfrak{K} \neq 0$, $\mathfrak{J}$. Then $\mathfrak{K} = \sum \mathfrak{K}_{ij}$, where $\mathfrak{K}_{ij} = \mathfrak{K} \cap \mathfrak{J}_{ij}$. Then $\mathfrak{K}_{ii} + \mathfrak{K}_{ij} + \mathfrak{K}_{ji}$ is an ideal in $\mathfrak{J}_{ii} + \mathfrak{J}_{ij} + \mathfrak{J}_{jj}$ and the latter has the form $\mathfrak{J}U_f$, f an idempotent, so it satisfies our axioms. Hence, by Lemma 3 and the remark following it, either $\mathfrak{J}_{ii} + \mathfrak{J}_{ij} + \mathfrak{J}_{jj}$ is simple or $\mathfrak{J}_{ij} = 0$. In the first case, either $\mathfrak{K}_{ii} + \mathfrak{K}_{ij} + \mathfrak{K}_{jj} = \mathfrak{J}_{ii} + \mathfrak{J}_{ij} + \mathfrak{J}_{jj}$ or $\mathfrak{K}_{ii} + \mathfrak{K}_{ij} + \mathfrak{K}_{jj} = 0$. In the second case $\mathfrak{K}_{ij} = 0$ and $\mathfrak{K}_{ii} + \mathfrak{K}_{jj} = 0$, $\mathfrak{J}_{ii}$, $\mathfrak{J}_{jj}$, or $\mathfrak{J}_{ii} + \mathfrak{J}_{jj}$. This implies that $\mathfrak{K} = \sum' \mathfrak{J}_{kl}$ a sum of Peirce spaces $\mathfrak{J}_{ij}$ such that if $\mathfrak{J}_{kl} \neq 0 \subseteq \mathfrak{K}$, then $\mathfrak{J}_{kk}$, $\mathfrak{J}_{ll} \subseteq \mathfrak{K}$. Let $\mathfrak{K}' = \sum'' \mathfrak{J}_{ij}$ the sum of Peirce spaces not contained in $\mathfrak{K}$. Then $\mathfrak{J} = \mathfrak{K} \oplus \mathfrak{K}'$ and we proceed to show that $\mathfrak{K}'$ is an ideal. If we define $\mathfrak{J}_{ji} = \mathfrak{J}_{ij}$ for $j > i$, then it is well known that we have the following multiplication table for the Peirce spaces: $\mathfrak{J}_{ij} \cdot \mathfrak{J}_{kl} = 0$ if $\{i,j\} \cap \{k,l\} = \phi$, $\mathfrak{J}_{ij} \cdot \mathfrak{J}_{jk} \subseteq \mathfrak{J}_{ik}$ if i, j, k are distinct, $\mathfrak{J}_{ii} \cdot \mathfrak{J}_{ij} = \mathfrak{J}_{ij}$ and $\mathfrak{J}_{ij}^{\cdot 2} \subseteq \mathfrak{J}_{ii} + \mathfrak{J}_{jj}$. Hence, the fact that $\mathfrak{K}'$ is an ideal will follow by showing that there is no index k such that there exists a nonzero $\mathfrak{J}_{kl} \subseteq \mathfrak{K}$ and a nonzero $\mathfrak{J}_{ki} \subseteq \mathfrak{K}'$. Suppose the contrary. Then $\mathfrak{J}_{kk} \subseteq \mathfrak{K}$ and $\mathfrak{J}_{ki} = \mathfrak{J}_{kk} \cdot \mathfrak{J}_{ki} \subseteq \mathfrak{K}$ which contradicts $\mathfrak{J}_{ki} \subseteq \mathfrak{K}'$. Thus, we have shown that every ideal in $\mathfrak{J}$ has a complementary ideal. This complete reducibility of the lattice of ideals implies that $\mathfrak{J}$ is a direct sum of minimal ideals. Since $\mathfrak{J}$ has a 1, $\mathfrak{J}$ is a direct sum of a finite number of minimal ideals. Hence, $\mathfrak{J} = \mathfrak{J}_1 \oplus \mathfrak{J}_2 \oplus \ldots \oplus \mathfrak{J}_s$, where $\mathfrak{J}_i$ is a minimal ideal in $\mathfrak{J}$. Then $\mathfrak{J}_i$ as algebra satisfies the axioms. Hence, if it is not simple, then $\mathfrak{J}_i = \mathfrak{J}_i' \oplus \mathfrak{J}_i''$ are nonzero ideals in $\mathfrak{J}_i$. Since $\mathfrak{J}_j \cdot \mathfrak{J}_i = 0$ if $i \neq j$, $\mathfrak{J}_i', \mathfrak{J}_i''$ are ideals in $\mathfrak{J}$. This contradicts the minimality of $\mathfrak{J}_i$. Hence every $\mathfrak{J}_i$ is a simple algebra.

Let $\mathfrak{J} = \sum_{i \leq j} \oplus \mathfrak{J}_{ij}$ be the Peirce decomposition of a Jordan algebra $\mathfrak{J}$ with 1 relative to the orthogonal idempotents $e_1, e_2, \ldots, e_n$ such that $\Sigma\, e_i = 1$. Let $k \neq l$ and let $a_{kl} \in \mathfrak{J}_{kl}$ $(= \mathfrak{J}_{lk})$ be invertible in $\mathfrak{J}_{kk} + \mathfrak{J}_{kl} + \mathfrak{J}_{ll}$. Then one sees easily that the inverse b_{kl} of a_{kl} in $\mathfrak{J}_{kk} + \mathfrak{J}_{kl} + \mathfrak{J}_{ll}$ is in $\mathfrak{J}_{kl}$. We can now prove

LEMMA 4. *Connectedness is an equivalence relation for the set of idempotents* $e_1, e_2, \ldots, e_n$ *(orthogonal with* $\sum e_i = 1$*).*

Proof: The relation is clearly reflexive and symmetric. Now assume i, j, k are distinct, e_i and e_j are connected, and e_j and e_k are connected. Then we have a_{ij}, $b_{ij} \in \mathfrak{J}_{ij}$ which are inverses in $\mathfrak{J}_{ii} + \mathfrak{J}_{ij} + \mathfrak{J}_{jj}$ and a_{jk}, $b_{jk} \in \mathfrak{J}_{jk}$ which are inverses in $\mathfrak{J}_{jj} + \mathfrak{J}_{jk} + \mathfrak{J}_{kk}$. Put $a_{ik} = 2a_{ij} \cdot a_{jk}$, $b_{ik} = 2b_{ij} \cdot b_{jk}$. We shall show that a_{ik} and b_{ik} are inverses in $\mathfrak{J}_{ii} + \mathfrak{J}_{ik} + \mathfrak{J}_{kk}$. For the proof we need the following identities for elements of the Peirce components: $(x_{ij} \cdot y_{ij}) \cdot z_{jk} = x_{ij} \cdot (y_{ij} \cdot z_{jk}) + y_{ij} \cdot (x_{ij} \cdot z_{jk})$, $((x_{ij} \cdot y_{jk}) \cdot z_{ki}) \cdot e_i = (x_{ij} \cdot (y_{jk} \cdot z_{ki})) \cdot e_i$, $(x_{ii} \cdot y_{ij}) \cdot z_{jk} = x_{ii} \cdot (y_{ij} \cdot z_{jk})$, where i, j, k are distinct x_{ij}, $y_{ij} \in \mathfrak{J}_{ij}$, $z_{jk} \in \mathfrak{J}_{jk}$, etc. These are obtained by making suitable substitutions in the basic multilinear identities for Jordan algebras. We shall also need the fact that if a and b are inverses in a subalgebra of $\mathfrak{J}$, then their multiplication R_a and R_b commute in $\mathfrak{J}$. This is well known if the subalgebra has the same identity as $\mathfrak{J}$ and the general case can be established using the Peirce decomposition relative to the identity element of the subalgebra. Now we have $a_{jk} \cdot b_{ik} = 2a_{jk} \cdot (b_{jk} \cdot b_{ij}) = a_{jk} \cdot (b_{jk} \cdot b_{ij}) + b_{jk} \cdot (a_{jk} \cdot b_{ij}) = (a_{jk} \cdot b_{jk}) \cdot b_{ij} = (e_j + e_k) \cdot b_{ij} = {}^1/_2 b_{ij}$ and $(a_{ik} \cdot b_{ik}) \cdot e_i = 2((a_{ij} \cdot a_{jk}) \cdot b_{ik}) \cdot e_i = 2(a_{ij} \cdot (a_{jk} \cdot b_{ik})) \cdot e_i = (a_{ij} \cdot b_{ij}) \cdot e_i = e_i$. By symmetry, $(a_{ik} \cdot b_{ik}) \cdot e_k = e_k$. Hence, $a_{ik} \cdot b_{ik} = e_i + e_k$. Next we have $b_{ij} \cdot a_{ik} = {}^1/_2 a_{jk}$ by the same type of argument used to prove $a_{jk} \cdot b_{ik} = {}^1/_2 b_{ij}$. Hence, $(a_{ik}^{\cdot 2} \cdot e_i) \cdot b_{ik} = 2(a_{ik}^{\cdot 2} \cdot e_i) \cdot (b_{ij} \cdot b_{jk}) = 2((a_{ik}^{\cdot 2} \cdot e_i) \cdot b_{ij}) \cdot b_{jk} = 2(a_{ik}^{\cdot 2} \cdot b_{ij}) \cdot b_{jk} = 4(a_{ik} \cdot (a_{ik} \cdot b_{ij})) \cdot b_{jk} = 2(a_{ik} \cdot a_{jk}) \cdot b_{jk} = (a_{ik} \cdot a_{jk}) \cdot b_{jk} + (a_{ik} \cdot b_{jk}) \cdot a_{jk} = a_{ik} \cdot (b_{jk} \cdot a_{jk}) = a_{ik} \cdot (e_j + e_k) = {}^1/_2 a_{ik}$. Similarly, $(a_{ik}^{\cdot 2} \cdot e_k) \cdot b_{ik} = {}^1/_2 a_{ik}$. Hence, $a_{ik}^{\cdot 2} \cdot b_{ik} = a_{ik}$ and a_{ik}, b_{ik} are inverses relative to $e_i + e_k$. This proves that connectedness is a transitive relation.

Let $\mathfrak{D}$ be an algebra with an identity element and an involution $d \rightarrow \bar{d}$, $\mathfrak{D}_n$ the algebra of $n \times n$ matrices with entries, γ a diagonal matrix with diagonal entries γ_i which are symmetric elements of the nucleus of $\mathfrak{D}$ having inverses in this nucleus. Let $\mathfrak{H}(\mathfrak{D}_n, \gamma)$ be the set elements $A \in \mathfrak{D}_n$ such that $\gamma^{-1} \bar{A}' \gamma = A$, where the prime denotes the transpose and if $A = (a_{ij})$, then $\bar{A} = (\bar{a}_{ij})$. Then $\mathfrak{H}(\mathfrak{D}_n, \gamma)$ is a subalgebra of $\mathfrak{D}_n^{+}$ and it is known that for $n = 3$, $\mathfrak{H}(\mathfrak{D}_n, \gamma)$ is Jordan if and only if $\mathfrak{D}$ is alternative and the symmetric elements of $\mathfrak{D}$ are in the nucleus, and for $n \geq 4$, $\mathfrak{H}(\mathfrak{D}_n, \gamma)$ is Jordan if and only if $\mathfrak{D}$ is associative.[4] In these cases we call $\mathfrak{H}(\mathfrak{D}_n, \gamma)$ a *Jordan matrix algebra* and $\mathfrak{D}$ and its involution a *coordinate algebra* for $\mathfrak{H}(\mathfrak{D}_n, \gamma)$. We can now state the

SECOND STRUCTURE THEOREM. *Let* $\mathfrak{J}$ *be a simple Jordan algebra satisfying axioms (i)–(iii). Then* $\mathfrak{J}$ *is one of the following types:* (1) *a division algebra,* (2) *a*

Jordan algebra whose identity is a sum of two connected orthogonal idempotents e_i such that $\mathfrak{J}U_{e_i}$ is a division algebra, (3) *a Jordan matrix algebra whose coordinate algebra is either a Cayley algebra over its center with standard involution, a split quaternion algebra over the center with standard involution, a direct sum of an associative division algebra and its opposite with involution exchanging the two factors, or an associative division algebra with involution.*

Proof: Let $\mathfrak{J} = \sum_{i \leq j} \oplus \mathfrak{J}_{ij}$ be the Peirce decomposition of $\mathfrak{J}$ relative to a set of orthogonal idempotents e_i such that $\sum_1^n e_i = 1$ and $\mathfrak{J}_{ii}$ is a division algebra (Theorem 3). Let $i \neq j$ and assume $\mathfrak{J}_{ij} \neq 0$. Then Lemma 3 and the fact that $\mathfrak{J}_{ii} + \mathfrak{J}_{ij} + \mathfrak{J}_{jj}$ satisfies the axioms imply that this is a simple algebra. Hence, $\mathfrak{J}_{ij}{}^{\cdot 2} \neq 0$ and e_i and e_j are connected by Lemma 2. Let $e_1, \ldots, e_m$ be the complete set of e_i connected with e_1 and set $\mathfrak{J}' = \sum_{k,l=1}^{m} \mathfrak{J}_{kl}$. Since connectedness is a transitive relation, every $\mathfrak{J}_{ki}$ with $k \leq m$ and $i > m$ is 0. This implies that $\mathfrak{J}'$ is an ideal. Hence, $\mathfrak{J}' = \mathfrak{J}$ and all the e_i are connected. If $n = 1$, $\mathfrak{J}$ is a division algebra and if $n = 2$, we have the second alternative. If $n \geq 3$, then by the Coordinatization Theorem for Jordan algebras, $\mathfrak{J}$ is a Jordan matrix algebra.[5] Also, since $\mathfrak{J}$ is simple, the coordinate algebra $\mathfrak{D}$ is a simple algebra with involution, that is, it has no ideals $\neq 0$, $\mathfrak{D}$ which are invariant under the involution. It follows that either $\mathfrak{D} = \Delta \oplus \Delta'$ where Δ is simple with involution exchanging the two factors, or $\mathfrak{D}$ itself is simple. In the first case, $\mathfrak{H}(\mathfrak{D})$ the subalgebra of $\mathfrak{D}^+$ of symmetric elements is isomorphic to Δ^+ and in both cases $\mathfrak{H}(\mathfrak{D})$ is isomorphic to $\mathfrak{J}U_{e_i}$ which is a Jordan division algebra. If $\mathfrak{D}$ is not associative, then the condition that the symmetric elements are in the nucleus rules out $\mathfrak{D} = \Delta \oplus \Delta'$. Hence, in this case $\mathfrak{D}$ is simple, so by a theorem of Kleinfeld's, $\mathfrak{D}$ is a Cayley algebra.[6] Moreover, the involution is a standard one since the symmetric elements are in the nucleus which, for Cayley algebras, coincides with the center. If $\mathfrak{D}$ is associative, then it is well known that an element a has an inverse in the Jordan algebra $\mathfrak{D}^+$ if and only if it has an inverse in $\mathfrak{D}$ in the usual sense. Hence, every nonzero element of $\mathfrak{H}(\mathfrak{D})$ has an inverse in $\mathfrak{D}$. If $\mathfrak{D} = \Delta \oplus \Delta'$, this implies that Δ is a division algebra. The remaining case: $\mathfrak{D}$ simple associative is settled by the following lemma which is due to J. M. Osborn, Jr. (oral communication).

LEMMA 4. *Let $\mathfrak{D}$ be a simple associative ring of characteristic $\neq 2$ containing an identity element and having an involution $\mathfrak{J}$ such that every nonzero symmetric element of $\mathfrak{D}$ has an inverse. Then either $\mathfrak{D}$ is a quaternion algebra over its center with standard involution or $\mathfrak{D}$ is a division algebra.*

Proof: We assume that $\mathfrak{D}$ is not a quaternion algebra with standard involution. Then, by a theorem of Herstein's, the symmetric elements of $\mathfrak{D}$ generate $\mathfrak{D}$.[7] Let $\mathfrak{H}$ and $\mathfrak{S}$ denote the sets of symmetric and skew elements, respectively, of $\mathfrak{D}$. Let $a \in \mathfrak{D}$ and write $a = h + s$, where $h \in \mathfrak{H}$, $s \in \mathfrak{S}$. If a is not invertible, then $aa^J = 0$ and $a^J a = 0$. Otherwise, since aa^J, $a^J a \in \mathfrak{H}$, the hypothesis on $\mathfrak{H}$ implies that one of aa^J, $a^J a$ is 0 and the other is invertible. By symmetry it is enough to suppose that $a^J a = 0$ and there exists an x such that $aa^J x = 1 = x aa^J$. Then $0 = a^J aa^J x = a^J$ so $aa^J = 0 = a^J a$, a contradiction. We now prove that if $h \neq 0$, then $a = h + s$

is invertible. Otherwise,

$$0 = {}^1/_2(aa^J + a^Ja) = {}^1/_2[(h + s)(h - s) + (h - s)(h + s)] = h^2 - s^2,$$

$$0 = {}^1/_2(aa^J - a^Ja) = {}^1/_2[(h + s)(h - s) - (h - s)(h + s)] = sh - hs.$$

Hence, $sh = hs$ and $h^2 = s^2$. Since $h \neq 0$ is in $\mathfrak{H}$, h^{-1} exists. Hence, if $c = h^{-1}s$, then $c^2 = h^{-1}sh^{-1}s = h^{-2}s^2 = 1$. Also, $c^J = -h^{-1}s = -c$. Let k be any element of $\mathfrak{H}$. Then $(1 + c)k(1 + c)^J = (1 + c)k(1 - c)$ and $(1 + c)^Jk(1 + c) = (1 - c)$ $k(1 + c)$. Also, if $b = (1 + c)k(1 + c)^J$, then $b^2 = (1 + c)k(1 - c)(1 + c)k(1 - c)$ $= 0$ since $(1 + c)(1 - c) = 1 - c^2 = 0$. Since $b \in \mathfrak{H}$, this gives $(1 + c)k(1 - c) =$ 0. Similarly, $(1 - c)k(1 + c) = 0$, so we have $k + ck - kc - ckc = 0$, $k - ck +$ $kc - ckc = 0$. Hence, $kc = ck$. By Herstein's theorem the symmetric elements generate $\mathfrak{D}$. Hence, c is in the center. Since $c^2 = 1$, $e = {}^1/_2(1 - c)$ is a central idempotent. Since c is skew, $c \neq \pm 1$ so $e \neq 0, 1$. This contradicts the simplicity of $\mathfrak{D}$. Hence, we have shown that if $a = h + s$, $h \in \mathfrak{H}$, $s \in \mathfrak{S}$, and $h \neq 0$, then a is invertible. Next, suppose $a = s$ is skew. If hs is not skew for some $h \in \mathfrak{H}$, then there exists a u such that $uhs = 1 = -shu^J$, so s is invertible. Hence, we may assume $(hs)^J = -hs$ for all $h \in \mathfrak{H}$. This gives $sh = hs$, so s commutes with every symmetric element. Then s is in the center and since the center is a field, either $s = 0$ or it has an inverse.

In the case[3] of the theorem it is easy to see that $\mathfrak{J}$ is either a reduced exceptional simple Jordan algebra over the center, or $\mathfrak{J}$ has the form $\mathfrak{A}^+$, $\mathfrak{A}$ simple Artinian, or $\mathfrak{J} = \mathfrak{H}(\mathfrak{A},J)$, where $\mathfrak{A}$ is simple Artinian with involution J. Conversely, one sees easily that these algebras satisfy the axioms. Clearly, any Jordan division algebra satisfies the axioms also. Further analysis of the remaining case[2] is desirable and we hope to undertake this in a later communication.

3. *Alternative Set of Axioms.*—Let $\mathfrak{J}$ be a Jordan algebra, $\mathfrak{J}' = \Phi 1 \oplus \mathfrak{J}$ the Jordan algebra obtained by adjoining an identity element 1 to $\mathfrak{J}$. If $b' = \beta 1 + b$, $b \in \mathfrak{J}$, then $b'^{\cdot 2} = \beta^2 1 + (2\beta b + b^{\cdot 2})$, so $b'^{\cdot 2} = 0$ implies $\beta = 0$ and $b' = b \in \mathfrak{J}$. It follows that $\mathfrak{J}'$ has no absolute zero divisors $\neq 0$ if $\mathfrak{J}$ has none. Next, assume $b'^{\cdot 2} = b'$. Then $\beta = 0$ or 1. In the first case, $b' = b \in \mathfrak{J}$ and in the second $b^{\cdot 2} =$ $-b$, so $f = -b$ is an idempotent and $b' = 1 - f$. If $e^{\cdot 2} = e \in \mathfrak{J}$, then $\mathfrak{J}_1'(e) =$ $\mathfrak{J}_1(e)$ since $\mathfrak{J}$ is an ideal in $\mathfrak{J}'$. If $e' = 1 - f, f^{\cdot 2} = f \in \mathfrak{J}$, then $\mathfrak{J}_1'(e') = \mathfrak{J}_0(f) + \Phi e'$ Clearly, the minimum condition holds for quadratic ideals of this form if and only if it holds for quadratic ideals of $\mathfrak{J}$ of the form $\mathfrak{J}_0(f), f^{\cdot 2} = f$.

Assume $\mathfrak{J}$ has no absolute zero divisors, so $\mathfrak{J}'$ has none. Let $\mathfrak{B}$ be a minimal quadratic ideal of $\mathfrak{J}$. By Theorem 1, $\mathfrak{B} = \mathfrak{J}U_b$ where $b \in \mathfrak{B}$ and $\mathfrak{B}$ is a subalgebra of $\mathfrak{J}$. Then $\mathfrak{J}'U_b = \mathfrak{J}U_b + \Phi b^{\cdot 2} = \mathfrak{B}$, so $\mathfrak{B}$ is a quadratic ideal of $\mathfrak{J}'$, which is clearly minimal. If $e' = 1 - f, f^{\cdot 2} = f \in \mathfrak{J}$, then $\mathfrak{J}_1'(e') = \mathfrak{J}_0(f) + \Phi e'$. Hence, if $\mathfrak{J}_0(f)$ contains a minimal quadratic ideal of $\mathfrak{J}$, then $\mathfrak{J}_1'(e')$ contains a minimal quadratic ideal of $\mathfrak{J}'$. If $\mathfrak{J}_0(f) = 0$, then $\mathfrak{J}_1'(e') = \Phi e'$ is a minimal quadratic ideal of $\mathfrak{J}'$.

We now state the following axiom on $\mathfrak{J}$.

(*iv*) The minimum condition holds for quadratic ideals of the form $\mathfrak{J}_0(e)$, $e^{\cdot 2} = e$, and if $\mathfrak{J}_0(e) \neq 0$, then $\mathfrak{J}_0(e)$ contains a minimal quadratic ideal.

THEOREM 4. *If $\mathfrak{J}$ satisfies (ii), (iii), and (iv), then $\mathfrak{J}$ has an identity element, so the structure theorems hold for $\mathfrak{J}$.*

Proof: The foregoing results show that $\mathfrak{J}' = \Phi 1 \oplus \mathfrak{J}$ satisfies (*i*), (*ii*), (*iii*). Hence, by the proof of the first structure theorem, every ideal in $\mathfrak{J}'$ has a complementary ideal. In particular, $\mathfrak{J}' = \mathfrak{J} \oplus \mathfrak{B}$, where $\mathfrak{B}$ is an ideal. Then $\mathfrak{J}$ has an identity element.

Axioms (*iii*) and (*iv*) are evidently satisfied if $\mathfrak{J}$ is finite dimensional over Φ. Also, it is quite easy to show in this case that (*ii*) is equivalent to the assumption that $\mathfrak{J}$ has no nil ideals. Therefore, our results give a new and improved derivation of Albert's structure theorems for finite dimensional semisimple Jordan algebras.

[1] See author's paper, "A coordinatization theorem for Jordan algebras," these PROCEEDINGS, **48,** 1154–1160 (1962); and the references in this paper.

[2] See ref. 1; also Jacobson, N., "A theorem on the structure of Jordan algebras," these PROCEEDINGS, **42,** 140–147 (1956); and McCrimmon, K., "Norms and non-commutative Jordan algebras," forthcoming in *Pacific J. Math.*

[3] "Jordan algebras of self-adjoint operators," *Mem. Am. Math. Soc.*, **53** (1965).

[4] See ref. 1, p. 1154, and the reference given there to a dissertation by Dallas Sasser.

[5] See ref. 1, p. 1158.

[6] "Simple alternative rings," *Ann. Math.*, **58,** 544–547 (1953).

[7] "Lie and Jordan systems in simple rings with involution," *Am. J. Math.*, **78,** 629–649 (1956).

Personal History and Commentary

1970–1980

I became President of the American Mathematical Society in January 1971. Previously, I had held a number of other positions in the Society and had served on a number of committees: Councillor, Trustee, Vice-President, President-Elect, member of the Editorial Boards of the *Bulletin* and the *Transactions*, member of the Colloquium Committee and the Committee to Select the Gibbs Lecturer, etc.

Political issues became an important part of the agendas of our national meetings in response to widespread dissatisfaction with the continuance of the war in Vietnam. A hotly debated issue was: Should the Society, whose primary purpose was that of fostering research express dissent with national policy on the war?

Among those who felt that we should keep aloof from politics was Zariski, who was President in 1969 and 1970. In a letter addressed to the Members of the Council and the Board of Trustees of the American Mathematical Society published in the October 1970 issue of the *Notices* (p. 869), he took a strong stand against political action by the society. Two excerpts from his letter are: "The main issue to me lies in the plain undisputable fact that the only thing the members of the Society have in common is that they are mathematicians"; and at the end of his letter, "I herewith serve notice, as a plain member fo the Society, that if the Society at any time votes to take a definite stand on a political issue, I will resign forthwith from the Society and I will do so regardless of whether I agree or do not agree with the political content of that stand."

There were voices also on the other side of this issue, and in fact, Zariski's letter was written in response to a petition by a group of mathematicians calling for "the immediate withdrawal of United States armed forces from Indochina" and urging "the AMS to take this position as a national stand."

The main business at the Business Meeting of the Annual Meeting of the Society at Atlantic City in January 1971 was the consideration of the following resolution that had been proposed by Professor Anatole Beck.

> Resolved that the Society shall poll its membership on the question: Are you in substantial agreement or disagreement with this statement: "I favor the prompt termination of American participation in the War in Southeast Asia, and the withdrawal of our troops with the greatest speed possible, limited only by the availability of transport and the safety of said troops regardless of the consequences for the present governments in Saigon and Pnom Penh."

This item was scheduled for this meeting as a result of a vote that had been taken at the Business Meeting of the Summer Meeting of the Society at Laramie. I had not attended that meeting, but I was in the chair at the Atlantic City Business Meeting.

After Professor Beck spoke at length for his motion, Professor Edwin Moise offered an amendment to the effect that instead of polling the mebership by mail, a vote should be taken at this meeting on the substance of Beck's resolution. He argued that if passed, this would have the same effect as a poll of the membership and would save the considerable expense of a mail vote. After some debate, Moise's amendment was defeated. Immediately after this, the question was called on Beck's motion. This was defeated by the close vote of 294 to 276.

The Summer meeting of the AMS for 1971 was held at the end of August at Pennsylvania State University. Shortly before this meeting, I received requests from Professors Anatole Beck and Mary Gray to address the Council at its meeting on August 31. At this time, the council operated under the rule that its meetings were closed except to persons who had been invited to attend. The requests by Beck and Gray brought this policy into question and after considerable debate it was decided by an 8 to 5 vote that the "Council meeting normally be open except when closed by the President or Chair or the majority vote of the Council." The meeting of the council was called at 5 P.M., recessed for dinner from 6:30 to 8 P.M., and adjouned at 11:05 P.M. Professor Gray addressed the council before the dinner recess and Professor Beck shortly before adjournment. Professor Gray spoke mainly on the nominating procedure for certain offices in the society. She regarded the one used by the council as undemocratic, and she suggested that the council adopt the rule that anyone proposed by a certain number of members (she suggested 100 or more) should automatically have his or her name included on the ballot. Some opposition was expressed to this procedure. It appeared that the issue could not be resolved at this meeting, so the council voted that the President appoint a committee to consider this proposal.

Nomination by petition for Vice-President and members-at-large of the Council became a reality soon after this meeting. The present rule is that anyone supported for one of these offices by 50 or more members would have his or her name placed on the ballot. On the other hand, failure of such a nominee to be elected after two tries would exclude the candidate from further consideration for a period of 10 years.

Professor Beck in his address stated that because both the AMS and the MAA (Mathematical Association of America) had decided not to consider a resolution on the war in Vietnam, he would not continue to work on this, although he hoped someone else would take it up. Instead, he wished now to talk about "alienation and a generational warfare" that was going on in the universities. He thought it important that this breach be healed, and he urged the council to do something about it. He had no concrete suggestion to offer. Because of the lateness of the hour, the subject was not pursued at this meeting.

The status of women in the profession had to be addressed by the Society. A Committee for Women with Cathleen Morawetz as Chairperson was appointed. One of the first orders of business of this committee was the following resolution that was voted at the Business Meeting of September 2, 1971, to be placed on the agenda of the next business meeting:

> Resolved that
>
> 1. The American Mathematical Society will work actively for equal opportunities for women in the following areas:
>   a. employment at all levels: this will include the search for and recruitment of qualified women;
>   b. advancement and tenure in academic positions;
>   c. admissions to graduate schools;
>   d. graduate and postdoctoral fellowships and assistantships; and
>   e. membership on advisory boards and panels; and
> 2. the Society will include more women on

a. Society programs and panels, including invited speakers and section chairman, and
b. Society committees and governing boards.

The Association of Women in Mathematics (AWM) was also founded in 1971. This has played an important role in making us aware of the problems of women in the profession.

The most serious problem before the Society in 1972 was the financial crisis of the *Mathematical Reviews* (MR). The deficit budgeted for MR for 1972 was $90,000, and the projected deficit for 1973 was approximately $290,000. At this juncture, there was a serious threat that we would have to discontinue publication of MR. I brought this to the attention of the membership in a letter published in the October 1972 *Notices*. This included an appeal for direct contributions to an MR fund. The total funds raised in this way were miniscule, but perhaps the appeal helped to make the membership and subscribers receptive to increases in dues and subscription prices that appeared to be the only practical way out of the financial difficulties. Also, at the suggestion of Ted Martin, the Treasurer, a Crisis Committee for MR headed by Ralph Boas was formed to make recommendations for alleviating the problem in the short run and suggesting a long-term solution. Another concern of MR at this time was the coverage o the field of applied mathematics that many felt was inadequate. A Committee on Mathematical Reviews Coverage, also headed by Ralph Boas, was studying this problem.

I think it is fair to say that the outcome of all this was a stronger MR that gave a much better coverage of applied mathematics. Moreover, this was achieved without sacrificing the quality of reviews of what appeared to be significant papers nor the quality of the physical make-up of the journal. MR continues to be the best mathematics review journal and is an outstanding achievement of the Society.

In November 1970, I attended an algebra conference in Rome. This was sponsored by the Istituto di Alta Matematica and was held November 23–26 at the mathematical institute of the University of Rome (Istituto Matematico Castelnuovo). The participants (in the order of their presentations) were R. Baer, P. M. Cohn, S. A. Amitsur, D. G. Northcott, A. Orsati, L. Lesieur, D. Rees, I. N. Herstein, P. Ribenboim, G. Mischler, N. Jacobson, P. Salmon, C. Procesi, and C. Faith. The organizers of the conference were Herstein and Procesi. The title of my talk was Connections Between Associative and Jordan Rings. This appears as [91] in this collection.

The visit to Rome turned out to be a prelude to a considerably longer stay in the spring of 1971. I was due to take a triennial leave during the second semester of 1970–1971, and we were planning to visit Rome and Israel during this period. In October 1970, I received a letter from Professor Lucio Lombardo Radici stating that he had heard that I was planning to spend some time in Rome and offering me a visiting professorship at the University of Rome during my stay. He was also going to invite Herstein, and he hoped that the two of us could organize a program in algebra between us. As I wrote him in my reply, I was unable to accept such a position under the terms of my triennial leave, but I offered to give some lectures during my visit to Rome, and I mentioned some possibilities for mini-courses I might give. The details of this could be worked out after we got settled in Rome.

We arrived in Rome in February 1971 after an Atlantic crossing on the *Michelangelo*. We had arranged to pick up a car we had purchased on our arrival at Genoa. From there we drove to Rome, where we spent a few days in a hotel until we found an apartment on Via Ripetta close to the Piaza del Popolo. This turned out to be a disaster because of the street noise, so after a month we gave up the apartment and moved to a small hotel, The Lord Byron, in a quiet and pleasant location in the Parioli section of Rome near the Museum of Modern Art.

I lectured once a week at the Istituto Matematico Castelnuovo. The audience was small, but included some excellent young people. One of these, Monica Barattieri, came to Yale the following year. The traffic in Rome was more than I could contend with alone, so on the days I lectured, Florie had to accompany me as navigator. Living in Rome gave us an opportunity to see at leisure its great cultural monuments. Even the location at Via Ripetta had the compensation that we were only a few steps away from the Church of Santa Maria del Popolo with its treasures, which included two masterpieces by Caravaggio.

During our stay in Rome, we took two trips for lectures. The first of these was to Genoa, where I was invited by our friend, Alessandro Figa-Talamanca, whom we had first met at Yale. Our second trip was for several days in London, where I had been invited for two lectures, one a University of London lecture and the second at Paul Cohn's seminar.

From mid-April to mid-May, we visited Israel (our third visit there), where I had a joint invitation from the Weizmann Institute and Hebrew University. During this time, Herstein was a professor at the former, and he was responsible for their invitation. The arrangements for our visit to Hebrew University were made by our friend Shaul Foguel, who had been a student at Yale. In addition to the lectures at the Weizmann Institute and Hebrew University, I also lectured at Tel Aviv University at the invitation of Harley Flanders, who had taken a professorship there, and at Haifa University. During our stay in Jerusalem, we also saw the Zelinskys, who were there for an extended stay.

During the month in Israel, we had several interesting trips: a trip with Herstein to Caesarea and Ako; a trip with Amitai Regev to Kibbutz Michmaroth, where he was born and where his parents still lived; a trip with Herstein to Nazareth, Tibereas, Capernaum, and Safad; and a trip with Amitsur to Masada and the caves of the Dead Sea Scrolls. In addition, we had several tours of Jerusalem under the expert guidance of Amitsur, who knew every stone of the city.

After our trip to Israel, we flew back to Rome, where we stayed for a short time. Then we returned to the United States on the *Rafaello*, which sailed from Naples. We drove from Rome to Naples and had arranged to have our car accompany us on the boat. On the drive to Naples, we stopped just outside of Rome to view a monument to the Holocaust that had been designed by Mirko (Basaldella), who was an important sculptor and whose widow, a sister of Yole (Mrs. Oscar) Zariski, we had visited to purchase a small bronze sculpture that he had designed. It was around noon when we stopped. The streets were deserted. Perhaps for this reason we were a bit careless and left our car with its windows down while we walked a few feet away to view Mirko's sculpture. During our brief absence from the car, two young men on a motorcycle swooped down on the car and drove off. After they disappeared in the distance, we suspected they had stolen something from the car but we did not realize it was my jacket until we had driven about halfway to Naples. This made our departure a hectic one since my passport, vaccination certificate, reading glasses, and boat tickets were in the inside pocket of the jacket that had been stolen.

A Conference on Lie Algebras and Related Topics was held at Ohio State University on the weekend of October 29–31, 1971. The organizers were Harry P. Allen, Robert H. Brown, Joseph C. Ferrar, and Hans Zassenhaus, all members of the Mathematics Department of Ohio State, whose chairman at the time was Arnold Ross. The conference was well attended. Sixty-five persons were listed as participants, of whom 17 presented papers. These were A. A. Albert, Richard E. Block, C. W. Curtis, John R. Faulkner, Robert Gordon, J. E. Humphreys, John B. Jacobs, Nathan Jacobson, K. McCrimmon, Kurt Meyberg, J. M. Osborn, Art Sagle, Richard D. Schafer, G.B. Seligman, Robert Wilson, D. Winter, and Hans Zassenhaus. The title of my talk was Remarks on Exceptional Lie Algebras. The results announced here were published in the monograph *Exceptional Lie Algebras*.

Albert's presentation at the conference was rather muddled, which was strange for him, as was his subdued manner. It was apparent to those of us who knew him that he was not well. This was indeed the case; he never recovered from this illness, and he died the following June.

In February 1972, I was elected to honorary membership of the London Mathematical Society. About the same time I learned of this honor in a letter from Professor C. A. Rogers of the University College, London, I received a telephone call from Florie's brother Al (Albert Dorfman, Professor of Biochemistry and Pediatrics at the University of Chicago) that he had just learned that I was to receive and honorary degree from the University of Chicago. A few days later I received a letter from President Edward H. Levi stating that they proposed to confer on me the honorary degree of Doctor of Science at one of several convocations that he listed. He requested that I choose one of these. I chose the Spring Convocation, which would occur June 9, 1972. The ceremony was truly impressive, especially the academic procession to and in Rockefeller Memorial Chapel. I was presented for the degree to the President by Saunders MacLane, who read a beautiful statement about me (that I believe he had written). Following the ceremony, the honorees and their wives and other guests had a luncheon with President and Mrs. Levi. The others who were awarded honorary degrees at the June 9 convocation were Edward Uhler Condon, John Philip Dawson, and Samuel Abraham Goudsmit. Condon had been an associate professor of mathematical physics at Princeton when I was a graduate student, and I knew him quite well in "the old days."

We arrived at Chicago on June 6, since I had accepted an invitation to give a lecture on June 7 in their series of lectures entitled *Contemporary Mathematics from a Historical Viewpoint*. When we arrived at Chicago, we learned that Albert had died early that morning. I dedicated my lecture on Jordan algebras to his memory. In 1974, I published an obituary article [92] on Albert in the *Bulletin of the American Mathemtical Society* (1972) and during October 21–27, 1976, I gave the first of the *Adrian Albert Memorial Lectures* (Groups and Lie Algebras Defined by Jordan Structures), an annual series of lectures dedicated to his memory. Adrian had been a close friend of both of us for many years. I had met him first in 1933 when we attended Weyl's lectures at the Institute in Princeton, and Florie had been his doctoral student when we met in 1941. I was also greatly indebted to him mathematically.

Soviet anti-Semitism became a serious concern of western mathematicians during the seventies. The specific abuses to which we objected were denial of travel to Jewish mathematicians who had been invited to address international meetings, especially the ICM's; denial of exit visas to those who wished to emigrate, usually on the highly suspect grounds that the applicant possessed "state secrets"; racial quotas for admission to study at the important mathematics centers; and difficulty for Jews to obtain advanced degrees. I was involved with these problems in a number of different capacities: as Vice-President and member of the Executive Committee of the IMU (International Mathematical Union) from 1972–1974; Chairman of the Mathematics Section of the National Academy of Sciences from 1973–1975; and Chairman of the Human Rights Committee of the AMS from 1975–1977. This committee was also concerned with human rights violations in other countries.

I accepted the position of Vice-President of the IMU to fill the vacancy that had been created by Albert's death in June 1972. The other Vice-President at the the time I served was L. S. Pontrjagin fo the USSR. K. Chandrasekharan of the ETH (Eidgenossische Technische Hochschule) was President and Henri Cartan was Past President. There was friction between Pontrjagin and other members of the EC (Excutive Committee) on the issue of non-participation of some invited speakers from the USSR at international meetings. Pontrjagin attempted to blunt the criticism of the USSR by pointing out that at conferences in Japan and Australia three each of Soviet mathematicians were present. In spite of Gorbachev's perestroika, the situation has not yet improved with regard to participation in the ICMs. A large number of Soviet invitees to the most recent ICM (Berkeley 1986) were unable to attend[14].

In late 1974, a number of mathematicians became aware of the plight of I. I. Piatetski-Shapiro. He had applied for an exit visa to go to Israel where his only son lived. He was refused on the grounds that he had worked on state secrets. As a consequence of his application, he was dismissed or resigned from the posts he held, and he could not publish his papers in Soviet journals. He made a public appeal for support in which he stated that he had been unemployed for more than a year and that he had requested the director of the institute where he had worked, M. V. Keldysh, who was also the President of the Academy of Sciences of the USSR, to review the work he had done and send the relevant information to the Soviet officials responsible for the issuance of exit visas. He appealed to the international community of mathematicians to ask Keldysh to vouch that he had never had access to classified information. He mentioned also that the only work the Academy of Science had offered him was occasional reviewing work, which would not pay him enough to live on.

In December of 1974, I received a letter from Professor Lars V. Ahlfors suggesting that the Mathematics Section of the National Academy of Science take some joint action on behalf of Piatetski-Shapiro. I asked him, G. D. Mostow, and J. Tate to draft a letter for Piatetski-Shapiro that I would circulate for signatures among the members of our section. Accompanying this letter was a Curriculum Vita of Piatetski-Shapiro (P.-S.) and a carefully researched report on his work written by Armand Borel of the Institute for Advanced Study. The fact that P.-S. was one of the top mathematicians of the USSR and indeed of the world was well known to many members of the Mathematics Section. The purpose of Borel's report was to make this known to all members. Forty-five out of a total of 59 members of the mathematics Section signed the Ahlfors–Mostow–Tate letter. Four others were in general agreement with the letter but had some objections that kept them from signing. One member strongly opposed the statement in the letter that asserted that P.-S. had not engaged in classified research. He wrote: "The evidence that Piatetski-Shapiro had never performed such research is

[14] See p. 56 for the record of attendance of invited Soviet speakers for the Vancouver ICM in 1974.

inevitably questionable since in such a complex case his simple statement made to an anonymous party must formally be subject to doubt. More generally, it is uncertain how far our country should go in telling a different country how they should handle problems about military secrecy."

Our letter was sent to President Podgorny of the USSR, to Keldysh, and to Ambassador Dobrynin. A letter to Brezhnev on behalf of Piatetski-Shapiro by a number of the leading German mathematicians was sent in January 1975 and another by leading French mathematicians was also sent to Brezhnev. This was published in *Le Monde*.

We tried to enlist the aid of the Academy itself on behalf of P.-S. To this end, Dan Mostow presented his case to the Council of the National Academy. We had hoped that President Handler of the Academy would write a letter to Keldysh in which P.-S. would be mentioned. In the end, we were rather disappointed in that the only action taken by the Academy itself was in private conversations by the Foreign Secretary, George S. Hammond, with several influential scientists of the USSR, including the Secretary General of the Academy of Sciences.

Piatetski-Shapiro was permitted to emigrate from the USSR early in 1976.

In the November 1978 *Notices of the American Mathematical Society*, a group of mathematicians who had recently emigrated from the Soviet Union published an article "The Situation in Soviet mathematics," in which they traced the history of discrimination against Jewish mathematicians and Jewish students who wished to study mathematics in the Soviet Union. The names of the authors were not revealed since they feared reprisals on persons close to them who had remained in the USSR. Instead, the article ended with a list of 16 American mathematicians who had signed the statement: "We believe the authors have made every effort to present a true picture of the situation." I was among the signers. A section of "The Situation ..." was devoted to admissions to institutions of higher learning especially to Mechmat (the mathematics department) of Moscow State University. The article noted that whereas in 1964 there were 84 out of 410 students admitted to Mechmat who were identified as Jews on their (internal) passports, there were only two to four Jews among 400 to 500 admitted in the years beginning with 1970. The mechanism for achieving this change was that of special "Jewish examinations" for Jewish students. Further details on this were given in an article by Boris Weisfeiler "(About non-admissions of Jews to universities in USSR)." Also, in an appendix to the book *It Seems I Am a Jew* by Grigory Freiman, the following comment by Academician Andrei Sakharov appears.

> To give such problems at admission examinations is, in my opinion, completely inadmissible. The problems are very complicated, require enormous concentration and accuracy in carrying out long computations, and presume knowledge and experience far beyond the level which is possible even for a very able high school graduate applying to a university. It is espcially inadmissible to offer such problems at oral examinations with the tense atmosphere, and with a time limitation of 20 minutes per problem.... A comparison with problems offered to others who were not destined to fail confirms the discriminating intentions of the examiners.

In 1979, A. I. Kostrikin was Chairman of the Department of Mathematics at Moscow State University, and he had the primary responsibility for admissions to Mechmat. I regarded him as a friend; I saw a good deal of him when I visited the Soviet Union in 1961, and I was his primary host when he visited Yale in 1967. George Seligman and I wrote him a letter in July 1979 that we had heard "of systematic use of separate entrance examinations and arbitrary practice of grading them, to exclude Jewish students from the study of mathematics at Moscow State University.... We urge you to disavow such practices, and to dissociate yourself from them."

The difficulty for Jews to obtain doctorates is well documented in "The Situation in Soviet Mathematics" and in other sources. I myself learned of a case that was so blatant that it caused a furor in the USSR itself and forced a reconsideration of the dissertation that was then accepted, within a year of its original rejection.

In 1978, L. S. Pontrjagin made himself famous (perhaps "infamous" is more appropriate) by an anti-Semitic attack on me in an autobiographical article he published in *Uspekhi Mathematicheskikh Nauk*. In describing his activities as a member of the EC of the IMU he wrote: "There was an attempt by Zionists to take the International Union of Mathmaticians into their hands. They attempted to raise N. Jacobson, a mediocre scientist but an aggressive Zionist, to the presidency. I managed to repel this attack."[15]. After consulting several mathematicians of high standing internationally, I decided to write some comments, in the form of a letter to *Uspekhi*, on Pontrjagin's account of the 1974 election of President of IMU and his bizarre attack on me. The editorial board refused to publish my letter, but I published it together with other relevant facts in the June 1980 *Notices of the American Mathematical Society*. Since it would be rather lengthy to duplicate or summarize this, I leave it to the reader to consult the *Notices* for this material.

I conclude this account of Soviet anti-Semitism with the following excerpt from a letter in English I received from a colleague in one of the socialist countries:

> There is one possibly "new" (old?) bio-ideological motivation I recently got to hear in a private discussion here with some mathematician to whom I announced your documents [my letter to the *Notices* on Pontrjagin] ... and who has close connections (during a long time) to leading mathematicians SU. He authentically reported to me about the following kind of ideology in SU: Since Jewish people in history come from South and since Southern people's children proceed quicklier in their development than Northern, the degree of difficulty of examination (mathematical olympiads etc.) of them had to be posed higher than for non-Jewish candidates—for otherwise in a short time all positions in SU would have to be occupied by Jewish mathematicians only.
>
> Since this motivation is unexpected strange as well as possibly of some use on occasion of further discussion, I thought it would be of some value to know about it. (Of course, our correspondence should officially not include or reflect these themes).

As I wrote my correspondent, it is interesting that the "theorem" discovered by the Russian was not new. It had been announced many years earlier by an eminent American mathematician who also believed that he had discovered an important corollary, namely, that Jews, because of their early development peter out early. Hence, it is unwise to hire Jews.

During the period 1973–1978, I participated in a number of conferences and institutes. These were (1) The Ring Theory Conference, March 29–31, 1973, at the University of Oklahoma at Norman; (2) The Winter School on Universal Algebras and the Theory of Radicals at Reinhardsbrünn, DDR (East Germany), January 26–February 9, 1974; (3) the Summer Institute of the Canadian Mathematical Society at Dalhousie University, Halifax, Nova Scotia, summer, 1975; (4) Scuola Matematica

[15] This is a translation of his statement as it appeared in *Russian Mathematical Surveys*, which is the English translation of *Uspekhi*.

Interuniversitarea, Cortona, Italy, July 10–August 6, 1977; and (5) the 18th Summer Research Institute of the Australian Mathematical Society, Canberra, January 9–February 17, 1978. In all of these, I lectured on various aspects of the theory of PI algebras (algebras with polynomial identity) in which there were a number of important new developments during this period.

The conference at the University of Oklahoma was organized by Bernard R. McDonald, Andy R. Magid, and Kirby C. Smith. I was one of the principal speakers; the other two were Robert Gilmer and Alex Rosenberg. The *Proceedings* of the conference published by Marcel Dekker lists 88 participants and the following list of speakers: J. W. Brewer, L. N. Childs, F. R. Demeyer, E. Graham Evans, Jr., Daniel J. Fieldhouse, Joe W. Fisher, Robert Gilmer, Melvin Hochster, Raymond T. Hoobler, N. Jacobson, Stephen McAdam, Thomas S. Shores, and Richard L. Tangeman. I gave two lectures, the first a historical survey of the theory of PI algebras and the second on central localization of PI algebras. The latter was a report on some results of Louis Rowen's dissertation. Other results of his dissertation were reported in the first lecture (see [93] and Rowen's paper, "Maximal quotients of semi-prime PI-algebras," *Transactions of the American Mathematical Society* **196**, 1974, pp. 127–135). At the conference, I met Craig Huneke and succeeded in persuading him to come to Yale for graduate study.

My correspondence on the Winter School at Reinhardsbrünn was with Professor Hans J. Hoehnke. I accepted his invitation to participate in the school from February 2 to February 9, and in my letter I mentioned that Florie would accompany me and that we wished to spend a few days in East Berlin after the Winter School. I also wrote that I wished to speak on some recent developments on PI -rings at this conference.

We flew from New York to East Berlin via Copenhagen. We were met at Schoenefeld (East Berlin) airport by Dr. Walter Romberg, who at this time was head of the Zentralblatt office in East Berlin. He assisted us with the formalities of entering the DDR (which were surprisingly simple), accompanied us to a hotel where we spent the night, and gave me an envelope with DDR money to get us started. Our expenses in the DDR were to be paid by the Akademie der Wissenschaften, which sponsored the Winter School. This was done by my receiving from time to time an envelope containing a substantial number of DDR marks. It turned out to be more than adequate to cover our expenses.

The next morning (February 2) Dr. Romberg picked us up in a car for the trip to Reinhardsbrünn. On the way we stopped at the historic city of Wittenberg. It was here that Martin Luther had posted his 95 theses on the church door on November 1, 1517. We arrived in Reinhardsbrünn in time for dinner, which we took with other participants at the Schloss hotel. The lectures were held at this hotel, which had been the palace of the Dukes of Thuringia. It was a beautiful building in an equally beautiful setting that included an 800-year-old linden tree. We stayed at a comfortable small hotel a short distance from the Schloss hotel.

The conference had a truly international character with participants from Australia, Canada, Czechoslovakia, East Germany, Hungary, Poland, Sweden, USA, USSR, and West Germany. Among these were Andrunakievic from Moldavia and L. A. Skorniakov from the USSR, whom I had met in Leningrad in 1961. Another participant with whom we became very friendly was Boris Schein from Saratov in the USSR. The second week was devoted to "main lectures by invited speakers." I gave two two-hour lectures on Some Recent Developments in the Theory of PI-Algebras [94]. These gave a survey of the main known results on these algebras, including results on the Jacobson radical and the most important application of these algebras: Amitsur's construction of division algebras that are not crossed products. A list of abstracts of the lectures was published in *Studien zur Algebra und ihre Anwendungen*, Band I, Akademie-Verlag, Berlin, 1976 (MR 56, #214).

There were several interesting excursions offered to the members of the conference. In general, this area of East Germany is very picturesque and has a number of important historic places (Eisenach, Wittenberg, Weimar) associated with some of Germany's greatest figures (Bach, Goethe, Luther, Schiller). On the other hand, one of our excursions included Buchenwald, one of Germany's greatest horrors. This has been turned into a museum with particular emphasis on the heroic communists who were murdered there. At the end of the tour, mention was made of the fact that many Jews had also been killed there.

For our return trip by train to East Berlin, we were assisted by a young man, Dr. Neumann. He helped us with our luggage and took us to our hotel. He was unhappy with the lack of freedom in the DDR and indicated his approval of Solzhenitsyn, who had just left the USSR at the time we were in East Germany. He also told us about a dissident poet, Reiner Kunze, whose small volume of poems, *Brief mit-blauem Siegel*, had been published, but whose publication had been stopped when the authorities became aware of its contents. The books also disappeared from the shelves of the bookstores. We visited a number of these hoping to find a copy but were unable to find a single one. When we mentioned this to Dr. Neumann, he gave Florie his own copy that had been autographed by the author. We found the bookstores strong on propaganda and weak on literature, even the great German classics. On the other hand, television was interesting if one turned off the highly slanted news broadcasts.

We spent four days in East Berlin. I lectured at Humboldt University and at the Academy on Some Groups and Lie Algebras Defined by Jordan Algebras. During our stay, we visited the Pergamon Museum, one of the greatest archaeological museums; the Berlin State Opera, where we saw a performance of *Der Freischütz* and were impressed with the formality of the dress of the audience; the Berliner Ensemble, a theater devoted to Brecht's plays, where we saw *Turandot*, a play based on text and notes Brecht had left. We also witnessed the changing of the guard before the tomb of the unknown soldier. This was preceded by a procession of goose-stepping soldiers with military bands that was witnessed by thousands of Berliners who lined the streets of the route. We also attended a pleasant party of mathematicians at Professor Hoehnke's apartment, and we had supper at Irene (Mrs. George) Seligman's parents' home.

We left the DDR on February 14 and spent one day in West Berlin on our way to Zurich. We found West Berlin very expensive, and we were under considerable pressure to see some of it in such a short time.

The Canadian Mathematical Society sponsored a Summer Institute at Dalhousie University in Halifax, Nova Scotia during the summer of 1975. I was invited to give some lectures at the institute, and we were its guests from July 27 to August 1. I lectured on PI-algebras, focusing on a new proof of an important theorem of Artin and Procesi that gives a characterization of Azumaya algebras of rank n^2 by identities for the matrix algebra $M_n(\mathbb{Z})$. William Schelter was on the faculty at Dalhousie at the time. He had a deep interest in PI-algebras and had done important work on these. Not long after my lecture, he discovered a much simpler proof of the Artin–Procesi theorem than the one I had presented, which was due to Amitsur and to Rowen. Schelter's improved proof was based on a judicious choice of the definition of an Azumaya algebra. This was published in his paper "Azumaya algebras and Artin's theorem," (*Journal of Algebra* **46**, 1977, pp. 303–304).

We drove to Halifax taking the overnight ferry from Portland, Maine, to Yarmouth, Nova Scotia. Schelter showed us around the city, which was pleasant since it was the provincial capital and was a center for Nova Scotia's gastronomic specialities of salmon and lobster. After our stay at Halifax, we drove through Acadia (Evangeline's country) to the Bay of Fundy. We spent a night on the road and returned to the United States by ferry from Yarmouth to Bar Harbor, Maine.

In July 1977, I gave a summer course at the Scuola Matematica Interuniversitaria at Cortona. There were three sections of the Scuola, and my course was in the Argomenti di Algebra non Commutativa in which the other lecturer was Claudio Procesi. The level of the courses was roughly that of a second- or third-year graduate course in the United States. The title of my course was Some Topics in the Theory of Algebras with Polynomial Identity. The outline I submitted to Figa-Talamanca, who was on the organizing committee of the Scuola, was: I. History of PI algebras; II. Review of general structure theory; III. Formal results on PI (Amitsur–Levitzki theorem, Razmyslov's skew identities); IV. Kaplansky–Amitsur–Posner theorem; V. Brief survey of Azumaya algebras; and VI. Artin–Procesi theorem. As texts for the course, I specified my monograph *PI-algebras, An Introduction* and Procesi's *Rings with Polynomial Indentity* published by Marcel Dekker. My monograph had been published in 1975 after I had given a course at Yale, whose main objective was Amitsur's construction of noncrossed product division algebras based on PI theory. Procesi lectured on invariant theory. His outline, which he admitted was overly ambitious, was: 1. Quadratic equations for the Grassmann variety; 2. Hodge standard tableaux, Rota double tableaux, Young tableaux, application to the representation theory of the classical groups in characteristic 0; 3. First and second fundamental theorems of invariant theory; 4. Applications: the invariants of matrices, cohomology of the classical groups and characteristic classes (by purely algebraic means); 5. Trace identities, classification of representations of algebras; 6. The infinite symmetric group and Mumford's conjecture; and 7. Inequalities in invariant theory.

The courses consisted of 20 1-hour lectures, 5 days a week (Monday through Friday) for 4 weeks (July 10–August 6). In addition, Procesi and I had a joint seminar that met several afternoons a week. In the seminar, the students reported on topics to fill in some of the details of the lectures. I had 12 students in my course, including Corrado deConcini and Elizabetta Strickland.

The lectures were held at the Palazzone, a palace that had been built for Cardinal Silvio Passerini around 1525. This had been donated to the Scuola Normale at Pisa by the heir of the family. We and a number of the participants of the Scuola lived in the Palazzone, others lived in a nearby hotel, the Oasi, which had been a convent. One or two families could not be accommodated in either of these two places and had to rent apartments in the center of Cortona, which was located about a kilometer from the Palazzone. The Palazzone had some fine frescoes, one of which was a ceiling that was designed by Luca Signorelli. He died in 1523 at the age of 73 as a result of a fall from the scaffolding that he was using to paint the ceiling. The work was completed by one of his pupils.

Cortona was an astonishing place. Its art treasures included 14 paintings by Signorelli, who is sometimes regarded as a precursor of Michelangelo, and 2 masterpieces by Fra Angelico. It also had a number of beautiful Renaissance buildings, especially around the central square, and several lovely churches in the vicinity. Also, the town had two outstanding restaurants and a number of excellent trattorias. All of this in a town of about twenty thousand that rated only one star in the Guide Michelin!

We had a pleasant routine at the Scuola: breakfast at the Palazzone, classes in the morning, and lunch at the Oasi hotel. For dinner, we walked to the center of Cortona to eat at one of its restaurants, usually Tonino's, which in addition to its fine food, had a wonderful view from its terrace. It was at this restaurant that the mayor of Cortona gave a sumptuous banquet for the Scuola. Also, it was here that many of the younger members of the Scuola would gather to have ice cream before returning to their rooms. We usually walked back to the Palazzone with some of the students.

Our weekends were free for short trips. Our first one was to Florence. We drove there with deConcinia and Strickland who returned the same day, while we stayed overnight in a hotel and returned the next day by train. This gave us an opportunity to view again some of the wonders of this city. On our second weekend, we rented a car and drove to Perugia for the day (Saturday). We had a chance meeting in the main square with the Larry Goldstein family. They were spending five weeks at Perugia where he was teaching at another summer school of the Scuola Matematica Interuniversitaria; this one at a somewhat lower level than the one at Cortona. For our third weekend, we rented a car for a two-day trip to Siena, San Gimignano, and Gubbio.

After our term at the summer school, we spent several days at Cortina d'Ampezzo in the Dolomites. The weather was poor most of the time, but it cleared on our last day when we were visiting the nearby resort of Misurina. This has a picturesque setting: a small lake almost completely surrounded by high peaks. It was a place where Bohnenblust, Walker, and I had spent a few days during our tour of Europe in 1935. From Cortina, we drove past Lake Garda on our way to Milan, where we took a plane to New York.

In November 1976, I received an invitation from Professor M. F. Newman, the Director of the 1978 Summer Research Institute (SRI) of the Australian Mathematical Society, to participate in the 18th SRI, which would be held at Australian National University (ANU) in Canberra, January 9–February 17. The subject of this SRI was "Algebra" in a broad sense that included number theory and applications. I was due to have a triennial leave the second semster of 1977–1978 so I could accept Newman's invitation.

There were more than 100 participants in the 1978 SRI of whom the following were from outside of Australia: A. Baker (England), T. Bisson (USA), B. Fischer (West Germany), G. H. Golub (USA), F. D. Jacobson (USA), N. Jacobson (USA), I. Kaplansky (USA), M. Lazard (France), D. J. McCaughan (New Zealand), I. Reiner (USA), J. C. Renaud (New Guinea), C. C. Sims (USA), A. H. Stone (USA), D. M. Stone (USA), M. Waldschmidt (France), and J. B. Wilker (Canada). There were invited general lectures and "splinter groups." Both covered a variety of topics in algebra.

I gave six lectures: two survey lectures on Jordan algebras and their generalizations and four lectures on PI algebras, the first of which was on the history of the subject. This was similar to an earlier lecture I had given at Norman, Oklahoma (see [94]). The three other lectures on PI algebras had the following titles: I. Razmyslov's Central Polynomial, II. The Artin-Procesi Theorem, and III. On Shirshov's Local Finiteness

Theorems (see [95]). These appeared in the Proceedings of the SRI, which was published as #697 in the Springer lecture notes series. This volume contains the list of participants and the program for the SRI.

We had a pleasant two-room apartment in University House on the campus of ANU. There were restaurants nearby where we usually met our friends the Kaplanskys and the Arnold Ross' who were visiting Canberra during part of our stay there. Canberra is a beautiful small city. It had been designed by a Chicago architect, Burley Griffin, whose plan for the capital city of Australia had won a competition in 1913. The result is a beautiful city that features a large artificial lake named after the architect and a lovely botanical garden that attracts many birds and wildfowl. The majority of the population is in government service. On weekends, the place was deserted and rather dull. This was relieved for us by two excursions to which we were invited by Newman and Professor B. H. Neumann and two weekend visits that were arranged by our friends, the Don Barnes's, who had spent the academic year of 1975–1976 at Yale. Our first excursion was with Newman and his young son, Peter, to Tidbinbilli, a nature reserve. If one is lucky, one can see koala bears in this reserve. We tramped around quite a lot on rough terrain but did not see any. On the other hand, we saw many kangaroos, wild turkeys, emus, etc. On another weekend, we took a full day's trip with the Neumanns. We saw quite a lot of the countryside and drove along the sea. In late January (midsummer), we spent three days at a lodge partly owned by the Barnes's at Snowy Mountain. This trip was organized by Don Barnes and G. E. Wall, who had the responsibility of bringing enough food for 3 days for 10 people. Besides Barnes and Wall, there were the B. Fitzpatricks from ANU, Fischer, the Jacobsons, the Kaplanskys, and Lazard. It rained most of the time we were there and even snowed a bit. Even so, we saw some interesting places: caves, the highest pass in the area, and an elaborate hydroelectric project. We spent the first weekend in February in Sydney as guests of the Barnes's. This was undoubtedly the high point of our visit to Australia: wonderful hosts, an exciting city with many beautiful buildings (especially the spectacular opera house), and a fine zoo, reached by a harbor trip. We heard Joan Sutherland in the *Merry Widow*. Her singing was still great, but she certainly did not look the part of a petite merry widow.

After the SRI, we vacationed for a while on the South Island of New Zealand. We rented a car for a drive around the island. The scenery was magnificent. Nevertheless, our assessment afterward was that we had found the Australian landscapes, animals, and birds more interesting because of their unique character.

After our trip to East Germany in the winter of 1974, we spent a month in Zurich, where I had been invited to be a visiting member of the ETH. I gave several lectures there and received a stipend that was enough to pay our rent at the Hotel zum Storchen. This is a beautiful small, hotel located on the Limmatquai with a fine restaurant where we dined frequently.

One March 7–8, there was a meeting of the EC of the IMU at the Storchen. The members of the EC at this time were the president (Chandrasekharan), Past President (Cartan), Vice-Presidents (Jacobson and Pontrjagin), Secretary (O. Frostman), and members M. F. Atiyah, Y. Kawada, N. H. Kuiper, M. Nicolescu, and E. Vesentini. Two of the main topics discussed at this meeting were the dues and the work of the Consultative Committee for the forthcoming ICM at Vancouver.

It was apparent at the previous meeting of the IMU, which was held in Frankfurt, June 1–2, 1973, that it was necessary to raise the dues paid by the member nations of the IMU in order to provide enough income simply to maintain the level of activity of the IMU. A Committee on Dues was appointed at the Frankfurt meeting with members Atiyah, Frostman, Jacobson (Convener), and Pontrjagin. We decided unanimously to adopt the Swiss franc for the definition of the dues (a switch from the US dollar) and over Pontrjagin's objection, we set the amount of the assessment per unit at 700SF. Also, I proposed the creation of a new adherence Group VI paying 13 units to include the United States and the Soviet Union. The numeral VI indicated the number of votes for each nation in this group. Atiyah favored this proposal, but Pontrjagin vehemently opposed. In a letter to the members of the committee, he wrote:

> Professor Jacobson's proposal to create a new Group VI with the USA and, perhaps the USSR as members, and with 13 units for dues is typical of certain circles in the USA which as it seems to me, try to win influence all over the world by means of dollar expansion. This is a dangerous way that might have already brought to the violent dollar inflation. Our Union should not follow this way. So I object to the creation of a new group VI.

The dues question was on the agenda of the meeting of the EC in Zurich. The EC resolved to recommend to the General Assembly that the unit of dues be set at 600SF, that the number of adhering groups be retained as I–V, but that the assessments for the last three groups be raised from 3, 5, and 8, to 4, 7, and 12, respectively.

I was a member of the Consultative Committee (CC) for the Vancouver ICM. The Chairman was Lars Hormander. The CC has the responsibility of determining the division of the scientific program into sections and appointing a Panel Chairman for each section. The Panel Chairman was instructed to name the other members of his panel and was provided with a list of suggested names. The chairman was free to choose from this list or from any other source. After their deliberations (conducted by mail), the panels were to provide the CC with an ordered (or partially ordered) list of possible speakers for the section as well as suggestions for the hour-long expository lectures. The CC used these lists to choose a list of speakers and alternates. This was passed on to the Organizing Committee, which issued the invitations to the speakers chosen by the CC. The CC could also decide whether or not to have short contributed papers for this ICM.

The CC had two meetings, one at Lund and a final meeting at Toronto. We decided to allow short communications of 15 minutes duration by any member of the Congress. At the Toronto meeting, we worked late on the evening prior to our scheduled day of departure to finish making the list of speakers, including the list of hour speakers. All that remained to be done the following morning was for each member to read the minutes of the meeting that would be available and to record his approval or disapproval. Things did not work out as smoothly as this. The next morning at our final meeting. Professor S. V. Jablonski of the Soviet Union stated that he wished it to be recorded in the minutes that he had reservations on the work of the CC. First, he stated that the division of the Congress into Sections was "unsuccessful ... subjects which were different in volume received almost an equal number of lectures (e.g., Algebra or Probability and Mathematical Statistics on the one hand, and Algebraic Groups or Operator Algebras on the other) ... The decision of the CC about speakers from the Soviet Union also in several cases, was unobjective, which was partly due to mistakes in the composition of the Panels." (At the meeting of the CC in Lund, Jablonski had tried unsuccessfully to block Gelfand's appointment as Panel Chairman to every panel for which his name had been proposed.) Jablonski also warned that "a situation could arise in which a number of speakers from the USSR would not, in fact, be supported by the official scientific bodies (the Academy of Sciences of the USSR and the National Committee of Soviet Mathematicians) and would not be included in the Soviet delegation to the Vancouver Congress."

The Chairman of the CC, speaking on behalf of the other members, responded to Jablonski's attack by affirming that the CC represented the international mathematical community with special emphasis on the needs of the host country and was definitely not a representation of the various national organizations. "The CC was guided by these principles in its deliberations. It strongly relied on the advice of the panels, which, to the best of our knowledge, were guided by the same principles."

The Jablonski matter was taken up at the EC meeting in Zurich. The President introduced the subject by stating that the role of the CC as affirmed by the Committee (minus Jablonski), was fully in keeping with established tradition, and was entirely in accord with the Working Paper given the Chairman and Members of the CC at the time of their appointment. An extensive discussion ensued. The gist of this was that Pontrjagin stated that the opinion of the Soviet National Committee was different from the CC in respect of some of the speakers invited to the Vancouver Congress, and it could happen that a small number, perhaps three or four, of the invited speakers from the USSR would be unable to attend. At one point, he said that the General Assembly could discuss the possibility of abolishing the CC altogether, its work being transferred to the EC and the National Committees. He was not demanding such a change, but he wanted to know the views of other members of the EC. After extensive discussion, the EC unanimously resolved not to recommend the abolition of the CC to the general assembly. The EC further resolved to send the following statement to all National Committees:

> The Executive Committee invites the active help of all National Committees in promoting international cooperation in mathematics to the fullest possible extent, and with that end in view, specifically requests that National Committees *should* do their utmost to ensure the participation in the International Congress at Vancouver (August 21–29, 1974) of the mathematicians from their respective countries who have been invited by the Organizing Committee of the Congress to give hour-addresses or specialized lectures, the selection of these mathematicians having been duly made by the Consultative Committee, following the advice of Panels of experts in the principal fields of mathematics, as resolved by the Sixth General Assembly of IMU, Menton, 1970. The Executive Committee wishes to point out that a more active functioning of the National Committees, especially in keeping the Executive Committee informed of mathematical developments, and related matters, in the countries concerned, could result in more effective cooperation within the Union.

I attended the meeting of the Assembly of the IMU at Harrison Hot Springs in Canada prior to the Congress at Vancouver. We also attended the Congress. The records show that considerably more than three or four of the invited speakers from the USSR were unable to attend. There were 31 speakers invited from the USSR and only 15 attended the Congress. None of the latter was Jewish. On the other hand, among the 16 invitees who did not attend $5\frac{1}{2}$ were Jewish, $3\frac{1}{2}$ of whom sent papers that were read by other members.

It is pleasant to be able to end this section on a lighter note: my career as Senior Marshal (SM) of Yale. I was asked by the Secretary, Henry Chauncey, to serve in this capacity for the academic year 1978–1979 and then again in 1979–1980. The main duty of the SM is to lead the Commencement procession wearing full academic regalia including a mace, the Havemeyer mace, a highly ornamental one made of wood and silver that is believed to have been an 18th century Dutch band leader's baton. For 1978–1979, there were two additional processions that I had to lead, one for the freshman assembly and one for the inauguration of our new President, A. Bartlett Giamatti. The Yale commencements, barring a torrential rain lasting the whole day, are always held in the courtyard of the Old Campus. The procession begins at the Noah Porter Gate, proceeds across Elm Street onto the New Haven Green, where it circles the oldest church in New Haven, the Center Church, and returns to the Old Campus. The following were my instructions as senior marshal for the 1979 commencement.

> At Noah Porter Gate line up behind 2 Campus Police, next to Procession Marshal, ahead of National Colors.
>
> March down Elm to College, up College to Phelps Archway.
>
> When Chief Marshal and Presidential Party are in proper position, the Procession Marshal and Senior Marshal lead Colors, the Band, and the rest of the procession through Phelps Archway. When Senior Marshal reaches Campus side of Archway, peel off to left with 4 flag bearers and take new position under direction of Old Campus Marshal.
>
> When signal is received by Old Campus Marshal for Procession Marshal, the Senior Marshal leads the Massed Banners down the Center Aisle (Aisle 5) to the Platform.
>
> When Senior Marshal arrives at Platform, turn to left and go up to seat, depositing mace in rack provided.
>
> Help with hooding of honorary degree recipients as directed.
>
> When Benediction has been pronounced, give signal to begin recessional.

The Procession Marshal at this time was George D. Vaill. My partner for the hooding ceremony was Professor Lloyd G. Reynolds, the Corporation Marshal. One of those who was honored at the 1979 Commencement was John G. Thompson, who was of the Yale Class of 1955 and had subsequently gone a long way to the Rouse Ball Professorship of Pure Mathematics at Cambridge and a 1970 Fields Medal.

The ceremony was duplicated with new actors in the 1980 Commencement. At this one, Danny Kaye was awarded the degree of Doctor of Humane Letters. The citation included his brilliance as an entertainer, but emphasized his humanitarian work for children through UNICEF. After he was hooded, I had to march back to my place on the platform, and I was followed by Danny Kaye, who sat on the same side as I. I was somewhat distracted by loud laughter from the audience. After the ceremony, I learned its cause: Kaye had been mimicking my walk as we returned to our seats. I saw him at the luncheon for the honorary degree recipients and told him that I thought he owed me a fee for acting as the straight man for his comedy. It was very much in keeping with the gaiety of Yale commencements that is provided by the graduating seniors who march in the procession with their residential colleges.

The papers I wrote in the period 1970–1980 are [74]–[77]. All are on quadratic Jordan algebras, and all except [76] are joint papers.

In [74], written jointly with McCrimmon, we considered a class of quadratic Jordan algebras $\mathfrak{J} = \mathrm{Jord}(Q, 1)$ defined by a quadratic form Q on the vector space $\mathfrak{J}$ over a field Φ such that Q has a base point, that is, an element $1 \in \mathfrak{J}$ such that $Q(1) = 1$. The definition of a (unital) quadratic Jordan algebra based on a distinguished element 1 and U operators U_a, $a \in \mathfrak{J}$, had been given by McCrimmon in the paper cited (on page 12). For Jord$(Q, 1)$, one defines $bU_a = Q(a, \bar{b})a - Q(a)\bar{b}$, where $Q(a, b) = Q(a + b) - Q(a) - Q(b)$, $T(a) = Q(a, 1)$, $\bar{b} = T(b)1 - b$. The algebras Jord$(Q, 1)$ can be characterized as the quadratic Jordan algebras that are generaically algebraic of degree two in the sense that there exists a linear function T and a quadratic form Q on $\mathfrak{J}$ such that $a^2 - T(a)a + Q(a)1 = 0$ and $a^3 - T(a)a^2 + Q(a)a = 0$ for all $a \in \mathfrak{J}$ and dim $\mathfrak{J} > 1$. Moreover, these equations are assumed to hold in $\mathfrak{J}_\rho$ for any extension field $_\rho$ of Φ. Here, $a^2 = 1U_a$, $a^3 = aU_a$ (and in general $a^{k+2} = a^k U_a$). We remark also that we must assume the cubic equation as well as the quadratic since it may not be a consequence of the latter (see [75]).

The concept of a special universal envelope of a quadratic Jordan algebra is defined as for (linear) Jordan algebras. For Jord$(Q, 1)$, this is the Clifford algebra $C(Q, 1)$, which is defined to be the algebra $\mathfrak{T}(\mathfrak{J})/\mathfrak{K}$, where $\mathfrak{T}(\mathfrak{J})$ is the tensor algebra defined by the vector space $\mathfrak{J}$ and $\mathfrak{K}$ is the ideal in $\mathfrak{T}(\mathfrak{J})$ generated by the elements $1_\Phi - 1$, $a \otimes a - T(a)a + Q(a)1$, $a \in \mathfrak{J}$, where 1_Φ is the identity element of the field Φ, 1 is that of $\mathfrak{J}$, and $T(a) = Q(a, 1)$. This can be reduced to ordinary Clifford algebras except where char $\Phi = 2$ and $T \neq 0$, so there exists a $d \in \mathfrak{J}$ such that $T(d) = 1$. Then, we can write $\mathfrak{J} = \Phi d \oplus \Phi 1 \oplus \mathfrak{B}$ where $T(v) = 0$ for $v \in \mathfrak{B}$. In this case we can define the usual Clifford algebra $\mathfrak{G} = C(\mathfrak{B}, Q)$. We have a derivation D *in* $\mathfrak{G}$ such that $vD = v + Q(v, d)1$, $v \in \mathfrak{B}$. Then, we can form the ring of differential polynomials $\mathfrak{C}[t, D]$ over $\mathfrak{C}$, where $tc = ct + cD$ for $c \in \mathfrak{C}$. Then $C(Q, 1) \simeq \mathfrak{C}[t, D]/(t^2 + \mathrm{t} + Q(d)1)$. The structure of Jord$(Q, 1)$ and of $C(Q, 1)$ were determined for Q nondegenerate.

The definition of $C(Q, 1)$ entails corresponding definitions of Clifford groups, orthogonal groups, etc., and these are related to Jordan concepts of structure groups, inner structure groups, etc.

The representation theoretic concepts of universal associative algebras for birepresentations of [46] and [52] were extended to quadratic Jordan algebras in McCrimmon's paper, "Representations of quadratic Jordan algebras," (*Transactions of the American Mathematical Society* **153**, 1971, pp. 279–305). For $\mathfrak{J} = $ Jord$(Q, 1)$, these became the meson algebra and its homomorphic image, which is the universal associative algebra of sums of commuting special representations.These are not separable even if Q is separable in the sense that Q is nondegenerate and remains so under field extensions. This follows from the fact proved in [74] that there exist sums of commuting special representations that are not completely reducible, which contrasts with the situation for linear (char $\Phi \neq 2$) Jordan algebras discussed in [52].

The paper [75], written jointly with J. Katz, is a contribution to the theory of generic norms. I had defined this concept in [63] for any finite dimensional strictly power associative algebra with 1 via the generic minimum polynomial. A more comprehensive study of these concepts was given in [69]. In a paper, "Generically algebraic algebras" (*Transactions of the American Mathematical Society* **127**, 1967, pp. 527–551), McCrimmon introduced the concept of a generically algebraic strictly power associative algebra that need not be finite dimensional, and he extended the theory of [69] to these algebras. In a second paper, "The Freudenthal-Springer-Tits constructions revisited" (*Transactions of the American Mathematical Society* **139**, 1970, pp. 293–314), he extended the theory to quadratic Jordan algebras satisfying a certain condition of strict power associativity that does not hold for all quadratic Jordan algebras. This condition was removed in [75] by defining the minimum polynomial of an algebraic element a of a (unital) quadratic Jordan algebra over a field to be the monic polynomial of least degree such that a is a root of both $\mu(\lambda)$ and $\lambda\mu(\lambda)$. It is necessary to assume both conditions, since the first does not imply the second if the characteristic is two. On the other hand, the two conditions that a is a root of $\mu(\lambda)$ and of $\lambda\mu(\lambda)$ imply that a is a root of $v(\lambda)\mu(\lambda)$ for every $v(\lambda)$. We defined generically algebraic quadratic Jordan algebra as had been done by McCrimmon, and we based the notion of a generic minimum polynomial on that of the minimum polynomial as just defined. A large part of the earlier results were carried over in [75]. We left open three questions on generic norms known to hold if the characteristic is not two. McCrimmon settled these in the paper, "The generic norm of an isotope of a Jordan algebra" ("quadratic Jordan algebra" in our terminology) that follows immediately after [75] in the same journal.

An alternative definition of generic norm for quadratic Jordan algebras has been given by T. Springer in his book, *Jordan Algebras and Algebraic Groups*, Ergebnisse der Mathematik und ihre Grenzgebiete, Band 75, 1973 Springer-Verlag. This is restricted to finite dimensional algebras. On the other hand, it can be extended to Jordan triple systems and Jordan pairs, as has been shown by Loos in *Jordan Pairs*, Springer Lecture Notes in Mathematics #460, Springer-Verlag 1975.

The definition of the structure group of a Jordan algebra appears to have been introduced by Koecher in the monograph, *Jordan Algebras and their Applications*, University of Minnesota Lecture Notes, 1962 (where the applications are to analysis, notably, to domains of positivity). One can generalize his definition to quadratic Jordan algebras: if $\mathscr{J} = (\mathscr{J}, U, 1)$ is a unital quadratic Jordan algebra over a commutative ring K, then the structure group Str $\mathscr{J}$ is the transformation group of bijective K-endomorphisms η of $\mathscr{J}$ for which there exists a bijective K-endomorphism η^* such that $U_{\eta(a)=\eta}U_a\eta^*$, $a \in \mathscr{J}$. In [76], I considered the problem of determining Str $\mathscr{J}$ for quadratic Jordan algebras of the form $\mathscr{J} = \mathscr{H}(\mathfrak{A}, J)$, where $(\mathfrak{A}, J)$ is an associative algebra with involution. The key to the results of [76] is an alternative definition of Str $\mathscr{J}$ as the set of isomorphisms of $\mathscr{J}$ into the various isotopes $\mathscr{J}^{(c)}$ of $\mathscr{J}$. The equivalence of the two definitions in the special case of Jordan algebras over fields of characteristic $\neq 2$ was given in *Structure and Representations of Jordan Algebras*, p. 59. The usefulness of the second definition stems from the following facts:

1. Any isomorphism of $\mathscr{J}_1$ onto $\mathscr{J}_2$ has a unique extension to an isomorphism of the special universal envelope ("universal associative algebra" in the terminology of [40]) $S(\mathscr{J}_1)$ onto $S(\mathscr{J}_2)$.
2. (McCrimmon): Let $(S(\mathscr{J}), \sigma_u)$ be a special universal envelope of $\mathscr{J}$, where σ_u is the map of $\mathscr{J}$ into $S(\mathscr{J})$ and let c be an invertible element of $\mathscr{J}$. Then $(S(\mathscr{J}), \sigma_u(c)_L\sigma_u)$, where $\sigma_u(c)_L$ is the left multiplication by $\sigma_u(c)$, is a special universal envelope of the isotope $\mathscr{J}^{(c)}$.

An associative algebra with involution $(\mathfrak{A}, J)$ is called perfect if $\mathfrak{A}$ and the injection map of $\mathscr{H} = \mathscr{H}(\mathfrak{A}, J)$ into $\mathfrak{A}$ constitute a special universal envelope for $\mathscr{H}$. For perfect $(\mathfrak{A}, J)$, I showed in Theorem 2.1 of [76] that if ζ is an automorphism of $\mathfrak{A}$ such that $\zeta J\zeta^{-1}J = I_d$ $(I_d x = dxd^{-1})$, then $d_L^{-1}\zeta$ restricted to $\mathscr{H}$ is in Str $\mathscr{H}$, and every element of Str $\mathscr{H}$ is obtained in this way. The condition on ζ is satisfied by any inner automorphism $x \to g^{-1}xg$, which implies that if γ is any invertible element of $\mathscr{C}$ and g is any invertible element of $\mathfrak{A}$, then $x \to \gamma(Jg)xg$ is in Str $\mathscr{H}$.

The main cases of $(\mathfrak{A}, J)$ considered in [76] are

I. $(\mathfrak{A}, J)$ is simple Artinian in the sense that $\mathfrak{A}$ is left or right Artinian and $\mathfrak{A}$ contains no ideal $\neq \mathfrak{A}$, 0 stabilized by J.

II. $(\mathfrak{A}, J)$ is finite dimensional central simple over a field Φ, that is, $\mathfrak{A}$ is finite dimensional over Φ, $\mathfrak{A}$ contains no ideal $\neq \mathfrak{A}$, 0 and $\mathscr{H}(\mathfrak{A}, J) \cap \mathscr{C} = \Phi$).

The results for I. are obtained by using McCrimmon's theorem and results on perfection given in [48] (or a generalization of these due to Martindale given also in [76]). The case II. is a special case of I. However, considerably more can be said here. In *Structure and Representations of Jordan Algebras*, p. 241f, I showed that if char $\Phi \neq 2$, Str $\mathscr{J}$ coincides with the group of norm similarities defined to be the bijective linear transformations η such that $N(\eta(a)) = \rho N(a)$, $\rho \in \Phi$. This applies to the algebras $\mathscr{H}(\mathfrak{A}, J)$, where $(\mathfrak{A}, J)$ is central simple and gives a determination of the groups of norm similarities from the results on structure groups (loc. cit., p. 247f). In [77], I obtained similar results in the general case of quadratic Jordan algebras. Special care is required to encompass the case char $\Phi = 2$. If $\mathfrak{A}$ is central simple and J is an involution in $\mathfrak{A}$ of symplectic type, then one has to replace $\mathscr{H}(\mathfrak{A}, J)$ by the outer ideal $\mathscr{H}(\mathfrak{A}, J)'$ generated by 1 (the smallest subspace of $\mathscr{H}(\mathfrak{A}, J)$ containing 1 and stabilized by the U_a, $a \in \mathscr{H}$). This is a subalgebra of $\mathscr{H}(\mathfrak{A}, J)$, which coincides with $\mathscr{H}(\mathfrak{A}, J)$ if char $\Phi \neq 2$. If $\mathfrak{A}$ is central simple and J is an orthogonal type involution in $\mathfrak{A}$, then the trace bilinear form is degenerate if char $\Phi = 2$. This causes a breakdown in the method I used to prove that the group of norm similarities is contained in the structure group. This gap in the theory was filled by William C. Waterhouse in "Symmetric determinants and Jordan norm similarities in Characteristic 2," (*Proceedings of the American Mathematical Society* **93**, 1985, pp. 538–589). His proof makes use of the theory of group schemes.

I developed the requisite definitions and results on generic norms for the quadratic Jordan algebras $\mathscr{H}(\mathfrak{A}, J)$ and $\mathscr{H}(\mathfrak{A}, J)'$ almost completely without reference to the general theory of [75] and McCrimmon's paper that followed it. This was done by case considerations of the various types of central simple $(\mathfrak{A}, J)$.

The last part of [76] is devoted to the study of the Lie algebra str $\mathscr{J}$, which in a formal sense is the Lie algebra of the group Str $\mathscr{J}$. The results here extend those given in [B7], pp. 253–258, for Lie algebras over fields of characteristic $\neq 2$.

Two aspects of associative ring theory form the background of [77], written jointly with McCrimmon and M. Parvathi: first, Ore's construction of a quotient division ring (see, e.g., *The Theory of Rings*, p. 31) and its generalization to construct the so-called classical quotient ring of an arbitrary ring (loc. cit., p. 118). The latter became important in Goldie's theorem giving a characterization of the class of rings that have semisimple Artinian classical quotient rings (see, e.g., *Structure of Rings*, p. 261). The second topic in associative ring theory that served as a model for a part of [77] was central localization, which had been used effectively by Rowen to obtain a sharpening and improved proof Posner's theorem for PI algebras (see *PI-Algebras: An Introduction*, p. 57).

Ore's construction was that of a quotient division ring for a noncommutative domain with common multiple property: for any nonzero a, b in the domain there exist nonzero a_1, b_1 such that $ab_1 = ba_1$. In [77], we attempted to do a similar thing for quadratic Jordan algebras. Such an algebra $\mathscr{J} = (\mathscr{J}, U, 1)$ is called a domain if every U_a, $a \neq 0$ is injective and a division ring if every U_a, $a \neq 0$, is bijective. The common multiple condition is that for nonzero a and b in $\mathscr{J}$ there exist a', $b' \in \mathscr{J}$ such that

$U_a a' = U_b b' \neq 0$, or, equivalently, $U_a \mathcal{J} \cap U_b \mathcal{J} \neq 0$ for the principal inner ideals $U_a \mathcal{J}$ and $U_b \mathcal{J}$. In *Structure and Representations of Jordan Algebras* (p. 426), I suggested that this condition for every $a, b \neq 0$ in a Jordan domain might be sufficient to obtain an imbedding in a Jordan division algebra.

We considered first an imbedding problem in a purely multiplicative setting. We defined a Jordan monad $\mathcal{M}$ as a triple $(\mathcal{M}, U, 1)$, where $\mathcal{M}$ is a set, 1 a distinguished element of $\mathcal{M}$, and U a map of $\mathcal{M}$ into the monad Mon $\mathcal{M}$ of maps of $\mathcal{M}$ into itself such that $U_1 = 1$ and $U_{U_a b} = U_a U_b U_a$. Any quadratic Jordan algebra defines a Jordan monad by forgetting the module structure, and if $\mathcal{J}$ is a domain, then the subset $\mathcal{J}^*$ of nonzero elements is a submonad of the monad $(\mathcal{J}, U, 1)$. An element a of a Jordan monad is cancellable (invertible) if U_a is injective (bijective) and $\mathcal{M}$ is a cancellable (division) Jordan monad if all its elements are cancellable (invertible). The pair of elements $a, b \in \mathcal{M}$ is said to have a common multiple if $U_a \mathcal{M} \cap U_b \mathcal{M} \neq \phi$. We proved that if $\mathcal{M}$ is a cancellable Jordan monad having the common multiple property for all (a, b), then $\mathcal{M}$ can be imbedded in a Jordan division monad $\mathcal{D}$, so that every element of $\mathcal{D}$ has the form $U_a^{-1} b$, a, $b \in \mathcal{M}$. This result was proved as a corollary of a localization theorem that gives conditions for the construction of a Jordan monad $\mathcal{S}^{-1}\mathcal{M}$ for $\mathcal{S}$ a submonad of $\mathcal{M}$. The proof of this makes use of a localization theorem for associative monoids.

We can apply the result on $\mathcal{S}^{-1}\mathcal{M}$ to $(\mathcal{J}^*)^{-1}\mathcal{J}$ for a quadratic Jordan domain $\mathcal{J}$ with the common multiple property on all $a, b \neq 0$. Also, one can introduce a Φ-module structure on $(\mathcal{J}^*)^{-1}\mathcal{J}$ if Φ is the base ring of $\mathcal{J}$. To show that $(\mathcal{J}^*)^*\mathcal{J}$ is a quadratic Jordan algebra, it is necessary to show that $U_{a,b} = U_{a+b} - U_a - U_b$ is bilinear in a and b. We were unable to do this without imposing some rather complicated conditions involving the set of structural pairs of $\mathcal{J}$, defined to be pairs of Φ-endomorphisms (M, M^*) of $\mathcal{J}$ such that $U_{Ma} = MU_a M^*$ and $U_{M^*a} = M^* U_a M$, $a \in \mathcal{J}$. As far as I know, the question raised on p. 427 of *Structure and Representations of Jordan Algebras* remains open. It should be added, however, that in view of the structure theory of quadratic Jordan algebras developed by Zel'manov and by McCrimmon and Zel'manov (see the paper, "The Structure of strongly prime Jordan Algebras," (*Advances in Mathematics* **69**, 1988, pp. 133–222), and the references in this paper) the question may have lost some of its interest since another quotient, the Martindale quotient of an associative algebra, appears to be the right one to use for the structure theory of special Jordan algebras.

We were more successful in seeking a Jordan analog of associative central localization. This was given in the first part of [77]. It was obtained by replacing the center used in the associative theory by the centroid, which is defined to be the set of Φ-endomorphisms γ of $\mathcal{J}$ such that $\gamma U_a = U_a \gamma$ and $U_{\gamma a} = \gamma^2 U_a$ for all $a \in \mathcal{J}$.

Journal of the Indian Math. Soc.
35 (1971) 1–45

QUADRATIC JORDAN ALGEBRAS OF QUADRATIC FORMS WITH BASE POINTS

By N. JACOBSON and K. McCRIMMON

[Received July 21, 1970]

THIS PAPER originated in some experimentation with algebraic structures now known as special quadratic Jordan algebras which the first named author undertook more than ten years ago. The algebras we shall consider in this paper and the related Clifford algebras appeared in this investigation, the results of which have not been published before. Comparatively recently the second-named author initiated the study of abstract quadratic Jordan algebras and succeeded in extending the results of Jacobson [1] on ordinary Jordan algebras to quadratic Jordan algebras (McCrimmon [1]). One of the classes of algebras occurring in the structure theory is the class which had appeared in Jacobson's earlier study. It is the purpose of this paper to give a detailed account of these algebras and the related Clifford algebras. Except for §6 our account is essentially self-contained and does not make use of the recently developed structure theory of quadratic Jordan algebras. The results of §6 constitute a contribution to the structure theory and make use of some of its results.

We briefly recall the basic notions concerning quadratic Jordan algebras in §1. In §2 we indicate the role that the particular algebras we shall consider — the algebras Jord $(Q, 1)$ defined by a quadratic form Q with base point 1 — play in the general theory: they are precisely the generically algebraic algebras of degree 2.

The Clifford algebra of quadratic forms with base points are introduced in §3, where it is shown that these are the special universal envelopes of Jordan algebras Jord $(Q, 1)$. The structure theory of these algebras for Q separable (in the sense that the extension of Q to $\mathfrak{J}_\Omega$ is non-degenerate for Ω the algebraic closure of the base field Φ) is given in § 7. Both in § 3 and § 7 we have adopted the point of view of reduction to the case of ordinary Clifford algebras. This is quite simple in all cases except that in which the characteristic char $\Phi = 2$ and there exists an element $x \in \mathfrak{J}$ such that $T(x) = Q(x, 1) \neq 0$. We refer to this case as the impure case and show that in this case the Clifford algebra with base point is a homomorphic image of an algebra of differential polynomials over an ordinary Clifford algebra. In all cases we show that if Q is separable then the Clifford algebra with base point is central simple, except when dim $\mathfrak{J}$ is even (finite) and $T \neq 0$, in which case it is either simple with center a separable quadratic field or a direct sum of two isomorphic central simple algebras according as the discriminant $[\delta] \neq 1$ or $= 1$ if char $\Phi \neq 2$, or the pseudo-discriminant $[\Delta] \neq 0$, or $= 0$ if char $\Phi = 2$. These results allow us in §8 to define Clifford and spin groups for a separable quadratic form with base point, and to establish their usual properties.

In §4 we consider the problem of characterizing the structure group and structure Lie algebra of Jord $(Q, 1)$ in terms of the form Q.

The structure theory for algebras of the form Jord $(Q, 1)$ is developed in §5, and in §6 we describe another way in which these algebras arise in the general theory of almost nil quadratic Jordan algebras.

In §9 we characterize the various Lie algebras associated with the algebra Jord $(Q, 1)$ – the structure algebra, inner structure algebra, derivation algebra, and Tits-Koecher algebra – in terms of the form Q.

In § 10 we show that, even for separable Q, the theory of birepresentations of Jord $(Q, 1)$ breaks down in characteristic 2. We show that the Clifford algebra with base point, regarded as a bimodule in the natural way, fails to be completely reducible, and consequently the meson algebra is not semisimple in characteristic 2.

1. Quadratic Jordan algebras. Let Φ be a field of any characteristic. We define a (*unital*) *quadratic Jordan algebra* over Φ to be a triple $(\mathfrak{J}, U, 1)$ where $\mathfrak{J}$ is a vector space over Φ, U is a mapping of $\mathfrak{J}$ into the (associative) endomorphism algebra $\mathrm{End}_\Phi\, \mathfrak{J}$, 1 a distinguished element of $\mathfrak{J}$ such that the following axioms hold:

*QJ*1. $U: a \to U_a$ is quadratic, that is, $U_{\alpha a} = \alpha^2 U_a$, $\alpha \in \Phi$, and $U_{a,b} \equiv U_{a+b} - U_a - U_b$ is Φ-bilinear in a and b

*QJ*2. $U_1 = 1$

*QJ*3. $U_a U_b U_a = U_{bU_a}$

*QJ*4. If $V_{a,b}$ is defined by $xV_{a,b} = aU_{x,b}$ then $U_b V_{a,b} = V_{b,a} U_b$

*QJ*5. If P is any extension field of Φ and U is the (unique) extension of the mapping U on $\mathfrak{J}$ to a quadratic mapping of $\mathfrak{J}_{\mathrm{P}} = \mathrm{P} \otimes \mathfrak{J}$ into $\mathrm{End}_{\mathrm{P}}\, \mathfrak{J}_{\mathrm{P}}$ then *QJ*3 and *QJ*4 hold for U on $\mathfrak{J}_{\mathrm{P}}$. (McCrimmon [1]. See also Jacobson [2]).

We remark that *QJ* 5 is equivalent to the validity in $\mathfrak{J}$ of a number of additional operator identities which are obtained by linearizing *QJ* 3 and *QJ* 4. These linearizations follow automatically if Φ has at least four elements. Hence in this case quadratic Jordan algebras can be defined by *QJ* 1 – *QJ* 4 alone.

If $\mathfrak{A}$ is a (unital) associative algebra, $\mathfrak{A}$ defines a quadratic Jordan algebra $\mathfrak{A}^{(q)} = (\mathfrak{A}, U, 1)$ where U_a is the mapping $x \to axa$ in $\mathfrak{A}$ and 1 is the unit of $\mathfrak{A}$. If Φ is of characteristic $\neq 2$ and $\mathfrak{J}$ is a (linear, unital) Jordan algebra in the usual sense with the multiplication $R_a: x \to x.a$ then $\mathfrak{J}$ defines a quadratic Jordan algebra $(\mathfrak{J}, U, 1)$ if we take $U_a = 2R_a^2 - R_{a^2}$ (see the reference above for this and other results quoted without proof in this section).

A *homomorphism* η of $(\mathfrak{J}, U, 1)$ into a second quadratic Jordan algebra $(\mathfrak{J}', U', 1')$ is a linear mapping of $\mathfrak{J}$ in $\mathfrak{J}'$ such that $(aU_b)^\eta = a^\eta U'_{b^\eta}$ and $1^\eta = 1'$. We shall often write $aU_b = bab$ (as in $\mathfrak{A}^{(q)}$). Then the first homomorphism condition is $(bab)^\eta = b^\eta a^\eta b^\eta$ (writing $cdc = dU'_c$ in $\mathfrak{J}'$). A *subalgebra* $\mathfrak{B}$ of $(\mathfrak{J}, U, 1)$ is a subspace of $\mathfrak{J}$ containing 1 and every $cbc = bU_c$, $b, c \in \mathfrak{B}$. An *inner* (*outer*) *ideal* $\mathfrak{B}$ is a subspace such that $xU_b \in \mathfrak{B}$ ($bU_x \in \mathfrak{B}$) for $x \in \mathfrak{J}$, $b \in \mathfrak{B}$. An *ideal* is a subset which is both an inner and an outer ideal. If $\mathfrak{B}$ is an ideal then the factor space $\bar{\mathfrak{J}} = \mathfrak{J}/\mathfrak{B}$ is a quadratic Jordan algebra $(\bar{\mathfrak{J}}, \bar{U}, \bar{1})$ in which $\bar{1} = 1 + \mathfrak{B}$ and $\bar{a}\bar{U}_{\bar{b}} = \overline{aU_b}$ where in general $\bar{x} = x + \mathfrak{B}$. A *derivation* D of $\mathfrak{J}$ is a linear mapping of $\mathfrak{J}$ into $\mathfrak{J}$ such that $1D = 0$ and $[U_a, D] = U_aD - DU_a = U_{a,aD}$, $a \in \mathfrak{J}$.

A quadratic Jordan algebra $\mathfrak{J}$ is *special* if there exists a monomorphism of $\mathfrak{J}$ into an algebra $\mathfrak{A}^{(q)}$, $\mathfrak{A}$ associative. A homomorphism σ of $\mathfrak{J}$ into an algebra $\mathfrak{A}^{(q)}$, $\mathfrak{A}$ associative, is called an *associative specialization* of $\mathfrak{J}$. Let $\mathfrak{T}(\mathfrak{J})$ be the tensor algebra over $\mathfrak{J}$, so $\mathfrak{T}(\mathfrak{J}) = \Phi \oplus \mathfrak{J} \oplus (\mathfrak{J} \otimes \mathfrak{J}) \oplus \dots$ with the bilinear product in $\mathfrak{T}(\mathfrak{J})$ such that

$$(x_1 \otimes \dots \otimes x_r)(x_{r+1} \otimes \dots \otimes x_{r+s}) = x_1 \otimes \dots \otimes x_{r+s} \tag{1}$$

and let $\mathfrak{R}$ be the ideal in $\mathfrak{T}(\mathfrak{J})$ generated by the elements

$$1_\Phi - 1 \ (1 \in \mathfrak{J}), \ aU_b - b \otimes a \otimes b, \ a, b \in \mathfrak{J}. \tag{2}$$

Put $S(\mathfrak{J}) = \mathfrak{T}(\mathfrak{J})/\mathfrak{R}$, $a^{\sigma_u} = a + \mathfrak{R}$, $a \in \mathfrak{J}$. Then σ_u is an associative specialization of $\mathfrak{J}$ in $S(\mathfrak{J})$ and $(S(\mathfrak{J}), \sigma_u)$ is universal in the sense that if σ is any associative specialization of $\mathfrak{J}$ in $\mathfrak{A}$ then there exists a unique homomorphism η of $S(\mathfrak{J})$ into $\mathfrak{A}$ (as associative algebras) such that $\sigma = \sigma_u \eta$. We call $(S(\mathfrak{J}), \sigma_u)$ a *special universal envelope* for $\mathfrak{J}$. This is uniquely determined in the usual strong sense that if $(S(\mathfrak{J}'), \sigma_u')$ is another pair having the properties stated for $(S(\mathfrak{J}), \sigma_u)$ then there is a unique isomorphism ζ of $S(\mathfrak{J})$ onto $S(\mathfrak{J})'$ such that $\sigma_u \zeta = \sigma_u'$. We have a uniquely determined involution J in $S(\mathfrak{J})$ such that $a^{\sigma_u J} = a^{\sigma_u}$, $a \in \mathfrak{J}$. We shall call this the *main involution* in $S(\mathfrak{J})$.

We define $a^0 = 1$, $a^1 = a$, $a^k = a^{k-2}\,U_a$ for $k > 1$. In $\mathfrak{A}^{(q)}$, $\mathfrak{A}$ associative, a^k as thus defined coincides with the associative powers a^k. As in this special case, one has $(a^k)^l = a^{kl}$. Also one defines $a \circ b = (a+b)^2 - a^2 - b^2$. In $\mathfrak{A}^{(q)}$ one has $a \circ b = ab + ba$. In general $a^k \circ a^l = 2\,a^{k+l}$. In the characteristic two case $\mathfrak{J}$ is a 2-Lie algebra (= restricted Lie algebra of characteristic two) relative to the composition $[a, b] = a \circ b$ and a^2. For, we have $[a, a] = 0$, $[[a, b], c] + [[b, c], a] + [[c, a], b] = 0$, $[a, b^2] = [[a, b], b]$ and $(a+b)^2 = a^2 + b^2 + [a, b]$.

For arbitrary $\mathfrak{J}$ one defines $\{a\ b\ c\} = bU_{a,c} = b(U_{a+c} - U_a - U_c)$ and one has the relation $\{a\ a\ b\} = aU_{a,b} = a^2 \circ b$. If λ is an indeterminate we have

$$\begin{aligned}(a + \lambda b)^3 &= (a + \lambda b)\, U_{a+\lambda b} = (a + \lambda b)\, (U_a + \lambda U_{a,b} + \lambda^2\, U_b) \\ &= a^3 + \lambda(bU_a + aU_{a,b}) + \lambda^2\, (aU_b + bU_{a,b}) + \lambda^3\, b^3 \\ &= a^3 + \lambda(bU_a + a^2 \circ b) + \lambda^2\, (aU_b + b^2 \circ a) + \lambda^3\, b^3.\end{aligned}$$

We shall denote the coefficient of λ in this as $\Delta_a^b\, x^3$ so

$$\Delta_a^b\, x^3 = bU_a + a^2 \circ b. \tag{3}$$

Similarly, one defines $\Delta_a^b\, x^2$ and $\Delta_a^b\, x$ and obtains

$$\Delta_a^b\, x^2 = a \circ b,\ \Delta_a^b\, x = b. \tag{4}$$

2. Generically algebraic quadratic Jordan algebras of degree two.

DEFINITION 1. *A quadratic Jordan algebra* $(\mathfrak{J}, U, 1)$ *will be called 'generically algebraic of degree two' if* $\mathfrak{J} \neq \Phi\, 1$ *and there exists a linear function* $T(x)$ *and a quadratic form* $Q(x)$ (*=quadratic mapping into* Φ) *on* $\mathfrak{J}$ *such that*

$$x^2 - T(x)\, x + Q(x)\, 1 = 0, \tag{5}$$

$$x^3 - T(x)\, x^2 + Q(x)\, x = 0, \tag{6}$$

hold for all x *in any* $\mathfrak{J}_{\mathrm{P}}$, P *any extension field of the base field* Φ. *It is understood here that* T *and* Q *in* $\mathfrak{J}_{\mathrm{P}}$ *are the extensions of the given* T *and* Q *to linear and quadratic functions respectively on* $\mathfrak{J}_{\mathrm{P}}$.

This notion is a special case of a general notion which has been introduced by McCrimmon in [2] and developed further by McCrimmon in [3]. Since everything we shall do can be developed without recourse to the general theory we need not recall the general results here.

Let $\mathrm{P} = \Phi(\lambda)$, λ an indeterminate, and consider the equations (5) and (6) for $x = a + \lambda b$, $a, b \in \mathfrak{J}$. Taking the coefficient of λ in these equations we obtain respectively

$$\Delta_a^b\, x^2 - T(a)\, \Delta_a^b\, x - (\Delta_a^b\, T)\, a + (\Delta_a^b\, Q)\, 1 = 0, \tag{7}$$

$$\Delta_a^b\, x^3 - T(a)\, \Delta_a^b\, x^2 - (\Delta_a^b\, T)\, a^2 + (\Delta_a^b\, Q)\, a + Q(a)\, \Delta_a^b\, x = 0, \tag{8}$$

where $\Delta_a^b\, T$ and $\Delta_a^b\, Q$ are the coefficients of λ in $T(a+\lambda b)$ and $Q(a+\lambda b)$. Using the formulas for the various functions we obtain

$$a \circ b - T(a)b - T(b)a + Q(a, b)\, 1 = 0 \tag{7'}$$

where $Q(a, b) = Q(a+b) - Q(a) - Q(b)$ and

$$bU_a + a^2 \circ b - T(a)(a \circ b) - T(b)a^2 + Q(a, b)\, a + Q(a)\, b = 0. \tag{8'}$$

Replacing a by $a + c$ in (8′) gives

$$bU_{a,c} + (a \circ c) \circ b - \sum_{(abc)} T(a) b \circ c + \sum_{(abc)} Q(a, b)\, c = 0$$

where $\sum\limits_{(abc)}$ denotes summation over the cyclic permutations of a, b, c. It is clear also that the coefficient of λ^2 in (6) for $x = a + \lambda b$ is obtained from the left hand side of (8′) by interchanging a and b, and the coefficients of the other powers of λ in (5) and (6) for $x = a + \lambda b$ are 0 by (5) and (6). It follows from this that if (5), (6), (7′), and (8′) hold for all $a, b \in \mathfrak{J}$ then these hold also in $\mathfrak{J}_{\mathrm{P}}$ for any extension field P of Φ. Hence these conditions on a, b in $\mathfrak{J}$ give an alternative definition of a generically algebraic quadratic Jordan algebra of degree two.

We observe next that if Φ has cardinality $|\Phi| > 2$ then (5) and (6) alone for a in $\mathfrak{J}$ imply (7′) and (8′) and so these can be used as alternatives to the conditions of Definition 1. It is clear that (7′) follows from (5) for any Φ. To see that (8′) is a consequence of (6) in $\mathfrak{J}$ we note that if a, b in $\mathfrak{J}$ and $\beta \in \Phi$ then the left hand side of (6) for $x = a + \beta b$ can be reduced by (6) to $\beta f(a, b) + \beta^2 f(b, a) = 0$ where $f(a, b)$ is the left hand side of (8′). Since we have two non-zero β's in Φ we obtain $f(a, b) = 0 = f(b, a)$, so (8′) holds. Hence (5), (6), (7′) and (8′) hold in $\mathfrak{J}$.

We shall now deduce from our relations a simple formula for bU_a. First we put $b = 1$ in (7′) and use the known formula $a \circ 1 = 2a$ to obtain $2a - T(a)\, 1 - T(1)\, a + Q(a, 1)\, 1 = 0$. Since $\mathfrak{J} \neq \Phi 1$ this gives $T(1) = 2$, $Q(a, 1) = T(a)$. Then if we put $x = 1$ in (5) we obtain $Q(1) = 1$. Thus we have

$$T(1) = 2,\ Q(1) = 1,\ Q(a, 1) = T(a). \tag{9}$$

By (5), (7′) and (8′) we obtain

$$bU_a = Q(a)\,b + T(a)\,T(b)\,a - Q(a, b)\,a - T(b)\,Q(a)\,1. \tag{10}$$

We can write this more compactly if we put

$$\bar{b} = T(b)\,1 - b. \tag{11}$$

Then (10) becomes

$$bU_a = Q(a, \bar{b})\,a - Q(a)\,\bar{b}. \tag{12}$$

We note also that the definition (11) and $T(1) = 2$ give

$$\bar{1} = 1, \bar{\bar{b}} = b, b \in \mathfrak{J}. \tag{13}$$

Conversely, suppose we have a vector space $\mathfrak{J}$ over Φ with, $\dim \mathfrak{J} > 1$, a quadratic form Q on $\mathfrak{J}$, a vector $1 \in \mathfrak{J}$ such that $Q(1) = 1$. Define $T(a) = Q(a, 1)$ where $Q(a, b) = Q(a + b) - Q(a) - Q(b)$ is the symmetric bilinear form associated with Q. Define $\bar{b}$ by (11) and the operator U_a by (12). We claim that $(\mathfrak{J}, U, 1)$, $U: a \to U_a$, is a quadratic Jordan algebra which is generically algebraic of degree two. We have $T(1) = Q(1, 1) = 2Q(1) = 2$ and direct verification gives (13) and

$$Q(\bar{b}) = Q(b) \tag{14}$$

which implies $Q(\bar{a}, \bar{b}) = Q(a, b)$ and $Q(a, \bar{b}) = Q(\bar{a}, b)$. Also $Q(bU_a) = Q(Q(a, \bar{b})\,a - Q(a)\,\bar{b}) = Q(a)\,Q(a, \bar{b})^2 - Q(a)\;Q(a, \bar{b})^2 + Q(a)^2\,Q(b)$. Hence Q satisfies

$$Q(bU_a) = Q(b)\,Q(a)^2. \tag{15}$$

We note that since the conditions on Q carry over to $\mathfrak{J}_{\mathrm{P}}$ for any extension field P of Φ it suffices to prove that $QJ1 - QJ4$ and (5) and (6) hold in $\mathfrak{J}$.

From the definition (12) of U_a we obtain for $U_{a,b} = U_{a+b} - U_a - U_b$ the formula

$$xU_{a,b} = Q(a, \bar{x})b + Q(b, \bar{x})\,a - Q(a, b)\bar{x}. \tag{16}$$

It is clear from this that $(a, b) \to U_{a,b}$ is a bilinear mapping into $\mathrm{End}_\Phi \mathfrak{J}$. Also it is clear from (12) that $U_{\alpha a} = \alpha^2 U_a$. Hence $QJ1$. Since $xU_1 = Q(1, \bar{x})1 - Q(1)\bar{x} = T(\bar{x})\,1 - \bar{x} = \bar{\bar{x}} = x$, $QJ2$ holds. It remains to verify $QJ3$ and $QJ4$. We have $xU_a = Q(a, \bar{x})\,a - Q(a)\,\bar{x}$ and, by (16) and $xV_{a,b} = aU_{x,b}$, we have

$$xV_{a,b} = Q(x, \bar{a})\,b + Q(b, \bar{a})\,x - Q(x, b)\bar{a}. \tag{17}$$

As preliminary to the proof of $QJ3 - QJ4$ we establish first the following formulas:

(i) $\overline{bU_a} = \bar{b}U_{\bar{a}}$

(ii) $Q(xU_a, y) = Q(x, yU_{\bar{a}})$

(iii) $\bar{a}U_a = Q(a)\,a$

(iv) $\bar{x}U_{\bar{a}}U_a = Q(a)^2\bar{x}$.

For (i) we have $\overline{bU_a} = Q(a, \bar{b})\,\bar{a} - Q(a)\,b = Q(\bar{a}, b)\,\bar{a} - Q(\bar{a})\,b = bU_{\bar{a}}$. For (ii), $Q(xU_a, y) = Q(a, \bar{x})Q(a, y) - Q(a)Q(\bar{x}, y)$, $Q(x, yU_{\bar{a}}) = Q(\bar{a}, \bar{y})Q(\bar{a}, x) - Q(\bar{a})Q(x, \bar{y})$. For (iii), $\bar{a}U_a = Q(a, a)a - Q(a)a = 2Q(a)a - Q(a)a = Q(a)a$. For (iv), $\bar{x}U_{\bar{a}}U_a = \overline{xU_a}\,U_a = Q(a, xU_a)\,a - Q(a)\,xU_a = Q(aU_{\bar{a}}, x)\,a - Q(a)\,xU_a = Q(\bar{a})Q(\bar{a}, x)a - Q(a)Q(a, \bar{x})\,a + Q(a)^2\,\bar{x}$. Now we have

$$\begin{aligned} xU_a\,U_b\,U_a &= (Q(b, \overline{xU_a})\,b - Q(b)\,\overline{xU_a})\,U_a \\ &= Q(b, \bar{x}U_a)\,bU_a - Q(b)\,\bar{x}U_{\bar{a}}\,U_a \\ &= Q(bU_a, \bar{x})\,bU_a - Q(b)\,Q(a)^2\,\bar{x} \\ &= xU_{bU_a} \end{aligned}$$

and

$$\begin{aligned} xV_{a,b}U_a &= (Q(x, \bar{a})b + (Q(b, \bar{a})x - Q(x, b)\bar{a})U_a \\ &= Q(x, \bar{a})Q(a, \bar{b})a - Q(x, \bar{a})Q(a)\,\bar{b} + Q(b, \bar{a})xU_a - \\ &\quad - Q(x, b)Q(a)a \\ &= Q(a, \bar{x})Q(a, \bar{b})\,a - Q(x, aU_{\bar{a}})\,\bar{b} + Q(a, \bar{b})\,xU_a - \\ &\quad - Q(a)Q(\bar{x}, \bar{b})a \\ &= Q(a, \bar{b})\,xU_a - Q(xU_a, a)\,\bar{b} + Q(xU_a, \bar{b})\,a \\ &= xU_aV_{b,a}. \end{aligned}$$

This completes the verification that $(\mathfrak{J}, U, 1)$ is a quadratic Jordan algebra.

Now we have $a^2 = 1U_a = Q(a, \bar{1})\, a - Q(a)\, \bar{1} = T(a)\, a - Q(a)\, 1$. Hence (5) holds. Also $a^3 = aU_a = (T(a)1 - \bar{a})U_a = T(a)\, a^2 - Q(a)a$. Thus (6) holds.

Our results may be summarized in the following

THEOREM 1. *Let* $\mathfrak{J}$ *be a vector space over* Φ, $\dim \mathfrak{J} > 1$, Q *a quadratic form on* $\mathfrak{J}$, 1 *an element of* $\mathfrak{J}$ *such that* $Q(1) = 1$. *Define* $T(a) = Q(a, 1)$, $\bar{a} = T(a)\, 1 - a$, $xU_a = Q(a, \bar{x})\, a - Q(a)\, \bar{x}$. *Then* $(\mathfrak{J}, U, 1)$ *is a quadratic Jordan algebra which is generically algebraic of degree two. Conversely, any such algebra is obtained from a quadratic form* Q *and a point* 1 *in the manner indicated.*

We shall now write $(\mathfrak{J}, U, 1) = \text{Jord}\ (Q, 1)$ and call this the *quadratic Jordan algebra of the quadratic form* Q *with base point* 1. Note that the only conditions we require are that $\dim \mathfrak{J} > 1$ and Q has a *base point,* that is, there is a vector 1 such that $Q(1) = 1$. If Q is a quadratic form $\neq 0$ then there exists a c such that $Q(c) \neq 0$. Then we can replace Q by $Q' = Q(c)^{-1} Q$ and obtain $Q', (c) = 1$ so c is a base point for Q'. In this way the construction is applicable to arbitrary quadratic forms $Q \neq 0$.

3. Clifford algebras for quadratic forms with base points. In this section we shall determine the special universal envelope of $\mathfrak{J} = \text{Jord}\ (Q, 1)$ and show that $\mathfrak{J}$ is a special quadratic Jordan algebra. Here Q is a quadratic form on $\mathfrak{J}$ with base point 1. We have $T(a) = Q(a, 1)$, $\bar{a} = T(a)1 - a$. If the characteristic char $\Phi \neq 2$, $a = \frac{1}{2}(a + \bar{a}) + \frac{1}{2}(a - \bar{a}) = \frac{1}{2}T(a)\, 1 + v$ where $v = \frac{1}{2}(a - \bar{a})$ satisfies $T(v) = 0$. Also $T(1) = Q(1, 1) = 2Q(1) = 2 \neq 0$. Thus $\mathfrak{J} = \Phi\, 1 \oplus \mathfrak{V}$ where $\mathfrak{V} = \{v \mid T(v) = 0\}$. If char $\Phi = 2$, $T(1) = 0$. In this case we distinguish two cases: $T \equiv 0$ and $T \not\equiv 0$. In the first case we can write $\mathfrak{J} = \Phi\, 1 \oplus \mathfrak{V}$ where $\mathfrak{V}$ is any subspace complementary to $\Phi 1$ and we shall have $T(v) = 0$, $v \in \mathfrak{V}$. In the second case we can choose a d such that $T(d) = 1$ and write $\mathfrak{J} = \Phi\, d \oplus \mathfrak{W}$, $\mathfrak{W} = \{w \mid T(w) = 0\}$. Then $1 \in \mathfrak{W}$ so we can write $\mathfrak{W} = \Phi 1 \oplus \mathfrak{V}$ and $T(v) = 0$, $v \in \mathfrak{V}$. In either the case char $\Phi \neq 2$ or the case $T = 0$ we have $\mathfrak{J} = \Phi 1 \oplus \mathfrak{V}$ where $T(\mathfrak{V}) = 0$. In these two cases we call Q *pure* and in the case in which $T = 0$ (so char $\Phi = 2$) we call Q *traceless.*

LEMMA. *Let σ be a linear mapping of $\mathfrak{J} = \text{Jord}(Q, 1)$ into an associative algebra $\mathfrak{A}$ (with 1). Then σ is an associative specialization if and only if $1^\sigma = 1$ and*

$$(x^\sigma)^2 - T(x)\, x^\sigma + Q(x)\, 1 = 0,\ x \in \mathfrak{J}. \tag{18}$$

If Q is pure, so $\mathfrak{J} = \Phi\, 1 \oplus \mathfrak{B}$ with $T(\mathfrak{B}) = 0$, then σ is an associative specialization if and only if $1^\sigma = 1$, $(v^\sigma)^2 = -Q(v)\, 1$, $v \in \mathfrak{B}$.

PROOF. In $\mathfrak{A}^{(q)}$ we have $x^2 = 1U_x$. Hence if σ is an associative specialization of $\mathfrak{J}$ in $\mathfrak{A}$ (= homomorphism of $\mathfrak{J}$ in $\mathfrak{A}^{(q)}$) then $1^\sigma = 1$ and (18) follows from the equation $x^2 - T(x)\, x + Q(x)\, 1 = 0$ in $\mathfrak{J}$. Conversely, assume σ satisfies $1^\sigma = 1$ and (18). Replacing x by $x + y$ in (18) we obtain

$$x^\sigma y^\sigma + y^\sigma x^\sigma - T(x)\, y^\sigma - T(y)\, x^\sigma + Q(x, y)\, 1 = 0. \tag{19}$$

If we multiply this on the right by x^σ and use (18) we obtain

$$x^\sigma y^\sigma x^\sigma = Q(x)\, y^\sigma + T(x)\, T(y)\, x^\sigma - Q(x, y)\, x^\sigma - T(y)\, Q(x) 1. \tag{20}$$

By this and (10) we have $(yU_x)^\sigma = x^\sigma y^\sigma x^\sigma = y^\sigma U_{x\sigma}$ in $\mathfrak{A}^{(q)}$. Hence σ is an associative specialization of $\mathfrak{J}$. This proves the first statement. Now suppose Q is pure so $\mathfrak{J} = \Phi\, 1 \oplus \mathfrak{B}$, $T(\mathfrak{B}) = 0$. Then (18) implies $(v^\sigma)^2 = -Q(v)\, 1$. Conversely, assume this and $1^\sigma = 1$ hold. Put $x = \alpha\, 1 + v$, $\alpha \in \Phi$, $v \in \mathfrak{B}$. Then $T(x) = 2\alpha$ and $Q(x) = \alpha^2 + Q(v)$. Also $x^\sigma = \alpha\, 1 + v^\sigma$, $(x^\sigma)^2 = \alpha^2\, 1 + 2\alpha v^\sigma - Q(v)\, 1$ so $(x^\sigma)^2 - T(x) x^\sigma + Q(x)\, 1 = 0$. Hence σ is an associative specialization by the first part of the lemma.

DEFINITION 2. *Let Q be a quadratic form on a vector space $\mathfrak{J}$ with base point 1 (that is, $Q(1) = 1$) and let $\mathfrak{T}(\mathfrak{J}) = \Phi \oplus \mathfrak{J} \oplus (\mathfrak{J} \otimes \mathfrak{J}) \oplus (\mathfrak{J} \otimes \mathfrak{J} \otimes \mathfrak{J}) \oplus \ldots$ be the tensor algebra defined by $\mathfrak{J}$. Then we define the Clifford algebra $C(\mathfrak{J}, Q, 1)$ of Q with base point 1 to be $C(\mathfrak{J}, Q, 1) = \mathfrak{T}(\mathfrak{J})/\mathfrak{K}$ where $\mathfrak{K}$ is the ideal in $\mathfrak{T}(\mathfrak{J})$ generated by*

$$1_\Phi - 1,\ x \otimes x - T(x)\, x + Q(x)\, 1,\ x \in \mathfrak{J}. \tag{21}$$

Let σ_u denote the mapping $x \longrightarrow x + \mathfrak{K}$ of $\mathfrak{J}$ in $C(\mathfrak{J}, Q, 1)$. Then we have

THEOREM 2. *$(C(\mathfrak{J}, Q, 1), \sigma_u)$ is a special universal envelope for $\mathfrak{J} = \text{Jord}(Q, 1)$.*

Proof. By (21), $1^{\sigma u} = 1_\Phi$ the unit element of $\mathfrak{T}(\mathfrak{J})$. Also $(x^{\sigma_u})^2 = (x + \mathfrak{K})^2 = x \otimes x + \mathfrak{K} = T(x)\, x^{\sigma_u} + Q(x)\, 1_\Phi$. Hence σ_u is an associative specialization of $\mathfrak{J}$ in $\mathfrak{T}(\mathfrak{J})/\mathfrak{K}$ by the lemma. Now let σ be any associative specialization of $\mathfrak{J}$ into an associative algebra $\mathfrak{A}$. Then $1^\sigma = 1$ and (18) holds in $\mathfrak{A}$. By the basic universal property of $\mathfrak{T}(\mathfrak{J})$ we have a homomorphism of $\mathfrak{T}(\mathfrak{J})$ into $\mathfrak{A}$ such that $x \to x^\sigma$, $x \in \mathfrak{J}$. Since $1^\sigma = 1$ and (18) holds the generators (21) of $\mathfrak{K}$ are in the kernel of this homomorphism. Hence we have the homomorphism η of $\mathfrak{T}(\mathfrak{J})/\mathfrak{K}$ into $\mathfrak{A}$ such that $(x + \mathfrak{K})^\eta = x^\sigma$ or $x^{\sigma_u \eta} = x^\sigma$. Also η is unique since the x^{σ_u}, $x \in \mathfrak{J}$, generate $C(\mathfrak{J}, Q, 1) = \mathfrak{T}(\mathfrak{J})/\mathfrak{K}$. Hence σ_u is a universal associative specialization.

We recall the definition of the usual Clifford algebra $C(\mathfrak{V}, Q)$ of a quadratic form Q on a vector space $\mathfrak{V}$ (without base point) as $\mathfrak{T}(\mathfrak{V})/\mathfrak{L}$ where $\mathfrak{L}$ is the ideal in the tensor algebra $\mathfrak{T}(\mathfrak{V})$ generated by the elements $x \otimes x - Q(x)\, 1$, $x \in \mathfrak{V}$. It is well known that the canonical mapping $\alpha\, 1 + x \to \alpha\, 1 + x + \mathfrak{L}$, $x \in \mathfrak{V}$, is an injective mapping of $\mathfrak{J} = \Phi 1 \oplus \mathfrak{V}$ into $C(\mathfrak{V}, Q)$. (Chevalley [1], p. 40 for the finite-dimensional case). Hence we may identify $\mathfrak{J}$ with its image in $C(\mathfrak{V}, Q)$. Then $C(\mathfrak{V}, Q)$ is generated by $\mathfrak{J}$. The following result can be derived from the second part of the foregoing lemma in the same way as Theorem 2.

Theorem 3. *Let Q be pure, so $\mathfrak{J} = \Phi 1 \oplus \mathfrak{V}$ where $T(\mathfrak{V}) = 0$. Then $C(\mathfrak{V}, -Q)$ together with the injection of $\mathfrak{J}$ in $C(\mathfrak{V}, -Q)$ constitute a special universal envelope for $\mathfrak{J} = Jord\ (Q, 1)$.*

We now consider the remaining case: characteristic $= 2$, $T \not\equiv 0$. Then $\mathfrak{J} = \Phi d \oplus \mathfrak{W}$ where $T(d) = 1$, $\mathfrak{W} = \{w \mid T(w) = 0\}$. Also $\mathfrak{W} = \Phi 1 \oplus \mathfrak{V}$. Let $C(\mathfrak{V}, Q)$ be the Clifford algebra of $Q (= -Q)$ on $\mathfrak{V}$ and identify $\mathfrak{W} = \Phi 1 \oplus \mathfrak{V}$ with the corresponding subspace of $C(\mathfrak{V}, Q)$. We have a derivation D' of $\mathfrak{T}(\mathfrak{V})$ such that

$$v \to v + Q(v, d)\, 1, \quad v \in \mathfrak{V}. \tag{22}$$

Since this maps $v \otimes v + Q(v)\, 1$ into 0, it maps the ideal $\mathfrak{L}$ defining $C(\mathfrak{V}, Q) = \mathfrak{T}(\mathfrak{V})/\mathfrak{L}$ into itself. Hence we have a derivation D of $C(\mathfrak{V}, Q)$ such that (22) holds. Since $vD^2 = vD$, $v \in \mathfrak{V}$, and D^2 is a derivation (since the characteristic is two), we have $D^2 = D$. We now write $\mathfrak{C} = C(\mathfrak{V}, Q)$ and we form the algebra $\mathfrak{C}[t, D]$ of differential polynomials $c_0 + c_1 t + c_2 t^2 + \dots$, $c_i \in \mathfrak{C}$, in an indeterminate t with coefficients in $\mathfrak{C}$ such that the commutation formula

$$ct + tc = cD, \; c \in \mathfrak{C} \tag{23}$$

is valid. This relation implies $ct^2 + t^2 c = cD$ so $c\,(t^2 + t) = (t^2 + t)c$. Since $t^2 + t$ also commutes with t it is clear that $t^2 + t$ is in the center of $\mathfrak{C}[t, D]$. Hence also

$$g(t) \equiv t^2 + t + Q(d)1 \tag{24}$$

is in the center. Let $(g(t))$ be the ideal in $\mathfrak{C}[t, D]$ generated by $g(t)$ and put $\mathfrak{D} = \mathfrak{C}[t, D]/(g(t))$. Let $f(t) \in \mathfrak{C}[t, D]$. Since the leading coefficient of $g(t)$ is 1 we can carry out the division algorithm and write $f(t) = q(t)\, g(t) + r(t)$ where $r(t) = c_0 + c_1 t$, $c_i \in \mathfrak{C}$. Also $r(t)$ is unique. Thus $f(t) + (g(t)) = r(t) + (g(t))$ and $c_0 + c_1 t + (g(t)) = 0$ in $\mathfrak{D}$ implies $c_0 = 0 = c_1$. Put $d = t + (g(t))$ (for a reason which will soon be clear) and identify $\mathfrak{C}$ with its image in $\mathfrak{D}$. Then $\mathfrak{D}$ consists of the elements $c_0 + c_1 d$, $c_i \in \mathfrak{C}$, and $c_0 + c_1 d = 0$ only if $c_0 = 0 = c_1$. Moreover, we have the following relations

$$cd + dc = cD, \; c \in \mathfrak{C}, \; d^2 + d + Q(d)\,1 = 0. \tag{25}$$

We can identify $\mathfrak{J} = \Phi d \oplus \Phi 1 \oplus \mathfrak{V}$ with the set of elements $\alpha 1 + \beta d + v$, $\alpha, \beta \in \Phi$, $v \in \mathfrak{V} \subseteq \mathfrak{C} = C(\mathfrak{V}, Q)$ of $\mathfrak{D}$. Using this identification we have

THEOREM 4. *Let* char $\Phi = 2$, $T \not\equiv 0$ *so* $\mathfrak{J} = \Phi d \oplus \Phi 1 \oplus \mathfrak{V}$ *as above. Let* $\mathfrak{C} = C(\mathfrak{V}, Q)$, $\mathfrak{D} = \mathfrak{C}[t, D]/(g(t))$ *and represent this as the set of elements* $c_0 + c_1 d$ *as indicated. Then* $\mathfrak{D}$ *and the injection of* $\mathfrak{J} = \Phi d \oplus \Phi 1 \oplus \mathfrak{V}$ *in* $\mathfrak{D}$ *constitute a special universal envelope for* $\mathfrak{J} = \text{Jord}\,(Q, 1)$.

PROOF. Put $x = \alpha 1 + \beta d + v$, $v \in \mathfrak{V}$. Then $T(x) = \beta$ and $Q(x) = \alpha^2 + \beta^2 Q(d) + Q(v) + \beta Q(v, d) + \alpha\beta$. Using (25) and (22) we have $x^2 = \alpha^2 1 + \beta^2 d^2 + v^2 + \beta(vD) = \alpha^2 1 + \beta^2(d + Q(d)\,1) + Q(v)\,1 + \beta(v + Q(v, d)\,1)$. Thus

$$\begin{aligned} x^2 + T(x)x + Q(x)\,1 &= \alpha^2 1 + \beta^2(d + Q(d)\,1) + Q(v)\,1 + \beta(v + Q(v, d)\,1) + \\ &\quad + \beta(\alpha 1 + \beta d + v) + \alpha^2 1 + \beta^2 Q(d)\,1 + \\ &\quad + Q(v)\,1 + \beta Q(v, d)\,1 + \alpha\beta\,1 \\ &= 0. \end{aligned}$$

Hence the injection of $\mathfrak{J}$ in $\mathfrak{D}$ is an associative specialization, so we have the homomorphism of $C(\mathfrak{J}, Q, 1)$ into $\mathfrak{D}$ such that $x^{\sigma_u} \longrightarrow x$, $x \in \mathfrak{J}$. We note next that for $v \in \mathfrak{B}$ we have $(v^{\sigma_u})^2 = Q(v)\,1$ so we have the homomorphism of $C(\mathfrak{B}, Q)$ into $C(\mathfrak{J}, Q, 1)$ such that $\alpha 1 + v \longrightarrow \alpha 1 + v^{\sigma_u}$. By (19) we have $v^{\sigma_u} d^{\sigma_u} + d^{\sigma_u} v^{\sigma_u} = v^{\sigma_u} + Q(d, v)\,1$. This implies that the inner derivation $y \longrightarrow [y, d^{\sigma_u}]$ in $C(\mathfrak{J}, Q, 1)$ maps the subalgebra generated by the v^{σ_u}, $v \in \mathfrak{B}$, into itself. Also we have a homomorphism of $\mathfrak{C}[t, D]$ into $C(\mathfrak{J}, Q, 1)$ such that $\alpha 1 + v \longrightarrow \alpha 1 + v^{\sigma_u}$, $t \longrightarrow d^{\sigma_u}$. Since we have the relation $(d^{\sigma_u})^2 + d^{\sigma_u} + Q(d)\,1 = 0$ in $C(\mathfrak{J}, Q, 1)$, our homomorphism maps $g(t)$ into 0. Thus we have a homomorphism of $\mathfrak{D} = \mathfrak{C}[t, D]/(g(t))$ into $C(\mathfrak{J}, Q, 1)$ such that $\alpha 1 + v \longrightarrow \alpha 1 + v^{\sigma_u}$, $d \longrightarrow d^{\sigma_u}$ and so $x = \alpha 1 + \beta d + v \longrightarrow x^{\sigma_u}$. Since the elements x generate $\mathfrak{D}$ and the x^{σ_u} generate $C(\mathfrak{J}, Q, 1)$ the homomorphism just defined is the inverse of the homomorphism we defined first of $C(\mathfrak{J}, Q, 1)$ into $\mathfrak{D}$. Hence these are isomorphisms and $\mathfrak{D}$ and the injection of $\mathfrak{J}$ constitute a special universal envelope.

We can now prove

THEOREM 5. *The canonical mapping* σ_u *of* $\mathfrak{J}$ *into* $C(\mathfrak{J}, Q, 1)$ *is injective. If* $\dim \mathfrak{J} = n$ *is finite then* $\dim C(\mathfrak{J}, Q, 1) = 2^{n-1}$.

PROOF. In the pure case : $\mathfrak{J} = \Phi 1 \oplus \mathfrak{B}$, $T(\mathfrak{B}) = 0$ this is clear since we have the imbedding of $\mathfrak{J}$ in $C(\mathfrak{B}, -Q)$ and so we have the isomorphism of $C(\mathfrak{J}, Q, 1)$ onto $C(\mathfrak{B}, -Q)$ such that $(\alpha 1 + v)^{\sigma_u} \longrightarrow \alpha 1 + v$. Moreover, it is well known that if $\dim \mathfrak{B} = n - 1$ then $\dim C(\mathfrak{B}, -Q) = 2^{n-1}$. Now suppose char $\Phi = 2$, $T \not\equiv 0$. Then we have the isomorphism of $C(\mathfrak{J}, Q, 1)$ onto $\mathfrak{D}$ such that $(\alpha 1 + \beta d + v)^{\sigma_u} \longrightarrow \alpha 1 + \beta d + v$. Hence σ_u is injective. If $\dim \mathfrak{J} = n$ then $\dim \mathfrak{B} = n - 2$ and $\dim \mathfrak{C} = 2^{n-2}$. Then $\dim \mathfrak{D} = 2^{n-1}$ since $\mathfrak{D} = \mathfrak{C} \oplus \mathfrak{C}d$.

Since σ_u is injective it is clear that $\mathfrak{J}$ is special. From now on we shall identify $\mathfrak{J}$ also with its image $\mathfrak{J}^{\sigma_u}$ in the special universal envelope $C(\mathfrak{J}, Q, 1)$. Also, in the pure case we identify $C(\mathfrak{J}, Q, 1)$ with $C(\mathfrak{B}, -Q)$ using the canonical isomorphism given by Theorems 2 and 3 for the identification. Finally, in the remaining case, we identify $C(\mathfrak{J}, Q, 1)$ with $\mathfrak{D} = \mathfrak{C}[t, D]/(g(t))$, $\mathfrak{C} = C(\mathfrak{B}, Q)$ as in Theorem 4.

4. Structure group and Lie algebra. Let η be a linear mapping of $\mathfrak{J} = \text{Jord}\,(Q, 1)$ into $\mathfrak{J}' = \text{Jord}\,(Q', 1')$ such that $1^\eta = 1'$ and $Q'(a^\eta) = Q(a)$, $a \in \mathfrak{J}$. Then $T'(a^\eta) = Q'(a^\eta, 1') = Q(a, 1) = T(a)$. Hence the formula (10) for bU_a shows that $(bU_a)^\eta = b^\eta U_{a^\eta}$. Thus η is a homomorphism of $\mathfrak{J}$ into $\mathfrak{J}'$. In particular, any linear isomorphism of $\mathfrak{J}$ onto $\mathfrak{J}'$ such that $1^\eta = 1'$, $Q'(a^\eta) = Q(a)$ is an isomorphism of Jord $(Q, 1)$ onto Jord $(Q', 1')$. Conversely, let η be an algebra isomorphism of Jord $(Q, 1)$ onto Jord $(Q', 1')$. Then $1^\eta = 1'$ and $a^2 = T(a)a + Q(a)\,1 = 0$ gives $(a^\eta)^2 - T(a)\,a^\eta + Q(a)\,1' = 0$. Also $(a^\eta)^2 - T'(a^\eta)\,a^\eta + Q'\,(a^\eta)\,1' = 0$. Then $[T(a) - T'(a^\eta)]\,a^\eta = [Q(a) - Q'\,(a^\eta)]\,1'$. If $T(a) \neq T'(a^\eta)$, $a^\eta = \alpha\,1'$, $\alpha \in \Phi$. Then $a = \alpha\,1$ and $Q(a) = \alpha^2 = Q'(a^\eta)$. If $T(a) = T'(a^\eta)$ then $Q(a) = Q'(a^\eta)$. Thus $Q(a) = Q'(a^\eta)$, $a \in \mathfrak{J}$, and we have proved

THEOREM 6. *A bijective linear mapping η of $\mathfrak{J} = \text{Jord}\,(Q, 1)$ into $\mathfrak{J}' = \text{Jord}\,(Q', 1')$ is an isomorphism of quadratic Jordan algebras if and only if $1^\eta = 1'$ and $Q'\,(a^\eta) = Q(a)$, $a \in \mathfrak{J}$.*

We have seen that $a \to \bar{a} = T(a)\,1 - a$ satisfies $\bar{1} = 1$, $Q(\bar{a}) = Q(a)$, $\bar{\bar{a}} = a$. Hence this mapping is an automorphism of period two, of Jord $(Q, 1)$.

We recall that an element of a quadratic Jordan algebra $\mathfrak{J}$ is called *invertible* if there exists a $b \in \mathfrak{J}$ such that $bU_a = a$, $b^2U_a = 1$. Then b is unique and is called the *inverse* of a. Also $U_aU_b = 1 = U_b\,U_a$. If $a \in$ Jord $(Q, 1)$ we have $\bar{a}U_a = Q(a)\,a$ and $\bar{a}^2U_a = 1U_{\bar{a}}U_a = Q(a)^2 1$ (as in the proof of Theorem 1). Hence if $Q(a) \neq 0$ then $Q(a)^{-1}\,\bar{a}$ satisfies the conditions for an inverse of a. If $Q\,(a) = 0$ we have $\bar{a}U_a = 0$, so U_a is not invertible. Then a is not invertible. Thus a is invertible if and only if $Q(a) \neq 0$, in which case the inverse is $Q(a)^{-1}\,\bar{a}$.

If u is an invertible element of a quadratic Jordan algebra $\mathfrak{J}$ then u determines another quadratic Jordan algebra $\mathfrak{J}^{(u)} = (\mathfrak{J}, U^{(u)}, 1^{(u)})$ where $U_a^{(u)} = U_uU_a$, $1^{(u)} = u^{-1}$. This is called the *u-isotope* of $\mathfrak{J}$. We now prove

THEOREM 7. *If u is an invertible element of* Jord $(Q, 1)$ *then the u-isotope of* Jord $(Q, 1)$ *is* Jord $(Q(u)Q, u^{-1})$.

PROOF. Clearly $Q(u)\,Q$ is a quadratic form and $Q(u)\,Q(u^{-1}) = Q(u)\,Q(Q(u)^{-1}\bar{u}) = Q(u)\,Q(u)^{-2}\,Q(\bar{u}) = 1$ so u^{-1} is a base point for $Q(u)Q$. Hence we can define the quadratic Jordan algebra Jord $(Q(u)Q,\, u^{-1})$ on $\mathfrak{J}$ whose unit is u^{-1}. The mapping corresponding to $x \to \bar{x}$ in Jord $(Q(u)\,Q,\, u^{-1})$ is $x \to x' = Q(u)\,Q(x,\, u^{-1})\, u^{-1} - x$. Since $u^{-1} = Q(u)^{-1}\bar{u}$ we have $x' = Q(u)^{-1}\,Q(x,\, \bar{u})\bar{u} - Q(u)^{-1}Q(u)x = Q(u)^{-1}\,(Q(x,\, \bar{u})\,\bar{u} - Q(u)x) = Q(u)^{-1}\,\overline{xU_u}$ (by (12)). Hence the U-operator U' in Jord $(Q\,(u)\,Q,\, u^{-1})$ is given by (12) as

$$\begin{aligned} xU'_a &= Q(u)\,Q(a,\, Q(u)^{-1}\,\overline{xU_u})\; a - Q(u)\,Q(a)\,Q(u)^{-1}\,\overline{xU_u} \\ &= Q(a,\, \overline{xU_u})\; a - Q(a)\,\overline{xU_u} \\ &= xU_uU_a = xU_a^{(u)}. \end{aligned}$$

Thus Jord $(Q\,(u)\,Q,\, u^{-1})$ has the same U-operator and unit as the u-isotope of Jord $(Q,\, 1)$ and so it coincides with this isotope.

An *isotopy* η of a quadratic Jordan algebra $(\mathfrak{J},\, U, 1)$ into $(\mathfrak{J}', U', 1')$ is an isomorphism of $\mathfrak{J}$ onto an isotope of $\mathfrak{J}'$. It is easily seen that the set of isotopies of $\mathfrak{J}$ onto $\mathfrak{J}$ is a group. This has been called the *structure group* by Koecher. It is easily seen that $\eta \in \operatorname{Str} \mathfrak{J}$, the structure group of $\mathfrak{J}$, if and only if η is bijective linear and there exists a bijective linear transformation η^* of $\mathfrak{J}$ such that $U_{a^\eta} = \eta^* U_a\, \eta$, $a \in \mathfrak{J}$. Putting $a = 1$ we find $\eta^* = U_{1^\eta}\, \eta^{-1}$. We now prove

THEOREM 8. *The structure group of* Jord $(Q,\, 1)$ *coincides with the group of similarities of* Q ($=$ *group of bijective linear mappings* η *such that* $Q(a^\eta) = \rho\, Q(a)$, $\rho \neq 0$ *in* Φ).

PROOF. By Theorem 6, if $\eta \in$ Str Jord $(Q,\, 1)$ then η is an isomorphism of Jord $(Q,\, 1)$ onto Jord$(Q(u)\, Q,\, u^{-1})$ for an invertible u. Then $Q(u)\, Q\, (a^\eta) = Q(a)$, $a \in \mathfrak{J}$. Hence $Q(a^\eta) = \rho\, Q(a)$, $\rho = Q(u)^{-1}$. Conversely, suppose η is bijective linear and $Q(a^\eta) = \rho Q(a)$, $a \in \mathfrak{J}$. Then $\rho = Q(1^\eta)$ and η is an isomorphism of Jord $(Q,\, 1)$ onto Jord $(Q(1^\eta)^{-1}Q,\, 1^\eta)$. Hence η is an isotopy of Jord $(Q,\, 1)$.

If $\eta \in$ Str Jord $(Q,\, 1)$ then $\eta^* = U_{1^\eta}\, \eta^{-1}$ and $xU_{1^\eta} = Q(1^\eta,\, \bar{x})\, 1^\eta - Q(1^\eta)\, \bar{x}$ so

$$x\,\eta^* = Q(1^\eta)\,[Q(1^\eta)^{-1}\,Q(1^\eta, \bar{x})\,1 - \bar{x}^{\eta-1}].$$

Also $Q(x^\eta) = Q(1^\eta)\,Q(x)$ so $Q(x^\eta, y^\eta) = Q(1^\eta)\,Q(x, y)$ and $Q(x^\eta, y) = Q(1^\eta)\,Q(x, y^{\eta-1})$. Then $T(\bar{x}^{\eta-1}) = Q(1, \bar{x}^{\eta-1}) = Q(1^\eta)^{-1}\,Q(1^\eta, \bar{x})$. Hence

$$x^{\eta^*} = Q(1^\eta)\,\overline{\bar{x}^{\eta-1}}. \tag{26}$$

The *structure Lie algebra* Strl $\mathfrak{J}$ of a quadratic Jordan algebra $(\mathfrak{J}, U, 1)$ is the set of linear transformations L of $\mathfrak{J}$ into $\mathfrak{J}$ for which there exists a linear L' satisfying

$$U_a L + L' U_a = U_{aL,a}\,,\; a \in \mathfrak{J}. \tag{27}$$

THEOREM 9. *A linear transformation L of $\mathfrak{J}$ = Jord $(Q, 1)$ is in* Strl $\mathfrak{J}$ *if and only if $Q(aL, a) = \rho\, Q(a)$, $a \in \mathfrak{J}$, where $\rho \in \Phi$.*

PROOF. Suppose $Q(aL,a) = \rho\, Q(a)$, $a \in \mathfrak{J}$. Then $Q(aL, b) + Q(a, bL) = \rho\, Q(a, b)$, $a, b \in \mathfrak{J}$, and by (16),

$$\begin{aligned}
\bar{x} U_{a,aL} &= Q(a, x)\, aL + Q(aL, x)\, a - Q(a, aL) x \\
&= Q(a, x)\, aL - Q(xL, a)\, a + \rho\, Q(a, x)\, a - \rho\, Q(a)\, x \\
&= [Q(a, x)\, aL - Q(a)\, xL] + [\rho\, Q(a, x)\, a - \rho\, Q(a)\, x - \\
&\quad - Q(x L, a) + Q(a)\, xL] \\
&= [Q(a, x)\, a - Q(a)\, x]L + (\rho \bar{x} - \overline{xL})\, U_a \\
&= \bar{x}(U_a L + L' U_a)
\end{aligned}$$

where $xL' = \rho\, x - \overline{\overline{x}L}$. Hence $L \in$ Strl $\mathfrak{J}$. Conversely, let $L \in$ Strl $\mathfrak{J}$, L' satisfy $U_{a,\,aL} = U_a L + L' U_a$. Then $L + L' = U_{1,\,1L}$ and since $1 U_{1,\,1L} = \{1\,1\,1L\} = 1^2 \circ 1L = 2(1L)$ we have $1L + 1L' = 2\,(1L)$ so $1L = 1L'$. We have

$$\begin{aligned}
1 U_{a,\,aL} &= T(a) aL + T(aL) a - Q(a, aL)\, 1 && \text{(by (16))} \\
1 U_a L + 1 L' U_a &= 1 U_a L + 1 L U_a \\
&= T(a) aL - Q(a)\, 1L + Q(a, \overline{1L}) a - Q(a)\, \overline{1L} && \text{(by (12))} \\
&= T(a) aL + Q(a, \overline{1L}) a - Q(a) T\,(1L)\, 1.
\end{aligned}$$

Since $1 U_{a,\,aL} = 1 U_a L + 1 L' U_a$ we have

$$[T(aL) - Q(a, \overline{1L})] a = [T(1L) Q(a) - Q(a,\, aL)]\, 1.$$

If $a \neq \alpha 1$ this implies $Q(a, aL) = \rho Q(a)$, $\rho = T(1L)$. If $a = \alpha 1$ then $Q(a, aL) = \alpha^2 T(1L)$ and $T(1L)Q(a) = \alpha^2 T(1L)$.

Hence for all a, $Q(a, aL) = \rho Q(a)$, $\rho = T(1L)$.

The derivations D of $\mathfrak{J}$ are the elements of Strl $\mathfrak{J}$ such that $1D = 0$. Thus we have the

COROLLARY. *A linear transformation D of $\mathfrak{J} =$ Jord $(Q, 1)$ is a derivation if and only if $1D = 0$ and $Q(a, aD) = 0$, $a \in \mathfrak{J}$.*

PROOF. The conditions $1D = 0$, $Q(a, aD) = \rho Q(a)$ give $\rho = 0$ so $Q(a, aD) = 0$.

5. Structure theory. An element z of a quadratic Jordan algebra is called an *absolute zero divisor* if $U_z = 0$. An element z is *nilpotent* if $z^n = 0$ for some n. An ideal $\mathfrak{N}$ is called *nil* if every $z \in \mathfrak{N}$ is nilpotent. Clearly if z is an absolute zero divisor then $z^2 = 1U_z = 0$ so z is nilpotent.

If Q is a quadratic form on $\mathfrak{J}$ *the radical of* Q, rad Q, is defined as $\{z \mid Q(z) = Q(a, z) = 0, a \in \mathfrak{J}\}$. Q is *non-degenerate* if rad $Q = 0$, rad Q is a subspace contained in $\mathfrak{J}^\perp = \{z \mid Q(a, z) = 0, a \in \mathfrak{J}\}$. We shall call $\mathfrak{J}^\perp$ the *bilinear radical* of Q. If char $\Phi \neq 2$, rad $Q = \mathfrak{J}^\perp$ since $Q(a, a) = 2Q(a)$. We define the *defect* of Q to be dimension of $\mathfrak{J}^\perp/\text{rad } Q$, so for finite dimensional $\mathfrak{J}$ we have def $Q = \dim \mathfrak{J}^\perp - \dim \text{rad } Q$. Clearly def $Q = 0$ if char $\Phi \neq 2$.

We now consider the structure of the quadratic Jordan algebras $\mathfrak{J} =$ Jord $(Q, 1)$. It will be convenient at times to regard $\mathfrak{J}$ as imbedded in $C(\mathfrak{J}, Q, 1)$. Then the powers of $a \in \mathfrak{J}$ as defined in the quadratic Jordan algebra coincide with the corresponding powers of a in $C(\mathfrak{J}, Q, 1)$. Also it is easily seen from the definitions that a and b are inverses in $\mathfrak{J}$ if and only if they are contained in $\mathfrak{J}$ and are inverses in $C(\mathfrak{J}, Q, 1)$.

THEOREM 10. *An element $z \in \mathfrak{J} =$ Jord $(Q, 1)$ is an absolute zero divisor if and only if $z \in$ rad Q. Moreover,* rad Q *is a nil ideal which contains every nil ideal of* Jord $(Q, 1)$.

PROOF. If $z \in \operatorname{rad} Q$ the formula for xU_z shows that z is an absolute zero divisor. Conversely, if z is an absolute zero divisor then $\bar{z}U_z = Q(z)\, z = 0$ so $Q(z) = 0$. Then $\bar{x}U_z = Q(x, z)z - Q(z)x = 0$ gives $Q(x, z) = 0$. Hence $z \in \operatorname{rad} Q$, so rad Q is the set of absolute zero divisors. It is clear from $U_{aU_b} = U_b U_a U_b$ that if z is an absolute zero divisor then zU_a is an absolute zero divisor. Since $aU_z = 0$ it follows that the subspace rad Q is an ideal, and clearly this is a nil ideal. Now let $\mathfrak{N}$ be any nil ideal of $\mathfrak{J}$ and let $z \in \mathfrak{N}$. Then z is nilpotent in $C(\mathfrak{J}, Q, 1)$ and z satisfies the equation $z^2 - T(z)\, z + Q(z)\, 1 = 0$ in $C(\mathfrak{J}, Q, 1)$. It follows that $z^2 = 0$ and $T(z) = Q(z) = 0$. If $a \in \mathfrak{J}$ then $z \circ a = T(z)\, a + T(a)\, z - Q(z, a)\, 1 = T(a) z - Q\,(z, a)\, 1$. Since $z \circ a$ and $z \in \mathfrak{N}$ but $1 \notin \mathfrak{N}$ we have $Q(a, z) = 0$, $a \in \mathfrak{J}$. Then $z \in \operatorname{rad} Q$ and $\mathfrak{N} \subseteq \operatorname{rad} Q$.

Let e be an idempotent element $\neq 0, 1$ in Jord $(Q, 1)$. Then $e^2 = e \neq 0, 1$ and $e^2 - T(e)\, e + Q(e)\, 1 = 0$ in $C\,(\mathfrak{J}, Q, 1)$. These imply $T(e) = 1$, $Q(e) = 0$. Conversely, if $T(e) = 1$, $Q(e) = 0$ then $e^2 = e$ and $e \neq 0, 1$ since $T(0) = 0$ and $Q(1) = 1$. We have $e = 1 - \bar{e}$ and $\bar{e}^2 = \bar{e} \neq 0, 1$. The idempotents e and $\bar{e}$ are orthogonal in $\mathfrak{J}$ in the sense that $e \circ \bar{e} = 0 = \bar{e}U_e = eU_{\bar{e}}$. The subspace $\Phi e + \Phi \bar{e}$ is non-isotropic relative to $Q(x, y)$ since $Q(e, e) = 0 = Q(\bar{e}, \bar{e})$ and $Q(e, \bar{e}) = Q(e, 1 - e) = T(e) = 1$. Hence $\mathfrak{J} = \Phi e \oplus \Phi \bar{e} \oplus \mathfrak{J}_{12}$ where $\mathfrak{J}_{12} = (\Phi e + \Phi \bar{e})^{\perp} = \{x \mid Q(x, e) = 0 = Q(x, \bar{e})\}$. We have $xU_e = Q(e, \bar{x})e \in \Phi e$ and $eU_e = e^3 = e$ so U_e is a projection on Φe. Similarly, $U_{\bar{e}}$ is a projection on $\Phi \bar{e}$. Also $xU_{e,\bar{e}} = Q(e, \bar{x})\, \bar{e} + Q(\bar{e}, \bar{x})\, e - \bar{x} = T(x)\, \bar{e} - Q\,(e, x)\, \bar{e} + T(x)\, e - Q(\bar{e}, x)\, \bar{e} - T(x)\, 1 + x = x - Q(e, x)\bar{e} - Q(\bar{e}, x)e$. Hence $U_{e,\bar{e}}$ is the projection on $\mathfrak{J}_{12} = (\Phi e + \Phi \bar{e})^{\perp}$ determined by $\mathfrak{J} = (\Phi e + \Phi \bar{e}) \oplus \mathfrak{J}_{12}$. If $x = \alpha e + \beta e + y$, $y \in \mathfrak{J}_{12}$ then $Q(x) = \alpha\beta + Q(y)$. Hence Q on $\mathfrak{J}$ is determined by its restriction to $\mathfrak{J}_{12}$.

We shall call $\mathfrak{J} =$ Jord $(Q, 1)$ *reduced* if it contains an idempotent $e \neq 0, 1$.

THEOREM 11. *If Q is non-degenerate then $\mathfrak{J} =$ Jord $(Q, 1)$ is of one of the following types*: (i) *a direct sum $\Phi e \oplus \Phi \bar{e}$ of two copies of the base field Φ*, (ii) *a quadratic Jordan division algebra*, (iii) *reduced* *and simple,* (iv) *not a division algebra yet simple of characteristic two with Q traceless.*

PROOF. Suppose first that $\mathfrak{J}$ is not simple, and let $\mathfrak{B}$ be an ideal $\neq 0$, $\neq \mathfrak{J}$, in $\mathfrak{J}$. Since $\mathfrak{B}$ contains no invertible elements, $Q(z) = 0$ for every $z \in \mathfrak{B}$. If $T(z) = 0$ for some $z \in \mathfrak{B}$ then $z \circ x = T(x)\, z - Q(x, z)\, 1 \in \mathfrak{B}$ and $1 \notin \mathfrak{B}$, which implies that $Q(x, z) = 0$, $x \in \mathfrak{J}$. Then $z \in \operatorname{rad} Q$ and $z = 0$ by the non-degeneracy of Q. It now follows that the linear mapping $z \to T(z)$ of $\mathfrak{B}$ into Φ is bijective. Hence $\mathfrak{B} = \Phi e$ where $T(e) = 1$. If $x \in \mathfrak{J}$ then $x \circ e = T(x)\, e + x - Q(x, e)\, 1$ and since $x \circ e \in \mathfrak{B} = \Phi e$ this shows that $x \in \Phi e + \Phi \bar{e}$. Hence $\mathfrak{J} = \Phi e \oplus \Phi \bar{e}$. From now on we assume $\mathfrak{J}$ simple. Suppose also that $\mathfrak{J}$ is not a division algebra and is not reduced. Then we have to show (iv) holds. Since $\mathfrak{J}$ is not a division algebra we have a $v \neq 0$ such that $Q(v) = 0$. If $T(v) = \beta \neq 0$, $e = \beta^{-1} v$ is an idempotent $\neq 0, 1$, contrary to hypothesis. Thus $T(v) = 0$ for every v such that $Q(v) = 0$. The non-degeneracy of Q implies the existence of a w such that $Q(v, w) = 1$. Also we may assume $Q(w) = 0$ since $Q(v, w + \alpha v) = 1$ and $Q(w + \alpha v) = Q(w) + \alpha$, so α can be chosen so that $Q(w) + \alpha = 0$. We now have v and w such that $Q(v) = 0 = Q(w)$, $T(v) = 0 = T(w)$, $Q(w, v) = 1$. Then $\mathfrak{J} = (\Phi v + \Phi w) \oplus (\Phi v + \Phi w)^{\perp}$. Suppose there exists an $x \in \mathfrak{J}$ such that $T(x) \neq 0$, and let y be the component of x in $(\Phi v + \Phi w)^{\perp}$. Then $T(y) \neq 0$ and we may assume that $T(y) = 1$. Put $z = y + v - Q(y)\, w$. Then $T(z) = T(y) = 1$ and $Q(z) = Q(y) - Q(y) = 0$. This contradicts the result established before that $Q(z) = 0$ implies $T(z) = 0$. Hence $T(x) = 0$ for all x. Then $T(1) = 2 = 0$ so char $\Phi = 2$. Moreover, Q is traceless since $T \equiv 0$.

It is clear that the classes (i) and (ii) specified in Theorem 11 are closed under isotopy, that is, if $\mathfrak{J}$ is of type (i) or (ii) then every isotope of $\mathfrak{J}$ is of the same type. The situation for types (iii) and (iv) is described in the following

THEOREM 12. *Let Q be non-degenerate with base point* 1. *Then* $\mathfrak{J} = \text{Jord}\,(Q, 1)$ *has an isotope which is reduced if and only if $\mathfrak{J}$ is not a division algebra (equivalently Q is not anisotropic). If* char $\Phi = 2$ *then $\mathfrak{J}$ has an isotope* $\widetilde{\mathfrak{J}} = \text{Jord}\,(\widetilde{Q}, \widetilde{1})$ *such that $\widetilde{Q}$ is traceless if and only if the bilinear radical of Q ($\mathfrak{J}^{\perp}$ relative to Q) $\neq 0$.*

PROOF. If $\mathfrak{J}$ is a division algebra so is every isotope and so no isotope is reduced. Conversely, suppose $\mathfrak{J}$ is not a division algebra. Then there exists a $v \neq 0$ with $Q(v) = 0$. Since Q is non-degenerate there is a u such that $Q(u, v) = 1$. Also we may assume $Q(u) \neq 0$ since u can be replaced by $u + \alpha v$, $\alpha \in \Phi$. Then u is invertible. Put $e = \bar{v}$. Then in the isotope $\tilde{\mathfrak{J}} = \mathfrak{J}^{(u)}$ we have $\tilde{Q}(e) = Q(u)\, Q(e) = Q(u)\, Q(v) = 0$ and

$$\tilde{T}(e) = \tilde{Q}(e, u^{-1}) = Q(u)\, Q(e, u^{-1}) = Q(u)\, Q(\bar{v}, Q(u)^{-1}\,\bar{u}) = 1.$$

Hence e is an idempotent $\neq 0$, u^{-1} in $\tilde{\mathfrak{J}}$ so $\tilde{\mathfrak{J}}$ is reduced. Now suppose char $\Phi = 2$. If $T \equiv 0$, $Q(1, x) = 0$ for all x so $1 \in \mathfrak{J}^{\perp}$. Since the bilinear radicals of Q and $\tilde{Q} = Q(u)\, Q$, $Q(u) \neq 0$ coincide, it is clear that if $\mathfrak{J}$ has an isotope $\tilde{\mathfrak{J}}$ with traceless $\tilde{Q}$ then $\mathfrak{J}^{\perp} \neq 0$. Conversely, if $u \neq 0$ is in $\mathfrak{J}^{\perp}$ then $Q(u) \neq 0$ since Q is non-degenerate. Then $\bar{u} = -u$ since $T(u) = Q(u, 1) = 0$, $u^{-1} = Q(u)^{-1}\,\bar{u} = -Q(u)^{-1}\,u$. Hence the unit u^{-1} of $\tilde{\mathfrak{J}} = \mathfrak{J}^{(u)}$ is in the bilinear radical so $\tilde{T}(x) = \tilde{Q}(u^{-1}, x) = 0$, $x \in \mathfrak{J}$. Thus $\tilde{Q}$ is traceless.

We now consider the inner and outer ideals of Jord $(Q, 1)$ (definitions in §1). We recall that a subspace $\mathfrak{B}$ is called *totally singular* if $Q(\mathfrak{B}) = 0$, that is, $Q\,(b) = 0$ for $b \in \mathfrak{B}$. We have

THEOREM 13. *The inner ideals of* $\mathfrak{J} =$ Jord $(Q, 1)$ *are* $\mathfrak{J}$ *and the totally singular subspaces of* $\mathfrak{J}$.

PROOF. Let $b \in \mathfrak{B}$ an inner ideal. Then $\mathfrak{B} \supseteq \mathfrak{J} U_b$. If $Q(b) \neq 0$, b is invertible so $\mathfrak{J} U_b = \mathfrak{J}$ and $\mathfrak{B} = \mathfrak{J}$. Hence if $\mathfrak{B} \neq \mathfrak{J}$ then $\mathfrak{B}$ is totally singular. Conversely, let $\mathfrak{B}$ be a totally singular subspace of $\mathfrak{J}$. Then $x U_b = Q(b, \bar{x})\, b \in \mathfrak{B}$ if $x \in \mathfrak{J}$, $b \in \mathfrak{B}$. Hence $\mathfrak{B}$ is an inner ideal.

We determine next the outer ideals of the simple Jord $(Q, 1)$. By Theorems 10 and 11 the simplicity condition is equivalent to: Q is non-degenerate and $\mathfrak{J} =$ Jord $(Q, 1)$ is not a direct sum of two copies of Φ.

THEOREM 14. *If* $\mathfrak{J} =$ Jord $(Q, 1)$ *is simple then the outer ideals of* $\mathfrak{J}$ *are* $\mathfrak{J}$ *and the subspaces of* $\mathfrak{J}^{\perp}$.

PROOF. Let $b \in \mathfrak{B}$, a subspace of $\mathfrak{J}^{\perp}$. Then $T(b) = Q(b, 1) = 0$ so $\bar{b} = -b$. Hence $bU_x = Q(x, \bar{b})x - Q(x)\bar{b} = Q(x)b \in \mathfrak{B}$ for all x and $\mathfrak{B}$ is an outer ideal. To prove the converse let $\mathfrak{B}$ be an outer ideal $\neq \mathfrak{J}$. Suppose $\mathfrak{B} \not\subset \mathfrak{J}^{\perp}$. Then $\bar{\mathfrak{B}} \not\subset \mathfrak{J}^{\perp}$ so we have $b_0 \in \mathfrak{B}$, $\bar{b}_0 \notin \mathfrak{J}^{\perp}$. For any x, $b_0 U_x = Q(x, \bar{b}_0)x - Q(x)\bar{b}_0 = Q(x, \bar{b}_0)x - Q(x)(T(b_0)1 - b_0) \in \mathfrak{B}$. Hence

$$Q(x, \bar{b}_0)x - Q(x)T(b_0)1 \in \mathfrak{B}, \; x \in \mathfrak{J}. \tag{28}$$

We now distinguish two cases :

I. Either $1 \in \mathfrak{B}$ or $T(b_0) = 0$. Then $Q(x, \bar{b}_0)x \in \mathfrak{B}$ so $x \in \mathfrak{B}$ for every x such that $Q(x, \bar{b}_0) \neq 0$. Since $\bar{b}_0 \notin \mathfrak{J}^{\perp}$ there exists x_0 such that $Q(x_0, \bar{b}_0) \neq 0$. Then $x_0 \in \mathfrak{B}$. If $y \in \mathfrak{J}$ satisfies $Q(y, \bar{b}_0) = 0$ then $x = x_0 + y$ satisfies $Q(x, \bar{b}_0) \neq 0$ so $x \in \mathfrak{B}$. Then $y \in \mathfrak{B}$. Then $\mathfrak{B} = \mathfrak{J}$ contrary to hypothesis.

II. $1 \notin \mathfrak{B}$ and $T(b_0) \neq 0$. By (28), if $x \in \mathfrak{B}$, $Q(x)T(b_0)1 \in \mathfrak{B}$. Hence $Q(x) = 0$. Thus $Q(\mathfrak{B}) = 0$ so $\mathfrak{B}$ is an inner ideal by Theorem 13. Hence $\mathfrak{B}$ is an ideal so $\mathfrak{B} = \mathfrak{J}$ or $\mathfrak{B} = 0$ by simplicity. These contradict $\mathfrak{B} \neq \mathfrak{J}$, $\mathfrak{B} \not\subset \mathfrak{J}^{\perp}$. Thus we must have either $\mathfrak{B} = \mathfrak{J}$ or $\mathfrak{B} \subseteq \mathfrak{J}^{\perp}$.

6. Almost nil quadratic Jordan algebras. It is easy to see that if $\mathfrak{N}_1$ and $\mathfrak{N}_2$ are nil ideals in a quadratic Jordan algebra $\mathfrak{J}$ then $\mathfrak{N}_1 + \mathfrak{N}_2$ is a nil ideal. It follows that there exists a nil ideal $\mathfrak{N}$ in $\mathfrak{J}$ such that $\mathfrak{N}$ contains every nil ideal and $\mathfrak{J}/\mathfrak{N}$ has no non-zero nil ideal. We call $\mathfrak{N}$ the *nil radical* of $\mathfrak{J}$. It is known that $\mathfrak{N}$ contains every absolute zero divisor of $\mathfrak{J}$ (McCrimmon [4] or Jacobson [2], § 3.3).

Let $\Phi[\lambda]$ be the polynomial algebra in the indeterminate λ and let $a \in \mathfrak{J}$. Then there exists a unique homomorphism of $\Phi[\lambda]^{(q)}$ into $\mathfrak{J}$ sending $\lambda \to a$. Let $\mathfrak{K}_a$ denote the kernel. Then we shall call $\mathfrak{J}$ *power special* if $\mathfrak{K}_a$ is an ideal in the associative algebra $\Phi[\lambda]$ for every $a \in \mathfrak{J}$.

LEMMA 1. *If $\mathfrak{J}$ contains no absolute zero divisors $\neq 0$ then $\mathfrak{J}$ is power special.*

PROOF. Let $a \in \mathfrak{J}$ and let $\mathfrak{K}_a$ be as above. Let $f(\lambda) \in \mathfrak{K}_a$ and $g(\lambda) = f(\lambda)\, h(\lambda)$. Then we have the known identity $U_{g(a)} = U_{f(a)}\, U_{h(a)}$ where, in general, $p(a)$ denotes the image of $p(\lambda)$ under the homomorphism mapping $\lambda \to a$ (McCrimmon [4] or Jacobson [2], §1.3). Since $f(a) = 0$ we have $U_{g(a)} = 0$ so $g(a) = 0$. Hence $g(\lambda) \in \mathfrak{K}_a$ so $\mathfrak{K}_a$ is an ideal in $\Phi\,[\lambda]$.

LEMMA 2. *If $\mathfrak{J}$ contains no absolute zero divisors $\neq 0$ then $\mathfrak{J}$ contains no $z \neq 0$ such that $z^2 = 0$ and $V_z = 0$.*

PROOF. We have the identities : $U_b U_{a,b} + U_{a,b} U_b = U_{b^2, a \circ b}$, $U_a U_b + U_b U_a + U_{a,b}^2 = U_{a^2,b^2} + U_{a \circ b}$, $U_a U_b U_c + U_c U_b U_a + U_{a,c} U_b U_{a,c} = U_{bU_a, bU_c} + U_{bU_{a,c}}$ (Jacobson [2], § 1.3). Hence if $z^2 = 0$ and $V_z = 0$ then $U_z\, U_{a,z} = -\, U_{a,z}\, U_z$, $U_a\, U_z + U_z\, U_a + U_{a,z}^2 = 0$, $U_{a,z}\, U_z\, U_{a,z} = U_{zU_{a,z^3}} + U_{zU_{a,z}}$ since $U_z^2 = U_{z^2} = 0$. Also since $z^3 = 0$ by Lemma 1, and $zU_{a,z} = z^2 \circ a$ (Jacobson loc. cit.), we have $U_{a,z}\, U_z U_{a,z} = 0$. Then $U_{aU_z} = U_z U_a U_z = -\, U_z^2 U_a - U_z U_{a,z}^2 = -\, U_{z^2} U_a + U_{a,z} U_z U_{a,z} = 0$. Since $\mathfrak{J}$ contains no absolute zero divisors this gives $aU_z = 0$ for all a and so $z = 0$.

We shall call a quadratic Jordan algebra $\mathfrak{J}$ *almost nil* if every $a \in \mathfrak{J}$ has the form $\alpha\, 1 + z$ where $\alpha \in \Phi$ and z is nilpotent. It is known that if $\mathfrak{J}$ is an almost nil (linear) Jordan algebra of characteristic $\neq 2$ then $\mathfrak{J} = \Phi\, 1 + \mathfrak{N}$ where $\mathfrak{N}$ is a nil ideal (Jacobson [3], p. 198). It is known that a quadratic Jordan algebra of characteristic $\neq 2$ becomes a (linear) Jordan algebra if one defines $a.b = \frac{1}{2} a \circ b$ (McCrimmon [1] or Jacobson [2], § 1.2). Hence if $\mathfrak{J}$ is an almost nil quadratic Jordan algebra of characteristic $\neq 2$ then $\mathfrak{J} = \Phi 1 + \mathfrak{N}$ where $\mathfrak{N}$ is a nil ideal. It is clear that $\mathfrak{N}$ is the nil radical of $\mathfrak{J}$ and that $\mathfrak{J} = \Phi 1 + \mathfrak{N}$, $\mathfrak{N}$ a nil ideal, if and only if the difference algebra of $\mathfrak{J}$ relative to its nil radical is one dimensional. In the characteristic 2 case an almost nil quadratic Jordan algebra may not have the structure since any traceless Jord $(Q, 1)$ such that $Q(a) \in \Phi^2$, $a \in \mathfrak{J}$, is almost nil. For, we have $a^2 = Q\,(a)\, 1$ so if $\alpha^2 = Q(a)$ then $a = \alpha 1 + z$ where $z^2 = 0$. If Q is non-degenerate then $\mathfrak{J}$ contains no nil ideal $\neq 0$, and since $\mathfrak{J} \neq \Phi\, 1$ by definition, the decomposition $\mathfrak{J} = \Phi\, 1 + \mathfrak{N}$, $\mathfrak{N}$ the nil radical, fails. More generally, it is clear that if $\mathfrak{N}$ is the nil radical of $\mathfrak{J}$ and $\mathfrak{J}/\mathfrak{N}$ = Jord $(Q,1)$, Q traceless and non-degenerate, $Q(a) \in \Phi^2$ for all $a \in$ Jord $(Q, 1)$, then $\mathfrak{J}$ is almost nil and $\mathfrak{J} \neq \Phi 1 + \mathfrak{N}$. We shall now show that for finite-dimensional $\mathfrak{J}$ over a field Φ with cardinality $|\Phi| > 2$ this is the only exception to $\mathfrak{J} = \Phi 1 + \mathfrak{N}$.

THEOREM 15. *Let $\mathfrak{J}$ be a finite dimensional quadratic Jordan algebra over Φ with $|\Phi| > 2$ and let $\mathfrak{N}$ be the nil radical. Then $\mathfrak{J}$ is almost nil if and only if either $\mathfrak{J}/\mathfrak{N}$ is one dimensional or $\mathfrak{J}/\mathfrak{N}$ =* Jord $(Q, 1)$ *where Q is non-degenerate and traceless, $Q(a) \in \Phi^2$, $a \in$* Jord $(Q, 1)$.

PROOF. The sufficiency is clear and the result quoted shows that the condition is necessary if char $\Phi \neq 2$. Hence we assume $\mathfrak{J}$ is finite-dimensional almost nil of characteristic two. Since we may replace $\mathfrak{J}$ by $\mathfrak{J}/\mathfrak{N}$, $\mathfrak{N}$ the nil radical, we may assume that $\mathfrak{J}$ contains no nil ideal $\neq 0$. We have noted in § 1 that $\mathfrak{J}$ is a restricted Lie algebra relative to $[a, b] = a \circ b$, $a^{[2]} = a^2$. Then ad $a = V_a$ in this Lie algebra. If $a \in \mathfrak{J}$, $a = \alpha 1 + z$, $\alpha \in \Phi$, z nilpotent. Since $x \circ 1 = 2x = 0$, ad $1 = 0$ so ad $a =$ ad z. Since z is nilpotent there exists a k such that $z^{2^k} = 0$. Then $(\text{ad } a)^{2^k} = (\text{ad } z)^{2^k} = \text{ad } z^{2^k} = 0$. It now follows from Engel's theorem that $\mathfrak{J}$ is a nilpotent Lie algebra. Next we note that the center $\mathfrak{C}$ of $\mathfrak{J}$ as Lie algebra, that is the set of c such that $V_c = 0$, is $\Phi 1$. Otherwise, since $c \in \mathfrak{C}$ implies $c^2 \in \mathfrak{C}$, $\mathfrak{C}$ contains a $z \neq 0$ such that $z^2 = 0$, contrary to Lemma 2. Hence $\mathfrak{C} = \Phi 1$ and since $\mathfrak{J}$ is a nilpotent Lie algebra either $\mathfrak{J} = \mathfrak{C} = \Phi 1$ or there exists a $d \notin \Phi 1$ such that $a \circ d = l(a) 1$ for every $a \in \mathfrak{J}$, where $l(a) \in \Phi$. Then l is a linear function on $\mathfrak{J}$. Assume this possibility. We have $V_{a^3} = V_{a^2} V_a - V_a U_a$ (Jacobson [2], § 1.3) $= V_a^3 + V_a U_a$ (char. 2). Since $dV_a^3 = (l(a) 1) V_a^2 = 0$ the foregoing identity gives $l(a^3) 1 = a^3 \circ d = (d \circ a) U_a = l(a) a^2$. Since $d \notin \mathfrak{C} = \Phi 1$ there exists a b such that $l(b) = 1$. Then for $\beta \neq l(a)$ we have $l(a + \beta b) \neq 0$ so $l(a + \beta b)(a + \beta b)^2 \in \Phi 1$ implies that $(a + \beta b)^2 \in \Phi 1$. Since $(a + \beta b)^2 = a^2 + \beta a \circ b + \beta^2 b^2$ and $b^2 \in \Phi 1$, now $a^2 + \beta a \circ b \in \Phi 1$ if $\beta \neq l(a)$. Since $|\Phi| > 2$ this implies that every $a^2 \in \Phi 1$. Thus $a^2 = Q(a) 1$, $a \in \mathfrak{J}$. It follows that Q is a quadratic form on $\mathfrak{J}$. Since $\mathfrak{J}$ contains no absolute zero divisors we have $a^3 = Q(a)\, a$ by Lemma 1. Hence by § 2, $\mathfrak{J}$ is a generically algebraic quadratic Jordan algebra of degree two so $\mathfrak{J} = \mathrm{Jord}\,(Q, 1)$. Since $\mathfrak{J}$ contains no nil ideal $\neq 0$, Q is non-degenerate and clearly Q is traceless. Since $a = \alpha 1 + z$, z nilpotent, and $a^2 = Q\,(a)\,1$ we have $z^2 = 0$ and $\alpha^2 = Q(a)$. Hence $\mathfrak{J}$ satisfies the conditions of the theorem.

Now suppose the base field is algebraically closed, $\mathfrak{J}$ is power special, and $a \in \mathfrak{J}$ is algebraic. Then the minimum polynomial $m_a(\lambda)$ of a, that is, the monic polynomial of least degree in $\mathfrak{K}_a$, has the form $\Pi\,(\lambda - \alpha_i)^{e_i}$, $\alpha_i \neq \alpha_j$ if $i \neq j$. Moreover, the subalgebra generated by a contains an idempotent $\neq 0$, 1 unless the minimum polynomial has the form $(\lambda - \alpha)^e$. In this case $z = a - \alpha\, 1$ is nilpotent and $a = \alpha 1 + z$.

Lemma 3. *Let $\mathfrak{J}$ be a quadratic Jordan algebra over a field, $\mathfrak{N}$ the nil radical of $\mathfrak{J}$. Then every idempotent of $\bar{\mathfrak{J}} = \mathfrak{J}/\mathfrak{N}$ can be lifted to an idempotent of $\mathfrak{J}$.*

Proof. If the characteristic $\neq 2$, $\mathfrak{J}$ is a (linear) Jordan algebra with $a.b = \frac{1}{2}\, a \circ b$. Then the result is known (Jacobson [3], p. 144). Now suppose char $\Phi = 2$. Let $a \in \mathfrak{J}$ satisfy $a^2 + a = z \in \mathfrak{N}$. Using the homomorphism of $\Phi\,[\lambda]^{(q)}$ into $\mathfrak{J}$ sending $\lambda \to a$ we obtain $a^m \circ a^n = 2a^{m+n} = 0$ and $(a^m)^n = a^{m+n}$. Hence $z^2 = (a^2 + a)^2 = a^4 + a^2$. Iteration gives $a^{2^{k+l}} + a^{2^k} = z^{2^k}$ so $e = a^{2^k}$ satisfies $e^2 = e$ if k is sufficiently large. Also $e = a^{2^k} \equiv a \pmod{\mathfrak{N}}$.

We can now prove

Theorem 16. *Let $\mathfrak{J}$ be a finite dimensional quadratic Jordan algebra over an algebraically closed field with nil radical $\mathfrak{N}$. Then the following conditions on $\mathfrak{J}$ are equivalent:* (1) *$\mathfrak{J}$ is almost nil,* (2) *$\mathfrak{J}$ contains no idempotents $\neq 0$, 1,* (3) *$\mathfrak{J}/\mathfrak{N}$ is either one dimensional or it is isomorphic to a traceless* Jord $(Q, 1)$.

Proof. The equivalence of (1) and (3) follows from Theorem 15 and algebraic closure of Φ. Now suppose z is nilpotent and $\alpha\, 1 + z = e$ is an idempotent. Then we have $(\alpha^2 - \alpha)\; 1 = (1 - 2\alpha)z - z^2$. Since the right hand side is nilpotent, $\alpha^2 = \alpha$, $\alpha = 0, 1$. Then $z^2 = \pm z$. Since z is nilpotent, raising both sides to a suitable 2^k-th power gives $z = 0$. Then $e = 0$ or 1. Thus if $\mathfrak{J}$ is almost nil it contains no idempotent $\neq 0, 1$. Conversely, suppose $\mathfrak{J}$ contains no idempotent $\neq 0,1$ and consider $\bar{\mathfrak{J}} = \mathfrak{J}/\mathfrak{N}$, $\mathfrak{N}$ the nil radical. By Lemma 3, $\bar{\mathfrak{J}}$ contains no idempotent $\neq 0, 1$. Also this algebra is power special so the minimum polynomial of every $\bar{a} \in \bar{\mathfrak{J}}$ is of the form $(\lambda - \alpha)^e$. Hence $\bar{\mathfrak{J}}$ and consequently $\mathfrak{J}$ is almost nil.

7. Structure of the Clifford algebras. $C(\mathfrak{J}, Q, 1)$. We shall call a quadratic form Q *separable* if the extension of Q to $\mathfrak{J}_\Omega$, Ω the algebraic closure of Φ, is non-degenerate. If char $\Phi \neq 2$ separability is equivalent to non-degeneracy and to non-degeneracy of the bilinear form $Q(x, y)$. If char $\Phi = 2$ the conditions for separability are: Q is non-degenerate and the defect def $Q = 0$ or 1. The first condition is clearly necessary. Now suppose it holds. Then def $Q = \dim \mathfrak{J}^\perp$ so our result will follow if we can show that Q is separable if and only if either $\mathfrak{J}^\perp = 0$ or $\mathfrak{J}^\perp$ is one dimensional. We have $\mathfrak{J}^\perp{}_\Omega = \mathfrak{J}_\Omega{}^\perp$ so if $\mathfrak{J}^\perp = 0$ then $\mathfrak{J}_\Omega{}^\perp = 0$ and Q is non-degenerate on $\mathfrak{J}_\Omega$. Now assume $\mathfrak{J}^\perp \neq 0$ and let $u_1, \dots, u_k$ be linearly independent elements of $\mathfrak{J}^\perp$. Since rad $Q = 0$, $Q(u_i) \neq 0$. The u_i are also linearly independent in $\mathfrak{J}_\Omega$ and if the $\xi_i \in \Omega$ then $Q(\Sigma_1^k \xi_i u_i) = \Sigma_1^k \xi_i^2 Q(u_i)$. It is clear that we can choose $x = \Sigma \xi_i u_i \neq 0$ in $\mathfrak{J}_\Omega$ so that $Q(x) = 0$ if and only if $k > 1$, and such an x belongs to rad Q_Ω. Thus in all cases (char $\Phi \neq 2$ or $= 2$) Q is separable if and only if Q is non-degenerate of defect 0 or 1.

We now suppose $(Q, 1)$ is a quadratic form with base point 1 and we consider the Clifford algebra $C(\mathfrak{J}, Q, 1)$ as defined in § 3. As usual we identify $\mathfrak{J}$ with the corresponding subspace of $C(\mathfrak{J}, Q, 1)$, so $\mathfrak{J}$ generates $C(\mathfrak{J}, Q, 1)$ and we have the following relations on the generators $x, y \in \mathfrak{J}$:

$$x^2 - T(x)x + Q(x)1 = 0 \tag{29}$$

$$xy + yx = T(x)y + T(y)x - Q(x, y)1 \tag{30}$$

$$yxy = Q(y, \bar{x})y - Q(y)\bar{x} \tag{31}$$

where $T(x) = Q(x, 1), \bar{x} = T(x)\,1 - x$. Let $z \in \text{rad}\, Q$ so $Q(z) = 0 = Q(z, x), x \in \mathfrak{J}$. Then $zxz = 0$, $x \in \mathfrak{J}$ and $xz = -zx + T(x)z = z\bar{x}$ Hence for $C = C(\mathfrak{J}, Q, 1)$ we have $Cz = zC$ and $zCz = 0$. Hence CzC. is a nilpotent ideal in C. It follows that if Q is not a separable then C_Ω has a nilpotent ideal $\neq 0$ so C is not a separable associative algebra in the usual sense. We shall now show that if Q is separable then either C is central simple, a direct sum of two central simple algebras, or simple with center a separable quadratic field. Then it will follow that in all cases C is separable.

These results are well known if $(Q, 1)$ is pure. Then C can be identified with $C(\mathfrak{B}, -Q)$ the usual Clifford algebra of the restriction of $-Q$ to $\mathfrak{B}$, $\mathfrak{B}$ a subspace such that $\mathfrak{J} = \Phi 1 \oplus \mathfrak{B}$, $T(\mathfrak{B}) = 0$. Suppose first $\dim \mathfrak{B} < \infty$ and char $\Phi \neq 2$. Then if $\dim \mathfrak{B} = 2\nu$ so $\dim \mathfrak{J} = 2\nu + 1$, it is well known that C is central simple (see Chevalley [1], § 2 . 2, or Jacobson [3], pp. 263-264 for this and the other results on Clifford algebras which will be assumed here without proof). If $\dim \mathfrak{B} = 2\nu - 1$, let $(v_1, v_2, \dots, v_{2\nu-1})$ be a basis for $\mathfrak{B}/\Phi$ and let $\delta = (-1)^{\nu\nu} 2 \det (Q(v_i, v_j))$. Then it is known that C is simple with center a quadratic field or a direct sum of two isomorphic central simple algebras according as δ is not or is a square in Φ.

We note next that our hypothesis that Q is separable implies that $\mathfrak{B}$ is not isotropic: $\mathfrak{B} \cap \mathfrak{B}^\perp = 0$ even in the characteristic two case. Otherwise, we have a $z \neq 0$ in $\mathfrak{B} \cap \mathfrak{B}^\perp$. Also in the characteristic two case 1 is orthogonal to $\mathfrak{B}$ (in $\mathfrak{J}$ relative to $Q(x, y)$). Since $\mathfrak{J} = \Phi 1 \oplus \mathfrak{B}$ this contradicts def $Q \leqslant 1$. This result implies that $\dim \mathfrak{B}$ is even if char $\Phi = 2$. Then it is known that C is central simple (Chevalley, loc. cit.).

Now assume $\dim \mathfrak{B} = \infty$, $(Q, 1)$ pure. Since $\mathfrak{B}$ is non-isotropic, any finite dimensional subspace of $\mathfrak{B}$ can be imbedded in an even (finite) dimensional non-isotropic subspace $\mathfrak{W}$. The subalgebra of $\mathfrak{C} = C(\mathfrak{B}, -Q)$ generated by $\mathfrak{W}$ is isomorphic to $C(\mathfrak{W}, -Q)$ which, by the results just indicated, is central simple. It follows that any finite subset of $\mathfrak{J}$ can be imbedded in a central simple subalgebra of $C(\mathfrak{J}, Q, 1) = C(\mathfrak{V}, -Q)$. This implies that $C(\mathfrak{J}, Q, 1)$ is central simple if $\dim \mathfrak{J} = \infty$, Q separable and $(Q, 1)$ pure.

It remains to consider the case char $\Phi = 2$, $(Q, 1)$ is not pure. Choose $d \in \mathfrak{J}$ so that $T(d) = 1$. Then $\Phi 1 + \Phi d$ is non-isotropic so $\mathfrak{J} = \Phi 1 \oplus \Phi d \oplus \mathfrak{V}$ where $\mathfrak{V} = (\Phi 1 + \Phi d)^{\perp}$ and $T(\mathfrak{V}) = 0$. Then $\mathfrak{J}^{\perp} = \mathfrak{V} \cap \mathfrak{V}^{\perp}$ so the restriction of Q to $\mathfrak{V}$ is separable and the defect of Q is the same as the defect of the restriction of Q to $\mathfrak{V}$. We showed in § 3 that if $\mathfrak{C} = C(\mathfrak{V}, Q)$ then $C(\mathfrak{J}, Q, 1) = \mathfrak{C}[t, D]/(g(t))$ where $\mathfrak{C}[t, D]$ is the algebra of differential polynomials determined by the derivation D in $\mathfrak{C}$ such that $vD = v$, $v \in \mathfrak{V}$, and $g(t) = t^2 + t + Q(d)1$. Since $[v, d] = vd + dv = v$ by (30), it is clear that D is the restriction to $\mathfrak{C}$ of the inner derivation $x \to [x, d]$ in C. We shall distinguish three cases: I. def $Q = 1$, II. def $Q = 0$, $\mathfrak{J}$ finite-dimensional, and III. def $Q = 0$, $\mathfrak{J}$ infinite-dimensional.

I. def $Q = 1$. Let $\mathfrak{J}^{\perp} = \Phi z$. Then $z \in \mathfrak{V}$ and $\mathfrak{V} = \Phi z \oplus \mathfrak{W}$ where $\mathfrak{W}$ is non-isotropic. We have $Q(z) \neq 0$. Also the subalgebra $\mathfrak{D}$ of $\mathfrak{C} = C(\mathfrak{V}, Q)$ generated by $\mathfrak{W}$ can be identified with $C(\mathfrak{W}, Q)$ and this is central simple. The element z is in the center of $\mathfrak{C}$ and every element of $\mathfrak{C}$ has the form $d_0 + d_1 z$, $d_i \in \mathfrak{D}$. Moreover, $d_0 + d_1 z = 0$ implies $d_0 = d_1 = 0$. Hence $\Phi[z]$ is the center of $\mathfrak{C}$ and $\mathfrak{C} = \Phi[z] \otimes \mathfrak{D}$. Since $\mathfrak{D}$ is central simple any non-zero ideal of $\mathfrak{C}$ contains a non-zero element of $\Phi[z]$. The isomorphism of $C = C(\mathfrak{J}, Q, 1)$ with $\mathfrak{C}[t, D]/(g(t))$ shows that the elements of C can be written in the form $c_0 + c_1 d$, $c_i \in C$, and $c_0 + c_1 d = 0$ implies $c_0 = c_1 = 0$. Also we have $[c, d] = cD$, $c \in \mathfrak{C}$, $d^2 + d + Q(d)1 = 0$. Let B be a non-zero ideal in C and let $\mathfrak{B} = \{b_1 \in \mathfrak{C} \mid c_0 + b_1 d \in B$ for some $c_0 \in \mathfrak{C}\}$. Then $\mathfrak{B}$ is a non-zero ideal in $\mathfrak{C}$ so $\mathfrak{B}$ contains an element $\alpha 1 + \beta z \neq 0$, $\alpha, \beta \in \Phi$. Then B contains an element of the form $(\alpha 1 + \beta z) d + c_0$, $c_0 \in \mathfrak{C}$. Since $z^2 = Q(z)1 \neq 0$, $[c, d] \in \mathfrak{C}$ if $c \in \mathfrak{C}$, and $[z, d] = z$, it follows that B contains an element of the form $d + c_0$, $c_0 \in \mathfrak{C}$. Taking the commutator with z shows that $z \in B$. Then $1 \in B$ so $B = C$. Thus C is simple. A similar argument can be used to show that the center of C is $\Phi 1$. Another way of seeing this is to note that if ρ is any extension field of Φ then $C_\rho C(\mathfrak{J}_\rho, Q, 1)$ is simple. This implies that C is central simple (Jacobson [4], p. 115). We remark also that in the case I if $\mathfrak{J}$ is finite-dimensional then dim $\mathfrak{J}$ is odd.

II. def $Q=0$, $\mathfrak{J}$ finite-dimensional. $\mathfrak{B}$ is non-isotropic so dim $\mathfrak{B}=2\nu-2$ and dim $\mathfrak{J}=2\nu$. $\mathfrak{B}$ has a symplectic base $(u_1, v_1, \ldots, u_{\nu-1}, v_{\nu-1})$ (that is, $Q(u_i, v_i)=1=Q(v_i, u_i)$ and all other $Q(-, -)$ determined by this base are 0). Then $(1, d, u_1, v_1, \ldots, u_{\nu-1}, \nu_{\nu-1})$ is a symplectic base for $\mathfrak{J}$. Put $\Delta=Q(d)+\Sigma Q(u_i)Q(v_i)=Q(d)\,Q(1)+\Sigma Q(u_i)\,Q(v_i)$, $w=\Sigma\, u_i v_i$, $z=d+w$. We have $[v, w]=v$ for $v\in\mathfrak{B}$ so $z=d+w$ commutes with every $v\in V$. Also $[w, d]=\Sigma\,(u_i\,D)v_i+\Sigma\, u_i\,(v_i\,D)=2\,\Sigma u_i\,v_i=0$. Hence $[z, d]=0$ and since $\mathfrak{B}$ and d generate C, z is in the center of C. Since $uv+vu=Q(u, v)\;1$, $u, v\in\mathfrak{B}$ by (30), $(u_i\,v_i)^2=u_i^2\,v_i^2+u_i\,v_i=Q(u_i)\,Q(v_i)\;1+u_i\,v_i$ and if $i\neq j$ then $[u_i\,v_i, u_j\,v_j]=0$. Hence $w^2=(\Sigma\,u_i\,v_i)^2=w+\Sigma Q(u_i)\,Q(v_i)\;1$. Then $z^2=(d+w)^2=d^2+w^2=d+Q\,(d)\;1+w+\Sigma\,Q(u_i)\,Q(v_i)\;1=z+\Delta 1$. As in the case I, C is a free left module over $\mathfrak{C}$ with base $(1, d)$. Hence also $(1, z)$ is a free base for C as left module over $\mathfrak{C}$. It follows that $C\cong\Phi[z]\otimes\mathfrak{C}$ and $(1, z)$ is a base for $\Phi[z]$ over Φ. Since $z^2=z+\Delta 1$, $\Phi[z]$ is a separable quadratic field over Φ or a direct sum of two copies of Φ according as Δ does not or does have the form $\beta^2+\beta$, $\beta\in\Phi$. Since $\mathfrak{C}$ is central simple it follows that C is simple with $\Phi\,[z]$ as its center or is a direct sum of two copies of $\mathfrak{C}$ according as $\Delta\neq\beta+\beta^2$ or $=\beta+\beta^2$, $\beta\in\Phi$.

III. def $Q=0$, $\mathfrak{B}$ infinite-dimensional. We claim that the derivation D in $\mathfrak{C}$ is not inner. On the contrary, suppose D is the inner derivation determined by an element $v\in\mathfrak{C}$. Then v is contained in the subalgebra $\mathfrak{D}$ generated by a finite dimensional non-isotropic subspace $\mathfrak{W}$ of $\mathfrak{B}$. Since $vD=v$, $v\in\mathfrak{B}$, D induces a derivation in $\mathfrak{D}$. If $(u_1, v_1, \ldots, u_\nu, v_\nu)$ is a symplectic base for $\mathfrak{W}$ then the argument in II shows that D in $\mathfrak{D}$ is the inner derivation $x\longrightarrow[x, \Sigma_1^\nu\, u_i v_i]$. Since $\mathfrak{D}$ is central simple, $v=\Sigma_1^\nu\, u_i\,v_i+\gamma\,1$, $\gamma\in\Phi$. We can now imbed $\mathfrak{W}$ in a $2\nu+2$-dimensional non-isotropic subspace $\mathfrak{W}'$ with symplectic base $(u_1, v_1, \ldots, u_{\nu+1}, v_{\nu+1})$. Then $v=\Sigma_1^{\nu+1}\,u_j\,v_j+\delta\,1$, $\delta\in\Phi$. This implies that $u_{\nu+1}\,v_{\nu+1}\in\Phi\,1$ which is impossible. Hence D is not inner. Now let B be a non-zero ideal in C and let $\mathfrak{B}=B\cap\mathfrak{C}$ so $\mathfrak{B}$ is an ideal in the simple algebra $\mathfrak{C}$.

Hence if $\mathfrak{B} \neq 0$ then $\mathfrak{B} = \mathfrak{C}$ contains 1 so $B = C$. Let $c_0 + c_1 d$ be a non-zero element of B, $c_i \in \mathfrak{C}$. If $c_1 = 0$, $c_0 = c_0 + c_1 d \in \mathfrak{B}$ so $B = C$. Hence suppose $c_1 \neq 0$. Then the fact that $\mathfrak{C}$ is simple implies that we may assume $c_1 = 1$ and we have $c_0 + d \in B$. If $c \in \mathfrak{C}$ we have $[c_0 + d, c] = [c_0, c] + cD \in \mathfrak{B}$ and since D is not inner c can be chosen so that this element is $\neq 0$. Then again $B = C$. Hence C is simple. Then the argument at the end of I shows also that C is central.

Before stating our results we recall the notions of discriminant and pseudo-discriminant of a quadratic form of defect 0 on an even (finite) dimensional vector space $\mathfrak{J}$. If char $\Phi \neq 2$ then $\mathfrak{J}$ has an orthogonal base $(u_1, u_2, \ldots, u_{2\nu})$ and $\delta = (-1)^\nu \Pi\, Q(u_i) \neq 0$. Let Φ^2 denote the subgroup of the multiplicative group of non-zero elements of Φ consisting of squares and let $[\delta] = \delta\,\Phi^2$ denote the coset of δ modulo Φ^2. Then $[\delta]$ is independent of the choice of the orthogonal base and this is called the *discriminant* of Q on $\mathfrak{J}$. If char $\Phi = 2$, let $(u_1, v_1, \ldots, u_\nu, v_\nu)$ be a symplectic base and let $\Delta = \Sigma_2^\nu Q(u_i)\, Q(v_i)$. Let $V(\Phi)$ denote the subgroup of the additive group of Φ consisting of the elements of the form $\beta^2 + \beta$, $\beta \in \Phi$, and let $[\Delta]$ now denote the coset $\Delta + V(\Phi)$. Then $[\Delta]$ is independent of the choice of the symplectic base and this is called the *pseudo-discriminant* of Q (*cf*. Dieudonné [1], p. 64). The results we have proved can be summarized in the following

THEOREM 17. *Let Q be a separable quadratic form on a (possibly infinite-dimensional) vector space $\mathfrak{J}/\Phi$ with base point* 1. *Then the Clifford algebra $C(\mathfrak{J}, Q, 1)$ is central simple except in the two cases*: dim $\mathfrak{J}$ *even (finite) and* char $\Phi \neq 2$ *or* dim $\mathfrak{J}$ *even*, $T \neq 0$, char $\Phi = 2$. *In these cases $C(\mathfrak{J}, Q, 1)$ is either simple with center a separable quadratic field or a direct sum of two isomorphic central simple algebras according as the discriminant* $[\delta] \neq 1$ *or* $= 1$ *if* char $\Phi \neq 2$ *or the psuedo-discriminant* $[\Delta] \neq 0$ *or* $= 0$ *if* char $\Phi = 2$, $T \not\equiv 0$.

8. Clifford groups and spin groups of quadratic forms with base points. In this section we extend the classical results on Clifford groups and spin groups to the case of quadratic forms with base points. Most of the results we shall give have been obtained in a different way by Faulkner [1]. We assume throughout that $\mathfrak{J}$ is finite dimensional and Q is separable so its defect is either 0 or 1.

We recall that the *inner structure group* Instr $\mathfrak{J}$ of a quadratic Jordan algebra $\mathfrak{J}$ is the set of mappings of the form $U_{a_1} U_{a_2} \cdots U_a$ where the a_i are invertible. This is a normal subgroup of the structure group Str $\mathfrak{J}$. For $\mathfrak{J} = \text{Jord}(Q, 1)$ the condition that a is invertible is $Q(a) \neq 0$, and if $\eta = U_{a_1} U_{a_2} \cdots U_{a_r}$ then $Q(x^\eta) = \Pi Q(a_i)^2 \, Q(x)$. Hence if $\eta \in$ Instr $\mathfrak{J}$, η is a similarity of ratio $\Pi \, Q(a_i)^2$. Thus if $0(\mathfrak{J}, Q)$ denotes the orthogonal group of $\mathfrak{J}$ relative to Q then $0(\mathfrak{J}, Q) \cap$ Instr $\mathfrak{J}$ is the set of mappings $U_{a_1} U_{a_2} \cdots U_{a_r}$ such that $\Pi Q(a_i)^2 = 1$. We recall that if $Q(a) \neq 0$ then the mapping

$$S_a \colon x \longrightarrow x - Q(a)^{-1} \, Q(a, x) \, a \tag{32}$$

is contained in $0(\mathfrak{J}, Q)$. This is called the *symmetry* determined by a. If $\alpha \neq 0$ in Φ then $S_a = S_{\alpha a}$. We have $S_1 \colon x \longrightarrow x - T(x) \, 1 = -\bar{x}$. Hence the formula for $U_a \colon x \longrightarrow Q\,(a, \bar{x})\, a - Q(a)\, \bar{x}$ for a invertible $(Q\,(a) \neq 0)$ can be written as

$$U_a = Q(a) S_1 S_a. \tag{33}$$

It is known that except in the case $|\Phi| = 2$, $\dim \mathfrak{J} = 4$, index $Q = 2$ every orthogonal transformation is a product of symmetries (Chevalley [1], p. 19 for the non-defective case, Tamagawa [1] for the general case). *We shall exclude the exceptional case from further consideration.* We recall the definition of the rotation group $0^+(\mathfrak{J}, Q)$ *for Q non-defective*: for char $\Phi \neq 2$ the subgroup of orthogonal transformations of determinant 1, for char $\Phi = 2$ the subgroup of orthogonal transformations having Dickson invariant 0 (Dieudonné [1], p. 64). $0^+(\mathfrak{J}, Q)$ is the set of products of an even number of symmetries. Also, for non-defective Q the *reduced orthogonal group* $0'(\mathfrak{J}, Q)$ is the subgroup of $0^+(\mathfrak{J}, Q)$ of $\eta = S_{a_1} S_{a_2} \cdots S_{a_{2k}}$ such that $\Pi_1^{2k} Q(a_i)$ is a square or, equivalently (in view of $S_{\alpha a} = S_a$), $\Pi_1^{2k} Q(a_i) = 1$.

THEOREM 18. *If Q is not defective, $0'(\mathfrak{J}, Q)$ coincides with the subgroup of* Instr $\mathfrak{J}$ *of $\eta = U_{a_1} U_{a_2} \cdots U_{a_r}$ such that $\Pi \, Q(a_i) = 1$. For char $\Phi = 2$ this is $0\,(\mathfrak{J}, Q) \cap$* Instr $\mathfrak{J}$.

Proof. Let $\eta = U_{a_1} U_{a_2} \dots U_{a_r}$ with $\prod_1^r Q(a_i) = 1$. Then, by (33), $\eta = S_1 S_{a_1} S_1 \dots S_{a_r}$ and $Q(1)Q(a_1) \dots Q(a_r) = 1$.
Conversely, let $\eta = S_{a_1} S_{a_2} \dots S_{a_{2k}}$ with $\prod_1^{2k} Q(a_i) = 1$. Since, as is easily verified, $S_a S_b S_a = S_a^{-1} S_b S_a = S_{bS_a}$,

$$\eta = (S_1 S_{\bar{a}_1} S_1) S_{a_2} (S_1 S_{\bar{a}_3} S_1) S_{a_4} \dots (S_1 S_{a_{2k-1}} S_1) S_{a_{2k}}$$

where $\bar{a}_i = -a_i S_1$. Thus

$$\begin{aligned}\eta &= (S_1 S_{\bar{a}_1}) (S_1 S_{a_2}) \dots (S_1 S_{\bar{a}_{2k-1}}) (S_1 S_{a_{2k}}) \\ &= Q(\bar{a}_1)^{-1} Q(a_2)^{-1} \dots Q(\bar{a}_{2k-1})^{-1} Q(a_{2k})^{-1} U_{\bar{a}_1} U_{a_2} \dots U_{\bar{a}_{2k-1}} U_{a_{2k}} \\ &= U_{\bar{a}_1} U_{a_2} \dots U_{\bar{a}_{2k-1}} U_{a_{2k}} \in \operatorname{Instr} \mathfrak{J}\end{aligned}$$

and $Q(\bar{a}_1) Q(a_2) \dots Q(\bar{a}_{2k-1}) Q(a_{2k}) = 1$.

From now on *we assume Q non-defective*. We write $0(\mathfrak{J}, Q, 1)$ for subgroup of $0(\mathfrak{J}, Q)$ such that $1^\eta = 1$. Since $\operatorname{Str} \mathfrak{J}$ coincides with the group of similarities of $\mathfrak{J}$ relative to Q and since $\operatorname{Aut} \mathfrak{J}$ is the subgroup of $\operatorname{Str} \mathfrak{J}$ of η satisfying $1^\eta = 1$, it is clear that $\operatorname{Aut} \mathfrak{J} = 0(\mathfrak{J}, Q, 1)$. We put $0^+(\mathfrak{J}, Q, 1) = 0^+(\mathfrak{J}, Q) \cap 0(\mathfrak{J}, Q, 1)$ and $0'(\mathfrak{J}, Q, 1) = 0'(\mathfrak{J}, Q) \cap 0(\mathfrak{J}, Q, 1)$. We call $0(\mathfrak{J}, Q, 1), 0^+(\mathfrak{J}, Q, 1), 0'(\mathfrak{J}, Q, 1)$ respectively the *orthogonal group of Q with base point* 1, *the rotation group of Q with base point* 1, *the reduced orthogonal group of Q with base point* 1. Clearly they are just the subgroups of the orthogonal, rotation, or reduced orthogonal group respectively fixing the element 1. If the characteristic $\neq 2$ and $\mathfrak{J} = \Phi 1 \oplus \mathfrak{B}$ for $T(\mathfrak{B}) = 0$ then the mapping sending η to its restriction to $\mathfrak{B}$ gives an isomorphism of $0(\mathfrak{J}, Q, 1)$, etc., with $0(\mathfrak{B}, Q)$, etc.

We define the *Clifford group of Q with base point* 1, $\Gamma(\mathfrak{J}, Q, 1)$, to be the group of invertible elements $a \in C(\mathfrak{J}, Q, 1)$ such that $a^{-1}\mathfrak{J}a = \mathfrak{J}$. The mapping $\chi_a : x \to a^{-1}xa$, $x \in \mathfrak{J}$, is an automorphism of $\mathfrak{J}$ and so it is contained in $0(\mathfrak{J}, Q, 1)$. The mapping $\chi : a \to \chi_a$ is a representation of $\Gamma(\mathfrak{J}, Q, 1)$ called the *vector representation*.

Theorem 19. *The image of the Clifford group $\Gamma(\mathfrak{J}, Q, 1)$ under the vector representation is either $0(\mathfrak{J}, Q, 1)$ or $0^+(\mathfrak{J}, Q, 1)$ according as $C(\mathfrak{J}, Q, 1)$ is central simple or not.*

PROOF. Suppose first that $C(\mathfrak{J}, Q, 1)$ is central simple and let $\eta \in \operatorname{Aut} \mathfrak{J} = 0(\mathfrak{J}, Q, 1)$. Since $C(\mathfrak{J}, Q, 1)$ is the special universal envelope for Jord $(Q, 1)$, η has a unique extension to an automorphism η of $C(\mathfrak{J}, Q, 1)$. Since $C(\mathfrak{J}, Q, 1)$ is central simple this is inner. Hence there exists an $a \in \Gamma(\mathfrak{J}, Q, 1)$ such that $\chi_a = \eta$. Next assume $C(\mathfrak{J}, Q, 1)$ is not central simple. In this case dim $\mathfrak{J} = 2\nu$ and we distinghish the two cases: I. char $\Phi \neq 2$, II. char $\Phi = 2$, $T \not\equiv 0$. In the first case the center of $C(\mathfrak{J}, Q, 1)$ is generated by $u = u_1 u_2 \dots u_{2\nu-1}$ where $(u_1, u_2, \dots, u_{2\nu-1})$ is an orthogonal base for the subspace $\mathfrak{V}$ of elements of trace zero. If $\eta \in 0(\mathfrak{J}, Q, 1)$, direct calculation shows that $u^\eta = (\det \eta)\, u$ so $u^\eta = u$ if and only if $\eta \in 0^+(\mathfrak{J}, Q, 1)$. Also $u^\eta = u$ if and only if the automorphism η of $C(\mathfrak{J}, Q, 1)$ is inner. In case II, $\mathfrak{J}$ has a symplectic base $(1, d, u_1, v_1, \dots, u_{\nu-1}, v_{\nu-1})$ such that $z = d + \Sigma\, u_i v_i$ generates the center of $C(\mathfrak{J}, Q, 1)$. Then $z^\eta = z$ or $z^\eta = z + 1$ according as $\eta \in 0^+(\mathfrak{J}, Q, 1)$ or $\notin 0^+(\mathfrak{J}, Q, 1)$. The result follows as in case I.

We define the *even Clifford algebra* $C^e(\mathfrak{J}, Q, 1)$ to be the subalgebra of $C(\mathfrak{J}, Q, 1)$ generated by the products $v_1 v_2$ *where* $T(v_i) = 0$. The structure of $C^e(\mathfrak{J}, Q, 1)$ for char $\Phi \neq 2$ is well known. If char $\Phi = 2$, $T(1) = 0$ so $C^e(\mathfrak{J}, Q, 1)$ is the subalgebra of $C(\mathfrak{J}, Q, 1)$ generated by the elements of trace 0. This is the subalgebra $\mathfrak{C} = C(\mathfrak{V}, Q)$ considered in the proof of Theorem 17 (under II). This algebra is central simple since we are assuming def $Q = 0$. We define the *even Clifford group* $\Gamma^e(\mathfrak{J}, Q, 1) = \Gamma(\mathfrak{J}, Q, 1) \cap C^e(\mathfrak{J}, Q, 1)$.

LEMMA. (Faulkner [1]). *Any* $\eta \in 0(\mathfrak{J}, Q, 1)$ *is a product of symmetries* S_v *with* $T(v) = 0$.

THEOREM 20. *Assume* def $Q = 0$. *Then we have the exact sequence* $1 \to \Phi^* \to \Gamma^e(\mathfrak{J}, Q, 1) \xrightarrow[\chi]{} 0^+(\mathfrak{J}, Q, 1) \to 1$ *where* Φ^* *is the multiplicative group of non-zero elements of* Φ. *Moreover,* $\Gamma^e(\mathfrak{J}, Q, 1) = \{v_1 v_2 \dots v_{2k} \mid T(v_i) = 0, Q(v_i) \neq 0\}$.

PROOF. If char $\Phi \neq 2$, $\mathfrak{J} = \Phi 1 \oplus \mathfrak{V}$ where $\mathfrak{V}$ is the set of elements of trace 0. Then $C^e(\mathfrak{J}, Q, 1)$ can be identified with the even Clifford algebra of $\mathfrak{V}$ relative to $-Q$, and $0(\mathfrak{J}, Q, 1)$ with the orthogonal group of $\mathfrak{V}$ relative to $-Q$. Then the theorem follows from standard results (Chevalley [1], p. 49 or Jacobson [3], p. 372). Now assume char $\Phi = 2$. Then $\mathfrak{J}$ has a symplectic base $(1, d, u_1, v_1, \ldots, u_{\nu-1}, v_{\nu-1})$ such that $z = d + \sum u_i v_i$ generates the center P of $C(\mathfrak{J}, Q, 1)$. Since $\dim P = 2$, if $P \cap C^e(\mathfrak{J}, Q, 1,) \neq \Phi$, $P \subseteq C^e = C^e(\mathfrak{J}, Q, 1)$. Now $(1, u_1, v_1, \ldots, u_{\nu-1}, v_{\nu-1})$ is a base for the subspace of elements of trace 0 so C^e is the subalgebra generated by these elements. Hence $\dim C^e = 2^{2(\nu-1)}$. On the other hand, if $P \subseteq C^e$ then $d \in C^e$ so $C^e = C(\mathfrak{J}, Q, 1)$, contrary to $\dim C = 2^{2\nu-1}$. Thus $C^e \cap P = \Phi$. It follows that the kernel of the vector representation χ of the even Clifford group $\Gamma^e = \Gamma^e(\mathfrak{J}, Q, 1)$ is Φ^*. By Theorem 19 the image of Γ^e under χ is contained in $0^+(\mathfrak{J}, Q, 1)$. Now let $\eta \in 0^+(\mathfrak{J}, Q, 1)$. By the Lemma, $\eta = S_{v_1} S_{v_2} \ldots S_{v2k}$, $T(v_i) = 0$. As in the proof of Theorem 18, $\eta = \alpha^{-1} U_{v_1} U_{v_2} \ldots U_{v2k}$ where $\alpha = \Pi_1^{2k} Q(v_i)$. Put $a = v_1 v_2 \ldots v_{2k}$. Then for $x \in \mathfrak{J}$, $x^\eta = \alpha^{-1} a^J x a$ where J is the main involution in $C(\mathfrak{J}, Q, 1)$. Since $1^\eta = 1$, $a^{-1} = \alpha^{-1} a^J$ and $x^\eta = a^{-1} x a$. Thus $a \in \Gamma^e$ and $\chi_a = \eta$. This completes the proof that $1 \to \Phi^* \to \Gamma^e(\mathfrak{J}, Q, 1) \underset{\chi}{\to} 0^+(\mathfrak{J}, Q, 1) \to 1$ is exact. The argument shows also that if $b \in \Gamma^e$ and $\chi_b = \eta$ then there exists an $a = v_1 v_2 \ldots v_{2k}$, $T(v_i) = 0$, in Γ^e such that $\chi_a = \eta$. Then $b = \beta v_1 v_2 \ldots v_{2k} = (\beta v_1) v_2 \ldots v_{2k}$ as required in the last statement of the theorem.

If $a \in \Gamma^e$, $a^J \in \Gamma^e$ so $aa^J \in \Gamma^e$. Since $a^{-1} x a = a^J x (a^J)^{-1}$, $x \in \mathfrak{J}$, aa^J is in the center. Hence, under the hypothesis of Theorem 20, $aa^J = \gamma(a) 1$, $\gamma(a) \in \Phi^*$. We have $\gamma(ab) = \gamma(a)\gamma(b)$ for $a, b \in \Gamma^e$ so if $a = v_1 v_2 \ldots v_{2k}$, $T(v_i) = 0$, then $\gamma(a) = \Pi_1^{2k} Q(a_i)$. We define the *spin group of Q with base point* 1 to be the subgroup Spin $(\mathfrak{J}, Q, 1)$ of Γ^e of those a such that $\gamma(a) = 1$.

Let $\eta \in 0^+(\mathfrak{J}, Q, 1)$ and let $a \in \Gamma^e$ satisfy $\chi_a = \eta$. Then a is determined up to a multiplier in Φ^* so $\gamma(a)$ is determined up to a multiplier in Φ^{*2}, the set of squares of the elements of Φ^*. Hence the coset $\sigma(\eta) \equiv \gamma(a) \Phi^{*2}$ is determined by η. We call this the *spinorial norm* of $\eta \in 0^+(\mathfrak{J}, Q, 1)$. Since

$$\eta = S_{v_1} S_{v_2} \ldots S_{v2k}, \; T(v_i) = 0, \; \sigma(\eta) = \prod_1^{2k} Q(v_i) \Phi^{*2}.$$

This shows that the spinorial norm as just defined is the restriction to $0^+(\mathfrak{J}, Q, 1)$ of the usual spinorial norm defined in $0(\mathfrak{J}, Q)$

(Dieudonné [1], p. 55 and p. 66). It is clear that $0'(\mathfrak{J}, Q, 1)$ is the subgroup of $0^+(\mathfrak{J}, Q, 1)$ of transformations of spinorial norm 1. This and Theorem 20 have the following consequence.

THEOREM 21. *Under the hypothesis of Theorem* 20 *we have an exact sequence* $1 \to \{1, -1\} \to \mathrm{Spin}(\mathfrak{J}, Q, 1) \xrightarrow[\chi]{} 0'(\mathfrak{J}, Q, 1) \longrightarrow 1$.

9. Related Lie algebras. We have seen in § 4 that the structure Lie algebra Strl $\mathfrak{J}$ for $\mathfrak{J} = \mathrm{Jord}(Q, 1)$ is the set of linear transformations L on $\mathfrak{J}$ such that $Q(xL, x) = \rho Q(x)$, $\rho \in \Phi$, $x \in \mathfrak{J}$. This implies

$$Q(xL, y) + Q(x, yL) = \rho Q(x, y) \tag{34}$$

or

$$Q(xL, y) + Q(x, yL') = 0,\ L' = L - \rho 1. \tag{35}$$

In $\mathfrak{J}$ we have

$$xV_{\bar{a},b} = Q(a, x)\, b - Q(b, x)\, a + Q(a, b)\, x \tag{36}$$

so if we define $S_{\bar{a},b}$ by $xS_{a,b} = Q(a, x)\, b - Q(b, x)\, a$, then

$$S_{a,b} = V_{\bar{a},b} - Q(a, b)\, 1. \tag{37}$$

We have

$$Q(xS_{a,b}, x) = 0 \tag{38}$$

$$Q(xV_{\bar{a},b}, x) = 2Q(a, b)\, Q(x) \tag{39}$$

so $S_{a,b} \in$ Strl $\mathfrak{J}$ with $\rho = 0$ and $V_{\bar{a},b} \in$ Strl $\mathfrak{J}$ with $\rho = 2Q(a, b)$. A straightforward verification using (34) and (35) shows that if $L \in$ Strl $\mathfrak{J}$ then

$$[V_{\bar{a},b}, L] = V_{\bar{a},bL} + V_{\overline{aL'},b} \tag{40}$$

$$[S_{a,b}, L] = S_{a,\, bL} + S_{aL',\, b}. \tag{41}$$

This shows that the set $V'_{\mathfrak{J},\mathfrak{J}}$ of mappings of the form $\alpha 1 + \Sigma V_{\bar{a}i,\, bi}$ (or, equivalently, the set $S'_{\mathfrak{J},\mathfrak{J}}$ of all $\alpha 1 + \Sigma S_{a_i,\, b_i}$) is an ideal in Strl $\mathfrak{J}$. We call this the *inner structure Lie algebra* of $\mathfrak{J}$, and denote it by Instr $\mathfrak{J}$. Taking $a = 1$ in (36) we obtain that $V_b : x \to T(x)b - Q(b, x)\, 1 + T(b)\, x$ belongs to Strl $\mathfrak{J}$, and for $T(v) = 0$

$$V_v = S_{1,v} : x \to T(x)v - Q(v, x)\, 1. \tag{42}$$

If $T(v_1) = 0 = T(v_2)$ then

$$[V_{v_1}, V_{v_2}] = 2S_{v_2, v_1} : x \to 2Q(v_2, x)\, v_1 - 2Q(v_1, x)\, v_2. \quad (43)$$

The scalar multiplier $\rho = \rho\,(L) = T\,(1L)$ in (34) defines a linear function on Strl $\mathfrak{J}$ which is 0 on the derived algebra [Strl $\mathfrak{J}$, Strl $\mathfrak{J}$]. For any quadratic form Q on a vector space $\mathfrak{V}$ we let $L\,(\mathfrak{V}, Q)$ denote the Lie algebra of transformations alternate with respect to Q,

$$L(\mathfrak{V}, Q) = \{\, L \in \mathrm{End}_\Phi\, \mathfrak{V} \mid Q\,(vL, v) = 0,\ v \in \mathfrak{V}\}. \quad (44)$$

Thus $L(\mathfrak{J}, Q)$ is the subset of those $L \in$ Strl $\mathfrak{J}$ for which $\rho\,(L) = 0$; it forms an ideal containing [Strl $\mathfrak{J}$, Strl $\mathfrak{J}$].

The structure of $L(\mathfrak{V}, Q)$ when $\mathfrak{V}$ is finite-dimensional and the bilinear form $Q\,(x, y)$ is non-degenerate is given in Jacobson [5], pp. 494-498. It is well known that in characteristic $\neq 2$ the mappings $S_{a,b} : v \to Q(a, v)\, b - Q\,(b, v)\, a$ as in (37) span $L\,(\mathfrak{V}, Q)$. This is also true in characteristic $= 2$. (If $(u_1, v_1 \ldots, u_\nu, v_\nu)$ is a symplectic base for $\mathfrak{V}$ and the linear transformation L satisfies $Q\,(vL, v) = 0$, $v \in \mathfrak{V}$, then $L = \Sigma_{i<j}\, Q\,(v_i\, L, v_j)\, S_{u_i,u_j} + \Sigma_{i<j}\, Q\,(u_i\, L, u_j)\, S_{v_i,v_j} + \Sigma_{i,j}$ $Q(u_i L, v_j)\, S_{v_i,u_j}$. In particular, $1 = \Sigma\, S_{v_i,u_i}$ so that in characteristic two $S_{\mathfrak{V},\mathfrak{V}} = S'_{\mathfrak{V},\mathfrak{V}} = V'_{\mathfrak{V},\mathfrak{V}}$). We write this as

$$L(\mathfrak{V}, Q) = S_{\mathfrak{V},\mathfrak{V}} \quad (45)$$

for finite-dimensional $\mathfrak{V}$ and non-degenerate $Q\,(x, y)$.

From now on we assume Q is separable and dim $\mathfrak{J} < \infty$, so Q is non-degenerate and def $Q = 0$ or 1. We distinguish three cases : I. char $\Phi \neq 2$, II. char $\Phi = 2$ and def $Q = 0$, III. char $\Phi = 2$ and def $Q = 1$.

I. char $\Phi \neq 2$. Since $Q(x1, x) = 2Q(x)$, $1 \in$ Strl $\mathfrak{J}$ but $1 \notin L(\mathfrak{J}, Q)$. Then Strl$\mathfrak{J} = \Phi 1 \oplus L(\mathfrak{J}, Q)$ where, since $Q\,(x, y)$ is non-degenerate, we have $L(\mathfrak{J}, Q) = S_{\mathfrak{J},\mathfrak{J}}$ by (45). If $\mathfrak{J} = \Phi 1 \oplus \mathfrak{V}$ for $T(\mathfrak{V}) = 0$ then by (42), (43) the elements $S_{1,v} = V_v$, $2S_{v_2,v_1} = [V_{v_1}, V_{v_2}]$ span $S_{\mathfrak{J},\mathfrak{J}}$, $v, v_1, v_2 \in \mathfrak{V}$. If $V(\mathfrak{V}) = \{V_v \mid v \in \mathfrak{V}\}$ then $S_{\mathfrak{J},\mathfrak{J}} = V(\mathfrak{V}) \oplus [V(\mathfrak{V}), V(\mathfrak{V})]$ where the sum is direct since $1V_v = 2v$, $1[V_{v_1}, V_{v_2}] = 2(v_1 V_{v_2} - v_2 V_{v_1}) = 0$. Thus

$$\mathrm{Strl}\,\mathfrak{J} = \mathrm{Instrl}\,\mathfrak{J} = \Phi\, 1 \oplus V(\mathfrak{V}) \oplus [V(\mathfrak{V}), V(\mathfrak{V})] \quad (46)$$

$$L(\mathfrak{J}, Q) = V(\mathfrak{V}) \oplus [V(\mathfrak{V}), V(\mathfrak{V})].$$

Since the derivation algebra Der $\mathfrak{J}$ is just the subset of $D \in$ Strl $\mathfrak{J}$ satisfying $1D = 0$, it is clear that

$$\text{Der}\,\mathfrak{J} = \text{Inder}\,\mathfrak{J} = [V(\mathfrak{B}),\ V(\mathfrak{B})] \tag{47}$$

where the *inner derivation algebra* is Inder $\mathfrak{J}$ = Der $\mathfrak{J} \cap$ Instrl $\mathfrak{J}$. We can give a more geometric description of Der $\mathfrak{J}$ as follows. If $\mathfrak{J} = \Phi 1 \oplus \mathfrak{B}$ then $1D = 0$ for any derivation D, and $\mathfrak{B}D \subseteq \mathfrak{B}$ because $\mathfrak{J}D \subset (\Phi 1)^{\perp}$ for any derivation: $0 = Q(1D, x) + Q(1, xD) = T(xD)$ by (34). Thus D is a derivation if and only if $1D = 0$ $\mathfrak{B}D \subseteq \mathfrak{B}$, and the restriction of D to $\mathfrak{B}$ satisfies $Q(vD, v) = 0$, $v \in \mathfrak{B}$. Thus the restriction mapping gives an isomorphism

$$\text{Der}\,\mathfrak{J} \simeq L(\mathfrak{B}, Q) \tag{48}$$

where the restriction of $Q(x, y)$ to $\mathfrak{B} = (\Phi 1)^{\perp}$ is again non-degenerate.

II. char $\Phi = 2$, def $Q = 0$. In characteristic two (no matter what the defect is) we have $1 \in L(\mathfrak{J}, Q)$ and $V_{\bar{a},b} \in L(\mathfrak{J}, Q)$ by (38), so Instrl $\mathfrak{J} \subseteq L(\mathfrak{J}, Q)$. Also all V_x, $x \in \mathfrak{J}$, are derivations since $V_x \in$ Strl $\mathfrak{J}$ and $1V_x = 2x = 0$, and by (40) these form an ideal $V(\mathfrak{J})$ in Der $\mathfrak{J}$: $[V_{\bar{a}}, D] = V_{\overline{aD}}$, $a \in \mathfrak{J}$. Moreover, by (42) the V_v for $T(v) = 0$ form an abelian subalgebra of $V(\mathfrak{J})$ which is an ideal in Der $\mathfrak{J}$ by our remark $T(xD) = 0$ in case I.

Now assume def $Q = 0$. The bilinear form $Q(x, y)$ is non-degenerate and dim $\mathfrak{J} = n = 2\nu$ is even. We may write $\mathfrak{J} = \Phi 1 \oplus \Phi d \oplus \mathfrak{B}$ for $T(d) = 1$, $\mathfrak{B} = (\Phi 1 + \Phi d)^{\perp}$. Choose a symplectic base $(u_1, v_1, \ldots, u_{\nu-1}, v_{\nu-1})$ for $\mathfrak{B}$. The linear transformation L_0 defined by $1L_0 = d$, $dL_0 = Q(d)\,1 + d$, $u_iL_0 = Q(u_i)\,v_i$, $v_iL_0 = Q(v_i)\,u_i + v_i$ is easily seen to belong to Strl $\mathfrak{J}$, and $\rho(L_0) = T(1L_0) = T(d) = 1$. Thus Strl $\mathfrak{J} = \Phi L_0 \oplus L(\mathfrak{J}, Q)$. Since $Q(x, y)$ is non-degenerate, by (45) $L(\mathfrak{J}, Q) = S_{\mathfrak{J},\mathfrak{J}}$. Since $\mathfrak{J} = \Phi 1 \oplus \Phi d \oplus \mathfrak{B}$ and $S_{x,y} = -S_{y,x}$, $S_{\mathfrak{J},\mathfrak{J}}$ is spanned by the transformations $S_{1,d}$, $S_{1,v}$ $(= V_v$ by (42)), $S_{d,v}$ and S_{v_1,v_2} for $v, v_1, v_2 \in \mathfrak{B}$. Hence

$$\text{Strl}\,\mathfrak{J} = \Phi L_0 \oplus \Phi S_{1,d} \oplus S_{d,\mathfrak{B}} \oplus V(\mathfrak{B}) \oplus S_{\mathfrak{B},\mathfrak{B}} \tag{49}$$

$$\text{Instrl}\,\mathfrak{J} = L(\mathfrak{J}, Q) = \Phi S_{1,d} \oplus S_{d,\mathfrak{B}} \oplus V(\mathfrak{B}) \oplus S_{\mathfrak{B},\mathfrak{B}}$$

where the sum is direct since $1L_0 = d$, $1S_{1,d} = 1$, $1S_{d,v} = v$, $1V_v = 1S_{v_1,v_2} = 0$ and $dL_0 = Q(d)\,1 + d$, $dS_{1,d} = d$, $dV_v = v$, $dS_{d,v} = dS_{v_1,v_2} = 0$. Since Der $\mathfrak{J}$ consists of the $D \in$ Strl $\mathfrak{J}$ for which $1D = 0$ we have

$$\text{Der}\,\mathfrak{J} = \text{Inder}\,\mathfrak{J} = V(\mathfrak{B}) \oplus S_{\mathfrak{B},\mathfrak{B}} \tag{50}$$

where, as we have seen, $V(\mathfrak{B})$ is an abelian ideal. Since $\Phi 1 + \Phi d$ is a hyperbolic subspace and $Q(x, y)$ is non-degenerate on $\mathfrak{J}$, the restriction of $Q(x, y)$ to $\mathfrak{B} = (\Phi 1 + \Phi\, d)^{\perp}$ is also non-degenerate. Therefore by (45) we have $S_{\mathfrak{B},\mathfrak{B}} \cong L(\mathfrak{B}, Q)$.

III. char $\Phi = 2$, def $Q = 1$. It is clear that the structure Lie algebra is unchanged in passing to an isotope. Hence, by Theorem 12, we may assume that $\mathfrak{J}^{\perp} = \Phi\, 1$, $T \equiv 0$ and write $\mathfrak{J} = \Phi\, 1 \oplus \mathfrak{B}$ where $\mathfrak{B}$ is a subspace. Then $Q\,(x, y)$ is non-degenerate on $\mathfrak{B}$ and dim $\mathfrak{J} = n = 2\nu + 1$, dim $\mathfrak{B} = 2\nu$. We have $\rho(L) = T(1L) = 0$ for all $L \in$ Strl $\mathfrak{J}$, so Strl $\mathfrak{J} = L\,(\mathfrak{J}, Q)$ (but $Q\,(x, y)$ is degenerate on $\mathfrak{J}$). If $L \in$ Strl $\mathfrak{J}$ has $1L = \alpha\, 1 + v$, $\alpha \in \Phi$, $v \in \mathfrak{B}$, then $0 = Q\,(1L, w) + Q\,(1, wL) = Q\,(v, w)$, $w \in \mathfrak{B}$, by (34) so by non-degeneracy on $\mathfrak{B}$ we have $v = 0$. Thus $D = L - \alpha\, 1$ satisfies $1D = 0$ and is a derivation. Hence Strl $\mathfrak{J} = \Phi\; 1 \oplus$ Der $\mathfrak{J}$. If D is a derivation and $vD = \lambda\,(v)\; 1 + vD_0$, $\lambda\,(v) \in \Phi$, $vD_0 \in \mathfrak{B}$, for $v \in \mathfrak{B}$ then $0 = Q\,(vD, v) = Q\,(vD_0, v)$ so that $D_0 \in L\,(\mathfrak{B}, Q)$ and λ is an arbitrary linear functional on $\mathfrak{B}$. By non-degeneracy on $\mathfrak{B}$, any linear functional has the form $\lambda\,(v) = Q\,(w, v)$ for some $w \in \mathfrak{B}$. Also $1V_w = 0$, $vV_w = v \circ w = Q\,(v, w)\; 1$, and by (45) $L(\mathfrak{B}, Q) = S_{\mathfrak{B},\mathfrak{B}}$ so we have

$$\begin{aligned} \text{Strl}\,\mathfrak{J} &= L(\mathfrak{J}, Q) = \Phi 1 \oplus V(\mathfrak{B}) \oplus S_{\mathfrak{B},\mathfrak{B}} \qquad (51)\\ \text{Der}\,\mathfrak{J} &= \text{Inder}\,\mathfrak{J} = V(\mathfrak{B}) \oplus S_{\mathfrak{B},\mathfrak{B}} \end{aligned}$$

where again $V(\mathfrak{B})$ is an abelian ideal and $S_{\mathfrak{B},\mathfrak{B}} = L(\mathfrak{B}, Q)$ for $Q(x, y)$ non-degenerate on $\mathfrak{B}$.

We summarize our results in

THEOREM 22. *Let Q be a separable quadratic form on a finite-dimensional vector space $\mathfrak{J}$ with base point* 1, $\mathfrak{J} =$ Jord $(Q, 1)$. *If* char $\Phi \neq 2$ *then* $\mathfrak{J} = \Phi 1 \oplus \mathfrak{B}$ *for* $T(\mathfrak{B}) = 0$, *and we have*

$$\begin{aligned} \text{Strl}\,\mathfrak{J} &= \text{Instrl}\,\mathfrak{J} = \Phi 1 \oplus V(\mathfrak{B}) \oplus [V(\mathfrak{B}), V(\mathfrak{B})]\\ \text{Der}\,\mathfrak{J} &= \text{Inder}\,\mathfrak{J} = [V(\mathfrak{B}), V(\mathfrak{B})] \qquad \text{(I)} \end{aligned}$$

where $V(\mathfrak{B}) \oplus [V(\mathfrak{B}), V(\mathfrak{B})] = S_{\mathfrak{J},\mathfrak{J}} = L(\mathfrak{J}, Q)$ and $[V(\mathfrak{B}), V(\mathfrak{B})] = S_{\mathfrak{B},\mathfrak{B}} \cong L(\mathfrak{B}, Q)$.

If char $\Phi = 2$, def $Q = 0$ *then* $\mathfrak{J} = \Phi 1 \oplus \Phi d \oplus \mathfrak{B}$ *for* $T(d) = 1$, $\mathfrak{B} = (\Phi 1 + \Phi d)^{\perp}$, *and we have*

$$\begin{aligned} \text{Strl}\,\mathfrak{J} &= \Phi L_0 \oplus \Phi S_{1,d} \oplus S_{d,\mathfrak{B}} \oplus V(\mathfrak{B}) \oplus S_{\mathfrak{B},\mathfrak{B}} \\ \text{Instrl}\,\mathfrak{J} &= L(\mathfrak{J}, Q) = \Phi S_{1,d} \oplus S_{d,\mathfrak{B}} \oplus V(\mathfrak{B}) \oplus S_{\mathfrak{B},\mathfrak{B}} \qquad \text{(II)} \\ \text{Der}\,\mathfrak{J} &= \text{Inder}\,\mathfrak{J} = V(\mathfrak{B}) \oplus S_{\mathfrak{B},\mathfrak{B}} \end{aligned}$$

where $V(\mathfrak{B})$ *is an abelian ideal in* Der $\mathfrak{J}$ *and* $S_{\mathfrak{B},\mathfrak{B}} \cong L(\mathfrak{B}, Q)$.

If char $\Phi = 2$, def $Q = 1$ *then some isotope of* $\mathfrak{J}$ *is traceless. In the traceless case we have* $\mathfrak{J} = \Phi 1 \oplus \mathfrak{B}$ *for* $T(\mathfrak{B}) = 0$, *and*

$$\begin{aligned} \text{Strl}\,\mathfrak{J} &= \text{Instrl}\,\mathfrak{J} = L(\mathfrak{J}, Q) = \Phi 1 \oplus V(\mathfrak{B}) \oplus S_{\mathfrak{B},\mathfrak{B}} \\ \text{Der}\,\mathfrak{J} &= \text{Inder}\,\mathfrak{J} = V(\mathfrak{B}) \oplus S_{\mathfrak{B},\mathfrak{B}} \qquad \text{(III)} \end{aligned}$$

where $V(\mathfrak{B})$ *is an abelian ideal in* Strl $\mathfrak{J}$ *and* $S_{\mathfrak{B},\mathfrak{B}} \cong L(\mathfrak{B}, Q)$.

Give any quadratic Jordan algebra $\mathfrak{J}$ we can form a Lie algebra on the vector space

$$T(\mathfrak{J}) = \mathfrak{J} \oplus \mathfrak{J}' \oplus \text{Strl}\,\mathfrak{J} \qquad (52)$$

where $\mathfrak{J}'$ is a vector space isomorphic to $\mathfrak{J}$ under a mapping $x \to x'$ and the Lie products satisfy

$$[x, y] = [x', y'] = 0 \qquad (53)$$

$$[L, M] = [L, M] \text{ (the usual product in Strl } \mathfrak{J}) \qquad (54)$$

$$[x, L] = xL \qquad (55)$$

$$[x', L] = (x\bar{L})' \text{ where } \bar{L} = L - V_{1L} \qquad (56)$$

$$[x, y'] = V_{y,x} \qquad (57)$$

where $x \in \mathfrak{J}$, $x' \in \mathfrak{J}'$, $L \in \text{Strl}\,\mathfrak{J}$. The Lie algebra $T(\mathfrak{J})$ is called the *Tits-Koecher algebra* of $\mathfrak{J}$ (*cf.* Jacobson [3], p. 324).

THEOREM 23. *The Tits-Koecher* $T(\mathfrak{J})$ *for* $\mathfrak{J} = \text{Jord}\,(Q, 1)$ *is canonically isomorphic to Strl* $\tilde{\mathfrak{J}}/\Phi\tilde{1}$ *where* $\tilde{\mathfrak{J}} = \text{Jord}\,(\tilde{Q}, 1)$ *for* Q *the quadratic form* $\tilde{Q}(\alpha_1 e_1 + \alpha_2 e_2 + x) = \alpha_1 \alpha_2 + Q(x)$ *on the vector space* $\tilde{\mathfrak{J}} = \Phi e_1 + \Phi e_2 + \mathfrak{J}$.

PROOF. Write an arbitrary $\tilde{L} \in \mathrm{End}_\Phi \tilde{\mathfrak{J}}$ as

$$x\tilde{L} = -\lambda_1(x)\, e_1 - \lambda_2(x)\, e_2 + xL$$

$$e_1\tilde{L} = \alpha_1 e_1 + \beta_1 e_2 + z_1$$

$$e_2\tilde{L} = \alpha_2 e_1 + \beta_2 e_2 + z_2$$

for $\lambda_i(x)$, α_i, $\beta_i \in \Phi$, $x, z_i \in \mathfrak{J}$ and $L \in \mathrm{End}_\Phi \mathfrak{J}$. The condition $\tilde{Q}(\tilde{x}\tilde{L}, \tilde{x}) = \rho(\tilde{L})\, \tilde{Q}(\tilde{x})$ that $\tilde{L} \in \mathrm{Strl}\, \tilde{\mathfrak{J}}$ reduces to

$$\tilde{Q}(x\tilde{L}, x) = \rho(\tilde{L})\, \tilde{Q}(x)$$

$$\tilde{Q}(e_i\tilde{L}, e_i) = 0$$

$$\tilde{Q}(x\tilde{L}, e_i) + \tilde{Q}(e_i\tilde{L}, x) = 0$$

$$\tilde{Q}(e_1\tilde{L}, e_2) + \tilde{Q}(e_2\tilde{L}, e_1) = \rho(\tilde{L})$$

and thus, by definition of $\tilde{Q}$, $\tilde{L}$ to

$$Q(xL, x) = \rho\,(\tilde{L})\, Q\,(x)$$

$$\beta_1 = \alpha_2 = 0,\ \alpha_1 + \beta_2 = \rho(\tilde{L})$$
$$-\lambda_2(x) + Q(z_1, x) = -\lambda_1(x) + Q(z_2, x) = 0$$

and therefore

$$x\tilde{L} = Q(z_2,\ x)\ e_1 + Q(z_1\, x)\ e_2 + xL$$
$$e_1\tilde{L} = \alpha_1 e_1 + z_1$$
$$e_2\tilde{L} = (\rho\,(L) - \alpha_1)\ e_2 + z_2$$

for arbitrary $L \in \mathrm{Strl}\,\mathfrak{J}$ (with multiplier $\rho\,(L) = \rho\,(\tilde{L})$, $z_1, z_2 \in \mathfrak{J}$ and $\alpha_1 \in \Phi$. Thus if we define the linear transformations $i(L), x_1, x_2, l$ in $\tilde{\mathfrak{J}}$ by

$$i(L)\colon y \to yL,\ e_1 \to 0,\ e_2 \to \rho(L)\, e_2 \qquad (L \in \mathrm{Strl}\,\mathfrak{J})$$

$$x_1\colon y \to Q(x, y)\, e_2,\ e_1 \to x,\ e_2 \to 0 \qquad (x \in \mathfrak{J})$$

$$x_2\colon y \to Q\,(x, y)\, e_1,\ e_1 \to 0,\ e_2 \to x$$

$$l\colon\ y \to 0,\ e_1 \to e_2,\ e_2 \to -e_2$$

these are contained in Strl $\tilde{\mathfrak{J}}$. Moreover, $L \to i(L)$, $x \to x_1$, $x \to x_2$ are injective linear mappings of Strl $\mathfrak{J}$, $\mathfrak{J}$ and $\mathfrak{J}$ into Strl $\tilde{\mathfrak{J}}$ and Strl $\tilde{\mathfrak{J}} = i\,(\text{Strl}\,\mathfrak{J}) \oplus \mathfrak{J}_1 \oplus \mathfrak{J}_2 \oplus \Phi\, l$, $\mathfrak{J}_i = \{x_i \mid x \in \mathfrak{J}\}$. Since the identity mapping $1 = 1_{\mathfrak{J}} \in$ Strl $\mathfrak{J}$ with $\rho\ (1) = 2$ and $i\,(1)$ sends $y \to y$, $e_1 \to 0$, $e_2 \to 2e_2$, $i(1) + l$ sends $y \to y$, $e_1 \to e_1$, $e_2 \to e_2$. Thus $i(1) + l = \tilde{1} = 1_{\tilde{\mathfrak{J}}}$. Hence

$$\text{Strl}\,\tilde{\mathfrak{J}} = \Phi\,\tilde{1} \oplus \mathfrak{J}_1 \oplus \mathfrak{J}_2 \oplus i\,(\text{Strl}\,\mathfrak{J}). \tag{58}$$

From the formulas above one computes directly the Lie products in Strl $\tilde{\mathfrak{J}}$ obtaining

$$[x_1, y_1] = 0 = [x_2, y_2] \tag{59}$$

$$[i\,(L), i\,(M)] = i\,([L, M]) \tag{60}$$

$$[x_1, i\,(L)] = (xL)_1 \tag{61}$$

$$[x_2, i\,(L)] = (xL')_2,\ L' = L - \rho(L)\,1 \tag{62}$$

$$[x_1, y_2] = i\,(S_{x,y}) + Q(x, y)\, l. \tag{63}$$

For (59), $x_1 y_1$ sends $\mathfrak{J} + \Phi\, e_2 \to 0$, $e_1 \to Q(y, x)\, e_2$. For (60), $i\,(L)\, i\,(M)$ coincides with LM on $\mathfrak{J}$ and sends $e_1 \to 0$ and $e_2 \to \rho\,(L)\, \rho\,(M)\, e_1$; recall $\rho\,([L, M]) = 0$. For (61), $x_1\, i\,(L)$ sends $y \to Q(x, y)\, e_1 \to Q(x, y)\ \rho\,(L)\, e_2$, sends $e_2 \to 0$ and $e_1 \to x \to xL$ while $i(L)\, x_1$ sends $y \to yL \to Q(x, yL)\, e_2$, sends $e_2 \to 0$ and $e_1 \to 0$; recall (34). For (62), $x_2\, i\,(L)$ sends $y \to Q\,(x, y)\, e_1 \to 0$, sends $e_1 \to 0$, $e_2 \to x \to xL$ while $i(L)\, x_2$ sends $y \to yL \to Q(x, yL)\, e_1$, $e_1 \to 0$, $e_2 \to \rho(L)\, e_2 \to \rho\,(L)\, x$; recall (35). For (63), $x_1\, y_2$ sends $z \to Q(x, z)\, e_2 \to Q(x, z)\, y$, $e_1 \to x \to Q(y, x)\, e_1$, $e_2 \to 0$ while $y_2\, x_1$ sends $z \to Q(y, z)\, e_1 \to Q(y, z)\, x$, $e_1 \to 0$, $e_2 \to y \to Q(x, y)\, e_2$; recall $S_{x,\,y} \in$ Strl $\mathfrak{J}$ has $\rho = 0$. Let η' be the linear mapping of Strl $\tilde{\mathfrak{J}}$ into $T(\mathfrak{J})$ defined by

$$\eta' : \alpha\,\tilde{1} + x_1 + y_2 + i(L) \to x - \bar{y}' + L$$

(where, as usual, $\bar{y} = T(y)\,1 - y$ in $\mathfrak{J}$). Clearly η' is surjective and $\Phi\,\tilde{1}$ is its kernel. We claim that η' is a homomorphism. This is clear for the products $[x_1, y_1]$, $[x_2, y_2]$, $[i\,(L), i\,(M)]$, $[x_1, i\,(L)]$ by comparing (59)–(61) with (53)–(55). It is clear also for any product $[\tilde{A}, \tilde{1}]$, $\tilde{A} \in$ Strl $\tilde{\mathfrak{J}}$, since $[\tilde{A}, \tilde{1}] = 0$. Now

$\overline{\bar{x}L'} = xL - T(x)(1L) - \rho(L)x - T(xL)1 + T(x)T(1L)1$ and $x\bar{L} = xL - xV_{1L} = xL - T(x)(1L) + Q(1L, x)1 - T(1L)x = xL - T(x)(1L) - Q(1, xL)1 + \rho(L)Q(1, x)1 - T(1L)x = xL - T(x)(1L) - T(1L)x - T(xL)1 + \rho(L)T(x)1$. Since $T(1L) = \rho(L)$ we have $(\overline{\bar{x}L'}) = x\bar{L}$. Then $[x_2, i(L)]^{\eta'} = (xL')_2^{\eta'} = -(\overline{\bar{x}L'})' = (-\bar{x}\bar{L})'$ and $[x_2^{\eta'}, i(L)^{\eta'}] = -[\bar{x}'L] = -(\bar{x}\bar{L})'$. Hence $[x_2, i(L)]^{\eta'} = [x_2^{\eta'}, i(L)^{\eta'}]$. Finally, $S(x, y) - Q(x, y)1_{\mathfrak{A}} = -S(y, x_{\mathfrak{J}} + Q(x, y)1) = -V_{y}^{1}, x$ (by (36)) $= V_{-\bar{y},x}$ so $[x_1, y_2]^{\eta'} = (i(S_{x,y}) + Q(x, y)l)^{\eta'} = (i(S_{x,y}) + Q(x, y)(1_{\mathcal{J}} - i(1_J))^{\eta'} = (i(S_{x,y} - Q(x, y)1_J)^{\eta'} = S_{x,y} - Q(x,y)1_{\mathcal{J}} = V_{-\bar{y},x} = [x_1^{\eta'}, y_2^{\eta'}]$. Hence η' is a homomorphism and $\eta : a + \Phi 1_{\mathcal{J}} \to a^{\eta'}$ is an isomorphism of $\operatorname{Str}\tilde{\mathfrak{J}}/\Phi 1_{\mathfrak{J}}$ onto $T(\mathfrak{J})$.

10. Birepresentations and the meson algebra. The quadratic analogue of an associative specialization of a quadratic Jordan algebra $\mathfrak{J}$ in an associative algebra $\mathfrak{A}$ is a *(unital) quadratic specialization*, a map μ of $\mathfrak{J}$ into $\mathfrak{A}$ such that

QS1. $\mu : x \longrightarrow \mu(x)$ is quadratic

QS2. $\mu(1) = 1$

QS3. $\mu(x)\,\mu(y)\,\mu(x) = \mu(yU_x)$

QS4. If $\nu(x, y)$ is defined by $\nu(x, y) = \nu(x)\,\nu(y) - \mu(x, y)$ for $\nu(x) = \mu(x, 1)$, $\mu(x, y) = \mu(x + y) - \mu(x) - \mu(y)$, then

$$\mu(x)\,\nu(y, x) = \mu(yU_x, x) = \nu(x, y)\,\mu(x)$$

QS5. If ρ is any extension field of Φ and μ is the (unique) extension of the mapping μ on $\mathfrak{J}$ to a quadratic mapping of $\mathfrak{J}_P$ into $\mathfrak{A}_P$ then QS3 and QS4 hold for μ on $\mathfrak{J}_P$. In the particular case where $\mathfrak{A} = \operatorname{End}_\Phi \mathfrak{M}$ for some vector space $\mathfrak{M}$ we speak of a *quadratic representation*; a *$\mathfrak{J}$-bimodule* is a vector space $\mathfrak{M}$ together with a quadratic representation of $\mathfrak{J}$ on $\mathfrak{M}$. For example, the *regular representation* $x \to U_x$ of $\mathfrak{J}$ in $\operatorname{End}_\Phi \mathfrak{J}$ defines the *regular bimodule* $\mathfrak{J}$ by comparing QS1-QS5 with QJ1-QJ5. Another way of obtaining quadratic specializations is to form $\mu(x) = \sigma(x)\,\tau(x)$ where σ and τ are commuting (linear) associative specializations of $\mathfrak{J}$ in $\mathfrak{A}$; such a quadratic specialization is called a *compound linear specialization.* For these and all other results relating to quadratic specializations we refer to McCrimmon [5].

The *universal quadratic envelope* of $\mathfrak{J}$ consists of an associative algebra UQE $(\mathfrak{J})$ together with a quadratic specialization $\mathfrak{J} \xrightarrow{u}$ UQE $(\mathfrak{J})$ which are universal in the sense that any quadratic specialization $\mathfrak{J} \xrightarrow{\mu} \mathfrak{A}$ factors uniquely through u, that is, there is a unique homomorphism η : UQE $(\mathfrak{J}) \to \mathfrak{A}$ of associative algebras such that $u\eta = \mu$. In a similar way we have a universal object UCE $(\mathfrak{J})$ for the compound linear specializations.

The universal quadratic envelope of $\mathfrak{J} =$ Jord $(Q, 1)$, Q a quadratic form with base point 1, is called the *meson algebra* of Q. In characteristic $\neq 2$ it is known (Jacobson [3], pp. 264-272) that for a non-degenerate quadratic form Q on a finite dimensional vector space we have UQE $(\mathfrak{J}) \cong$ UCE $(\mathfrak{J})$ and the meson algebra is a finite dimensional semisimple associative algebra. This implies that all birepresentations of $\mathfrak{J}$ (or all $\mathfrak{J}$-bimodules) are completely reducible.

We now wish to show that these results fail in characteristic $= 2$. In this case UQE $(\mathfrak{J})$ and UCE $(\mathfrak{J})$ are finite dimensional if $\mathfrak{J}$ is, but they are no longer semisimple if dim $\mathfrak{J} > 2$. To prove this we need only find a bimodule, given by a compound linear specialization, which is not completely reducible.

For the rest of this section we assume Q is a separable quadratic form on a finite dimensional vector space $\mathfrak{J}$ over a field Φ of characteristic two. First suppose that Q has defect 1. We claim the regular bimodule $\mathfrak{J} =$ Jord $(Q, 1)$ is not completely reducible. The sub-bimodules of $\mathfrak{J}$ are just the outer ideals, and by Theorem 14 the outer ideals are just $\mathfrak{J}$ and the subspaces of $\mathfrak{J}^{\perp}$. Since Q is non-degenerate and has defect 1, $\mathfrak{J}^{\perp}$ is one-dimensional. Thus $\mathfrak{J}^{\perp}$ is the unique irreducible sub-bimodule, and as dim $\mathfrak{J} > 1$ we see that the socle ($=$ sum of all irreducible sub-bimodules) of $\mathfrak{J}$ is $\mathfrak{J}^{\perp}$ and $\mathfrak{J}$ is not completely reducible.

Now consider the more difficult case where Q has defect 0. Then by Theorem 14 the regular bimodule $\mathfrak{J}$ is irreducible since $Q(x, y)$ is non-degenerate, $\mathfrak{J}^{\perp} = 0$. However, we will prove that the Clifford bimodule $\mathfrak{M} = C(\mathfrak{J}, Q, 1)$ is not completely reducible, indeed that the only irreducible sub-bimodules of $\mathfrak{M}$ have the form $\omega\mathfrak{J}$ for ω in the center Ω of $C(\mathfrak{J}\, Q, 1)$ (thus the socle of $\mathfrak{M}$ is $\Omega\mathfrak{J} < \mathfrak{M}$). Here $C(\mathfrak{J}, Q, 1)^{(q)}$ is a quadratic Jordan algebra containing $\mathfrak{J}$ as a subalgebra, so the mapping $a \rightarrow U_a$, $a \in \mathfrak{J}$, U the U-operator in $C(\mathfrak{J}, Q, 1)^{(q)}$, defines a quadratic specialization and turns $C(\mathfrak{J}, Q, 1)$ into a $\mathfrak{J}$-bimodule. Since $U_a = r_a l_a$ for r_a, l_a the right, left multiplications in the associative algebra $C(\mathfrak{J}, Q, 1)$ determined by $a \in \mathfrak{J}$, we see $a \rightarrow U_a$ is a compound linear specialization. We need only show any non-zero sub-bimodule $\mathfrak{N}$ of $\mathfrak{M}$ contains some $\omega\,\mathfrak{J}$ for $\omega \in \Omega$, and since $\omega\,\mathfrak{J}$ is irreducible if $\mathfrak{J}$ is; it will even be enough to show $\mathfrak{N} \cap \omega\mathfrak{J} \neq 0$.

We write $\mathfrak{J} = \Phi 1 + \Phi d + \mathfrak{V}$ for $T(d) = 1$, $\dim \mathfrak{V} = 2\nu$ even. By the proof of Theorem 17 we know $C(\mathfrak{J}, Q, 1) = \mathfrak{C}_{\Omega} = \mathfrak{C} + \mathfrak{C}z$ for $\mathfrak{C} = C(\mathfrak{V}, Q)$ and $\Omega = \Phi\,[z]$ the center of $C(\mathfrak{J}, Q, 1)$. First consider the case where $\mathfrak{N} \subseteq \mathfrak{C}$ is not necessarily a $\mathfrak{J}$ sub-bimodule, but merely a subspace invariant under all V_v, $v \in \mathfrak{V}$. Relative to a fixed symplectic base $(u_1, v_1, \ldots, u_\nu, v_\nu)$ for $\mathfrak{V}$, the monomials $u_{i_1} \ldots u_{i_r} v_{j_1} \ldots v_{j_s}$ $(i_1 < \ldots < i_r,\ j_1 < \ldots < j_s)$ form a base for $\mathfrak{C} = C(\mathfrak{V}, Q)$; call $r + s$ the *length* of such a monomial, and let the length of a general element of $C(\mathfrak{V}, Q)$ be the greatest length of any monomial appearing in the expression of that element in terms of the above basis. Let $n \in \mathfrak{N}$ have minimal length $r + s$; if we can show $r + s \leqslant 1$ then $n \in \Phi 1 + \mathfrak{V} \subseteq \mathfrak{J}$ and we will be done : $\mathfrak{N} \cap \mathfrak{J} \neq 0$. Suppose on the contrary that $r + s > 1$, say $s \geqslant 1$. Now $u_j \circ (u_{i_1} \ldots \ldots u_{i_r} v_{j_1} \ldots v_{j_s}) = 0$ if $j \neq j_1, \ldots, j_s$ and $= u_{i_1} \ldots u_{i_r} v_{j_1} \ldots \hat{v}_{j_1} \ldots v_{j_s}$ if $j = j_k$ (where the circumflex denotes omission); here we have used the fact that the characteristic is two. If $u_{i_1} \ldots u_{i_r} v_{j_1} \ldots v_{j_s}$ is a monomial of maximal length in $n \in \mathfrak{N}$ then $u_{i_1} \ldots \hat{v}_{j_k} \ldots v_{j_s} \neq 0$ is a monomial of maximal length in $u_j \circ n$, so that $u_j \circ n \in \mathfrak{N}$ has length $< r + s$. This contradicts the minimality of $r + s$, so we must have $r + s \leqslant 1$ as desired.

Next consider the case where $\mathfrak{N}$ is a $\mathfrak{J}$-bimodule and $\mathfrak{N} \cap (\mathfrak{J} + \mathfrak{J}z) \neq 0$. If $\mathfrak{N} \cap \mathfrak{J} \neq 0$ or $\mathfrak{N} \cap \mathfrak{J}z \neq 0$ we are done (taking $\omega = 1$ or $\omega = z$). Otherwise $\mathfrak{J}' = \{x' \in \mathfrak{J} \mid x + x'z \in \mathfrak{N}$ for some $x \in \mathfrak{J}\}$ and $\mathfrak{J}'' = \{x'' \in \mathfrak{J} \mid x'' + xz \in \mathfrak{N}$ for some $x \in \mathfrak{J}\}$ are non-zero sub-bimodules of $\mathfrak{J}$, hence $\mathfrak{J} = \mathfrak{J}' = \mathfrak{J}''$ by irreducibility of $\mathfrak{J}$. Thus for each $x \in \mathfrak{J}$ there is a *unique* $x^\eta \in \mathfrak{J}$ with $x + x^\eta z \in \mathfrak{N}$ (because $\mathfrak{N} \cap \mathfrak{J} = \mathfrak{N} \cap \mathfrak{J}z = 0$). The map $\eta : \mathfrak{J} \to \mathfrak{J}$ is an isomorphism of $\mathfrak{J}$-bimodules, which implies $x^\eta = \alpha x$ for some $\alpha \in \Phi$. (It is enough if this is true of the extension $\eta : \mathfrak{J}_P \to \mathfrak{J}_P$, P the algebraic closure of Φ, since $Q(x, y)$ remains non-degenerate and $\mathfrak{J}_P$ irreducible by Theorem 14. By Schur's Lemma a module homomorphism of $\mathfrak{J}_P$ is either zero or an isomorphism, and if P is an eigenvalue of η, $\rho \in P$, then $\eta - \rho 1$ is not an isomorphism so $\eta = \rho 1$ is a scalar multiplication). Thus $\mathfrak{N}$ contains all $x + \alpha xz$ and $\mathfrak{N} \supseteq \omega \mathfrak{J}$ for $\omega = 1 + \alpha z$.

Finally, consider the general case of a $\mathfrak{J}$-bimodule $\mathfrak{N} \subseteq \mathfrak{M} = \mathfrak{C} + \mathfrak{C}z$. If $\mathfrak{N} \subseteq \mathfrak{C}z$ we are done by the first case, otherwise $\mathfrak{N}' = \{c' \in \mathfrak{C} \mid c' + cz \in \mathfrak{N}$ for some $c \in \mathfrak{C}\}$ is a non-zero subspace of $\mathfrak{C}$ invariant under all V_v, so by the first case $\mathfrak{N}' \cap \mathfrak{J} \neq 0$. Therefore $x + cz \in \mathfrak{N}$ for some $0 \neq x \in \mathfrak{J}$, $c \in \mathfrak{C}$. If $c = 0$ then $\mathfrak{N} \cap \mathfrak{J} \neq 0$, so we may assume $c \neq 0$. Then $\mathfrak{N}'' = \{c \in \mathfrak{C} \mid x + cz \in \mathfrak{N}$ for some $x \in \mathfrak{J}\}$ is a non-zero subspace of $\mathfrak{C}$ invariant under all V_v, so again $\mathfrak{N}'' \cap \mathfrak{J} \neq 0$ and $x + yz \in \mathfrak{N}$ for some $x, y \in \mathfrak{J}$ with $y \neq 0$. This means $\mathfrak{N} \cap (\mathfrak{J} + \mathfrak{J}z) \neq 0$, so by the previous case $\mathfrak{N}$ contains some $\omega \mathfrak{J}$.

Thus in all cases we have bimodules for $\mathfrak{J} = \mathrm{Jord}(Q, 1)$ determined by compound linear specializations which are not completely reducible. Consequently the meson algebra UQE ($\mathfrak{J}$) and the algebra UCE ($\mathfrak{J}$) cannot be semisimple for separable quadratic forms Q in characteristic two.

REFERENCES

C. CHEVALLEY (1): *The Algebraic Theory of Spinors*, Columbia University Press, New York, 1954.

J. DIEUDONNÉ (1): Pseudo-discriminant and Dickson invariant, *Pac. J. Math.*, 5 (1955), 907-910.

J. FAULKNER (1): Octonion Planes Defined by Quadratic Jordan Algebras, *Memoirs Amer. Math. Soc.*

N. JACOBSON (1): Structure theory for a class of Jordan algebras, *Proc. Nat. Acad. Sci. U.S.A.*, 55 (1966), 243-251.

N. JACOBSON (2): Structure Theory of Quadratic Jordan Algebras, *Lecture notes*, Tata Institute, Bombay, 1970.

N. JACOBSON (3): Structure and Representations of Jordan Algebras, Colloq. Publ. Vol. 39, *Amer. Math. Soc.*, Providence, 1968.

N. JACOBSON (4): Structure of Rings, Colloq. Publ. Vol. 37, *Amer. Math. Soc.*, Providence, 1964 (revised ed.).

N. JACOBSON (5): Classes of restricted Lie algebras of characteristic p. I, *Amer. J. Math.*, 63 (1941), 481-515.

K. MCCRIMMON (1): A general theory of Jordan rings, *Proc. Nat. Acad. Sci. U.S.A.*, 56 (1966), 1072-1079.

K. MCCRIMMON (2): Generically algebraic algebras, *Trans. Amer. Math. Soc.*, 127 (1967), 527-551.

K. MCCRIMMON (3): The Freudenthal-Springer-Tits constructions revisited, *Trans. Amer. Math. Soc.*, 148 (1970), 293-314.

K. MCCRIMMON (4): The radical of a Jordan algebra, *Proc. Nat. Acad. Sci. U.S.A.*, 59 (1969), 671-678.

K. MCCRIMMON (5): Representations of quadratic Jordan algebras, *Trans. Amer. Math. Soc.*, 153 (1971), 279-305.

T. TAMAGAWA (1): *Mimeographed notes on quadratic forms*, Yale University, 1969.

Yate University
New Haven, Conn.

SCRIPTA MATHEMATICA, Volume XXIX, No. 3–4

GENERICALLY ALGEBRAIC QUADRATIC JORDAN ALGEBRAS

BY N. JACOBSON AND J. KATZ[1]

In memory of Adrian Albert

The notions of generic minimum polynomial, trace and norm of a finite dimensional strictly power associative algebra were introduced by Jacobson in [1] and [2]. These were generalized by McCrimmon in [8] to generically algebraic strictly power associative algebras. In [11] McCrimmon extended the theory to quadratic Jordan algebras. However, he made use of a restriction to strict power associativity of such algebras which amounts to assuming that the subalgebras generated by single elements of any extension algebra $\mathfrak{J}^P$ are special. In this note we shall remove this restriction from the theory.

1. Algebraic elements: Throughout this paper we deal exclusively with algebras over an infinite field Φ. For such a field we can define a *unital quadratic Jordan* algebra as a triple $(\mathfrak{J}, U, 1)$ where $\mathfrak{J}$ is a Φ-vector space, 1 a distinguished element of $\mathfrak{J}$ and U is a mapping $a \to U_a$ of $\mathfrak{J}$ into $\mathrm{End}_\Phi \mathfrak{J}$ satisfying the following axioms:

QJ1. *U is a quadratic mapping, that is, $U_{\alpha a} = \alpha^2 U_a$ for $\alpha \in \Phi$ and the mapping $(a, b) \to U_{a,b} \equiv U_{a+b} - U_a - U_b$ is Φ-bilinear.*

QJ2. $U_1 = 1$.

QJ3. $U_{aU_b} = U_b U_a U_b$.

QJ4. *If $V_{a,b}$ is defined by $xV_{a,b} = aU_{x,b}$ then $U_b V_{a,b} = V_{b,a} U_b$.*

The notion of a quadratic Jordan algebra and the basic results on these algebras are due to McCrimmon [7]. A fairly comprehensive exposition of these is given in Jacobson [4]. The reader is referred to this for results we shall state without proofs. We recall that if $\mathfrak{A}$ is an associative algebra with unit 1, $\mathfrak{A}$ gives rise to a quadratic Jordan algebra $\mathfrak{A}^{(q)} = (\mathfrak{A}, U, 1)$ where U is defined by $xU_a = axa$. A subalgebra of an $\mathfrak{A}^{(q)}$ is called a *special* quadratic Jordan algebra.

Powers in a quadratic Jordan algebra $\mathfrak{J}$ are defined inductively: $a^0 = 1, a^1 = a$ and $a^{n+2} = a^n U_a$. An element $a \in \mathfrak{J}$ is *invertible* if there exists a $b \in \mathfrak{J}$ such that $bU_a = a$, $b^2 U_a = 1$; such a b is uniquely determined and is denoted as a^{-1}. It is

Received by the editors, November 26, 1971.

[1]Research supported in part by NSF GP-29081 and NSF GP-19960.

well known that a is invertible if and only if U_a is bijective and that $a^{-1}=aU_a^{-1}$. Moreover, bU_a is invertible if and only if a and b are invertible and $(bU_a)^{-1}=b^{-1}U_{a^{-1}}$.

If $a\in\mathfrak{J}$ we define a Φ-linear mapping ν_a of $\Phi[\lambda]^{(q)}$, λ an indeterminate, into $\mathfrak{J}$ such that $\lambda^i\to a^i$, $i=0,1,2,\ldots$. We write $f(\lambda)\nu_a=f(a)$. It is easily seen that ν_a is a homomorphism of quadratic Jordan algebras. This is an immediate consequence of the following identities:

$$a^iU_{a^j}=a^{2j+i},\qquad a^iU_{a^j,a^k}=2a^{i+j+k}. \tag{1}$$

These can either be proved directly, or, preferably, they can be deduced from the weak MacDonald theorem ([4], pp. 1.24—1.26). It follows that the image $\Phi[a]=\Phi[\lambda]\nu_a$ is a subalgebra of $\mathfrak{J}$. If $f(\lambda),g(\lambda)\in\Phi[\lambda]$ we shall find it suggestive to write $f(a)g(a)$ for $(fg)(a)$ or $(f(\lambda)g(\lambda))\nu_a$. In particular, we have $a^kf(a)=(\lambda^kf)(a)$ so if $f(\lambda)=\alpha_0+\alpha_1\lambda+\cdots+\alpha_m\lambda^m$ then $a^kf(a)=\alpha_0a^k+\alpha_1a^{k+1}+\cdots+\alpha_ma^{m+k}$. By definition, $f(a)^2g(a)=(g(\lambda)f(\lambda)^2)\nu_a=(g(\lambda)U_{f(\lambda)})\nu_a=g(a)U_{f(a)}$. Thus $f(a)^2g(a)$ coincides with the Jordan product $g(a)U_{f(a)}$ in $\mathfrak{J}$. Let $\mathfrak{K}_a$ be the kernel of ν_a so $\mathfrak{K}_a$ is an ideal in $\Phi[\lambda]^{(q)}$. We call a *algebraic* if ν_a is not a monomorphism, that is, $\mathfrak{K}_a\neq 0$. Suppose this is the case and let $f(\lambda)\neq 0$ be in $\mathfrak{K}_a$. Then, since $\mathfrak{K}_a$ is an ideal in $\Phi[\lambda]^{(q)}$, $g(\lambda)f(\lambda)^2=g(\lambda)U_{f(\lambda)}\in\mathfrak{K}_a$ for every $g(\lambda)\in\Phi[\lambda]$. Thus $\mathfrak{K}_a$ contains the principal ideal $(f(\lambda)^2)$ of $\Phi[\lambda]$. It follows that $\mathfrak{K}_a$ contains a non-zero ideal $\mathfrak{K}_a^*$ of the associative algebra $\Phi[\lambda]$ such that $\mathfrak{K}_a^*$ contains every ideal of $\Phi[\lambda]$ contained in $\mathfrak{K}_a$. $\mathfrak{K}_a^*$ is a principal ideal and has a unique monic generator $\mu_a(\lambda)$. We shall call $\mu_a(\lambda)$ *the minimum polynomial* of the algebraic element a. This is the monic polynomial of least degree such that $f(a)\mu_a(a)=0$ for all $f(\lambda)$. Since $g(a)=0=ag(a)$ implies $a^2g(a)=g(a)U_a=0$, $a^3g(a)=ag(a)U_a=0$ etc. it is clear that $\mu_a(\lambda)$ can be characterized as the monic polynomial of least degree satisfying

$$\mu_a(a)=0=a\mu_a(a). \tag{2}$$

It is well known that if the characteristic is $\neq 2$ then a quadratic Jordan algebra is the same thing as a Jordan algebra. In this case $\mathfrak{K}_a=\mathfrak{K}_a{}^*$ for every a and $\mu_a(\lambda)$ is the polynomial of least degree such that $\mu_a(a)=0$. Even in characteristic two fairly mild restrictions on $\mathfrak{J}$ insure that $\mathfrak{K}_a=\mathfrak{K}_a{}^*$ for every a. For example, this is the case if $\mathfrak{J}$ has no absolute zero divisors $\neq 0$, that is, elements $z\neq 0$ such that $U_z=0$. However, it may happen that $\mathfrak{K}_a\neq\mathfrak{K}_a{}^*$. For example, this will be the case in $\mathfrak{J}=\Phi[\lambda]^{(q)}/\mathfrak{K}$ where Φ is of characteristic two and $\mathfrak{K}$ is the ideal in $\Phi[\lambda]^{(q)}$ which is the Φ-space spanned by $\lambda^2,\lambda^i,i\geqslant 4$. Then if $a=\lambda+\mathfrak{K}$, $a^2=0$ but $a^3\neq 0$. The element a is algebraic with minimum polynomial λ^4.

We now consider the associative algebra $\Phi[\lambda]/\mathfrak{K}_a{}^*$ determined by the algebraic element a. Put $\lambda_a=\lambda+\mathfrak{K}_a{}^*$ so $\Phi[\lambda]/\mathfrak{K}_a{}^*=\Phi[\lambda_a]$. Since $\mathfrak{K}_a{}^*=(\mu_a(\lambda))$ the minimum polynomial $\mu_a(\lambda)$ of a coincides with the minimum polynomial of λ_a in

$\Phi[\lambda_a]$ and hence with the characteristic polynomial $\det(\lambda 1 - \lambda_{aR})$ of the multiplication λ_{aR} in $\Phi[\lambda_a]$ determined by λ_a. We have the homomorphism $\nu_a{}^*$ of $\Phi[\lambda_a]^{(q)}$ into $\mathfrak{J}$ sending $f(\lambda_a)\to f(a)$. The kernel of $\nu_a{}^*$ is $\mathfrak{K}_a/\mathfrak{K}_a{}^*$. Since $f(\lambda)\in\mathfrak{K}_a$ implies $f(\lambda)^2\in\mathfrak{K}_a{}^*$ it is clear that $\mathfrak{K}_a/\mathfrak{K}_a{}^*$ is a nil ideal. Since every $f(\lambda_a)$ is algebraic, every element $f(a)\in\Phi[a]$ is algebraic. Moreover, if the degree of the minimum polynomial $\mu_{f(a)}(\lambda)$ of $f(a)$ is $m=\deg\mu_a(\lambda)$ then

$$\mu_{f(a)}(\lambda)=\det(\lambda 1 - f(\lambda_a)_R) \tag{3}$$

where $f(\lambda_a)_R$ is the multiplication in $\Phi[\lambda_a]$ determined by $f(\lambda_a)$. This is clear since if we denote the polynomial on the right by $g(\lambda)$ then $\deg g(\lambda)=m$ and $g(f(a)) =0=f(a)g(f(a))$.

LEMMA 1. *An element $a\in\mathfrak{J}$ is invertible if there exists a polynomial $f(\lambda)$ such that $f(0)\neq 0$ and $f(a)=0$. If a is algebraic then the following conditions are equivalent:* (i) *a is invertible in $\mathfrak{J}$,* (ii) *$\mu_a(0)\neq 0$ for the minimum polynomial $\mu_a(\lambda)$,* (iii) *there exists an element $g(a)$ in $\Phi[a]$ such that $ag(a)=1$,* (iv) *λ_a is invertible in $\Phi[\lambda_a]$ (or $\Phi[\lambda_a]^{(q)}$). In these cases $a^{-1}\in\Phi[a]$.*

Proof. If $f(a)=0$ and $f(0)\neq 0$ we have a relation of the form $1=\alpha_1 a+\alpha_2 a^2+\cdots +\alpha_n a^n$, $\alpha_i\in\Phi$. Squaring gives a relation of the form $1=\beta_1 a^2+\cdots+\beta_m a^m$, $m=2n$, which can be written as $1=(\beta_1 1+\cdots+\beta_n a^{m-2})U_a$. Hence $1\in\mathfrak{J}U_a$. It is known that this implies that a is invertible. Now suppose a is algebraic with minimum polynomial $\mu_a(\lambda)$. If $\mu_a(0)\neq 0$ the result just proved shows that a is invertible. Now suppose $\mu_a(0)=0$ so $\mu_a(\lambda)=\lambda g(\lambda)$. Since $ag(a)=0$ and $\deg g(\lambda) <\deg\mu_a(\lambda)$ we have $g(a)\neq 0$. On the other hand, $a\mu_a(a)=0$ gives $g(a)U_a=0$. Hence U_a is not invertible so a is not invertible. This proves the equivalence of (i) and (ii). The equivalence of (iv) with these conditions is clear also since $\mu_a(\lambda)$ is the minimum polynomial of λ_a and, as is well known, λ_a is invertible in $\Phi[\lambda_a]$ if and only if $\mu_a(0)\neq 0$. If λ_a is invertible we have a polynomial $g(\lambda)$ such that $\lambda_a g(\lambda_a)=1$. Applying $\nu_a{}^*$ we obtain $ag(a)=1$. Moreover, since $g(\lambda_a)$ is the inverse of λ_a in $\Phi[\lambda_a]^{(q)}$ it is clear that $g(a)$ is the inverse of a in $\Phi[a]$. Conversely, if we have a polynomial $g(\lambda)$ such that $ag(a)=1$ then the polynomial $f(\lambda)=\lambda g(\lambda) -1$ satisfies $f(a)=0$, $f(0)\neq 0$ so a is invertible. This proves the equivalence of (iii) with the other conditions and shows also that $a^{-1}\in\Phi[a]$.

LEMMA 2. *An element $a\in\mathfrak{J}$ is algebraic if and only if U_a is an algebraic linear transformation. More precisely, if $g(\lambda)\in\Phi[\lambda]$ satisfies $g(U_a)=0$ then $f(\lambda)=g(\lambda^2)$ satisfies $f(a)=0=af(a)$. On the other hand, if $f(a)=0=af(a)$ for $f(\lambda)=\lambda^n- \alpha_1\lambda^{n-1}+\alpha_2\lambda^{n-2}-\cdots+(-1)^n\alpha_n$ then there exist n^2 polynomials $b_j(p_1,\ldots,p_n)$ with integer coefficients in indeterminates $p_1,\ldots,p_n$ such that if $\beta_j=b_j(\alpha_1,\ldots,\alpha_n)$ and $g(\lambda)=\lambda^{n^2}-\beta_1\lambda^{n^2-1}+\beta_2\lambda^{n^2-2}-\cdots$ then $g(U_a)=0$.*

Proof. The first part follows by applying $g(U_a)$ to 1 and to a. Next assume $f(\lambda)$ is a monic polynomial in $\Phi[\lambda]$ such that $f(a)=0=af(a)$. Suppose that $f(\lambda)=\prod_1^n(\lambda-\rho_i)$ in a splitting field and define $g(\lambda)=\prod_{i,j}(\lambda-\rho_i\rho_j)$. If the x_i, $1\leqslant i\leqslant n$, are indeterminates then the coefficients of $\prod_{i,j}(\lambda - x_i x_j)$ are symmetric polynomials with integer coefficients in the x_i. Hence $\Pi(\lambda - x_i x_j) = \lambda^{n^2} - b_1(p_1,\ldots,p_n)\lambda^{n^2-1} + b_2(p_1,\ldots,p_n)\lambda^{n^2-2} - \cdots$ where the b_i are polynomials with integer coefficients in the elementary symmetric polynomials $p_1 = \sum x_i$, $p_2 = \sum_{i<j} x_i x_j, \ldots$. Then $g(\lambda) = \lambda^{n^2} - b_1(\alpha_1,\ldots,\alpha_n)\lambda^{n^2-1} + b_2(\alpha_1,\ldots,\alpha_n)\lambda^{n^2-2}\ldots$. It remains to prove that $g(U_a) = 0$. For this purpose we introduce the free quadratic Jordan algebra $FQJ(x,y)$ on two generators x,y over Φ. Let $\mathfrak{K}$ be the ideal in $FQJ(x,y)$ generated by $f(x)$ and $xf(x)$. By a result of McCrimmon's (corollary on p. 769 of [12]) one knows that $\mathfrak{J}' \equiv FQJ(x,y)/\mathfrak{K}$ is special so we may suppose that $\mathfrak{J}'$ is a subalgebra of an algebra $\mathfrak{A}^{(q)}$, $\mathfrak{A}$ associative. Put $x' = x + \mathfrak{K}$, $y' = y + \mathfrak{K}$ and consider $U_{x'}$ in $\mathfrak{A}^{(q)}$. We have $U_{x'} = x'_L x'_R = x'_R x'_L$ where x'_L and x'_R denote the left and right multiplications in $\mathfrak{A}$ determined by x'. Since $f(x') = 0$ we have $f(x'_L) = 0 = f(x'_R)$. By passing to an extension field if necessary we may assume that Φ contains the roots ρ_i of $f(\lambda)$. To prove that $g(U_{x'}) = 0$ is equivalent to proving that $\prod_{i,j}(\lambda x'_L - \rho_i\rho_j 1)$ becomes 0 on putting $\lambda = x'_R$. We note first that division of $g_i(\lambda) \equiv \prod_j(\lambda x'_L - \rho_i\rho_j 1)$ by $x - \rho_i 1$ in $\Phi[x'_L][\lambda]$ gives the following remainder: $g_i(\rho_i) = \prod_j(\rho_i x'_L - \rho_i\rho_j 1) = \rho_i^n \prod_j(x'_L - \rho_j 1) = \rho_i^n f(x'_l) = 0$. Hence $g_i(\lambda)$ is divisible by $\lambda - \rho_i 1$ and $g(\lambda x'_L) = \prod_{i,j}(\lambda x'_L - \rho_i\rho_j 1)$ is divisible by $f(\lambda) = \Pi(\lambda - \rho_i 1)$. Then $g(x'_R x'_l) = 0$ follows from $f(x'_R) = 0$. We now observe that we have a homomorphism of $\mathfrak{J}'$ into $\mathfrak{J}$ sending $x' \to a$ and y' into any $b \in \mathfrak{J}$. Since $y' g(U_{x'}) = 0$ we obtain $bg(U_a) = 0$. Since this holds for all $b \in \mathfrak{J}$ we have $g(U_a) = 0$.

2. Generically algebraic quadratic Jordan algebras: McCrimmon in [8] has defined the Zariski topology, polynomials and rational mappings for arbitrary vector spaces over a field and extended some of the basic results of the finite dimensional case to these. He has defined the *Zariski topology* on a vector space V as the topology in which a set O is open if and only if $O_\alpha = O \cap V_\alpha$ is open in V_α for every finite dimensional subspace V_α of V. A mapping $F\colon V \to \Phi$ is a *polynomial* (*rational*) *function* if every restriction $F_\alpha = F|V_\alpha$ to finite dimensional V_α is a polynomial (rational) function in the usual sense. We denote the ring of polynomial functions on V by $\mathfrak{P}(V)$ and the field of rational functions (defined on open subsets with the obvious identification) by $\mathfrak{R}(V)$. A polynomial function which is homogeneous of degree $n = 1,2,\ldots$ is called a *form of degree n*. One has the unique factorization theorem for forms: If Q is a form on V, x_0 a point with $Q(x_0) \neq 0$ then Q has a unique factorization as $Q = Q(x_0)\Pi Q_i^{q_i}$ where the Q_i are irreducible (forms) such that $Q_i(x_0) = 1$. Also one has the *Hilbert Nullstellensatz:* If Φ is algebraically closed and P and Q are forms on vector space V/Φ such that $P(x) = 0$ implies $Q(x) = 0$ then P is composed of the irreducible factors of Q. The machinery of differential calculus of rational functions also carries over to this setting.

DEFINITION 1. *A quadratic Jordan algebra* $\mathfrak{J}$ *is* generically algebraic *if there exists a monic polynomial* $f_x(\lambda)\in\mathfrak{P}(\mathfrak{J})[\lambda]$ *such that* $f_x(x)=0$ *for all* $x\in\mathfrak{J}$.

If $\mathfrak{J}$ is generically algebraic then the argument of the last section shows that there exists a unique monic polynomial $m_x(\lambda)\in\mathfrak{R}(\mathfrak{J})[\lambda]$ of least degree such that $m_x(x)=0=xm_x(x)$ for all x's on an open subset of $\mathfrak{J}$ (in the Zariski topology). Since $g(\lambda)=f_x(\lambda)^2\in\mathfrak{P}(\mathfrak{J})[\lambda]$ and $g(x)=xg(x)=0$ for all x it follows as in [8], p. 533 that $m_x(\lambda)|g(\lambda)$ and that $m_x(\lambda)\in\mathfrak{P}(\mathfrak{J})[\lambda]$. The polynomial $m_x(\lambda)$ is called the *generic minimum polynomial* of the generically algebraic quadratic Jordan algebra $\mathfrak{J}$ and the degree m of $m_x(\lambda)$ in λ is called the *degree* of $\mathfrak{J}$.

We shall now give some instances of generically algebraic quadratic Jordan algebras. First we prove

THEOREM 1. $\mathfrak{J}$ *is generically algebraic if and only if there exists a monic polynomial* $g_x(\lambda)\in\mathfrak{P}(\mathfrak{J})[\lambda]$ *such that* $g_x(U_x)=0$ *for all* $x\in\mathfrak{J}$.

Proof. As in the proof of Lemma 2, $g_x(U_x)=0$ implies $f_x(x)=0=xf_x(x)$ for $x\in\mathfrak{J}$ if $f_x(\lambda)=g_x(\lambda^2)$. Conversely, suppose $f_x(x)=0=xf_x(x)$ for $f_x(\lambda)=\lambda^n-\alpha_1(x)\lambda^{n-1}+\cdots+(-1)^n\alpha_n(x)$ where $\alpha_i(x)\in\mathfrak{P}(\mathfrak{J})$. Then Lemma 2 gives a polynomial $g_x(\lambda)=\lambda^{n^2}-\beta_1(x)\lambda^{n^2-1}+\cdots$ where the $\beta_i(x)$ are polynomials with integer coefficients in the $\alpha_j(x)$ such that $g_x(U_x)=0$.

Now let $\mathfrak{J}$ be a finite dimensional quadratic Jordan algebra and let $g_x(\lambda)=\det(\lambda 1-U_x)$ the characteristic polynomial of U_x. Since $x\to U_x$ is a polynomial mapping it is clear that $g_x(\lambda)$ is a monic polynomial contained in $\mathfrak{P}(\mathfrak{J})[\lambda]$. Since $g_x(U_x)=0$ by the Hamilton-Cayley theorem we have the following consequence of Theorem 1.

THEOREM 2. *Any finite dimensional quadratic Jordan algebra is generically algebraic.*

We give next two important instances of generically algebraic quadratic Jordan algebras which need not be finite dimensional.

First, let $\mathfrak{J}$ be a vector space over Φ equipped with a quadratic form Q and a base point 1, that is, we have $Q(1)=1$. Let $Q(x,y)=Q(x+y)-Q(x)-Q(y)$ be the bilinearization of Q and put $T(x)=Q(x,1)$. Then one can verify that if we define U by

$$yU_x=Q(x)y+T(y)T(x)x-Q(x,y)x-Q(x)T(y)1$$

one obtains a quadratic Jordan algebra with 1 as unit (Jacobson and McCrimmon [5]). We denote this as Jord$(Q,1)$. If $\dim\mathfrak{J}>1$ then $\mathfrak{J}$ is generically algebraic with

$$m_x(\lambda)=\lambda^2-T(x)\lambda+Q(x)1$$

as generic minimum polynomial. Moreover, it can be proved that conversely any generically algebraic quadratic Jordan algebra of degree two has the form Jord$(Q,1)$, $(Q,1)$ a quadratic form with base point (*loc. cit.*).

Another important class of generically algebraic quadratic Jordan algebras are those defined by a certain class of cubic forms (McCrimmon [10] and [11]). Here one begins with a vector space $\mathfrak{J}/\Phi$, a cubic form N, an element $1\in\mathfrak{J}$ and a quadratic mapping $x\to x^{\#}$ in $\mathfrak{J}$ satisfying the following conditions:

(i) $$x^{\#\#}=N(x)x\,.$$

(ii) If $T(x,y)=(\Delta_1^xN)(\Delta_1^yN)-\Delta_1^x(\Delta^yN)$ where Δ_y^xF is the directional derivative of F at y in the direction x, then

$$T(x^{\#},y)=\Delta_x^yN.$$

(iii) If $x\times y=(x+y)^{\#}-x^{\#}-y^{\#}$ and $T(y)=T(y,1)$ then

$$1\times y=T(y)1-y.$$

(iv) $$1^{\#}=1.$$

(v) $$N(1)=1.$$

Then one obtains a quadratic Jordan algebra with unit 1 by defining

$$yU_x=T(x,y)x-x^{\#}\times y^{\#}$$

We denote this Jordan algebra as Jord$(N,\#,1)$. It is known that

$$x^3-T(x)x^2+S(x)x-N(x)1=0$$

for $S(x)=T(x^{\#})$, $x^{\#}=x^2-T(x)x+S(x)1$, $x^{\#}\times x=[S(x)T(x)-N(x)]1-S(x)x-T(x)x^{\#}$, $2S(x)=T(x)^2-T(x^2)$. Using these formulas and $N(x)x=x^{\#\#}=(x^2-T(x)x+S(x)1)$ one can verify that

$$x^4-T(x)x^3+S(x)x^2-N(x)x=0.$$

Hence Jord$(N,\#,1)$ is generically algebraic of degree $\leqslant 3$. After we have shown that the generic minimum polynomial satisfies the usual properties, one can use McCrimmon's argument in [11] to prove that conversely any generically algebraic quadratic Jordan algebra of degree three has the form Jord$(N,\#,1)$.

Clearly any subalgebra of a generically algebraic quadratic Jordan algebra is generically algebraic. Also any homomorphic image $\overline{\mathfrak{J}}$ of a generically algebraic quadratic Jordan algebra $\mathfrak{J}$ is generically algebraic. For, if ν is a homomorphism of $\mathfrak{J}$ onto $\overline{\mathfrak{J}}$ then there exists a linear mapping μ of $\overline{\mathfrak{J}}$ into $\mathfrak{J}$ such that $\mu\nu=1_{\overline{\mathfrak{J}}}$ (by the existence of a complementary space of ker ν in $\mathfrak{J}$). Then if P is a polynomial function on $\mathfrak{J}$, $\bar{x}\to P(\bar{x}^{\mu})$ is a polynomial function $\overline{P}$ on $\overline{\mathfrak{J}}$. If

$x^m - m_1(x)x^{m-1} + \cdots + (-1)^m m_m(x) = 0$ for $m_i \in \mathfrak{P}(\mathfrak{J})$ then $(\bar{x}^\mu)^m - \overline{m}_1(\bar{x})^\mu \times (\bar{x}^\mu)^{m-1} + \cdots = 0$. Applying ν we obtain $\bar{x}^m - \overline{m}_1(\bar{x})\bar{x}^{m-1} + \cdots = 0$ which shows that $\overline{\mathfrak{J}}$ is generically algebraic.

If P is an extension field of the base field Φ and $\mathfrak{J}/\Phi$ is generically algebraic then $\mathfrak{J}^{\mathrm{P}}/\mathrm{P}$ is generically algebraic. Moreover, the generic minimum polynomial in $\mathfrak{J}$ is the restriction of that in $\mathfrak{J}^{\mathrm{P}}$ if $\mathfrak{J}$ is identified in the usual way with a subset of $\mathfrak{J}^{\mathrm{P}}$.

3. Basic properties of the generic minimum polynomial: We write the generic minimum polynomial $m_x(\lambda)$ of the generically algebraic quadratic Jordan algebra $\mathfrak{J}$ as

$$m_x(\lambda) = \sum_0^m m_i(x)\lambda^i, \qquad m_m(x) = 1 \tag{4}$$

and we define the *generic trace* $t(x) = -m_{m-1}(x)$, and the *generic norm* $n(x) = (-1)^m m_0(x)$. Also we define the *adjoint* $x^\#$ by

$$x^\# = (-1)^{m-1} \sum_1^m m_i(x) x^{i-1} \tag{5}$$

so $m_x(x) = 0 = x m_x(x)$ is equivalent to

$$xx^\# = n(x)1, \qquad x^2 x^\# = n(x)x. \tag{6}$$

We note first that the polynomial function m_i is a form of degree $m-i$. For, if $\alpha \neq 0$ is in Φ then the relations $m_x(x) = 0 = x m_x(x)$ imply $m_x^{(\alpha)}(x) = 0 = x m_x^{(\alpha)}(x)$ for $m_x^{(\alpha)}(\lambda) = \alpha^{-m} m_{\alpha x}(\alpha\lambda) = \lambda^m + \alpha^{-1} m_{m-1}(\alpha x)\lambda^{m-1} + \alpha^{-2} m_{m-2}(\alpha x)\lambda^{m-2} + \cdots$. Hence $m_i(\alpha x) = \alpha^{m-i} m_i(x)$.

It is clear that if $\mu_x(\lambda)$ is the minimum polynomial of x then $\mu_x(\lambda) | m_x(\lambda)$. An element $a \in \mathfrak{J}$ will be called *general* if $\mu_a(\lambda) = m_a(\lambda)$. We shall now show that the set of these elements is a Zariski open subset of $\mathfrak{J}$. Let $\{ \mu_\alpha | \alpha \in I \}$ be a base for $\mathfrak{J}/\Phi$. For $k = 0, 1, 2, \ldots$ there exist forms $\sigma_{k\alpha}$ of degree k such that

$$x^k = \sum_\alpha \sigma_{k\alpha}(x)\mu_\alpha$$

where for any given x only a finite number of $\sigma_{k\alpha}(x) \neq 0$. Then $m_x(x) = 0 = x m_x(x)$ if and only if

$$\sum_0^m m_{m-i}(x)\sigma_{m-i,\alpha}(x) = 0 = \sum_0^m m_{m-i}(x)\sigma_{m-i+1,\alpha}(x)$$

for all $\alpha \in I$. The system of linear equations

$$\sum_0^{m-1} y_i \sigma_{m-i,\alpha}(x) = 0$$
$$\sum_0^{m-1} y_i \sigma_{m-i+1,\alpha}(x) = 0, \qquad \alpha \in I \tag{7}$$

has only the trivial solution $y_i = 0$ in $\Re\,\Im$ since otherwise there would exist an $f_x(\lambda) \neq 0$ in $\Re\,\Im\,[\lambda]$ of degree less than m such that $f_x(x) = 0 = x f_x(x)$ contradicting the fact that $m_x(\lambda)$ divides every such $f_x(\lambda)$. Hence the m-columned matrix of the coefficients of (7) contains non-zero m-rowed minors. Let $\{M_\beta(x) | \beta \in K\}$ be the set of these non-zero m-rowed minors. It is easily seen that $\deg \mu_a(\lambda) < m$ if and only if $M_\beta(a) = 0$ for all $\beta \in K$. Thus the set of general a coincides with the open set defined by the set of polynomial inequations $M_\beta(a) \neq 0$, $\beta \in K$. We use this result in the proof of the following.

Theorem 2. *The generic minimum polynomial* $m_x(\lambda) = \sum_0^m m_i(x)\lambda^i$, *generic trace* $t = -m_{m-1}$ *and generic norm* $n = (-1)^m m_0$ *have the following properties:*

(i) $n(f(x)g(x)) = n(f(x))n(g(x))$ *for any* $f(\lambda), g(\lambda) \in \Phi[\lambda]$.

(ii) $n(\lambda 1 - x) = m_x(\lambda)$ *if* $n(\lambda 1 - x)$ *is the generic norm as defined in the extension algebra* $\Im^{\Phi(\lambda)}$.

(iii) x *is invertible if and only if* $n(x) \neq 0$, *in which case*

$$x^{-1} = n(x)^{-1} x^{\#} \tag{8}$$

(iv) $m_a(\lambda)$ *and the minimum polynomial* $\mu_a(\lambda)$ *have the same irreducible factors in* $\Phi[\lambda]$.

(v) $m_i(1) = (-1)^{m-i}\binom{m}{i}$

(vi) z *is nilpotent if and only if* $m_z(\lambda) = \lambda^m$

(vii) $m_x(\lambda)$ *is invariant under isomorphism*

(viii) *if* D *is a derivation then* $\Delta_a^{aD} m_i = 0$

(ix) $\Delta_a^1 m_i = -(i+1)m_{i+1}(a)$

Proof. It suffices to prove (i) and (ii) for $x = a$ general. For such an a we consider the mappings $(f(\lambda), g(\lambda)) \to f(a)$, $(f(\lambda), g(\lambda)) \to g(a)$, $(f(\lambda), g(\lambda)) \to f(a)g(a)$ of $\Phi[\lambda] \times \Phi[\lambda]$ into $\Phi[a]$. These are polynomial mappings and the set of $(f(\lambda), g(\lambda))$ such that $f(a)$, $g(a)$ and $f(a)g(a)$ are general is a non-vacuous open subset of $\Phi[\lambda] \times \Phi[\lambda]$. It therefore suffices to prove $n(f(a)g(a)) = n(f(a))n(g(a))$ for such pairs of polynomials. Then we have: $\mu_a(\lambda)=m_a(\lambda)$, $\mu_{f(a)}(\lambda)=m_{f(a)}(\lambda)$, $\mu_{g(a)}(\lambda)=m_{g(a)}(\lambda)$, $\mu_{f(a)g(a)}(\lambda)=m_{f(a)g(a)}(\lambda)$. Let $\lambda_a=\lambda+(\mu_a(\lambda))$ in $\Phi[\lambda_a]=\Phi[\lambda]/(m_a(\lambda))$ as in Section 1. Then (3) implies that $n(f(a))=\det f(\lambda_a)_R$, $n(g(a))=\det g(\lambda_a)_R$, $n(f(a)g(a))=\det(f(\lambda_a)g(\lambda_a))_R$. Hence $n(f(a)g(a))=n(f(a))n(g(a))$ follows from the homomorphism property of R and the multiplicative property of determinants. This proves (i). (ii) follows in a similar manner if we observe that if a is general in $\mathfrak{J}$ then $\lambda 1-a$ is general in $\mathfrak{J}^{\Phi(\lambda)}$. (iii): If $n(a)\neq 0$, then $n(a)^{-1}a^{\#}a=1$ and a is invertible by Lemma 1. To prove the converse we note first that the multiplicative property of n given in (i) implies that either $n(1)=1$ or $n(1)=0$. In the second case $n(x)=0$ for all x. This carries over to $\mathfrak{J}^{\Phi(\lambda)}$ and gives $m_x(\lambda)=n(\lambda 1-x)=0$ contrary to $m_x(\lambda)=\lambda^m+\cdots\neq 0$. Now if a is invertible we have a polynomial $f(\lambda)$ such that $af(a)=1$. Then $n(a)n(f(a))=n(1)=1$ so $n(a)\neq 0$. It remains to establish the formula (8) for the inverse. Let $\mu_a(\lambda)$ be the minimum polynomial of a and let $\lambda_a=\lambda+(\mu_a(\lambda))$ in $\Phi[\lambda]/(\mu_a(\lambda))$ as before. Then $\mu_a(\lambda)|m_a(\lambda)$ so $m_a(\lambda_a)=0$. Hence if $n(a)\neq 0$ and we put $p(\lambda_a)=n(a)^{-1}(-1)^{m-}\sum_1^m m_i(a)\lambda_a^{i-1}$ then $\lambda_a p(\lambda_a)=1$ so $p(\lambda_a)$ is the inverse of λ_a in $\Phi[\lambda_a]$. Hence $p(\lambda_a)$ is the inverse of λ_a in $\Phi[\lambda_a]^{(q)}$. Applying the homomorphism ν_a^* we see that $n(a)^{-1}a^{\#}=a^{-1}$ in $\mathfrak{J}$. It suffices to prove (iv) in the case in which Φ is algebraically closed. Then (iii) and Lemma 1 show that $\lambda|\mu_a(\lambda)$ if and only if $\lambda|m_a(\lambda)$. It is clear that for any $\rho\in\Phi$, $\mu_a(\lambda+\rho 1)=\mu_{a-\rho 1}(\lambda)$ and $m_x(\lambda+\rho 1)=m_{x-\rho 1}(\lambda)$ so $m_a(\lambda+\rho 1)=m_{a-\rho 1}(\lambda)$. Thus $\lambda|m_a(\lambda+\rho 1)$ if and only if $\lambda|\mu_a(\lambda+\rho 1)$ and so $\lambda-\rho 1|m_a(\lambda)$ if and only if $\lambda-\rho 1|\mu_a(\lambda)$. This proves (iv). To establish (v) we observe that $\mu_1(\lambda)=\lambda-1$ so, by (iv), $m_a(\lambda)=(\lambda-1)^m$. This implies (v). An element z is called nilpotent if there exists an integer r such that $z^r=0$. Then $z^{2r}=0=z^{2r+1}$ and so $\mu_z(\lambda)|\lambda^{2r}$. Then $\mu_z(\lambda)$ is a power of λ and the same thing is true of $m_z(\lambda)$, by (iv). Hence $m_z(\lambda)=\lambda^m$. Conversely, if $m_z(\lambda)=\lambda^m$ then $z^m=0$. (vii) is clear. To prove (viii) we recall that a derivation in $\mathfrak{J}$ is a linear mapping D such that $1D=0$ and $[U_a,D]=U_{a,aD}$ for every $a\in\mathfrak{J}$ and the directional derivative of a rational function $\Delta_a^b f$ is the coefficient of the indeterminate ξ in $f(a+\xi b)$. For any constant function $c\,(x\to c)$ we have $\Delta_a^b c=0$ and for the identity function $u\,(x\to x)$ we have $\Delta_a^b u=b$. Then for $u^n\,(x\to x^n)$ one has the recursion formula $\Delta_a^b u^{k+2}=(\Delta_a^b u^k)U_a+a^kU_{a,b}$. Also one has $1D=0$, $a^{k+2}D=a^kU_aD=a^kDU_a+a^kU_{a,aD}$. Hence $\Delta_a^{aD}u^k=a^kD$, $k=0,1,2,\ldots$, so if we apply D to the relations $m_a(a)=0=am_a(a)$ and Δ_a^{aD} to the functional relation $m_u(u)=0=um_u(u)$ and subtract the resulting relations we obtain

$$\sum_0^{m-1}(\Delta_a^{aD}m_i)a^i=0=\sum_1^m(\Delta_a^{aD}m_i)a^{i+1}.$$

Assuming a general these relations imply that $\Delta_a^{aD}m_i=0$. Then $\Delta_a^{aD}m_i=0$ holds for arbitrary a by density. (ix) Since $m_{a+\xi 1}(\lambda)=m_a(\lambda-\xi 1)$, $\Delta_a^1 m_i$ is the coefficient of $\xi\lambda^i$ in $m_a(\lambda-\xi 1)=\sum m_j(a)(\lambda-\xi 1)^j$. Hence $\Delta_a^1 m_i=-(i+1)m_{i+1}(a)$.

4. Properties of the generic norm: We have seen that the generic norm n of a generically algebraic algebra $\mathfrak{J}$ of degree m is a form of degree m and $n(1)=1$. Hence n has a unique factorization as $n=\prod n_j$ where the n_j are irreducible forms which are *normalized* in the sense that $n_j(1)=1$ (McCrimmon [8]). The key facts for establishing the properties of n are that an element $a\in\mathfrak{J}$ is invertible if and only if $n(a)\neq 0$, and if p and q are forms on a vector space V which have no common factors in $\mathfrak{P}(V)$ then their extensions have no common factors in $\mathfrak{P}(V^{\mathrm{P}})$ for any field extension P of Φ. The latter result follows from the finite dimensional case which is equivalent to the analogous result for homogeneous polynomials in $\Phi[x_1,\ldots,x_n]$, x_i indeterminates. We consider first the relation between the generic norm on $\mathfrak{J}$ and related algebras.

THEOREM 3. (1) *If $\mathfrak{B}$ is a subalgebra of $\mathfrak{J}$ then the restriction $n|\mathfrak{B}$ of the generic norm on $\mathfrak{J}$ to $\mathfrak{B}$ has the same irreducible factors as the generic norm $n_{\mathfrak{B}}$ of $\mathfrak{B}$. Moreover, $n_{\mathfrak{B}}|(n|\mathfrak{B})$ and $n_{\mathfrak{B}}=n|\mathfrak{B}$ if and only if $\mathfrak{B}$ contains a general element of $\mathfrak{J}$.* (2) *Let $\mathfrak{J}=\mathfrak{J}_1\oplus\mathfrak{J}_2$ where the $\mathfrak{J}_i$ are ideals and let n_i denote the generic norm on $\mathfrak{J}_i$. Then $n(x)=n_1(x_1)n_2(x_2)$, $x_i\in\mathfrak{J}_i$.* (3) *Let $\mathfrak{N}$ be a nil ideal in $\mathfrak{J}$, $\bar{\mathfrak{J}}=\mathfrak{J}/\mathfrak{N}$ and let $\bar{n}$ be the generic norm on $\bar{\mathfrak{J}}$, n the generic norm on $\mathfrak{J}$. We can regard $\bar{n}$ as a polynomial function on $\mathfrak{J}$ by defining $\bar{n}(x)=\bar{n}(\bar{x})$, $\bar{x}=x+\mathfrak{N}$. Then n and $\bar{n}$ have the same irreducible factors in $\mathfrak{P}(\mathfrak{J})$. Also $n(\bar{x})\equiv n(x)$ defines a polynomial function on $\bar{\mathfrak{J}}$, $\bar{n}|n$ in $\mathfrak{P}(\bar{\mathfrak{J}})$ and n and $\bar{n}$ have the same irreducible factors in $\mathfrak{P}(\bar{\mathfrak{J}})$.*

Proof. The proofs of (1) and (2) are as usual (McCrimmon [8], Jacobson [3]). To prove (3) we observe that the polynomial mapping $x\rightarrow x^m$ is 0 on $\mathfrak{N}$. Hence this holds also for the ideal $\mathfrak{N}^{\Omega}$, Ω the algebraic closure of Φ. Hence $\mathfrak{N}^{\Omega}$ is a nil ideal in $\mathfrak{J}^{\Omega}$. Also n and $\bar{n}$ have the same irreducible factors in $\mathfrak{P}(\mathfrak{J})$ if and only if their extensions n and $\bar{n}$ have the same irreducible factors in $\mathfrak{P}(\mathfrak{J}^{\Omega})$. Thus to prove the first assertion we may assume the base field is algebraically closed. We now note that x is invertible in $\mathfrak{J}$ if and only if $\bar{x}$ is invertible in $\bar{\mathfrak{J}}$. Clearly x invertible implies $\bar{x}$ invertible. Conversely if $\bar{x}$ is invertible we have a $y\in\mathfrak{J}$ such that $yU_x=1-z$ where $z\in\mathfrak{N}$. Then $b=1-z$ is invertible by Lemma 1 (iii) since $(1-z)(1+z+z^2+\cdots)=1$. Thus yU_x is invertible and hence x and y are invertible. The norm condition for invertibility now implies that $n(x)=0$ if and only if $n(x)=\bar{n}(\bar{x})=0$. Then, by the *Hilbert Nullstellensatz*, n and $\bar{n}$ have the same irreducible factors in $\mathfrak{P}(\mathfrak{J})$. We prove next that $n(x+z)=n(x)$ for $x\in\mathfrak{J}$, $z\in\mathfrak{N}$. Again we may assume the base field is algebraically closed. Also the result just proved shows that $n(x)=0$ if and only if $n(x+z)=0$. It follows that for any z the polynomial functions n and $n^{(z)}$ defined by $n^{(z)}(x)=n(x+z)$ have the same irreducible factors. Let the distinct ones of these be $n_1,n_2,\ldots,n_r$. Then for each z we have a permutation $\pi(z)$, $i\rightarrow i'$, of $1,2,\ldots,r$ such that $n_i(x+z)=n_{i'}(x)$, $x\in\mathfrak{J}$. The mapping $z\rightarrow\pi(z)$ is a homomorphism of the additive group $(\mathfrak{N},+)$ of $\mathfrak{N}$ into the symmetric group S_r. The inverse image of a particular π in this homomorphism is a closed subset of $\mathfrak{N}$, by definition of the Zariski topology on $\mathfrak{N}$. Since $\mathfrak{N}$ is connected in the Zariski topology the kernel of $z\rightarrow\pi(z)$ is 0, so we have $n_i(x+z)=n_i(x)$ for all i and $z\in\mathfrak{N}$. Then $n(x+z)=n(x)$, $z\in\mathfrak{J}$. Using the formula $n(\lambda 1-x)=m_x(\lambda)$ established in Theorem 2 we see also that $m_x(\lambda)=m_{x+z}(\lambda)$. It follows that the mapping $\bar{x}\to m_x(\lambda)$ is single-valued. Hence if $m_x(\lambda)=\sum m_i(x)\lambda^i$ then we have well defined mappings of $\bar{\mathfrak{J}}$ into Φ given by $m_i(\bar{x})=m_i(x)$. The proof of the fact that $\bar{\mathfrak{J}}$ is generically algebraic (p.221) shows that the mappings $\bar{x}\to m_i(\bar{x})$ are polynomial functions and we have $\sum m_i(\bar{x})\bar{x}^i=0=\sum m_i(\bar{x})\bar{x}^{i+1}$. Hence $m_{\bar{x}}$ (λ) is a factor of $\sum m_i(\bar{x})\lambda^i$ (as just defined) in $\mathfrak{P}(\bar{\mathfrak{J}})[\lambda]$. It follows that the generic norm $\bar{n}$ on $\bar{\mathfrak{J}}$ is a factor of n where $n(\bar{x})=n(x)$ in $\mathfrak{P}(\bar{\mathfrak{J}})$. The proof that n and $\bar{n}$ have the same irreducible factors in $\mathfrak{P}(\bar{\mathfrak{J}})$ is the same as that given for n and $\bar{n}$ on $\mathfrak{P}(\mathfrak{J})$.

As in the usual theories we also have the following characterizations of the normalized irreducible factors of the generic norm.

THEOREM 4. *If $\mathfrak{J}$ is a generically algebraic quadratic Jordan algebra and $n_i(x)$ is an irreducible factor of the generic norm, then*

$$n_i(f(a)g(a))=n_i(f(a))n_i(g(a)) \qquad f(\lambda),g(\lambda)\in\Phi[\lambda]. \tag{9}$$

$$n_i(bU_a)=n_i(b)n_i(a)^2, \qquad a,b\in\mathfrak{J}. \tag{10}$$

Conversely, if Q is a normalized form satisfying either

$$Q(aa^{\#})=Q(a)Q(a^{\#}) \tag{9'}$$

or

$$Q(bU_a)=Q(b)Q(a)^2, \qquad a,b\in\mathfrak{J}, \tag{10'}$$

then Q is a product of normalized irreducible factors of n.

The usual proofs of this carry over (*loc. cit.*).We note also the following useful consequences of the multiplicative properties given in the last theorem.

THEOREM 5. *We have*

$$n(a^{\#})=n(a)^{m-1} \tag{11}$$

$$a^{\#\#}=n(a)^{m-2}a \tag{12}$$

$$(bU_a)^{\#}=b^{\#}U_{a^{\#}} \tag{13}$$

m the degree of the algebra.

Proof. We have $n(a)n(a^{\#})=n(n(a)1)=n(a)^m n(1)=n(a)^m$. Hence (11) holds for invertible a. Since the invertible elements form an open, hence dense, set in the Zariski topology we have (11) for all a. It is sufficient also to prove (12) and (13) for invertible a and b. Then $a^{-1}=n(a)^{-1}a^{\#}$ and $a=n(a)(a^{\#})^{-1}$. Also replacing a

by $a^{\#}$ in the first relation we get $(a^{\#})^{-1}=n(a^{\#})^{-1}a^{\#\#}$. If we equate the two values for $(a^{\#})^{-1}$ and use (11) we obtain (12). To obtain (13) we note that $(bU_a)^{\#}=n(bU_a)(bU_a)^{-1}=n(b)n(a)^2b^{-1}U_{a^{-1}}=n(b)n(a)^2n(b)^{-1}b^{\#}U_{n(a)^{-1}a^{\#}}=b^{\#}U_{a^{\#}}$.

5. Generic trace form: For a generically algebraic quadratic Jordan algebra $\mathfrak{J}$ one defines the *generic trace form* by

$$\begin{aligned} t(a,b) &= -\Delta_1^b\Delta_x^a \log n \equiv -\Delta_1^b(n^{-1}\Delta_x^a n) \\ &= (\Delta_1^a n)(\Delta_1^b n) - \Delta_1^b\Delta_x^a n. \end{aligned} \tag{14}$$

The first formulation indicates the origin of this concept in analysis and differential geometry (see Koecher [6]); the second is the one which has a meaning in our purely algebraic context. It follows from well-known properties of the directional derivative ([3] p. 217) that $t(a,b)$ is a symmetric bilinear form. Since $\Delta_1^a n$ is the coefficient of λ in $n(1+\lambda a)=\lambda^m n(\lambda^{-1}1+a)$, we have $\Delta_1^a n=t(a)$. Since n is homogeneous of degree m, $\Delta_x^a n$ is homogeneous of degree $m-1$. Hence by Euler's differential equation for homogeneous forms we have $\Delta_1^1 n=m$ and $\Delta_1^1\Delta_x^a n=(m-1)t(a)$. Thus

$$t(a,1)=t(a) \tag{15}$$

We shall now prove

THEOREM 6. *For a generically algebraic quadratic Jordan algebra we have*

$$t(f(a)g(a),1)=t(f(a),g(a)) \tag{16}$$

for $f(\lambda), g(\lambda)\in\Phi[\lambda]$, $a\in\mathfrak{J}$.

Proof. It suffices to prove this for a general. Then the generic norm of $f(a)$ in $\mathfrak{J}$ and $\Phi[a]$ are identical, by Theorem 3(1). Hence it suffices to prove the result for $\mathfrak{J}=\Phi[a]$. Here $\Phi[a]\cong\Phi[\lambda_a]^{(q)}/\mathfrak{N}$ where $\lambda_a=\lambda+(m_a(\lambda))$ and $\mathfrak{N}$ is a nil ideal. Moreover, it is clear that $\Phi[\lambda_a]^{(q)}$ and $\Phi[a]$ both have degree m. Hence, by Theorem 3(3), the generic norm of $f(a)$ in $\Phi[a]$ coincides with the generic norm of $f(\lambda_a)$ in $\Phi[\lambda_a]$. This reduces the proof to the case of an algebra of the form $\mathfrak{J}=\mathfrak{A}^{(q)}$, $\mathfrak{A}$ associative, with generic norm n satisfying $n(bc)=n(b)n(c)$.[2] We now consider this case. Let b be invertible in $\mathfrak{J}$ (or equivalently $\mathfrak{A}$). Since Δ_b^a is the coefficient of λ in $n(b+\lambda a)=n(b(1+\lambda b^{-1}a))=n(b)n(1+\lambda b^{-1}a)$ we have

$$\Delta_b^a n=n(b)t(b^{-1}a)=t(b^{\#}a). \tag{17}$$

The second formula is valid for all b by density. Now $\Delta_x^a\log n=t(x^{-1}a)$ and since $\Delta_c^b x^{-1}=-c^{-1}bc^{-1}$ we have

$$t(a,b)=-\Delta_1^b\Delta_x^a\log n=t(ba) \tag{18}$$

[2] This is clear for $\Phi[\lambda_a]^{(q)}$ since the generic norm in this case coincides with the determinant of the right multiplication. Moreover, the multiplicative property has been proved in general by McCrimmon in [8], p. 536.

The required relation (16) is an immediate consequence of this.

For further results on the generic norm and trace form we refer the reader to the following paper by McCrimmon who, at least for the time being, will have the last word on this subject.

REFERENCES

1. N. Jacobson,"Some groups of linear transformations defined by Jordan algebras I", *J. Reine Angew. Math.* **201** (1959), 178–195.
2. N. Jacobson, "Generic norm of an algebra", *Osaka Math. J.* **15** (1963), 25–50.
3. N. Jacobson, "Structure and Representatons of Jordan Algebras", *Amer. Math. Soc. Colloq. Publ.,* Vol. 39, Amer. Math. Soc., Providence, Rhode Island, 1968.
4. N. Jacobson, *Lectures on Quadratic Jordan Algebras,* Tata Institute, Bombay, 1970.
5. N. Jacobson and K. McCrimmon, "Quadratic Jordan algebras of quadratic forms with base points", *J. Indian Math. Soc.* **35** (1971), 1–45.
6. M. Koecher, *Jordan Algebras and Their Applications,* University of Minnesota, Minneapolis, Minn., 1962.
7. K. McCrimmon, "A general theory of Jordan rings", *Proc. Nat. Acad. Sci. U.S.A.* **56** (1966), 1072–1079.
8. K. McCrimmon, "Generically algebraic algebras", *Trans. Amer. Math. Soc.* **127** (1967), 527–551.
9. K. McCrimmon, "The radical of a Jordan algebra", *Proc. Nat. Acad. Sci. U.S.A.* **59** (1969), 671–678.
10. K. McCrimmon, "The Freudenthal-Springer-Tits constructions of exceptional Jordan algebras", *Trans. Amer. Math. Soc.* **139** (1969), 495–510.
11. K. McCrimmon, "The Freudenthal-Springer-Tits constructions revisited", *Trans. Amer. Math. Soc.* **148** (1970), 293–314.
12. K. McCrimmon, "Speciality of quadratic Jordan algebras", *Pac. J. Math.* **36** (1971), 763–776.

YALE UNIVERSITY
UNIVERSITY OF MISSOURI AT ST. LOUIS

Reprinted from ADVANCES IN MATHEMATICS
All Rights Reserved by Academic Press, New York and London

Vol. 20, No. 2, May 1976
Printed in Belgium

Structure Groups and Lie Algebras of Jordan Algebras of Symmetric Elements of Associative Algebras with Involution*

N. JACOBSON

Department of Mathematics, Yale University, New Haven, Connecticut 06520

Let $\mathscr{A}$ be an associative algebra with involution J and let $\mathscr{H}(\mathscr{A}, J)$ denote the set of J-symmetric elements of $\mathscr{A}$. The set $\mathscr{H}(\mathscr{A}, J)$ can be endowed with a structure based on its module structure, the unit element 1 and the binary composition $(a, b) \to aba$ which is linear in b and quadratic in a. This is an instance of a (unital) quadratic Jordan algebra in the sense of McCrimmon ([22]). For such an algebra one can define a group, the structure group, which has been defined by Koecher [20] to be the group of module automorphisms η of the algebra for which there exists a module automorphism η^* such that

$$U_{\eta(a)} = \eta U_a \eta^*$$

where $U_a b = aba$ in the given quadratic Jordan algebra. An alternative definition of the structure group can be based on a concept of isotopy which is due to the author [9]. According to this alternative definition the structure group is the group of isomorphisms of a quadratic Jordan algebra onto its various isotopes.

Associated with any quadratic Jordan algebra one has a special universal envelope. This is an associative algebra that is analogous to the universal enveloping algebra of a Lie algebra. The method we shall give for determining the structure group of quadratic Jordan algebras of the form $\mathscr{H}(\mathscr{A}, J)$ will be based on our definition of the structure group in terms of isotopes and a result due to McCrimmon (see Section 3) giving the relation between the special universal envelope of a quadratic Jordan algebra and an isotope of such an algebra. The special universal

* This is an expanded version of the author's Retiring Presidential Address presented at the Annual Meeting of the Society, January 16, 1974 with the title, "Some Groups and Lie Algebras Defined by Jordan Algebras." A sketch of some of the results of this paper was given in the author's paper [13]. The research was supported in part by National Science Foundation Grant No. NSF GP 33591X2.

envelope for certain classes of quadratic Jordan algebras $\mathscr{H}(\mathscr{A}, J)$ were determined by Jacobson and Rickart in [18] and these results were generalized by Martindale in [26]. A somewhat modified version of Martindale's theorem will be given below. This combined with McCrimmon's theorem will give quite explicit results in a number of important cases of quadratic Jordan algebras of the form $\mathscr{H}(\mathscr{A}, J)$.

The cases we shall treat in detail in this paper are the algebras $\mathscr{H}(\mathscr{A}, J)$ where $(\mathscr{A}, J)$ is simple Artinian with involution (in Section 6) and where $(\mathscr{A}, J)$ is finite dimensional central simple (in Section 7). In the latter case the structure group coincides with the group of bijective linear transformations having the generic norm as seminvariant. These groups were determined by the author in the characteristic $\neq 2$ case in [11] (see also [10]) using an argument on Galois descent. The present method does not require this and is also applicable to a number of cases in characteristic two. These results constitute an extensive generalization of a classical theorem of Frobenius [4] determining the group of bijective linear transformations η of the space of $n \times n$ matrices over a field such that

$$\det \eta(a) = \rho \det a. \tag{1}$$

Frobenius showed that these maps have one of the two forms

$$x \to bxc \qquad \text{or} \qquad x \to b^t xc \tag{2}$$

where b and c are invertible matrices and ${}^t x$ denotes the transpose of x.

In this paper we have sought to emphasize the associative aspects of the theory. In line with this we have given an account of the theory of generic norms for the algebras $\mathscr{H}(\mathscr{A}, J)$, $(\mathscr{A}, J)$ finite dimensional central simple that is almost completely independent of the general theory developed in [15] and [25].

The results on structure groups have analogues for Lie algebras. One obtains a definition of the structure Lie algebra using a standard method for defining the Lie algebra of an algebraic group (see, for example, [28]). This makes use of quadratic Jordan algebras over commutative rings of dual numbers and gives a justification for studying algebras over commutative rings as we shall do in this paper.

We remark finally that structure groups and Lie algebras play an important role in the applications of the Jordan theory For example, these arise as groups of bijective linear transformations which stabilize homogeneous domains of positivity [20]. Also the structure groups of the so-called exceptional simple Jordan algebras are forms of Lie groups of type E_6 [3, 9, 10].

1. Jordan Algebras and Quadratic Jordan Algebras

One of the main purposes of the Jordan theory is to provide an adequate framework for dealing in an intrinsic fashion with the sets of symmetric elements of associative algebras with involution. The Jordan algebras defined by such sets are the ones we shall consider in this paper. The types of problems that will concern us are interesting also for other classes of Jordan algebras, notably, exceptional ones, which are intimately related with some of the exceptional Lie groups, and Jordan algebras defined by quadratic forms, which are related to Clifford algebras. These have been treated elsewhere [16, 23, 24, 27] and will therefore be omitted from the present discussion.

We deal with algebras in the most general sense: algebras over a commutative ring k—always assumed to have a unit. k-modules are assumed to be left and unital, that is, $1x = x$ for all x. Associative algebras are assumed to have units, subalgebras contain the unit and homomorphisms map 1 into 1. By an *associative algebra with involution* we mean a pair $(\mathscr{A}, J)$ where $\mathscr{A}$ is an associative algebra and J is an involution in $\mathscr{A}$, that is, J is a map $a \to a^*$ of $\mathscr{A}$ into itself such that

$$\begin{gathered}(a + b)^* = a^* + b^*, \qquad (\alpha a)^* = \alpha a^*, \qquad \alpha \in k \\ (ab)^* = b^*a^*, \qquad a^{**} = a.\end{gathered} \tag{3}$$

The simplest nontrivial examples are the algebras $\mathscr{A} = M_n(k)$ of $n \times n$ matrices over k with the involution $A \to {}^tA$, and the algebra of quaternions with J as the conjugation $1 \to \bar{a}$. By a *homomorphism* η of an algebra with involution $(\mathscr{A}, J)$ into a second one we mean a homomorphism of associative algebras which respects the involutions in the sense that

$$\eta(a^*) = (\eta(a))^*. \tag{4}$$

The class of associative algebras with involutions with homomorphisms as morphisms is a category.

Let $\mathscr{H}(\mathscr{A}, J)$ denote the subset of $\mathscr{A}$ of elements that are *symmetric* in the sense that $h^* = h$. It is clear that $\mathscr{H}(\mathscr{A}, J)$ is a k-submodule of $\mathscr{A}$. The Jordan theory arises when one seeks to endow $\mathscr{H}(\mathscr{A}, J)$ with a richer algebraic structure than that of k-module. If k contains an element $\frac{1}{2}$ such that $\frac{1}{2} + \frac{1}{2} = 1$ one achieves this objective by introducing the *Jordan product*

$$a \cdot b = \tfrac{1}{2}(ab + ba). \tag{5}$$

This is bilinear and it satisfies the following two identities:

$$a \cdot b = b \cdot a \tag{6}$$

$$a^2 \cdot (b \cdot a) = (a^2 \cdot b) \cdot a. \tag{7}$$

Also 1 acts as unit for the Jordan product:

$$1 \cdot a = a. \tag{8}$$

These observations lead to the definition of a (*unital*) *Jordan algebra* over a commutative ring k containing $\frac{1}{2}$ as a k-module equipped with a bilinear product $a \cdot b$ satisfying (6) and (7) and an element 1 satisfying (8).

Now suppose k is arbitrary. Then the Jordan product $a \cdot b$ is unavailable. Experience has shown that in place of this one should base the Jordan theory on the product, aba which is linear in b and quadratic in a. It is clear that if $a, b \in \mathscr{H}(\mathscr{A}, J)$ as defined before then $aba \in \mathscr{H}(\mathscr{A}, J)$. Moreover, if $k \ni \frac{1}{2}$ then we can express the Jordan product in terms of aba and 1 by

$$a \cdot b = \tfrac{1}{2}(ab + ba) = \tfrac{1}{2}(ab1 + 1ba) = \tfrac{1}{2}(a + 1)\, b(a + 1) - \tfrac{1}{2}aba - \tfrac{1}{2}1b1.$$

What are the basic properties of the product aba that may serve as the axioms for an abstract theory? To formulate these it is somewhat simpler to work with the map $U_a : x \to axa$ in place of the product aba. Evidently U_a is a k-endomorphism of $\mathscr{H}(\mathscr{A}, J)$. Also the map

$$U: a \to U_a$$

is *quadratic* in the sense that 1) $U_{\alpha a} = \alpha^2 U_a$, $a \in k$, 2) $U_{a,b} \equiv U_{a+b} - U_a - U_b$ is k-bilinear. We now single out two identities on the U-map:

$$U_a U_b U_a = U_{U_a b} \tag{9}$$

If $V_{a,b}$ is the operator defined by $V_{a,b}x = U_{a,x}b$ then

$$V_{a,b} U_a = U_a V_{b,a}. \tag{10}$$

To verify these, we note that $U_aU_bU_ax = a(b(axa)b)a$ and $U_{aba}x = (aba)\,x(aba)$. Hence (9) holds. Also the definition of $U_{a,b}$ gives $U_{a,b}x = axb + bxa$ from which we obtain $V_{a,b}x = abx + xba$. Hence $V_{a,b}U_ax = ab(axa) + (axa)\,ba$ and $U_aV_{b,a} = a(bax + xab)a$. Then (10) holds.

It turns out that to obtain a reasonable theory we require a bit more than the foregoing conditions (9) and (10), namely, certain linearizations of these conditions. These will be satisfied automatically if the base ring is a field with at least four elements, and they can be assured for an arbitrary base ring by assuming that (9) and (10) are maintained on "extension" of the base ring k.

By *an extension* of the base ring k we shall mean any commutative associative k-algebra K. By *the extension* M^K *of a* k*-module* M we shall mean $K \otimes_k M$. This has a K-module structure in which $\alpha(\beta \otimes x) = \alpha\beta \otimes x$ if $\alpha, \beta \in K$, $x \in M$, and we have the natural map ν: $x \to 1 \otimes x$ of M into M^K, which need not be injective. Now it is easily seen that if Q is a quadratic map of a k-module M into a second one N, then there exists a unique quadratic map Q^K of M^K into N^K such that

$$\begin{array}{ccc} M & \xrightarrow{Q} & N \\ \downarrow & & \downarrow \\ M^K & \xrightarrow[Q^K]{} & N^K \end{array} \qquad (11)$$

is commutative. For a proof of this we refer to the monograph, Jacobson [12], which will serve as a general reference for results we shall state without proof or explicit reference. We recall also the well-known result on tensor products that if L is a commutative algebra over K then $(M^K)^L$ may be identified with M^L where L is regarded as an algebra over k by defining $al = (a1)l$ for $a \in K$, $l \in L$. Then $(Q^K)^L = Q^L$.

Now suppose $N = \text{End}\, M$ regarded as a k-module in the usual way. Then we can follow the map ν of End M into (End $M)^K$ by the map of this module into End M^K such that $1 \otimes_k A$ is sent into the K-endomorphism $1 \otimes' A$ of M^K such that $(1 \otimes' A)(1 \otimes x) = 1 \otimes Ax$. Let $\tilde{\nu}$ denote the composite of ν and this map. Then given a quadratic map U of M into End M, there exists a unique quadratic map U^K of M^K into End M^K such that we have the commutative diagram

$$\begin{array}{ccc} M & \xrightarrow{U} & \text{End}\, M \\ \nu\downarrow & & \downarrow\tilde{\nu} \\ M^K & \xrightarrow[Q^K]{} & \text{End}\, M^K \end{array} \qquad (12)$$

We are now ready to give the definition of a *(unital) quadratic Jordan algebra*, which is due to McCrimmon [22]:

DEFINITION 1. Let k be an arbitrary commutative associative ring. Then a *quadratic Jordan algebra* over k is a triple $(\mathcal{J}, U, 1)$ where $\mathcal{J}$ is a k-module, 1 a distinguished element of $\mathcal{J}$, and U is a map of $\mathcal{J}$ into $\mathrm{End}_k\, \mathcal{J}$ such that

$QJ1$. U is quadratic

$QJ2$. $U_1 = 1$ (the identity map on J)

$QJ3$. $U_a U_b U_a = U_{U_a b}$ (the "fundamental formula")

$QJ4$. If $U_{a,b} = U_{a+b} - U_a - U_b$ and $V_{a,b}$ is defined by $V_{a,b}x = U_{a,x}b$ then

$$V_{a,b}U_a = U_a V_{b,a} .$$

$QJ5$. If K is any extension of the base ring k and U^K is the quadratic map of $\mathcal{J}^K$ into End $\mathcal{J}^K$ such that (12) is commutative then U^K satisfies $QJ3$ and $QJ4$.

We remark that $QJ4$ states that $U_{a,U_a x}b = U_a U_{b,x}a$. Since the right hand side is symmetric in b and x so is the left. Hence we obtain the following addendum to $QJ4$:

$$QJ4' \cdot V_{a,b}U_a = U_a V_{b,a} = U_{a,U_a b} .$$

It is clear from $QJ5$ and the transitivity of extension of the base ring that if $(\mathcal{J}, U, 1)$ is a quadratic Jordan algebra over k then $(\mathcal{J}^K, U^K, 1 \otimes 1)$ is a quadratic Jordan algebra over K for any extension K of k. If we apply $QJ5$ to the polynomial ring $K = k[t]$ in an indeterminate t, and use the method of equating coefficients of powers of t (which can be done since $k[t]$ is k-free) we can deduce from $QJ3$ and $QJ4$ for $\mathcal{J}^K$ four additional identities $QJ6$-$QJ9$. In turn these can be used to derive $QJ5$. In this way one obtains a set of intrinsic conditions $QJ1$-$QJ4$, $QJ6$-$QJ9$ defining quadratic Jordan algebras. However, because of the rather complicated form of $QJ6$-$QJ9$ it is generally preferable to work with the first definition.

A *homomorphism* η of a quadratic Jordan algebra $(\mathcal{J}, U, 1)$ into a second one $(\mathcal{J}', U', 1')$ is a k-module homomorphism of $\mathcal{J}$ into $\mathcal{J}'$ such that

$$\eta(1) = 1', \qquad \eta U_a = U_{\eta(a)}\eta, \qquad a \in \mathcal{J}. \tag{13}$$

If we write axa for $U_a x$, the second condition is $\eta\,(axa) = \eta(a)\,\eta(x)\,\eta(a)$. An immediate consequence of the second condition is $\eta U_{a,b} = U_{\eta(a),\eta(b)}\eta$ for all $a, b \in \mathscr{J}$. On the other hand, suppose $G = \{u_i\}$ is a set of generators for $\mathscr{J}$ as k-module and η is a module homomorphism of $\mathscr{J}$ into $\mathscr{J}'$ such that

$$\eta(1) = 1', \qquad \eta U_{u_i} = U_{\eta(u_i)}\eta, \qquad \eta U_{u_i,u_j} = U_{\eta(u_i),\eta(u_j)}\eta$$

for all $u_i, u_j \in G$. Then it is clear that η is a homomorphism of quadratic Jordan algebras of $\mathscr{J}$ into $\mathscr{J}'$. It follows that if η is a homomorphism of $\mathscr{J}$ into $\mathscr{J}'$ and K is an extension of k then the K homomorphism η^K of $\mathscr{J}^K$ into $\mathscr{J}'^K$ such that $\eta^K(1 \otimes x) = 1 \otimes \eta(x)$ is a homomorphism of quadratic Jordan algebras.

If $\mathscr{A}$ is an associative algebra, $\mathscr{A}$ gives rise to a quadratic Jordan algebra $\mathscr{A}^+ = (\mathscr{A}, U, 1)$ where 1 is the unit of $\mathscr{A}$ and U_a is the map $x \to axa$. It is clear that $U\colon a \to U_a$ is a quadratic map of $\mathscr{A}$ into End $\mathscr{A}$ (as k-module). We have $U_1 = 1$ and we have seen that $QJ3$ and $QJ4$ hold. Also $QJ5$ is clear since $U^K_{a'}$ is $x' \to a'xa'$ in $\mathscr{A}^K$.

If $(\mathscr{A}, J)$ is an associative algebra with involution, then the set $\mathscr{H}(\mathscr{A}, J)$ of symmetric elements ($h^* = h$) is a *subalgebra* of $\mathscr{A}^+$ in the sense that it is a k-submodule containing 1 and every $U_a b$ for $a, b \in \mathscr{H} = \mathscr{H}(\mathscr{A}, J)$. There is a useful trick which permits viewing the quadratic Jordan algebras $\mathscr{A}^+$ as algebras of symmetric elements. Given an associative algebra $\mathscr{A}$ we form the algebra $\mathscr{B} = \mathscr{A} \oplus \mathscr{A}^0$ where $\mathscr{A}^0$ is the opposite algebra. Then $\mathscr{B}$ is the set of pairs (x, y), $x, y \in \mathscr{A}$, with the usual k-module structure and the multiplication

$$(x_1, y_1)(x_2, y_2) = (x_1x_2, y_2y_1).$$

The "exchange" map $E\colon (x, y) \to (y, x)$ is an involution in $\mathscr{B}$ and $\mathscr{H}(\mathscr{B}, E)$ is the set of elements (x, x). It is clear that $x \to (x, x)$ is an isomorphism of $\mathscr{A}^+$ onto $\mathscr{H}(\mathscr{B}, E)$.

The roles of one sided ideals of the associative theory are played by inner and outer ideals in quadratic Jordan algebras. A k-submodule $\mathscr{K}$ of $\mathscr{J}$ is called an *inner* (*outer*) *ideal* if bxb (xbx) $\in \mathscr{K}$ fore very $b \in \mathscr{K}$, $x \in \mathscr{J}$. Unlike the situation in the associative theory the two Jordan concepts differ sharply from each other since aba is linear in one of its arguments and quadratic in the other. A subset $\mathscr{K}$ which is both an inner and an outer ideal is called an *ideal* in $\mathscr{J}$.

The concept of a quadratic Jordan algebra reduces to that of a Jordan algebra if $k \ni \frac{1}{2}$. If $\mathcal{J}$ is a Jordan algebra, then $\mathcal{J}$ defines a quadratic Jordan algebra $(\mathcal{J}, U, 1)$ in which $U_a = 2L_a{}^2 - L_{a^2}$ where L_a is $x \to a \cdot x$ and $a^2 = a \cdot a$. On the other hand, if $(\mathcal{J}, U, 1)$ is a quadratic Jordan algebra over $k \ni \frac{1}{2}$ we obtain a Jordan algebra in which $a \cdot b = \frac{1}{2}a \circ b$ and $a \circ b = V_a b$, $V_a = U_{a,1}$ (which turns out to be also $V_{a,1}$ and $V_{1,a}$). The two constructions are inverses and homomorphisms mean the same things for the two types of structures. All of this will seem plausible if we observe that for $\mathscr{A}^+$, $\mathscr{A}$ associative, then

$$U_a b = aba = \tfrac{1}{2}\{a(ab + ba) + (ab + ba)\,a - (a^2b + ba^2)\}$$

$$= 2a \cdot (a \cdot b) - a^2 \cdot b = (2L_a{}^2 - L_{a^2})b$$

and $a \circ b = ab + ba$ so $a \cdot b = \frac{1}{2}a \circ b$.

2. Structure Group. Isotopy

The definition (13) of a homomorphism of a quadratic Jordan algebra and the so-called fundamental formula $U_{U_b a} = U_b U_a U_b$ suggest the consideration of k-endomorphisms η of a quadratic Jordan algebra $\mathcal{J}$ such that there exists a k-endomorphism η^* satisfying

$$U_{\eta(a)} = \eta U_a \eta^*, \qquad a \in \mathcal{J}. \tag{14}$$

It is clear that if the pairs (η, η^*), (ζ, ζ^*) have this property, then so does $(\eta\zeta, \zeta^*\eta^*)$. Also if η and η^* are bijective, then $(\eta^{-1}, (\eta^*)^{-1})$ satisfies (14). Hence the set of k-module automorphisms of $\mathcal{J}$ for which a k-module automorphism η^* exists satisfying (14) is a transformation group. This group has been introduced by Koecher (in a more special situation, [20] and [2]) and called the *structure group* of $\mathcal{J}$. We shall denote it as Str $\mathcal{J}$.

Using (13) it is clear that any automorphism is contained in Str $\mathcal{J}$. Hence the group of automorphisms Aut $\mathcal{J} \subset$ Str $\mathcal{J}$. It is immediate from (14) that Aut $\mathcal{J}$ can be characterized as the isotropy group of the element 1 in Str $\mathcal{J}$, that is, the subgroup of Str $\mathcal{J}$ of η such that $\eta(1) = 1$. It is clear also that for an automorphism η we have $\eta^* = \eta^{-1}$.

We proceed to define some other elements of Str $\mathcal{J}$. We recall that an element c of a quadratic Jordan algebra $(\mathcal{J}, U, 1)$ is called *invertible* if there exists a d in $\mathcal{J}$ such that $U_c d = c$, $U_c d^2 = 1$ where $d^2 = U_d 1$. If $\mathcal{J} = \mathscr{A}^+$, $\mathscr{A}$ associative, this reads $cdc = c$, $cd^2c = 1$ which is equivalent to c being invertible in $\mathscr{A}$ with $d = c^{-1}$. It is easy to show that c is invertible if and only if U_c is bijective and if and only if 1 is in the range of U_c. The *inverse* d is uniquely determined as $d = U_c^{-1}c$. Denoting the inverse of c as c^{-1} we have $U_c^{-1} = U_{c^{-1}}$ and c^{-1} is invertible with $(c^{-1})^{-1} = c$. Also $cdc = U_c d$ is invertible if and only if c and d are invertible, in which case $(cdc)^{-1} = c^{-1}d^{-1}c^{-1}$. The formula $U_{U_c a} = U_c U_a U_c$ for invertible c shows that $U_c \in \operatorname{Str} \mathscr{J}$ with $U_c{}^* = U_c$. The subgroup of Str $\mathscr{J}$ generated by these maps is called the *inner structure group* and will be denoted as Instr $\mathscr{J}$. Since $U_c^{-1} = U_{c^{-1}}$ it is clear that Instr $\mathscr{J}$ is the set of products $U_{c_1} \cdots U_{c_r}$, c_i invertible in $\mathscr{J}$. By (14), we have $\eta^* = \eta^{-1}U_{\eta(1)}$ and $U_{\eta(1)} = \eta\eta^*$ is invertible. Hence $\eta(1)$ is invertible. Moreover, we have $U_{\eta(a)} = \eta U_a \eta^* = \eta U_a \eta^{-1} U_{\eta(1)}$ for any $\eta \in \operatorname{Str} \mathscr{J}$. If c is invertible $U_{\eta(c)} = \eta U_c \eta^{-1} U_{\eta(1)}$ implies that $\eta(c)$ is invertible. Then $\eta U_c \eta^{-1} = U_{\eta(c)} U_{\eta(1)}$ implies that Instr $\mathscr{J}$ is a normal subgroup of Str $\mathscr{J}$.

The formula $\eta^* = \eta^{-1}U_{\eta(1)}$ now shows that if $\eta \in \operatorname{Str} \mathscr{J}$ then $\eta^* \in \operatorname{Str} \mathscr{J}$. If $\zeta \in \operatorname{Str} \mathscr{J}$ then $\eta\zeta \in \operatorname{Str} \mathscr{J}$ and $(\eta\zeta)(\eta\zeta)^* = U_{\eta\zeta(1)} = \eta U_{\zeta(1)}\eta^* = \eta\zeta\zeta^*\eta^*$. Hence $(\eta\zeta)^* = \zeta^*\eta^*$. Since $(\eta^*)^* = (\eta^{-1}U_{\eta(1)})^* = U_{\eta(1)}(\eta^*)^{-1} = U_{\eta(1)}U_{\eta(1)}^{-1}\eta = \eta$ we see that $\eta \to \eta^*$ is an antiautomorphism of Str $\mathscr{J}$ with square the identity map. Then the map $\eta \to \hat{\eta} \equiv (\eta^*)^{-1} = (\eta^{-1})^*$ is an automorphism of Str $\mathscr{J}$ with square the identity. Since $\eta^* = \eta^{-1}U_{\eta(1)}$, $\hat{\eta} = U_{\eta(1)}^{-1}\eta$ and since Aut $\mathscr{J}$ is the isotropy subgroup of 1 in Str $\mathscr{J}$ it is clear that Aut $\mathscr{J}$ is contained in the subgroup of Str $\mathscr{J}$ of elements fixed under the automorphism $\eta \to \hat{\eta}$.

Now suppose $\mathscr{J} = \mathscr{A}^+$, $\mathscr{A}$ associative. Then it is clear that any automorphism or anti-automorphism of $\mathscr{A}$ is an automorphism of $\mathscr{A}^+$, and there are a number of results in the literature that assert that for certain $\mathscr{A}$ all the automorphisms of $\mathscr{A}^+$ are obtained in this way. The earliest result of this kind is that this holds for finite dimensional central simple algebras. This was proved by Ancochea [1] for characteristic $\neq 2$ and by Kaplansky [19] for arbitrary characteristic. A similar result for arbitrary division algebras is due to Hua [7] and this was extended to rings without zero divisors $\neq 0$ by Jacobson and Rickart [17]. The most general result of this sort is due to Herstein [5] and [6]: If $\mathscr{A}$ is a prime algebra then any automorphism of $\mathscr{A}^+$ is either an automorphism or anti-automorphism of $\mathscr{A}$.

We have seen that if $\eta \in \operatorname{Str} \mathscr{J}$ then $\eta(1)$ is invertible. On the other hand, if c is an invertible element of $\mathscr{A}$ then the multiplications c_L : $x \to cx$ and $c_R : x \to xc$ are k-automorphisms of $\mathscr{A}$ and, since $U_{c_L a}x = caxca$ and $c_L U_a c_R x = caxca$, $c_L \in \operatorname{Str} \mathscr{A}^+$ and $c_L{}^* = c_R$. It is clear from these results and the fact that Aut $\mathscr{A}^+$ is the isotropy subgroup of 1 in Str $\mathscr{A}^+$ that

$$\operatorname{Str} \mathscr{A}^+ = \mathscr{A}_L \operatorname{Aut} \mathscr{A}^+ \tag{15}$$

where $\mathscr{A}_L$ is the set of left multiplication c_L, c invertible in $\mathscr{A}$. Hence, by Herstein's theorem, we have

THEOREM 1. *If $\mathscr{A}$ is a prime associative algebra then* Str $\mathscr{A}^+$ *is the set of maps $c_L\eta$ where c is invertible in $\mathscr{A}$ and η is either an automorphism or anti-automorphism of $\mathscr{A}$.*

The problem of determining the structure groups of quadratic Jordan algebras $\mathscr{H}(\mathscr{A}, J)$ where $(\mathscr{A}, J)$ is an associative algebra with involution is considerably more difficult than the one we have just considered. We shall now give an alternative definition of Str $\mathscr{J}$ which will permit us to solve this in a number of important cases.

We shall base the second definition on a concept of isotopy which we proceed to define. This has an associative background which we consider first. Let $\mathscr{A}$ be an associative algebra, c an invertible element of $\mathscr{A}$. We define a c-product in $\mathscr{A}$ by $x_c y = xcy$. This is associative and c^{-1} acts as unit. Hence we have a new associative algebra $\mathscr{A}^{(c)}$ which is the k-module $\mathscr{A}$ with the c-multiplication and unit c^{-1}. We call this the *c-isotope* of $\mathscr{A}$. The left multiplication c_L is an isomorphism of $\mathscr{A}^{(c)}$ onto $\mathscr{A}$ since this is a k-module automorphism sending c^{-1} into 1 and satisfying $c_L(x_c y) = cxcy = (c_L x)(c_L y)$. The U-map defined by a in $\mathscr{A}^{(c)}$ is $U^{(c)} = U_a U_c$ since $U_a^{(c)}x = a_c x_c a = acxca = U_a U_c x$. Then $\mathscr{A}^{(c)+}$ is $(\mathscr{A}, UU_c, c^{-1})$.

We now try to do the same thing for arbitrary quadratic Jordan algebras. Let c be an invertible element of $\mathscr{J} = (\mathscr{J}, U_j\, 1)$. Put $U_a^{(c)} = U_a U_c$, $a \in \mathscr{J}$. It is easily seen that

$$\mathscr{J}^{(c)} = (\mathscr{J}, U^{(c)}, c^{-1}) \tag{16}$$

is a quadratic Jordan algebra. We call this the *c-isotope* of $\mathscr{J}$. Since d is invertible in $\mathscr{J}$ if and only if U_d is bijective, d is invertible in $\mathscr{J}^{(c)}$ if and only if $U_d^{(c)} = U_d U_c$ is bijective, hence if and only if d is invertible in $\mathscr{J}^{(c)}$. In this case we can form the d-isotope of the c-isotope $\mathscr{J}^{(c)}$. We denote this as $\mathscr{J}^{(c)(d)}$ and observe that $U_a^{(c)(d)} = U_a^{(c)} U_d^{(c)} = U_a U_c U_d U_c = U_a U_{U_c d}$. Also $U_c d$ is invertible. It follows that $\mathscr{J}^{(c)(d)} = \mathscr{J}^{(cdc)}$. Clearly $\mathscr{J}^{(1)} = \mathscr{J}$ and it is easily seen that $\mathscr{J}$ is the $c^{-2} = (c^{-1})^2$-isotope of $\mathscr{J}^{(c)}$.

Unlike the situation for associative algebras, it is easy to give examples of isotopes of Jordan algebras which are not isomorphic [11, p. 61].

We now give the following

DEFINITION 2. Let $(\mathscr{J}, U, 1)$ and $(\mathscr{J}', U', 1')$ be quadratic Jordan algebras. Then a map η of $\mathscr{J}$ into $\mathscr{J}'$ is called *an isotopy* if η is an isomorphism of $\mathscr{J}$ onto some isotope $\mathscr{J}'^{(c')}$ of $\mathscr{J}'$. If such a map exists then $\mathscr{J}$ and $\mathscr{J}'$ are called *isotopic* (or *isotopes*).

Using $\eta = 1$ we see that $\mathscr{J}^{(c)}$ and $\mathscr{J}$ are isotopic. It is clear also that isomorphic algebras are isotopic. The definition of isotopy gives $\eta(1) = c'^{-1}$ and $\eta U_a = U'^{(c')}_{\eta(a)}\eta$ (cf. (13)). Hence we have $\eta U_a = U'_{\eta(a)}U'_{c'}\eta$ and so

$$U'_{\eta(a)} = \eta U_a \eta^*, \qquad a \in \mathscr{J}, \tag{17}$$

where $\eta^* = \eta^{-1}(U'_{c'})^{-1}$. Conversely, suppose η is a k-module isomorphism of $\mathscr{J}$ onto $\mathscr{J}'$ such that there exists a k-module isomorphism η^* of $\mathscr{J}'$ onto $\mathscr{J}$ satisfying (17). Then c is invertible in $\mathscr{J}$ if and only if $\eta(c)$ is invertible in $\mathscr{J}'$. In particular, $\eta(1)$ is invertible in $\mathscr{J}'$. Write $c' = \eta(1)^{-1}$. Then $\eta\eta^* = U'_{c'^{-1}} = (U'_{c'})^{-1}$ and $\eta^* = \eta^{-1}U'_{c'^{-1}}$. Hence $\eta U_a = U'_{\eta(a)}U'_{c'}\eta = U'^{(c')}_{\eta(a)}\eta'$ and so η is an isomorphism of $\mathscr{J}$ onto the c'-isotope $\mathscr{J}'$. We therefore have the following alternative definition of isotopy: A map η of $\mathscr{J}$ into $\mathscr{J}'$ is an isotopy if and only if η is a k-module isomorphism of $\mathscr{J}$ onto $\mathscr{J}'$ such that there exists a k-module isomorphism η^* of $\mathscr{J}'$ onto $\mathscr{J}$ satisfying (17). This gives the following alternative definition of Koecher's structure group.

DEFINITION 3. The *structure group* of $\mathscr{J} = (\mathscr{J}, U, 1)$ is the group of isomorphisms of $\mathscr{J}$ onto the various isotopes of $\mathscr{J}$.

It is this definition that we shall exploit in considering the problem of determining Str $\mathscr{H}(\mathscr{A}, J)$, $(\mathscr{A}, J)$ an associative algebra with involution.

3. Special Universal Envelopes

We have a functor from the category of associative algebras to the category of quadratic Jordan algebras, which maps an associative algebra $\mathscr{A}$ into the quadratic Jordan algebra $\mathscr{A}^+$ and a homomorphism of $\mathscr{A}$ into the same map of $\mathscr{A}^+$, which is a homomorphism of this quadratic

Jordan algebra. This functor has an adjoint which we proceed to define. For this purpose we introduce the following terminology. If $(\mathscr{J}, U, 1)$ is a quadratic Jordan algebra then we call a homomorphism of $\mathscr{J}$ into an algebra $\mathscr{A}^+$, $\mathscr{A}$ associative, an *associative specialization* of $\mathscr{J}$ in $\mathscr{A}$. We define a *special universal envelope* of $\mathscr{J}$ to be a pair $(S(\mathscr{J}), \sigma_u)$ where $S(\mathscr{J})$ is an associative algebra and σ_u is an associative specialization of $\mathscr{J}$ in $S(\mathscr{J})$ such that if σ is any associative specialization of $\mathscr{J}$ then there exists a unique homomorphism ζ (of associative algebras) such that

(18)

is commutative. If such a universal envelope exists it is unique up to isomorphism in the strong sense that if $(S(\mathscr{J})', \sigma_u')$ is another one then there exists a unique isomorphism ζ of $S(\mathscr{J})$ onto $S(\mathscr{J})'$ such that $\zeta\sigma_u = \sigma_u'$. One can easily construct a special universal envelope in a manner similar to the construction of the universal enveloping algebra of a Lie algebra or the Clifford algebra of a quadratic form. Given $\mathscr{J}$, consider the tensor algebra defined by $\mathscr{J}$ as k-module:

$$T(\mathscr{J}) = k \oplus \mathscr{J} \oplus (\mathscr{J} \otimes \mathscr{J}) \oplus \cdots \oplus (\mathscr{J} \otimes \cdots \otimes \mathscr{J}) \oplus \cdots.$$

where all the tensor products are taken with respect to k. We have the obvious associative multiplication and unit in $T(\mathscr{J})$ ($1 \in k$ is the unit and $(x_1 \otimes \cdots \otimes x_r)(x_{r+1} \otimes \cdots \otimes x_s) = x_1 \otimes \cdots \otimes x_s$, $x_i \in \mathscr{J}$). We now factor out the ideal $\mathscr{K}$ of $T(\mathscr{J})$ generated by $1_{\mathscr{J}} - 1$, where $1_{\mathscr{J}}$ now denotes the unit of $\mathscr{J}$ and 1 is the unit of k, and all the elements of the form $a \otimes b \otimes a - U_a b$, $a, b \in \mathscr{J}$. Put $S(\mathscr{J}) = T(\mathscr{J})/\mathscr{K}$ and $\sigma_u(a) = a + \mathscr{K}$, $a \in \mathscr{J}$. Then it is readily seen that $(S(\mathscr{J}), \sigma_u)$ is a special universal envelope for $(\mathscr{J}, U, 1)$.

$S(\mathscr{J})$ is generated by the image $\sigma_u(\mathscr{J})$ and $S(\mathscr{J})$ has a unique involution π called its *main involution* which is characterized by the property that $\pi(\sigma_u(a)) = \sigma_u(a)$, if $a \in \mathscr{J}$. In other words, we have $\sigma_u \eta \mathscr{J}) \subset \mathscr{H}(S(\mathscr{J}), \pi)$ and π is determined by this property. We have the usual functorial property of $(S(\mathscr{J}), \sigma_u)$ that if η is a homomorphism of $(\mathscr{J}, U, 1)$ into $(\mathscr{J}', U', 1')$ and $(S(\mathscr{J}), \sigma_u)$ and $(S(\mathscr{J}'), \sigma_u')$ are special universal envelopes for $\mathscr{J}$ and $\mathscr{J}'$ respectively, then we have a unique homomorphism of associative algebras such that

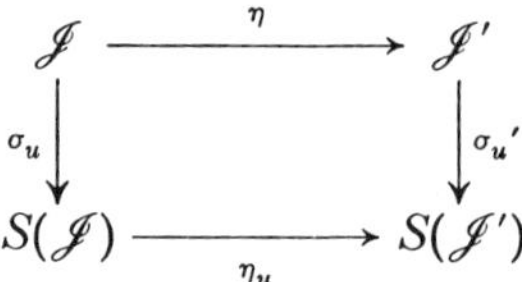

is commutative. In fact, η_u is a homomorphism of associative algebra with involution of $(S(\mathscr{J}), \pi)$ into $(S(\mathscr{J}'), \pi')$, π' the main involution in $S(\mathscr{J})$. The maps $\mathscr{J} \to S(\mathscr{J})$, $\eta \to \eta_u$ define the adjoint functor of the functor $\mathscr{A} \to \mathscr{A}^+$, $\zeta \to \zeta$ from the category of associative algebras to that of quadratic Jordan algebras. It is clear from functorial considerations that if η is an isomorphism of $\mathscr{J}$ onto $\mathscr{J}'$ then η_u is an isomorphism of $S(\mathscr{J})$ onto $S(\mathscr{J}')$.

A quadratic Jordan algebra is called *special* if it is isomorphic to a subalgebra of some $\mathscr{A}^+$, $\mathscr{A}$ associative. Evidently, $\mathscr{J}$ is special if it has an injective associative specialization. It follows that $\mathscr{J}$ is special if and only if the associative specialization σ_u of $\mathscr{J}$ in $S(\mathscr{J})$ is injective. We have seen that $\sigma_u(\mathscr{J}) \subset \mathscr{H}(S(\mathscr{J}), \pi)$; $\mathscr{J}$ will be called *reflexive* if $\sigma_u(\mathscr{J}) = \mathscr{H}(S(\mathscr{J}), \pi)$. On the other hand, we shall call an associative algebra with involution *perfect* if $\mathscr{A}$ and the injection map of $\mathscr{H}(\mathscr{A}, J)$ in $\mathscr{A}$ constitute a special universal envelope of $\mathscr{H}(\mathscr{A}, J)$. This is equivalent to the condition that any associative specialization has a unique extension to a homomorphism of the associative algebra $\mathscr{A}$. It is clear that in this case $\mathscr{H}(\mathscr{A}, J)$ is special and reflexive. It is clear also that if $\mathscr{J}$ is special and reflexive then $(S(\mathscr{J}), \pi)$ is perfect.

We shall obtain next a determination of a special universal envelope of an isotope of $\mathscr{J}$ in terms of $(S(\mathscr{J}), \sigma_u)$. We observe first that if η is a homomorphism of $(\mathscr{J}, U, 1)$ into $(\mathscr{J}', U', 1')$ and c is an invertible element of $\mathscr{J}$ then $c' = \eta(c)$ is invertible in $\mathscr{J}'$ and $\eta(c^{-1}) = c'^{-1}$. Moreover, if $x, y \in \mathscr{J}$ then

$$\eta(U_x^{(c)}y) = \eta(U_x U_c y) = U_{\eta(x)} U_{\eta(c)} \eta(y) = U_{\eta(x)}^{(c')} \eta(y).$$

Hence η is a homomorphism of $\mathscr{J}^{(c)}$ into $\mathscr{J}'^{(c)'}$, $c' = \eta(c)$. We recall also that if d is an invertible element of an associative algebra $\mathscr{A}$ then the d-isotope of $\mathscr{A}^+$ is $\mathscr{A}^{(d)+}$ where $\mathscr{A}^{(d)}$ is the d-isotope of the associative algebra $\mathscr{A}$. Combining these two results we see that if σ is an associative specialization of $\mathscr{J}$ in $\mathscr{A}$ then σ is an associative specialization of $\mathscr{J}^{(c)}$ in

$\mathscr{A}^{(d)}$, $d = \sigma(c)$. We can now prove the following result which is due to McCrimmon (unpublished).

PROPOSITION 1. *Let* $(S(\mathscr{J}), \sigma_u)$ *be a special universal envelope for* $\mathscr{J} = (\mathscr{J}, U, 1)$, *c an invertible element of* $\mathscr{J}$. *Then* $(S(\mathscr{J}), \sigma_u(c)_L\sigma_u)$ *is a special universal envelope for the isotope* $\mathscr{J}^{(c)}$.

Proof. Let σ be an associative specialization of $\mathscr{J}^{(c)}$ in $\mathscr{A}$. Then σ is an associative specialization of $\mathscr{J} = \mathscr{J}^{(c)(c^{-2})}$ in $\mathscr{A}^{(\sigma(c^{-2}))}$. Hence there is a unique homomorphism ζ of $S(\mathscr{J})$ into $\mathscr{A}^{(\sigma(c^{-2}))}$ such that $\sigma = \zeta\sigma_u$. Moreover, ζ is a homomorphism of the associative isotope $S(\mathscr{J})^{(\sigma_u(c))}$ into $\mathscr{A}^{(\sigma(c^{-2})),(\zeta\sigma_u(c))} = \mathscr{A}^{(\sigma(c^{-2})),(\sigma(c))} = \mathscr{A}^{(\sigma(c^{-2})\sigma(c)\sigma(c^{-2}))}$. Since σ is a homomorphism of $\mathscr{J}^{(c)}$ into $\mathscr{A}^+$ and c^{-1} is the unit of $\mathscr{J}^{(c)}$ we have $\sigma(c^{-1}) = 1$. Also

$$\begin{aligned}\sigma(c^{-2})\,\sigma(c)\,\sigma(c^{-2}) &= \sigma(U^{(c)}_{c^{-2}}c) = \sigma(U_{c^{-2}}U_c c) = \sigma(U_c^{-2}U_c c)\\ &= \sigma(U_c^{-1}c) = \sigma(c^{-1}) = 1.\end{aligned}$$

Thus $\mathscr{A}^{(\sigma(c^{-2})\sigma(c)\sigma(c^{-2}))} = \mathscr{A}$ and ζ is a homomorphism of $S(\mathscr{J})^{(\sigma_u(c))}$ into $\mathscr{A}$. We recall that $\sigma_u(c)_L$ is an isomorphism of $S(\mathscr{J})^{(\sigma_u(c))}$ onto $S(\mathscr{J})$. Since σ_u is an associative specialization of $\mathscr{J}$ in $S(\mathscr{J})$ it is also an associative specialization of $\mathscr{J}^{(c)}$ in $S(\mathscr{J})^{(\sigma_u(c))}$. Hence $\sigma_u(c)_L\sigma_u$ is an associative specialization of $\mathscr{J}^{(c)}$ in $S(\mathscr{J})$. We have

$$\sigma = \zeta\sigma_u = \zeta'\sigma_u(c)_L\sigma_u$$

where $\zeta' = \zeta\sigma_u(c)_L^{-1}$ is a homomorphism of $S(\mathscr{J})$ into $\mathscr{A}$. To show that ζ' is unique it suffices to show that $\sigma_u(c)\,\sigma_u(\mathscr{J})$ generates $S(\mathscr{J})$. Now $\sigma_u(c)\,\sigma_u(\mathscr{J})$ contains $\sigma_u(c) = \sigma_u(c)\,\sigma_u(1)$ and it contains $\sigma_u(c)\,\sigma_u(U_{c^{-1}}a) = \sigma_u(c)\,\sigma_u(c)^{-1}\,\sigma_u(a)\,\sigma_u(c)^{-1} = \sigma_u(a)\,\sigma_u(c)^{-1}$ for every $a \in \mathscr{J}$. Then $\sigma_u(c)\,\sigma_u(\mathscr{J})$ contains $\sigma_u(\mathscr{J})$ and since $\sigma_u(\mathscr{J})$ generates $S(\mathscr{J})$ so does $\sigma_u(c)\,\sigma_u(\mathscr{J})$. We have therefore proved that given an associative specialization σ of $\mathscr{J}^{(c)}$ in $\mathscr{A}$ there exists a unique homomorphism ζ' of $S(\mathscr{J})$ into $\mathscr{A}$ such that $\sigma = \zeta'\sigma_u(c)_L\sigma_u$. Hence $(S(\mathscr{J}), \sigma_u(c)_L\sigma_u)$ is a special universal envelope for $\mathscr{J}^{(c)}$.

4. Applications to Structure Groups

Let $\eta \in \operatorname{Str} \mathscr{J}$, so η is an isomorphism of $\mathscr{J}$ onto the isotope $\mathscr{J}^{(c)}$ where $c = \eta(1)^{-1}$. Since $(S(\mathscr{J}), \sigma_u(c)_L\sigma_u)$ is a special universal envelope for $\mathscr{J}^{(c)}$ we have a unique automorphism η_u of $S(\mathscr{J})$ such that

$$\begin{array}{ccc} \mathscr{J} & \xrightarrow{\eta} & \mathscr{J}^{(c)} \\ \downarrow{\scriptstyle \sigma_u} & & \downarrow{\scriptstyle \sigma_u(c)_L \sigma_u} \\ S(\mathscr{J}) & \xrightarrow[\eta_u]{} & S(\mathscr{J}) \end{array}$$

is commutative. Hence if we put

$$d = \sigma_u(c) = \sigma_u(\eta(1)^{-1}) = ((\sigma_u\eta)(1))^{-1} \tag{19}$$

we have

$$(\eta_u\sigma_u)(x) = d((\sigma_u\eta)(x))), \qquad x \in \mathscr{J}. \tag{20}$$

Now apply the main involution π to this equation. Since $\pi\sigma_u(x) = \sigma_u(x)$, $x \in \mathscr{J}$, this gives

$$(\pi\eta_u)(\sigma_u(x)) = ((\sigma_u\eta)(x))\, d = d^{-1}((\eta_u\sigma_u)(x))\, d = d^{-1}((\eta_u\pi)(\sigma_u(x))d.$$

Thus for the two anti-automorphisms $\pi\eta_u$ and $\eta_u\pi$ and the inner automorphism $I_d : a \to dad^{-1}$ of $S(\mathscr{J})$ we have $(\eta_u\pi)(y) = (I_d\pi\eta_u)(y)$ for all $y \in \sigma_u(\mathscr{J})$. Since $\sigma_u(\mathscr{J})$ generates $S(\mathscr{J})$ we have the relation

$$(\eta_u\,, \pi) \equiv \eta_u\pi\eta_u^{-1}\pi = I_d\,. \tag{21}$$

Next suppose $(\mathscr{A}, J)$ is an associative algebra with involution and let ζ be an automorphism of $\mathscr{A}$ such that $(\zeta, J) = \zeta J\zeta^{-1}J$ is an inner automorphism by an invertible element d of $\mathscr{H} = \mathscr{H}(\mathscr{A}, J)$:

$$(\zeta, J) = I_d\,, \qquad d \in \mathscr{H}. \tag{22}$$

Then we claim that the map

$$\eta = d_L^{-1}\zeta \mid \mathscr{H} \tag{23}$$

that is, the restriction to $\mathscr{H}$ of $x \to d^{-1}\zeta(x)$, is in Str $\mathscr{H}$. First, if $x \in \mathscr{H}$, then $\eta(x)^* = (d^{-1}\zeta(x))^* = \zeta(x)^*d^{-1} = d^{-1}I_dJ\zeta(x) = d^{-1}\,\zeta J(x) = d^{-1}\,\zeta(x) = \eta(x)$. Hence $\eta(x) \in \mathscr{H}$. A similar calculation shows that $\zeta^{-1}(dx) \in \mathscr{H}$ for $x \in \mathscr{H}$. Clearly, the maps $x \to d^{-1}\,\zeta(x)$ and $x \to \zeta^{-1}(dx)$ are inverses. Hence η is a k-automorphism of $\mathscr{H}$.

Moreover, if $x, y \in \mathscr{H}$ then

$$\begin{aligned}\eta U_y \eta^{-1} U_{d^{-1}} x &= \eta U_y \eta^{-1}\, d^{-1} x\, d^{-1} = \eta U_y \zeta^{-1}(x\, d^{-1}) \\ &= \eta(y\zeta^{-1}(x\, d^{-1})y) = d^{-1}\zeta(y\zeta^{-1}(x\, d^{-1})y) \\ &= d^{-1}\zeta(y)\, x\, d^{-1}\zeta(y) = \eta(y)\, x \eta(y) = U_{\eta(y)} x.\end{aligned}$$

Thus $U_{\eta(y)} = \eta U_y \eta^*$ where $\eta^* = \eta^{-1} U_{d^{-1}}$. This is the condition that $\eta \in \operatorname{Str} \mathscr{H}$. Note also that $\hat{\eta} = (\eta^*)^{-1} = U_d \eta = d_R \zeta$. Hence for η as in (23) we have

$$\eta^* = \zeta^{-1}\, d_R^{-1} \mid \mathscr{H}. \tag{24}$$

We remark that the element d in (22) can be replaced by $d' = \gamma d$ where γ is any invertible element in $\mathscr{H} \cap \mathscr{C}$, $\mathscr{C}$ the center of $\mathscr{A}$. Moreover, this is the only alteration that can be made in d, that is, if (ζ, d) and (ζ, d'), $d' \in \mathscr{H}$, satisfy the stated conditions then $d' = \gamma d$ where γ is an invertible element of $\mathscr{H} \cap \mathscr{C}$. Replacing d by d' changes η to $\gamma_L^{-1}\eta$. In particular, we see that if γ is any invertible element of $\mathscr{H} \cap \mathscr{C}$ then

$$x \to \gamma x \tag{25}$$

is in $\operatorname{Str} \mathscr{J}$.

As a special case we now take $\zeta = I_{g^{-1}} : a \to g^{-1}\, ag$, g any invertible element of $\mathscr{A}$. Then

$$(\zeta J \zeta^{-1} J)a = g^{-1}(g^{-1})^*\, ag^* g$$

and (22) holds for $\zeta = I_{g^{-1}}$ and any element $\gamma g^* g$ where γ is an invertible element of $\mathscr{H} \cap \mathscr{C}$. This gives the map

$$x \to \gamma g^* x g \tag{26}$$

in $\operatorname{Str} \mathscr{H}$.

Now suppose $\mathscr{J} = (\mathscr{J}, U, 1)$ is a special Jordan algebra. Then we can identify $\mathscr{J}$ with the subalgebra $\sigma_u(\mathscr{J})$ of $S(\mathscr{J})^+$ and so we may assume σ_u is the injection map of $\mathscr{J}$ in $S(\mathscr{J})$. In this case we see that any $\eta \in \operatorname{Str} \mathscr{J}$ has the form $x \to d^{-1}\, \zeta(x)$ where ζ and d satisfy the stated conditions. On the other hand, if d is any invertible element of $\mathscr{J}$ and ζ is an automorphism of $S(\mathscr{J})$ such that $(\zeta, \pi) = I_d$ and $d_L^{-1}\zeta$ maps $\mathscr{J}$ onto $\mathscr{J}$ then $d_L^{-1}\, \zeta \mid \mathscr{J}$ is in $\operatorname{Str} \mathscr{J}$. For, we have seen that $\eta = d_L^{-1}\, \zeta \mid \mathscr{H}(S(\mathscr{J}), \pi)$ is in $\operatorname{Str} \mathscr{H}$ and since $\eta^* = \eta^{-1} U_{d^{-1}}$ maps the subalgebra $\mathscr{J}$ of $\mathscr{H}$ onto itself, $\eta \mid \mathscr{J}$ is in $\operatorname{Str} \mathscr{J}$. Hence $d_L^{-1}\, \zeta \mid \mathscr{J}$ is in Str $\mathscr{J}$. If $\mathscr{J}$ is reflexive as well as special, so $\mathscr{J} = \mathscr{H}(S(\mathscr{J}, \pi)$, then Str $\mathscr{J}$ is the set of maps $d_L^{-1} \zeta \mid J$ where d is an invertible element of $\mathscr{J}$ and ζ is an automorphism of $S(\mathscr{J})$ such that $(\zeta, \pi) = I_d$. It is somewhat more convenient to state this last result in terms of perfect associative algebras with involution. We summarize these results in

THEOREM 2.1. *Let $\mathscr{J}$ be a special quadratic Jordan algebra and identify $\mathscr{J}$ with its image in $S(\mathscr{J})$. Then* Str $\mathscr{J}$ *is the set of maps $d_L^{-1} \zeta \mid \mathscr{J}$ such that d is an invertible element of $\mathscr{J}$ and ζ is an automorphism of $S(\mathscr{J})$ such that $(\zeta, \pi) = I_d$ and $d_L^{-1}\zeta$ maps $\mathscr{J}$ onto $\mathscr{J}$.* 2. *If $(\mathscr{A}, J)$ is a perfect associative algebra with involution then* Str $\mathscr{H}$ *for $\mathscr{H} = \mathscr{H}(\mathscr{A}, J)$ is the set of maps $d_L^{-1} \zeta \mid \mathscr{H}$ where ζ is an automorphism of $\mathscr{A}$ and d is an invertible element of $\mathscr{H}$ such that $\zeta J \zeta^{-1} J = I_d$.*

We also have the following

COROLLARY. 1. *Let $\mathscr{J}$ and $S(\mathscr{J})$ be as in Theorem 2 and assume that every automorphism of $S(\mathscr{J})$ is inner. Then* Str $\mathscr{J}$ *is the set of maps η: $x \rightarrow \gamma\pi(g)\, xg$ where g is an invertible element of $S(\mathscr{J})$, γ an invertible element of $\mathscr{C} \cap \mathscr{J}$, $\mathscr{C}$ the center of $S(\mathscr{J})$ and η maps $\mathscr{J}$ onto $\mathscr{J}$.* 2. *Let $(\mathscr{A}, J)$ be a perfect associative algebra with involution such that every automorphism of $\mathscr{A}$ is inner. Then* Str $\mathscr{H}$ *for $\mathscr{H} = \mathscr{H}(\mathscr{A}, J)$ is the set of maps $x \rightarrow \gamma g^* xg$ where g is an invertible element of $\mathscr{A}$ and γ is an invertible element of $\mathscr{C} \cap \mathscr{H}$, $\mathscr{C}$ the center of $\mathscr{A}$.*

We shall now derive the most comprehensive result on perfection of an associative algebra with involution. This is due to Martindale in [26] and generalizes an earlier result by the present author and Rickart [18]. We shall formulate this in a somewhat different fashion than Martindale, in that our main hypothesis will be in the form of a purely Jordan condition on $\mathscr{H}(\mathscr{A}, J)$.[1]

5. MARTINDALE'S THEOREM

We consider first the analogue for quadratic Jordan algebras of the classical two sided Peirce decomposition of an associative algebra $\mathscr{A}$. An element e of a quadratic Jordan algebra $\mathscr{J}$ is called *idempotent* if $e^2(= U_e 1) = e$ and the idempotents e and f are *orthogonal* if $U_e f =$

[1] A similar result has been obtained independently by B. D. Smith in [29].

$U_f e = e \circ f = 0$. If $\mathscr{J} = \mathscr{A}^+$ then idempotent means idempotent in $\mathscr{A}$ and orthogonality amounts to the conditions: $efe = fef = ef + fe = 0$ which are immediately seen to be equivalent to orthogonality in $\mathscr{A}(ef = 0 = fe)$. Let $e_1, e_2, \ldots, e_n$ be orthogonal idempotents in $\mathscr{J}$ such that $1 = \sum e_i$. Then it is easily seen that the $n(n+1)/2$ endomorphisms U_{e_i}, $U_{e_i e_j}$, $i \neq j$, are orthogonal idempotents whose sum is the identity map. Hence we obtain the *Peirce decomposition*

$$\mathscr{J} = \bigoplus_{i \leqslant j} \mathscr{J}_{ij}, \mathscr{J}_{ii} = U_{e_i}\mathscr{J}, \mathscr{J}_{ij} = U_{e_i, e_j}\mathscr{J}, \qquad i \neq i. \tag{27}$$

If $\mathscr{J} = \mathscr{A}^+$ the e_i are orthogonal idempotents in $\mathscr{A}$, and we have the two sided Peirce decomposition $\mathscr{A} = \bigoplus \mathscr{A}_{ij}$, $\mathscr{A}_{ij} = e_i \mathscr{A} e_j$. Since $\mathscr{A}_{ii} = U_{e_i}\mathscr{A} = e_i \mathscr{A} e_i$ and $\mathscr{A}_{ij} = U_{e_j, e_i}\mathscr{A} = \{e_i a e_j + e_j a e_i \mid a \in \mathscr{A}\}$ for $i \neq j$ it is clear that in this case

$$\mathscr{J}_{ii} = \mathscr{A}_{ii}, \qquad \mathscr{J}_{ij} = \mathscr{A}_{ij} + \mathscr{A}_{ji}. \tag{28}$$

Suppose next that we have an associative algebra with involution $(\mathscr{A}, \mathscr{J})$ and we have n orthogonal idempotents e_i, $1 \leqslant i \leqslant n$, in $\mathscr{H} = \mathscr{H}(\mathscr{A}, J)$ such that $\sum e_i = 1$. Then we have the Peirce decomposition $\mathscr{H} = \bigoplus_{i \leqslant j} \mathscr{H}_{ij}$ and $\mathscr{H}_{ii} = \mathscr{A}_{ii} \cap \mathscr{H}$, $\mathscr{H}_{ij} = (\mathscr{A}_{ij} + \mathscr{A}_{ji}) \cap \mathscr{H}$ where $\mathscr{A}_{ij} = e_i \mathscr{A} e_j$. Since $e_i{}^* = e_i$, $\mathscr{A}_{ij} = \mathscr{A}_{ji}$. It follows that

$$\mathscr{H}_{ij} = \{a_{ij} + a_{ij}^* \mid a_{ij} \in \mathscr{A}_{ij}\} \qquad \text{if} \quad i \neq i. \tag{29}$$

We shall now assume that for every i, j, $i \neq j$, e_i and e_j are *interconnected* in the sense that there exist elements $h_j{}^\lambda \in \mathscr{H}_{jj}$ and elements $h_{ij}^\lambda \in \mathscr{H}_{ij}$ such that

$$e_i = \sum_\lambda U_{h_{ij}^\lambda} h_j{}^\lambda. \tag{30}$$

Writing $h_{ij}^\lambda = a_{ij}^\lambda + a_{ij}^{\lambda *}$ we obtain $U_{h_{ij}^\lambda} h_j{}^\lambda = (a_{ij}^\lambda + a_{ij}^{\lambda *})\, h_j{}^\lambda (a_{ij} + a_{ij}^{\lambda *}) = a_{ij}^\lambda h_j{}^\lambda a_{ij}^{\lambda *}$. Hence the interconnectedness condition is equivalent to the existence of $a_{ij}^\lambda \in \mathscr{A}_{ij}$ and $h_j{}^\lambda \in \mathscr{H} \cap \mathscr{A}_{jj}$ such that

$$e_i = \sum_\lambda a_{ij}^\lambda h_j{}^\lambda a_{ij}^{\lambda *} \tag{31}$$

for all $i \neq j$. In the foregoing calculations we have used the standard multiplication properties of the two sided Peirce components $\mathscr{A}_{ij}$:

$$\mathscr{A}_{ij}\mathscr{A}_{jk} \subset \mathscr{A}_{ik}, \qquad \mathscr{A}_{ij}\mathscr{A}_{lk} = 0 \qquad \text{if} \quad j \neq l. \tag{32}$$

Now (31) implies that $e_i \in \mathscr{A}_{ij}\dot{\mathscr{A}}_{jj}\mathscr{A}_{ji} \subset \mathscr{A}_{ij}\mathscr{A}_{ji}$ so $\mathscr{A}_{ii} = e_i\mathscr{A}_{ii} \subset \mathscr{A}_{ii}\mathscr{A}_{ij}\mathscr{A}_{ji} \subset \mathscr{A}_{ij}\mathscr{A}_{ji}$. Hence $\mathscr{A}_{ij}\mathscr{A}_{ji} = \mathscr{A}_{ii}$ and $\mathscr{A}_{ik} = e_i\mathscr{A}_{ik} \subset \mathscr{A}_{ii}\mathscr{A}_{ik} = \mathscr{A}_{ii}\mathscr{A}_{ji}\mathscr{A}_{ik} \subset \mathscr{A}_{ii}\mathscr{A}_{jk}$. Hence the first relation in (32) can be replaced by

$$\mathscr{A}_{ij}\mathscr{A}_{jk} = \mathscr{A}_{ik} \tag{33}$$

when we have interconnected $e_i \in \mathscr{H}$.

We can now state

MARTINDALE'S THEOREM. *Let $(\mathscr{A}, J)$ be an associative algebra with involution such that $\mathscr{H} = \mathscr{H}(\mathscr{A}, J)$ contains $n \geqslant 2$ interconnected orthogonal idempotents e_i such that $\sum e_i = 1$. Then $(\mathscr{A}, J)$ is perfect if either $n \geqslant 3$ or $n = 2$ and $\mathscr{H}_i = \mathscr{H} \cap e_i \mathscr{A} e_i$ generates $e_i \mathscr{A} e_i$ for $i = 1, 2$.*

We shall sketch the proof. We have to show that if σ is an associative specialization of $\mathscr{H}$ into an associative algebra $\mathscr{B}$ then σ has a unique extension to a homomorphism ζ of $\mathscr{A}$ into $\mathscr{B}$. Now $\mathscr{H}$ generates $\mathscr{A}$, since the subalgebra $\mathscr{A}'$ generated by $\mathscr{H}$ contains every $a_{ij} = e_i(a_{ij} + a_{ij}^*)$, $a_{ij} \in \mathscr{A}_{ij}$ for $i \neq j$. Thus $\mathscr{A}' \supset \mathscr{A}_{ij}$, $i \neq j$ and hence $\mathscr{A}' \supset \mathscr{A}_{ii} = \mathscr{A}_{ij}\mathscr{A}_{ji}$ so $\mathscr{A}' = \mathscr{A}$. It is now clear that if an extension ζ exists then it is unique. We now proceed to construct an extension ζ. We observe first that if $f_i = \sigma(e_i)$ then the f_i are orthogonal idempotents of $\mathscr{B}^+$ with $\sum f_i = 1$. Hence these are orthogonal idempotents in $\mathscr{B}$ with $\sum f_i = 1$ and we have the associative Peirce decomposition

$$\mathscr{B} = \bigoplus \mathscr{B}_{ij}, \qquad \mathscr{B}_{ij} = f_i\mathscr{B}f_j. \tag{34}$$

Evidently σ maps $\mathscr{H}_{ii} = U_{e_i}\mathscr{H}$ into $U_{f_i}\mathscr{B} = \mathscr{B}_{ii}$ and $\mathscr{H}_{ii} = U_{e_i,e_j}\mathscr{H}$ into $U_{f_i,f_j}\mathscr{B} = \mathscr{B}_{ij} + \mathscr{B}_{ji}$ for $i \neq j$. Let $x_{ij} \in \mathscr{A}_{ij}$, $i \neq j$. Then $x_{ij} + x_{ij}^* \in \mathscr{H}_{ij}$ and $\sigma(x_{ij} + x_{ij}^*) = y_{ij} + y_{ji}$ where $y_{ij} \in \mathscr{B}_{ij}$, $y_{ji} \in \mathscr{B}_{ji}$ are uniquely determined. Hence we can define a map $\zeta_{ij} : \mathscr{A}_{ij} \to \mathscr{B}_{ij}$ by $\zeta_{ij}(x_{ij}) = y_{ij}$. This is a k-homomorphism and

$$\sigma(x_{ij} + x_{ij}^*) = \zeta_{ij}(x_{ij}) + \zeta_{ji}(x_{ij}^*), \qquad i \neq i. \tag{35}$$

One can prove next by straightforward calculations the following relations for $x_{ij} \in \mathscr{A}_{ij}$, $y_i \in \mathscr{H}_{ii}$ etc. with $i, j, k \neq$:

$$\sigma(x_{ij}x_{ji} + x_{ji}^*x_{ij}^*) = \zeta_{ij}(x_{ij})\,\zeta_{ji}(x_{ji}) + \zeta_{ij}(x_{ji}^*)\,\zeta_{ji}(x_{ij}^*)$$

$$\zeta_{ik}(x_{ij}x_{jk}) = \zeta_{ij}(x_{ij})\,\zeta_{jk}(x_{jk}) \tag{37}$$

$$\zeta_{ij}(y_ix_{ij}) = \sigma(y_i)\,\zeta_{ij}(x_{ij}), \qquad \zeta_{ij}(x_{ij}y_j) = \zeta_{ij}(x_{ij})\,\sigma(y_j) \tag{38}$$

$$\sigma(x_{ij}x_{ij}^*) = \zeta_{ij}(x_{ij})\,\zeta_{ji}(x_{ij}^*). \tag{39}$$

The first of these is obtained by applying σ to the equation

$$(x_{ij} + x_{ij}^*) \circ (x_{ji} + x_{ji}^*) = (x_{ij}x_{ji} + x_{ji}^*x_{ij}^*) + (x_{ji}x_{ij} + x_{ij}^*x_{ji}^*)$$

and using (35). The others are obtained in a similar fashion from the relations $(x_{ij} + x_{ij}^*) \circ (x_{jk} + x_{jk}^*) = x_{ij}x_{jk} + x_{jk}^*x_{ij}^*$, $y_i \circ (x_{ij} + x_{ij}^*) = y_ix_{ij} + x_{ij}^*y_i$, $y_j \circ (x_{ij} + x_{ij}^*) = x_{ij}y_j + y_jx_{ij}^*$, $(x_{ij} + x_{ij})\, e_j(x_{ij} + x_{ij}^*) = x_{ij}x_{ij}^*$. Also, applying σ to (30) with $h_{ij}^\lambda = a_{ij}^\lambda$ yields

$$f_i = \sum_\lambda \zeta_{ij}(a_{ij}^\lambda)\, \sigma(h_j{}^\lambda)\, \zeta_{ji}(a_{ij}^{\lambda *}). \tag{40}$$

We now assume $n \geqslant 3$ and we proceed to define a k-homomorphism ζ_{ii} of $\mathscr{A}_{ii}$ into $\mathscr{B}_{ii}$. For this purpose we choose $j \neq i$ and use the fact that $\mathscr{A}_{ii} = \mathscr{A}_{ij}\mathscr{A}_{ji}$ to write any element of $\mathscr{A}_{ii}$ as $\sum_\mu x_{ij}^\mu x_{ji}^\mu$, $x_{ij}^\mu \in \mathscr{A}_{ij}$, $x_{ji}^\mu \in \mathscr{A}_{ji}$. Then we define $\zeta_{ii}(\sum x_{ij}^\mu x_{ji}^\mu) = \sum \zeta_{ij}(x_{ij}^\mu)\, \zeta_{ji}(x_{ji}^\mu)$. To see that this is well defined we have to show that if $\sum x_{ij}^\mu x_{ji}^\mu = 0$ then

$$\sum \zeta_{ij}(x_{ij}^\mu)\, \zeta_{ji}(x_{ji}^\mu) = 0.$$

For this, we multiply $\sum x_{ij}^\mu x_{ji}^\mu$ on the right by $a_{ik} \in \mathscr{A}_{ik}$ where $k \neq i, j$ and apply ζ_{ik}. By (37) we obtain

$$\sum_\mu \zeta_{ij}(x_{ij}^\mu)\, \zeta_{ji}(x_{ji}^\mu)\, \zeta_{ik}(a_{ik}) = 0$$

for all $a_{ik} \in \mathscr{A}_{ik}$. Then if we use (40) with j replaced by k we obtain $\sum_\mu \zeta_{ij}(x_{ij}^\mu)\, \zeta_{ji}(x_{ji}^\mu) f_i = 0$ and since $\sum \zeta_{ij}(x_{ij}^\mu)\, \zeta_{ji}(x_{ji}^\mu) \in \mathscr{B}_{ii}$ this implies $\sum \zeta_{ij}(x_{ij}^\mu)\, \zeta_{ji}(x_{ji}^\mu) = 0$. We now have our k-homomorphism ζ_{ii} of $\mathscr{A}_{ii}$ into $\mathscr{B}_{ii}$ and it is clear from its definition that

$$\zeta_{ii}(x_{ij}x_{ji}) = \zeta_{ij}(x_{ij})\, \zeta_{ji}(x_{ji}) \tag{41}$$

for the $j \neq i$ we have chosen. We show next that ζ_{ii} is independent of the choice of $j \neq i$. Let $k \neq i, j$ and let ζ_{ii}' be defined by k so that we have $\zeta_{ii}'(x_{ik}x_{ki}) = \zeta_{ik}(x_{ik})\, \zeta_{ki}(x_{ki})$. We have $\zeta_{ii}(x_{ij}x_{jk}x_{ki}) = \zeta_{ij}(x_{ij})\, \zeta_{ji}(x_{jk}x_{ki}) = \zeta_{ij}(x_{ij})\, \zeta_{jk}(x_{jk})\, \zeta_{ki}(x_{ki})$ and $\zeta_{ii}'(x_{ij}x_{jk}x_{ki}) = \zeta_{ik}(x_{ij}x_{jk})\, \zeta_{ki}(x_{ki}) = \zeta_{ij}(x_{ij})\, \zeta_{jk}(x_{jk})\, \zeta_{ki}(x_{ki})$. Since $\mathscr{A}_{ii} = \mathscr{A}_{ij}\mathscr{A}_{jk}\mathscr{A}_{ki}$ this implies that $\zeta_{ii} = \zeta_{ii}'$ and hence we have (41) for all i, j.

Now let ζ be the k-homomorphism of $\mathscr{A}$ into $\mathscr{B}$ such that $\zeta \mid \mathscr{A}_{ij} = \zeta_{ij}$ for all i, j. Then $\zeta(x_{ij}x_{ji}) = \zeta(x_{ij})\, \zeta(x_{ji})$ if $i \neq j$ and $\zeta(x_{ij}x_{jk}) = \zeta(x_{ij})\, \zeta(x_{jk})$ if $i, j, k \neq$. Using the fact that $\mathscr{A}_{ii} = \mathscr{A}_{ik}\mathscr{A}_{ki}$ one sees that $\zeta(x_{ii}x_{ij}) = \zeta(x_{ii})\, \zeta(x_{ij})$, $\zeta(x_{ij}x_{jj}) = \zeta(x_{ij})\, \zeta(x_{jj})$ and $\zeta(x_{ii}y_{ii}) = \zeta(x_{ii})\, \zeta(y_{ii})$. It is clear from these relations and $\mathscr{A}_{ij}\mathscr{A}_{kl} = 0$, $\mathscr{B}_{ij}\mathscr{B}_{kl} = 0$ if $j \neq k$ that ζ is multiplicative. Now let $y_i \in \mathscr{H}_{ii}$. Then $\sigma(y_i) = \sigma(y_i) f_i = \sum \sigma(y_i) \times \zeta_{ij}(a_{ij}^\lambda)\, \sigma(h_j{}^\lambda)\, \zeta_{ji}(a_{ii}^{\lambda *}) = \sum_\lambda \zeta_{ij}(y_i a_{ij}^\lambda)\, \zeta_{ji}(h_j{}^\lambda a_{ij}^{\lambda *}) = \zeta(\sum_\lambda y_i a_{ij}^\lambda h_j{}^\lambda a_{ij}^{\lambda *}) =$

$\zeta(y_i e_i) = \zeta(y_i)$. $\zeta(x_{ij} + x_{ij}^*) = \zeta_{ij}(x_{ij}) + \zeta_{ji}(x_{ij}^*) = \sigma(x_{ij} + x_{ij}^*)$ we see that ζ coincides with σ on $\mathscr{H}$. Then $\zeta(1) = 1$ and ζ is a homomorphism of $\mathscr{H}$ which extends σ.

Now assume $n = 2$ and $\mathscr{H}_{ii}$ generates $\mathscr{A}_{ii}$ for $i = 1, 2$. Suppose we have a relation of the form $\sum m = 0$ where each m is a product $k_1 k_2 \cdots k_r$ where all the k's are in $\mathscr{H}_{ii}$. Let $m' = \sigma(k_1)\,\sigma(k_2) \cdots \sigma(k_j)$. Then for any $a_{ij} \in \mathscr{A}_{ij}$ we have $\sum m' \zeta_{ij}(a_{ij}) = \sum \zeta_{ij}(k_1 \cdots k_r a_{ij})$ (by repeated application of (38)) $= \zeta_{ij}((\sum m) a_{ij}) = 0$. Then it follows from (40) that $\sum m' f_i = 0$ so $\sum m' = 0$. This implies that we have a homomorphism ζ_{ii} of $\mathscr{A}_{ij}$ as algebra with unit e_i into $\mathscr{B}_{ii}$ as algebra with unit f_i such that ζ_{ii} coincides with σ on $\mathscr{H}_{ii}$. Now let ζ be the k-homomorphism of $\mathscr{A}$ into $\mathscr{B}$ which is ζ_{ii} on $\mathscr{H}_{ii}$ and ζ_{ij} on $\mathscr{A}_{ij}$ for $i \neq j$. Then it is easily seen that ζ is a homomorphism of $\mathscr{A}$ into $\mathscr{B}$ which extends σ.

There are two important cases of Martindale's theorem which we shall now consider. First, let $(\mathscr{D}, j)$ be an associative algebra with involution j: $d \to \bar{d}$ and let $M_n(\mathscr{D})$ be the algebra of $n \times n$ matrices with entries in $\mathscr{D}$. Let c_i, $1 \leqslant i \leqslant n$, be invertible elements of $\mathscr{H}(\mathscr{D}, j)$ and put $C = \operatorname{diag}\{c_1, c_2, \ldots, c_n\}$ the diagonal matrix with diagonal entries c_i. Then C defines a *canonical* involution $J = J_C$ in $\mathscr{A} \equiv M_n(\mathscr{D})$ which is $X \to X^* = C\,{}^t\bar{X} C^{-1}$ where ${}^t\bar{X}$ denotes transpose of $\bar{X} = (\bar{x}_{ij})$ for $X = (x_{ij})$. If e_{ij} is the usual (i, j)-matrix unit then $e_{ij}^* = c_j c_i^{-1} e_{ji}$. Hence $e_i = e_{ii}$ and $e_{ij} + c_j c_i^{-1} e_{ji} \in \mathscr{H}$. The e_i are orthogonal idempotents such that $\sum e_i = 1$ and if $i \neq j$ then $e_{ij} \in \mathscr{A}_{ij} = e_i \mathscr{A} e_j$ and $e_i = e_{ij} e_j e_{ij}^*$. Hence the e_i are interconnected and so Martindale's theorem applies if $n \geqslant 2$. It is easily seen that the supplementary condition for $n = 2$ in the theorem is satisfied if and only if $\mathscr{D}$ is generated by $\mathscr{H}(\mathscr{D}, j)$.

Now let $\mathscr{A}$ be an associative algebra, let $\mathscr{A}^0$ denote the opposite algebra and put $\mathscr{B} = \mathscr{A} \oplus \mathscr{A}^0$. We have the exchange involution E: $(x, y) \to (y, x)$ in $\mathscr{B}$ and $\mathscr{H}(\mathscr{B}, E)$ is isomorphic to $\mathscr{A}^+$. It is easy to translate the hypotheses on $\mathscr{H}(\mathscr{A}, J)$ of Martindale's theorem to conditions on $\mathscr{A}$. These are that $\mathscr{A}$ contain $n \geqslant 2$ orthogonal e_i such that $\sum e_i = 1$ and that $e_i \in \mathscr{A}_{ij}\mathscr{A}_{ji}$ for every $i \neq j$—which is the interconnectedness condition for the idempotents (e_i, e_i) of $\mathscr{H}(\mathscr{B}, E)$. One can also formulate a condition on $\mathscr{A}$ which is equivalent to the supplementary condition in the $n = 2$ case of the theorem. We shall be interested primarily in the case in which $\mathscr{A} = M_n(\mathscr{D})$ the algebra of $n \times n$ matrices with entries in an algebra $\mathscr{D}$. This has the orthogonal idempotents $e_i = e_{ii}$ and $e_i \in \mathscr{A}_{ij}\mathscr{A}_{ji}$. If we apply the theorem for $n \geqslant 3$ we see that if σ is an associative specialization of $\mathscr{A}^+$ then the map $(x, x) \to \sigma(x)$ of $\mathscr{A} \oplus \mathscr{A}^0$ can be extended in one and only one way to a homomorphism ζ of $\mathscr{A} \oplus \mathscr{A}^0$. This result was first proved by Jacobson and Rickart in [1] who showed that it is valied also for the algebra $\mathscr{A} = M_2(\mathscr{D})$.

We remark finally that we can extend Martindale's theorem to outer ideals containing 1 in quadratic Jordan algebras of the form $\mathscr{H}(\mathscr{A}, J)$. It is easily seen that if $k \ni \frac{1}{2}$ (so we are in the situation of ordinary Jordan algebras) then an outer ideal is necessarily an ideal. Hence any outer ideal containing 1 coincides with $\mathscr{J}$. This need not be the case if $k \not\ni \frac{1}{2}$, and in this case, the outer ideals containing 1 play an important role in the structure theory. If $\mathscr{K}$ is an outer ideal containing 1 in $\mathscr{J}$ then $\mathscr{K}$ contains every idempotent $e = U_e 1$ and if $e_1, e_2, \ldots, e_n$ are orthogonal idempotents in $\mathscr{J}$ such that $\sum e_i = 1$ then $\mathscr{K}$ contains every Peirce submodule $\mathscr{J}_{ij}$ for $i \neq j$, since for $a_{ij} \in \mathscr{J}_{ij}$, $i \neq j$, $U_{a_{ij}, e_i} e_i = a_{ij}$. It follows that $\mathscr{K}$ has the form $\sum_{i \neq j} \mathscr{J}_{ij} + \sum_i \mathscr{K}_{ii}$ where $\mathscr{K}_{ii} = \mathscr{K} \cap \mathscr{J}_{ii}$. The proof of Martindale's theorem carries over verbatim to prove the following extension: Let $(\mathscr{A}, J)$ be an associative algebra with involution, $\mathscr{K}$ an outer ideal in $H(A, J)$ containing 1 and $n \geqslant 2$ interconnected (in $\mathscr{K}$) orthogonal idempotents e_i such that $\sum e_i = 1$. Assume also that if $n = 2$ then $\mathscr{K}_{ii} = \mathscr{K} \cap e_i \mathscr{A} e_i$ generates $e_i \mathscr{A} e_i$, $i = 1, 2$. Then any homomorphism of $\mathscr{K}$ into a quadratic Jordan algebra $\mathscr{B}^+$, $\mathscr{B}$ associative, has a unique extension to a homomorphism of $\mathscr{A}$ into $\mathscr{B}$. In other words, $\mathscr{K}$ and the injection map of $\mathscr{K}$ in $\mathscr{A}$ constitute a special universal algebra for $\mathscr{K}$, which is evidently a subalgebra of $\mathscr{H}(\mathscr{A}, J)$ and of $\mathscr{A}^+$.

6. Simple Artinian Algebras with Involution

An algebra with involution $(\mathscr{A}, J)$ is *simple* if $\mathscr{A}$ contains no ideal $\mathscr{B} \neq 0, \neq A$ which is stabilized by $\mathscr{J}$. It is easily seen that $(\mathscr{A}, J)$ is simple if and only if either $\mathscr{A}$ is simple or $\mathscr{A} = \mathscr{B} \oplus \mathscr{B}^*$ where $\mathscr{B}$ is a simple ideal. $(\mathscr{A}, J)$ is called (left) *artinian* if $\mathscr{A}$ is (left) artinian. The structure of simple artinian algebras with involution is known. There are three types: I $\mathscr{B} \oplus \mathscr{B}^0$ where $\mathscr{B} = \mathrm{End}_\Delta V$, V an n-dimensional vector space over a division algebra Δ, J the exchange involution. II $\mathscr{A} = \mathrm{End}_\Delta V$ where (Δ, j) is a division algebra with involution and J is the adjoint map $A \to A^*$ in $\mathrm{End}_\Delta V$ relative to a non-degenerate diagonalizable hermitian form h. (This will be explained more fully below.) III $\mathscr{A} = \mathrm{End}_\Phi V$ where Φ is a field and J is the adjoint map relative to a non-degenerate alternate bilinear form f.

It is not difficult to see that in all three cases, $(\mathscr{A}, J)$ is isomorphic to a matrix algebra $M_n(\mathscr{D})$ with a canonical involution. For I, we have $\mathscr{D} = \Delta \oplus \Delta^0$ with exchange involution j. For II, $\mathscr{D} = \Delta$ and j is an involution in Δ. For III, $\mathscr{D} = M_2(\Phi)$ with the involution

$$a \to \begin{pmatrix} 0 & 1 \\ -1 & 0 \end{pmatrix} t_a \begin{pmatrix} 0 & -1 \\ 1 & 0 \end{pmatrix}.$$

The results of the last section show that $(\mathscr{A}, j)$ is perfect if $n \geqslant 2$ with no further condition in case I and in cases II and III provided that $\mathscr{H}(\mathscr{D}, j)$ generates $\mathscr{D}$ with $n = 2$. In case III the form of the involution shows that $\mathscr{H}(\mathscr{D}, j)$ generates $\mathscr{D} = M_2(\Phi)$ if and only if char $\Phi = 2$. In case I, $\mathscr{H}(\mathscr{A}, J) = \mathscr{B}^+$, $\mathscr{B} = \operatorname{End}_\Delta V$. Since $\mathscr{B}$ is simple it is prime. Hence, by Theorem 1, Str $\mathscr{B}^+$ is the set of maps $c_L\zeta$ where c is an invertible element of $\mathscr{B}$ and ζ is either an automorphism or anti-automorphism of $\mathscr{B}$.

We now consider case II. Here we have a division algebra with involution (Δ, j) where j is $a \to \bar{a}$ and we have a non-degenerate hermitian form $h(x, y)$ on an n-dimensional vector space V over Δ. The conditions on h are that $h(x, y)$ is bi-additive, $h(ax, y) = ah(x, y)$, $h(x, ay) = h(x, y)\bar{a}$, $h(y, x) = \overline{h(x, y)}$ and $h(x, z) = 0$ for all x implies $z = 0$. We are assuming also that V has an orthogonal base $(u_1, \ldots, u_n)$, that is, $h(u_i, u_j) = 0$ for $i \neq j$.

Let S be a semilinear transformation of V corresponding to the automorphism σ of Δ, that is, $S(x \times y) = Sx + Sy$, $S(ax) = \sigma(a)\, Sx$ for $x, y \in V$, $a \in \Delta$. For a fixed y, the map

$$x \to \sigma^{-1}h(Sx, y)$$

is a linear function on V. Hence there exists a unique vector $y' \in V$ such that $\sigma^{-1}h(Sx, y) = h(x, y')$ for all $x \in V$. We have the map S^*: $y \to y'$ such that for all $x, y \in V$

$$h(Sx, y) = \sigma h(x, S^*y). \tag{42}$$

S^* is called the *adjoint* of S relative to h. It follows directly from the defining property (42) that S^* is σ^*-semilinear where $\sigma^* = j\sigma^{-1}j$. If S' is a second σ-semilinear transformation then so is $S + S'$ and $(S + S')^* = S^* + (S')^*$. If T is a τ-semilinear transformation then ST is a $\sigma\tau$-semilinear transformation and $(ST)^* = T^*S^*$. If a_L denotes the scalar multiplication $x \to ax$ then a_L is semilinear with associated automorphism I_a in Δ and (42) implies that $a_L{}^* = \bar{a}_L$. These results imply that the map J: $A \to A^*$ is an involution in $\mathscr{A} = \operatorname{End}_\Delta V$. The pair $(\mathscr{A}, J)$ is the algebra with involution we have to consider in case

II. As we have indicated this is perfect if $n \geqslant 3$ and also if $n = 2$ provided that $\mathscr{H}(\Delta, j)$ generates Δ.

The automorphisms of $\mathscr{A} = \mathrm{End}_\Delta V$ have the form $A \to SAS^{-1}$ where S is a semilinear automorphism of V. S is determined to within a scalar multiplication $a_L : x \to ax$, $a \in \Delta$. For ζ: $A \to SAS^{-1}$ the map $(\zeta, J) = \zeta J \zeta^{-1} J$ is $A \to SS^*A(SS^*)^{-1}$. If this has the form $A \to HAH^{-1}$ where $H \in \mathscr{H}(\mathscr{A}, J)$ then $SS^* = a_L H$ ($= Ha_L$ since H is linear) for some a in Δ. Since $(a_L H)^* = Ha_L{}^* = H\bar{a}_L = \bar{a}_L H$ and $(a_L H)^* = (SS^*)^* = SS^* = a_L H$ we have $\bar{a} = a$. Also if S is σ-semilinear then SS^* is $\sigma\sigma^* = \sigma j \sigma^{-1} j = (\sigma, j)$-semilinear. Hence we have the condition $(\sigma, j) = I_a$ where $a \in \mathscr{H}(\Delta, j)$. These remarks, the condition for perfection and Theorem 2 yield the following

THEOREM 3. *Let (Δ, j) be a division algebra with involution, $h(x, y)$ a non-degenerate diagonalizable hermitian form (associated with j) on an n-dimensional vector space V over Δ and let J be the adjoint map $A \to A^*$ of $\mathscr{A} = \mathrm{End}_\Delta V$ defined by h. Assume either $n \geqslant 3$ or $n = 2$ and $\mathscr{H}(\Delta, j)$ generates Δ. Then* Str $\mathscr{H}$ *for $\mathscr{H} = \mathscr{H}(\mathscr{A}, J)$ is the set of maps*

$$A \to a_L T^* A T \tag{43}$$

where $a \in \mathscr{H}(\Delta, j)$ and T is a σ^{-1}-semilinear automorphism of V such that $(\sigma, j) = I_a$.

If $\Delta = \Phi$ a field and $j = 1$ then $\mathscr{H}(\Phi, j) = \Phi$ so that supplementary condition for $n = 2$ is satisfied. Here a can be taken to be any non-zero element of Φ and T any semilinear automorphism of V.

The case III can be handled in the same way. The result one obtains in this case is

THEOREM 4. *Let Φ be a field and let $f(x, y)$ be a nondegenerate alternate form on a $2n$-dimensional vector space V over Φ. Let J: $A \to A^*$ be the adjoint map in $\mathscr{A} = \mathrm{End}_\Phi V$ relative to f. Assume $n \geqslant 2$ and $n \geqslant 3$ if* char $\Phi \neq 2$. *Then* Str $\mathscr{H}$ *for $\mathscr{H} = \mathscr{H}(\mathscr{A}, J)$ is the set of maps*

$$A \to \gamma T^* A T \tag{44}$$

where γ is any non-zero element of Φ and T is any semilinear automorphism of V.

7. Finite Dimensional Central Simple Algebras with Involution

We define the *center of an algebra* with *involution* $(\mathscr{A}, J)$ as $\mathscr{H}(\mathscr{A}, J) \cap \mathscr{C}$, $\mathscr{C}$ the center of $\mathscr{A}$. $(\mathscr{A}, J)$ is called *central* if its center is $k1$, k the base ring. If $(\mathscr{A}, J)$ is simple its center is a field Φ and if $(\mathscr{A}, J)$ is central simple it is natural to replace k by Φ and regard $\mathscr{A}$ as algebra over Φ. We shall now do this and we assume, moreover, that $\mathscr{A}$ is finite dimensional over Φ. Then $(\mathscr{A}, J)$ is simple Artinian so the results of Section 6 apply. However, we can obtain more complete results by applying a field extension argument, as we shall show.

It is well known that if $(\mathscr{A}, J)$ is central simple over the field Φ and P is an extension field of Φ then $(\mathscr{A}^P, J^P)$ is a central simple algebra with involution. In particular, this holds if $P = \bar{\Phi}$ the algebraic closure of Φ. In this case if $\mathscr{A}$ is finite dimensional then $(\bar{\mathscr{A}}, \bar{J})$ for $\bar{\mathscr{A}} = \mathscr{A}^{\bar{\Phi}}$, $\bar{J} = J^{\bar{\Phi}}$ is isomorphic to one of the following algebras with involution: (1) $M_n(\bar{\Phi})$, $\bar{J}$: $a \to {}^t a$ (the transpose involution), (2) $M_{2n}(\bar{\Phi})$, $\bar{J}$: $a \to s({}^t a)s^{-1}$ where $s = \operatorname{diag}\{q, q, \ldots, q\}$, $q = \left(\begin{smallmatrix} 0 & 1 \\ -1 & 0 \end{smallmatrix}\right)$, (3) $M_n(\bar{\Phi}) \oplus M_n(\bar{\Phi})$ with $\bar{J}$: $(a, {}^t b) \to ({}^t a, b)$. In cases (1) and (2), J is of *first kind* in the sense that the center of $(\mathscr{A}, J)$ is the center of $\mathscr{A}$. In (3) J is of the *second kind* in that the center of $(\mathscr{A}, J)$ is a proper subset of that of $\mathscr{A}$. In fact, the latter is two dimensional over Φ. No two of the algebras $(\bar{\mathscr{A}}, \bar{J})$ we have listed are isomorphic. Accordingly, we say that $(\mathscr{A}, J)$ is of first kind and *orthogonal type* if $(\bar{\mathscr{A}}, \bar{J})$ is as in (1) and $(\mathscr{A}, J)$ is of first kind and *symplectic type* if $(\bar{\mathscr{A}}, \bar{J})$ is as in (2). In all cases n is called the *degree* of $(\mathscr{A}, J)$. It is clear that the dimensionality, $[\mathscr{H}(\bar{\mathscr{A}}, \bar{J}): \bar{\Phi}] = n(n+1)/2$ in case (1) and $= n^2$ in case (3). In case (2) the condition that $a \in \mathscr{H}(\bar{\mathscr{A}}, \bar{J})$ is that $s({}^t a)s^{-1} = a$ which is equivalent to ${}^t(as) = -(as)$. Hence in this case $[\mathscr{H}(\bar{\mathscr{A}}, \bar{J}): \bar{\Phi}] = n(n-1)/2$ or $n(n+1)/2$ according as the characteristic char $\Phi \neq 2$ or $= 2$. It is clear that $\mathscr{H}(\mathscr{A}, J)^{\bar{\Phi}}$ can be identified with $\mathscr{H}(\bar{\mathscr{A}}, \bar{J})$. Hence $[\mathscr{H}(\mathscr{A}, J): \Phi] = [\mathscr{H}(\bar{\mathscr{A}}, \bar{J}): \bar{\Phi}]$ and so we have the following table for $[\mathscr{H}(\mathscr{A}, J): \Phi]$:

$n(n+1)/2$ if J is of orthogonal type

$n(n-1)/2$ if J is of symplectic type and char $\Phi \neq 2$

$n(n+1)/2$ if J is of symplectic type and char $\Phi = 2$

n^2 if J is of second kind.

It is interesting to determine also the outer ideal of $\mathscr{H}(\mathscr{A}, J)$ generated by 1. We denote this as $\mathscr{H}(\mathscr{A}, J)'$. It is easily seen that $\mathscr{H}(\mathscr{A}, J)'^{\bar{\Phi}}$ can be identified with $\mathscr{H}(\bar{\mathscr{A}}, \bar{J})'$. It follows by applying the remark on outer ideals at the end of Section 5 that in all cases except that in which char $\bar{\Phi} = 2$ and the involution is symplectic, $\mathscr{H}(\bar{\mathscr{A}}, \bar{J})' = \mathscr{H}(\bar{\mathscr{A}}, \bar{J})$. On the other hand, if the involution is symplectic and char $\bar{\Phi} = 2$ then it is easily seen that $\mathscr{H}(\bar{\mathscr{A}}, \bar{J})'$ is the set of matrices a such that as is alternate, that is, ${}^t(as) = as$ and the diagonal elements of as are 0. Then $[\mathscr{H}(\bar{\mathscr{A}}, \bar{J})' : \bar{\Phi}] = n(n-1)/2$. It follows that $\mathscr{H}(\mathscr{A}, J)' = \mathscr{H}(\mathscr{A}, J)$ in all cases except that in which char $\Phi = 2$ and J is of symplectic type. Then $[\mathscr{H}(\mathscr{A}, J)' : \Phi] = n(n-1)/2$ while $[\mathscr{H}(\mathscr{A}, J) : \Phi] = n(n+1)/2$.

To determine the structure groups of the quadratic Jordan algebras $\mathscr{H}(\mathscr{A}, J)$ and $\mathscr{H}(\mathscr{A}, J)'$ we are considering we need to see what happens to a special universal envelope under extension of the base ring of a quadratic Jordan algebra. For the present we need this only for algebras over fields and extensions that are fields. In Section 9 we shall need the results also for certain ring extensions. Accordingly, we now consider the general case: $\mathscr{J} = (\mathscr{J}, U, 1)$ is a quadratic Jordan algebra over the commutative ring k and K is an extension of k, that is, K is a commutative associative algebra over k. If η is a homomorphism of a k-module M then η^K is the K-homomorphism of M^K such that $\eta^K(1 \otimes x) = 1 \otimes \eta(x)$. We now prove

PROPOSITION 2. *Let $(S(\mathscr{J}), \sigma_u)$ be a special universal envelope for $\mathscr{J}$ and let K be an extension of k. Then $(S(\mathscr{J})^K, \sigma_u{}^K)$ is a special universal envelope for $\mathscr{J}^K$.*

Proof. Let σ' be an associative specialization of $\mathscr{J}^K$ into the associative algebra $\mathscr{A}'$ over K. We can regard $\mathscr{A}'$ as an associative algebra over k by defining $\alpha a' = (\alpha 1)a'$ for $\alpha \in k$. If ν is the map $x \to 1 \otimes x$ of $\mathscr{J}$ into $\mathscr{J}^K$, $\sigma = \sigma'\nu$ is an associative specialization of $\mathscr{J}$ into $\mathscr{A}'$ as k-algebra. Hence we have a k-algebra homomorphism ζ of $S(\mathscr{J})$ into $\mathscr{A}'$ such that $\zeta\sigma_u = \sigma$. This defines a k-algebra homomorphism ζ' of $S(\mathscr{J})^K$ into $\mathscr{A}'$ such that $\zeta'(\alpha' \otimes a) = \alpha'\zeta(a)$, $\alpha' \in K$, $a \in S(\mathscr{J})$. Now $\sigma_u{}^K$ is an associative specialization of $\mathscr{J}^K$ into $S(\mathscr{J})^K$ and we have $\zeta'\sigma_u{}^K(1 \otimes x) = \zeta'(1 \otimes \sigma_u(x)) = \zeta\sigma_u(x) = \sigma(x) = \sigma'(1 \otimes x)$. Hence $\sigma' = \zeta'\sigma_u{}^K$. We know that $\sigma_u(\mathscr{J})$ generates $S(\mathscr{J})$ and this implies that $\sigma_u{}^K(\mathscr{J}^K)$ generates $S(\mathscr{J})^K$. It follows that ζ' is unique so $(S(\mathscr{J})^K, \sigma_u{}^K)$ satisfies the definition of a special universal envelope for $\mathscr{J}^K$.

If $(\mathscr{A}, J)$ is an associative algebra with involution then $(\mathscr{A}^K, J^K)$ is an associative algebra with involution over K. If $h \in \mathscr{H}(\mathscr{A}, J)$ then $1 \otimes h \in \mathscr{H}(\mathscr{A}^K, J^K)$. In particular, the elements $1 \otimes \sigma_u{}^K(x)$, $x \in J$, are contained in $\mathscr{H}(S(\mathscr{J})^K, \pi^K)$, π the main involution of $S(\mathscr{J})$. Then

$\pi(J^K) \subset \mathscr{H}(S(\mathscr{J})^K, \pi^K)$ from which it follows that π^K is the main involution in $S(\mathscr{J})^K$.

The interesting case for the applications is that in which K is k-free so K has a base $\{\rho_\lambda\}$ over k. Then if M is a k-module, the elements of M^K can be written as finite sums $\sum \rho_\lambda \otimes x_\lambda$, $x_\lambda \in M$, and $\sum \rho_\lambda \otimes x_\lambda = 0$ if and only if every $x_\lambda = 0$. If N is a submodule of M then the set of elements $\sum \rho_\lambda \otimes y_\lambda$, $y_\lambda \in N$, is a submodule of M^K which can be identified with N^K. The map $N \to N^K$ thus defined is injective. We use these remarks to prove

PROPOSITION 3. *Let K be k-free. Then $\mathscr{J}$ is special (reflexive) if and only if $\mathscr{J}^K$ is special (reflexive). Also if $(\mathscr{A}, J)$ is an associative algebra with involution, then $(\mathscr{A}, J)$ is perfect if and only if $(\mathscr{A}^K, J^K)$ is perfect.*

Proof. The kernel of $\sigma_u{}^K$ is $(\ker \sigma_u)^K$. Hence $\sigma_u{}^K$ is injective if and only if σ_u is injective. Accordingly, $\mathscr{J}^K$ is special if and only if J is special. If $(\mathscr{A}, J)$ is an associative algebra with involution, the involution J^K in $\mathscr{A}^K$ is $\sum \rho_\lambda \otimes a_\lambda \to \sum \rho_\lambda \otimes a_\lambda{}^*$ (J: $a \to a^*$). Hence $\mathscr{H}(\mathscr{A}^K, J^K)$ is the set of elements $\sum \rho_\lambda \otimes h_\lambda$, $h_\lambda \in \mathscr{H}(\mathscr{A}, J)$, and this set can be identified with $\mathscr{H}(\mathscr{A}, J)^K$. Since the main involution in $S(\mathscr{J}^K) = S(\mathscr{J})^K$ is π^K it follows that $\sigma_u(\mathscr{J}) = \mathscr{H}(\mathscr{A}, J)$ if and only if $\sigma_u(\mathscr{J}^K) = \mathscr{H}(\mathscr{A}^K, J^K)$. Thus $\mathscr{J}$ is relexive if and only if $\mathscr{J}^K$ is reflexive. Again let $(\mathscr{A}, J)$ be an associative algebra with involution and let ζ be the homomorphism of $S(\mathscr{H}(\mathscr{A}, J))$ into $\mathscr{A}$ such that $\zeta(\sigma_u(x)) = x$, $x \in \mathscr{H}(\mathscr{A}, J)$. Then $(\mathscr{A}, J)$ is perfect if and only if ζ is an isomorphism. The conditions for this are: (1) $\ker \zeta = 0$, (2) $\mathscr{H}(\mathscr{A}, J)$ generates $\mathscr{A}$. We have seen that the special universal envelope of $\mathscr{H}(\mathscr{A}, J)^K = \mathscr{H}(\mathscr{A}^K, J^K)$ is $(S(\mathscr{H}(\mathscr{A}, J))^K, \sigma_u{}^K)$. Moreover, ζ^K is the homomorphism of $S(\mathscr{H})^K$ into $\mathscr{A}^K$ such that $\zeta^K(\sigma_u{}^K(1 \otimes x)) = \zeta^K(1 \otimes \sigma_u(x)) = 1 \otimes \zeta\sigma_u(x) = 1 \otimes x$, $x \in \mathscr{H}$. Hence $\zeta^K\sigma_u{}^K(x') = x'$ for $x' \in \mathscr{H}(\mathscr{A}^K, J^K)$. Since $\ker \zeta^K = (\ker \zeta)^K$, ζ is injective if and only if ζ^K is injective. Also if $\mathscr{B}$ is the subalgebra of $\mathscr{A}$ generated by $\mathscr{H}$ then the subalgebra of $\mathscr{A}^K$ generated by $\mathscr{H}^K = \mathscr{H}(\mathscr{A}^K)$ is $\mathscr{B}^K$. Hence $\mathscr{H}(\mathscr{A}^K)$ generates $\mathscr{A}^K$ if and only if $\mathscr{H}(\mathscr{A})$ generates $\mathscr{A}$. Thus $(\mathscr{A}, J)$ is perfect if and only if $(\mathscr{A}^K, J^K)$ is perfect.

We shall now apply this result to the finite dimensional central simple algebra with involution $(\mathscr{A}, J)$. Consider the algebras $(\bar{\mathscr{A}}, \bar{J})$ obtained by extending the base field Φ to its algebraic closure $\bar{\Phi}$. Applying Martindale's theorem and the earlier result of Jacobson and Rickart on the algebras $M_2(\bar{\Phi})^+$ we see that if the degree is > 1 then $(\bar{\mathscr{A}}, \bar{J})$ is perfect unless the degree is two, and the involution is of symplectic type. Hence, by Proposition 3, we have

THEOREM 5. *Let $(\mathscr{A}, J)$ be a finite dimensional central simple associative algebra with involution and assume the degree > 1. Then $(\mathscr{A}, J)$ is perfect in all cases except that in which the degree is two and the involution is of symplectic type.*

We shall now apply this result and Theorem 2 and its corollary to obtain the structure groups of the corresponding quadratic Jordan algebras $\mathscr{H}(\mathscr{A}, J)$. We consider first the case of central simple algebras of first kind. Here $\mathscr{A}$ is finite dimensional central simple over Φ, and $\mathscr{J}$ is an involution in $\mathscr{A}$ over Φ. The automorphisms of $\mathscr{A}$ are all inner. The corollary to Theorem 2 now yields the following result.

THEOREM 6. *Let $\mathscr{A}$ be a finite dimensional central simple associative algebra over a field Φ such that $\mathscr{A}$ over Φ has an involution J. Assume that the degree of $(\mathscr{A}, J)$ is > 1 and in the degree two case that $(\mathscr{A}, J)$ is of orthogonal type. Then* Str $\mathscr{H}$, $\mathscr{H} = H(A, J)$ *is the set of maps*

$$x \to \gamma g^* x g$$

where g is an invertible element of $\mathscr{A}$ and $\gamma \neq 0$ is in Φ.

In treating the case in which J is of second kind we shall distinguish the two subcases in which $\mathscr{A}$ is not simple and in which $\mathscr{A}$ is simple. In the first we have $\mathscr{A} = \mathscr{B} \oplus \mathscr{B}^0$ with the involution $(a, b) \to (b, a)$. The automorphisms of $\mathscr{A}$ are all inner unless $\mathscr{B}$ has an anti-automorphism μ. In the latter case the automorphisms are either inner or products of inner automorphisms by the automorphism $(a, b) \to (\mu(b), \mu(a))$. The quadratic Jordan algebra $\mathscr{H}(\mathscr{A}, J)$ is isomorphic to $\mathscr{B}^+$. We prefer to state the results in terms of the algebra $\mathscr{B}^+$ and changing the notation from $\mathscr{B}$ to $\mathscr{A}$, our results yield the following

THEOREM 7. *Let $\mathscr{A}$ be a finite dimensional central simple associative algebra, $\mathscr{A}^+$ the corresponding quadratic Jordan algebra. Then* Str $\mathscr{A}^+$ *consists of the maps*

$$x \to gxh$$

where g, h are invertible elements of $\mathscr{A}$, unless $\mathscr{A}$ has an anti-automorphism μ. In the latter case Str $\mathscr{A}^+$ *consists of the maps $x \to gxh$, and the maps $x \to g\mu(x)h$ where g and h are invertible elements of $\mathscr{A}$.*

Now assume $(\mathscr{A}, J)$ is finite dimensional central simple of second kind and $\mathscr{A}$ is simple. The center of $\mathscr{A}$ is a quadratic extension field P of the base field Φ and the restriction of J to P is an automorphism

$\neq 1$. It may happen that every automorphism of $\mathscr{A}$ is the identity on P, in which case, the automorphisms of $\mathscr{A}$ are all inner. On the other hand, suppose $\mathscr{A}$ has an outer automorphism μ. We claim that μ satisfies the condition of Theorem 2, namely $\mu J \mu^{-1} J$ is an inner automorphism by an invertible element contained in $\mathscr{H}(\mathscr{A}, J)$. Since μ and J induce the same automorphism in P it is clear that $\mu J \mu^{-1} J$ is P-linear and hence is an inner automorphism I_d. Thus we have for $a \in \mathscr{A}$ that $(\mu J \mu^{-1} J)a = dad^{-1}$. Then $(J\mu J \mu^{-1})a = (\mu J \mu^{-1} J)^{-1} a = d^{-1}ad$. On the other hand, $(\mu J \mu^{-1} J)(a^*) = d(a^*)d^{-1}$, and operating with J on both sides of this relation, gives $(J\mu J\mu^{-1})a = (d^*)^{-1} a(d^*)$. It follows that there exists a $\delta \in P$ such that $d^* = \delta d$. Then $d = \delta^* d^* = \delta\delta^* d$ so $\delta^*\delta = 1$. Hence by Hilbert's Satz 90 there exists a $\gamma \neq 0$ in P such that $\delta = \gamma(\gamma^*)^{-1}$ and $d^* = \delta d = \gamma(\gamma^*)^{-1} d$. Hence $(\gamma d)^* = \gamma d$. Thus, since we can replace d by γd, we may assume that $d \in \mathscr{H}$ and so μ satisfies the condition of Theorem 2. This result and the Corollary to Theorem 2 imply

THEOREM 8. *Let $\mathscr{A}$ be a finite dimensional simple associative algebra with center P a quadratic extension of the base field Φ and an involution of second kind. Then* Str $\mathscr{H}(\mathscr{A}, J)$ *is the set of maps*

$$x \to \gamma g^* x g$$

where g is an invertible element of $\mathscr{A}$ and γ is a nonzero element of P if all automorphisms of $\mathscr{A}$ over Φ are inner. On the other hand, if $\mathscr{A}$ has outer automorphisms and μ is a fixed one of these then Str $\mathscr{H}(\mathscr{A}, J)$ *consists of the foregoing maps and the products of these by the map $x \to d^{-1}\mu(x)$ where d is chosen in $\mathscr{H}$ such that $\mu J \mu^{-1} J = I_d$.*

We consider also the quadratic Jordan algebra $\mathscr{H}(\mathscr{A}, J)'$ over Φ of characteristic two, J symplectic. We need to show that if $a \in \mathscr{H}(\mathscr{A}, J)'$ and $g \in \mathscr{A}$ then $g^*ag \in \mathscr{H}(\mathscr{A}, J)'$. It suffices to prove this for the algebra obtained by extending the base field to its algebraic closure. Then if s is as before, $g^* = s({}^tg)s^{-1}$ and the condition that $a \in \mathscr{H}(\bar{\mathscr{A}}, \bar{J})'$ is that as is alternate. Then

$$g^*ag = s({}^tg)\, s^{-1}ag = s({}^tg)\, s^{-1}(as)\, s^{-1}g$$

and $(g^*ag)s = {}^tu(as)u$ for $u = s^{-1}gs$. As is well known, as alternate implies ${}^tu(as)u$ alternate. Hence $g^*ag \in \mathscr{H}(\bar{\mathscr{A}}, \bar{J})'$. We can now apply Theorem 2 and the remark at the end of Section 5 on Martindale's theorem to show that Theorem 6 is valid also for $\mathscr{H}(\mathscr{A}, J)'$, char $\Phi = 2$, J symplectic.

8. Generic Norm in $\mathscr{H}(\mathscr{A}, J)$ and $\mathscr{H}(\mathscr{A}, J)'$, $(\mathscr{A}, J)$ Finite Dimensional Central Simple. Norm Similarities

The theory of generic minimum polynomials, traces and norms for quadratic Jordan algebras has been developed in [15] and [25]. This is based on the definition of the power of an element in a quadratic Jordan algebra $\mathscr{J}$ by the rules that $a^0 = 1$, $a^1 = a$ and $a^n = U_a a^{n-2}$ if $n > 1$. If $\mathscr{J}$ is a subalgebra of $\mathscr{A}^+$, $\mathscr{A}$ associative, then a^n in $\mathscr{J}$ is the same thing as a^n in the associative algebra $\mathscr{A}$. The theory of generic minimum polynomials for a finite dimensional $\mathscr{J}$ is analogous to that of generic minimum polynomials for power associative algebras, which in turn extends classical results on associative algebras. One can largely bypass the general theory in considering the algebras $\mathscr{H}(\mathscr{A}, J)$ and $\mathscr{H}(\mathscr{A}, J)'$ for $(\mathscr{A}, J)$ finite dimensional central simple by a direct determination of the generic minimum polynomials for the various cases of $(\mathscr{A}, J)$. We shall follow this approach here and for the sake of simplicity we assume the base field is infinite. This will enable us to use the Zariski topology for finite dimensional vector spaces and elementary results on polynomial functions on vector spaces and more generally rational mappings between vector spaces.

We shall require the following result whose proof is staightforward and is given in [14, p. 44].

Lemma. *Let V be a finite dimensional vector space over an infinite field Φ, P an extension field of Φ, $V' = V^P$. Identify V with the subset of V' of elements $1 \otimes v$, $v \in V$. Then:* (i) *if O' is a non-vacuous open subset of V' (in the Zariski topology) then $O = O' \cap V$ is a nonvacuous open subset of V,* (ii) *if f is a polynomial function on V' such that $f(v) \in \Phi$ for all v in an open subset of V then $f \mid V$ is a polynomial function on V.*

Again let $(\mathscr{A}, J)$ be a finite dimensional central simple associative algebra with involution and let n be the degree of $(\mathscr{A}, J)$ as defined in Section 7. We now define the degree of $\mathscr{H}(\mathscr{A}, J)$ and $\mathscr{H}(\mathscr{A}, J)'$ to be n in all cases except $\mathscr{H}(\mathscr{A}, J)$ of characteristic two with J symplectic, in which case, we define the *degree* to be $2n$. We have the following

Theorem 9. *Let $(\mathscr{A}, J)$ be a finite dimensional central simple associative algebra with involution over an infinite field Φ and let m denote the degree of $\mathscr{H}(\mathscr{A}, J)$ (or $\mathscr{H}(\mathscr{A}, J)'$). Then:*

1. *There exist uniquely determined polynomial functions $T, \ldots, N$ on $\mathscr{H}(\mathscr{A}, J)$ $(\mathscr{H}(\mathscr{A}, J)')$ such that*

$$a^m - T(a)\, a^{m-1} + \cdots + (-1)^m N(a)\, 1 = 0 \tag{45}$$

for all $a \in \mathscr{H} = \mathscr{H}(\mathscr{A}, J)$ (or $\mathscr{H}'$). The polynomial

$$\chi_a(\lambda) = \lambda^m - T(a)\, \lambda^{m-1} + \cdots + (-1)^m N(a) \tag{46}$$

$\in \Phi[\lambda]$, *λ an indeterminate, is called the* generic minimum polynomial *of a, $T(a)$ and $N(a)$ are respectively the* generic trace *and* generic norm *of a. We have*

$$\chi_a(\lambda) = N(\lambda 1 - a) \tag{47}$$

where $\lambda 1 - a \in \mathscr{H}^{\Phi(\lambda)}$ $(\mathscr{H}'^{\Phi(\lambda)})$.

2. *The generic trace function T is linear on $\mathscr{H}(\mathscr{H}')$ and the generic norm function N is homogeneous of degree m. Moreover, $T(1) = m$, $N(1) = 1$ and if $a, c \in \mathscr{H}(\mathscr{H}')$ and $b \in \mathscr{A}$ then*

$$N(b^*ab) = N(b^*b)\, N(a) \tag{48}$$

$$N(U_c a) = N(c)^2\, N(a). \tag{49}$$

3. *An element $a \in \mathscr{H}(\mathscr{H}')$ is invertible in $\mathscr{H}(\mathscr{H}')$ if and only if $N(a) \neq 0$.*

Proof. We consider first the case (3) (of Section 7): $\bar{\mathscr{A}} = M_n(\bar{\Phi}) \oplus M_n(\bar{\Phi})$, $\bar{J}$: $(a, {}^t b) \to ({}^t b, a)$. Here $\mathscr{H}(\bar{\mathscr{A}}, \bar{J})$ is the set of elements $a' = (a, {}^t a)$. Let $\chi_a'(\lambda)$ be the characteristic polynomial of the matrix a and write this as $\chi_a'(\lambda) = \lambda^n - T(a')\lambda^{n-1} + \cdots + (-1)^n N(a')$. Then $T, \ldots, N$ are polynomial functions on $\bar{\mathscr{H}} = \mathscr{H}(\bar{\mathscr{A}}, \bar{J})$ and $a'^n - T(a')a'^{n-1} + \cdots + (-1)^n N(a')1 = 0$ follows from the Hamilton–Cayley theorem. It is clear that there exist a' such that $\chi_a'(\lambda)$ is the minimum polynomial of a' (as element of $\bar{\mathscr{A}}$). The condition that a' have this property is that $1, a', \ldots, a'^{m-1}$ are linearly independent. It is easily seen that this is an open subset of $\bar{\mathscr{H}}$. Hence it is a nonvacuous open subset O' of $\bar{\mathscr{H}}$ and by the lemma, $O = O' \cap \mathscr{H}$ is a nonvacuous open subset of $\mathscr{H}$. Since the minimum polynomial is unchanged on extension of the base field it is clear that if $a \in O$, $\chi_a(\lambda)$ is the minimum polynomial of a. Hence $\chi_a[\lambda] \in \Phi[\lambda]$ and so, by the lemma, the restrictions of $T, \ldots, N$ to H are polynomial functions. We have (45) and the other assertions of the theorem follow from well known properties of characteristic polynomials, traces and norms of matrices.

Next assume we have case (1): $\bar{\mathscr{A}} = M_n(\bar{\Phi})$ with the transpose involution. For $a \in \bar{\mathscr{H}}$ let $\chi_a(\lambda)$ be the characteristic polynomial of a. Since there exist symmetric matrices whose minimum polynomials have degree n the argument in the previous case carries over verbatim.

Now assume we have case (2): $\bar{\mathscr{A}} = M_{2n}(\bar{\Phi})$, $\bar{J}$: $a \to s({}^t a)s^{-1}$, $s = \operatorname{diag}\{q, \ldots, q\}$, $q = \begin{pmatrix} 0 & 1 \\ -1 & 0 \end{pmatrix}$. If we partition the $2n \times 2n$ matrices into n blocks of 2×2 matrices we can identify $M_{2n}(\bar{\Phi})$ with $M_n(\mathscr{D})$ where $\mathscr{D} = M_2(\bar{\Phi})$. The involution $\bar{J}$ is $(a_{ij}) \to {}^t(a_{ij})$ where $\bar{a}_{ij} \in M_2(\bar{\Phi})$ and $\bar{a}_{ij} = q({}^t a_{ij})q^{-1}$. Then $\bar{a}_{ij} = (\operatorname{tr} a_{ij})1 - a_{ij}$. The condition that $a \in \bar{\mathscr{H}}$ is that

$$a = \begin{pmatrix} a_1 & a_{12} & \cdots & a_{1n} \\ \bar{a}_{12} & a_2 & \cdots & a_{2n} \\ \cdot & \cdot & \cdots & \cdot \\ \bar{a}_{1n} & \bar{a}_{2n} & \cdots & a_n \end{pmatrix} \tag{50}$$

where $\bar{a}_i = a_i$. If char $\bar{\Phi} \neq 2$, the condition $\bar{a}_i = a_i$ is satisfied if and only if $a_i = \alpha_i 1$, $\alpha_i \in \Phi$. On the other hand, if char $\bar{\Phi} = 2$ then $\bar{a}_i = a_i$ if and only if the trace tr $a_i = 0$. Also in this case (which is the only one in which $\mathscr{H}(\mathscr{A}, J) \neq \mathscr{H}(\mathscr{A}, J)'$), $\bar{\mathscr{H}}'$ is the set of matrices (50) in which $a_i = \alpha_i 1$, $\alpha_i \in \bar{\Phi}$. We now consider simultaneously the two cases: $\bar{\mathscr{H}}$, char $\Phi \neq 2$ and $\bar{\mathscr{H}}'$, char $\Phi = 2$. In both cases we define $\chi_a(\lambda) = Pf(s\lambda - as)$ the Pfaffian of the alternate matrix $s\lambda - as$. Then the argument in [11, p. 231] shows that $\chi_a(a) = 0$. Moreover, if we take the a_{ij} in (50) to be 0 and $a_i = \alpha_i 1$ with distinct α_i it is easy to verify that $\chi_a(\lambda)$ is the minimum polynomial of a. The results then follow as in the first case from properties of the Pfaffian. Now let char $\bar{\Phi} = 2$ and consider $\bar{\mathscr{H}}$. In this case we take $\chi_a(\lambda)$ to be the characteristic polynomial of the matrix a as in (50) where the condition on the a_i is tr $a_i = 0$. If we take the $a_{ij} = 0$ for $i \neq j$ then $\chi_a(\lambda) = \prod_1^n (\lambda^2 + \det a_i)$. By a suitable choice of the det a_i we shall have that $\chi_a(\lambda)$ is the minimum polynomial of a. Then the results follow as in the first case.

We shall call an element $a \in \mathscr{H}(\mathscr{H}')$ *regular* if $\chi_a(\lambda)$ has distinct roots in $\bar{\Phi}$. It is easy to see that the regular elements of $\mathscr{H}$ and $\mathscr{H}'$ form a non-vacuous open subset of $\mathscr{H}$ or $\mathscr{H}'$ in all case except $\mathscr{H}$, char $\Phi = 2$, J symplectic. In this case, if $a \in \mathscr{H}$ then $a^2 \in \mathscr{H}'$ and hence a^2 is a root of $Pf(s\lambda - a^2 s)$. Thus a is a root of $Pf(s\lambda^2 - a^2 s)$. It follows that

$$\chi_a(\lambda) = Pf(s\lambda^2 - a^2 s) \tag{51}$$

and so $\chi_a(\lambda)' = 0$. Hence no regular elements exist in this case.

We shall now need to use directional derivatives of rational maps and logarithmic derivatives of rational functions for finite dimensional vector spaces over infinite fields. We refer the reader to [11] and [30] for the definitions and the properties we shall need. If F is a rational map we write $\Delta_b{}^a F$ for the directional derivation of F at b in the direction a.

We now define the *generic trace bilinear* form on $\mathscr{H}(\mathscr{A}, J)$ $(\mathscr{H}(\mathscr{A}, J)')$ by

$$T(a, b) = -\Delta_1{}^a \Delta^b \log N. \tag{52}$$

This is a symmetric bilinear form and we have

PROPOSITION 4. *If* char $\Phi \neq 2$ *the generic trace form is nondegenerate in all cases. If* char $\Phi = 2$ *the generic trace form is nondegenerate for* $\mathscr{H}(\mathscr{A}, J)$, *$J$ of second kind and for* $\mathscr{H}(\mathscr{A}, J)'$, *$J$ symplectic (of first kind). In the remaining cases the generic trace form is degenerate. In all cases we have*

$$-\Delta_c{}^a \Delta^b \log N = T(U_c^{-1}a, b) \tag{53}$$

for all a, b and invertible c.

We shall give the proof for involutions of second kind and sketch it for the remaining cases. By extending the base field to its algebraic closure and using the isomorphism of $\bar{\mathscr{H}}$ with $M_n(\bar{\Phi})^+$ it suffices to prove the result for $M_n(\Phi)^+$, Φ an infinite field with $N(a) = \det a$, $\chi_a(\lambda) = \det(\lambda 1 - a)$. Then $T(a) = \operatorname{tr} a$. We calculate first $\Delta_c{}^a N$ for arbitrary a and invertible c. This is the coefficient of λ in $N(c + \lambda a) = N(c)\, N(1 + \lambda b)$, $b = c^{-1}a$. Now $N(\lambda 1 - b) = \lambda^n - T(b)\lambda^{n-1} + \cdots$ so $N(\lambda^{-1}1 - b) = \lambda^{-n} - T(b)\lambda^{-(n-1)} + \cdots$ and $N(1 - \lambda b) = \lambda^n N(\lambda^{-1}1 - b) = 1 - T(b)\lambda + \cdots$. Hence

$$\Delta_c{}^a N = N(c)\, T(c^{-1}a) \tag{54}$$

if $N(c) \neq 0$ and

$$\Delta_c{}^a \log N = T(c^{-1}a). \tag{55}$$

Next we use the formula

$$\Delta_c{}^b x^{-1} = -c^{-1}bc^{-1} \tag{56}$$

which is easy to prove. From (55) and (56) we can obtain

$$-\Delta_c{}^b \Delta^a \log N = T((U_c^{-1}b)a). \tag{57}$$

If we take $c = 1$ we obtain

$$T(a, b) = T(ab). \tag{58}$$

It is well known that $T(ab) = \operatorname{tr} ab$ is a nondegenerate symmetric bilinear form on $M_n(\Phi)$. The formula (53) is an immediate consequence of (57) and (58).

If the characteristic is $\neq 2$ the results in the other cases follow easily from these results for $M_n(\Phi)$. It suffices to consider the Jordan algebra of symmetric matrices in $M_n(\Phi)$ and of symplectic symmetric matrices in $M_{2n}(\Phi)$. In the first case the generic norm of an element is the same as its generic norm in $M_n(\Phi)^+$. If a is symmetric and b is skew then $T(ab) = T(a \cdot b) = 0$ since $a \cdot b = \frac{1}{2}(ab + ba)$ is skew. Hence the space of symmetric matrices is orthogonal to the space of skew symmetric matrices and since the sum of these two spaces is $M_n(\Phi)$ it follows that $T(a, b) = T(ab)$ is nondegenerate in the Jordan algebra of symmetric matrices. Also (53) is clear since it holds in $M_n(\Phi)^+$. A similar argument applies for the subalgebra $\mathscr{H}$ of $M_{2n}(\Phi)$ of symplectic symmetric matrices. If we take $s = \operatorname{diag}\{q, \dots, q\}$, $q = \left(\begin{smallmatrix} 0 & 1 \\ -1 & 0 \end{smallmatrix}\right)$ as before then $N(a) = Pf(sa)$ and $N(a)^2 = \det(sa) = \det a$ which is the generic norm of a in $M_{2n}(\Phi)$. Then we have the relation $T_1(a, b) = \frac{1}{2}T_2(a, b)$ where T_1 and T_2 are, respectively, the generic trace form for $\mathscr{H}$ and for $M_{2n}(\Phi)^+$. The results follow from this relation as in the case of symmetric matrices.

Now let char $\Phi = 2$, J an involution of first kind. The generic norm of $a \in \mathscr{H}(\mathscr{A}, J)$ is the same as its generic norm as element of $\mathscr{A}^+$. Hence $T(a, b) = \operatorname{tr}(ab)$. Taking $\mathscr{A} = M_n(\Phi)$, J of orthogonal type we can verify that if a has diagonal elements all 0 then $T(a, b) = 0$ for all $b \in \mathscr{H}(\mathscr{A}, J)$. Also if $\mathscr{A} = M_{2n}(\Phi)$ and J is of symplectic type then any a of the form (50) in which the $a_{ij} = 0$ for $i \neq j$ and $a_i = \alpha_i 1$, $\alpha_i \in \Phi$ satisfies $T(a, b) = 0$ for all $b \in \mathscr{H}(\mathscr{A}, J)$. It follows that the generic trace bilinear form is degenerate on $\mathscr{H}(\mathscr{A}, J)$ in both cases. Also (57) follows from this equation for $M_n(\Phi)$. Finally we have to consider the case of $\mathscr{H}(\mathscr{A}, J)'$, char $\Phi = 2$, J of symplectic type. For this one case we shall draw on some of the known general results on trace bilinear forms. We note first that one has the formula $T(x, y) = T(x)\, T(y) - S(x, y)$ where $S(x, y)$ is a symmetric bilinear form obtained by linearizing a quadratic form. Hence in characteristic two, $S(x, x) = 0$ and $T(x, x) = T(x)^2$. Now in the case we are considering there exist a such that $T(a) \neq 0$. In fact, $\mathscr{H}(\mathscr{A}, J)'$ contains n non-zero orthogonal idempotents e_i such that $\sum e_i = 1$ and if we take $a = \sum \rho_i e_i$ with distinct ρ_i and $\sum \rho_i \neq 0$, then the minimum polynomial of a is $\prod (\lambda - \rho_i)$ and this is the same as $\chi_a(\lambda)$. Then $T(a) \neq 0$. We now see that $T(x, y) \neq 0$. To prove nondegeneracy we observe that the radical of $T(x, y)$, that is, the subspace of z such that $T(x, z) = 0$ for all x, is invariant under Str $\mathscr{H}'$ since we have $T(\eta x, y) = T(x, \eta^* y)$ for $\eta \in \text{Str}\, \mathscr{H}'$. It is easily seen using elementary results on alternate matrices that the set of $\eta: x \to g^* x g$, g invertible, acts irreducibly on H'. It follows that the radical of $T(x, y)$ is either the whole space or 0 and since $T(x, y) \neq 0$ it must be 0. Then $T(x, y)$ is nondegenerate. Finally we note that the relation (57) is readily deduced from the relation $\Delta_c^a \log N = T(c^{-1}, a)$ for arbitrary a and invertible c (see Eq. (17) of McCrimmon [25]).

We shall now consider the problem of determining the group of bijective linear transformations η of $\mathscr{H}(\mathscr{A}, J)$ or $\mathscr{H}(\mathscr{A}, J)$ having the generic norm as semi-invariant, that is, satisfying

$$N(\eta(a)) = \rho N(a) \tag{59}$$

for a nonzero ρ in Φ and all a in the quadratic Jordan algebra. An η satisfying this condition is called a *norm similarity*. The element ρ is called the *multiplier* of η. Since $N(1) = 1$ we have $\rho = N(u)$, $u = \eta(1)$. We have the following:

THEOREM 10. *Let $(\mathscr{A}, J)$ be a finite dimensional central simple associative algebra with involution over an infinite field Φ. Then the group of norm similarities of $\mathscr{H}(\mathscr{A}, J)$ (or $\mathscr{H}(\mathscr{A}, J)'$) coincides with the structure group if* char $\Phi \neq 2$. *The same result holds for* char $\Phi = 2$ *in the two cases: $\mathscr{H}(\mathscr{A}, J)$, J of second kind and $\mathscr{H}(\mathscr{A}, J)'$, J of symplectic type.*

Proof. The fact that the structure group is contained in the group of norm similarities is a general result which has been proved recently by McCrimmon [25]. For the algebras we are considering it follows also from Theorem 9, part 2 and the determination of the structure groups which we made in the last section. It remains to show that any norm similarity is contained in the structure group. For this we need the nondegeneracy of the trace bilinear from which was established in Proposition 4 for the cases covered by the statement of the theorem. Let η be a norm similarity and let $a, b, c \in \mathscr{H}(\mathscr{H}')$ where $N(c) \neq 0$. Then $N(\eta(c)) \neq 0$ and c and $\eta(c)$ are invertible. Using the chain rule for differentiation we obtain from (53) that

$$-\Delta_c^b \Delta^a \log N\eta = T(U_{\eta(c)}^{-1}(\eta(b)), \eta(a)).$$

Since $N\eta = \rho N$ this implies that

$$T(U_{\eta(c)}^{-1}(\eta(b)), \eta(a)) = T(U_c^{-1}(b), a). \tag{60}$$

Let η^* denote the adjoint of η relative to $T(x, y)$. Then (60) gives the relation $\eta^* U_{\eta(c)}^{-1}\eta = U_c^{-1}$ and taking inverses we obtain

$$U_{\eta(c)} = \eta U_c \eta^* \tag{61}$$

for all c such that $N(c) \neq 0$. Since the set of these c is open in the Zariski topology and (61) is a relation between polynomial maps, it follows that it holds for all c. Hence η is in the structure group. This completes the proof.

This result and our determination of the structure groups give explicit determinations of the groups of norm similarities. These constitute an extensive generalization of the results of Frobenius mentioned in the Introduction. The results we have obtained leave open the question of determination of the groups of norm similarities for the algebra $\mathscr{H}(\mathscr{A}, J)$, J of first kind, char $\Phi = 2$. As in the other cases these groups contain the structure groups.

9. Structure Algebras

We consider again an arbitrary quadratic Jordan algebra $\mathscr{J} = (\mathscr{J}, U, 1)$. We shall now define by the standard method of formal algebraic groups the Lie algebra str J of the structure group Str $\mathscr{J}$. For this purpose one introduces the algebra Λ of dual numbers over the base ring k. This has the base $(1, \delta)$ over k with multiplication defined by $\delta^2 = 0$. Consider the quadratic Jordan algebra $\mathscr{J}^\Lambda$ over Λ. The natural map $x \rightarrow 1 \otimes x$ of $\mathscr{J}$ into $\mathscr{J}^\Lambda$ is injective; hence one can identify $\mathscr{J}$ with its image in $\mathscr{J}^\Lambda$. Then the elements of $\mathscr{J}^\Lambda$ can be written in one and only one way in the form $x + \delta y$, $x, y \in \mathscr{J}$. Any k-endomorphism H of $\mathscr{J}$ determines a Λ-endomorphism $1 + \delta H$ of $\mathscr{J}^\Lambda$ such that $(1 + \delta H)x = x + \delta(Hx)$, $x \in \mathscr{J}$, and consequently $(1 + \delta H)(x + \delta y) = x + \delta(Hx + y)$. It is clear that $1 + \delta H$ is invertible with $(1 + \delta H^{-1} = 1 - \delta H$. Now define

$$\text{str } J = \{H \in \text{End}_k \mathscr{J} \mid 1 + \delta H \in \text{Str } \mathscr{J}^\Lambda\}. \tag{62}$$

Then it is easily shown that str $\mathscr{J}$ is a Lie algebra of endomorphisms of $\mathscr{J}$ and Str $\mathscr{J}$ acts on str $\mathscr{J}$ by the "adjoint" action

$$\eta H = \eta H \eta^{-1} \tag{63}$$

if $\eta \in \operatorname{Str} \mathscr{J}$ and $H \in \operatorname{str} \mathscr{J}$. This action is one by automorphisms of the Lie algebra str $\mathscr{J}$. The Lie algebra str $\mathscr{J}$ is called the *structure* (*Lie*) *algebra of* $\mathscr{J}$.

We proceed to derive an explicit condition that an endomorphism $H \in \operatorname{str} \mathscr{J}$. We note first that for any $H \in \operatorname{End}_k \mathscr{J}$, $(1 + \delta H)1 = 1 + \delta(H1)$ is invertible in $\mathscr{J}^{\Delta}$ with inverse $1 - \delta(H1)$. More generally, it is easily seen that $a + \delta b$, $a, b \in \mathscr{J}$ is invertible in $\mathscr{J}^{\Delta}$ if and only if a is invertible in $\mathscr{J}$. We have for any $x = a + \delta b$ in $\mathscr{J}^{\Delta}$

$$U_{(1+\delta H)x} = U_{a+\delta(Ha+b)} = U_a + \delta(U_{a,Ha} + U_{a,b})$$

and $U_{1+\delta(H1)} = 1 + \delta U_{1,H1} = 1 + \delta V_{H1}$ (since $U_{1,x} = V_x$ generally). Hence

$$\begin{aligned}(1 + \delta H)\, U_x(1 - \delta H)\, U_{1+\delta(H1)} &= (1 + \delta H)\, U_x(1 - \delta H)(1 + \delta V_{H1}) \\ &= U_x + \delta[H, U_x] + U_x V_{H1} = U_a + \delta(U_{a,b} + [H, U_a] + U_a V_{H1})\end{aligned}$$

Now the condition that $\eta = 1 + \delta H \in \operatorname{Str} \mathscr{J}^{\Delta}$ is that $U_{\eta(x)} = \eta U_x \eta^{-1} U_{\eta(1)}$ (cf. (17)) so the foregoing formula shows that $\eta = a + \delta H \in \operatorname{Str} \mathscr{J}^{\Delta}$ if and only if

$$U_{a,Ha} = [HU_a] + U_a V_{H1} \tag{64}$$

for all $a \in \mathscr{J}$. Hence this is the condition that $H \in \operatorname{str} \mathscr{J}$. We can write this also as

$$U_{a,Ha} = HU_a - U_a\bar{H} \tag{65}$$

where

$$\bar{H} = H - V_{H1}.$$

As in the case of Str $\mathscr{J}$ one sees that str $\mathscr{J}$ can also be characterized as the set of k-endomorphisms H of $\mathscr{J}$ for which there exists a k-endomorphism $\bar{H}$ such that (65) holds. Then necessarily $\bar{H}$ is given by (66).

One can define also a Lie algebra analogue of the inner structure group. We claim that for any $b, c \in \mathscr{J}$, $V_{b,c} \in \operatorname{str} \mathscr{J}$ and that for $H = V_{b,c}$

we have $\bar{H} = -V_{c,b}$ to satisfy (65). This amounts to the identity

$$U_{a,V_{b,c}a} = V_{b,c}U_a + U_aV_{c,b}\,. \tag{67}$$

Before giving a formal proof of this we sketch an argument which will perhaps indicate why (67) holds. Let c be invertible in $\mathcal{J}$. Then $c^{-1} + \delta b$ is invertible in $\mathcal{J}^{\Lambda}$ and $U_{c^{-1}+\delta b}U_c \in \mathrm{Instr}\,\mathcal{J}^{\Lambda}$. This map is $1 + \delta U_{c^{-1},b}U_c$. It is easy to see that this coincides with $1 + \delta V_{b,c}$. Hence $V_{b,c} \in \mathrm{str}\,\mathcal{J}$. Then (65) holds for $H = V_{b,c}$ with $\bar{H} = V_{b,c} - V_{b\circ c}$ and since it is easily seen that $V_{b,c} + V_{c,b} = V_{b\circ c}$, we have (67) for arbitrary b and invertible c. In may important situations (e.g. finite dimensional algebras over a field) one can argue that the validity of (67) for arbitary b and invertible c implies its validity for all b, c. However, instead of pursuing this line of argument further we shall now give a direct proof of (67) for all b, c which is due to McCrimmon. It is useful to write $\{abc\} \equiv U_{a,c}b = V_{a,b}c$ and $aba = U_ab$. We refer to the identities in Jacobson [2, pp. 1.18–1.23] noting that these need to be translated from the right hand notation for operators used in the reference to the left hand notation employed here. We have the identity

$$V_{U_ba,a} = V_{b,U_ab} \tag{68}$$

(QJ 31 of Jacobson [2], p. 1.22). Linearization of this with respect to a gives

$$V_{b,U_{a,d}b} = V_{U_ba,d} + V_{U_bd,a}\,. \tag{69}$$

We now employ the identity $U_{b,c}V_{a,b} + U_bV_{a,c} = V_{b,a}U_{b,c} + V_{c,a}U_b$ (QJ9 on p. 1.17) which can be written out as

$$\{b\{abd\}c\} + b\{acd\}b = \{ba\{bdc\}\} + \{ca(bdb)\}.$$

Taking c as operand this becomes

$$V_{b,U_{a,d}b} + U_bU_{a,d} = V_{b,a}V_{b,d} + V_{U_bd,a}\,. \tag{70}$$

By (69) and (70) we have

$$V_{U_ba,d} + U_bU_{a,d} = V_{b,a}V_{b,d} \tag{71}$$

or

$$\{(bab)\,dc\} + b\{acd\}b = \{ba\{bdc\}\}$$

which give the required relation: $V_{c,d}U_b + U_bV_{d,c} = U_{b,V_{c,d}b}\,.$

We now see that

$$\text{instr}\,\mathscr{J} = \left\{\sum V_{b_i,c_i} \mid b_i\,,\, c_i \in J\right\} \tag{72}$$

is a k-submodule of str $\mathscr{J}$. We claim that this is an ideal. To see this we observe that (65) implies that $U_{a,Hb} + U_{Ha,b} = HU_{a,b} - U_{a,b}\bar{H}$. Using the definition $V_{b,c}x = U_{b,x}c$ we see that this is equivalent to

$$[H, V_{a,c}] = V_{Ha,c} + V_{a,Hc} \tag{73}$$

which evidently implies that instr $\mathscr{J}$ is an ideal in str $\mathscr{J}$. We shall call this the *inner structure* (*Lie*) *algebra* of $\mathscr{J}$.

Since $V_a = V_{a,1} \in \text{instr}\,\mathscr{J}$ it is clear from (66) that if $H \in \text{str}\,\mathscr{J}$ (instr $\mathscr{J}$) then $\bar{H} \in \text{str}\,\mathscr{J}$ (instr $\mathscr{J}$). It is easily seen also that str $\mathscr{J} =$ str $\mathscr{J}^{(c)}$ and instr $\mathscr{J} =$ instr $\mathscr{J}^{(c)}$ for any isotope $\mathscr{J}^{(c)}$ of $\mathscr{J}$.

One defines a *derivation* D of a quadratic Jordan algebra $\mathscr{J}$ as a k-endomorphism of $\mathscr{J}$ such that

$$D1 = 0, [D, U_a] = U_{a,Da}\,. \tag{74}$$

It is immediate that D is a derivation if and only if $1 + \delta D \in \text{Aut}\,\mathscr{J}^{\Lambda}$. Hence the set Der $\mathscr{J}$ (or aut $\mathscr{J}$) of derivations is a Lie algebra and Aut $\mathscr{J}$ acts on Der $\mathscr{J}$ by $\eta D = \eta D \eta^{-1}$. It is clear also that Der $\mathscr{J} \subset$ str $\mathscr{J}$ and Der $\mathscr{J}$ is the subalgebra of str $\mathscr{J}$ of endomorphisms D such that $D1 = 0$. The intersection Der $\mathscr{J} \cap$ instr $\mathscr{J}$ is an ideal in Der $\mathscr{J}$ whose elements are called *inner derivations*. Since instr $\mathscr{J} = \{\sum V_{b_i,c_i}\}$ it is immediate that the inner derivations are the maps $\sum V_{b_i,c_i}$ such that $\sum b_i \circ c_i = 0$.

After all of these formal results we now consider the analogues for str $\mathscr{J}$ of our results connecting Str $\mathscr{J}$ with automorphisms of its special universal envelope $(S(\mathscr{J}), \sigma_u)$. We recall that $(S(\mathscr{J})^{\Lambda}, \sigma_u{}^{\Lambda})$ is a special universal envelope for $\mathscr{J}^{\Lambda}$ and we can identify $S(\mathscr{J})$ with the corresponding subset of $S(\mathscr{J})^{\Lambda}$ (Proposition 3). Then $S(\mathscr{J})^{\Lambda}$ is the set of elements $a + \delta b$, $a, b \in S(\mathscr{J})$ and $a + \delta b = 0$ implies $a = b = 0$. Now let $H \in \text{str}\,\mathscr{J}$, so $\eta = 1 + \delta H \in \text{Str}\,\mathscr{J}^{\Lambda}$. Then, as in (20), we have an automorphism η_u of $S\mathscr{J}^{\Lambda}$ such that

$$(\eta_u \sigma_u)(x) = d((\sigma_u{}^{\Lambda}\eta)(x)), \qquad x \in \mathscr{J}$$

where $d = \sigma_u{}^{\Lambda}(c)$, $c = \eta(1)^{-1}$. Since $\eta(1) = 1 + \delta(H1)$, $c = 1 - \delta(H1)$.

Also $\sigma_u{}^A(x) = \sigma_u(x)$ and $(\sigma_u{}^A\eta)(x) = \sigma_u(x) + \delta\sigma_u H(x)$. Hence we have

$$\begin{aligned}\eta_u(\sigma_u(x)) &= (1 - \delta\sigma_u(H1))(\sigma_u(x) + \delta\sigma_u(Hx)) \\ &= \sigma_u(x) + \delta(\sigma_u(Hx) - \sigma_u(H1)\,\sigma_u(x)).\end{aligned} \tag{75}$$

Since the elements $\sigma_u(x)$, $x \in \mathscr{J}$, generate $S(\mathscr{J})$ and η_u is an automorphism, the foregoing formula implies that we have a k-endomorphism D of $S(\mathscr{J})$ such that

$$\eta_u(a) = a + \delta(Da) \tag{76}$$

for all $a \in S(\mathscr{J})$. It follows that D is a derivation in $S(\mathscr{J})$, and comparing (75) and (76), we obtain

$$\sigma_u(Hx) = \sigma_u(H1)\,\sigma_u(x) + D\sigma_u(x) \tag{77}$$

for $x \in \mathscr{J}$. We now observe that if D is a derivation in $S(\mathscr{J})$ then so is $\pi D\pi$, π the main involution of $S(\mathscr{J})$. Hence $\pi D\pi - D$ is a derivation of $S(\mathscr{J})$ and

$$D\sigma_u(x) + \sigma_u(H1)\,\sigma_u(x) = \sigma_u(Hx) = \pi\sigma_u(Hx) = \pi\, D\sigma_u(x) + \sigma_u(x)\,\sigma_u(H1).$$

Thus

$$(\pi\, D\pi - D)\,\sigma_u(x) = [h, \sigma_u(x)]$$

where $h = \sigma_u(H1)$. Since the $\sigma_u(x)$ generate $S(\mathscr{J})$ we have

$$(\pi\, D\pi - D)a = [h, a] \tag{78}$$

for all $a \in S(\mathscr{J})$. Thus $\pi D\pi - D$ is the inner derivation $i_h : a \to [ha]$ determined by $h = \sigma_u(H1)$ or by $h - \gamma$ for any $\gamma \in \mathscr{C} \cap \mathscr{H}$, $\mathscr{C}$ the center of $\mathscr{A}$. Next let $(\mathscr{A}, J)$ be any associative algebra with involution and assume D is a derivation of $\mathscr{A}$ such that the derivation $JDJ - D$ is an inner derivation i_h by an element $h \in \mathscr{H}(\mathscr{A}, J)$. Consider the map

$$H = D + h_L \tag{79}$$

of $\mathscr{A}$. If $x \in \mathscr{H}(\mathscr{A}, J)$ we have

$$\begin{aligned}(Hx)^* &= (Dx + hx)^* = (Dx)^* + xh = (Dx^*)^* + xh \\ &= Dx + hx - xh + xh = Dx + hx = Hx.\end{aligned}$$

Hence $Hx \in \mathscr{H}(\mathscr{A}, J)$. Also $H1 = h$ and direct verification shows that (65) holds for $\bar{H} = H - V_{H1}$. Hence the restriction of H to $\mathscr{H}(\mathscr{A}, J)$ is in str $\mathscr{H}(\mathscr{A}, J)$.

Any inner derivation $D = i_d$ satisfies the foregoing condition since $Dx = [d, x]$ implies $JDJx = [-d^*, x]$ so $JDJ = D = i_h$ where $h = -(d + d^*) \in \mathscr{H}(\mathscr{A}, J)$. The corresponding element of str $\mathscr{H}$ is $x \to dx - xd - (d + d^*)x = b^*x + xb$ with $b = -d$. Since h can be replaced by $h - \gamma$, $\gamma \in \mathscr{C} \cap \mathscr{H}$ it follows that the elements of str $\mathscr{H}$ determined from inner derivations by our procedure are the maps

$$x \to b^*x + xb + \gamma x \tag{80}$$

where $b \in \mathscr{A}$ and $\gamma \in \mathscr{C} \cap \mathscr{H}$. If k contains $\frac{1}{2}$ we can write $\gamma = \frac{1}{2}\gamma + \frac{1}{2}\gamma = \frac{1}{2}\gamma^* + \frac{1}{2}\gamma$ and then absorb this into the part of (80) involving b. Thus we need consider only the maps $x \to b^*x + xb$ in this case.

We shall now apply these results to the problem of determining the structure algebra of a special quadratic Jordan algebra and in particular for algebras $\mathscr{H}(\mathscr{A}, J)$ where $(\mathscr{A}, J)$ is a perfect associative algebra with involution. The main result which is the analogue of Theorem 2 is:

THEOREM 11. 1. *Let $\mathscr{J}$ be a special quadratic Jordan algebra and identify J with its image in $S(\mathscr{J})$. Then* str $\mathscr{J}$ *is the set of maps $(D + h_L) \mid \mathscr{J}$ where D is a derivation of $S(\mathscr{J})$ and h is an element of $\mathscr{J}$ such that $\pi D \pi - D = i_h$ and $D + h_L$ stabilizes $\mathscr{J}$.* 2. *If $(\mathscr{A}, J)$ is a perfect associative algebra with involution then* str $\mathscr{H}$ *for $\mathscr{H} = \mathscr{H}(\mathscr{A}, J)$ is the set of maps $(D + h_L) \mid \mathscr{H}$ where D is a derivation of $\mathscr{A}$ and h is an element of $\mathscr{H}$ such that $JDJ - D = i_L$.*

The proof is clear. We also have the

COROLLARY 1. *Let $\mathscr{J}$, $S(\mathscr{J})$ be as in Theorem* 11 *and assume that every derivation of $S(\mathscr{J})$ is inner. Then* str $\mathscr{J}$ *is the set of maps $x \to b^*x + xb + \gamma x$, $b \in S(\mathscr{J})$, $\gamma \in \mathscr{C} \cap \mathscr{J}$, which stabilize $\mathscr{J}$.* 2. *Let $(\mathscr{A}, J)$ be a perfect associative algebra with involution such that every derivation of $\mathscr{A}$ is inner. Then* str $\mathscr{H}$ *is the set of maps $x \to b^*x + xb + \gamma x$, $b \in \mathscr{A}$, $\gamma \in \mathscr{C} \cap \mathscr{H}$.*

We are now in a position to carry over all our results on structure groups, except those based on Herstein's theorem, to structure algebras. Herstein's theorem can not be applied since $\mathscr{A}^{\Delta}$ is not prime for any

$\mathscr{A} \neq 0$. However, we can apply Theorem 11 to obtain the following result on str $\mathscr{A}^+$, $\mathscr{A}$ associative.

THEOREM 12. *Let $\mathscr{A}$ be an associative algebra such that $\mathscr{B} = \mathscr{A} \oplus \mathscr{A}^0$ with the exchange involution E* (see $S1$) *is perfect. Then* str $\mathscr{A}^+$ *is the set of maps $D + h_L$ where D is a derivation of the associative algebra $\mathscr{A}$ and $h \in \mathscr{A}$.*

Proof. Any derivation of $\mathscr{B}$ maps the ideals $\mathscr{A}$ and $\mathscr{A}^0$ into themselves since $\mathscr{A}^2 = \mathscr{A}$ and $\mathscr{A}^{02} = \mathscr{A}^0$. Hence a derivation $\tilde{D}$ of $\mathscr{B}$ has the form $(a, b) \to (Da, D'b)$ where D is a derivation of $\mathscr{A}$ and D' is one of $\mathscr{A}^0$. Then $E\tilde{D}E - \tilde{D}$ has the form $(a, b) \to ((D' - D)a, (D - D')b)$. The condition in Theorem 11 is that this has the form $(a, b) \to [(h, h), (a, b)]$ for some $h \in \mathscr{A}$ and this holds if and only if $D' = D + i_h$. The corresponding elements of str $\mathscr{H}(\mathscr{B}, E)$ given by Theorem 11 then give the maps $D + h_L$ of $\mathscr{A}^+ \cong \mathscr{H}(\mathscr{B}, E)$.

We now consider the simple artinian algebras with involution. The preceding result is applicable to the case $\mathscr{B} = \mathscr{A} \oplus \mathscr{A}^0$ with exchange involution provided $\mathscr{A}$ is not a division algebra. The derivations of $\mathscr{A} = \mathrm{End}_\Delta V$ where V is a finite dimensional vector space are known. These have the form $A \to [D, A]$ where D is a differential transformation of V, that is, V is an endomorphism of V for which there exists a derivation d of Δ such that $D(ax) = a(Dx) + d(a)x$ for $a \in \Delta$, $x \in V$ [8, p. 87]. We consider next the analog of Theorem 3 which gives Str $\mathscr{H}$ for $\mathscr{H} = \mathscr{H}(\mathscr{A}, J)$ where J is the adjoint map of $\mathscr{A} = \mathrm{End}_\Delta V$ given by a diagonalizable hermitian form. We refer to the statement of this theorem for the procise conditions we require. Let D be a differential transformation in V with associated derivation d. For a fixed vector $y \in V$, the map

$$x \to h(Dx, y) - d(h(x, y))$$

is a linear function on V. Hence there is a unique $y' \in V$ such that $h(Dx, y) - d(h(x, y)) = h(x, y')$ for all x. Then we have the map $D^*: y \to y'$ such that

$$h(Dx, y) - d(h(x, y)) = h(x, D^*y) \tag{81}$$

for all $x, y \in V$. D^* is called the *adjoint of D relative to h.* One can check that D^* is a $d^* = -jdj$-differential transformation of V. Also a_L: $x \to ax$ is a differential transformation whose associated derivation is $i_a : b \to [a, b]$. We can now state the following analogue of Theorem 3:

THEOREM 13. *Let* (Δ, j), h, $\mathscr{A}$, $\mathscr{H}$ *etc. be as in Theorem* 3. *Then* str $\mathscr{H}$ *is the set of maps*

$$A \to D^*A + AD + a_L A$$

where D *is a* d*-differential transformation,* $a \in \mathscr{H}(\Delta, j)$ *and* $jdj - d = i_a$.

The proof is a straightforward application of Theorem 12 and the foregoing results. We leave the details to the read3r. Similarly we can obtain the analog of Theorem 4.

We now consider the problem of determining the structure algebras of the quadratic Jordan algebra $\mathscr{H}(\mathscr{A}, J)$ where $(\mathscr{A}, J)$ is finite dimensional central simple over a field Φ. It is well known that the derivations of $\mathscr{A}$ are all inner. Assuming the degree is > 1 in all cases and > 2 if J is of symplectic type we can use the Corollary to Theorem 11 to derive the following results.

THEOREM 14. *Let* $\mathscr{A}$ *be a finite dimensional central simple associative algebra,* $\mathscr{A}^+$ *the corresponding quadratic Jordan algebra. Then* str $\mathscr{A}^+$ *is the set of maps*

$$x \to ax + xb$$

where $a, b \in \mathscr{A}$.

THEOREM 15. *Let* $\mathscr{A}$ *be either a finite dimensional central simple associative algebra over* Φ *with an involution* J *of first kind or a finite dimensional simple associative algebra with center a quadratic extension field* P *of the base field* Φ *and an involution* J *of second kind. Then* str $\mathscr{H}$ *for* $\mathscr{H} = \mathscr{H}(\mathscr{A}, J)$ *(or* $\mathscr{H}(\mathscr{A}, J)'$*) are the maps*

$$x \to b^*x + xb + \gamma x$$

where $b \in \mathscr{A}$, $\gamma \in \Phi$.

We can also obtain analogues of the results on the groups of norm similarities of the algebras $\mathscr{H}(\mathscr{A}, J)$ for $(\mathscr{A}, J)$ finite dimensional central simple with involution. The Lie algebra analogous to the group of norm similarities is comprised of the set of linear transformations H such that

$$\Delta_a^{Ha} N = \rho N \tag{82}$$

for some $\rho \in \Phi$ and all $a \in \mathscr{H}$ (or $\mathscr{H}'$) (see Jacobson [7], p. 220). It can be shown that under the hypotheses of Theorem 10 this Lie algebra coincides with str $\mathscr{H}$ (str $\mathscr{H}'$).

References

1. G. Ancochea, On semi-automorphisms of division algebras, *Ann. Math.* **48** (1947), 147–154.
2. H. Braun and M. Koecher, Jordan-algebren, Springer-Verlag, Berlin, 1966.
3. C. Chevalley and R. D. Schafer, The exceptional Lie algebras F_4 and E_6, *Proc. Nat. Acad. Sci. U.S.A.* **36** (1950), 137–141.
4. G. Frobenius, Über die Darstellung der endlichen Gruppen durch lineare Substitutionen, *S.-B. Berlin Akad.* 1897, 994–1015.
5. I. N. Herstein, Jordan homomorphisms, *Trans. Amer. Math. Soc.* **81** (1956), 331–341.
6. I. N. Herstein, "Topics in Ring Theory," University of Chicago Press, Chicago, Ill., 1969.
7. L. K. Hua, On the automorphisms of a *s*-field, *Proc. Nat. Acad. Sci.* **35** (1949), 386–389.
8. N. Jacobson, "Structure of Rings," Colloq. Publ. 37, Amer. Math. Soc., Providence, 1956; Rev. Ed. 1964.
9. N. Jacobson, "Exceptional Lie Algebras," Mimeographed notes, Yale University 1958, published in augmented from by Marcel Dekker, 1971.
10. N. Jacobson, Some groups of linear transformations defined by Jordan algebras I, *J. Reine Angew. Math.* **201** (1959), 178–195, III, *J. Reine Angew. Math.* **207** (1961), 61–85.
11. N. Jacobson, "Structure and Representations of Jordan Algebras," Colloq. Publ. 39, Amer. Math. Soc., Providence, R.I., 1968.
12. N. Jacobson, "Lectures on Quadratic Jordan Algebras," Tata Institute of Fundamental Research, Bombay, 1969.
13. N. Jacobson, Connections between associative and Jordan rings, *Symposia Math. Ist. Nazionale di Alta Mat.* **8** (1972), 261–268.
14. N. Jacobson, "*PI*-Algebras. An Introduction," Lecture Notes in Mathematics, Springer-Verlag, Berlin, 1975.
15. N. Jacobson and J. Katz, Generically algebraic quadratic Jordan algebras, *Scripta Math.* **29** (1973), 215–227.
16. N. Jacobson and K. McCrimmon, Quadratic Jordan algebras of quadratic forms with base points, *J. Indian Math. Soc.* **35** (1971), 1–45.
17. N. Jacobson and C. E. Rickart, Jordan homomorphisms of rings, *Trans. Amer. Math. Soc.* **69** (1950), 479–502.
18. N. Jacobson and C. E. Rickart, Homomorphisms of Jordan rings of self-adjoint elements, *Trans. Amer. Math. Soc.* **72** (1952), 310–322.
19. I. Kaplansky, Semi-automorphisms of rings, *Duke Math. J.* **14** (1947), 37–50.
20. M. Koecher, "Jordan Algebras and their Applications," University of Minnesota Lecture Notes, Minneapolis, 1962.
21. O. Loos, "Jordan Pairs," Lecture Notes in Mathematics, Springer-Verlag, Berlin, 1975.

22. K. McCrimmon, A general theory of Jordan rings, *Proc. Nat. Acad. Sci. U.S.A.* **56** (1966), 1072–1079.
23. K. McCrimmon, The Freudenthal–Springer–Tits construction of exceptional Jordan algebras, *Trans. Amer. Math. Soc.* **139** (1969), 495–510.
24. K. McCrimmon, The Freudenthal–Springer–Tits constructions revisited, *Trans. Amer. Math. Soc.* **148** (1970), 293–314.
25. K. McCrimmon, The generic norm of an isotope of a Jordan algebra, *Scripta Math.* **29** (1975), 229–236.
26. W. S. Martindale, III, Jordan homomorphisms of symmetric elements of a ring with involution, *J. Algebra* **5** (1967), 232–249.
27. M. Racine, A note on quadratic Jordan algebras of degree 3, *Trans. Amer. Math. Soc.* **164** (1972), 93–103.
28. J. P. Serre, Lie Algebras and Lie Groups, Harvard University Lecture Notes, Benjamin, 1965.
29. B. D. Smith, Jordan rings with polynomial identities, Doctoral dissertation, University of Virginia, 1975.
30. T. A. Springer, Jordan Algebras and Algebraic Groups, Ergebnisse der Math. 75, Springer-Verlag, Berlin, 1973.

Printed by the St Catherine Press Ltd., Tempelhof 37, Bruges, Belgium.

COMMUNICATIONS IN ALGEBRA, 6(9), 911-958 (1978)

LOCALIZATION OF JORDAN ALGEBRAS

N. Jacobson, K. McCrimmon and M. Parvathi

Yale University
New Haven, Connecticut 06520

University of Virginia
Charlottesville, Virgina 22903

Ramanujan Institute, University of Madras
Madras, India

Introduction

A number of years ago the first named author gave a definition of Jordan domain and division algebras and suggested that it would be interesting to obtain imbeddings of Jordan domains in division algebras by constructing algebras of quotients in a manner analogous to Ore's construction of division rings from associative domains satisfying a common multiple condition (see Jacobson [1], p. 426). There is a natural analogue of Ore's condition in the Jordan case, namely, given any two non-zero elements a and b there exist a', b' such that $U_a a' = U_b b' \neq 0$, where U_a is the U-operator determined by a (see §1 for definitions). In a special Jordan algebra, $U_a a' = aa'a$ in terms of the associative multiplication, and it is easy to see that for a Jordan algebra A^+ obtained from an associative algebra A the Jordan common multiple condition is equivalent to the existence of common left and common right multiples in A. Also, as in the associative case, the Jordan condition has a formulation in terms of ideals, namely, the inner ideals $U_a J$ and $U_b J$ have a non-zero intersection.

Aside from the interest in constructing Jordan division rings the study of Jordan quotients has been motivated by the problem of obtaining an analogue for Jordan algebras of Goldie's characterization of the associative rings that have classical quotient rings that are semi-simple artinian. The corresponding problem for Jordan algebras is a natural one in view of the Jacobson-McCrimmon structure theory of semi-simple Jordan algebras satisfying minimum conditions for inner ideals (see Jacobson [1] and [2]).

One of the most important tools in commutative algebra is localization with respect to submonoids of the multiplicative monoid. The nearest relative of this in non-commutative (associative) algebra is central localization. This has played an important role in the recent simplification of the theory of algebras satisfying a polynomial identity (see Jacobson [3]). There is a natural analogue of central localization for Jordan algebras: centroid localization. We consider this first in §1 of the present paper. After this we study the problem of quotients or localization with respect to subsets of a Jordan algebra closed under the product $U_a b$. The chief difficulty we shall encounter in developing a theory of localization for Jordan algebras is the establishment of the bilinear character of $U_{a,b} = U_{a+b} - U_a - U_b$ for the extended operator U_a. It seems to be advisable to separate off this problem and we shall do so by first developing a purely multiplicative theory in which we drop the module structure. This leads to the definition of Jordan monads analogous to associative monoids. We define this in §2 as a set M together with a distinguished element $1 \in M$ and a map $U: a \to U_a$ of M into the monoid $\mathrm{Mon}(M)$ of maps of M into itself such that $U_1 = 1$, $U_{U_a b} = U_a U_b U_a$. An element $a \in M$ is called cancellable (invertible) if U_a is injective (bijective) and M is a cancellation (division) monad if every $a \in M$ is cancellable (invertible). We show that if M is a cancellation monad satisfying the element common multiple property then M can be embedded in a division monad. More generally we consider localization of an arbitrary monad with respect to a submonad S. To define this we require cancellation and commom multiple conditions on the submonoid $\underline{S}$ generated by the U_s $(s \in S)$ in the submonoid $\underline{M}$ of $\mathrm{Mon}(M)$ generated by the U_a $(a \in M)$. This leads to the study of localization of an associative monoid $\underline{M}$ and its action on a set X with respect to a submonoid $\underline{S}$ of $\underline{M}$. We consider this in §3 and apply the results to Jordan monads in §4. In §5 we impose module conditions that reduce the verification of identities defining Jordan algebras to a manageable form. In §6 we obtain a localization theory for Jordan algebras based on some fairly complicated common multiple properties. We show that in important situations arising from associative algebras these conditions are satisfied. It remains to be seen whether or not our localization theory is adequate for attacking the problems of constructing Jordan division algebras and in developing a Goldie theory for jordan algebras (cf. Britten [1], [2], Montgomery [1]).

1. Centroid Localization

Our primary interest in this paper will be in (unital) quadratic Jordan algebras over a commutative ring Φ (with unit). We recall that such an algebra is a triple $(J,U,1)$ where J is a (unital) Φ-module, U is a map of J into $\mathrm{End}_\Phi J$ and 1 is a distinguished element of J such that the following conditions hold:

QJ1. U is a quadratic map: $U_{\alpha a} = \alpha^2 U_a$ if $\alpha \in \Phi$, $a \in J$ and $U_{a,b} \equiv U_{a+b} - U_a - U_b$ is Φ-bilinear in a and b.

QJ2. $U_1 = 1$.

QJ3. $U_a U_b U_a = U_{U_a b}$ (the "fundamental" formula).

QJ4. If $V_{a,b}$ is defined by $V_{a,b}x = U_{a,x}b$ then

$$U_a V_{b,a} = V_{a,b} U_a.$$

QJ5. If Ψ is a commutative associative algebra over Φ and U is the "extension" of U to a quadratic map of $J^\Psi = \Psi \otimes_\Phi J$ into $\mathrm{End}_\Psi J^\Psi$ then QJ3 and QJ4 are valid in J^Ψ.

It can be shown that QJ5 is equivalent to the following additional identities that are obtained by linearizing QJ3 and QJ4:

QJ3'. $U_{U_a c, U_{a,b} c} = U_a U_c U_{a,b} + U_{a,b} U_c U_a$.

QJ3''. $U_{U_a c, U_b c} + U_{U_{a,b} c} = U_a U_c U_b + U_b U_c U_a + U_{a,b} U_c U_{a,b}$.

QJ4'. $U_a V_{b,c} + U_{a,c} V_{b,a} = V_{c,b} U_a + V_{a,b} U_{a,c}$.

The foregoing definition is due to McCrimmon (McCrimmon [1]). An adequate reference for the results we shall state without proofs or explicit references is Jacobson [2].

The quadratic theory is equivalent to the more familiar linear theory if $2 = 1 + 1$ is invertible in Φ (briefly $\Phi \ni \frac{1}{2}$). In this case one defines a linear Jordan algebra over Φ as a non-associative algebra with unit in the usual sense whose product composition $a.b$ satisfies the following identities:

J1. $a.b = b.a$

J2. $(a^2.b).a = a^2.(b.a)$, $a^2 = a.a$.

The definition of a linear Jordan algebra can also be formulated in terms of a map L of J into $\mathrm{End}_\Phi J$. On the other hand, the definition of a quadratic Jordan algebra could also be formulated in terms of a binary product $(a,b) \to aba$ that is assumed to be linear in b and quadratic in a. If $\Phi \ni \frac{1}{2}$, any quadratic Jordan algebra becomes a linear Jordan algebra by defining $a.b = \frac{1}{2} U_{a,b}1$, and any linear Jordan algebra becomes a quadratic Jordan algebra if we define

$$U_a = 2L_a{}^2 - L_{a^2} \tag{1}$$

where L_a is $x \to a.x$.

The simplest type of localization of associative algebras is central localization. This plays an important role in recent developments in the theory of associative algebras satisfying polynomial identities (PI-algebras). This suggests that it would be worthwhile to consider a corresponding theory for quadratic Jordan algebras.

The natural analogue in the Jordan theory of central localization of associative algebras appears to be centroid localization.

We recall that the centroid $\Gamma(J)$ of a quadratic Jordan algebra J (over Φ) is the set of Φ-endomorphisms γ of J such that

$$(2) \qquad \gamma U_a = U_a\gamma, \qquad U_{\gamma a} = \gamma^2 U_a, \qquad a \in J.$$

Under fairly mild non-degeneracy conditions the centroid turns out to be a commutative algebra of endomorphisms of J (McCrimmon [3]). For example, this is the case if the quadratic map U is non-degenerate in the sense that the only z such that $U_z = 0 = U_{z,a}$ for all a is $z = 0$. It is readily seen that this holds if $\frac{1}{2} \in \Phi$. Then the condition $\gamma U_a = U_a\gamma$ for all a is equivalent to $\gamma L_a = L_a\gamma$ for all a and the latter implies $L_{\gamma a} = \gamma L_a$ which in turn implies the second condition in (2). Hence this condition is superfluous in the case of linear Jordan algebras. It is readily seen also that in the linear case the centroid can be identified with the center, defined to be the set of elements $c \in J$ such that $(a.c).b = a.(c.b)$. We have an isomorphism of the center onto the centroid sending $c \to L_c$ (Jacobson [1], p. 206).

Now let Δ be a commutative algebra of endomorphisms of the module J contained in the centroid $\Gamma(J)$. We can regard J as a Δ-module in the natural way. Moreover, the condition $\delta U_a = U_a\delta$ for $\delta \in \Delta$, $a \in J$ shows that $U_a \in \mathrm{End}_\Delta J$ and so U can be regarded as a map of J into $\mathrm{End}_\Delta J$. Since δ commutes with every $U_{a,b} = U_{a+b} - U_a - U_b$ and with every $V_a \equiv U_{a,1}$, δ commutes with every $V_{a,b} = V_aV_b - U_{a,b}$. Hence $U_{a,\delta b}x = V_{a,x}\delta b = \delta V_{a,x}b = \delta U_{a,b}x$. Thus $U_{a,\delta b} = \delta U_{a,b}$ and $U_{\delta a,b} = U_{b,\delta a} = \delta U_{b,a} = \delta U_{a,b}$. It follows that U is a quadratic map of J into $\mathrm{End}_\Delta J$ and $(J,U,1)$ is a quadratic Jordan algebra over Δ. In particular, if Γ itself is a commutative algebra of endomorphisms then J can be regarded as a quadratic Jordan algebra over Γ.

Let $\underline{S}$ be a submonoid of the multiplicative monoid of Δ (that is, $S \subset \Delta$, $\underline{S}$ is closed under multiplication and $\underline{S}$ contains the identity map). By the standard theory of localization of commutative algebras we can form the commutative algebra $\underline{S}^{-1}\Delta$ and the module $\underline{S}^{-1}J$, which is canonically isomorphic to $\underline{S}^{-1}\Delta \otimes_{\Delta} J$. Hence we have a quadratic Jordan structure on the $\underline{S}^{-1}\Delta$-module $\underline{S}^{-1}J$ whose U map is the "extension" of the U map on J to $\underline{S}^{-1}J$. We shall call $\underline{S}^{-1}J$ the Δ-localization of J at $\underline{S}$. We have the canonical map $\nu_{\underline{S}}$: $a \to 1^{-1}a$ of J into $\underline{S}^{-1}J$ whose kernel is the set of elements $b \in J$ for which there exists an $s \in \underline{S}$ such that $sb = 0$. Thus $\underline{S}^{-1}J = 0$ if $\underline{S}$ contains 0 and $\nu_{\underline{S}}$ is injective if no element of $\underline{S}$ annihilates a non-zero element of J. We remark also that $\underline{S}^{-1}J$ can be regarded as a quadratic Jordan algebra over Δ and over Φ if we define $\delta(s^{-1}a) = s^{-1}(\delta a)$, $\alpha(s^{-1}a) = s^{-1}(\alpha a)$ for $\delta \in \Delta$, $\alpha \in \Phi$.

The foregoing situation can be realized by beginning with any subset X of Γ whose elements commute and taking $\underline{S}$ to be the submonoid of $\mathrm{End}_{\Phi}(J)$ generated by X, Δ the subalgebra of $\mathrm{End}_{\Phi}(J)$ generated by X. We note that if $\gamma_1, \gamma_2 \in \Gamma$ and $\gamma_1\gamma_2 = \gamma_2\gamma_1$ then $U_{\gamma_1\gamma_2 a} = \gamma_1^2 U_{\gamma_2 a} = \gamma_1^2\gamma_2^2 U_a = (\gamma_1\gamma_2)^2 U_a$. It follows that $\gamma_1\gamma_2 \in \Gamma$. Also $U_{(\gamma_1+\gamma_2)a} = U_{\gamma_1 a+\gamma_2 a} = U_{\gamma_1 a} + U_{\gamma_1 a,\gamma_2 a} + U_{\gamma_2 a}$. As before, $U_{\gamma_1 a,b} = \gamma_1 U_{a,b}$ and $U_{b,\gamma_2 a} = \gamma_2 U_{b,a}$. Hence $U_{(\gamma_1+\gamma_2)a}$

$= \gamma_1{}^2U_a + \gamma_1\gamma_2U_{a,a} + \gamma_2{}^2U_a = \gamma_1{}^2U_a + 2\gamma_1\gamma_2U_a + \gamma_2{}^2U_a = (\gamma_1 + \gamma_2)^2U_a$. It follows that $\gamma_1 + \gamma_2 \in \Gamma$. Also $\alpha\gamma \in \Gamma$ if $\alpha \in \Phi$ and $\gamma \in \Gamma$. These results imply that if X is a subset of Γ of elements that commute then the subalgebra Δ (submonoid $\underline{S}$) of $\mathrm{End}_\Phi(J)$ generated by X is contained in Γ. Hence we can form the Δ-localization of J at $\underline{S}$.

The most important instance of localization of the type we have defined is that in which the centroid Γ is a commutative algebra of endomorphisms of J. Then $\underline{S}$ can be taken to be any submonoid of Γ and the Γ-localization of J will be called the <u>centroid localization</u> of J at $\underline{S}$.

We now consider some special cases of these constructions.

1. <u>Quadratic Jordan algebras without 2-torsion</u>. Let J be an arbitrary Jordan algebra and let $X = \{2 \cdot 1\}$ where $2 \cdot 1$ is the endomorphism $x \to 2x$ of J. This is contained in Γ and it generates the algebra of endomorphisms $\Phi 1$ and the monoid of endomorphisms $\underline{S} = \{2^k \cdot 1,\ k = 0,1,2,\ldots\}$. We can form the Φ-localization $\underline{S}^{-1}J$ of J at $\underline{S}$. This is a quadratic Jordan algebra over $\underline{S}^{-1}\Phi$ which is a commutative ring containing $\frac{1}{2}$. Hence $\underline{S}^{-1}J$ can be regarded as a linear Jordan algebra over $\underline{S}^{-1}\Phi$. Now suppose J has no 2-torsion. Then the canonical map $x \to 1^{-1}x$ of J into $\underline{S}^{-1}J$ is injective. In this sense we have an imbedding of J in a linear Jordan algebra.

2. <u>Prime rings</u>. A quadratic Jordan algebra J is called <u>prime</u> (<u>semi-prime</u>) if there exist no non-zero ideals $\mathcal{B}$ and $\mathcal{C}$

such that $U_{\mathcal{B}}\mathcal{C} = 0$ $(U_{\mathcal{B}}\mathcal{B} = 0)$ (Tsai [1]). Here $U_{\mathcal{B}}\mathcal{C}$ is the Φ-submodule generated by all $U_b c$, $b \in \mathcal{B}$, $c \in \mathcal{C}$. In any quadratic Jordan algebra, $\operatorname{rad} U = \{z \in J \mid U_z = 0 = U_{a,z},\ a \in J\}$ is an ideal. It is clear that rad U is an inner ideal and the fact that rad U is an outer ideal follows from the identity $U_{U_b z,c} = V_{b,z}U_{b,c} - U_b V_{z,c}$ which implies that $U_b z \in \operatorname{rad} U$ for every $z \in \operatorname{rad} U$ and every $b \in J$. Evidently $U_{\operatorname{rad} U}(\operatorname{rad} U) = 0$. Hence $\operatorname{rad} U = 0$ if J is semi-prime. Then $\Gamma(J)$ is a commutative algebra of endomorphisms.

Now suppose that J is prime. Then Γ is a domain and $\gamma a = 0$ for $\gamma \in \Gamma$, $a \in J$, implies either $\gamma = 0$ or $a = 0$. The first of these assertions is an immediate consequence of the fact that γJ is an ideal for any $\gamma \in \Gamma$. To see the second, we let $N_\gamma = \{a \in J \mid \gamma a = 0\}$. We claim that if $N_\gamma \neq 0$ then N_γ contains a non-zero ideal of J. Evidently N_γ is an outer ideal, so if N_γ is an inner ideal, then it is an ideal. Otherwise, we have a $b \in J$, $a \in N_\gamma$ such that $\gamma U_a b \neq 0$. Since $\gamma^2 U_a b = U_{\gamma a} b = 0$, $\gamma U_a b \in N_\gamma$. Also $U_{\gamma U_a b} = \gamma^2 U_{U_a b} = \gamma^2 U_a U_b U_a = U_{\gamma a} U_b U_a = 0$. Thus $\gamma U_a b$ is an absolute zero divisor. It follows that the ideal $\mathcal{B}$ generated by $\gamma U_a b$ is the set of Φ linear combinations of elements of the form $U_{c_1} \cdots U_{c_r}(\gamma U_a b)$, $c_i \in J$ (McCrimmon [2]). Hence $\gamma u = 0$ for every $u \in \mathcal{B}$ and so N_γ contains the non-zero ideal $\mathcal{B}$. Now if J is prime and $\gamma a = 0$ for $\gamma \in \Gamma$ and $a \neq 0$ then $N_\gamma \neq 0$ and hence $U_{\gamma J}\mathcal{B} = 0$ for a non-zero ideal $\mathcal{B}$. Since J is prime and γJ is an ideal, this implies that $\gamma J = 0$, that is, $\gamma = 0$.

We now let $\underline{S} = \Gamma^*$ the submonoid of non-zero elements of Γ. Then $\underline{S}^{-1}\Gamma$ is the field of fractions of the domain Γ and the canonical map of $\mathcal{J}$ into $\tilde{\mathcal{J}} = \underline{S}^{-1}\mathcal{J}$ is injective. If $\tilde{\mathcal{B}}$ is a non-zero ideal in $\tilde{\mathcal{J}}$ then $\mathcal{B} = \tilde{\mathcal{B}} \cap \mathcal{J}$ is a non-zero ideal in $\mathcal{J}$. It follows that $\tilde{\mathcal{J}}$ is prime. Thus $\tilde{\mathcal{J}}$ is a prime algebra over a field.

3. Orders in finite dimensional algebras. Let $\mathcal{J}$ be a finite dimensional quadratic Jordan algebra over a field Φ which is the field of fractions of a Dedekind domain D. Let $\mathcal{O}$ be a D-order in $\mathcal{J}$ (in the sense of Knebusch [1] and Racine [1]), that is, $\mathcal{O}$ is a D-subalgebra of $\mathcal{J}$ that is finitely generated as D-module and satisfies $\Phi\mathcal{O} = \mathcal{J}$. We can identify D with a subring of the centroid $\Gamma(\mathcal{O})$. Let D^* be the monoid of non-zero elements of D so $D^{*-1}D = \Phi$. We can identify $\mathcal{J}$ with the D-localization $D^{*-1}\mathcal{O}$ of $\mathcal{O}$ at D^*.

The elements of $\mathcal{O}$ have generic minimum polynomials in $D[\lambda]$ (not merely in $\Phi[\lambda]$; see Jacobson and Katz [1] for the definition of the generic minimum polynomial), and $a \in \mathcal{O}$ has injective U_a in $\mathcal{O}$ if and only if the generic norm $N(a) \neq 0$, in which case a^{-1} exists in $\mathcal{J}$. Any element of $\mathcal{J}$ has the form $\gamma^{-1}a$, $\gamma \in D^*$, $a \in \mathcal{O}$; this can be written as $\gamma^{-2}(\gamma a) = U_{(\gamma 1)^{-1}}\gamma a = U_b{}^{-1}c$ for $b,c \in \mathcal{O}$.

2. Jordan Monads

Our main concern in this paper will be with an analogue for quadratic Jordan algebras of the classical localization of associative rings which had its origin in Ore's construction of the division ring of fractions of a non-commutative domain with common multiple property. We shall encounter some difficulties in carrying out this program and we shall have to impose some restrictions to overcome some of these, notably, to insure that the U-maps in the extensions we shall construct are quadratic. A purely multiplicative aspect of the theory analogous to localization of monoids relative to submonoids satisfying some common multiple conditions can be developed quite smoothly. We shall separate off this part of the theory as a study of a Jordan analogue of (associative) monoids that we shall call Jordan monads. The results on these will then be applied to quadratic Jordan algebras.

We define a _Jordan monad_ to be a triple $(M,U,1)$ where M is a set, 1 a distinguished element of M (the _unit element_), and U is a map $a \to U_a$ of M into the monoid $\text{Mon}\, M$ of transformations of M into itself satisfying the identities:

JM1. $U_1 = 1$ $(= 1_M,$ the identity map)

JM2. $U_{U_a b} = U_a U_b U_a$.

We could also formulate the definition in terms of the distinguished element 1 and a binary composition $(a,b) \to aba$ satisfying: JM1'. $1a1 = a$, JM2'. $(aba)c(aba) = a(b(aca)b)a$.

Examples. 1. If $(J,U,1)$ is a quadratic Jordan algebra, we obtain the _multiplicative monad_ of J by discarding the module structure. Thus we regard J simply as a set with distinguished element 1 and map U: $a \to U_a$ of J into $\text{Mon}\, J$. Then JM1 and JM2 are the conditions QJ2 and QJ3 of a quadratic Jordan algebra.

2. Let $\underline{M}$ be a monoid. Let 1 be the given unit of $\underline{M}$ and define $U_a b = aba$. This defines the Jordan monad $\underline{M}^+ = (\underline{M},U,1)$ of the monoid $\underline{M}$.

3. Let $\underline{G}$ be a group, $\underline{I}$ the set of involutions in $\underline{G}$, that is, the elements a such that $a^2 = 1$. Evidently $1 \in \underline{I}$ and if $a,b \in I$ then $aba = aba^{-1} \in \underline{I}$. Hence $\underline{I}$ is a submonad (in the obvious sense) of the Jordan monad $\underline{G}^+$.

We define powers in a Jordan monad M by $a^0 = 1$, $a^1 = a$, $a^{n+2} = U_a a^n$ for $n \geq 0$. Repeated use of JM2 gives $U_{a^n} = U_a{}^n$, $n \geq 0$, and induction on n gives $(a^m)^n = a^{mn}$.

An element a M is <u>cancellable</u> if U_a is injective and a is <u>invertible</u> if U_a is bijective. As for quadratic Jordan algebras we have the

<u>Theorem on Inverses</u>. The following conditions on an element a of a Jordan monad are equivalent:

(1) a is invertible

(2) U_a is surjective

(3) $1 \in U_a M$

(4) there exists a b such that $U_a b = a$ and $U_a b^2 = 1$.

When a is invertible then the element b is uniquely determined by a as $U_a{}^{-1} a$. Denoting this as a^{-1} we have that a^{-1} is invertible, $U_{a^{-1}} = U_a{}^{-1}$ and $(a^{-1})^{-1} = a$. In general, $U_a b$ is invertible if and only if a and b are invertible, in which case, $(U_a b)^{-1} = U_{a^{-1}} b^{-1}$. A power a^n, $n > 1$, is invertible if and only if a is invertible, in which case $(a^n)^{-1} = a^{-n} = (a^{-1})^n$ and $a^m a^n = a^{mn}$ holds for all integral m and n.

Proof. (1) ⇒ (2) ⇒ (3) is obvious; (3) ⇒ (1) since if $U_a b = 1$ then $1 = U_1 = U_{U_a b} = U_a U_b U_a$, which shows that U_a is invertible. (4) ⇒ (3) is clear, and (1) ⇒ (4) since we have a b such that $U_a b = a$ which gives $U_a U_b U_a = U_a$ and $U_b U_a = 1 = U_a U_b$ so $U_b = U_a^{-1}$. Then $U_a b^2 = U_a U_b 1 = 1$. This proves the equivalence of (1)-(4).

If b satisfies (4) then $b = U_a^{-1} a$ is uniquely determined; we had $U_b = U_a^{-1}$, hence $b = a^{-1}$ is invertible and $(a^{-1})^{-1} = U^{-1}_{a^{-1}} a^{-1} = U_a a^{-1} = a$.

Now $U_a b$ invertible ⇔ $U_{U_a b} = U_a U_b U_a$ invertible ⇔ U_a and U_b invertible ⇔ a and b invertible. In this case $(U_a b)^{-1} = U_{U_a b}{}^{-1} U_a b = U_a{}^{-1} U_b{}^{-1} U_a{}^{-1} U_a b = U_a{}^{-1} U_b{}^{-1} b = U_a{}^{-1} b^{-1}$.

Since $U_{a^n} = U_a{}^n$, a^n for $n \geq 0$ is invertible if and only if a is invertible. Since $U_{a^n}{}^{-1} = (U_a{}^{-1})^n$ induction on n gives $U_{a^n}{}^{-1} a^n = (a^{-1})^n$; hence $(a^n)^{-1} = (a^{-1})^n$. It is easy to see that $(a^m)^n = a^{mn}$ for $m,n \in \mathbb{Z}$ follows from the special case in which $m \geq 0$, $n \geq 0$. □

A Jordan monad is called a <u>cancellation</u> (<u>division</u>) monad if all of its elements are cancellable (invertible). Jordan division monads are equivalent to the abstract symmetric spaces studied by Loos [1] in terms of the products $S(x)y = U_x y^{-1}$. A submonad S of a Jordan monad M is a <u>cancellable submonad</u> if every $s \in S$ is cancellable in M.

We define the <u>multiplication monoid</u> $\underline{M}(M)$ of M to be the submonoid of maps on M generated by the U_a, $a \in M$. More generally, if S is a submonad of M then we define $\underline{M}_M(S)$ to be the submonoid of maps on M generated by the U_s, $s \in S$ (acting in M). This consists of the products $U_{s_1} U_{s_2} \ldots U_{s_r}$, $s_i \in S$. If $x, a_i \in M$ we can iterate JM2 to obtain

$$(2.1) \qquad U_{U_{a_1} \ldots U_{a_r} x} = U_{a_1} \ldots U_{a_r} U_x U_{a_r} \ldots U_{a_1}.$$

A pair of maps (T, T^*) of M into M will be called a <u>structural pair</u> if

$$(2.2) \qquad U_{Tx} = T U_x T^*, \qquad U_{T^* x} = T^* U_x T$$

for all $x \in M$ (cf. Meyberg [1], p.126). The formula (2.1) shows that if we put $T = U_{a_1} \ldots U_{a_r}$, $T^* = U_{a_r} \ldots U_{a_1}$ then (T, T^*) is a structural pair. A structural pair of this form will be called <u>inner</u>. Evidently, if (T, T^*) is a structural pair, then so is (T^*, T) and if (T_1, T_1^*) and (T_2, T_2^*) are then so is $(T_1 T_2, T_2^* T_1^*)$. If we put $x = 1$ in (2.2) we obtain

$$(2.3) \qquad U_{T1} = TT^*, \qquad U_{T^* 1} = T^* T.$$

Hence the terms T, T^* are factors of U-maps.

We may have $T = U_{a_1} \ldots U_{a_r} = U_{b_1} \ldots U_{b_s}$ but $T_1^* = U_{a_r} \ldots U_{a_1} \neq U_{b_s} \ldots U_{b_1} = T_2^*$. An example of this sort can be obtained by starting with the free monad generated by x_1, x_2, x_3, x_4 and forming the quotient M relative to the congruence relation generated by all the pairs of the form $(x_1x_2xx_2x_1, x_3x_4xx_4x_3)$, x arbitrary. Then if $\bar{x}$ denotes the congruence class of x we have $U_{\bar{x}_1}U_{\bar{x}_2} = U_{\bar{x}_3}U_{\bar{x}_4}$ in $\underline{M}(M)$ and it is readily seen that $U_{\bar{x}_2}U_{\bar{x}_1} \neq U_{\bar{x}_4}U_{\bar{x}_3}$. If M is a Jordan monad in which any relation of the form $U_{a_1} \ldots U_{a_r} = U_{b_1} \ldots U_{b_s}$ implies $U_{a_r} \ldots U_{a_1} = U_{b_s} \ldots U_{b_1}$ then we have a unique involution $T \to T^*$ in $\underline{M}(M)$ such that $U_a^* = U_a$ for $a \in M$. In this case (T,T^*) is a structural pair for any $T \in \underline{M}(M)$.

In the sequel we shall be interested in submonads S of M such that for every $s \in S$, U_s is left and right cancellable in the multiplication monoid: $U_sA_1 = U_sA_2$ for $A_i \in \underline{M}(M)$ implies $A_1 = A_2$ and $A_1U_s = A_2U_s$ implies $A_1 = A_2$. If S is a cancellable submonad of M then every U_s is injective and hence is left cancellable in $\underline{M}(M)$. We have the following

2.4. Proposition. Let S be a submonad of the Jordan monad M such that every U_s, $s \in S$, is left cancellable in $\underline{M} = \underline{M}(M)$. Then there is a unique involution $T \to T^*$ in $\underline{M}_M(S)$ such that (T,T^*) is a structural pair for every $T \in \underline{M}_M(S)$. The elements of $\underline{M}_M(S)$ are right cancellable in $\underline{M}_M(S)$ (though perhaps not in $\underline{M}$).

Proof. Every $T = U_{s_1} \dots U_{s_r}$, $s_i \in S$ is left cancellable in $\underline{M}$. If $T^* = U_{s_r} \dots U_{s_1}$ then (T,T^*) is a structural pair. Now suppose (T,T_1^*) and (T,T_2^*) are structural pairs. Then $TT_1^* = U_{T(1)} = TT_2^*$; hence $T_1^* = T_2^*$ and T^* is the only element of $\underline{M}_M(S)$ such that (T,T^*) is a structural pair. Since $(T_1T_2)^* = T_2^*T_1^*$, $T^{**} = T$ and $1^* = 1$, the map $T \to T^*$ is an involution in $\underline{M}_M(S)$. Since every element of $\underline{M}_M(S)$ is left cancellable in $\underline{M}_M(S)$, application of the involution shows that every element of $\underline{M}_M(S)$ is right cancellable in $\underline{M}_M(S)$. □

The hypothesis of Proposition 2.4 is satisfied if S is a cancellable submonad of M. In particular it holds with $S = M$ if M is a cancellation monad. Hence we have

2.5. Corollary. Let M be a cancellation Jordan monad. Then every element of the multiplication monoid $\underline{M}$ is left and right cancellable in $\underline{M}$ and there exists a unique involution $T \to T^*$ in $\underline{M}$ such that (T,T^*) is a structural pair for every $T \in \underline{M}$. □

3. Localization of Monoids

In the next section we shall consider localization of a Jordan monad M with respect to a submonad S. The method we shall employ will be based on localization of (associative) monoids applied to the multiplication monoid $\underline{M}(M)$ and its submonoid $\underline{M}_M(S)$. We shall now develop the associative theory in a form suitable for application to Jordan monads.

Let $\underline{M}$ be a monoid, $\underline{S}$ a submonoid such that $\underline{S}$ is right cancellable in $\underline{M}$: $a_1s = a_2s$ for $a_i \in \underline{M}$, $s \in \underline{S}$, implies $a_1 = a_2$. We assume also that $\underline{S}$ has the <u>left Ore</u> (or <u>left common multiple</u>) <u>properties</u> that given $s \in \underline{S}$, $a \in \underline{M}$, there exist $s' \in \underline{S}$, $a' \in \underline{M}$ such that $a's = s'a$ and, moreover, if $a \in \underline{S}$ then a' can be chosen in $\underline{S}$ to fulfill this condition.

We suppose also that we have an action of $\underline{M}$ on a set X (that is, $ax \in M$ is defined for $a \in \underline{M}$, $x \in X$ so that $1x = x$ and $(ab)x = a(bx)$ for $a,b \in \underline{M}$, $x \in X$). We shall define a set $\underline{S}^{-1}X$ of fractions $s^{-1}x$, $s \in \underline{S}$, $x \in X$ on which we have a natural action of $\underline{M}$. To define $\underline{S}^{-1}X$ we consider the product set $\underline{S} \times X$ and we introduce a relation $\sim$ in this set by declaring that $(s_1,x_1) \sim (s_2,x_2)$ if there exist $t_1,t_2 \in \underline{S}$ such that

$$(t_1s_1,t_1x_1) = (t_2s_2,t_2x_2). \tag{3.1}$$

Evidently $\sim$ is reflexive and symmetric. Also it is transitive since if we have $t_i \in \underline{S}$ such that $(t_1s_1,t_1x_1) = (t_2s_2,t_2x_2)$ and $t_j' \in \underline{S}$ such that $(t_2's_2,t_2'x_2) = (t_3's_3,t_3'x_3)$ then there exist $r,r' \in \underline{S}$ such that $rt_2 = r't_2'$ and then $(rt_1s_1,rt_1x_1) = (rt_2s_2,rt_2x_2) = (r't_2's_2,r't_2'x_2) = (r't_3's_3,r't_3'x_3)$. Thus $(s_1,x_1) \sim (s_3,x_3)$ follows from $(s_1,x_1) \sim (s_2,x_2)$ and $(s_2,x_2) \sim (s_3,x_3)$. We denote the quotient set of $\underline{S} \times X$ relative to $\sim$ as $\underline{S}^{-1}X$ and denote the equivalence class of (s,x) as $s^{-1}x$.

Let $a \in \underline{M}$ and choose $s' \in \underline{S}$, $a' \in \underline{M}$ so that $a's = s'a$. Then we put

$$a(s^{-1}x) = s'^{-1}(a'x). \tag{3.2}$$

We have to show that this is single valued. First, we show independence of the choice of the common multiple: Let $a''s = s''a$. Then we have $t', t'' \in \underline{S}$ such that $t's' = t''s''$ and hence $t''a''s = t''s''a = t's'a = t'a's$. Since s is right cancellable this gives $t''a'' = t'a'$. Hence $(t's', t'a'x) = (t''s'', t''a''x)$ so $s'^{-1}(a'x) = s''^{-1}(a''x)$. Next we show independence of the choice of representative (s,x) of $s^{-1}x$. From the definition of $\sim$ it suffices to show that the right hand side of (3.2) is unchanged if we replace (s,x) by (ts,tx), $t \in \underline{S}$. Let $a'' \in \underline{M}$, $s'' \in \underline{S}$ satisfy $a''ts = s''a$. Then the new right hand side of (3.2) is $(s'')^{-1}(a''tx)$. On the other hand, the relation $(a''t)s = s''a$ and the independence of (3.2) of the choice of common multiple gives $(s'')^{-1}(a''tx) = (s')^{-1}(a'x)$.

We claim that (3.2) defines an action of $\underline{M}$ on $\underline{S}^{-1}X$. For, let $a,b \in \underline{M}$ and determine $s' \in \underline{S}$, $b' \in \underline{M}$ so that $b's = s'b$. Then $b(s^{-1}x) = (s')^{-1}(b'x)$. Next determine $a'' \in \underline{M}$, $s'' \in \underline{S}$ so that $a''s' = s''a$. Then

$$a(b(s^{-1}x)) = a((s')^{-1}b'x)) = (s'')^{-1}(a''b'x).$$

Since $s''(ab) = a''s'b = (a''b')s$ we have $(ab)s^{-1}x = (s'')^{-1}(a''b'x)$. Hence $a(b(s^{-1}x)) = (ab)(s^{-1}x)$. It is clear also that $1(s^{-1}x) = s^{-1}x$ and so we do have an action of $\underline{M}$ on $\underline{S}^{-1}X$. In other words, if $\lambda(a)$ denotes the map $s^{-1}x \to a(s^{-1}x)$ then

$$\lambda(ab) = \lambda(a)\lambda(b), \qquad \lambda(1) = 1_{\underline{S}^{-1}X}. \tag{3.3}$$

If $r,t \in \underline{S}$ then $(st)^{-1}x = (rst)^{-1}(rx)$; hence $s^{-1}x \to (st)^{-1}x$ is a well-defined transformation of $\underline{S}^{-1}X$. Moreover, this is the inverse of $\lambda(t)$. For, $\lambda(t)$ is $s^{-1}x \to t(s^{-1}x) = (s')^{-1}(t'x)$ if $t's = s't$, $s',t' \in S$, and $(s't)^{-1}(t'x) = (t's)^{-1}(t'x) = s^{-1}x$ and $t((st)^{-1}x) = s^{-1}x$ since $1(st) = st$.

We have the "regular" action of $\underline{M}$ on $\underline{M}$ by left multiplication. Taking $X = \underline{M}$ we obtain the set $\underline{S}^{-1}\underline{M}$ on which the action of $\underline{M}$ is defined by (3.2). In this case we denote the map $s^{-1}x \to a(s^{-1}x)$ by L_a and we define a product in $\underline{S}^{-1}\underline{M}$ by

$$(3.4) \qquad (s_1{}^{-1}a_1)(s_2{}^{-1}a_2) = L_{s_1}{}^{-1}L_{a_1}(s_2{}^{-1}a_2).$$

This is well-defined since if $t \in \underline{S}$ then $L_{ts_1}{}^{-1}L_{ta_1} = L_{s_1}{}^{-1}L_t{}^{-1}L_tL_{a_1} = L_{s_1}{}^{-1}L_{a_1}$ (by (3.3)). If $t' \in \underline{S}$, $a' \in \underline{M}$ satisfy $t'a_1 = a's_2$ then (3.4) and the formula for $L_{s_1}{}^{-1}$ give $(s_1{}^{-1}a_1)(s_2{}^{-1}a_2) = (t's_1)^{-1}(a'a_2)$. Hence if we denote the map $s^{-1}a \to (s_1{}^{-1}a_1)(s^{-1}a)$ in $\underline{S}^{-1}\underline{M}$ by $L_{s_1{}^{-1}a_1}$ then $L_{s_1{}^{-1}a_1} = L_{s_1}{}^{-1}L_{a_1}$ and

$$\begin{aligned} L_{(s_1{}^{-1}a_1)(s_2{}^{-1}a_2)} &= L_{(t's_1)^{-1}(a'a_2)} = L_{t's_1}{}^{-1}L_{a'a_2} \\ &= L_{s_1}{}^{-1}L_{t'}{}^{-1}L_{a'}L_{a_2} \text{ (by (3.3))} = L_{s_1}{}^{-1}L_{a_1}L_{s_2}{}^{-1}L_a \\ &\text{(by } L_{t'}L_{a_1} = L_{a'}L_{s_2} \text{ by (3.3))} = L_{s_1{}^{-1}a_1}L_{s_2{}^{-1}a_2}. \end{aligned}$$

This is equivalent to the associative law of multiplication in $\underline{S}^{-1}\underline{M}$. Also if we put $1 = 1^{-1}1 \in \underline{S}^{-1}\underline{M}$ then (3.4) shows that this is the unit in $\underline{S}^{-1}\underline{M}$; hence $\underline{S}^{-1}\underline{M}$ is a monoid.

We consider again an arbitrary set X on which $\underline{M}$ acts and the set $\underline{S}^{-1}X$ of fractions $s^{-1}x$. We define

$$(s^{-1}a)(t^{-1}x) = \lambda(s)^{-1}\lambda(a)(t^{-1}x) \tag{3.5}$$

for $s,t \in \underline{S}$, $a \in \underline{M}$, $x \in X$. As before this is well-defined and $1(t^{-1}x) = t^{-1}x$. The foregoing calculations with the L's replaced by the λ's show that (3.5) defines an action of $\underline{S}^{-1}\underline{M}$ on $\underline{S}^{-1}X$.

We have the natural maps i: $a \to \bar{a} = 1^{-1}a$ and ι: $x \to \bar{x} = 1^{-1}x$ of $\underline{M}$ into $\underline{S}^{-1}\underline{M}$ and of X into $\underline{S}^{-1}X$. The first is a homomorphism of monoids and $\bar{a}\bar{x} = \overline{ax}$. Moreover, $a(s^{-1}x) = \bar{a}(s^{-1}x)$ for the actions of $\underline{M}$ and $\underline{S}^{-1}\underline{M}$ on $\underline{S}^{-1}X$. Every element of $\underline{S}^{-1}X$ has the form $\bar{s}^{-1}\bar{x}$ for $s \in \underline{S}$, $x \in X$. These assertions follow directly from the definitions. The definition of $\underline{S}^{-1}X$ shows also that we have $\bar{x} = \bar{y}$ if and only if there is an $s \in \underline{S}$ such that $sx = sy$. Hence ι is injective if and only if every $s \in \underline{S}$ is cancellable on X. In particular, i is a monomorphism if and only if $\underline{S}$ is left cancellable in $\underline{M}$ (as well as right cancellable, which we assumed at the start).

We summarize our results in the

3.6. Localization Theorem for Monoids. Let $\underline{M}$ be a monoid, $\underline{S}$ a submonoid satisfying the following conditions:

1. Every element of $\underline{S}$ is right cancellable in $\underline{M}$

2. $\underline{S}$ satisfies the left Ore conditions: Given s_1, s_2, $s \in \underline{S}$, $a \in \underline{M}$, there exist t_1, t_2, $t \in \underline{S}$, $b \in \underline{M}$ such that

$$t_1s_1 = t_2s_2, \qquad ta = bs.$$

Then we can construct a monoid $\underline{S}^{-1}\underline{M}$ and a homomorphism $i\colon a \to \bar{a}$ of $\underline{M}$ into $\underline{S}^{-1}\underline{M}$ such that every $\bar{s}$, $s \in \underline{S}$, is invertible in $\underline{S}^{-1}\underline{M}$ and every element of $\underline{S}^{-1}\underline{M}$ has the form $\bar{s}^{-1}\bar{a}$, $s \in \underline{S}$, $a \in \underline{M}$.

Moreover, given a set X on which $\underline{M}$ acts, we can construct a set $\underline{S}^{-1}X$ on which $\underline{M}$ and $\underline{S}^{-1}\underline{M}$ act and a map ι: $x \to \bar{x}$ of X into $\underline{S}^{-1}X$ such that every element of $\underline{S}^{-1}X$ has the form $\bar{s}^{-1}\bar{x}$, $s \in \underline{S}$, $s \in X$ and

$$(3.7) \qquad \overline{ax} = \bar{a}\bar{x}, \qquad a(\bar{s}^{-1}\bar{x}) = \bar{a}(\bar{s}^{-1}\bar{x})$$

hold for $a \in \underline{M}$, $s \in \underline{S}$, $x \in X$.

The homomorphism i is a monomorphism if and only if every element of $\underline{S}$ is left cancellable in $\underline{M}$ and ι is injective if and only if every element of $\underline{S}$ is (left) cancellable on X ($sx = sy \Rightarrow x = y$). □

Remark. The hypothesis that $\underline{S}$ is right cancellable in $\underline{M}$ was used in only one place: to show that (3.2) is independent of the choice of a' and s'. It would be preferable if we could avoid the use of right cancellability in these considerations.

4. Localization of Jordan Monads

We shall now consider the analogue fcr Jordan monads of the localization of associative monoids. Let M be a Jordan monad, S a submonad. We seek to construct a Jordan monad $S^{-1}M$ and a homomorphism $i\colon a \to \bar{a}$ of M into $S^{-1}M$ such that the elements $\bar{s}$,

$s \in S$, are invertible in $S^{-1}M$ and every element of $S^{-1}M$ has the form $U_{\bar{s}}^{-1}\bar{a}$, $s \in S$, $a \in M$. The procedure for defining $S^{-1}M$ will be based on associative localization. Let $\underline{M} = \underline{M}(M)$ be the multiplication monoid of M and let $\underline{S} = M_M(S)$ the submonoid generated by the maps U_s, $s \in S$. Under suitable conditions we shall construct the monoid $\underline{S}^{-1}\underline{M}$, the set $\underline{S}^{-1}M$ and the action of the first of these on the second as in the last section. Then we shall define a map U of $\underline{S}^{-1}M$ into its monoid of maps to obtain the Jordan monad $S^{-1}M$.

We require the following conditions on S, or, more precisely, on the submonoid $\underline{S} = \underline{M}_M(S)$ of $\underline{M}$.

1. (Operator cancellation). The maps U_s, $s \in S$, are left and right cancellable in $\underline{M}$.

2. (Operator common multiple property). Let s_1, s_2, $s \in S$, $a \in M$. Then there exist structural pairs (S_1, S_1^*), (S_2, S_2^*), (S, S^*), (A, A^*) with $S_1, S_2, S, S_1^*, S_2^*, S^* \in \underline{S}$ and $A, A^* \in \underline{M}$ such that

$$(4.1) \qquad S_1 U_{s_1} = S_2 U_{s_2}, \qquad U_{s_1} S_1^* = U_{s_2} S_2^*$$

$$(4.2) \qquad S U_a = A U_s, \qquad U_a S^* = U_s A^*.$$

Remark: We can always assume S is a U-operator U_t, replacing S by $S' = S^* S = U_{S^*(1)} = S'^*$ and A by $A' = S^* A$, $A'^* = A^* S$.

By Corollary 2.5, if M is a cancellation Jordan monad (every U_a is injective) then any submonad S of M satisfies the operator cancellation condition 1. By Proposition 2.4, we have the involution $S \to S^*$ in $\underline{S}$ such that (S, S^*) is a structural pair. Evidently this implies that the two conditions in (4.1) are equivalent and the same is true of those in (4.2). Moreover, in a cancellation monad, condition

(4.1) is equivalent to the <u>element common multiple property</u> that given s_1, s_2 there exist $t_1, t_2 \in S$ such that $U_{s_1} t_1 = U_{s_2} t_2$. For this gives $U_{s_1} U_{t_1} U_{s_1} = U_{s_2} U_{t_2} U_{s_2}$ and so we have (4.1) with $S_1 = U_{s_1} U_{t_1}$ and $S_2 = U_{s_2} U_{t_2}$. On the other hand, if we have the second condition in (4.1) then we have $U_{s_1} t_1 = U_{s_2} t_2$ for $t_1 = S_1^* 1$, $t_2 = S_2^* 1$. Evidently (4.2) is superfluous if $S = M$. Hence we have

4.3. <u>Proposition</u>. If M is a cancellation Jordan monad then M automatically satisfies the operator cancellation condition 1, and satisfies the operator common multiplication condition 2, if and only if it has the element common multiple property. □

We shall refer to the two first conditions in (4.1) and (4.2) taken together as the <u>left operator common multiple conditions</u> and the two second conditions in (4.1) and (4.2) as the <u>right operator common multiple conditions</u>. The importance of these conditions is that they imply the usual left (right) Ore property for the submonoid $\underline{S}$ of the monoid $\underline{M}$.

4.4. <u>Proposition</u>. Assume that the submonad S of M satisfies the left operator common multiple conditions (4.1) and (4.2). Then the left Ore conditions are satisfied for the submonoid $\underline{S}$ of the multiplication monoid $\underline{M}$ of M.

Proof. We have to show that given $S_1, S_2, S \in \underline{S}$, $A \in \underline{M}$, there exist $T_1, T_2, T \in \underline{S}$, $B \in \underline{M}$ such that $T_1 S_1 = T_2 S_2$ and $TA = BS$. The proofs in the two cases are similar, so we shall give only the one for the second case. We write $A = U_{a_1} U_{a_2} \ldots U_{a_n}$

and we shall prove the result by induction on n. For the case $n = 1$, $A = U_a$, we write $S = U_{s_1} \dots U_{s_m}$, $s_i \in S$, $S^* = U_{s_m} \dots U_{s_1}$. Then (S,S^*) is a structural pair, so $S^*S = U_s$, $s = S^*1 \in S$. By the hypothesis, we have $A' \in \underline{M}$, $T' \in \underline{S}$ such that $A'U_s = T'U_a$. Then $A'S^*S = A'U_s = T'U_a$ and so $B = A'S^*$, $T = T'$ satisfy the condition. Now assume the result for $n-1$. Then there exist $B' \in \underline{M}$, $T' \in \underline{S}$ such that $B'S = T'U_{a_2} \dots U_{a_n}$, and, by the case $n = 1$, there exist $B'' \in \underline{M}$, $T'' \in \underline{S}$ such that $T''U_{a_1} = B''T'$. Hence

$$T''A = T''U_{a_1} \dots U_{a_n} = B''T'U_{a_2} \dots U_{a_n} = B''B'S.$$

Thus $T = T''$, $B = B''B'$ satisfy the condition. □

Evidently the cancellation condition 1. implies that every element of $\underline{S}$ is right cancellable in $\underline{M}$. The right cancellability and Proposition 4.4 imply that we can construct as in the last section the monoid $\underline{S}^{-1}\underline{M}$ and the set $\underline{S}^{-1}M$ on which $\underline{M}$ and $\underline{S}^{-1}\underline{M}$ act with the properties stated in Proposition 3.6. The elements of $\underline{S}^{-1}M$ can be written as $S^{-1}x$, $S \in \underline{S}$, $x \in M$, and $S^{-1}x = T^{-1}y$ if and only if there exist $T_1,T_2 \in \underline{S}$ such that $(T_1S,T_1x) = (T_2T,T_2y)$. By Proposition 2.4 we have the involution $S \to S^*$ in $\underline{S}$ such that (S,S^*) is a structural pair. Then $S^{-1}x = (S^*S)^{-1}(S^*x) = U_s^{-1}y$ where $s = S^*1 \in S$ and $y = S^*x$. To avoid confusion with the meaning of $U_s^{-1}y$ in M, we now denote the element $U_s^{-1}y$ of $\underline{S}^{-1}M$ by $s^{-1}y$. Thus every element of $\underline{S}^{-1}M$ can be written in the form

$$s^{-1}y, \quad s \in S, \quad y \in M.$$

Accordingly, we denote $\underline{S}^{-1}M$ by $S^{-1}M$. As in the last section, we denote the map $S^{-1}x \to A(S^{-1}x)$, $A \in \underline{M}$, by $\lambda(A)$. Then $\lambda(1) = 1_{S^{-1}M}$, $\lambda(AB) = \lambda(A)\lambda(B)$ and $\lambda(S)$ is invertible if $S \in \underline{S}$.

We wish to introduce a U-operator on $S^{-1}M$ to make this a Jordan monad in which $U_{\bar{x}} = \lambda(U_x)$ for $\bar{x} = 1^{-1}x$. This and the condition JM2 require that $U_{s^{-1}x} = \lambda(U_s)^{-1}\lambda(U_x)\lambda(U_s)^{-1}$. It follows that we would have to have

$$U_{S^{-1}x} = \lambda(S)^{-1}\lambda(U_x)\lambda(S^*)^{-1}. \tag{4.5}$$

We need to show that the right hand side is unaltered if we replace (S,x) by (TS,Tx), $T \in \underline{S}$. This follows from

$$\begin{aligned}\lambda(TS)^{-1}\lambda(U_{Tx})\lambda((TS)^*)^{-1} &= \lambda(TS)^{-1}\lambda(TU_xT^*)\lambda((TS)^*)^{-1}\\ &= \lambda(S)^{-1}\lambda(T)^{-1}\lambda(T)\lambda(U_x)\lambda(T^*)\lambda(T^*)^{-1}\lambda(S^*)^{-1}\\ &= \lambda(S)^{-1}\lambda(U_x)\lambda(S^*)^{-1}\end{aligned}$$

We wish to show that $(S^{-1}M,U,1)$ where $1 = \bar{1} = 1^{-1}1 \in S^{-1}M$ is a Jordan monad. The condition JM1 is clear since $U_{\bar{1}} = \lambda(U_1) = \lambda(1) = 1_{S^{-1}M}$. To prove JM2 it suffices to show that

$$U_{\lambda(U_y)(s^{-1}x)} = \lambda(U_y)U_{s^{-1}x}\lambda(U_y). \tag{4.6}$$

For (4.5) gives $U_{\bar{y}} = \lambda(U_y)$ so (4.6) is equivalent to $U_{U_{\bar{y}}p} = U_{\bar{y}}U_pU_{\bar{y}}$ for $p = s^{-1}x$. If this holds, then $U_{U_{\bar{t}}p} = U_{\bar{t}}U_pU_{\bar{t}}$ for $t \in S$ gives $U_p = U_{\bar{t}}^{-1}U_{U_{\bar{t}}p}U_{\bar{t}}^{-1}$. By (4.5), $U_{t^{-1}y} = U_{\bar{t}}^{-1}U_{\bar{y}}U_{\bar{t}}^{-1}$. Hence

$$U_{U_{t^{-1}y}p} = U_{U_{\bar{t}}^{-1}U_{\bar{y}}U_{\bar{t}}^{-1}p} = U_{\bar{t}}^{-1}U_{U_{\bar{y}}U_{\bar{t}}^{-1}p}U_{\bar{t}}^{-1}$$

$$= U_{\bar{t}}^{-1}U_{\bar{y}}U_{U_{\bar{t}}^{-1}p}U_{\bar{y}}U_{\bar{t}}^{-1} = U_{\bar{t}}^{-1}U_{\bar{y}}U_{\bar{t}}^{-1}U_pU_{\bar{t}}^{-1}U_{\bar{y}}U_{\bar{t}}^{-1}$$

$$= U_{t^{-1}y}U_pU_{t^{-1}y}.$$

Since every element of $S^{-1}M$ has the form $t^{-1}y$, this will prove JM2.

To prove (4.6) we use the common multiple property 2. to obtain structural pairs (Y,Y^*), $Y \in \underline{M}$, (S,S^*), $S \in \underline{S}$, such that $SU_y = YU_s$, $U_yS^* = U_sY^*$. Then $\lambda(S)\lambda(U_y) = \lambda(Y)\lambda(U_s)$, $\lambda(U_y)\lambda(S^*) = \lambda(U_s)\lambda(Y^*)$ and hence $\lambda(U_y)\lambda(U_s)^{-1} = \lambda(S)^{-1}\lambda(Y)$ and $\lambda(U_s)^{-1}\lambda(U_y) = \lambda(Y^*)\lambda(S^*)^{-1}$. Hence

$$\lambda(U_y)(s^{-1}x) = U_y(U_s^{-1}x) = S^{-1}(Yx) \quad \text{(by definition (3.3))}$$

and

$$U_{\lambda(U_y)s^{-1}x} = U_{S^{-1}(Yx)} = \lambda(S)^{-1}\lambda(U_{Yx})\lambda(S^*)^{-1} \quad \text{(by definition (4.5))}$$

$$= \lambda(S)^{-1}\lambda(YU_xY^*)\lambda(S^*)^{-1}$$

$$= \lambda(S)^{-1}\lambda(Y)\lambda(U_x)\lambda(Y^*)\lambda(S^*)^{-1}$$

$$= \lambda(U_y)\lambda(U_s)^{-1}\lambda(U_x)\lambda(U_s)^{-1}\lambda(U_y)$$

$$= \lambda(U_y)U_{s^{-1}x}\lambda(U_y) \qquad \text{(by (4.5))}.$$

This establishes (4.6) and proves that $(S^{-1}M,U,1)$ is a Jordan monad.

We now consider the maps $x \to \bar{x} = 1^{-1}x$ of M into $S^{-1}M$ and $A \to \bar{A} = 1^{-1}A$ of $\underline{M}$ into $\underline{S}^{-1}\underline{M}$. By the localization theorem for monoids (3.8), we have $\overline{U_x y} = \overline{U_x}\,\overline{y} = U_{\bar{x}}\bar{y} = \lambda(U_x)\bar{y} = U_{\bar{x}}\bar{y}$. Also $\bar{1} = 1 \in S^{-1}M$. Hence $x \to \bar{x}$ is a homomorphism of M into $S^{-1}M$. If $s \in S$ then $U_{\bar{s}} = \lambda(U_s)$ is invertible. Hence every $\bar{s}$, $s \in S$, is invertible in $S^{-1}M$. We have seen that every element of $S^{-1}M$ can be written as $s^{-1}y$ which is $U_s{}^{-1}y = \lambda(U_s)^{-1}\bar{y} = U_{\bar{s}}{}^{-1}\bar{y}$. We have therefore shown that every element of $S^{-1}M$ has the form $U_{\bar{s}}{}^{-1}\bar{y}$, $s \in S$, $y \in M$.

The associative result shows also that $x \to \bar{x}$ is injective, hence a monomorphism, if and only if every element of $\underline{S}$ is cancellable on M. This is the case if and only if the submonad S is cancellable in M in the sense that every U_s, $s \in S$, is injective.

Our results prove the

4.7. Localization Theorem for Jordan Monads. Let M be a Jordan monad, S a submonad satisfying the operator cancellation and common multiple properties 1 and 2. Then we can construct a Jordan monad $S^{-1}M$ and a homomorphism ι: $x \to \bar{x}$ of M into $S^{-1}M$ such that every $\bar{s}, s \in S$, is invertible in $S^{-1}M$ and every element of $S^{-1}M$ has the form $U_{\bar{s}}{}^{-1}\bar{x}$, $s \in S$, $x \in M$. The homomorphism ι is a monomorphism if and only if S is a cancellable submonad of M. □

By Corollary 2.5 and Proposition 4.3 the foregoing hypotheses are satisfied for $S = M$ if M is a cancellable monad with the element common multiple property. Hence we have

4.8. Corollary. Let M be a cancellable Jordan monad satisfying the element common multiple condition. Then M can be embedded in a Jordan division monad $\mathcal{D}$ so that every element of $\mathcal{D}$ has the form $U_a^{-1}b$, $a,b \in M$. □

5. Some Characterizations of Quadratic Jordan Algebras

There are two types of difficulties that one encounters in seeking to apply the localization theory of Jordan monads to Jordan algebras: 1) the verification of the identities QJ4, QJ3', QJ3", and QJ4' and 2) the proof that the map U is quadratic. The first of these is readily overcome by imposing some mild conditions on the additive group and module structure of J as Φ-module. These will permit the replacement of QJ3', QJ3", QJ4' by a single condition QJ7 that is comparatively easy to verify. The result establishing the equivalence of the two sets of conditions defining quadratic Jordan algebras -- in the presence of suitable module conditions -- is an extension of a theorem on Jordan algebras due to Koecher [1]. We shall now consider this extension.

We shall now call a triple $(J,U,1)$ a quadratic algebra over the commutative (and associative) ring Φ if J is a Φ-module, 1 is a distinguished element of J and U is a quadratic map of J into $\mathrm{End}_\Phi J$ such that $U_1 = 1$. Thus our assumptions are QJ1 and QJ2 defining quadratic Jordan algebras. We shall obtain some characterizations of quadratic Jordan algebras among quadratic algebras by imposing restrictions on the Φ-module J. We begin by proving

5.1 Proposition. Let $(J,U,1)$ be a quadratic algebra satisfying QJ3 and QJ4 such that the base ring Φ contains distinct elements λ_1,λ_2 and λ_3 such that λ_i and $\lambda_i - \lambda_j$, $i \neq j$, are cancellable on J. Then $(J,U,1)$ is a quadratic Jordan algebra. In particular this holds if Φ is a field with more than three elements or if the additive group of J has no 2 or 3 torsions.

Proof. We have to show that QJ3', QJ3" and QJ4' hold for U. Applying QJ3 to a replaced by $\lambda a + b$, $\lambda \in \Phi$, gives

$$0 = U_{U_{\lambda a+b}c} - U_{\lambda a+b} U_c U_{\lambda a+b}$$

Since $U_{\lambda a+b} = \lambda^2 U_a + \lambda U_{a,b} + U_b$ and $U_{U_{\lambda a+b}c} = U_{\lambda^2 U_a c + \lambda U_{a,b}c + U_b c}$ $= \lambda^4 U_{U_a c} + \lambda^3 U_{U_a c, U_{a,b}c} + \lambda^2 (U_{U_{a,b}c} + U_{U_a c, U_b c}) + \lambda U_{U_{a,b}c, U_b c} + U_{U_b c}$ we obtain

$$(5.2)\quad A(b,c) + \lambda B(b,a,c) + \lambda^2 C(a,b,c) + \lambda^3 B(a,b,c) + \lambda^4 A(a,c) = 0$$

where

$$A(a,c) = U_{U_a c} - U_a U_c U_a = 0$$

$$B(a,b,c) = U_{U_a c, U_{a,b}c} - U_a U_c U_{a,b} - U_{a,b} U_c U_a$$

$$C(a,b,c) = U_{U_a c, U_b c} + U_{U_{a,b}c} - U_a U_c U_b - U_b U_c U_a - U_{a,b} U_c U_{a,b}.$$

Since we can cancel the λ_i, we obtain by a Vandermonde determinant argument $\prod_{i<j=1}^{3} (\lambda_i - \lambda_j) B(a,b,c) = 0$ and $\prod_{i<j=1}^{3} (\lambda_i - \lambda_j) C(a,b,c)$. Since every $\lambda_i - \lambda_j$ is cancellable this gives $B(a,b,c) = 0$ and $C(a,b,c) = 0$. The first relation is QJ3' and the second is QJ3".

A similar argument applied to QJ4 gives QJ4'. Evidently the hypothesis on the λ_i is satisfied if Φ is a field with more than three elements. It is clear also that if J has no 2 or 3-torsion then $\lambda_1 = 1$, $\lambda_2 = 2$, $\lambda_3 = 3$ (in Φ) satisfy the conditions on the λ_i so in this case we have QJ', QJ3" and QJ4'. □

Thus under rather mild restrictions on J as Φ-module the axioms QJ3 and QJ4 alone imply that J is a quadratic Jordan algebra. We shall show next that QJ4 can be replaced by simpler quadratic conditions.

5.4. Proposition. Let $(J,U,1)$ be a quadratic algebra such that QJ3 and the condition on J as Φ-module given in Proposition 5.1 hold. Then the condition QJ4 on U is equivalent to:

QJ6. $$\{aab\} = a^2 \circ b$$

where $\{abc\} = U_{a,c}b = V_{a,b}c$, $a^2 = U_a 1$, $a \circ b = U_{a,b}1$. Moreover, if J has no 2-torsion then QJ6 is equivalent to

QJ7. $$1 \circ a = 2a.$$

Proof. By Proposition 5.1, if QJ4 holds then $(J,U,1)$ is a quadratic Jordan algebra. Then QJ6 is a known identity for such algebras. Conversely, assume QJ6. Define $V_a = U_{a,1}$ so $V_a b = a \circ b$. Then the operator form of QJ6 is

(i) $$V_{a,a} = V_{a^2}.$$

Linearization of QJ6 is automatic since this is quadratic in a;

this gives $\{acb\} + \{cab\} = (a \circ c) \circ b$ and interpreting this as an operator relation on c we obtain

$$\text{(ii)} \qquad V_{b,a} = V_b V_a - U_{a,b}.$$

Putting $b = a$ and using (i) gives

$$\text{(iii)} \qquad 2U_a = V_a^{\,2} - V_{a^2}.$$

If we put $a = c = 1$ in QJ3' (which is valid by the proof of Proposition 5.1) we obtain $U_{1,\{11b\}} = 2U_{1,b}$ which acting on 1 gives $\{11\{11b\}\} = 2\{11b\}$. Since $\{11b\} = 1 \circ b$ we have $1 \circ (1 \circ b) = 2(1 \circ b)$ or $V_1^{\,2} = 2V_1$. On the other hand, $a = b = 1$ in (ii) gives $V_{1,1} = V_1^{\,2} - 2(1_J)$ and so by (i) we have $V_1 = V_{1,1} = 2(1_J)$. Then by (ii) we have

$$\text{(iv)} \qquad V_{1,a} = V_a = V_{a,1} = U_{a,1}.$$

The relation QJ4 will follow if we can show that

$$\text{(v)} \qquad U_{a,a \circ b} = U_a V_b + V_b U_a$$

$$\text{(vi)} \qquad U_{a^2,a \circ b} = U_a U_{a,b} + U_{a,b} U_a$$

$$\text{(vii)} \qquad U_{a^2,a} = U_a V_a = V_a U_a$$

$$\text{(viii)} \qquad U_{a^2,b} = U_{a,b} V_a - V_b U_a = V_a U_{a,b} - U_a V_b.$$

For then we shall have

$U_a V_{b,a} - V_{a,b} U_a = U_a(V_b V_a - U_{a,b}) - (V_{a \circ b} - V_b V_a + U_{a,b}) U_a$ (by (ii) and $V_{a,b} + V_{b,a} = V_{a \circ b}$, which is the linearization of (i)) $=$ $(U_a V_b + V_b U_a) V_a - (U_a U_{a,b} + U_{a,b} U_a) - V_{a \circ b} U_a$ (by (vii)) $=$

$U_{a,a\circ b}V_a - U_{a^2,a\circ b} - V_{a\circ b}U_a$ (by (v) and (vi)) $= 0$ (by (viii) with b replaced by $a\circ b$).

The first two of (v)-(viii) are easy: The first of these is obtained by putting $a = 1$ in QJ3', the second by putting $c = 1$ in QJ3'. Moreover, if we have (vii), then replacing a by $a + \lambda b$, $\lambda \in \Phi$, in this relation gives

$$\lambda(U_{a^2,b} + U_{a\circ b,a}) + \lambda^2(U_{b^2,a} + U_{a\circ b,b}) =$$
$$\lambda(U_aV_b + U_{a,b}V_a) + \lambda^2(U_bV_a + U_{a,b}V_b) =$$
$$\lambda(V_bU_a + V_aU_{a,b}) + \lambda^2(V_aU_b + V_bU_{a,b}).$$

It follows that $U_{a^2,b} + U_{a\circ b,a} = U_aV_b + U_{a,b}V_a = V_bU_a + V_aU_{a,b}$. Subtracting (v) then gives (viii).

We proceed to derive (vii). Puting $b = 1$ in (vi) gives

$$\text{(ix)} \qquad 2U_{a^2,a} = U_aV_a + V_aU_a$$

and applying (vi) to 1 gives $a^2\circ(a\circ b) = U_a(a\circ b) + \{aa^2b\}$ which has the operator form $V_{a^2}V_a = U_aV_a + V_{a,a^2}$. Then, by (ii), we obtain

$$\text{(x)} \qquad U_{a,a^2} - U_aV_a = V_aV_{a^2} - V_{a^2}V_a.$$

Linearizing (iii) gives $2U_{a,b} = V_aV_b + V_bV_a - V_{a\circ b}$; hence $2U_{a,a^2} = V_aV_{a^2} + V_{a^2}V_a - V_{a\circ a^2} = V_aV_{a^2} + V_{a^2}V_a - 2V_{a^3}$ where $a^3 = U_aa$ and hence $2a^3 = 2U_aa = (V_a^2 - V_{a^2})a = a\circ(a\circ a) - a^2\circ a = 2a\circ a^2 - a^2\circ a = a^2\circ a$. Applying (v) to a gives $\{a\,a\,a\circ b\} = U_a(a\circ b) + b\circ a^3$; hence, by (i), $a^2\circ(a\circ b) = U_a(a\circ b) + b\circ a^3$. In operator form this is $V_{a^3} =$

$V_{a^2}V_a - U_aV_a$. Hence $2U_{a,a^2} = V_aV_{a^2} + V_{a^2}V_a - 2V_{a^2}V_a + 2U_aV_a$ and

$$\text{(xi)} \qquad 2(U_{a,a^2} - U_aV_a) \doteq V_aV_{a^2} - V_{a^2}V_a.$$

Comparing (x) and (xi) gives $U_{a,a^2} = U_aV_a$ and subtracting this from (ix) gives $U_{a,a^2} = V_aU_a$. Hence the required relation (vii) holds.

The condition QJ7, which is equivalent to $V_1 = 2(1_J)$ holds in any quadratic Jordan algebra (as we showed above). We shall now show that conversely this condition together with those we are assuming throughout (QJ3 and the hypothesis on J as Φ-module) and the absence of 2-torsion imply QJ6, hence, that J is a quadratic Jordan algebra. First, we put $b = c = 1$ in QJ3" to obtain $U_{a^2,1} + U_{a \circ 1} = 2U_a + U_{a,1}^2$. By QJ7 we obtain

$$\text{(xii)} \qquad 2U_a = U_{a,1}^2 - U_{a^2,1}$$

which linearizes to

$$\text{(xiii)} \qquad 2U_{a,b} = U_{a,1}U_{b,1} + U_{b,1}U_{a,1} - U_{a \circ b,1}.$$

Next we put $a = 1$ in QJ3' and linearize the c to obtain $U_{c,U_{1,b}d} + U_{d,U_{1,b}c} = U_{c,d}U_{1,b} + U_{1,b}U_{c,d}$. Then putting $d = 1$ gives $U_{c,2b} + U_{1,U_{1,b}c} = U_{c,1}U_{1,b} + U_{1,b}U_{c,1}$. Thus $2U_{b,c} + U_{1,\{1cb\}} = U_{1,b}U_{1,c} + U_{1,c}U_{1,b}$. Comparing this with (xiii) shows that $U_{1,\{1ba\}} = U_{a \circ b,1}$. Then acting on 1 gives $2a \circ b = 2\{1ba\}$ and since J has no 2-torsion we have

$$\text{(xiv)} \qquad a \circ b = \{ab1\}$$

which gives $V_a = U_{a,1}$. Now put $c = 1$ in QJ3' to obtain $U_{a^2,b^2} + U_{a\circ b} = U_aU_b + U_bU_a + U_{a,b}^2$ which linearizes to

$U_{a^2,b\circ c} + U_{a\circ b,a\circ c} = U_aU_{b,c} + U_{b,c}U_a + U_{a,b}U_{a,c} + U_{a,c}U_{a,b}$.

Putting $c = 1$ in this gives $U_{a^2,b\circ 1} + U_{a\circ b,a\circ 1} =$ $U_aU_{b,1} + U_{b,1}U_a + U_{a,b}U_{a,1} + U_{a,1}U_{a,b}$, hence $2U_{a^2,b} + 2U_{a,a\circ b}$ $= U_aV_b + V_bU_a + U_{a,b}V_a + V_aU_{a,b}$. Applying this to 1 gives $2a^2\circ b + 2a\circ(a\circ b) = 2U_ab + b\circ a^2 + 2\{aab\} + a\circ(a\circ b)$

Thus we have

$$\text{(xv)} \qquad 2U_a = V_{a^2} + V_a^2 - 2V_{a,a}.$$

Comparing this with (xii) gives $2V_{a,a} = 2V_{a^2}$. Hence $V_{a,a} = V_{a^2}$ which is QJ6. □

Our results reduce the verification of the Jordan axioms for a quadratic algebra to QJ3 and QJ7, provided that there exist $\lambda_1,\lambda_2,\lambda_3$ in Φ such that λ_i and every $\lambda_i - \lambda_j$, $i \neq j$ are cancellable on $\mathcal{J}$ and $\mathcal{J}$ has no 2-torsion. We remark that, as we saw in section 1, the latter condition implies that $\mathcal{J}$ can be embedded in a linear Jordan algebra. Thus in effect we shall be dealing with linear Jordan algebras from the quadratic point of view.

6. Localization of Jordan Algebras

We now consider a quadratic Jordan algebra $(\mathcal{J},U,1)$ and a submonad $\mathcal{S}$ of the multiplicative monad of $\mathcal{J}$. Thus $\mathcal{S}$ is a subset of $\mathcal{J}$ containing 1 and closed under the composition U_ab. If we assume the operator cancellation and common multiple conditions 1 and 2 on S then we can construct the quotient monad $S^{-1}J$ as in section 4 so that we have the properties stated in the localization theorem for Jordan monads.

We use the Φ-module structure of J to endow $S^{-1}J$ with a Φ-module structure in which

$$\alpha(S^{-1}x) = S^{-1}(\alpha x)$$
$$(6.1)\qquad S^{-1}x_1 + S^{-1}x_2 = S^{-1}(x_1+x_2)$$

for $\alpha \in \Phi$, the x's $\in J$, $S \in \underline{S} = M_M(S)$. If S is replaced by TS and x by Tx, $T \in \underline{S}$ then the first right hand side of (6.1) becomes $(TS)^{-1}(\alpha Tx) = (TS)^{-1}(T\alpha x)$ (since T is a Φ-endomorphism) $= S^{-1}(\alpha x)$. Hence (6.1) gives a well-defined action of α on $S^{-1}J$. Similarly, $(TS)^{-1}(Tx_1 + Tx_2) = (TS)^{-1}(T(x_1+x_2)) = S^{-1}(x_1+x_2)$ which implies that the second part of (6.1) defines an addition composition in $S^{-1}J$. It is straightforward to verify that $S^{-1}J$ becomes a Φ-module with the addition and Φ-action defined by (6.1). If $A \in \underline{M}$ then $A(S^{-1}x) = T^{-1}(Bx)$ where $T \in \underline{S}$, $B \in \underline{M}$ and $TA = BS$. Hence $A(\alpha S^{-1}x) = AS^{-1}(\alpha x) = T^{-1}B(\alpha x) = T^{-1}\alpha(Bx)$ (since B is a Φ-endomorphism of J) $= \alpha T^{-1}(Bx) = \alpha A(S^{-1}x)$. Similarly $A(S^{-1}x_1+S^{-1}x_2) = AS^{-1}x_1 + AS^{-1}x_2$. Thus the action of A in $S^{-1}J$ is a Φ-endomorphism. We have defined $U_{S^{-1}x} = S^{-1}U_x(S^*)^{-1}$; hence, it is clear that $U_{S^{-1}x}$ is a Φ-endomorphism of J. Thus U is a map of $S^{-1}J$ into $\mathrm{End}_\Phi S^{-1}J$. It is clear also from definition (6.1) that ι is a Φ-homomorphism of J into $S^{-1}J$. We also have QJ2 and the fundamental formula QJ3 in $S^{-1}J$.

We now impose the conditions on J as Φ-module given in Proposition 5.1 and assume also that J has no 2-torsion. It is readily seen that if $\lambda \in \Phi$ is cancellable on J then λ is cancellable on $S^{-1}J$. Hence our conditions carry over to the Φ-module $S^{-1}J$. It follows that to prove that $(S^{-1}J,U,1)$ is a quadratic Jordan algebra it suffices to show that the map $S^{-1}a \to U_{S^{-1}a}$ is quadratic and QJ3 and QJ7 hold (Proposition 5.4). Since we have seen that QJ3 holds, it remains to prove QJ7 and the quadratic character of the map U. By far the more difficult of the two is the quadratic character of U. We consider this first.

By the definition (4.5), $U_{S^{-1}x} = \lambda(S)^{-1}\lambda(U_x)\lambda(S^*)^{-1}$. Hence for $\rho \in \Phi$, $U_{\rho(S^{-1}x)} = U_{S^{-1}(\rho x)} = \lambda(S)^{-1}\lambda(U_{\rho x})\lambda(S^*)^{-1} =$ $\lambda(S)^{-1}\lambda(U_{\rho 1}U_x)\lambda(S^*)^{-1} = \lambda(S)^{-1}\lambda(U_{\rho 1})\lambda(U_x)\lambda(S^*)^{-1} =$ $\rho^2\lambda(S)^{-1}\lambda(U_x)\lambda(S^*)^{-1} = \rho^2 U_{S^{-1}x}$ since, as is readily verified $\lambda(U_{\rho 1}) = \rho^2 1$. Thus U is homogeneous of degree two. It remains to show that for $p,q \in S^{-1}J$, $U_{p,q} = U_{p+q} - U_p - U_q$ is Φ-bilinear. Since this is symmetric in p and q it suffices to prove that

$\lambda(\Delta U(\rho,p_1,p_2,p_3))$

$$\Delta U(\rho,p_1,p_2,p_3) = U_{\rho p_1+p_2,p_3} - \rho U_{p_1,p_3} - U_{p_2,p_3}$$

$$= U_{\rho p_1+p_2+p_3} - U_{\rho p_1+p_2} - U_{p_3} - \rho U_{p_1+p_3} + \rho U_{p_1} + \rho U_{p_3} - U_{p_2+p_3} + U_{p_2} + U_{p_3}$$

$$= U_{\rho p_1+p_2+p_3} - U_{\rho p_1+p_2} - \rho U_{p_1+p_3} - U_{p_2+p_3} + \rho U_{p_1} + U_{p_2} + \rho U_{p_3}$$

is the 0 map on $S^{-1}J$ for any $\rho \in \Phi$, $p_1,p_2,p_3 \in S^{-1}J$. As is readily seen, we can write $p_i = T^{-1}x_i$ for $x_i \in J$, $T \in \underline{S}$.

Since $U_{T^{-1}(\Sigma\alpha_i x_i)} = \lambda(T)^{-1}U_{\Sigma\alpha_i x_i}\lambda(T^*)^{-1}$ it suffices to show that for all $s \in S$, $a \in J$,

$$\Delta U(\rho,x_1,x_2,x_3)U_s^{-1}a = 0.$$

We do not know whether our hypotheses assure this. All that we shall be able to do is to prove the result under the assumption of two further conditions. These will permit us to write $U_{\Sigma\alpha_i x_i}U_s^{-1}a = U_t^{-1}(M_\alpha a)$ where $\alpha = (\alpha_1,\alpha_2,\alpha_3)$ and M_α is a quadratic function of α in the sense that $M_\alpha = \sum_{i\leq j} \alpha_i\alpha_j M_{ij}$ where $M_{ij} \in \mathrm{End}_\Phi J$. It will follow that $\Delta U(\rho,x_1,x_2,x_3)U_s^{-1}a = U_t^{-1}(M_{(\rho,1,1)} - M_{(\rho,1,0)} - \rho M_{(1,0,1)} - M_{(0,1,1)} + \rho M_{(1,0,0)} + M_{(0,1,0)} + \rho M_{(0,0,1)})$. Then the formula for M_α will show that the parentheses is 0 and hence we shall have the required result that $U_{\Sigma\alpha_i x_i}U_s^{-1}a = 0$.

The conditions we shall impose pertain to the set Str J of <u>structural pairs for</u> J defined to be pairs of Φ-endomorphisms (M,M^*) of J such that $U_{Ma} = MU_aM^*$ and $U_{M^*a} = M^*U_aM$ for all $a \in J$. Evidently these are structural pairs for the multiplicative monad of J and they have the properties we noted in section 2. The conditions we require are:

1'. S is cancellable in Str J in the sense that if $(M_1,M_1^*),(M_2,M_2^*) \in$ Str J and $M_1U_s = M_2U_s$ and $U_sM_1^* = U_sM_2^*$ then $M_1 = M_2$ and $M_1^* = M_2^*$.

2'. Given $x_1,x_2,x_3 \in J$, $s \in S$, there exist $t \in S$ and structural pairs $(M_{x_i},M_{x_i}^*)$, $(M_{x_i+x_j},M_{x_i+x_j}^*)$, $i \neq j = 1,2,3$ such that

(a) $U_tU_{x_i} = M_{x_i}U_s, \quad U_{x_i}U_t = U_sM_{x_i}^*$

$U_tU_{x_i+x_j} = M_{x_i+x_j}U_s, \quad U_{x_i+x_j}U_t = U_sM_{x_i+x_j}^*$

(b) For all $\alpha_i \in \Phi$ and $\alpha = (\alpha_1,\alpha_2,\alpha_3)$, (M_α,M_α^*) is a structural pair if

$$M_\alpha = \Sigma\alpha_i^2M_{x_i} + \sum_{i<j} \alpha_i\alpha_j(M_{x_i+x_j}-M_{x_i}-M_{x_j})$$

$$(6.2)\qquad M_\alpha^* = \Sigma\alpha_i^2M_{x_i}^* + \sum_{i<j} \alpha_i\alpha_j(M_{x_i+x_j}^*-M_{x_i}^*-M_{x_j}^*).$$

These conditions imply that if $x(\alpha) = \Sigma\alpha_ix_i$ then

$$(6.3)\qquad U_tU_{x(\alpha)} = M_\alpha U_s, \quad U_{x(\alpha)}U_t = U_sM_\alpha^*.$$

Thus we have common left and right multiples of U_s and $U_{x(\alpha)}$ where the multipliers M_α and M_α^* given by (6.2) are quadratic functions of $\alpha = (\alpha_1,\alpha_2,\alpha_3)$ and $(M_\alpha,M_\alpha^*) \in \mathrm{Str}\, J$. It would be simpler to assume that the M's are contained in the multiplication monoid. However, the condition 2' would then be too restrictive.

If we put $M_{ii} = M_{x_i}$ and $M_{ij} = M_{x_i+x_j} - M_{x_i} - M_{x_j}$, for $i \neq j$, then we have $M_\alpha = \sum_{i\le j} \alpha_i\alpha_jM_{ij}$. Hence if $M_{\alpha,\beta} = M_{\alpha+\beta} - M_\alpha - M_\beta$ for $\beta = (\beta_1,\beta_2,\beta_3)$ then $M_{\alpha,\beta} = \sum_{i\le j} (\alpha_i\beta_j + \alpha_j\beta_i)M_{ij}$ is bilinear in α and β. A calculation like the one used to obtain $\Delta U(\rho,p_1,p_2,p_3)$ taken in reverse order shows that

$$
\begin{aligned}
(6.4)\quad \Delta M(\rho) &\equiv M_{(\rho,1,1)} - M_{(\rho,1,0)} - \rho M_{(1,0,1)} - M_{(0,1,1)} \\
&\qquad + \rho M_{(1,0,0)} + M_{(0,1,0)} + \rho M_{(0,0,1)} \\
&= M_{(\rho,1,0),(0,0,1)} - \rho M_{(1,0,0),(0,0,1)} - M_{(0,1,0),(0,0,1)} \\
&= 0.
\end{aligned}
$$

In our main theorem 6.9, we shall use conditions 1' and 2' (along with our other conditions) to prove that the map U in $S^{-1}J$ is quadratic. Before doing this we indicate that, while restrictive, these conditions are satisfied in some interesting cases.

We note first that if U_s is injective and (M_1, M_1^*), (M_2, M_2^*) are structural pairs such that $M_1U_s = M_2U_s$ and $U_sM_1^* = U_sM_2^*$ then $M_1 = M_2$ and $M_1^* = M_2^*$ provided there is an $a \in J$ such that $M_1(a)$ is cancellable. For the injectivity of U_s and $U_s(M_1^* - M_2^*) = 0$ imply $M_1^* = M_2^*$. Then $M_1^*M_1 = U_{M_1^*(1)}$ $U_{M_2^*(1)} = M_2^*M_2$ so $M_1^*(M_1 - M_2) = 0$. Hence $M_1U_aM_1^*(M_1 - M_2) = 0$ and $U_{M_1(a)}(M_1 - M_2) = 0$. Since $U_{M_1(a)}$ is injective this gives $M_1 = M_2$. A consequence of this is

6.5. <u>Proposition</u>. If J is a Jordan domain then 1' holds for any submonad S of J not containing 0.

Proof. Every U_s, $s \in S$, is injective so the foregoing result shows that 1' holds unless $M_1 = M_2 = 0$. Then $U_{M_i^*(a)} = M_i^*U_aM_i = 0$, $i = 1,2$, and $M_i^*(x) = 0$ for all x. Hence $M_1^* = M_2^*$ so the result holds also in this case. □

We consider next two instances arising from associative algebras in which the quadratic common multiple property 2' holds.

6.6. Proposition. Let $\underline{A}$ be an associative algebra, $\underline{S}$ a submonoid of the multiplication monoid of $\underline{A}$ satisfying the left and right Ore conditions in $\underline{A}$. Then the quadratic Jordan algebra $\underline{A}^+ = (\underline{A},U,1)$ satisfies 2'.

Proof. Let $x_1,x_2,x_3 \in \underline{A}$, $s \in S$. By the hypothesis, there exist $a_i,a_i' \in \underline{A}$, $s_i,s_i' \in \underline{S}$ such that

$$s_ix_i = a_is, \qquad x_is_i' = sa_i'.$$

We can further find a common left multiple $t_0 = t_1s_1 = t_2s_2 = t_3s_3$ where the $t_i \in \underline{S}$. This can be done by determining $t_1',t_2' \in S$ such that $t = t_1's_1 = t_2's_2$ and then determining $t'',t_3 \in S$ so that $t_0 = t''t = t_3s_3$. This gives $t_0 = t_1s_1 = t_2s_2 = t_3s_3$ for $t_1 = t''t_1'$, $t_2 = t''t_2'$. Similarly, there exist $s_i' \in \underline{S}$ so that $t_0' = s_1't_1' = s_2't_2' = s_3't_3'$. Then $t = t_0't_0 \in \underline{S}$ satisfies $tx_i = t_0't_0x_i = t_0't_is_ix_i = t_0't_ia_is$, $x_it = x_it_0't_0 = x_is_i't_i't_0 = sa_i't_i't_0$; hence

$$t\,x_i = y_is, \qquad x_it = sy_i'$$

for $y_i = t_0't_ia_i$, $y_i' = a_i't_i't_0$. If $c \in \underline{A}$ then (L_c,R_c) is a structural pair for $\underline{A}^+$ since $L_cU_aR_c = L_cL_aR_aR_c = L_{ca}R_{ca} = U_{ca} = U_{L_ca}$ and similarly, $R_cU_aL_c = U_{R_ca}$. It follows that $(L_cR_d,L_dR_c) \in \mathrm{Str}\,\underline{A}^+$ for any $c,d \in \underline{A}$. For $\alpha = (\alpha_1,\alpha_2,\alpha_3)$, put $x(\alpha) = \Sigma\alpha_ix_i$, $y(\alpha) = \Sigma\alpha_iy_i$, $y'(\alpha) = \Sigma\alpha_iy_i'$ and $M_\alpha = L_{y(\alpha)}R_{y'(\alpha)}$, $M_\alpha^* = L_{y'(\alpha)}R_{y(\alpha)}$. Then $tx(\alpha) = y(\alpha)s$, $x(\alpha)t = sy'(\alpha)$ so

$$U_tU_{x(\alpha)}a = tx(\alpha)ax(\alpha)t = y(\alpha)sasy'(\alpha) = M_\alpha U_sa$$

$$U_{x(\alpha)}U_ta = x(\alpha)tatx(\alpha) = sy'(\alpha)ay(\alpha)s = U_sM_\alpha^*a.$$

Thus $U_tU_{x(\alpha)} = M_\alpha U_s$ and $U_{x(\alpha)}U_t = U_sM_\alpha^*$. Hence 2'a holds if we take $M_{x_1} = M_{(1,0,0)}$, $M_{x_2} = M_{(0,1,0)}$, $M_{x_3} = M_{(0,0,1)}$, $M_{x_1+x_2} = M_{(1,1,0)}$, $M_{x_1+x_3} = M_{(1,0,1)}$, $M_{x_2+x_3} = M_{(0,1,1)}$. Moreover, $M_\alpha = L_{y(\alpha)}R_{y'(\alpha)} = \sum_{i,j} \alpha_i\alpha_j L_{y_i}R_{y_j'} = \sum_i \alpha_i^2 L_{y_i}R_{y_i'} + \sum_{i<j} \alpha_i\alpha_j(L_{y_i}R_{y_j'} + L_{y_j}R_{y_i'}) = \Sigma\alpha_i^2 M_{x_i} + \sum_{i<j} \alpha_i\alpha_j(M_{x_i+x_j} - M_{x_i} - M_{x_j})$ since $M_{x_i} = L_{y_i}R_{y_i'}$ and $M_{x_i+x_j} = L_{y_i+y_j}R_{y_i'+y_j'}$. Similarly, $M_\alpha^* = \Sigma\alpha_i^2 M_{x_i}^* + \sum_{i<j} \alpha_i\alpha_j(M_{x_i+x_j}^* - M_{x_i}^* - M_{x_j}^*)$. Hence 2'b holds. □

A second important example in which 2' holds is given in

6.7. <u>Proposition</u>. Let $(\underline{A},*)$ be an associative algebra with involution, $H(\underline{A},*)$ the subalgebra of $\underline{A}^+$ of $*$-symmetric elements, $\underline{T}$ a submonoid of $\underline{A}$ closed under the involution and satisfying the left and right Ore conditions relative to $\underline{A}$. Put $\underline{S} = H(\underline{A},*) \cap \underline{T}$. Then $\underline{S}$ is a submonoid of $H(\underline{A},*)$ satisfying 2' in $H(\underline{A},*)$.

Proof. Given $x_1,x_2,x_3 \in H(\underline{A},*)$, $s \in \underline{S}$, we can, as in the proof of 6.6, determine $\tilde{t} \in \underline{T}$, $z_i \in \underline{A}$ so that $\tilde{t}x_i = z_is$. Then $t = \tilde{t}^*\tilde{t} \in \underline{S}$ and $tx_i = \tilde{t}^*\tilde{t}x_i = \tilde{t}^*z_is = y_is$ for $y_i = \tilde{t}^*z_i \in \underline{A}$. Applying $*$ we obtain $x_it = sy_i^*$. For $x(\alpha) = \Sigma\alpha_ix_i$, $y(\alpha) = \Sigma\alpha_iy_i$ we have

$$tx(\alpha) = y(\alpha)s, \qquad x(\alpha)t = sy(\alpha)^*.$$

If $b \in A$ then $(L_bR_{b^*}, L_{b^*}R_b) \in \text{Str } \underline{A}^+$ and $L_bR_{b^*}$ and $L_{b^*}R_b$ stabilize $H(\underline{A},*)$. It follows that the pair of restrictions of these maps is contained in $\text{Str } H(\underline{A},*)$. We denote these again as

$(L_{b^*}R_b, L_b R_{b^*})$. Hence if we put $M_\alpha = L_{y(\alpha)}R_{y(\alpha)^*}$ and $M_\alpha^* = L_{y(\alpha)^*}R_{y(\alpha)}$, then $(M_\alpha, M_\alpha^*) \in \operatorname{Str}(H(\underline{A},*))$ and

$$U_t U_{x(\alpha)} a = tx(\alpha)ax(\alpha)t = y(\alpha)sasy(\alpha)^* = M_\alpha U_s a$$

$$U_{x(\alpha)} U_t a = x(\alpha)tatx(\alpha) = sy(\alpha)^* ay(\alpha)s = U_s M_\alpha^* a.$$

It follows as in the proof of (6.6) that 2' is satisfied for $\underline{S}$ in $H(\underline{A},*)$. □

Before proceeding to the main theorem, we establish a property of common multiples that will enable us to express the action of $\underline{M}$ on $\underline{S}^{-1}J$ in terms of common multiples in $\operatorname{Str} J$. This is given in

6.8. Lemma. Let M be a Jordan monad, S a submonad satisfying the conditions 1 and 2 of section 4 and condition 1' above. Let $x \in M$, $s,t \in S$ and let (M,M^*) be a structural pair for M such that

$$U_t U_x = MU_s, \qquad U_x U_t = U_s M^*.$$

Then for all $a \in M$ we have

$$U_x(U_s^{-1} a) = U_t^{-1}(Ma)$$

in $S^{-1}M$.

Proof. By the operator common multiple property 4.2, there exists a $t' \in S$ and an inner structural pair (A,A^*) such that

$$U_{t'} U_x = AU_s, \qquad U_x U_{t'} = U_s A^*.$$

Also there exist $T_1, T_2 \in \underline{S} = M_M(S)$ such that

$$T_1 U_{t'} = T_2 U_{t'}, \qquad U_{t'} T_1^* = U_t T_2^*.$$

Then

$$T_1 A U_s = T_1 U_{t'} U_x = T_2 U_t U_x = T_2 M U_s$$

and similarly $U_s A^* T_1^* = U_s M^* T_2^*$. Hence, by 1', $T_1 A = T_2 M$ and $A^* T_1^* = M^* T_2^*$. Then for any $a \in M$ the definition of the action of $\underline{M}$ on $S^{-1}M$ gives

$$U_x(U_s^{-1}a) = U_{t'}^{-1}(Aa) = (T_1 U_{t'})^{-1}(T_1 Aa) = (T_2 U_t)^{-1}(T_2 Ma) = U_t^{-1}(Ma)$$

as required. □

We are now ready to state our main result.

6.9. Localization Theorem for Jordan Algebras. Let $(J,U,1)$ be a quadratic Jordan algebra over a ring Φ such that Φ contains three element $\lambda_1, \lambda_2, \lambda_3$ such that every λ_i and $\lambda_i - \lambda_j$, $i \neq j$, are cancellable on J and J has no 2-torsion. Let S be a submonad of the multiplicative monad of J satisfying the following conditions:

1. The maps U_s, $s \in S$, are left and right cancellable in the multiplication monoid $\underline{M} = \underline{M}(J)$ of J.

1'. The maps U_s, $s \in S$, are cancellable for structural pairs in the sense that if (T_1, T_1^*), (T_2, T_2^*) are structural pairs such that $T_1 U_s = T_2 U_s$ and $U_s T_1^* = U_s T_2^*$ then $T_1 = T_2$ and $T_1^* = T_2^*$.

2. (Operator common multiple property in $\underline{M}$): Given $s_1, s_2, s \in S$, $a \in J$, then there exist **inner** structural pairs (S_1, S_1^*) (S_2, S_2^*), (A, A^*) where the S's $\in \underline{M}_J(S)$ and the A's $\in \underline{M}$ such that

$$S_1 U_{s_1} = S_2 U_{s_2}, \qquad U_{s_1} S_1^* = U_{s_2} S_2^*$$

$$SU_a = AU_s, \qquad U_a S^* = U_s A^*.$$

2'. (The quadratic common multiple property in $\mathrm{Str}\, J$): Given $x_1, x_2, x_3 \in J$, $s \in S$, there exist $t \in S$, and structural pairs $(M_{x_i}, M_{x_i}^*)$, $(M_{x_i+x_j}, M_{x_i+x_j}^*)$, $i \neq j = 1,2,3$, such that $U_t U_{x_i} = M_{x_i} U_s$, $U_{x_i} U_t = U_s M_{x_i}^*$, $U_t U_{x_i+x_j} = M_{x_i+x_j} U_s$, $U_{x_i+x_j} U_t = U_s M_{x_i+x_j}^*$ and for all $\alpha_i \in \Phi$, $(M_\alpha, M_\alpha^*) \in \mathrm{Str}\, J$ if $M_\alpha = \Sigma \alpha_i^2 M_{x_i} + \sum_{i<j} \alpha_i \alpha_j (M_{x_i+x_j} - M_{x_i} - M_{x_j})$ and $M_\alpha^* = \Sigma \alpha_i^2 M_{x_i}^* + \sum_{i<j} \alpha_i \alpha_j (M_{x_i+x_j}^* - M_{x_i}^* - M_{x_j}^*)$.

Then we can construct a Jordan algebra $(S^{-1}J, U, 1)$ and a homomorphism ι: $a \to \bar{a}$ of J into $S^{-1}J$ such that every $\bar{s}$, $s \in S$, is invertible in $S^{-1}J$ and every element of $S^{-1}J$ has the form $U_{\bar{s}}^{-1} \bar{a}$, $s \in S$, $a \in J$. Moreover, ι is a monomorphism if and only if every U_s, $s \in S$, is injective.

Proof. As we showed above, the conditions 1 and 2 permit the construction of the Jordan monad $S^{-1}J$ as in 4.7 and we can endow this with a Φ-module structure satisfying (6.1). To prove $(S^{-1}J, U, 1)$ is a quadratic Jordan algebra it suffices to prove that U is quadratic and QJ7 holds. The proof of the first of these has been reduced to showing that $\Delta U(\rho,x_1,x_2,x_3)U_s^{-1}a = 0$ for all $\rho \in \Phi$, $a,x_i \in J$, $s \in S$. By the quadratic common multiple property we can determine $t \in S$ and structural pairs $(M_{x_i},M_{x_i}^*)$, $i \neq j$, so that 2' holds. By Lemma 6.8, we have $\Delta U(\rho,x_1,x_2,x_3)U_s^{-1}a = U_t^{-1}(\Delta M(\rho))a = 0$ by (6.4). This proves U is quadratic. Next we shall prove QJ7: $1 \circ a = 2a$ where $1 \circ a = U_{a,1}1$. As in the proof of Proposition 5.1, we can linearize QJ3 to obtain QJ3': $U_{U_ac,U_{a,b}c} = U_aU_cU_{a,b} + U_{a,b}U_cU_a$ in $S^{-1}J$. If we put $a = c = 1$ in this, we obtain $U_{1,b\circ 1} = 2U_{1,b}$. Hence if $d = 1\circ b-2b$ then $U_{1,d} = 0$ in $S^{-1}J$. We now use the homomorphism ι: $x \to \bar{x} = 1^{-1}x$ of J into $S^{-1}J$. This is not only a homomorphism of monads as in 4.7, but by (6.1), ι is a homomorphism of quadratic algebras. Hence the image $\bar{J} = \iota J$ is a quadratic Jordan algebra. Now write $d = U_{\bar{s}}^{-1}\bar{z}$ for $z \in J$, $s \in S$, and use QJ3 to obtain $0 = U_{\bar{s}}U_{1,d}U_{\bar{s}} = U_{\bar{s}}U_{1,U_{\bar{s}}^{-1}\bar{z}}U_{\bar{s}} = U_{U_{\bar{s}}1,\bar{z}}$. Hence we have $U_{\bar{c},\bar{z}} = 0$ for $\bar{c} = \bar{s}^2$ in $\bar{J}$ and $\bar{c}$ is cancellable. Now because $\bar{J}$ is a quadratic Jordan algebra we know $U_{\bar{c}}(2\bar{z}) = 2U_{\bar{c}}\bar{z} = (V_{\bar{c}}^2 - V_{\bar{c}^2})\bar{z} = V_{\bar{c}}U_{\bar{c},\bar{z}}1 - U_{\bar{c},\bar{z}}\bar{c} = 0$ since $U_{\bar{c},\bar{z}} = 0$. Since $\bar{c}$ is cancellable and $\bar{J}$ has no 2-torsion (see the remark above that if λ is cancellable on J then it is cancellable on $S^{-1}J$) it follows that $\bar{z} = 0$ and hence $d = 0$ and $1 \circ b = 2b$. This completes the proof that $S^{-1}J$ is a quadratic Jordan algebra. It remains to remark that the statements on ι follows from 4.7. □

In the case of a Jordan domain, we can apply the localization theorem to obtain an embedding in a Jordan division algebra.

Moreover, the hypotheses of 6.9 can be simplified in this case. We note first that the hypothesis on the λ_i implies that the centroid Γ contains three distinct elements $x \to \lambda_i x$. On the other hand, if J is a prime algebra then its centroid is a domain whose elements are cancellable on J, and J can be regarded as an algebra over Γ. Hence if $|\Gamma| \geq 3$ then the hypothesis on the λ_i in 6.9 is fulfilled if we regard J as an algebra over Γ. In particular, this holds if J is a domain. We note next that if J is a domain and for non-zero $s_i \in J$ there exists a $t \neq 0$ in J and a structural pair (M,M^*) such that $U_tU_{s_1} = MU_{s_2}$, $U_{s_1}U_t = U_{s_2}M^*$ then there exists another $t \neq 0$ and an inner structural pair satisfying these conditions. For we have $U_{s_1}U_t{}^2U_{s_1} = U_{s_2}M^*MU_{s_2} = U_{s_2}U_{M^*1}U_{s_2}$. Then

$$U_{U_t{}^2 s_1}U_{s_1} = U_t{}^2U_{s_1}U_t{}^2U_{s_1} = U_t{}^2U_{s_2}U_{M*1}U_{s_2}$$

$$U_{s_1}U_{U_t{}^2 s_1} = U_{s_1}U_t{}^2U_{s_1}U_t{}^2 = U_{s_2}U_{M*1}U_{s_2}U_t{}^2$$

and so if we replace t by $U_t{}^2 s_1$ then this element and the inner structural pair $(U_t{}^2U_{s_2}U_{M*1}, U_{M*1}U_{s_2}U_t{}^2)$ satisfy the conditions. Thus th quadratic common multiple property 2' of 6.9 implies the operator common multiple property 2 of 6.9. We have seen also that conditions 1 and 1' o 6.9 are satisfied in the domain case by any submonad not containing 0 (Propositions 2.4 and 6.5). We can now prove the

6.10. Imbedding Theorem for Jordan Domains. Let $(J,U,1)$ be a quadratic Jordan domain over a commutative ring Φ such that J has no 2-torsion and $|\Gamma| \geq 3$ for the centroid Γ of J. Assume the quadratic common multiple property (2' of 6.9) for the submonad S of non-zero elements. Then J can be imbedded in a Jordan division algebra $S^{-1}J$ whose elements have the form $U_s^{-1}a$, $a,s \in J$, $s \neq 0$.

Proof. We treat J as an algebra over Γ. The foregoing remarks show that the theorem holds for J as algebra over Γ. Since $x \to \alpha x$ is in Γ for every $\alpha \in \Phi$ it follows that the theorem holds for J over Φ. □

We remark that the condition $|\Gamma| \geq 3$ in 6.10 will hold if J has no 3-torsion. Hence the theorem holds in particular if J has no 2-torsion and no 3-torsion and the quadratic common multiple property holds.

References

Daniel J. Britten

[1] On prime Jordan rings H(R) with chain conditions, J. of Algebra 27 (1973), 414-421.

[2] Goldie-like conditions on Jordan matrix rings, Trans. Amer. Math. Soc. 192 (1974), 87-98.

N. Jacobson

[1] Structure and Representations of Jordan Algebras, Colloq. Publ. 39, Amer. Math. Soc., Providence, R.I., 1968.

[2] Lectures on Quadratic Jordan Algebras, Tata Institute of Fundamental Research, Bombay, 1969.

[3] PI-Algebras, An Introduction, Springer-Verlag Lecture Notes 441, Berlin-Heidelberg-New York, 1975.

N. Jacobson and J. Katz

[1] Generically algebraic Jordan algebras, Scripta Math. 29 (1973), 215-227.

M. Knebusch

[1] Der Begriff der Ordung einer Jordan-algebra, Abh. Math. Sem. Hamburg 28 (1965), 168-184.

M. Koecher

[1] Eine characterisierung der Jordan-Algebras, Math. Ann. 148 (1962), 244-256.

O. Loos

[1] Symmetric Spaces I, W. A. Benjamin, New York, 1969.

K. McCrimmon

[1] A general theory of Jordan rings, Proc. Nat. Acad. Sci. U.S.A. 56 (1966), 1072-1079.

[2] The radical of a Jordan algebra, Proc. Nat. Acad. Sci. U.S.A. 62 (1969), 671-678.

[3] The Freudenthal-Springer-Tits construction revisited, Trans. Amer. Math. Soc. 148 (1970), 293-314.

K. Meyberg

[1] Lectures on algebras and triple systems, University of Virginia, 1972.

S. Montgomery

[1] Rings of quotients for a class of special Jordan rings, J. of Algebra 31 (1974), 154-165.

M. L. Racine

[1] The arithmetic of quadratic Jordan algebras, Mem. Amer. Math. Soc. 136 (1973).

C. Tsai

[1] The prime radical in a Jordan ring, Proc. Amer. Math. Soc. 19 (1968), 1171-1175.

Received: February 1977

Personal History and Commentary

1980–1988

In the Spring of 1976, a 10-member delegation of mathematicians was sent by the National Academy of Sciences to visit the People's Republic of China (PRC) as part of an exchange program that had recently been established. The year before, Saunders MacLane and I had suggested such a delegation to the chairman of the Committee on Scholarly Communication with the People's Republic of China at the April meeting of the Academy. I was asked to be a member of our delegation, but declined since the invitations did not include spouses, and I did not wish to go without Florie. By 1979, the exchange program between the US National Academy of Sciences and the PRC had gained momentum in both directions. The time appeared ripe for us to apply for a visit. During 1979, I had some correspondence both oral and written with Professor L. K. Hua who was chairman of the Institute of Mathematics of the Academia Sinica of the PRC, and with Professor Hsio Fu Tuan, chairman of the Department of Mathematics of Peking University, on the subject of a visit by Florie and myself to the PRC. Both Hua and Tuan were old friends whom I had known in the forties when they were in the United States. I had last seen Hua in Leningrad in 1961 when we were delegates to the 4th All-Union Congress of Mathematicians of the USSR, and I had last seen Tuan when he visited me at Johns Hopkins in 1946. In the meantime, Hua had become one of the top scientists of the PRC (Vice Chairman of the Academia Sinica, and member of the Praesidium of the Chinese Parliament).

In February 1980, Florie and I received an invitation to be guests of the Academia Sinica for a period of two weeks in June. We fixed the date of our visit to begin June 1. We planned to enter and to leave the PRC (Canton) by rail from Hong Kong. I had also accepted an invitation to be a delegate and give an invited address at a meeting in Hong Kong of the Southeast Asian Mathematical Society scheduled for June 16–20, 1980. My travel expenses for the visit to the PRC were paid by Yale. Our stay in the PRC stretched to seventeen days, nine in Beijing and eight in visits to Shanghai, Soochow, Hangchow and Canton.

In Beijing we stayed at the Friendship Hotel, a complex of rooms and apartments that had been built by the Russians and used by them until the Sino-Soviet break. We had a comfortable suite of rooms, approximately air-conditioned, and the food in the dining room for Foreign Experts was quite good. Our meals were interspersed with numerous banquets where the food and good fellowship, including speeches by our hosts (that required responses from me), were extraordinary. We had a guide and interpreter, Mr. Shi-Hsueh Chu, assigned to us, as well as a car with chauffeur. Mr Chu took us with the car to our lectures and to cultural events, for example, the Peking Opera, and he accompanied us on our trips until the very end of our stay, when he put us on the train at Canton for Hong Kong. He was an extremely pleasant person. During the time we were in Beijing, the Chilean dictator Pinochet was there on a state visit. We found it amusing to compare his treatment with ours. The only obvious differences we could observe were that his car was black while ours was grey, and there were streamers across the main streets around Tien-An-Min Square welcoming General Pinochet as representative of the Chilean People to the PRC. (This is a good example of the realism of the Deng Xiau-Ping regime that had recently succeeded the "Cultural" Revolution.)

I gave six lectures in Beijing: two lectures at the Mathematical Institute of Academia Sinica entitled Survey of Structure Theory of Jordan Algebras; a lecture at Peking University, History of PI-Algebras; a lecture at the Mathematical Institute of Academia Sinica, PI-Algebras; a survey lecture at Peking Normal University on finite dimensional division algebras; lecture at the Mathematical Institute of Academia Sinica, Finite Dimensional Division Algebras and PI-algebras.

The mathematical background and knowledge of English of my audience varied widely. Ideally, I should have made available English texts ahead of time so that these could have been translated into Chinese for distribution to the audience. I did give Professor Zhu-Xian Wan handwritten copies of two lectures on Jordan algebras. He was the member of the Institute who arranged our program in Beijing. During the time I was lecturing at Peking Normal University, Florie met with a large group of the faculty for an informal disucssion of secondary school mathematics education in the United States. They were especially interested in hearing about SMSG (School Mathematics Study Group), which Florie had been involved in as a writer.

All our lectures were scheduled for mornings, which left the afternoons free for sight-seeing. Our itinerary included Tien-An-Min Square and Mao's mausoleum, the Forbidden City (or Palace Museum), the Summer Palace, the Temple of Heaven, the Fragrant Hill Park, the Great Wall and Ming Tombs. Professor Wan accompanied us on a number of our trips. Also, Dr. Hsi-Tun Hsu, who had received his doctorate at Yale in 1955 with Hille, took the trip with us to the Great Wall.

On June 11, we departed for Shanghai. I gave a lecture at Fudan University, An Introduction to PI-Algebras. This was translated sentence by sentence by one of the mathematicians who had some familiarity with the subject. While I was lecturing, Florie browsed in the library of the university and came across a copy of the Chinese translation of my *Lie Algebras*. I knew that the Chinese had translated the first two volumes of *Lectures in Abstract Algebras*, but I was unaware of the existence of a Chinese translation of *Lie Algebras*. We tried to get a copy. This turned out to be impossible since all existing copies had been destroyed by their owners during the Cultural Revolution. Possession of such foreign books might have led to serious reprisals by the Red guards. As we were leaving Shanghai, I was given the library copy of *Lie Algebras* as a present from the president of Fudan University. Fortunately, I was able to reciprocate by giving them a copy of my *Basic Algebra II* that had just been published in the United States.

We had some very interesting sight-seeing in Shanghai: visits to a large agricultural commune, visit to the Bund, and the Children's palace, where we saw a lovely dance recital. Perhaps the most amazing performance we witnessed during our trip was that of the Shanghai acrobats. Although the light was not quite adequate, we took some wonderful stills and movies of this performance.

After Shanghai, we visited Soochow, Hangchow, Canton and then took the train from Canton to Hong Kong. We arrived June 17, a day late for the start of the 5th Biennial General Meeting and Conference of the Southeast Asian Mathematical Society. Altogether there were 160 mathematicians from 17 countries at the Conference, including a delegation from the PRC headed by Hua. The title of my address at the Conference was Survey of Jordan Structure Theory [96]. After the Conference, we returned to the United States by way of San Francisco.

In August of 1980, shortly before my "official" birthday, I received from Herstein a beautifully bound copy of the February and March issue of the *Journal of Algebra*. On the Table of Contents page of the February issue, there appears the caption, "This issue is Dedicated to Nathan Jacobson on his 70th birthday." On the next page, there is a photograph of me that had been taken and processed by Florie. Apparently there were more papers submitted than could be accomodated in a single issue; the overflow was published in the March issue. Herstein had been the editor and the special bound copy had been prepared by Academic Press to be presented to me by him. What a lovely birthday present! I responded by writing a brief note of thanks to each of the contributors with a quotation from Browning's *Rabbi Ben Ezra* ("Grow old along with me..."). The following is the list of contributors: Marshall Osborn, Michel Racine, Charles Rickart, Craig Huneke, Joe Ferrar, Earl Taft, Bob Wilson, Jim Humphreys, Ed Formanek, Charlie Curtis, David Saltman, Yitz Herstein, Shimshon Amitsur, Lance Small, John Faulkner, Tonny Springer, David Winter, Georgia Benkart, Harry Allen, D. S. Trushin, Bruno Harris, George Seligman, Irving Kaplansky, Ferdinand Veldkamp, Max Koecher, Goro Azumaya, Daya-Nand Verma, Louis Rowen, Jacques Tits, Kevin McCrimmon, Edward Cline, Brian Parshall, Leonard Scott, Leonard Chastkofsky, Walter Feit, Paul Cohn, Warren Dicks.

I was barely over the euphoria of the 70th birthday when I learned of a conference in algebra honoring my retirement from Yale that was scheduled for June 2–5, 1981. The members of the Organizing Committee for the *Conference in Algebra* were Charles Curtis, Kevin McCrimmon, George Seligman, Earl Taft and Maria Wonenburger, all former students, and support was provided by the National Science Foundation. More than 70 mathematicians attended the conference, including 24 of my former Ph.D. students. One of these was Charles Carroll, Jr., who had been my one Ph.D. student at the University of North Carolina.

There were seven 50-minute addresses at the *Conference* given by Shimshon Amitsur, Maurice Auslander, Ed Formanek, Jim Humphreys, Tonny Springer, Paul Cohn and Moss Sweedler. The first five of these gave general survey lectures; Cohn's and Sweedler's talks were more narrowly focused on comparatively recent research. Sweedler reported on joint work with Darrell Haile and Richard Larson. Shorter talks (25 or 15 minutes) were given (in order of presentation) by David Saltman, Murray Schacher, Murray Gerstenhaber, Georgia Benkart, Richard Block, Robert Wilson, Michael Artin, John Faulkner, Robert Bix, Michel Racine, Holger Petersson, Leslie Hogben, Kevin McCrimmon, Bill Lister, Louis Rowen, William Schelter, Lance Small, Wallace Martindale, Bruno Harris, Craig Huneke, Dan Zelinsky, Saul Steinberg, Julius Zelmanowitz, Darrell Haile, Irving Kaplansky, Ronald Infante, Earl Taft, Herbert Kreimer, Saunders MacLane, David Winter, Piotr Blass. The program covered the following topics: Ring Theory; Lie, Jordan, and Nonassociative Systems, Galois Theory. All of these had been of special interest to me over the years. The program was very successful in achieving the stated aim of the Conference of presenting the current status of these fields and of suggesting directions for future research.

Midway in the Conference, on Wednesday evening, there was a gala banquet in the President's Room of Woolsey Hall. In addition to the participants of the Conference and members of the department and their spouses, a number of old friends attended the banquet and cocktail party that preceded it. Among these were Elsie Begle from Palo Alto, Gudrun Ore, Marshall and Vyla Stone, and Rabbi Robert Goldberg, Emeritus Rabbi of Temple Mishkan Israel. After dinner there were several short talks, beginning with an introduction by George Seligman, who acted as MC for the evening. The first speakers were Alice and Dick Schafer, whose friendship dated far back and, in Florie's case, to graduate student days at Eckhart Hall in Chicago. Another speaker was Charlie Carroll who recalled the period before the outbreak of World War II, when he was working with me at North Carolina. Rabbi Goldberg, who had been an inspiration to us for many years, especially because of his espousal of unpopular social and political causes, gave a witty extemporaneous talk in which he alluded to my efforts on human rights. Finally, Saunders MacLane read a poem he had written. The first couplet read:

"We've gathered here in festive state
A special guest to celebrate."

On Thursday, Florie gave a luncheon for the "women mathematicians and mathematicians' women." It was *not* catered.

Thursday evening was very special: a program of cello and piano sonatas by Louis Rowen and Leonard Gillman. They played four sonatas: Sonata in D Major (BMV 1028) by J. S. Bach, Sonata in E minor OP. 38 by Brahms, Sonata in A major OP. 69 by Beethoven, and Sonata in G minor OP. 19 by Rachmaninoff. Louis had come to Yale in 1969 to study mathematics with me and the cello with a master cellist and teacher, Aldo Parisot, of the Yale School of Music. During his first three years at Yale, it was a toss-up as to whether he would become a mathematician or a cellist. In the end, he chose mathematics after a brief try at being a professional musician as cellist with the Jerusalem Philharmonic. Lenny Gillman had graduated from the Juilliard School of Music before becoming a mathematician. Their concert was of professional caliber in spite of the fact that they had almost no time for rehearsals.

The proceedings of the Conference in Algebra, called *Algebraists' Homage*, was published as Volume 13 of the American Mathematical Society series *Contemporary Mathematics*. This contains the papers presented at the conference and several additional papers written by Barbara Osofsky, Carl Faith, Walter Feit and Greg Zuckerman, and Tsuneo Tamagawa. George Seligman, who, together with Shimshon Amitsur and David Saltman, was an editor of *Algebraists' Homage*, presented me with a bound copy that had the title *Algebraists Homage to Nathan Jacobson* on its cover. The introduction stating the theme of the conference was written by George, and he also managed to get the signatures of all the contributors, which appear on the first page of my bound copy.

In 1977, the Mathematics Department of the University of Arkansas at Fayetteville, under the chairmanship of Bernard L. Madison, initiated a Spring Lecture Series. These lectures were to serve as bases for monographs to be published in the *University of Arkansas Lecture Notes in Mathematics*. I was asked to give the fifth of this series of lectures in the spring of 1981. My five lectures on Jordan Structure Theory gave an outline of the subject that included the remarkable results of Effim Zel'manov on prime Jordan algebras as extended by McCrimmon to the quadratic case and Zel'manov's results on Jordan division algebras. The monograph giving complete proofs was published in the Arkansas series in 1981 under the title *Structure Theory of Jordan Algebras*. The five lectures I gave were supplemented by two by McCrimmon on Generalized Jordan Structures (triples and pairs) and single lectures by Robert Bix, Harry F. Smith, John Faulkner, Josef Dorfmeister and Erhard Neher on non-associative systems and their applications. Thus the Spring Lecture Series became a conference that attracted a number of algebraists working in non-associative algebra.

A comparatively recent addition to the faculty at Fayetteville was an old friend from another world, Boris Schein of the Soviet Union, whom we had met in 1974 in East Germany. We enjoyed seeing him again and meeting his family.

My retirement from Yale began July 1981. This, of course, marked a discontinuity in my career—a comparatively minor one, since we had decided to remain in New Haven except for visits of no more than a semester's duration elsewhere. I was allowed to keep my office, which I used mainly for correspondence connected with editorial work on the *Journal of Algebra* and *Advances in Mathematics*. I also continued on our algebra grant, which paid my travel expenses to conferences, and I continued to participate in the algebra seminar. Florie took an early retirement from Albertus Magnus College in 1983. She had been teaching there since 1955 and had been chairperson since 1958. She received many honors on her retirement, including the award of an honorary degree of Doctor of Humane Letters from Albertus Magnus College.

In August 1980, I received a very friendly letter from Beno Eckmann of the ETH (Eidgenössische Technische Hochschule) inviting me to be a guest of their Mathematics Research Institute for an indefinite period with no duties except an occasional seminar lecture. The stipend for the guest membership would be adequate to cover living expenses and a portion of the travel costs. An alternative he proposed was that I visit during one semester and give a "postgraduate course" of my own choosing for their advanced students. I chose the latter alternative for the period of October 15, 1981 to February 1, 1982, and offered one of the following two subjects for the course: (1) Structure Theory of Jordan algebras (2) Finite Dimensional Associative Division Algebras. They chose (2), which fit in very well with my own plans since I had gone back to this subject, which had been my first love in mathematics, and was just at the beginning of a joint writing project with David Saltman on these algebras.

We decided to go to Zurich by way of Athens since this would give us a bit of a vacation and, oddly enough, cost slightly less than the direct trip to Zurich. We flew to Paris where we spent four days and then to Greece where we stayed for ten days. I had written Spiros Zervos of our plans and through him we were invited to stay at an apartment belonging to Dr. "Rosie" Voreadou, who had obtained her doctorate with MacLane at Chicago. She looked after us during our stay in Athens, and, at Zervos' suggestion, she accompanied us to Crete for two days and to Santorini for a day. We traveled to these islands on regularly scheduled commercial steamers, which were cheap and so attracted many young back-packing tourists. Rosie also accompanied us on a trip to the important Byzantine center of Mistras near ancient Sparta. On our last day in Athens, I gave a two-hour survey lecture on Jordan structure theory.

We had found Athens noisy and polluted. The charming tavernas with their folk music and dancing that we remembered from our 1965 visit had given way to discos and rock music. This was not the case in Crete, where the old traditions, such as dancing by male couples, were still maintained.

We arrived in Zurich on October 1 and were quickly settled in a small apartment on Tal Strasse that we had rented for three months. For our last month, we moved to another apartment since the owner of the one on Tal Strasse returned from the United States to spend some months in Zurich. Our apartment had a balcony that over-looked a small canal, and it was easily accessible to the ETH Centrum by street car or cable car. It was also very close to Bahnhofstrasse, the main shopping and promenade street of Zurich. We enjoyed window-shopping and people-watching on this street, but at times we felt a bit stifled by the upper-class bourgeois atmosphere. At such times, it was an easy matter to cross the Limmat River to the Niederdorf, where the hippies concentrated and where one could hear street musicians from many parts of the world. There were quite a few seedy characters in this area, but we never felt threatened since muggings were unheard of.

I gave a two-hour lecture once a week on Finite Dimensional Associative Division Algebras. I prepared a set of lecture notes consisting of two chapters: I. Skew Polynomials and Division Algebras and II. General Structure Theory of Central Simple Algebras. These were typed and duplicated by the department and made available to the students and others. The two auditors who appeared most interested in my lectures were Professor Albert Knus of the ETH and Dr. K. Hoechsmann of Heidelberg, who had been a student of Eckmann's and was visiting the ETH. A question raised by Knus after one of my lectures led to a paper of mine [79]. I also gave a survey talk on Jordan algebras at the colloquium.

During our stay in Zurich, I lectured at a number of mathematics centers. I went alone to Freiburg in October at the invitation of Otto Kegel. In November, we visited Innsbruck, Münster and Hagen, where our hosts were Loos, Koecher and Petersson, respectively. From December 12 to December 16, we were the guests of the Tits in Paris. Since I was to lecture at the Collège de France, we were housed in a visitors suite of the Collège. George Seligman was at the IHES on a trienniel leave from Yale at this time, and they invited us to a concert by Jean Pierre Rampal. I gave two lectures at the Collège on Zel'manov's work determining the Jordan division algebras. Lecturers at the Collège are asked to sign a register that has been kept since almost the founding of the Collège in 1530. Also, they receive a beautiful bronze medallion on which is engraved the lecturer's name and the year of his lecture.

After my lectures, there was a reception that overlapped a bit with another reception at the Institute. The latter preceded a "Séance Solonnel" of the Academy of Sciences. Tits, as a member, invited us to attend. The main business of the "Séance" was the award of honors beginning with foreign memberships and ending with book awards to brillant high school students of science. After these, there was a rather long, largely philosophical address by the Secretary. It was a memorable visit, not the least because December 13 was the day Polish General Jaruzelski staged a coup to snuff out the incipient democracy represented by the Solidarity movement.

After we returned to Zurich, we began thinking about a place to visit during the Christmas recess from the ETH. We considered several possibilities (the high Alps, Morocco), and we decided to go to Istanbul. We had corresponded with our Yale friends, the Feza Gurseys, who came from Istanbul, and they referred us to a former student of his, Dr. Mehmet Erbudak, who was Turkish and who held a position at the ETH. We had several conferences with him on where we should stay and what we should see in Istanbul. At one of these, we mentioned that we were interested in having some contact with the Jewish community of Istanbul, which we knew was a declining remnant of what had once been a relatively large and important minority in the Ottoman Empire. Dr. Erbudak was acquainted with a member of the ETH, Professor Mahir Sayer, who was Jewish and came from Istanbul. He suggested that we try to see him before we left for Turkey. We managed to do this. He immediately phoned his wife's parents, the Lazaro Frankos, and gave them the name of the hotel (the Étap) where we would be staying. When we arrived at the hotel, we found a bouquet of roses for Florie together with a note inviting us to their house for the following Tuesday evening to meet some of the leaders of the Jewish community. It turned out that these included the Grand Rabbi of Turkey. It was a wonderful evening. The conversation was mostly in French except when we found it difficult to tell some of our favorite stories in French, or when the president of the Bnai Brith, who was a marvelous story teller, switched to English to avoid the embarrassment of telling a story that the Grand Rabbi (who did not understand English) might not approve of. We saw the Frankos and their friends a number of times during our stay in Istanbul. We also visited the University of the Bosphorous, where we had lunch with Professor F. Kortel and other members of the mathematics department, and on another occassion we had dinner with his family at his home.

Istanbul is a magnificent city because of its location and its Byzantine and Ottoman past. One should record, however, two negatives: (1) It was easy to get sick on the food, and we did. (2) The city was under martial law when we were there. Our friends assured us that this was an improvement over the political assassinations and near-anarchy that preceded the military take-over. Nevertheless, it was clear that one had to be careful of what one said in earshot of passersby.

We moved to our second apartment shortly after we returned to Zurich from Istanbul. On January 15, Florie returned to the United States to resume her teaching at Albertus Magnus. On my next to last weekend, I visited the area around the Jungfrau. I returned home on February 1. We had enjoyed our term at the ETH. We saw a good deal of the Eckmanns and enjoyed meeting other members of the department. Unfortunately, we saw rather little of our friends, the Chandrasekharans. His health had deteriorated since we had seen him last at the Vancouver ICM.

In the spring of 1982, we attended a Symposium Sponsored by the Association for Women in Mathematics in Honor of Emmy Noether's 100th Birthday. This was held at Bryn Mawr College, March 17–19. The speakers were asked to present talks on subjects related to Emmy Noether's work and nearly all complied with this request. The speakers were Nathan Jacobson, Richard Swan, David Mumford, Judith Sally, Olga Taussky-Todd, Michele Vergne, Karen Uhlenbeck, Walter Feit, Armand Borel. The title of my talk was Brauer Factor Sets, Noether Factor Sets, and Crossed Products. Besides the mathematical talks, there was a panel discussion on Emmy Noether in Erlangen, Göttingen, and Bryn Mawr. The participants for this panel were Gottfried Noether, who was a nephew of Emmy Noether's, and whose address had been written jointly with his wife Emiliana; Grace Quinn, and Ruth McKee, who had been doctoral students of Emmy's at Bryn Mawr; Olga Taussky-Todd, who had been a postdoctoral fellow at Bryn Mawr working with Emmy; and Marguerite Lehr, who had been a junior faculty colleague of Emmy's at Bryn Mawr. A proceedings of the Symposium entitled *Emmy Noether in Bryn Mawr* was published by Springer-Verlag in 1983.

On the last morning of the Symposium (March 19), there was a dedication of a plaque marking the grave of Emmy Noether. This is located in the cloisters of Thomas Hall (the old library building) and bears the inscription: E.N. 1882–1935. The dedication was a Quaker service led by Professor F. Cunningham, who was chairman of the department at Bryn Mawr.

In the summer of 1982, the Bolyai Janos Mathematical Society organized an international conference on the Theory of Radicals in Eger, Hungary, which is located approximately 60 miles northwest of Budapest. The dates for the conference were August 1–7. The organizing committee was headed by R. Wiegandt as Chairman and L. Marki as Secretary. I had received a note in February from Professor Marki that, according to Professor L. Bokut, who was a member of the Organizing Committee, there was a good chance that Effim Zel'manov would attend the conference. This was its main attraction for me, especially since I had not worked on radical theory of rings in many years. Also, since the conference was scheduled to take place just before the 1982 Oberwolfach meeting on Jordan algebras that we were going to attend, we decided to accept the invitation for the conference in Eger.

This conference had a truly international character with participants from Australia, Austria, Belgium, Bulgaria, Canada, Czechoslovakia, England, Egypt, Finland, East Germany, West Germany, Hungary, Israel, Italy, the Netherlands, Scotland, the United States, the Soviet Union, and Vietnam. Thirteen Russians attended, most of them, including Zel'manov, from Novosibirsk.

I did not contribute a paper, but I chaired one of the sessions. This was one of the most interesting ones, whose speakers (in order of presentations) were E. I. Zel'manov, K. McCrimmon, S. A. Amitsur, L. A. Bokut, A. E. Zalesski, E. N. Kuzmin, Y. P. Razmyslov.

In addition to the formal program, Kevin McCrimmon, Michel Racine and I heard some private lectures by Zel'manov on his work on structure theory of Jordan algebras that had been published shortly before the conference in *Algebra i Logika*.

On Wednesday afternoon, we had a bus excursion with a short walk in the Bükk mountains, which are close to Eger. On Thursday evening, the group visited a winery where we had a tasting of the local wines in a wine cellar attached to the winery. It was an interesting evening, especially because of a lively conversation I had with one of the Soviet delegates on the subject of Soviet anti-Semitism. As I recall, this was the gist of our conversation:

He introduced the subject by saying that he felt that I had been unwise in writing my letter to the *Notices of the American Mathematical Society* on Pontrjagin's personal attack on me (see, p. 49), since this might have made it more difficult to have fruitful scientific exchanges between the United States and the Soviet Union. On the other hand, I and most western scientists believe that we must not compromise established principles for such exchanges. From here we moved to the general question of anti-Semitism. My Soviet colleague insisted that this did not exist in the Soviet Union since it was illegal and since shortly after the Reveolution, Lenin had delivered a speech that was an eloquent denunciation of anti-Semitism. On my part, I pointed out some examples of anti-Semitic acts that had been perpetrated in recent times in the Soviet Union and I observed that anti-Zionism was allowed and even sponsored by the government. The line between anti-Semitism and anti-Zionism was a very thin one that rapidly disappeared when it served the purposes of the regime. Our discussion lasted for quite a while.

On Friday night, there was a farewell reception for the members of the conference. Paul Erdös came from Budapest to attend this reception. We spent one day in Budapest on our way to Oberwolfach, and he invited us to dinner and a brief sight-seeing trip by taxi.

We were disappointed that none of our Soviet colleagues were permitted to attend the Oberwolfach conference. This would have been easy since the dates for Oberwolfach were August 8–14, and the distance between the two sites was quite short.

After returning to the United States, I accepted an invitation from the University of Indiana to give some lectures in their *Mathematics Distinguished Lecture Series*. I visited there for a week and gave four lectures under the general title Jordan Norms and Associative Algebras with the following titles for the individual lectures: I. Survey of Structure Theory, II–IV. Application of Jordan Norms to Associative Algebras. This was a very enjoyable visit with a number of old friends: Goro Azumaya, Darrell Haile, Paul Halmos, Maria Wonenburger, Max Zorn.

In the autumn of 1983, we visited all three Chinas: Hong Kong, the People's Republic of China and Taiwan. We used Hong Kong as a point of entry to the PRC and as a connecting link between the PRC and Taiwan. We were in Hong Kong from September 4 to September 10. During this time, we visited with King Fai Lai, and I had some informal discussions on algebra with some of his students at the Chinese University of Hong Kong. A severe typhoon struck the area while we were there (eventually graded as 10, which is maximal intensity). As a consequence, our flight had to be postponed until September 10, when we flew to Shanghai. Our primary host for this second visit to the PRC was the head of the Mathematics Department at Nanjing University, Professor Peh-Lsun Cheo, who met us with a colleague at the airport in Shanghai and travelled with us on the train to Nanjing.

I had accepted an invitation to be a guest lecturer at Nanjing for eight weeks and had decided to give a course entitled Finite Dimensional Division Algebras. They wanted me to give three lectures a week each two hours long so as to be roughly equivalent to one of their standard courses of three hours a week for 18 weeks. I succeeded in getting this reduced to three double periods of 45 minutes separated by a 15-minute (tea) break. In addition to these lectures, I gave several lectures indicating some other active areas of algebra (e.g., the theory of algebras with polynomial identity), and I met with the students once a week to answer questions on the material and for informal discussions.

There were 38 students in my class coming from 20 different institutions in eastern China. These were either advanced graduate students or faculty. Only five were women, who appeared to huddle together in isolation from the male students. The largest number from a single institution, five, came from Wuhan Normal University. Several years earlier, six students from WNU and their professor, Ji-sun Sum, had corresponded with me about my books, *Basic Algebra I* and *II*, and they were anxious for us to visit WNU during our stay in Nanjing. This proved to be impossible until our next trip to China in 1986.

To overcome the language barrier, I gave the class a set of my lecture notes at the end of each lecture. The students also recorded the lectures on tape recorders. They met on the off-days between lectures to translate the text. Shortly after we returned to the United States, I received a copy of the Chinese translation of my lectures, One of the students, Mr. Guo Li, obtained a simpler proof of one of the results I presented. Subsequently, I switched to his proof (which will appear in the forthcoming book by Saltman and myself). Another student, Mr. Zheng Yumei, showed me a little paper that he had written giving a characterization of Ore polynomial rings. Unfortunately, I had anticipated him by about fifty years in [2].

Florie was interested in studying Chinese during our stay in Nanjing. Unfortunately, the beginning course for foreign students was of second year level, so that she required private instruction. It took a while to negotiate a deal in which she exchanged such lessons for a course in English that she taught at the Institute for Radio and Telecommunications. Florie enjoyed this experience, which also gave us a contact with another group. They were very hospitable and took us on a very interesting excursion near Nanjing. Florie was also interested in learning more about secondary education in China. It proved rather difficult for her to visit any of the schools. However, she did have a chance to discuss secondary school education both in China and the United

States with the group from Wuhan on the afternoons when I met with the students for informal discussions of my course.

We lived in the Foreign Experts Building on the campus of the University of Nanjing and took our meals at the canteen for foreign students. We ate with the other "experts." Their academic backgrounds were interesting and varied. Through our daily contacts, we became good friends. This compensated for the rather poor accommodations—a single room with bath and hot water and the poor food we had at the canteen. Also, the sanitary conditions were poor, and as a result I was sick for a long stretch and had to be treated by a doctor. She appeared to be competent and prescribed a large number of pills each with its own function. Also, both of us suffered from colds during our stay in Nanjing.

Several times a week we escaped the canteen food by dining at the new Jinling Hotel located a short distance from the campus. This hotel was the tallest building in the PRC, 36 stories, with a revolving cocktail lounge on the top floor. The Chinese, unless they were entertaining foreign visitors, were not allowed to enter. It was a special treat for the four students (our "gang of four," who had been assigned by Professor Cheo to look after us) to be invited to have a drink in the cocktail lounge with us.

From our vantage point on the campus, we had an excellent view of student life in a Chinese university. We had arrived in Nanjing early enough to observe the military training by the PLA (People's Liberation Army) that every student, male or female, was obliged to undergo prior to the beginning of classes. It should be mentioned that all the students were fully subsidized and lived on the campus. The dormitory rooms appeared to be about 15×15 feet, and each accommodated eight students who slept in four double bunk beds. The students took their meals by lining up in the canteen with their rice bowls to have these filled with rice and something else, usually a couple of vegetable dishes. They took these to their rooms to eat. To relieve the monotony and to get a good home-cooked meal, many of the students who lived near Nanjing cycled home for week-ends. At about 4 P.M. every day, all academic activities gave way to sports or, more precisely, non-competitive play at volleyball, tennis, soccer, basketball, etc. This lasted about an hour, until it was time to line up for the evening meal. At a central location on the campus, there were bulletin boards for posting the daily newspapers, including one in English, and for posting CP announcements. There were also chalk-boards for the use of the students to express grievances and occasionally satire. The sexes were strictly segregated and apparently there was no dating or partying. One exception to this occured at a Halloween masquerade party given by American students to which a few of the Chinese students were invited. The mixed dancing that made its appearance a few years later in China was a cultural shock to the Chinese at this time.

Nanjing has many wonderful sights: the Yangtze River and its bridge, which was finished in 1968 during the "Cultural" Revolution; Muchou Lake; the tomb of the first Ming emperor; the mausoleum of Sun-Yat-sen; the palace of the Taiping ruler, which was later the offices of Sun-Yat-sen and still later Chiang Kai-shek's residence during the period when Nanjing was the capital of Nationalist China. We visited all of these. We were taken to see the Sun Yat-sen mausoleum by Professor Cheo and his wife. During this trip, we also visited a Buddhist temple that had been built by the first Ming emperor. The other sites we visited with some of Cheo's students or on exursions that had been arranged by the Foreign Office of the University for its foreign visitors. The students took us to the Yangtze River Bridge that we crossed to see the sculptures of heroic figures that had been erected at the end of the bridge. The students were rather disdainful of this example of socialist realism in art.

Perhaps the most interesting sights for us were simply the crowded streets, especially the Sun Yat-sen Boulevard, which was on one of the boundaries of the campus. Until quite late at night, this was crowded with pedestrians and cyclists, as well as an occasional taxi or bus. On Sundays, people aired their quilts on the streets. The threadbare condition of many of these was a forceful reminder of the poverty of much of the population. However, there were signs of progress over 1980: colorful clothing, leather shoes, and availability of fresh fruits and vegetables in the privately owned stores that had mushroomed since our last visit.

We had a shocking experience one afternoon when we were returning from a walk. As we were about to enter the gate to the campus, we heard sirens, and then we saw a motorcade of motorcycles and a police car carrying condemned persons to the soccer stadium for public display to a full stadium before they were taken off to be executed. We were not completely surprised by this since the event had been announced by posters around the city with the names of the convicted persons together with statements of their alleged crimes. We consulted a colleague who was teaching English and American law at the university, and we learned that the crimes that could be punished by death included embezzlement, grand larceny and counter-revolutionary activity, as well as murder and rape. A grim addendum to the process was the appearance of red crosses through the names of the condemned persons shortly after their executions. The event was also portrayed in photographs on the bulletin boards of the university as a warning to the students.

A happier event in which we participated was the Mid-Autumn Festival that was celebrated in 1983 on September 21. For us, this was an evening of entertainment at the Xiang Liang Sang Park, where we sat at tables and consumed much fruit, nuts, and "moon-cakes," which were specially made for this occasion, while we waited for the moon to rise. After this, we had a long speech by the Waiban (the head of the Foreign Office of the University). This was followed by a response by one of the visitors and then some pleasant musical entertainment by a visiting Peruvian professor and his young daughter. We enjoyed the evening very much, in spite of the fact that I was coughing a good deal of the time.

Professor Cheo was away during a good part of our stay in Nanjing. He was conferring with other algebraists on the development of algebra in the PRC and the introduction of Ph.D. programs in mathematics at some of the universities, including Nanjing.

My course at Nanjing was recessed from September 24 to October 6, when we visited Beijing and Xian. We had been invited by Professor Hua to be guests of the Institute of Mathematics at Beijing for a week during our stay at Nanjing, and we also had an invitation to visit Northwest University at Xian. It turned out to be most convenient for Professor Hua for us to begin our visit in Beijing on September 24.

We were met at the airport in Beijing by Professor V. G. Deng of the University, Assistant Chairman Lu, and another member of the Institute, and we were taken to the Friendship Hotel, where we had stayed on our last visit to Beijing. The fine new highway from the airport and the numerous new buildings that loomed over us on our ride were striking evidence that much had been achieved in Deng Xiau-Ping's program of modernization. We arrived at the hotel too late for lunch. However, we were met there by Mr. Chu, who had been our guide in 1980, and he had a picnic lunch for us. We were delighted that he had been assigned to look after us again.

At the hotel, we ran into the Garrett Birkhoffs, who were on the third week of an eight week tour of China. We had supper with them and saw them several times at breakfast before they departed.

I gave only two lectures at Beijing, so we had ample time for sight-seeing. The autumn weather was almost perfect, clear and cool with no rain except one evening. We re-visited a number of the great sights we had seen three years earlier—all except the Great Wall. Besides these, we scrambled around Coal Hill, which overlooks the Forbidden City, to obtain some panoramic views of this enormous complex. We visited a newly opened Lama temple. On other occasions, we visited the Marco Polo bridge and the magnificent new Fragrant Hill Hotel, which had been designed by I. M. Pei. At our request, Mr. Chu arranged for us to visit a "neighborhood"—a well-defined social and political entity. We were introduced to several families, including the block leaders, and we visited briefly in their apartments. Usually, three generations occupied an apartment. We also met a lady artist living alone in a single room. She had been a costume designer for the Peking Opera but had been obliged to retire for health reasons. We purchased a small silk screen painting from her, which unfortunately was spoiled by a framer in Nanjing who had mounted it and rolled it up while it was still wet.

My first lecture was on Tuesday afternoon at the Institute. That evening we were invited to a banquet by Hua. We were taken to the restaurant in a taxi by Mr. Chu, and when we arrived there we saw Hua drive up in a long black chauffered limousine, which was assigned to him as a top government official. He greeted us warmly—hugs and kisses for Florie (unusual for the Chinese). There were nine of us at dinner, including Tuan, Wan and Chu. After dinner, Hua invited Florie and me and Chu to his apartment for tea and fruits. His hostess on this occasion was his daughter-in-law, Dr. Koh, who lived with him with her family. We had met her in the United States. We did not stay long since Hua had another caller waiting to see him. He had his chauffeur drive us home in the limousine.

A few days later, we dined with the Wans in their apartment in a brand new building that belonged to the Institute. Two of Wan's students dined with us. One of these, Xiu Xiau Xi was applying for admission and financial support at Yale. I had a chance to talk to him after dinner and was impressed by his background. He is now working for his doctorate at Yale.

October 1 is National Day in China—the equivalent of our Fourth of July. It marks the establishement of the communist government by a proclamation by Mao on October 1, 1949, in Tien-An-Min Square. This is the center of the celebration in Beijing. We joined the crowds there with Chu and his nine-year-old daughter, who called us grandmother and grandfather because of our gray hair. She also came with her father the next day when he took us to the airport for a flight to Xian.

It rained constantly during the four days we were in Xian. In spite of this, the archeological sights more than made up for the discomfort of sight-seeing under an umbrella. While we were in Xian, there was a showing on prime time national televison of a six-part series on the early life of Hua. Florie watched all of these, but I was unable to see anything more than some glimpses of the program since I needed the time to think about the lecture that I would give on the last day of our visit. There were some physical problems associated with this lecture: a poor room with only a portable blackboard and terrible chalk. However, the translator appeared to have a good understanding of the material. After my talk, the chairman of the department presented me with a small terra cotta replica of one of the life-size terra cotta soldiers that were being excavated from the tomb near Xian of the first Qin emperor (circa 200 B.C.). We had, of course, visited this excavation site as well as other important sights of Xian: the 6000-year-old Banpo village, the Wild Goose Pagodas, the Xian Hot Spring, where Chiang Kai-shek had been captured during the famous "Xian incident" of 1936.

We returned to Nanjing October 2 and remained there until November 12, when we departed for Hong Kong. We had written our friend, King Lai, to reserve us a room in a luxury hotel since we felt the need of R and R after nine weeks in the PRC. By pulling some strings, he was able to get us into the Regents, which is one of the most luxurious hotels in Hong Kong. We enjoyed the culinary delights of this unique city, as well as the sight-seeing, especially around the harbor on a junk that was owned by two of King Lai's friends. We left Hong Kong November 15 for a month's visit to Taiwan.

The trip to Taiwan had been planned prior to our visit to Nanjing. In 1981, I had mentioned to two of our Taiwanese students, Dr. Jing Yu and Dr. Ying Cheng, that I was interested in visiting Taiwan. They relayed this information to Professor Liang-Chi Tsao, the chairman of the Department of Mathematics at the National Taiwan University, who then wrote me that they would be happy to have me visit their department for a period between one month to one year. For several reasons, it appeared that the earliest most convenient time for us would be in the autumn of 1983. In my correspondence with Professor Tsao, I mentioned Florie's background and suggested that she could give some lectures on mathematics education, which was her strong interest at the time, and that she was anxious to consult with people in Taiwan with similar interests. After some additional correspondence on details, I accepted an offer of a visiting position at the National Taiwan University from November 15 to December 15, 1983, and Florie accepted a similar offer at Taiwan Normal University, both located in Taipei. Florie's position gave us an entrée to an excellent two-bedroom apartment (without cooking facilities) in the Union of Taiwan Normal, which was within easy walking distance of both institutions. This was also a convenient location for restaurants. One of the best was located in the basement of the Union. The food in Taiwan was extremely good and we enjoyed some fabulous banquets there.

Florie's duties began almost immediately. She was asked to give a lecture two days after we arrived and to supply a manuscript for the lecture the day before it was given. A few days later, she took a trip to a mountain village south of Taipei to observe a school. Also, she was a delegate to the First Asian-Pacific Conference on Science Education, which was held at the Ambassador Hotel in Taipei from November 24 to 28. This was a lavish affair at which all the foreign delegates and their families were fully subsidized and were treated to a post-conference excursion by plane to one of the most scenic places on the island. Unfortunately, I was unable to take advantage of the dinners, and we had to skip the excursion since I had a bout with food poisoning that lasted for several days.

I gave a mini-course of eight lectures on Generic Norm of Algebras. I lectured twice a week, one lecture at the Mathematics Department building and another at the Institute for Mathematics, which was located opposite the mathematics building. The topic of generic norms turned out to be a good choice for these lectures since the concept was a familiar one to several of the auditors, who had received their doctorates at Chicago with Walter Baily. One of my auditors was Huah Chu, who had translated *Basic Algebra I* and had sent me some corrections and suggestions for improvements that were incorporated in the second edition of this book.

Form December 4 to December 7, we visited Tainan and Kao-Hsuing. I had been invited to lecture at the National Chung-Kung University at Tainan and at the Kao-Hsiung Teachers College at Kao-Hsuing. I gave the same lecture at both places with the title, Some Applications of a Theorem of Hua Loo-Kung. I chose this since it could be presented in a relatively elementary fashion and since I thought it might be intriguing for Taiwanese audiences to hear about an important contribution of a famous Chinese mathematician who was very likely known to them. During our visits, we enjoyed the hospitality of both departments and also of Professor and Mrs. King N. Wu, the parents-in-law of Ying Cheng. They gave a splended banquet for us, and he took us sight-seeing for an afternoon at Tainan and, together with Professor Li, the Chairman at Tainan, on a day's excursion to the National Scenic Area near Kao-Hsiung and the coastal region at the southerm tip of the island.

Taipei also had some wonderful places to visit. These included the Lungshan temple and the Sun Yat-sen Memorial Park. This was especially colorful on week-ends, when families came to fly kites and cycle on all manner of cycles from one to three wheelers. We found the Chung Cheng (Chiang Kai-shek Memorial Hall) impressive but somewhat incongruous in its obvious imitation of the Lincoln Memorial in Washington. The most remarkable tourist attraction in Taiwan is undoubtedly the National Palace Museum in Taipei. This contains the greatest collection of Chinese art, much of which was formerly housed in the Palace Museum in Beijing. This had been transported from the mainland when Chiang Kai-shek's followers fled to Taiwan.

We returned to the United States from Taipei on December 16.

During January 1984, I was lecturer for two weeks in the Distinguished Lecture Series of the University of Texas. This provided me an opportunity to work with David Saltman, my co-author of *Finite Dimensional Division Algebras*. I gave several lectures on Jordan norms and associative algebras and a colloquium survey lecture on Jordan algebras. Florie gave a Pi Mu Epsilon lecture on Women, Creativity and Mathematics. This represented the theme of a course, Women and Creativity, that she had given at Albertus Magnus College jointly with a colleague from the English Department. On our way home from Austin, we stopped off at the University of Oklahoma at Stillwater where I gave a Colloquium talk on Jordan Norms.

From February 15 to April 15, 1984, we visited the University of Virginia. My position there was funded by the University of Virginia Center of Advanced Studies. Thanks to Kevin McCrimmon, we were quickly settled in an excellent small apartment close to the university. We felt very much at home in Charlottesville, where we had many friends. I gave a series of lectures on projective varieties defined by Jordan algebras and a colloquium lecture on the history of PI algebras. The subject of projective varieties was close to the interests of McCrimmon and John Faulkner. We had some lively discussions on this subject; their contributions to its development are noted in [81].

During our visit to Charlottesville, we took a side trip to North Carolina State University at Raleigh, where I gave two lectures: (1) Jordan Algebras and Mathematical Physics and (2) Recent Developments on Jordan Algebras.

A conference, Algebra in its Variety, honoring Irving Kaplansky, was held at the

University of Chicago from July 13 to July 15, 1984. Kaplansky was leaving Chicago to become the Director of the Mathematical Sciences Research Institute at Berkeley. The speakers at the conference, in the order of their presentations, were I. N. Herstein, Nathan Jacobson, Robert Moody, Richard Block, Peter Freund, Richard Kadison, Alex Rosenberg, Wolmer Vasconcelos, Donald Ornstein, Melvin Hochster, Hyman Bass, Graham Evans and Richard Swan. My talk was on Generic Reducing Fields for Jordan Algebras. This was published as "Some Projective Varieties Defined by Jordan Algebras" [81], with a dedication to Kaplansky "with whom I have shared many common mathematical interests over the years." (A voluminous correspondence attests to this fact.)

From September 15 to October 15, I was a Visiting Professor at Pennsylvania State University. The mathematicians there whose interests were close to mine were Ed Formanek, William Waterhouse and L. N. Vaserstein. Some of my work overlapped with Waterhouse's, as I noted in the commentary on [80]. Also, Formanek's work on PI algebras was of great interest to me. I gave a series of six lectures on Generic Splitting Fields and a colloquium lecture on Some Projective Varieties Defined by Associative Algebras at Penn State.

While at University Park, we made a quick trip to Pittsburgh to visit our old friend Bob Walker, who had been a student at Princeton at the same time (1930–1934) as I. He had retired from Cornell and was living with his sister in their home in Pittsburgh.

The year 1986 was a busy year of travel for us: major trips to Israel and China and a brief visit to Ottawa.

We were in Israel for four weeks beginning February 16. This was our fourth visit. I had been invited to lecture at the Wladimir Schreiber Institute for Mathematics of Tel Aviv University by Piatetski-Shapiro and at the Hebrew University of Jerusalem by Amitsur. After I accepted these invitations, I was also asked by Louis Rowen to lecture at his university, Bar Ilan.

On our way to Israel, we spent ten days as tourists in Egypt—in Cairo and on a boat trip up the Nile from Luxor to the Aswan Dam. We flew to Ben Gurion airport from Cairo. The security precautions in Cairo were impressive, but these caused a considerable delay in our flight. Rowen, who met us at the airport, had waited many hours before we finally arrived and were able to pick up our luggage at around 2 A.M. We spent our first night (or what was left of it) at the Rowens and were driven the next day to Jerusalem, where we stayed at the Maiersdorf Guest House of Hebrew University on Mount Scopus.

We spent two weeks in Jerusalem, where I gave two hour and a half lectures on the main campus of Hebrew University on *Generic Splitting Fields*. This is a subject that had been intitiated by Amitsur. The point of view of my lectures was that of Chapter 3 of *Finite Dimensional Division Algebras*.

Under Amitsur's expert guidance, we revisited some of the important sights in Jerusalem that we had seen on earlier visits. Besides these, he took us to some places we had not seen before: the Memorial to John F. Kennedy; Pater Noster Church, which has inscribed on its walls the Lord's Prayer in more than 40 languages; the Garden Tomb, which some religious Protestants believe was Jesus' sepulcher; and the Cardo, a beautiful colonnaded street that dated originally from late Byzantine times and had been built on an earlier broad Roman street. Our old friends, the Henry Wallmans', live in a beautiful apartment in the Old City very close to the Cardo. We had dinner and a wonderful evening with them at their apartment. We also enjoyed the hospitality of a number of friends in Jerusalem: the Amitsurs, the Foguels, the Lubotzkys, and the Smiths (Josie Shamash and Stephen Smith).

After Jerusalem, we spent two weeks at Tel Aviv, where we stayed at the Ramat Aviv Hotel. I gave three lectures each at both of my host institutions in the Tel Aviv area. My first lecture at Tel Aviv University was on Free Special and Free Jordan Algebras. After this, I gave two lectures on A Survey of the Structure Theory of Jordan Algebras. At Bar Ilan, I gave a lecture on Generic Norms and two lectures on Some Projective Varieties Defined by Jordan Algebras.

We were taken by the Rowens to visit the archaeological sites of Megiddo and Beit She'arim. Megiddo dates back to ancient Egypt when it was a strategic point on the route from Egypt to Assyria. In Solomon's time (circa 1000 B.C.), it had been a stronghold in northern Palestine. Bei She'arim had been an important religious center, especially in the second century A.D.

We had a pleasant Saturday afternoon and evening with the Piatetski-Shapiros. They took us for a drive to Yafo along a route near the shore. We stopped to take walks on some of the beaches along the way, and we spent some time at Yafo itself. This is an attractive tourist spot containing a number of art and jewelry shops and a number of seafood restaurants owned by Arabs. The Piatetski-Shapiros knew a special seafood restaurant near Yafo to which they took us. It was Arab-owned and was excellent.

Florie has a cousin who lives with his family at Kibbutz Hahotrim, which is located near Haifa. They had come to Israel from Argentina primarily because of a strong interest in the kibbutz movement. We spent an afternoon with them at the kibbutz. It was pleasant to visit a kibbutz again, but it was disturbing to hear Florie's cousin describe his experience shortly after Israel's invasion of Lebanon when, as cultural leader of the kibbutz, he was obliged to pass on to the other members the official justification of Israel's action. He found it impossible to do so since he did not believe it himself and, as he put it, "the words stuck in my throat."

We returned to the United States on March 16.

During the latter part of August and the beginning of September 1986, Michel Racine and Erhard Neher organized a Workshop on Jordan Structures at Ottawa. The participants came at various times during this period, some for only a few days and some for the duration of the workshop. We attended from August 18 to August 22. During our stay, we heard talks by Bruce Allison, Kevin McCrimmon, B. Russo, R. D. Schafter and I. P. Shestakov. Allison's was espcially interesting to me since it gave an extension of a classical theorem of Alberts' on tensor products of two quaternion division algebras that I had considered in [79] (see the commentary on this paper). I gave a talk on projective varieties and Jordan algebras.

In the autumn of 1986, we had our third visit to the People's Republic of China. The dates were September 17–October 25, and this time we planned to fly to Beijing from Tokyo, to travel southward, and to return by way of Hong Kong. We had decided to spend four days in Tokyo. I wrote Iwahori of our plans and asked him to suggest a hotel. He reserved a room for us in a hotel across the street of the Grand Kabuki Theater in the Ginza section of the city. He also invited me to give a lecture at the university. At noon, prior to my lecture, which was scheduled from 4 to 5 in the afternoon, we were treated to a luncheon at a restaurant in the Hilltop Hotel that specialized in French haute cuisine. Besides ourselves, those attending the luncheon were our old friends, Ianaga, Iwahori, Kawada and Yosida. Also present was Professor H. Komatsu, whom we met for the first time. It was typical of the changes that had occurred since out last visit in 1969 that the Hilltop Hotel, which had been a modest small hotel then had become a high-rise luxury hotel in 1986. I lectured on Projective Varieties Defined by Jordan Algebras. The day after the lecture we spent with the Iwahoris. We had lunch and supper at their apartment, along with several of his students, and they took us to see some of the sections of the city that had sprung up since our last visit, most notably, the great concentration of high-rise office buildings and luxury hotels that constitute the Shinjuku area. We took in an aftenoon performance of Kabuki, and we enjoyed some splendid (and expensive) Japanese dinners in the Ginza area.

Our primary host for this visit to the PRC was the Graduate School of the Academia Sinica. The overall planning for the visit was done by Professor Zhe-xian Wan, and details were arranged by Mr. Guary Hua, who was the youngest son of Professor L. K. Hua. Professor Hua had died in 1985 of a heart attack immediately following a lecture he had given in Tokyo. He had been chronically ill for a number of years and had been hospitalized on several of his visits abroad. We were pleased that his son would look after us in Beijing and would arrange for our visits elsewhere in China.

We were met at the airport by Professor Wan and Mr. Hua, and we were taken to the Friendship Hotel. Again, we were amazed at the number of new buildings, especially apartments, that had been built or were under construction. As on our earlier visits, we met many interesting "foreign experts" in the dining room of the hotel.

I gave six lectures at Beijing: two lectures on Involutorial Division Algebras of Degree 8, three lectures on Projective Varieties Defined by Jordan Algebras one lecture on Norm Similarities of Algebras. Florie gave two lectures on mathematics education: Problems and Solutions in Mathematics Education and Secondary School Preparation in Mathematics for College and the Workplace. Our lectures were given at a number of different institutions in Beijing: the university, the Graduate School of Academia Sinica, Beijing Normal University, and the Beijing Mathematical Congress. We were treated to a number of banquets by these institutions. One of these was hosted by Tuan and another by Wan. We also had dinner at the Wans with the Hermans of Penn State.

Basic Algebra I had been translated by a team of algebraists from Shanghai Normal College headed by Jone-Wen Cohn. A second team headed by Professor Fu-Zu Zhu of East China Normal University, also located in Shanghai, had been assigned the task of translating *Basic Algebra II*. During our stay in Beijing, Professor Zhu visited me at the Freindship Hotel to ask about misprints and errors that needed correcting in this text. I informed him that I was preparing a new edition of *BA II*, and I said that I would send him a copy of the new material when ready. In turn, he offered suggestions for improvements that will be incorporated in the new edition. Altogether, we had a fruitful exchange on the subject of *BA II*.

We visited the Great Wall again and some of the other sights we had seen on our first two visits to the PRC. We also visited several places that were new to us: Yuan Ming Yuan Park, which had been the summer palace of the Qing emperors and had been destroyed by the English and French during the Boxer Rebellion; a new park containing a complex of gardens and houses designed to recall the setting described in the Chinese classic, *The Dream of the Red Chamber* (or *The Story of the Stone*); and the Temple of Azure Clouds in the Fragrant Hill suburb of Beijing.

On October 1 we flew to Xian, where we stayed in the guest house of Shanxi Normal University. Our host was Dr. Zum-man Li, a former student of Professor Wan's, whom we had met in 1983 at a dinner at the Wans'. I gave a lecture on Generalization of the Characteristic Polynomial of a Matrix on October 3. This was followed by a lecture in the same room by Florie on Problems and Programs in the State of Connecticut on Mathematics Education. October 4 was a day of sight-seeing, when we revisited the great archaeological sights that we had seen in 1983—this time in fair weather.

We flew to Chongqing on October 5. We were met at the airport by a delegation from Xinan Normal University, which is located in Bebei, an hour-and-a-half's bus ride from Chongqing. We stayed at the guest house of the university. On October 6–7, we were taken on an excursion to Dazu, a recently opened site of magnificent Buddhist stone figures dating from the tenth century. We also had some interesting stops on the return trip to Bebei. I gave a two hour lecture on Brauer Groups on the morning of October 9, and in the afternoon Florie lectured on Teaching Mathematics and Problem Solving. Both lectures were translated by Professor Guang-xian Zhang, who had just returned from a stay of two years at Yale. The next afternoon we were driven to Chongqing for sight-seeing and dinner before boarding a ship for a two-day trip down the Yangtze River.

Our accomodations were the best available on this ship—a private room with two beds. The ship was primarily designed to transport large numbers of people inexpensively and hence in minimal comfort. Most slept on the decks and brought their own food; the ship provided hot water for tea. The standard of cleanliness left much to be desired. However, the scenery, especially the passage through the Three Gorges of the river, was magnificent. As often happens at this time of the year, the weather was cold and rainy. In spite of this, we had wonderful views; the narrowness of the canyon brought us so close to its granite walls; it seemed that we could reach out and touch them. There were a number of stops at ports along the way to drop off and pick up passengers and freight. At the end of the trip, just before we reached Wuhan, the ship negotiated a lock to a lower level at the great dam on the river.

I had a special relation with Hubei University in Wuhan (formerly Wuhan Teachers College). In 1980, I had received a letter signed by six graduate students of WTC and their professor, Sun Ji-sun, asking me to send them a copy of *Basic Algebra II*, which was unavailable at the time in the PRC. They had studied *BA I* and were anxious to continue with *BA II*. I sent them a copy that arrived several months later. On receiving this, they expressed their gratitude by sending me a beautiful scroll painting of plum blossoms that they had commisioned for me. On the left margin, there is an inscription in Chinese characters that in English begins as "Venerable Professor Jacobson" and ends with "an ancient Chinese poem by Su Shi. Dan-Mu Meng-Xi at the age of eighty-two in Wuhan 1981." As mentioned earlier, five of my students at Nanjing had come from WTC. Also, Professor Sun had come to see me at Nanjing to try to persuade me to make a quick trip to Wuhan when I was at Nanjing. The schedule proposed for this appeared unrealistic, so I declined the invitation.

It was natural that we would receive a red-carpet treatment at Hubei University. Besides the usual sight-seeing: the bridge across the Yangtze (that antedates the one at Nanjing), a children's park that overlooks the river and several other parks, the Yellow Crane Pagoda, etc., we had a number of special treats. One of these was a dance given in our honor that featured western ball-room dancing (somewhat dated). At a special ceremony, we received from the president of Hubei University attractively bound letters of appointment "to act as honourable Professor in the Department of Mathematics." We were also given an unusual piece of handicraft: a finely carved Pekinese dog with puppy and ball in light wood, mounted on a piece of dark wood.

We worked hard at Hubei, both in lectures and conferences with students. I gave two lectures on Some Topics in the Theory of PI-Algebras. I had suggested this topic as a possibility for research when I lectured at Nanjing, and this had been pursued in a seminar at Hubei. Several of the students showed me their research results. One of them, Mr. Chang Qing, had a couple of nice results that I thought were publishable. He is now working for a doctorate with Formanek at Penn State.

At the same time I gave my second lecture (October 17) to about 40 auditors, Florie was lecturing on Secondary School Preparation in Mathematics, Problems and Solutions in another auditorium to 150–200. Both of our lectures were translated and there was a lively question and answer session following hers.

On October 18, we took an over-night train to Guilin. This is in one of the most scenic areas of China, a favorite of scroll painters and poets. We were in Guilin for five nights, the first in a guest house that was quite poor and then in a very comfortable hotel in the city. We were fortunate with the weather for our two main sight-seeing trips: the boat trip on the Li River after a bus ride to reach a point where the river is deep enough for the tourist boats, and the bus trip to the Ling Canal. Both of these trips were spectacular, and the second was also interesting historically since the canal had been ordered by the first Qin emperor some 2200 years ago to facilitate the unification of China under his rule. I gave three lectures at Guangxi Normal University. These repeated lectures I had given earlier on our trip. I also had a question and answer session on graduate work in mathematics in the United States. Florie repeated one of her earlier lectures on mathematics education, again to a large crowd.

On October 25, we flew to Canton, where we spent the night in an excellent hotel. The next day we flew to Hong Kong and we returned to the United States on October 27.

We attended the Annual Meeting of the AMS in January 1987 at San Antonio. I am not very fond of large meetings and have generally avoided these unless I have had a specific reason for attending. In this case, the reason was the Special Session on Brauer Groups and Galois Theory that was organized by Burton Fein, David Saltman and Murray Schacher. The participants were Lindsay Childs, Frank Demeyer, Burton Fein, Walter Feit, Gary Greenfield, Darrell Haile, Raymond Hoobler, Bill Jacob, Nathan Jacobson, David Leep, Patrick Morandi and Leonard Stern. The title of my talk was Jordan Norms and Associative Algebras. Besides the talks in the *Special Session* one of the invited addresses of the meeting was Saltman's Noether's Problem, Brauer Groups and Galois Theory, which was on the subject matter of our *Special Session.*

The Department of Mathematics of the University of Chicago held a conference in honor of Herstein, March 25–27, 1987. Yitz had been at Chicago for twenty-five years, and he had had a great impact there, especially on a large number of students. The speakers at the conference (in order of their lectures) were Nathan Jacobson, Lance Small, Toby Stafford, Edward Formanek, Michael Artin, Murray Schacher, Susan Montgomery, S. A. Amitsur, Irving Kaplansky, Wallace Martindale, David Saltman and Claudio Procesi. The title of my talk was Jordan Algebras of Real Symmetric Matrices. I published this as [82] and dedicated it "To Yitz."

From April 4 to April 22, 1988, I had an appointment as Van Vleck Distinguished

Visiting Professor of Mathematics at Wesleyan University. It was an easy commute for us to Wesleyan, especially since Florie had been making this journey several times a week for three academic years as Visiting Scholar at Wesleyan. We, and especially Florie, were well acquainted with the Wesleyan department, especially with Jim Reid, its chairman. I gave a series of four lectures on Splitting Fields. These began at the very beginning of the subject and ended with the latest results on splitting fields of central simple algebras that are consequences of a deep theorem of Merkurjev and Suslin.

The eighth Oberwolfach conference on Jordan Algebras and Related Topics occured August 8–14, 1988. The first had been held in August 1967. Some of us had attended all eight of these conferences. The one of 1988 was especially interesting because of the presence of a large number of Soviet mathematicians. The previous conference on Jordan algebras, etc., in 1985 had been attended by three Soviet mathematicians: Yuri Medvedev, I. P. Shestakov, and Effim Zel'manov, all from Novosibirsk. These three were present also at the 1988 meeting and in addition the following attended: L. A. Borsuk and V. G. Skosirskii from Novosibirsk, Dzumadildaev from Alma Ata and Sh. A. Ayupov from Tashkent. The total of seven Soviet mathemticians amounted to about one seventh of the number of participants of the conference.

The Novosibirsk school of non-associative, non-Lie algebras is perhaps the strongest in the the world in this area. They are especially interested in free algebras and in varieties of algebras. They presented some of their most recent results on these.

Another outstanding contribution presented at this conference was the work of Georgia Benkart, Marshall Osborn and Helmut Strade that they believe will lead to the determination of all simple Lie algebras of finite dimensionality over any algebraically closed field of characteristic p.

There were many other interesting papers. My own contribution was a lecture on the paper [82], in which I indicated two directions that might be interesting to pursue in connection with the theory developed in this paper.

The papers I published during 1970–1978 are [78]–[82], all of which deal with finite dimensional (quadratic) Jordan algebras and their applications to associative algebras.

In [78], I proved a Jordan algebra analogue of an earlier result ([5], p. 221) on two-element generation of (fintie dimensional) central simple associative algebras. The Jordan algebras considered in [78] have one of the following forms: (1) A^+, where A is central simple associative; (2) $H(A,j)$, where A is central simple with involution j; (3) $H(A,j)$, where A is simple with center a quadratic field over F and j is an involution of second kind.

If $(S(H), \sigma_u)$ is special universal envelope of the Jordan algebra H, then one defines the squared special universal envelope $S''(H)$ to be the subalgebra of $S(H) \otimes S(H)$ generated by the element $\sigma_u(a) \otimes \sigma_u(a)$, $a \in H$ ([52], p. 14, and McCrimmon, "Representation of quadratic Jordan algebras," *Transactions of the American Mathematical Society* **153**, 1971, pp. 279–305). If H is a subalgebra of K, then K is an $S(H)''$ module in which $(\sigma_u(a) \otimes \sigma_u(a))x = U_a x$. If H is generated by a single element a, then $S(H)$ is the associative algebra $F[a]$. The main result of [78] is that if a is an element of any one of the algebras, now denoted as H, in (1), (2), or (3) and a is separable if char $F = 2$, then there exists a second element b in H, such that $H = F[a]''b \oplus M$, where M is an $F[a]''$ submodule of H. In the special case in which $H = H(A,j)$, where A is central simple and j is an involution of orthogonal type, we have the corollary that if the degree of A

is n and the degree of the minimum polynomial of a is n, then there exists an element $b \in H$ such that the elements $a^i b a^j + a^j b a^i$, $1 \leq i < j \leq n$, constitute a base for H.

It is a well-known result of Albert's that if $\mathscr{A}$ is a central simple associative algebra over a field Φ then $\mathscr{A}$ has an involution if and only if $[\mathscr{A}]^2 = 1$ for the class $[\mathscr{A}]$ in the Brauer group. In this case, if the degree of $\mathscr{A}$ is odd, then all its involutions are of orthogonal type, and if n is even, $\mathscr{A}$ has involutions of both orthogonal and symplectic types. If J is an involution in $\mathscr{A}$, we let N_J denote the generic (or reduced) norms in the (quadratic) Jordan algebra $\mathscr{H} = \mathscr{H}(\mathscr{A}, J)$. The type of question considered in [79] is to what extent does N_J determine the structure of $\mathscr{A}$? The first result on this is the easy one that if char $\Phi \neq 2$, then N_J is anisotropic ($N_J(a) \neq 0$ for $0 \neq a \in \mathscr{H}$) if and only if either $\mathscr{A}$ is a division algebra or $\mathscr{A} = M_2(\Phi)$ with J of symplectic type.

The main result of [79] is that if char $\Phi \neq 2$ and $\mathscr{A}_i$, $i = 1, 2$, is central simple with involution J_i and both of the J_i are of the same type, then $\mathscr{A}_1 \cong \mathscr{A}_2$ if and only if the forms N_{J_i} are similar (that is, there exists a bijective linear map η of $\mathscr{A}_1$ onto $\mathscr{A}_2$ and a nonzero element $\rho \in \Phi$ such that $N_1(a) = \rho N_2(\eta a)$ for $a \in \mathscr{H}(\mathscr{A}_1, J_1)$). Similar results hold if char $\Phi = 2$, if we replace the Jordan algebra $\mathscr{H}(\mathscr{A}, J)$ by the outer ideal $\mathscr{H}(\mathscr{A}, J)'$ in $\mathscr{H}(\mathscr{A}, J)$ generated by 1. For the sake of simplicity, we stick to the case char $\Phi \neq 2$.

Another well-known result of Albert's is that if $\mathscr{A}$ is involutorial central simple of degree 4 then $\mathscr{A} = \mathscr{A}_1 \otimes \mathscr{A}_2$, where the $\mathscr{A}_i$ are quaternion algebras. Let $\mathscr{A}_i'$ denote the subspace of $\mathscr{A}_i$ of elements of trace 0 and let $Q(a_1 + a_2) = N(a_1) - N(a_2)$ for $a_i \in \mathscr{A}'$, where N is the usual norm form on the quaternion algebra $\mathscr{A}_i$. The quadratic form Q in six variables is called an Albert form on $\mathscr{A}$. In [79], I obtained the following results on Albert forms: (1) any two Albert forms of $\mathscr{A}$ are similar, (2) two involutorial central simple algebras of degree 4 are isomorphic if and only if their Albert forms are similar, (3) if Q is an Albert form of $\mathscr{A}$, then the index of $\mathscr{A}$ is 4, 2 or 1 according as the Witt index of Q is 0, 1, or 3. Statement (1) answered a question that had been raised by M. A. Knus and (3) extended a result of Albert's. A consequence of this is the char $\Phi \neq 2$ case of a theorem due independently to Albert and L. H. Sah stating that the tensor product of two quaternion division algebras is not a division algebra if and only if they have a common quadratic subfield. This result was extended to composition division algebras of characteristic 0 by B. Allison in "Structurable division algebras and type 1 Lie Algebras", (*Canadian Mathematical Society Conference* **5**, 1986, pp. 139–156). On the other hand, it has been shown by J. P. Tignol and A. Wadsworth that there exist central division algebras $\mathscr{A}_i$, $i = 1, 2$ of degree 3, such that $\mathscr{A}_1 \otimes \mathscr{A}_2$ is not a division algebra but the $\mathscr{A}_i$ do not have a common cubic subfield ("Totally ramified valuations in finite dimensional division algebras," *Transactions of the American Mathematical Society* **302**, 1987, 223–250).

In the original (uncorrected) version of [79], the definition and the theorem on Albert quadratic forms was stated without restriction on the characteristic. It was pointed out to me by Professor Tignol that the proof is not valid if char $\Phi = 2$. Another definition of Albert quadratic forms that works for all characteristics appears in the paper "The Albert quadratic form for an algebra of degree four" by Pasquale Mammone and Daniel B. Shapiro, forthcoming in the *Proceedings of the American Mathematical Society*. In this paper, the authors give purely quadratic form proofs of the theorem on Albert quadratic forms in both cases char $\Phi \neq 2$ and char $\Phi = 2$.

In the last part of [79], I considered generic norm fields defined by the generic norm N_J, where $\mathscr{A}$ is central simple with involution J. If $(u_1, \ldots, u_m)$ is a base for $\mathscr{H}(\mathscr{A}, J)$

and $\xi_1, \ldots, \xi_m$ are indeterminates then $N_J(\Sigma\, \xi_i u_i)$ is irreducible in $\Phi[\xi_1, \ldots, \xi_m]$ [69]. The field of fractions of the domain $\Phi[\xi_1, \ldots, \xi_m]/N_J(\Sigma\, \xi_i u_i)$ is called a generic norm field of $(\mathscr{A}, J)$. Using methods that had been used by Heuser and by Saltman for generic norm fields of central simple algebras (without involution), I showed that if J is of orthogonal type, then the generic norm field determined by N_J is a generic splitting field for $\mathscr{A}$ in the sense of Amitsur. If $\mathscr{A}$ is central simple of degree n over Φ and s is a divisor of n, then I called an extension field E of Φ a $1/s$-splitting field of $\mathscr{A}$ if the extension algebra $\mathscr{A}^E$ has index $\leq s$. Based on this concept, one can define generic $1/s$-splitting fields. I showed that if J is of symplectic type, then the generic norm field defined by J is a generic $1/2$-splitting field for $\mathscr{A}$.

W. C. Waterhouse in the paper "Twisted forms of the determinant," (*Journal of Algebra* **86**, 1984, pp.60—75), considered the problem of twisted forms of the determinant function det defined on $M_n(\Phi)$, where Φ is a field. This is the problem of determining the n^2-dimensional vector spaces V over Φ and the polynomial functions N on V such that for a suitable extension $\bar{\Phi}$ of the base field (which can always be taken to be the algebraic closure of Φ) the extension of N to $V^{\bar{\Phi}}$ is equivalent to the determinant function on $M_n(\bar{\Phi})$. He also considered the same problem for equivalence "up to scalar," that is, in which equivalence is replaced by similarity of forms. Waterhouse based his results on group schemes.

In [80], I considered a natural extension of Waterhouse's problem to generic norms of Jordan algebras and obtained results like his for separable Jordan algebras. My formulation was slightly different from his and was based on the following definitions. Let $\mathscr{J}$ be a finite dimensional (quadratic with unit) Jordan algebra over a field $\bar{\Phi}$, $\bar{N}$ the generic norm function on $\bar{\mathscr{J}}$, Φ a subfield of $\bar{\Phi}$. Then a Φ-form of $(\bar{\mathscr{J}}, N)$ is a Φ-subspace $\mathscr{J}$ of $\bar{\mathscr{J}}$ such that $\bar{\Phi}\mathscr{J} \cong \bar{\Phi} \otimes_{\Phi} \mathscr{J}$ (canonically) and $\bar{N}(\mathscr{J}) \subset \Phi$. In a similar manner, I defined a Φ-form of $(\bar{\mathscr{J}}, N, 1)$ as a Φ-form $\mathscr{J}$ of $(\bar{\mathscr{J}}, N)$ such that $1 \in \mathscr{J}$. Also, one has the concept that had been considered extensively in the literature (e.g., in [9] or [40]) of an algebra Φ-form, in which the condition on the subspace $\mathscr{J}$ is a Φ subalgebra of $\bar{\mathscr{J}}$ such that $\bar{\Phi}\,\mathscr{J} = \bar{\mathscr{J}}$.

The main result of [80] is that if $\bar{\Phi}$ is the algebraic closure of Φ and $\bar{\mathscr{J}}$ is separable over $\bar{\Phi}$, then any Φ-form $\mathscr{J}$ of $(\bar{\mathscr{J}}, \bar{N}, 1)$ is an algebra Φ-form, and any Φ-form of $(\bar{\mathscr{J}}, \bar{N})$ is an algebra Φ-form of an isotope of $\mathscr{J}$. This applies in particular to the $\mathscr{J}$ that are simple over $\bar{\Phi}$. The simple Jordan algebras over $\bar{\Phi}$ are known; hence, one can list the cases in which $\bar{\mathscr{J}}$ is simple. Among these is $\bar{\mathscr{J}} = M_n(\bar{\Phi})^+$. Here, $\bar{N} = \det$. Also, the algebra forms of $\bar{\mathscr{J}}$ are known [40]. These are either the Jordan algebras $\mathscr{A}^+$, where $\mathscr{A}$ is central, or the $\mathscr{H}(\mathscr{A}, J)$, where $\mathscr{A}$ is simple and J is an involution of second kind. In either case, N is the generic norm. This is the case considered by Waterhouse. Other cases of simple $\bar{\mathscr{J}}$ are $\bar{\mathscr{J}} = \mathscr{H}(M_n(\bar{\Phi}, t)$ and $\mathscr{H}(M_{2n}(\bar{\Phi}), t_s)$, where t is the transposed involution and t_s is the symplectic involution. We also have the exceptional simple Jordan algebra $\mathscr{H}(M_3(\mathbb{O}),^*)$ where $\mathbb{O}$ is an octonion algebra. I also consider one case in which $\bar{\mathscr{J}}$ is not simple, namely, $\bar{\mathscr{J}} = \bar{\Phi} \oplus \cdots \oplus \bar{\Phi}$.

The paper [81] deals with finite dimensional unital quadratic Jordan algebras $\mathscr{J}$ over an arbitrary field F. To state the results in this generality requires a number of technical definitions. For this reason, I shall restrict the discussion here to the characteristic $\neq 2$ case. Then the algebras can be considered as ordinary Jordan algebras defined by a bilinear product satisfying the usual conditions.

Following Faulkner and McCrimmon, one calls an element c of $\mathscr{J}/F$ reduced if $U_c\mathscr{J} \subset F_c(U_c = 2L_c^2 - L_{c^2})$. Let $\rho(\mathscr{J})$ denote the set of reduced elements of $\mathscr{J}$. It is readily seen that $\rho(\mathcal{J}) = \rho(\mathcal{J}')$ if $\mathcal{J}$ and $\mathcal{J}'$ are isotopes and $F\rho(\mathcal{J})$ is an ideal in $\mathcal{J}$. Hence, if $\mathcal{J}$ is simple, then either $F\rho(\mathcal{J}) = \mathcal{J}$ or $F\rho(\mathcal{J}) = 0$. In the former case, $\mathcal{J}$ is called reduced. The reduced simple Jordan algebras can be determined (cf. *Structure and Representations of Jordan Algebras*, p. 203). This applies to the case in which F is algebraically closed since any simple $\mathcal{J}$ over such a field is reduced. The main results of [81] were first derived using the determination of the reduced simple Jordan algebras. Subsequently, Faulkner gave uniform proofs of two of these that did not use this determination. These are the ones given in [81].

Let $(u_1, \ldots, u_n)$ be a base for $\mathcal{J}/F$, $\xi_1, \ldots, \xi_n$ indeterminates and let $x = \Sigma\, \xi_i u_i \in \mathcal{J}_{F(\xi)}$, $F(\xi) = F(\xi_1, \ldots, \xi_n)$. We have the U-multiplication table

$$U_{u_i} u_k = \sum_i \gamma_{ik\ell} u_\ell, \qquad U_{u_i, u_j} u_k = \sum_i \delta_{ijk\ell} u_\ell,$$

where the γ's and δ's are in F. Then, $U_x u_k = \Sigma_\ell \zeta_{k\ell} u_\ell$, where $\zeta_{k\ell} = \Sigma_i \gamma_{ik\ell} \xi_i^2 + \Sigma_{i<j} \delta_{ijk\ell} \xi_i \xi_j$. But $f_{jk\ell} = \xi_j \zeta_{k\ell} - \xi_\ell \zeta_{kj}$, so $f_{jk\ell} \in F[\xi] = F[\xi_1, \ldots, \xi_n]$. These homogeneous polynomials define a projective F variety $\mathscr{V}_{\mathcal{J}}$, whose points in $\mathbb{P}\mathcal{J}_E$ for E an extension field of F are the one-dimensional subspaces Ea, where a is reduced in $\mathcal{J}_E$.

I showed that if $\mathcal{J}$ is simple over an algebraically closed field, then the algebraic component of 1 of the structure group of $\mathcal{J}$ acts transitively on $\rho(\mathcal{J})\backslash\{0\}$, and used this to prove the $\mathscr{V}_{\mathcal{J}}$ is irreducible for any central simple $\mathcal{J}$, or, equivalently, the coordinate algebra $F[\mathscr{V}_{\mathcal{J}}]$ is a domain. Hence one can define the field of fractions $F(\mathscr{V}_{\mathcal{J}})$ and the field $PF(\mathscr{V}_{\mathcal{J}})$ of homogeneous rational functions of degree 0 on $\mathscr{V}_{\mathcal{J}}$. If $\mathcal{J}$ is reduced and simple, then $\mathscr{V}_{\mathcal{J}}$ is birationally equivalent to a projective space $\mathbb{P}F^{(r)}$, where the dimensionality r can be determined in the following way, following Faulkner: r is the dimensionality of the half-space $\mathcal{J}_{1/2}(e)$ for any idempotent $e \in \mathcal{J}$ such that $\mathcal{J}_1(e) = Fe$. The existence of such idempotents in a reduced simple Jordan algebra is readily proved. For any central simple $\mathcal{J}$, the dimensionality of $\mathscr{V}_{\mathcal{J}}$ is the same as that of $\mathscr{V}_{\mathcal{J}_{\bar{F}}}$, $\bar{F}$ the algebraic closure of F. Hence Faulkner's result gives a determination of the dimensionality of $\mathscr{V}_{\mathcal{J}}$ for any central simple $\mathcal{J}$.

If $\mathcal{J}$ is central simple Jordan over F, an extension field E/F is called a reducing field of $\mathcal{J}$ if $\mathcal{J}_E$ is reduced. This Jordan concept is analogous to that of a splitting field of a central simple associative algebra; and one has results in the Jordan case that are similar to those for associative algebras. For example, the algebraic closure $\bar{F}$ of F is a reducing field for any central simple $\mathcal{J}$. Also, I showed in [81] that any central simple $\mathcal{J}$ has a finite dimensional separable reducing field, and if E is a separable field such that $[E{:}F]$ is the degree of $\mathcal{J}$ and E^+ is a subalgebra of $\mathcal{J}$, then E is a reducing field.

An analogue of Amitsur's well-known concept of generic splitting field of a central simple associative algebra is the concept of generic reducing field: E/F is a generic reducing field of $\mathcal{J}$ (central simple) if it has the property that E'/F is a reducing field of $\mathcal{J}$ if and only if there exists an F place on E to E'. If follows from the definition of the projective variety $\mathscr{V}_{\mathcal{J}}$ that E/F is a reducing field of $\mathcal{J}$ if and only if $\mathscr{V}_{\mathcal{J}}$ contains an E-rational point. This can be used to show that the field $PF(\mathscr{V}_{\mathcal{J}})$ is a generic splitting field for $\mathcal{J}$. Also, we have the criterion that $\mathcal{J}$ is reduced if and only if $PF(\mathscr{V}_{\mathcal{J}})$ is rational (= purely transcendental) over F.

Another class of projective varieties considered in [81] is the class of norm hypersurfaces (cf. Sec. 4 of [79]). For a Jordan algebra $\mathcal{J}$, one defines $\mathscr{N}_{\mathcal{J}}$ to be the projective variety defined by the homogeneous polynomial $N(x)$ where N is the generic norm. This is an irreducible variety if $\mathcal{J}$ is simple, since in this case $N(x)$ is an irreducible polynomial (see [8]). Then we have the field $PF(\mathcal{N}_{\mathcal{J}})$ (also denoted as $PNF(\mathcal{J})$ in [81]) of homogeneous rational functions of degree 0 on $\mathcal{N}_{\mathcal{J}}$. If $\mathcal{J}$ is reduced simple, then $\mathcal{N}_{\mathcal{J}}$ is birationally equivalent to a projective space $\mathbb{P}F^{(n-2)}$ where $n = \dim \mathcal{J}$. The proof of this given by Faulkner is based on a set of norm equations defining the variety $\mathcal{V}_{\mathcal{J}}$ for central simple $\mathcal{J}$. Let m be the degree of $\mathcal{J}$. Then we can write $N(a + \lambda x) = \Sigma_0^m N_i(a, x)\lambda^i$, where $N_i(a, x)$ is a homogeneous polynomial of degree m-i in the ξ's. Then the set of polynomials $N_i(a, x)$ for all $a \in \mathcal{J}$ is a defining set of polynomials for the variety $\mathcal{V}_{\mathcal{J}}$.

I showed also that $PF(\mathcal{N}_{\mathcal{J}})$ is a generic reducing field for a central simple $\mathcal{J}$, and $\mathcal{J}$ is reduced if and only if $PF(\mathcal{N}_{\mathcal{J}})$ is rational over F.

The results on reducing fields of central simple Jordan algebras are applicable to the study of central simple associative algebras with involution. One can list the central simple $(\mathcal{A}, J)$: A_I, $\mathcal{A} = \mathcal{B} \oplus \mathcal{B}^{op}$, $\mathcal{B}$ cental simple, J the exchange involution; A_{II}, $\mathcal{A}$ simple with center a separable quadratic extension with J an involution of second kind; B, $\mathcal{A}$ central simple, J of orthogonal type; and C, $\mathcal{A}$ central simple, J of symplectic type. In all cases $H(\mathcal{A}, J)$ is central simple. I gave a number of results relating reducing fields of $H(\mathcal{A}, J)$ and splitting fields and $\frac{1}{2}$-splitting fields of $\mathcal{A}$. For example, for the algebras A_I, $H(\mathcal{A}, J) \cong \mathcal{B}^+$, and in this case we have that the fields $PF(\mathcal{V}_{\mathcal{J}})$ and $PF(\mathcal{N}_{\mathcal{J}})$, $\mathcal{J} = \mathcal{B}^+$, are generic splitting fields for $\mathcal{B}$. The case of the fields $PF(\mathcal{N}_{\mathcal{J}})$ had been considered earlier in [79]. I shall not give the details of the results here.

After I had finished [81], I learned of the paper by Holger Petersson, "Generic reducing fields for Jordan pairs" in *Transactions of the American Mathematical Society* **285**, 1984, pp. 825–843, that overlaps substantially with [81]. The points of view of the two papers are quite different. Two notable results of Petersson's papers that are not contained in [81] are (1) $\mathcal{N}_{\mathcal{J}}$ is birationally equivalent to the product of $\mathcal{V}_{\mathcal{J}}$ and a projective space, and (2) if $\mathcal{B}_{\mathcal{A}}$ denotes the Brauer–Severi variety of the central simple associative algebra $\mathcal{A}$, then $\mathcal{B}_{\mathcal{A}} \times \mathcal{B}_{\mathcal{A}^0}$ and $\mathcal{V}_{\mathcal{A}^+}$ are isomorphic projective varieties.

Paper [82] originated in a question that was communicated to me by J. D. Malley of the National Institutes of Health: Can every Jordan algebra $J = \mathbb{R}1 \oplus V$ of a positive definite symmetric bilinear form on a finite dimensional vector space V over $\mathbb{R}$ be realized as a Jordan algebra of real symmetric matrices? More generally, in connection with a problem in statistics (see reference 4 in [82]), Malley was interested in knowing what kind of Jordan algebras of real symmetric matrices existed. Any such algebra is a subalgebra of a formally real Jordan algebra and hence is formally real. These algebras were determined by Jordan, v. Neumann and Wigner in 1934 to be direct sums of ideals of one of the following types:

I. $H(M_n(C),*)$ the Jordan algebra of *-symmetric matrices over C where C is (1) $\mathbb{R}$, (2) $\mathbb{C}$, 3 $\mathbb{H}$ (Hamilton's quaternion algebra), or (4) $\mathbb{O}$ (the octonion algebra of Cayley and Graves). In all cases, * is the hermitian involution, the multiplication is the usual $a \cdot b = \frac{1}{2}(ab + ba)$, and $n \leq 3$ if $C = \mathbb{O}$.

II. Jordan algebras $\mathbb{R}1 \oplus V$ of positive definite symmetric bilinear forms on real vector spaces.

It was shown by Albert that $H(M_3(\mathbb{O}),*)$ is not special so it cannot be realized as a Jordan algebra of real symmetric matrices. On the other hand, it is easy to see that the other Jordan algebras in I can be so realized. Hence it remains to consider this question for the Jordan algebras in II.

Let $Q(x) = f(x, x)$ for the symmetric bilinear form defining $J = \mathbb{R}1 \oplus V$ and let

$C(V, Q)$ be the Clifford algbera defined by Q. We have the canonical imbedding of J into $C(V, Q)$. If π is the main involution, then $J \subset H(C(V, Q), \pi)$ is the canonical imbedding of J in $C(V, Q)$. Hence, if we have a monomorphism of algebras with involution of $(C(V, Q), \pi)$ in an $(M_m(\mathbb{R}), t)$, then we obtain by restriction an imbedding of J in $H(M_m(\mathbb{R}), t)$. I verified directly that we have the following isomorphisms of algebras with involution: (1) $(C(V, Q), \pi) \cong (\mathbb{R} \oplus \mathbb{R}, 1_{\mathbb{R} \oplus \mathbb{R}})$ if $\dim V = 1$; (2) $(C(V, Q), \pi) \cong (M_2(\mathbb{R}), t)$ if $\dim V = 2$; (3) $(C(V, Q), \pi) \cong (M_2(\mathbb{C}), *)$ if $\dim V = 3$; (4) $(C(V, Q), \pi) \cong (M_2(\mathbb{H}), *)$ if $\dim V = 4$. I used this information and a standard induction procedure of the theory of Clifford algebras to obtain similar isomorphisms of $(C(V, Q), \pi)$ for any $\dim V$. This gives imbeddings of $H(C(V, Q), \pi)$ and hence of $\mathbb{R}1 \oplus V$ into Jordan algebras of real symmetric matrices. I showed that the representation in real symmetric matrices given for the $H(M_n(C), *)$ with $n \leq 3$ and the ones given for the $\mathbb{R}1 \oplus V$ are irreducible. This gave a determination of all the irreducible Jordan algebras of real symmetric matrices. I showed also in Theorem 3 of [82] that similarity of representations of a Jordan algebra over $\mathbb{R}$ implies orthogonal similarity.

These results determine effectively the orbits under the orthogonal group $O(n)$ acting by similarity on Jordan algebras of real $n \times n$ symmetric matrices for any n.

Hokkaido Mathematical Journal Vol. 10 (1981) p. 333-342

Bimodule structure of certain Jordan algebras relative to subalgebras with one generator

By N. JACOBSON*

Dedicated to Goro Azumaya on his sixtieth birthday

(Received October 17, 1980)

Throughout this paper "algebra" will mean finite dimensional algebra with unit over a field F and, unless otherwise indicated, "algebra" without modifier will mean associative algebra. An algebra is called a *Frobenius algebra* if there exists a non-degenerate associative bilinear form $f(x, y)$ on A, where associativity means that

$$f(ab, c)=f(a, bc) \tag{0.1}$$

for a, b, $c\in A$. This condition is readily seen to be equivalent to: A contains a hyperplane that contains no non-zero one sided ideal.

A number of years ago we proved the following result on generation of central simple algebras.

0.1. THEOREM. *Let A be a central simple algebra of degree n, C a commutative Frobenius subalgebra of n dimensions. Then A contains an element b such that $A=CbC$ (Jacobson* [1]).

It is well known that an algebra with a single generator is Frobenius (see e.g. Jacobson [1], p. 219). Hence we have the following consequence of this theorem.

0.2. COROLLARY. *Let A be a central simple algebra of degree n, a an element of A such that $[F[a]:F]=n$. Then A contains an element b such that $A=F[a]bF[a]$.*

The proof of Theorem 0.1 is based on the following facts:

1. The tensor product of Frobenius algebras is Frobenius. 2. If C is a commutative Frobenius algebra then any faithful representation of C contains the regular representation as a direct component. 3. If B is a subalgebra of a central simple algebra then A regarded as a bimodule for B in the natural way can be regarded as a faithful module for $B\otimes B^{op}$. This follows from the fact that A is faithful as $A\otimes A^{op}$ module which in turn follows from the simplicity of $A\otimes A^{op}$.

* This research was partially supported by the National Science Foundation grant MCS 79-04473.

In this note we propose to investigate to what extent the analogue of the above Corollary is valid for special Jordan algebras.

1. We recall that a subspace H of an algebra A is a special Jordan algebra if H contains 1 and H is closed under the composition aba and hence under $abc+cba$. If the characteristic is $\neq 2$ this is equivalent to closure under $a.b=\frac{1}{2}(ab+ba)$. Associated with H we have a special universal envelope $S(H)$, which is analogous to the universal enveloping algebra of a Lie algebra. We recall the definition. First, we define an associative specialization σ of H into the algebra B as a linear map such that $\sigma(1)=1$, $\sigma(aba)=\sigma(a)\sigma(b)\sigma(a)$. Then a special universal envelope is a pair $(S(H),\sigma_u)$ where $S(H)$ is an algebra and σ_u is an associative specialization of H into $S(H)$ such that if σ is any associative specialization of H into B then we have a unique (associative algebra) homomorphism η of $S(H)$ into B such that

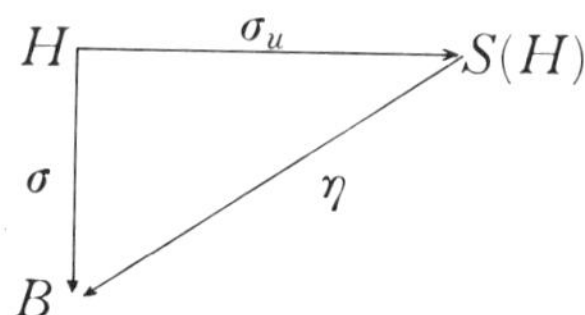

is commutative. The pair $(S(H),\sigma_u)$ is unique in the usual strong sense of uniqueness of universals in category theory. The existence of $S(H),\sigma)$ is easily proved. We refer the reader to Jacobson [2] or [3] for this and proofs of other properties which we shall state. Among these we note that $S(H)$ is generated by $\sigma_u(H)$ and $S(H)$ has a (unique) in- volution ι such that $\iota\sigma_u(a)=\sigma_u(a)$, $a\in H$. Unlike the situation for universal enveloping algebras of Lie algebras, $S(H)$ is finite dimensional for finite dimensional H.

Let H be a subalgebra of a special Jordan algebra K with ambient associative algebra B. If $a, b\in H$ we write $U_K(a)$ for the linear map $x\mapsto axa$ in K and $U_K(a,b)=U_K(a+b)-U_K(a)-U_K(b)$. This is $x\mapsto axb+bxa$. If $K=H$ we drop the subscripts K and write $U(a)$ and $U(a,b)$. We now define the *squared special Jordan algebra* $S''(H)$ to be the subalgebra of $S(H)\otimes_F S(H)$ generated by the elements $\sigma(a)\otimes\sigma(a)$, $a\in H$. We shall now show that K has a left *$S''(H)$-module* structure in which

$$(1.1)\qquad \left(\sigma(a)\otimes\sigma(a)\right)x=U_K(a)\,x\,,\qquad a\in H,\ x\in K.$$

First, we have the homomorphism of $S(H)$ into B such that $\sigma(a)\mapsto a$, $a\in H$. Combining this with $b\mapsto b_L$ where b_L is $x\mapsto bx$ we obtain the homomorphism of $S(H)$ into $\mathrm{End}_F\,B$ such that $\sigma(a)\mapsto a_L$. Similarly, since we have the involution ι of $S(H)$ such that $\iota\sigma_u(a)=\sigma_u(a)$, we have the homomorphism of $S(H)$ into $\mathrm{End}_F B$ such that $\sigma(a)\mapsto a_R$. Since left multiplications commute with right multiplications we obtain a homomorphism of $S(H)\otimes S(H)$ into $\mathrm{End}_F B$ such that $\sigma(a)\otimes\sigma(b)\mapsto a_L b_R$. Then B is a $S(H)\otimes S(H)$ module in which $(\sigma(a)\otimes(b))\, x=axb$, $x\in B$. Hence B is an $S''(H)$-module in which $(\sigma(a)\otimes\sigma(a))\, x=axa$. Since H is a subalgebra of K, K is a submodule of B as $S''(H)$-module. Then K is an $S''(H)$-module in which we have (2).

It is clear from the definition that $S''(H)$ is contained in the subalgebra of $S(H)\otimes S(H)$ of elements fixed under the exchange automorphism such that $c\otimes d\mapsto d\otimes c$, c, $d\in S(H)$. In important cases $S''(H)$ coincides with this subalgebra. This is the case if H is the subalgebra of K generated by a single element a. For, it follows from the power formulas $U(a)\,1=a^2$, $U(a)\,a^k=a^{k+2}$ that the subalgebra generated by a is $F[a]$. It is readily seen that the special universal envelope of $F[a]$ is $F[a]$ together with the identity map, so $F[a]''$ is the subalgebra of $F[a]\otimes F[a]$ generated by the elements $b\otimes b$, $b\in F[a]$. It is readily seen also that this is the subalgebra of fixed points under the exchange automorphism. If the dimensionality $[F[a]:F]=m$ then $(1, a, \cdots, a^{m-1})$ is a base for $F[a]$ and $(a^i\otimes a^i,\ a^i\otimes a^j+a^j\otimes a^i,\ 0\leq i<j\leq m-1)$ is a base for $F[a]''$.

We recall that an element e of H is idempotent if $e^2=e$ and e and f are orthogonal idempotents if $e\circ f\equiv ef+fe=U_e f=U_f e=0$. These Jordan relations are equivalent to the associative relations $ef=0=fe$ in the ambient associativealgebra A. Let $\{e_1, \cdots, e_k\}$ be a set of non-zero orthogonal idempotents that are supplementary in the sense that $\Sigma e_i=1$. Then the operations $U(e_i)$, $U(e_i, e_j)$, $i<j$, are orthogonal idempotent endomorphisms such that $\Sigma U(e_i)+\Sigma U(e_i, e_j)=1$. Hence these give a Peirce decomposition of H as

$$H=\bigoplus_{i\leq j} H_{ij} \tag{1.2}$$

where $H_{ii}=U(e_i)\,H$, $H_{ij}=U(e_i, e_j)\,H$ for $i<j$. (Jacobson [4], p. 2. 1 ff.). Since the e_i form a su form a supplementary set of orthogonal idempotents in A we have the two-sided Peirce decomposition $A=\bigoplus A_{ij}$ where $A_{ij}=e_i A e_j$. Then

$$H_{ii}=H\cap A_{ii}\,, \qquad H_{ij}=H\cap(A_{ij}+A_{ji})\,, \qquad i<j \tag{1.4}$$

Evidently H_{ii} is a special Jordan algebra with unit e_i.

We shall now derive two results on Peirce decompositions of H that are valid in the abstract case (cf. Jacobson [2], p. 105 or McCrimmon [1], p. 294).

1.5. Proposition. *We have a module action of* $S(H_{ii})\otimes S(H_{jj})$ *on* H_{ij}, $i<j$, *in which*

(1. 6) $$\left(\sigma_u(h_{ii})\otimes\sigma_u(h_{jj})\right) h_{ij} = U(h_{ii}, h_{jj})\, h_{ij}\,.$$

PROOF. Let $h_{ii}=e_i a e_i$, $h_{ij}=e_i x e_j+e_j x e_i$, $a, x\in H$. Then

$$U(h_{ii}, e_j)\, h_{ij} = e_i a e_i\,(e_i x e_j+e_j x e_i)\, e_j+ \\ e_j(e_i x e_j+e_j x e_i)\, e_i a e_i = e_i a e_i x e_j+e_j x e_i a e_i\,.$$

It follows that if h'_{ii} is a second element of H_{ii} then

$$U(h_{ii}, e_j)\, U(h'_{ii}, e_j)\, U(h_{ii}, e_j)\, h_{ij} \\ = U\left(U(h_{ii})\, h'_{ii}, e_j\right) h_{ij}\,.$$

Also $U(e_i, e_j)\, h_{ij}=h_{ij}$. Hence we have a homomorphism of $S(H_i)$ into $\mathrm{End}_F\, H_{ij}$ such that $\sigma_u(h_{ii})\mapsto U(h_{ii}, e_j)|H_{ij}$. Similarly, we have a homomorphism of $S(H_j)$ into $\mathrm{End}_F\, H_{ij}$ such that $\sigma_u(h_{jj})\mapsto U(e_i, h_{jj})|H_{ij}$. Next we can verify that

$$U(h_{ii}, e_j)\, U(e_i, h_{jj})\, h_{ij} = U(h_{ii}, h_{jj})\, h_{ij} \\ = U(e_i, h_{jj})\, U(h_{ii}, e_j)\, h_{ij}\,.$$

It follows that we have a homomorphism of $S(H_{ii})\otimes S(H_{jj})$ into $\mathrm{End}_F\, H_{ij}$ such that $\sigma_u(h_{ii})\otimes\sigma_u(h_{jj})\mapsto U(h_{ii}, h_{jj})|H_{ij}$. Hence we have a module action of $S(H_i)\otimes S(H_j)$ on H_{ij} for which (1. 6) holds. □

We now suppose that H is a subalgebra of the Jordan algebra K with ambient algebra B containing A as a subalgebra. Suppose $H=H_1\oplus H_2 \oplus\cdots\oplus H_s$ where H_i is an ideal in H (H_i is a subspace and $U(h_i)\, h$ and $U(h)\, h_i\in H_i$ for $h_i\in H_i$, $h\in H$). We have $1=\Sigma 1_i$, $1_i\in H_i$ and the 1_i are orthogonal idempotents. The decomposition $H=\oplus H_i$ is the Peirce decomposition of H relative to this set of idempotents. Let E be the ring of endomorphisms in K generated by the $U_K(h)$, $h\in H$. Then E contains the supplementary set $\Sigma=\{U_K(1_i), U_K(1_i, 1_j), i<1)\}$ of orthogonal idempotent operators. If $h_i\in H_i$ then $U_K(1_i)\, U_K(h_i)\, U_K(1_i)=U_K(U_K(1_i)\, h_i)=U_K(h_i)$. Hence $U_K(1_i)U_K(h_i)=U_K(h_i)=U_K(h_i)\, U_K(1_i)$ and $U_K(1_j)\, U_K(h_i)=0=U_K(h_i)\, U_K(1_j)$ for $j\neq i$, $U_K(1_j, 1_k)\, U_K(h_i)=0=U_K(h_i)\, U_K(1_j, 1_k)$ for $j<k$. Direct verification shows also that $U_K(1_i, 1_j)\, U_K(h_i, h_j) = U_K(h_i, h_j) = U_K(h_i, h_j)\, U_K(1_i, 1_j)$ for $h_i\in H_i$, $h_j\in H_j$, $i<j$. It now follows that multiplication of $U_K(h_i, h_j)$ on either side by any idempotent operator $\neq U_K(1_i, 1_j)$ in the set Σ gives 0. We use these results to prove

1. 7. PROPOSITION. *The idempotents $U_K(1_i)$, $U_K(1_i, 1_j)$ are central in E so $E=\bigoplus_{i\leq j} E_{ij}$ where $E_{ii}=U_K(1_i)\, E$, $E_{ij}=U_K(1_i, 1_j)\, E$, $i<j$, are ideals. Moreover, $U_K(h'_i)\in E_{ii}$ and $U_K(h_i, h_j)\in E_{ij}$ if $h_i\in H_i$, $h_j\in H_j$ and $i<j$.*

PROOF. Any $h\in H$ can be written as Σh_i, $h_i\in H_i$. Since $U_K(h)$ is a quadratic function of h we have $U_K(h)=\Sigma U_K(h_i)+\sum_{i<j}U_K(h_i, h_j)$. Thus E is generated by the $U_K(h_i)$ and the $U_K(h_i, h_j)$. Since these commute with the idempotents $U_K(1_i)$, $U_K(1_i, 1_j)$, these idempotents are central and we have $E=\bigoplus_{i<j}E_{ij}$ when the E_{ij} are as defined in the statement of the proposition. These are ideals and since $U_K(h_i)=U_K(1_i)\,U_K(h_i)$, $U_K(h_i)\in E_{ii}$. Similarly $U_K(h_i, h_j)\in E_{ij}$. □

2. The special Jordan algebras we shall consider in this paper are the following: 1. A^+ where A is central simple assocative and A^+ is obtained from A by replacing the associative product by the composition $U(a)\,b=aba$. 2. The subalgebra $H(A,j)$ of A^+ of j-symmetric elements $(j(h)=h)$ of A^+ where A is central simple with involution j. 3. $H(A,j)$ where A is simple with center a separable quadratic extension of the base field and j is an involution of second kind. All of these Jordan algebras are central simple. Under extension of the base field F to its algebraic closure $\bar{F}$ the Jordan algebras in 1. and 3. become $M_n(\bar{F})^+$ and the algebras in 2. become either $H(M_n)$ the Jordan algebra of $n\times n$ symmetric matrices or $H(M_n, s)$ the Jordan algebra of $n\times n$ (n even) symplectic symmetric matrices. In the first case j is said to be of *orthogonal type* and in the second of *symplectic type*.

If a is an element of a Jordan algebra H we call the degree of the minimum polynomial of a the *degree* of a. The maximum degree of the elements of H is called the *degree* of H. For the algebras we have listed the degrees in all cases are n except in the case $H(A,j)$ where j is of symplectic type, in which case the degree is $\nu=n/2$.

From now on H will denote one of the central simple Jordan algebras we have listed. In this section we shall show that H is a faithful $F[a]''$-module for any a that is *separable* in the sense that its minimum polynomial has distinct roots and the same result holds for all a if char $F\neq 2$. The statement that H is a faithful $F[a]''$-module is equivalent to the following: If $[F[a]:F]=m$ so $(1, a, \cdots, a^{m-1})$ is a base for $F[a]/F$ then the $m(m+1)/2$ linear transformations $U(a^i)$, $U(a^i, a^j)$, $0\leq i<j\leq m-1$, are linearly independent. Since we can extend F to its algebraic closure $\bar{F}$ and replace H by $H_{\bar{F}}$, it suffices to prove this for F algebraically closed.

2.1. LEMMA. *If e and f are non-zero orthogonal idempotents in H then $U(e,f)\neq 0$.*

PROOF. Since $e+f$ is idempotent and U_eH is a Jordan algebra of the same type as H we may assume $e+f=1$. Also we may assume F is algebraically closed. Then we can we can write $e=\sum_1^m e_i$, $f=\sum_{m+1}^n e_j$ where $\{e_k|1\leq k\leq n\}$ is a supplementary set of orthogonal idempotents $\neq 0$. Then we have the Peirce decomposition $\oplus H_{kl}$, rèlative to $\{e_k\}$. Since F is algebraically closed $H_{kk}=Fe_k$ $(\neq 0)$. Since H is simple every $H_{kl}\neq 0$ (Jacobson [4], p. 3. 26). Then the orthogonal idempotent operators $U(e_k)$ and $U(e_k, e_l)$ are $\neq 0$ and so these are linearly independent. Hence $U(e,f)=\sum_{i\leq m, j>m} U(e_i, e_j)\neq 0$. □

2. 2. THEOREM. *Let H be one of the central simple Jordan algebras listed above and let a be a separable element of H. Then H is faithful as $F[a]''$-module.*

PROOF. We may assume F algebraically closed. Then $F[a]=Fe_1\oplus \cdots \oplus Fe_m$ where the e_i are orthogonal idempotents and $a=\Sigma\alpha_i e_i$ where the α_i are distinct. Then every $U(e_i)$, $U(e_i, e_j)$ is non-zero so these are linearly independent for $1\leq i<j\leq m$. Hence the $U(a^i)$, $U(a^i, a^j)$ for $0\leq i<j\leq m-1$ are linearly independent. □

We now assume char $F\neq 2$ and we shall prove

2. 3. THEOREM. *Let H be as before and assume char $F\neq 2$. Then H is a faithful $F[a]''$-module for every $a\in H$.*

PROOF. We assume first that $a=z$ is nilpotent with minimum polynomial λ^m. Then it is known (Jacobson [5]) that z can be imbedded in a subalgebra that is a direct sum of ideals H_i where H_i can be identified with the Jordan algebra of $m_i\times m_i$ matrices that are j_i-symmetric relative to the involution

$$j_i : x\mapsto M_i({}^t x)\, M_i^{-1} \tag{2. 4}$$

where

$$M_i = \sum_{k+l=m_i+1} e_{kl} = M_i^{-1} \tag{2. 5}$$

(e_{kl} the matrix with 1 in the (k, l)-position, 0's elsewhere). Moreover, we may suppose $m_1=m$ and the component z_i of z in H_i is

$$z_i = \sum_1^{m_i-1} e_{k,\, k+1} \tag{2. 6}$$

To prove the operators $U(z^i, z^j)$, $0\leq i\leq j\leq m-1$, $(U(z^j, z^i)=2U(z^i))$ are linearly independent it suffiices to show this is the case for their restrictions to H_1. Hence we may assume H is the Jordan algebra of $m\times m$ matrices h such that $M({}^t h)\, M=h$ where $M=\sum_{k+l=m+1} e_{kl}$ and $z=\sum_1^{m-1} e_{k,k+1}$. Now H contains

(2.7) $$h_{kl} = e_{kl} + Me_{lk}M = e_{kl} + e_{m+1-l,m+1-k}$$

and

(2.8) $$(h_{kl} | 1 \leq k,\ l \leq m,\ k+1 \leq m+1)$$

is a base for H. We have

(2.9) $$U(z^i, z^j)\, h_{kl} = h_{k-i,l+j} + h_{k-j,l+i}$$

where $h_{u,v}=0$ unless the (u, v) satisfy the inequalities for (k, l) given in (2.8).

These formulas show that every $U(z^i, z^j)$, $0 \leq i \leq j \leq m-1$, $\neq 0$ and that no two of the matrices of these linear transformations relative to the base (2.8) (taken in some order) have non-zero entries in the same position. Hence the matrices are linearly independent and the operators $U(z^i, z^j)$ are linearly independent.

Now let a be arbitrary and assume F is algebraically closed. Then $F[a]$ contains a supplementary set of orthogonal idempotents $\{e_i | 1 \leq i \leq s\}$ such that $a_i = U(e_i)\, a = \alpha_i e_i + z_i$ where $\alpha_i \in F$ and $z_i \in U(e_i)\, F[a]$ has minimum polynomial λ^{m_i}. Then

(2.10) $$(e_1, z_1, \cdots, z_1^{m_1-1};\ e_2, z_2, \cdots, z_2^{m_2-1}, \cdots;\ \cdots z_s^{m_s-1})$$

is a base for $F[a]$. It auffices to prove the linear independence of the set $\{U(x, y)\}$ where x and y are chosen in (2.10) and $x \leq y$ in the order displayed in (2.10). By Proposition 1.7 and the result just proved on nilpotent elements, it suffices to show that for every $i \neq j$ the operators $U(z_i^k, z_j^l)$, $0 \leq k \leq m_i - 1$, $0 \leq l \leq m_j - 1$ are linearly independent. By replacing H by $U(e_i + e_j)\, H$ we may assume $s=2$ and write $e=e_i$, $f=e_j$, $z=z_i$, $w=z_j$. If we apply the imbedding theorem for nilpotent elements to the algebras $U(e)\, H$ and $U(f)\, H$ we see that there exists a subalgebra of H of the form $\bigoplus_1^t H_k$ where H_k is an ideal in this subalgebra that can be identified with the algebra of $m_k \times m_k$ symmetric matrices and $z \in H_1 + \cdots + H_r$, $w \in H_{r+1} + \cdots + H_t$. The result on the linear independence of the $U(z^k, w^l)$ will follow if we can show that if $(z_1, \cdots, z_u)$ is a base for $H_1 + \cdots + H_r$ and $(w_1, \cdots, w_v)$ is a base for $H_{r+1} + \cdots + H_t$ then the set of operators $\{U(z_p, w_q)\}$ are linearly independent. By using Proposition 1.7 we may assume $t=2$. Now it is known that the special universal envelope of the algebra $HM_n(F)$ of $n \times n$ symmetric matrices is $M_n(F)$ (Jacobson [2], p. 134). This applies to H_i and implies that $S(H_1) \otimes S(H_2)$ is simple. Since $U(e,f)\, H \neq 0$ (Lemma 2.1), this is a faithful module for $S(H_1) \otimes S(H_2)$ via (1.6). Then the operators $U(z_p, w_q)$ are linearly independent. □

2.11. Remark. The proof of Theorem 2.3 is considerably more

complicated than the proof of the corresponding result in the associative case. The latter follows from the fact that if A is central simple associative then $A^e = A \otimes A^{op}$ is simple and hence A is a faithful A^e-module. On the other hand, if H is central simple Jordan then H is generally not faithful as $S''(H)$-module. For example, let $H = H(A, j)$ where A is central simple of degree ≥ 3 and j is an involution in A. Assume char $F \neq 2$. Then $S''(H) \cong M_{n(n-1)/2}(F) \oplus M_{n(n+1)/2}(F)$ (Jacobson [2], p. 273). Since $H(A, j)$ is simple it is irreducible as $S''(H)$-module. Since $S''(H)$ is not simple, H is not faithful as $S''(H)$-module.

3. We prove next a result due to Azumaya (1956. unpublished).

3.1. Theorem. *Let B be the subalgebra of $F[a] \otimes F[a]$ of fixed elements under the automorphism ε such that $b \otimes c \mapsto c \otimes b$. Then B is a Frobenius algebra if and only if either a is separable or char $F \neq 2$.*

Proof. We recall first the well known result that an algebra A over F is Frobenius if and only if $A_{\bar{F}}$ is Frobenius for $\bar{F}$ the algebraic closure of F. Also it is clear that the exchange automorphism $\bar{\varepsilon}$ in $\bar{F}[a] \otimes \bar{F}[a]$ is the extension of ε in $F[a] \otimes F[a]$ and that the subalgebra of $\bar{\varepsilon}$-fixed elements is $B_{\bar{F}}$. Hence it suffice to prove the theorem for an algebraically closed base field. In this case we have $F[a] = F[a_1] \oplus \cdots \oplus F[a_r]$ where $a_i = \alpha_i e_i + z_i$, $\alpha_i \in F$, e_i the unit of $F[a_i]$ and z_i has minimum polynomial λ^{m_i}. Then $F[a] \otimes F[a] \cong \bigoplus_{i,j} (F[a_i] \otimes F[a_j])$. The automorphism ε interchanges $F[a_i] \otimes F[a_j]$ and $F[a_j] \otimes F[a_i]$. It follows that $B \cong \bigoplus_{i<j} (F[a_i] \otimes F[a_j]) \oplus (\bigoplus B_i)$ where B_i is the subalgebra of ε-fixed points of $F[a_i] \times F[a_i]$. Since the direct sum of algebras is Frobenius if and only if every component is Frobenius and the tensor product of Frobenius algebras is Frobenius, it suffices to prove the theorem for the case $r = 1$.

We now suppose $a = \alpha 1 + z$ where the minimum polynomial of z is λ^m. By replacing a by z we may assume a is nilpotent with minimum polynomial λ^m. If $m = 1$, $a = 0$ and $F[a] = F$. Then $B = F1$ is Frobenius. This and the preceding reductions show that B is Frobenius if a is separable.

Now assume a is nilpotent with minimum polynomial λ^m, $m > 1$. Then B has the base

$$(1\,;\ a^i \otimes a^i,\ 1 \leq i \leq m-1\,;\ a^i \otimes a^j + a^j \otimes a^i,\ 0 \leq i < j \leq m-1)\,. \tag{3.2}$$

These elements $\neq 1$ are nilpotent and hence span a nilpotent ideal N. We have $B/N \cong F$. Hence B is a local ring with N as its unique maximal ideal. It follows that B is Frobenius if and only if it has a unique minimal ideal (see e.g. Jacobson [1], p. 219). Since $B/N \cong F$ a B-module is completely reducible if and only if it is annihilated by N and irreducible B-modules are one dimensional. Hence the socle of B (= sum of the minimal ideals of B) is the annihilator M of N in B and B is Frobenius if and only if M is one dimensional.

Now N is generated as an ideal by the elements $a\otimes a$ and $1\otimes a^i+a^i\otimes 1$, $1\leq i\leq m-1$, since

$$(a\otimes a)^i = a^i\otimes a^i$$
$$a^i\otimes a^j+a^j\otimes a^i=(a\otimes a)^i(1\otimes a^{j-i}+a^{j-i}\otimes 1)$$

if $1\leq i<j\leq m-1$. We determine first the annihilator M_1 of $a\otimes a$. For this purpose we multiply the base (3. 2) by $a\otimes a$. This multiplication gives 0 for the base elements $a^{m-1}\otimes a^{m-1}$ and $a^i\otimes a^{m-1}+a^{m-1}\otimes a^i$. Otherwise, we obtain distinct elements of the base (3. 2). It follows that M_1 is the subspace spanned by $a^{m-1}\otimes a^{m-1}$ and $a^i\otimes a^{m-1}+a^{m-1}\otimes a^i$, $0\leq i<m-1$. We now multiply these elements by $1\otimes a+a\otimes 1$ to obtain successively 0 and $a^{i+1}\otimes a^{m-1}+a^{m-1}\otimes a^{i+1}$. It follows that the annihilator of $a\otimes a$ and $a\otimes 1+1\otimes a$ is spanned by $a^{m-1}\otimes a^{m-1}$ if char $F\neq 2$ and by $a^{m-1}\otimes a^{m-1}$ and $a^{m-2}\otimes a^{m-1}+a^{m-1}\otimes a^{m-2}$ if char $F=2$. Now

$$1\otimes a^i+a^i\otimes 1\equiv(1\otimes a+a\otimes 1)^i \mod \big((a\otimes a)\, B\big).$$

Hence $a^{m-1}\otimes a^{m-1}\in M$ for any F and $a^{m-2}\otimes a^{m-1}+a^{m-1}\otimes a^{m-2}\in M$ if char $F=2$. Thus

$$M=F(a^{m-1}\otimes a^{m-1}) \qquad \text{if char } F\neq 2$$
$$M=F(a^{m-1}\otimes a^{m-1})+F(a^{m-2}\otimes a^{m-1}+a^{m-1}\otimes a^{m-2}) \qquad \text{if char } F=2$$

and B is Frobenius if and only if char $F\neq 2$. □

4. We can now derive our main results on the central simple Jordan algebras listed at the beginning of 2.

4. 1. THEOREM. *Let H be a Jordan algebra in the following list: 1. A^+, A central simple associative, 2. $H(A,j)$ where A is central simple and j is an involution, 3. $H(A,j)$ where A is simple with center a separable quadratic extension of the base field and j is an involution of second kind. Let $a\in H$ be of degree m and suppose a is separable if char $F=2$. Then there exists an element b in H such that $F[a]''b$ has the base a^iba^i, $a^iba^j+a^jba^i$ where $0\leq i<j\leq n-1$ and $H=F[a]''b\oplus M$, M an $F[a]''$-submodule of H.*

PROOF. The results of the last two sections show that $F[a]''$ is a Frobenius algebra and H is a faithful module for $F[a]''$. Hence $H=P\oplus M$ where P and M are $F[a]''$-submodules and $p\cong F[a]''$. Let b denote the image of 1 in a $F[a]''$ isomorphism of $F[a]''$ onto P. Since the elements $a_i \otimes a^i$, $a^i \otimes a^j + a^j \otimes a^i$ form a base for $F[a]''$ it follows that the elements $a^i ba^i$, $a^i ba^j + a^j ba^i$, $0 \leq i < j \leq m-1$, form a base for P.

The dimensionality of $F[a]''b$ is $m(m+1)/2$. This coincides with $[H:F]$ if and only if $H=H(A, j)$ where A is central simple and j is an involution of orthogonal type, and a has maximal degree. The positive part of this result can be stated as

4.2. Corollary. *Let $H=H(A, j)$ where A is central simple and j is an involution of orthogonal type in A. Let a be an element of H of degree n where $[A:F]=n^2$ and assume a is separable if char $F=2$. Then there exists a $b \in H$ such that*

$$(a^i ba^i,\ a^i ba^j + a^j ba^i,\ 0 \leq i < j \leq n-1) \tag{4.3}$$

is a base for H over F.

4.4. Remarks. The foregoing theorem and corollary appear to constitute an adequate Jordan analogue for Corollary 0.2. One is tempted to define a Frobenius Jordan algebra as a Jordan algebra that possesses a non-degenerate symmetric associative bilinear form. If char $F \neq 2$ the analogue of a commutative associative algebra is a Jordan algebra that is associative in the multiplication $a.b = \frac{1}{2} U(a, b)\, 1$. This leads to the question of the valdity of an analogue of Theorem 0.1 in which commutative Frobenius subalgebras are replaced by associative Frobenius subalgebras of H (for char $F \neq 2$).

References

[1] N. Jacobson: Generation of separable and central simple algebras, Journal de Math. v. 36 (1957), 217–227.

[2] N. Jacobson: Structure and Representation of Jordan Algebras, AMS Colloq. Publ., 1968.

[3] N. Jacobson: Structure groups and Lie algebras of Jordan algebras of symmetric elements of associative algebras with involution, Advances in Mathematics, v. 20 (1976), 106–150.

[4] N. Jacobson: Lectures on Quadratic Jordan Algebras, Tata Institute Lecture Notes, 1969.

[5] N. Jacobson: Nilpotent elements in Semi-simple Jordan algebras, Math. Annalen. v. 136 (1958), 375–386.

[6] K. McCrimmon: Representations of quadratic Jordan Algebras, Trans. AMS v. 153 (1971), 279–305.

Yale University

Reprinted with corrections from Advances in Mathematics
Vol. 48, No. 2, May 1983

Some Applications of Jordan Norms to Involutorial Simple Associative Algebras

(*corrected*)

N. JACOBSON*

Department of Mathematics, Yale University, New Haven, Connecticut 06520

The following result on reduced norms of central simple associative algebras has been proved by the author:

THEOREM. *Let $\mathscr{A}$ and $\mathscr{B}$ be central simple associative algebras of the same degree. Then the reduced norm forms on $\mathscr{A}$ and $\mathscr{B}$ are equivalent if and only if $\mathscr{A}$ and $\mathscr{B}$ are either isomorphic or anti-isomorphic.*

This was first proved by the author in [6] with the restriction that char $\Phi \neq 2, 3$ for Φ the base field. The restriction char $\Phi \neq 3$ and char $\Phi \neq 2$ were removed in [7] and [9], respectively (see also [16]). We also have the obvious result that $\mathscr{A}$ is a division algebra if and only if the reduced norm is anisotropic ($N(a) \neq 0$ if $a \neq 0$).

These results apply in particular to central simple algebras that are involutorial. However, as we shall show in this paper, these results can be improved by replacing N by a norm N_J of degree n, the degree of $\mathscr{A}$, in $n(n+1)/2$ variables, and if n is even by a form N_J of degree $n/2$ in $n(n-1)/2$ variables. The special case in which $n = 2$ gives well-known results on quaternion algebras and the case $n = 4$ gives an improvement of a result of Albert's and provides an answer to a question that had been raised (orally) by Professor Max Knus.

The reduced norm of a central simple algebra $\mathscr{A}$ defines a norm surface $N = 0$ whose function field is a generic splitting field for $\mathscr{A}$ (Heuser [5], Saltman [13]). In a similar manner we can define a function field of the surface $N_J = 0$. This is a generic splitting field for $\mathscr{A}$ if J is an involution of orthogonal type and is generic for fields E such that $\mathscr{A}^E \cong M_n(\mathscr{D})$, where $\mathscr{D}$ is a quaternion algebra if J is of symplectic type. The last section of the paper is devoted to the study of these fields.

* This research was partially supported by National Science Foundation Grant MCS 79-05018.

0001-8708/83 $7.50

1. Forms Associated with an Involutorial Central Simple Associative Algebra

This note is an addendum to the author's paper [9], especially to Sections 7 and 8 of that paper. We follow the notations and freely use the results of [9]. Let $\mathscr{A}$ be a finite-dimensional central simple associative algebra over the field Φ such that $\{\mathscr{A}\}^2 = 1$ in the Brauer group $\mathrm{Br}(\Phi)$. It is a well-known result of Albert's that $\mathscr{A}$ has an involution J, that is, an anti-automorphism of $\mathscr{A}/\Phi$ such that $J^2 = 1_{\mathscr{A}}$ (Albert [2, p. 161]). The converse is clear since $\{\mathscr{A}\}^{-1} = \{\mathscr{A}^0\}$, $\mathscr{A}^0$, the opposite algebra of $\mathscr{A}$. The involution J defines the (unital quadratic) Jordan algebra $\mathscr{H}(\mathscr{A}, J)$. The basic Jordan composition here is $U_a b = aba$, $a, b \in \mathscr{H}(\mathscr{A}, J)$ and $\mathscr{H}(\mathscr{A}, J)$ contains 1. If $\bar{\Phi}$ is the algebraic closure of Φ, then the extension algebra $\mathscr{A}^{\bar{\Phi}} = \bar{\Phi} \otimes_{\Phi} \mathscr{A} = M_n(\bar{\Phi})$ and J extends to an involution $J^{\bar{\Phi}}$ in $M_n(\Phi)$. Relative to a suitable choice of the matrix base $\{e_{ij} \mid 1 \leqslant i, j \leqslant n\}$, $J^{\bar{\Phi}}$ is either the transpose involution $a \rightsquigarrow {}^t a$ or the symplectic involution $a \rightsquigarrow s({}^t a) s^{-1}$, where

$$s = \mathrm{diag}\{q, q, \ldots, q\}, \qquad q = \begin{pmatrix} 0 & 1 \\ -1 & 0 \end{pmatrix}. \tag{1.1}$$

In the first case J is said to be of *orthogonal type* and in the second case J is of *symplectic type*. We have the following table of dimensionalities:

$$\begin{aligned} [\mathscr{H}(\mathscr{A}, J) : \Phi] &= n(n+1)/2 && \text{if } J \text{ is of orthogonal type,} \\ &= n(n-1)/2 && \text{if } J \text{ is of symplectic type} \\ & && \text{char } \Phi \neq 2, \\ &= n(n+1)/2 && \text{if } J \text{ is of symplectic type and} \\ & && \text{char } \Phi = 2, \end{aligned} \tag{1.2}$$

where n is the degree of $\mathscr{A}$ ($[\mathscr{A} : \Phi] = n^2$). In the case of char $\Phi = 2$ and J is of symplectic type it is generally more interesting to replace $\mathscr{H}(\mathscr{A}, J)$ by the Jordan algebra $\mathscr{H}(\mathscr{A}, J)'$ defined to be the outer ideal in $\mathscr{H}(\mathscr{A}, J)$ generated by 1. We have $\mathscr{H}(\mathscr{A}, J)'^{\bar{\Phi}} = \mathscr{H}(\mathscr{A}^{\bar{\Phi}}, J^{\bar{\Phi}})'$ and this is the set of matrices a such that as is alternate, that is ${}^t(as) = as$ and the diagonal elements of as are 0. We have

$$[\mathscr{H}(\mathscr{A}, J)' : \Phi] = n(n-1)/2. \tag{1.3}$$

In the remainder of the paper $\mathscr{H}(\mathscr{A}, J)'$ will denote the outer ideal generated by 1 in $\mathscr{H}(\mathscr{A}, J)$, J of symplectic type and char $\Phi = 2$.

The vector space $\mathscr{K}(\mathscr{A}, J)$ of J-skew elements of $\mathscr{A}$ is a Lie algebra with composition $[ab] = ab - ba$, which is restricted if char $\Phi = p \neq 0$. If char $\Phi \neq 2$, then $\mathscr{A} = \mathscr{H}(\mathscr{A}, J) \oplus \mathscr{K}(\mathscr{A}, J)$. If u is an invertible element in $\mathscr{H}(\mathscr{A}, J) \cup \mathscr{K}(\mathscr{A}, J)$, then $J_u \equiv i_u J$, where i_u is the inner automorphism $a \rightsquigarrow uau^{-1}$, is an involution in $\mathscr{A}$. It follows from the Skolem–Noether theorem that every involution in $\mathscr{A}$ is of this form. If $u \in \mathscr{H}(\mathscr{A}, J)$, then

$$\mathscr{H}(\mathscr{A}, J_u) = u\mathscr{H}(\mathscr{A}, J) \tag{1.4}$$

and the Jordan algebra $\mathscr{H}(\mathscr{A}, J_u)$ is isotopic to $\mathscr{H}(\mathscr{A}, J)$. In fact, the map $u_L : a \rightsquigarrow ua$ is an isomorphism of the u-isotope $\mathscr{H}(\mathscr{A}, J)^{(u)}$ of $\mathscr{H}(\mathscr{A}, J)$ onto $\mathscr{H}(\mathscr{A}, J_u)$. The same remarks apply to the Jordan algebra $\mathscr{H}(\mathscr{A}, J)'$ and any invertible element $u \in \mathscr{H}(\mathscr{A}, J)'$. If $u \in \mathscr{K}(\mathscr{A}, J)$ is invertible, then $\mathscr{H}(\mathscr{A}, J_u) = u\mathscr{K}(\mathscr{A}, J)$. Hence if char $\Phi \neq 2$, then J is of orthogonal type if and only if J_u is of symplectic type. If u_1 and u_2 are invertible elements of $\mathscr{H}(\mathscr{A}, J)$, then $\mathscr{H}(\mathscr{A}, J_{u_1})$ and $\mathscr{H}(\mathscr{A}, J_{u_2})$ are isotopic since $u_2 u_1^{-1} \in \mathscr{H}(\mathscr{A}, J_{u_1})$ and $(J_{u_1})_{u_2 u_1^{-1}} = J_{u_2}$.

We recall that a Jordan algebra $\mathscr{J}$ is a division algebra if every $a \neq 0$ in $\mathscr{J}$ is invertible ($\equiv U_a$ is invertible in $\mathrm{End}_\Phi \mathscr{J}$). If $\mathscr{A}$ is an associative division algebra then $\mathscr{H}(\mathscr{A}, J)$ is a Jordan division algebra.

1.5. Proposition. *Assume the degree $n > 1$ (that is, $\mathscr{A} \neq \Phi$). Then $\mathscr{H}(\mathscr{A}, J)(\mathscr{H}(\mathscr{A}, J)')$ is a division algebra if and only if either $\mathscr{A}$ is a division algebra or $\mathscr{A} = M_2(\Phi)$ and J is of symplectic type.*

Proof. We may replace $(\mathscr{A}, J)$ by an isomorphic pair $(\mathscr{B}, K)$, where $(\mathscr{B}, K) \cong (\mathscr{A}, J)$ means that there exists an isomorphism η of $\mathscr{A}$ onto $\mathscr{B}$ such that $\eta J = K\eta$. We use the structure theorem that either $(\mathscr{A}, J) \cong (M_m(\mathscr{D}), \bar{t}_c)$, where $\mathscr{D}$ is a division algebra with an involution $d \rightsquigarrow \bar{d}$ and $\bar{t}_c$ is the involution $a \rightsquigarrow c({}^t\bar{a})\, c^{-1}$, $\bar{a} = (\bar{a}_{ij})$ for $a = (a_{ij})$ and $c = \mathrm{diag}\{c_1, c_2, \ldots, c_m\}$, where $c_i = \bar{c}_i$ is invertible, or $(\mathscr{A}, J) \cong (M_m(\mathscr{D}), t_s)$, where $\mathscr{D} = M_2(\Phi)$ and t_s is $a \rightsquigarrow s({}^t a)\, s^{-1}$, s as in (1.1) (see, for example, Jacobson [8, p. 0.11]). In either case $\mathscr{H}(\mathscr{A}, J)$ and $\mathscr{H}(\mathscr{A}, J)'$ contain m orthogonal idempotents $\neq 0$. Hence, if $\mathscr{H}(\mathscr{A}, J)$ and $\mathscr{H}(\mathscr{A}, J)'$ are division algebras, then $m = 1$. Then either $\mathscr{A}$ is a division algebra or $\mathscr{A} = M_2(\Phi)$ with J symplectic. Conversely, if $(\mathscr{A}, J)$ is of either of these forms, then $\mathscr{H}(\mathscr{A}, J)$ is a division algebra. ∎

From now on it is convenient to assume Φ infinite. Let N_J denote the generic norm defined on the Jordan algebra $\mathscr{H}(\mathscr{A}, J)$, and if char $\Phi = 2$ and J is of symplectic type, let N_J' denote the generic norm on $\mathscr{H}(\mathscr{A}, J)'$. For any characteristic, if J is of orthogonal type, then N_J is the restriction to $\mathscr{H}(\mathscr{A}, J)$ of the reduced (or generic) norm N on $\mathscr{A}$. Hence N_J is a form of degree $n = \deg \mathscr{A}$. If char $\Phi \neq 2$ and J is of symplectic type, then $\deg N_J = \frac{1}{2}n$. In this case, an exact determination of N_J is the following: If $\bar{\Phi}$ is the algebraic closure of Φ, then we have an imbedding of $\mathscr{A}$ in $M_n(\bar{\Phi})$ so that the elements $a \in \mathscr{H}(\mathscr{A}, J)$ satisfy the condition $s({}^t a) s^{-1} = a$. Then ${}^t(as) = -as$ and $N_J(a)$ is the Pfaffian $\mathrm{Pf}(as)$. If char $\Phi = 2$ and J is of symplectic type, then N_J is the restriction of the reduced norm to $\mathscr{H}(\mathscr{A}, J)$ and N'_J has degree $n/2$ and its determination is the same as that of N_J for char $\neq 2$ and J symplectic.

We can now determine the isotopy classes of Jordan algebras $\mathscr{H}(\mathscr{A}, J)$ and $\mathscr{H}(\mathscr{A}, J)'$. First, if deg $\mathscr{A} = n$ is odd, then there are no symplectic involutions. Hence all the $\mathscr{H}(\mathscr{A}, J)$ are isotopic. Next, suppose n is even and char $\Phi \neq 2$. In this case there exist symplectic involutions. For, let J be of orthogonal type. Then the subset of $\mathscr{H}(\mathscr{A}, J)$ of invertible elements is given by the condition $N(u) \neq 0$. Hence this is a Zariski open subset of $\mathscr{H}(\mathscr{A}, J)$. Since the corresponding subset of $\mathscr{A}^{\bar{\Phi}}$ is not vacuous, there exist invertible u in $\mathscr{H}(\mathscr{A}, J)$ and we have a symplectic involution J_u. Thus if n is even and char $\Phi \neq 2$, then there exist two isotopy classes of algebras $\mathscr{H}(\mathscr{A}, J)$. The same argument applies in the case n even and char $\Phi = 2$. In addition, in this case we have the isotopy class of algebras $\mathscr{H}(\mathscr{A}, J)'$, J of symplectic type.

1.6. Definition. Let f and g be polynomial functions on the vector spaces V/Φ and W/Φ, respectively. Then f and g are *similar* ($f \sim g$) if there exists a bijective linear map η of V into W and a nonzero $\rho \in \Phi$ such that $g(\eta(x)) = \rho f(x)$, $x \in V$.

1.7. Proposition. *Let J and K be involutions of orthogonal (symplectic) type in $\mathscr{A}$. Then N_J and N_K (defined on $\mathscr{H}(\mathscr{A}, J)$ and $\mathscr{H}(\mathscr{A}, K)$, respectively) are similar. The same holds for N'_J and N'_K in the characteristic two case for J and K symplectic.*

Proof. If J and K are of orthogonal type, then $K = i_u J$ and $\mathscr{H}(\mathscr{A}, K) = u\mathscr{H}(\mathscr{A}, J)$ for an invertible element $u \in \mathscr{H}(\mathscr{A}, J)$. Then $\eta = u_L$ is a bijective linear map of $\mathscr{H}(\mathscr{A}, J)$ onto $\mathscr{H}(\mathscr{A}, K)$. Now N_J and N_K are the restrictions to $\mathscr{H}(\mathscr{A}, J)$ and $\mathscr{H}(\mathscr{A}, K)$, respectively, of the reduced norm N on $\mathscr{A}$ and this function is multiplicative. Hence $N_K(\eta(a)) = N(ua) = N(u)\,N(a) = \rho N(a)$ for $\rho = N(u)$. Thus $N_J \sim N_K$. Next let J and K be of symplectic type. Then $N_J(a)^2 = N(a)$, $N_K(b)^2 = N(b)$ for the reduced norm N. Again we have an invertible $u \in \mathscr{H}(\mathscr{A}, J)$ such that $K = i_u J$ and $\eta = u_L$ is a bijective linear map of $\mathscr{H}(\mathscr{A}, J)$ onto $\mathscr{H}(\mathscr{A}, K)$. We have $N_K(\eta(a))^2 = \rho N_J(a)^2$, $a \in \mathscr{H}(\mathscr{A}, J)$. Since $N_J(1) = 1$, this gives $\rho = \tau^2$, where $\tau = N_K(\eta(1)$. Then $N_K(\eta(a))^2 = \tau^2 N_J(a)$ and $N_K(\eta(a)) = \pm\tau N_J(a)$, $a \in \mathscr{H}(\mathscr{A}, J)$. Thus the polynomial function $a \rightsquigarrow [N_K(\eta(a)) + \tau N_J(a)][N_K(\eta(a) - \tau N_J(a)] = 0$. It follows that either $N_K(\eta(a)) = \tau N_J(a)$ or $N_K(\eta(a)) = -\tau N_J(a)$ for all $a \in \mathscr{H}(\mathscr{A}, J)$. Hence $N_K \sim N_J$. A similar argument applies to N'_J and N'_K. ∎

2. Norm Conditions for a Division Algebra and for Isomorphism

We recall that an element $a \in \mathscr{H}(\mathscr{A}, J)$ $(\mathscr{H}(\mathscr{A}, J)')$ is invertible if and only if $N_J(a) \neq 0$ $(N_J'(a) \neq 0)$. Hence $\mathscr{H}(\mathscr{A}, J)$ is a Jordan division algebra if and only if the form N_J (N_J') is *anisotropic* in the sense that $N_J(a) \neq 0$ $(N_J'(a) \neq 0)$ for every $a \neq 0$ in $\mathscr{H}(\mathscr{A}, J)$ $(\mathscr{H}(\mathscr{A}, J)')$. The following result is therefore an immediate consequence of Proposition 1.5.

2.1. Theorem. *Assume* $\deg \mathscr{A} > 1$. *Then* $N_J(N_{J'})$ *is anisotropic if and only if either* $\mathscr{A}$ *is a division algebra or* $\mathscr{A} = M_2(\Phi)$ *and* J *is of symplectic type.*

We recall the definition of the generic trace bilinear form in $\mathscr{H}(\mathscr{A}, J)$ as

$$T_J(a, b) = -\Delta_1^a \Delta^b \log N_J = (\Delta_1^a N_J)(\Delta_1^b N_J) - \Delta_1^a \Delta^b N_J, \tag{2.2}$$

where $\Delta_b^a F$ is the directional derivative at a in the direction b of the rational function F. We have a similar definition of $T_J'(a, b)$ defined by N_J'. Then T_J is a nondegenerate symmetric bilinear form if char $F \neq 2$, but not otherwise. However, T_J' is nondegenerate. From now on we discard the degenerate cases. We have the important formula

$$\Delta_c^a \Delta^b \log N_J = T_J(U_c^{-1} a, b) \tag{2.3}$$

for all a, b and invertible c and we have a similar formula for T_J'.

2.4. Theorem. *Let* $\mathscr{A}_i$, $i = 1, 2$, *be an involutorial central simple algebra of degree* $n > 1$, J_1 *an involution in* $\mathscr{A}_i$ *of orthogonal (symplectic) type.*

(i) *If* char $\Phi \neq 2$, *we assume* $n > 2$ *in the symplectic case. Then* $\mathscr{A}_1 \cong \mathscr{A}_2$ *if and only if* $N_{J_1} \sim N_{J_2}$.

(ii) *If* char $\Phi = 2$, *we assume* $n > 2$ *and* J_i *is of symplectic type. Then* $\mathscr{A}_1 \cong \mathscr{A}_2$ *if and only if* $N_{J_1}' \sim N_{J_2}'$.

Proof. (i) The proof is similar to that of [5, Theorem 7, p. 244] (see also the proof of [7, Theorem 10]). Suppose $N_{J_1} \sim N_{J_2}$ and let η be a bijective linear map of $\mathscr{H}(\mathscr{A}_1, J_1)$ onto $\mathscr{H}(\mathscr{A}_2, J_2)$ and ρ a nonzero element of Φ such that

$$N_{J_2}(\eta(x)) = \rho N_{J_1}(x), \qquad x \in \mathscr{H}(\mathscr{A}_1, J_1). \tag{2.5}$$

Then x is invertible in $\mathscr{H}(\mathscr{A}_1, J_1)$ if and only if $\eta(x)$ is invertible in $\mathscr{H}(\mathscr{A}_2, J_2)$. From (2.3) and (2.5) we can obtain

$$T_{J_1}(U_c^{-1}b, a) = T_{J_2}(U_{\eta(c)}^{-1}\eta(b), \eta(a)) \tag{2.6}$$

for all a, b, and invertible c in $\mathscr{H}(\mathscr{A}_1, J_1)$. Since T_{J_1} and T_{J_2} are nondegenerate, we have a bijective linear map η^* of $\mathscr{H}(\mathscr{A}_2, J_2)$ onto $\mathscr{H}(\mathscr{A}_1, J_1)$ such that for $y_1 \in \mathscr{H}(\mathscr{A}_1, J_1)$, $x_2 \in \mathscr{H}(\mathscr{A}_2, J_2)$,

$$T_{J_1}(\eta^*(x_2), y_1) = T_{J_2}(x_2, \eta(y_1)). \tag{2.7}$$

Then by (2.6) and the nondegeneracy of T_{J_1} we obtain

$$U_{\eta(c)} = \eta U_c \eta^* \tag{2.8}$$

for all c with $N_{J_1}(c) \neq 0$. It follows that this holds for all c. This implies that η is an isomorphism of $\mathscr{H}(\mathscr{A}_1, J_1)$ onto an isotope of $\mathscr{H}(\mathscr{A}_2, J_2)$ [7, p. 116]. We shall show that this implies that $\mathscr{A}_1 \cong \mathscr{A}_2$. Suppose first that either the J_i are of orthogonal type or they are of symplectic type and $n > 4$. Then our hypotheses imply that $\mathscr{A}_1$ and the injection map constitute a special universal envelope for $\mathscr{H}(\mathscr{A}_1, J_1)$. By Proposition 1, p. 118 of [9], $\mathscr{A}_2$ and a suitable map constitute a special universal envelope for the isotope of $\mathscr{H}(\mathscr{A}_2, J_2)$. Hence the isomorphism of $\mathscr{H}(\mathscr{A}_1, J_1)$ onto the isotope of $\mathscr{H}(\mathscr{A}_2, J_2)$ can be extended to an isomorphism of $\mathscr{A}_1$ onto $\mathscr{A}_2$. Thus $\mathscr{A}_1 \cong \mathscr{A}_2$.

It remains to consider the case in which $n = 4$ and the J_i are of symplectic type. Then every element a of $\mathscr{H}(\mathscr{A}_i, J_i)$ is a root of an equation $\lambda^2 - T_J(a)\lambda + N_J(a) = 0$ and hence $\mathfrak{J}_i = \mathscr{H}(\mathscr{A}_i, J_i)$ is the Jordan algebra Jord$(N_{J_i}, 1)$ of the quadratic form N_{J_i} with basepoint 1 ([18], p. 5f). The special universal algebra of Jord$(N_{J_i}, 1)$ is the Clifford algebra $C_i = C(\mathfrak{J}_i, N_{J_i}, 1)$ and the canonical map of $\mathfrak{J}_i$ into C_i. We shall identify $\mathfrak{J}_i$ with its image in C_i. We have $[C_i : \Phi] = 2^5$ and either C_i is simple or it is a direct sum of two isomorphic central simple algebras of degree 4. By the universal mapping property of the special universal envelope we have a homomorphism of C_1 into $\mathscr{A}_1$ and this is an isomorphism if C_1 is simple. Since $[C_1 : \Phi] = 32$ and $[\mathscr{A}_1 : \Phi] = 16$ this is impossible. Hence C_1 is a direct sum of two central simple algebras of degree 4 and these are isomorphic to $\mathscr{A}_1$. Similarly, C_2 is isomorphic to a direct sum of two copies of $\mathscr{A}_2$ and the same is true of the special universal envelope of any isotope of $\mathfrak{J}_2$. Since $\mathfrak{J}_1$ is isomorphic to an isotope of $\mathfrak{J}_2$ it follows that C_1 is isomorphic to C_2. Then $\mathscr{A}_1 \cong \mathscr{A}_2$. Conversely, suppose $\mathscr{A}_1 \cong \mathscr{A}_2$ and let η be an isomorphism of $\mathscr{A}_1$ onto $\mathscr{A}_2$. Then $\eta J_1 \eta^{-1}$ is an involution in $\mathscr{A}_2$ of the same type as J_1 and η is a similarity of N_{J_1} and $N_{\eta J_1 \eta^{-1}}$. Thus $N_{J_1} \sim N_{\eta J_1 \eta^{-1}}$ and, by Proposition 1.7, $N_{\eta J_1 \eta^{-1}} \sim N_{J_2}$. Hence $N_{J_1} \sim N_{J_2}$.

(ii) The proof is identical to that of (i). ∎

3. Involutorial Central Simple Algebras of Degrees Two and Four

The central simple algebras of degree two are quaternion algebras. Any such algebra $\mathscr{A}$ has the *standard involution* J: $a \rightsquigarrow \bar{a}$ such that $a + \bar{a} = T(a)$, $a\bar{a} = N(a) = \bar{a}a$, where T and N are the reduced trace and norm on $\mathscr{A}$. The involution J is of symplectic type since $\mathscr{H}(\mathscr{A}, J) = \Phi$. Since $i_\alpha = 1$ if $\alpha \in \Phi$, J is the only involution of symplectic type in $\mathscr{A}$. Let $\mathscr{A}' = \{a \mid T(a) = 0\}$, so $\mathscr{A}'$ is three-dimensional and coincides with $\mathscr{K}(\mathscr{A}, J)$ $(=\mathscr{H}(\mathscr{A}, J))$ if char $\Phi = 2$). Let u be an invertible element of $\mathscr{K}(\mathscr{A}, J)$, $u \notin \Phi$. Then J_u is of orthogonal type and $\mathscr{H}(\mathscr{A}, J_u) = u\mathscr{A}'$. These remarks and Theorems 2.1 and 2.4 give the following well-known results:

3.1. Theorem. *Let $\mathscr{A}$ be a quaternion algebra over Φ, $\mathscr{A}'$ the subspace of $\mathscr{A}$ of elements of reduced trace* 0. *Then*

(i) *$\mathscr{A}$ is a division algebra if and only if $N \mid \mathscr{A}'$ is anisotropic.*

(ii) *If* char $\Phi \neq 2$ *and $\mathscr{A}_i$, $i = 1, 2$, is a quaternion algebra over Φ then $\mathscr{A}_1 \cong \mathscr{A}_2$ if and only if $N|\mathscr{A}_1' \sim N \mid \mathscr{A}_2'$.*

Proof. (i) Choose $u \in \mathscr{A}'$, $u \notin \Phi$. Then J_u is of orthogonal type and $\mathscr{H}(\mathscr{A}, J_u) = u\mathscr{A}'$. Hence $N \mid \mathscr{A}' \sim N_{J_u}$ and $N \mid \mathscr{A}'$ is anisotropic if and only if N_{J_u} is anisotropic. The result now follows from Theorem 2.1.

(ii) This follows in the same way by replacing $N \mid \mathscr{A}_1'$ and $N \mid \mathscr{A}_2'$ by similar forms $N_{J_{u_1}}$ and $N_{J_{u_2}}$, where J_{u_i} is of orthogonal type, and then applying Theorem 2.4. ∎

For the remainder of this section we assume char $F \neq 2$.

Now let $\mathscr{A}$ be involutorial central simple of degree 4. By a theorem of Albert's, $\mathscr{A} = \mathscr{A}_1 \otimes_\Phi \mathscr{A}_2$, where $\mathscr{A}_i$ is a quaternion algebra (Albert [2] or Racine [10]). Let J_i be the standard involution in $\mathscr{A}_i$ and let u_1 be an element of $\mathscr{A}_1' = \mathscr{K}(\mathscr{A}_1, J_1)$, $u_1 \notin \Phi$. Then J_{1u_1} is of orthogonal type and hence $J = J_{1u_1} \otimes J_2$ is of symplectic type. If $a_i \in \mathscr{A}_i'$, then

$$l(a_1, a_2) = u_1 \otimes a_2 + u_1 a_1 \otimes 1 \in \mathscr{H}(\mathscr{A}, J) \tag{3.2}$$

and the set

$$\mathscr{L} = \{l(a_1, a_2) \mid a_i \in \mathscr{A}_i'\} \tag{3.3}$$

is a six-dimensional subspace of $\mathscr{H}(\mathscr{A}, J)$. Hence $\mathscr{L} = \mathscr{H}(\mathscr{A}, J)$ if char $\Phi \neq 2$.

Since a_1, u_1, and $a_1 + u_1 \in \mathscr{A}_1'$, $-N(u_1 + a_1) = (u_1 + a_1)^2 = u_1^2 + u_1 a_1 + a_1 u_1 + a_1^2 = -N(u_1) - N(u_2) + u_1 a_1 + a_1 u_1$, where N is the reduced norm in $\mathscr{A}_1$. Thus

$$u_1 a_1 + a_1 u_1 = N(u_1) + N(a_1) - N(u_1 + a_1).$$

On the other hand, a calculation with matrices of trace 0 shows that $N(u_1) + N(a_1) - N(u_1 + a_1) = T(u_1 a_1)$, T the reduced traces in $\mathscr{A}_1$. Hence

$$u_1 a_1 + a_1 u_1 = T(u_1 a_1). \tag{3.4}$$

Now we have

$$(u_1 a_1)^2 = T(u_1 a_1)\, u_1 a_1 - N(u_1 a_1). \tag{3.5}$$

Hence

$$\begin{aligned} l(a_1, a_2)^2 &= (u_1 \otimes a_2 + u_1 a_1 \otimes 1)^2 \\ &= N(u_1)\, N(a_2) + u_1^2 a_1 \otimes a_2 + u_1 a_1 u_1 \otimes a_2 + (u_1 a_1)^2 \otimes 1 \\ &= N(u_1)\, N(a_2) + T(u_1 a_1)\, u_1 \otimes a_2 + T(u_1 a_1)\, u_1 a_1 \otimes 1 \qquad (3.6) \\ &\quad - N(u_1 a_1) \qquad ((3.4) \text{ and } (3.5)) \\ &= T(u_1 a_1)\, l(a_1, a_2) - N(u_1)(N(a_1) - N(a_2)). \end{aligned}$$

Hence $l(a_1, a_2)$ is a root of

$$\lambda^2 - T(ua_1)\lambda + N(u_1)((N(a_1) - N(a_2)). \tag{3.7}$$

It follows that

$$N_J(l(a_1, a_2)) = N(u_1)((N(a_1) - N(a_2)) \tag{3.8}$$

if char $\Phi \neq 2$, where N_J is the generic norm in $\mathscr{H}(\mathscr{A}, J)$.

3.10. Definition. Let $\mathscr{A}$ be involutorial central simple of degree 4 and $\mathscr{A} = \mathscr{A}_1 \otimes_\Phi \mathscr{A}_2$ a factorization of $\mathscr{A}$ as a tensor product of quaternion algebras. Then the quadratic form on $\mathscr{A}' \equiv \mathscr{A}_1' \oplus \mathscr{A}_2'$ defined by

$$Q(a_1 + a_2) = N(a_1) - N(a_2) \tag{3.9}$$

for $a_i \in \mathscr{A}_i'$ will be called an *Albert quadratic form* for $\mathscr{A}$.

Then we have

3.12. Theorem. (i) *Any two Albert quadratic forms for an involutorial central simple algebra of degree* 4 *are similar.*

(ii) *Two involutorial central simple algebras of degree* 4 *are isomorphic if and only if their Albert quadratic forms are similar.*

(iii). *If Q is an Albert quadratic form for $\mathscr{A}$, then the index of $\mathscr{A}$ is* 4, 2, *or* 1 *according as the Witt index of Q is* 0, 1, *or* 3.

Remarks. Part (iii) extends a result of Albert's in [1]. A similar result going in the direction of quadratic form to algebra of degree 4 via Clifford algebras has been given by Tamagawa in [12]. See also Seligman [11, p. 340]. Part (ii) gives an affirmative answer to a question which had been asked of the author by Professor Max Knus.

Proof. (i) We have shown that the Albert quadratic form Q associated with a factorization $\mathscr{A} = \mathscr{A}_1 \otimes \mathscr{A}_2$ is similar to a Jordan norm form N_J or N_J' where J is an involution of symplectic type in $\mathscr{A}$. Since the N_J and N_J' determined by two such involutions are similar, any two Albert quadratic forms are similar.

(ii) This is an immediate consequence of the argument in (i) and Theorem 2.4.

(iii) By Theorem 2.1, $\mathscr{A}$ is a division algebra if and only if N_J is anisotropic. Since $N_J \sim Q$ or $\mathscr{A}$ is a division algebra if and only if Q is anisotropic. Thus $\mathscr{A}$ has index 4 if and only if the Witt index of Q is 0. Next, suppose $\mathscr{A}$ has index 1 so $\mathscr{A} \cong M_4(\Phi)$. Then $\mathscr{A} = \mathscr{A}_1 \otimes \mathscr{A}_2$, where $\mathscr{A}_1 \cong \mathscr{A}_2$. Then it is clear that the Witt index of Q is 3. Finally, suppose $\mathscr{A}$ has index 2. Then $\mathscr{A} = M_2(\Phi) \otimes \mathscr{D}$, where $\mathscr{D}$ is a division algebra. Then $\mathscr{A}'$ is an orthogonal direct sum of $M_2(\Phi)'$ and $\mathscr{D}'$ relative to Q. It is readily seen that the $Q|M_2(\Phi)'$ is of Witt index 1. Since $\mathscr{D}$ is a division algebra, $Q|\mathscr{D}'$ is anisotropic. It follows that Q has index 1. ∎

Theorem 3.12(iii) has the following consequence.

3.13 Corollary (Albert–Sah). *If $\mathscr{A}_i$, $i = 1, 2$, is a quaternion division algebra, then $\mathscr{A}_1 \otimes_\Phi \mathscr{A}_2$ is not a division algebra if and only if the $\mathscr{A}_i$ contain isomorphic quadratic subfields.*

Proof. The condition is sufficient that $\mathscr{A}_1 \otimes \mathscr{A}_2$ is not a division algebra since the tensor product of isomorphic extension fields $E_i/\Phi \neq \Phi$ is never a field. Now suppose $\mathscr{A}_1 \otimes \mathscr{A}_2$ is not a division algebra. Then the Albert quadratic form determined by this factorization is not anisotropic. Hence we have an element $a = a_1 + a_2 \neq 0$, $a_i \in \mathscr{A}_i'$ such that $N(a_1) = N(a_2)$. Since $N|\mathscr{A}_i'$ is anisotropic, $N(a_i) \neq 0$. Since $T(a_i) = 0$ and $a_i^2 - T(a_i)\,a_i + N(a_i) = 0$, we have $\Phi(a_1) \cong \Phi(a_2)$. ∎

4. Generic Norm Fields of an Involutorial Central Simple Algebra

The results we shall derive in this section are valid also for $\mathscr{H}(\mathscr{A}, J)'$, J symplectic, Φ of characteristic two. However, for the sake of simplicity of exposition we shall assume char $\Phi \neq 2$.

Let $(u_1, \ldots, u_m)$ be a base for $\mathscr{H}(\mathscr{A}, J)$, $\xi_1, \ldots, \xi_m$ indeterminates, and let

$P = \Phi(\xi_1, \ldots, \xi_m)$. We have the generic element

$$x = \sum_1^m \xi_i u_i \tag{4.1}$$

in $\mathscr{H}(\mathscr{A}, J)^P$ whose minimum polynomial can be written as

$$m_x(\lambda) = \lambda^r - \tau_1(\xi_1, \ldots, \xi_m)\,\lambda^{r-1} + \cdots \\ + (-1)^r\,\tau_r(\xi_1, \ldots, \xi_m), \tag{4.2}$$

where τ_i is a homogeneous polynomial of degree i in the ξ's. Here $m = n(n+1)/2$ or $n(n-1)/2$ and $r = n$ or $n/2$ according as J is of orthogonal or symplectic type ($n = \deg \mathscr{A}$). We have

$$N_J(x) = \tau_r(\xi_1, \ldots, \xi_m) \tag{4.3}$$

and

$$m_x(\lambda) = N_J(\lambda 1 - x). \tag{4.4}$$

Let $(v_1, \ldots, v_m)$ be a second base for $\mathscr{H}(\mathscr{A}, J)$, $v_i = \sum_{j=1}^m \alpha_{ji} u_j$ and put $y = \sum \xi_i v_i = \sum \xi'_i u_j$, where $\xi'_j = \sum \alpha_{ji} \xi_i$. Then $N_J(y) = \tau_r(\xi'_1, \ldots, \xi'_m)$ and $m_y(\lambda) = N_J(\lambda 1 - y) = \lambda^r - \tau_1(\xi'_1, \ldots, \xi'_m)\,\lambda^{r-1} + \cdots + (-1)^r\,\tau_r(\xi'_1, \ldots, \xi'_m)$. Hence $m_y(\lambda)$ and $N_J(y)$ are obtained by making a linear change of variables in $m_x(\lambda)$ and $N_J(x)$, respectively. We recall also that if u is an invertible element of $\mathscr{H}(\mathscr{A}, J)$ and $J_u = i_u J$, then $\mathscr{H}(\mathscr{A}, J_u) = u\mathscr{H}(\mathscr{A}, J)$. Hence $(uu_1, \ldots, uu_n)$ is a base for $\mathscr{H}(\mathscr{A}, J_u)$ and $N_J(\sum \xi_i uu_i)$ is a multiple of $N_J(x)$.

The polynomial $N_J(x)$ is irreducible in $\Phi[\xi_1, \ldots, \xi_m]$ ([6, Theorem 7]). This implies that $m_x(\lambda)$ is irreducible in $\Phi[\xi_1, \ldots, \xi_m, \lambda]$ and hence in $\Phi(\xi_1, \ldots, \xi_m)[\lambda]$. Also the discriminant of $m_x(\lambda)$ as a polynomial in λ is $\neq 0$. It follows that

$$\Phi(\xi_1, \ldots, \xi_m, x) = \Phi(\xi_1, \ldots, \xi_m)[x] \tag{4.5}$$

is a separable algebraic extension of $\Phi(\xi_1, \ldots, \xi_m)$ and hence $\Phi(\xi_1, \ldots, \xi_m, x)$ is separable over Φ.

Since $N_J(x) = \tau_r(\xi_1, \ldots, \xi_m)$ is irreducible in $\Phi[\xi_1, \ldots, \xi_m]$,

$$\Phi[\xi_1, \ldots, \xi_m]/(\tau_r(\xi_1, \ldots, \xi_m)) \tag{4.6}$$

is a domain. Now we have

$$\Phi[\xi_1, \ldots, \xi_m]/(\tau_r(\xi_1, \ldots, \xi_m)) = \Phi[\bar{\xi}_1, \ldots, \bar{\xi}_m],$$

where $\bar{\xi}_i = \xi_i + (\tau_r(\xi_1, \ldots, \xi_m))$ and we can form the field of fractions

$\Phi(\bar{\xi}_1, \dots, \bar{\xi}_m)$ which we shall call a *generic norm field* of the Jordan algebra $\mathscr{H}(\mathscr{A}, J)$. It is readily seen from the remarks we made before that generic norm fields determined by different bases for $\mathscr{H}(\mathscr{A}, J)$ and by different involutions of the same type are isomorphic. Now we have the following analogue of a result of Heuser ([5, Satz 2]):

4.7 PROPOSITION. *The field $\Phi(\xi_1, \dots, \xi_m, x)$ is isomorphic to a simple transcendental extension of $\Phi(\bar{\xi}_1, \dots, \bar{\xi}_m)$.*

Proof. Let $1 = \sum \alpha_i u_i$, $\alpha_i \in \Phi$, and put

$$y_i = \alpha_i x - \xi_i 1, \qquad 1 \leqslant i \leqslant m. \tag{4.8}$$

Then $\Phi(\xi_1, \dots, \xi_m, x) = \Phi(y_1, \dots, y_m, x)$ and

$$\begin{aligned} m_x(\lambda) = N_J(\lambda 1 - x) = N_J \left(\sum (\alpha_i \lambda - \xi_i)\, u_i \right) \\ = \tau_r(\alpha_1 \lambda - \xi_1, \dots, \alpha_m \lambda - \xi_m). \end{aligned} \tag{4.9}$$

Since $m_x(x) = 0$,

$$\tau_r(y_1, \dots, y_m) = 0. \tag{4.10}$$

Hence $\operatorname{tr\,deg}_\Phi \Phi(y_1, \dots, y_m) \leqslant m - 1$ and $\operatorname{tr\,deg}_\Phi \Phi(\xi_1, \dots, \xi_m, x) = \operatorname{tr\,deg}_\Phi \Phi(y_1, \dots, y_m, x) \leqslant m$. Since the ξ_i are algebraically independent, $\operatorname{tr\,deg}_\Phi \Phi(\xi_1, \dots, \xi_m, x) = m$. It follows that $\operatorname{tr\,deg}_\Phi \Phi(y_1, \dots, y_m) = m - 1$ and $\Phi(y_1, \dots, y_m, x)$ is a simple transcendental extension of $\Phi(y_1, \dots, y_m)$. Now consider the homomorphism ν of $\Phi[\eta_1, \dots, \eta_m]/\Phi$, η_i indeterminates, onto $\Phi[y_1, \dots, y_m]$ such that $\eta_i \rightsquigarrow y_i$, $1 \leqslant i \leqslant m$. By (4.10), $\ker \nu$ contains the prime ideal $(\tau_r(\eta_1, \dots, \eta_m))$. Since $\operatorname{tr\,deg}_\Phi \Phi(y_1, \dots, y_m) = m - 1$, $\ker \nu = (\tau_r(\eta_1, \dots, \eta_m))$ (see, e.g., [17, p. 193]). Hence

$$\Phi[y_1, \dots, y_m] \cong \Phi[\eta_1, \dots, \eta_m]/(\tau_r(\eta_1, \dots, \eta_r))$$

and $\Phi(y_1, \dots, y_m) \cong \Phi(\bar{\xi}_1, \dots, \bar{\xi}_m)$. Thus $\Phi(\xi_1, \dots, \xi_m, x)$ is isomorphic to a simple transcendental extension of $\Phi(\bar{\xi}_1, \dots, \bar{\xi}_m)$. ∎

If E is an extension field of Φ, then $(u_1, \dots, u_m)$ is a base for $\mathscr{A}^E$. The minimum polynomial and generic norm of $x = \sum \xi_i u_i$ are unchanged on passing from $\mathscr{A}$ to $\mathscr{A}^E$. Hence $\tau_r(\xi_1, \dots, \xi_m)$ is irreducible for every extension field E/Φ, which means that $N_J(x) = \tau_r(\xi_1, \dots, \xi_m)$ is absolutely irreducible. The generic norm field determined by x (for $((\mathscr{H}, J)^E)$ is the field of fractions of $E[\xi_1, \dots, \xi_m]/(\tau_r(\xi_1, \dots, \xi_m)) \cong E \otimes \Phi[\xi_1, \dots, \xi_m]/(\tau_r(\xi_1, \dots, \xi_m))$. Hence $\Phi(\bar{\xi}_1, \dots, \bar{\xi}_m)$ is isomorphic to a subfield of $E(\bar{\xi}_1, \dots, \bar{\xi}_m)$. If $[E : \Phi] < \infty$, then $E \otimes_\Phi \Phi(\bar{\xi}_1, \dots, \bar{\xi}_m)$ is the field of fractions of $E[\xi_1, \dots, \xi_m]/(\tau_r(\xi_1, \dots, \xi_m))$. This implies that $\Phi(\bar{\xi}_1, \dots, \bar{\xi}_m)/\Phi$ contains no finite-dimensional subfield and hence Φ is algebraically closed in $\Phi(\bar{\xi}_1, \dots, \bar{\xi}_m)$. We observe also that since

$\Phi(\xi_1,\ldots,\xi_m,x)$ is separable over Φ, $\Phi(\bar{\xi}_1,\ldots,\bar{\xi}_m)$ which is isomorphic to a subfield of $\Phi(\xi_1,\ldots,\xi_m,x)/\Phi$ is separable. Now an extension field E/Φ is called a regular extension of Φ in the sense of Weil if E/Φ is separable and Φ is algebraically closed in E. Hence we have

4.11. THEOREM. *Any generic norm field $\Phi(\bar{\xi}_1,\ldots,\bar{\xi}_m)$ is a regular extension of Φ.*

We recall the definition of a Φ-place from an extension field E/Φ to an extension field E'/Φ: a homomorphism $\mathscr{P}$ of a Φ-subalgebra R into E'/Φ such that if $a \in E$, $\notin R$, then $a^{-1} \in R$ and $\mathscr{P}(a^{-1}) = 0$. We also have

4.12. DEFINITION. Let E_i, $i = 1, 2$, be an extension field of Φ. Then E_1 and E_2 are *Φ-place equivalent* if there exists a Φ-place from E_1/Φ to E_2/Φ and a Φ-place from E_2/Φ to E_1/Φ.

The following result is readily established by induction on m using the fact that the composite of Φ-places is a Φ-place:

4.13. PROPOSITION. *If the ξ_i, $1 \leqslant i \leqslant m$, are indeterminates, then Φ and $\Phi(\xi_1,\ldots,\xi_m)$ are Φ-place equivalent.*

We omit the proof.

We also have the following result [4, p. 34; 11, p. 428]:

4.14. PROPOSITION. *Let E/Φ be a splitting field for a central simple algebra $\mathscr{A}/\Phi$ and suppose there is a Φ-place from E/Φ to E'/Φ. Then E'/Φ is a splitting field for $\mathscr{A}$.*

We can now prove the following result on involutions of orthogonal type:

4.15. PROPOSITION. *If J is of orthogonal type, then any generic norm field $\Phi(\bar{\xi}_1,\ldots,\bar{\xi}_m)$ determined by $N_J(x)$ is a splitting field for $\mathscr{A}$.*

Proof. If $x = \sum \xi_i u_i$, then $m_x(\lambda)$ is of degree $r = n$. Hence $\Phi(\xi_1,\ldots,\xi_m,x)$ is a maximal subfied of $\mathscr{A}^P$, $P = \Phi(\xi_1,\ldots,\xi_m)$. Hence this is a splitting field for $\mathscr{A}^P$ and for $\mathscr{A}$. Since $\Phi(\xi_1,\ldots,\xi_m,x)$ is a simple transcendental extension of $\Phi(\bar{\xi}_1,\ldots,\bar{\xi}_m)$, we have a Φ-place from $\Phi(\xi_1,\ldots,\xi_m,x)$ to $\Phi(\bar{\xi}_1,\ldots,\bar{\xi}_m)$. Hence, by 4.14, $\Phi(\bar{\xi}_1,\ldots,\bar{\xi}_m)$ is a splitting field for $\mathscr{A}$. ∎

We shall require also the following well-known result:

4.16. PROPOSITION. (i) *If $\mathscr{D}$ is a finite-dimensional division algebra over Φ and $\xi_1,\ldots,\xi_m$ are indeterminates, then $\mathscr{D}^{\Phi(\xi_1,\ldots,\xi_m)}$ is a division algebra.*

(ii) *The canonical map $\{\mathscr{A}\} \rightsquigarrow \{\mathscr{A}^{\Phi(\xi_1,\ldots,\xi_m)}\}$ of* Br Φ *into* Br$(\Phi(\xi_1,\ldots,\xi_m))$ *is injective.*

Proof. (i) The algebra $\mathscr{D}^{\Phi(\xi_1,\ldots,\xi_m)}$ contains the subalgebra $\mathscr{D}^{\Phi[\xi_1,\ldots,\xi_m]} \cong \mathscr{D}[\xi_1,\ldots,\xi_m]$ and this is a domain. Every element of $\mathscr{D}^{\Phi(\xi_1,\ldots,\xi_m)}$ can be written in th form $\varphi(\xi_1,\ldots,\xi_m)^{-1}f$ when $\varphi \in \Phi[\xi_1,\ldots,\xi_m]$ and $f \in \mathscr{D}^{\Phi[\xi_1,\ldots,\xi_m]}$. It follows that $\mathscr{D}^{\Phi(\xi_1,\ldots,\xi_m)}$ is a domain and since it is finite dimensional over $\Phi(\xi_1,\ldots,\xi_m)$, it is a division algebra.

(ii) This is an immediate consequence of (i). ∎

Our first main result on generic norm fields of $\mathscr{H}(\mathscr{A},J)$ is

4.17. Theorem. *If J is of orthogonal type, then $\mathscr{A} \sim 1$ if and only if the generic norm field $\Phi(\bar{\xi}_1,\ldots,\bar{\xi}_m)$ of N_J is a rational function field.*

Proof. Suppose first that $\mathscr{A} \sim 1$ so $\mathscr{A} = M_n(\Phi)$. We may assume $J = t$ the transpose involution and the base for $\mathscr{H}(\mathscr{A},J)$ is $\{e_{ii}, e_{ij}+e_{ji}, i<j=1,\ldots,n\}$. Then $x = \sum \xi_{ii}e_{ii} + \sum_{i<j}\xi_{ij}(e_{ij}+e_{ji})$ with ξ_{ii}, ξ_{ij} indeterminates. Then $N_J(x) = \det x$, which is of first degree in ξ_{ii}. It follows that $\bar{\xi}_{ii}$, $i>1$, $\bar{\xi}_{ij}$, $i<j$, is a transcendency base for the generic norm field $\Phi(\bar{\xi}_{ii}, \bar{\xi}_{ij})$. Conversely, assume $\Phi(\bar{\xi}_1,\ldots,\bar{\xi}_m)$ is rational. By Proposition 4.15, $\Phi(\bar{\xi}_1,\ldots,\bar{\xi}_m)$ is a splitting field for $\mathscr{A}$. Hence, by Proposition 4.16, $\mathscr{A} \sim 1$. ∎

4.18. Definition (Amitsur). An extension field E/Φ is a *generic splitting field* for a central simple algebra $\mathscr{A}/\Phi$ if it has the property that an extension field E'/Φ is a splitting field for $\mathscr{A}$ if and only if there exists a Φ-place from E/Φ to E'/Φ.

Since there exists a Φ-place from E/Φ to E/Φ, the definition implies that any generic splitting field is a splitting field.

We can now prove the main theorem on generic norm field for involutions of orthogonal type.

4.19. Theorem. *Let $\mathscr{A}$ be an involutorial central simple algebra over Φ, J an involution of orthogonal type in $\mathscr{A}$. Then any generic norm field determined by an $N_J(x)$ is a generic splitting field for $\mathscr{A}/\Phi$.*

Proof. Let $\Phi(\bar{\xi}_1,\ldots,\bar{\xi}_m)$ be the generic norm field determined by $x = \sum \xi_i u_i$, where $(u_1,\ldots,u_m)$ is a base for $\mathscr{H}(\mathscr{A},J)$. We have seen that $\Phi(\bar{\xi}_1,\ldots,\bar{\xi}_m)$ is a splitting field for $\mathscr{A}$. Hence if there exists a Φ-place from $\Phi(\bar{\xi}_1,\ldots,\bar{\xi}_m)$ to E/Φ, then E is a splitting field for $\mathscr{A}$. Conversely, suppose E is a splitting field for $\mathscr{A}$. Then $\mathscr{A}^E = M_n(E)$ and hence the generic norm field $E(\bar{\xi}_1,\ldots,\bar{\xi}_m)$ of $\mathscr{A}^E$ is rational. Then there exists an E-place from $E(\bar{\xi}_1,\ldots,\bar{\xi}_m)$ to E. On the other hand, $\Phi(\bar{\xi}_1,\ldots,\bar{\xi}_m)$ is isomorphic to a subfield of $E(\bar{\xi}_1,\ldots,\bar{\xi}_m)$. If we identify $\Phi(\bar{\xi}_1,\ldots,\bar{\xi}_m)$ with the corresponding subfield of $E(\bar{\xi}_1,\ldots,\bar{\xi}_m)$, then the restriction of $\mathscr{P}$ to $\Phi(\bar{\xi}_1,\ldots,\bar{\xi}_m)$ is a Φ-place from $\Phi(\bar{\xi}_1,\ldots,\bar{\xi}_m)$ to E. ∎

We now consider involutions J of symplectic type. Let $\Phi(\bar{\xi}_1,\ldots,\bar{\xi}_m)$ be a generic norm field determined by such an involution. We shall show that $\Phi(\bar{\xi}_1,\ldots,\bar{\xi}_m)$ is a $\frac{1}{2}$-splitting field for $\mathscr{A}$ in the sense of the following

4.20. Definition. If $\mathscr{A}$ is central simple over Φ, an extension field E/Φ is a $(1/s)$-*splitting field for* $\mathscr{A}$ if $\mathscr{A}^E$ has index $\leqslant s$ or, equivalently, $\mathscr{A}^E$ has a splitting field of degree $\leqslant s$.

If $s \mid n$ and E is a subfield of $\mathscr{A}$ such that $[E\colon \Phi] = n/s$, then $\mathscr{A}^E$ has the same index as the centralizer $C_{\mathscr{A}}(E)$, which is central simple over E. Moreover, $[C_{\mathscr{A}}(E)\colon \Phi][E\colon \Phi] = [\mathscr{A}\colon \Phi] = n^2$, which implies that $[C_{\mathscr{A}}(E)\colon E] = s^2$. Then the index of $C_{\mathscr{A}}(E) \leqslant s$. Thus if $\mathscr{A}$ has a subfield of degree $\leqslant n/s$, then E is a $(1/s)$-splitting field for $\mathscr{A}$.

4.21. Proposition. *If $\mathscr{A}$ is involutorial and J is an involution of symplectic type in $\mathscr{A}$, then the generic norm field $\Phi(\bar{\xi}_1,\ldots,\bar{\xi}_m)$ determined by N_J is a $\frac{1}{2}$-splitting field for $\mathscr{A}$.*

Proof. The field $\Phi(\bar{\xi}_1,\ldots,\bar{\xi}_m)$ is a subfield of degree $r = n/2$ of $\mathscr{A}^{\Phi(\xi_1,\ldots,\xi_m)}$ and for $\mathscr{A}$. Now $\Phi(\xi_1,\ldots,\xi_m, x)$ is a simple transcendental extension of $\Phi(\bar{\xi}_1,\ldots,\bar{\xi}_m)$. Hence, by Proposition 4.16, $\mathscr{A},\ldots, \mathscr{A}^{\Phi(\xi_1,\ldots,\xi_m,x)}$ and $\mathscr{A}^{\Phi(\bar{\xi}_1,\ldots,\bar{\xi}_m)}$ have the same index $\leqslant 2$. Then $\Phi(\bar{\xi}_1,\ldots,\bar{\xi}_m)$ is a $\frac{1}{2}$-splitting for $\mathscr{A}$.

We prove next an analogue for involutions of symplectic type of Theorem 4.17.

4.22. Theorem. *Let $\mathscr{A}$ be involutorial, J an involution of symplectic type, and let $\Phi(\bar{\xi}_1,\ldots,\bar{\xi}_m)$ be a generic norm field determined by N_J. Then the index of $\mathscr{A} \leqslant 2$ if and only if $\Phi(\bar{\xi}_1,\ldots,\bar{\xi}_m)$ is rational.*

Proof. Suppose first that $\mathscr{A} = M_r(\mathscr{D})$, where $\mathscr{D}$ is a quaternion algebra (possibly split) and J is of symplectic type. To encompass the case char $\Phi = 2$ we assume $\mathscr{D}$ is generated by u and v such that

$$u^2 = u + \alpha, \qquad vu = (1-u)v, \qquad v^2 = \beta, \tag{4.23}$$

where $\alpha, \beta \in \Phi$, $4\alpha^2 + 1 \neq 0$, $\beta \neq 0$. Here $\Phi[u]$ has the automorphism $a \rightsquigarrow \bar{a}$, where $\bar{u} = 1 - u$, which extends to the involution in $\mathscr{D}$ such that $\bar{v} = -v$. This is the standard involution $\mathscr{D}$. We have a representation of $\mathscr{D}$ in $M_2(\Phi[v])$ such that if $a, b \in \Phi[u]$, then

$$d = a + bv \rightsquigarrow \begin{pmatrix} a & b \\ \beta\bar{b} & \bar{a} \end{pmatrix}. \tag{4.24}$$

We have

$$\bar{d} = \bar{a} - bv \rightsquigarrow \begin{pmatrix} 0 & 1 \\ -1 & 0 \end{pmatrix} ({}^t d) \begin{pmatrix} 0 & -1 \\ 1 & 0 \end{pmatrix}.$$

Now consider $\mathscr{A} = M_r(\mathscr{D})$. This has the involution $\bar{t}$: $(d_{ij}) \rightsquigarrow {}^t(\bar{d}_{ij})$. Since $[\mathscr{H}(\mathscr{A}, \bar{t}): \Phi] = r(2r-1)$ and $\deg \mathscr{A} = 2r$, $\bar{t}$ is of symplectic type. We have the representation of $M_r(\mathscr{D})$ in $M_{2r}(\Phi[u])$ such that if $d_{ij} = a_{ij} + b_{ij}v$, where $a_{ij}, b_{ij} \in \Phi[u]$, then

$$(d_{ij}) \rightsquigarrow \left(\begin{pmatrix} a_{ij} & b_{ij} \\ \beta\bar{b}_{ij} & \bar{a}_{ij} \end{pmatrix} \right). \tag{4.25}$$

Now $(d_{ij}) \in \mathscr{H}(\mathscr{A}, \bar{t})$ if and only if $\bar{d}_{ii} = d_{ii} = \alpha_{ii} \in \Phi$ and $d_{ji} = \bar{d}_{ij}$ for $i < j$. The corresponding matrix in $M_{2r}(\Phi[u])$ has the form

$$A = \begin{bmatrix} \alpha_{11} & 0 & a_{12} & b_{12} & \cdots \\ 0 & \alpha_{11} & \beta\bar{b}_{12} & \bar{a}_{12} & \cdots \\ \bar{a}_{12} & -b_{12} & \alpha_{22} & 0 & \cdots \\ -\beta\bar{b}_{12} & a_{12} & 0 & \alpha_{22} & \cdots \\ \vdots & \vdots & \vdots & \vdots & \end{bmatrix}. \tag{4.26}$$

The condition defining this matrix is $s({}^tA)\, s^{-1} = A$, where

$$s = \operatorname{diag} \left\{ \begin{pmatrix} 0 & 1 \\ -1 & 0 \end{pmatrix}, \ldots, \begin{pmatrix} 0 & 1 \\ -1 & 0 \end{pmatrix} \right\}. \tag{4.27a}$$

This is equivalent to As, is alternate, and hence $N_{\bar{t}}(A) = \operatorname{Pf}(As)$. Now

$$As = \begin{bmatrix} 0 & \alpha_{11} & -b_{12} & a_{12} & \cdots \\ -\alpha_{11} & 0 & -\bar{a}_{12} & \beta\bar{b}_{12} & \cdots \\ b_{12} & \bar{a}_{12} & 0 & \alpha_{22} & \cdots \\ -a_{12} & -\beta\bar{b}_{12} & -\alpha_{22} & 0 & \cdots \\ \vdots & \vdots & \vdots & & \end{bmatrix}. \tag{4.27b}$$

It follows that $N_t(A) = \operatorname{Pf}(As)$ is linear in α_{11}. Hence if we choose as base for $\mathscr{H}(\mathscr{A}, \bar{t})$: $\{e_{ii}, d_k e_{ij} + \bar{d}_k e_{ji}\}$, where $i < j$, and (d_1, d_2, d_3, d_4) is a base for $\mathscr{D}$, then $N_{\bar{t}}(x)$ for $x = \sum \xi_{ii} e_{ii} + \sum_{i<j} \xi_{kij}\, (d_k e_{ij} + \bar{d}_k e_{ji})$, the ξ indeterminates, then the $\bar{\xi}_{ii}$, $i > 1$, $\bar{\xi}_{kij}$, are algebraically independent and $\Phi(\bar{\xi}_{ii}, i > 1, \bar{\xi}_{kij})$ is a generic norm field for $\mathscr{H}(\mathscr{A}, J)$. Hence this field is rational.

Conversely, suppose the generic norm field $\Phi(\bar{\xi}_1, \ldots, \bar{\xi}_m)$ determined by N_J is rational. Since $\Phi(\bar{\xi}_1, \ldots, \bar{\xi}_m)$ is a $\frac{1}{2}$-splitting field, $\mathscr{A}^{\Phi(\bar{\xi}_1, \ldots, \bar{\xi}_m)} = M_r(\bar{\mathscr{D}})$, where $\bar{\mathscr{D}}$ is a quaternion algebra over $\Phi(\bar{\xi}_1,\ldots,\bar{\xi}_m)$. It follows from Proposition 4.16 that the index of $\mathscr{A} \leqslant 2$. ∎

4.28. Definition. An extension field E/Φ is a *generic* $(1/s)$*-splitting field* for a central simple algebra $\mathscr{A}/\Phi$ if it has the following property: An extension field E'/Φ is a $(1/s)$-splitting field if and only if there exists a Φ-place from E/Φ to E'/Φ.

We can now prove our main result for generic norm fields determined by involutions of symplectic type.

4.29. Theorem. *If J is of symplectic type, then any generic norm field determined by an* $N_J(x)$ *is a generic* $\frac{1}{2}$*-splitting field for* $\mathscr{A}$.

Proof. If $\Phi(\bar{\xi}_1,\ldots,\bar{\xi}_m)$ is a generic norm field determined by $N_J(x)$, then $\mathscr{A}^{\Phi(\bar{\xi}_1,\ldots,\bar{\xi}_m)} = M_r(\mathscr{D})$, where $\mathscr{D}$ is a quaternion algebra. It is readily seen that $\mathscr{D}$ is a division algebra. Hence if $\bar{\xi}_{m+1} \in \mathscr{D}$, $\notin \Phi(\bar{\xi}_1,\ldots,\bar{\xi}_m)$, then $\Phi(\bar{\xi}_1,\ldots,\bar{\xi}_{m+1})$ is a splitting field for $\mathscr{A}$. Now let E be an extension field of Φ such that there exists a Φ-place $\mathscr{P}$ from $\Phi(\bar{\xi}_1,\ldots,\bar{\xi}_m)$ to E. Then $\mathscr{P}$ can be extended to a Φ-place $\mathscr{P}'$ from $\Phi(\bar{\xi}_1,\ldots,\bar{\xi}_{m+1})$ into algebraic extension E' of E. Let 0, P $(0', P')$ be the valuation ring and maximal ideal associated with $\mathscr{P}(\mathscr{P}')$. Then $0'/P'$ is an extension field of $0/P$ and since $[\Phi(\bar{\xi}_1,\ldots,\bar{\xi}_{m+1}):\Phi(\bar{\xi}_1,\ldots,\bar{\xi}_m)] = 2$, $[0'/P' : 0/P] \leqslant 2$. Now $0/P$ is isomorphic to a subfield of E. Hence we have a Φ-place from $\Phi(\bar{\xi}_1,\ldots,\bar{\xi}_{m+1})$ into either E or a quadratic field extension E' of E. Thus either E or E' is a splitting field for $\mathscr{A}$ and E is a $\frac{1}{2}$-splitting field for $\mathscr{A}$.

Conversely assume that E is a $\frac{1}{2}$-splitting field for $\mathscr{A}$. Then the argument used in the proof of Theorem 4.19 shows that there exists a Φ-place from $\Phi(\bar{\xi}_1,\ldots,\bar{\xi}_{m+1})$ to E. ∎

References

1. A. A. Albert, On the Wedderburn norm condition for cyclic algebras, *Bull. Amer. Math. Soc.* **37** (1931), 301–312.
2. A. A. Albert, "Structure of Algebras," Amer. Math. Soc. Colloq. Publ. XXIV, Providence, R. I., 1939.
3. A. A. Albert, Tensor products of quaternion algebras, *Proc. Amer. Math. Soc.* **35** (1972), 65–66.
4. S. A. Amitsur, Generic splitting fields of central simple algebras, *Ann.of Math.* **62** (1955), 8–43.
5. A. Heuser, Über den Funktionenkörper der Norm-fläche..., *J. für Math.* **301** (1978), 105–113.
6. N. Jacobson, Generic norm of an algebra, *Osaka J. Math.* **15** (1963), 25–50.
7. N. Jacobson, "Structure and Representations of Jordan Algebras," Amer. Math. Soc. Colloq. Publ. XXXIX, Providence, R.I., 1968.

8. N. Jacobson, "Lecture on Quadratic Jordan Algebras," Tata Institute of Fundamental Research, Bombay, 1969.
9. N. Jacobson, Structure groups and Lie algebras of Jordan algebras of symmetric elements of associative algebras with involution, *Advan. in Math.* **20** (1976), 106–150.
10. M. L. Racine, A simple proof of a theorem of Albert, *Proc. Amer. Math. Soc.* **43** (1974), 487–488.
11. P. Roquette, On the Galois cohomology of the projective linear group, *Math. Ann.* **150** (1963), 411–439.
12. L.-H. Sah, Symmetric bilinear forms and quadratic forms, *J. Algebra* **20** (1972), 144–160.
13. D. Saltman, Norm polynomials and algebras, *J. Algebra* **62** (1980), 333–345.
14. G. Seligman, "Rational Methods in Lie Algebras," Marcel Dekker Lecture Notes, New York, 1976.
15. T. Tamagawa, On quadratic forms and Pfaffians, *J. Fac. Sci. Univ. Tokyo Sect. IA Math.* **24**, 213–219.
16. W. C. Waterhouse, Linear maps preserving reduced norms, *Linear Algebra Appl.* **43** (1982), 197–200.
17. O. Zariski and P. Samuel, "Commutative Algebra," Vol. II, Springer-Verlag, New York, 1960.
18. N. Jacobson and K. McCrimmon, "Quadratic Jordan algebras of quadratic forms with base points", J. Indian Math. Soc. 35, 1–45.

Printed by the St. Catherine Press Ltd., Tempelhof 41, Bruges, Belgium

Reprinted from JOURNAL OF ALGEBRA
All Rights Reserved by Academic Press, New York and London

Vol. 86, No. 1, January 1984
Printed in Belgium

Forms of the Generic Norm of a Separable Jordan Algebra

N. JACOBSON*

Department of Mathematics, Yale University, New Haven, Connecticut 06520

Communicated by the Editors

Received September 5, 1982

TO MAX KOECHER ON HIS 60TH BIRTHDAY

The purpose of this note is to show that the type of problem considered by Professor Waterhouse in the preceding paper has its natural setting in the theory of generic norms of finite-dimensional Jordan algebras. Using this approach the results can be obtained on forms of the determinant and considerably more by invoking well-known results on Jordon algebras. The most general class of Jordan algebras that shall be considered is that of the separable ones. The key fact, which is easily established, is that forms of the generic norm of such an algebra are algebra forms of isotopes of the algebra. The algebra forms of the important algebras in our class are well known. Perhaps the most important of these were determined in 1949 by F. D. Jacobson and the author by a method based on special universal envelopes. This circumvented the use of Galois descent that had been used earlier for the analogous problems for Lie algebras by W. Landherr and by the author. Another known Jordan result is that similarity of generic norms of separable Jordan algebras holds if and only if the algebras are isotopic. Thus the determination of the forms and conditions for similarity can be derived from the Jordan theory.

1. BACKGROUND

In order to encompass the case of characteristic two we shall deal with (unital) quadratic Jordan algebras as defined by McCrimmon [10]. General references for these are the monographs [4, 6]. We shall be interested only in algebras over an infinite field Φ. For such a base ring we can define a (unital quadratic) Jordan algebra to be a triple $(\mathscr{J}, U, 1)$, where $\mathscr{J}$ is a vector space over Φ, $1 \in \mathscr{J}$ and U is a quadratic map of $\mathscr{J}$ into $\mathrm{End}_\Phi \mathscr{J}$ satisfying

$QJ1 \qquad U_1 = 1_{\mathscr{J}},$

$QJ2 \qquad U_a U_b U_a = U_{U_a b},$

* This research was partially supported by National Science Foundation Grant MCS79-04473.

76

0021-8693/84 $3.00

$QJ3$ If $U_{a,b} = U_{a+b} - U_a - U_b$ and $V_{a,b}$ is defined by

$$V_{a,b}x = U_{a,x}b, \text{ then } U_a V_{b,a} = V_{a,b} U_a.$$

The prototype of such an algebra is the Jordan algebra $\mathscr{A}^+$ obtained from an associative algebra $\mathscr{A}$ over Φ by taking 1 to be the unit of $\mathscr{A}$ and defining U_a by $U_a x = axa$. If char $\Phi \neq 2$ the concept of Jordan algebra as we have defined it is equivalent to the original one defined by Jordan (see [4]).

We recall a few key concepts and results that we shall need. First, we define the powers of $a \in \mathscr{J}$ inductively by $a^0 = 1$, $a^1 = a$, $a^{n+2} = U_a a^n$. Also we define invertibility of an element a and the inverse by: U_a is invertible in $\mathrm{End}_\Phi \mathscr{J}$ and $a^{-1} = U_a^{-1} a$. For a to be invertible it suffices that $1 \in U_a(\mathscr{J})$. In $\mathscr{A}^+$, $\mathscr{A}$ associative, the Jordan powers of a are the same as the associative powers and a is invertible in $\mathscr{A}^+$ if and only if it is invertible in $\mathscr{A}$. Then a^{-1} in $\mathscr{A}^+$ is $\mathscr{A}^{-1}$ in $\mathscr{A}$.

If u is an invertible element of a Jordan algebra $\mathscr{J}$, then u defines a new Jordan algebra $\mathscr{J}^{(u)} = (\mathscr{J}, UU_u, u^{-1})$. This is called the u-isotope of $\mathscr{J}$. If E is an extension field of $\mathscr{J}$ the quadratic map U has a unique extension U to a quadratic map of $\mathscr{J}_E = E \otimes_\Phi \mathscr{J}$ into $\mathrm{End}_E \mathscr{J}_E$. This defines the extension Jordan algebra $(\mathscr{J}_E, U, 1)$. For $\mathscr{A}$ associative we have $(\mathscr{A}_E)^+ = (\mathscr{A}^+)_E$.

Now let $\mathscr{J}$ be finite dimensional. If $a \in \mathscr{J}$ and $f(\lambda) \in \Phi[\lambda]$, then the meaning of $f(a)$ is clear. We also write $af(a)$ for $g(a)$, where $g(\lambda) = \lambda f(\lambda)$. In characteristic two it may happen then that $f(a) = 0$ but $af(a) \neq 0$. On the other hand, there exists a unique monic polynomial $\mu_a(\lambda)$ of least degree such that $\mu_a(a) = 0$ and $a\mu_a(a) = 0$ (Jacobson and Katz [8]). $\mu_a(\lambda)$ is called the minimum polynomial of a. Now let $(u_1,\ldots, u_n)$ be a base for $\mathscr{J}/\Phi$ and let $\xi_1,\ldots, \xi_n$ be indeterminates, $P = \Phi(\xi_1,\ldots, \xi_n)$ the rational function field in the ξ_i. Let $\tilde{\mathscr{J}} = \mathscr{J}_P$ and let $x = \sum \xi_i u_i$. We call x a generic element of $\mathscr{J}$. Let

$$m_x(\lambda) = \lambda^m - \tau_1(\xi_1,\ldots, \xi_n)\, \lambda^{m-1} + \cdots + (-1)^m\, \tau_m(\xi_1,\ldots, \xi_n) \tag{1}$$

be the minimum polynomial of x. It can be shown that $\tau_i(\xi_1,\ldots, \xi_n)$ is homogeneous polynomial in the ξ's. Hence if $a = \sum \alpha_i u_i$, $\alpha_i \in \Phi$, then we can specialize $\xi_i \rightsquigarrow \alpha_i$ to obtain the monic polynomial $m_a(\lambda) = \lambda^m - \tau_1(a)\, \lambda^{m-1} + \cdots + (-1)^m\, \tau_m(a)$, where $\tau_i(a) = \tau_i(\alpha_i,\ldots, \alpha_n)$. This is called the generic minimum polynomial of a. It is easy to see that $m_a(a) = 0$ and that $m_a(\lambda)$ is independent of the choice of the base and is unchanged in passing from $\mathscr{J}$ to $\mathscr{J}_E$ for any extension field E/Φ. The element $N(a) = \tau_m(a)$ is called the generic norm of a. We have $N(a) = 1$ and $m_a(\lambda) = N(\lambda 1 - a)$. We define the adjoint $a^{\#}$ of a by

$$a^{\#} = (-1)^{m-1} \sum (-1)^i\, \tau_i(a)\, a^{m-i-1} \tag{2}$$

and the trace bilinear form of N by

$$T(a, b) = (\Delta_1^a N)(\Delta_1^a N) - \Delta_1^a \Delta_x^b N, \tag{3}$$

where $\Delta_b^a F$ for a polynomial map F is defined to be the coefficient of λ in $F(b + a\lambda)$. The following formula has been proved by McCrimmon [13]:

$$T(a^{\#}, b) = \Delta_a^b N. \tag{4}$$

An element a is invertible if and only if $N(a) \neq 0$. Then $a^{-1} = N(a)^{-1} a^{\#}$. We also have Hua's identity: If a, b and $a + b^{-1}$ are invertible, then $a + U_a b$ is invertible and

$$(a + U_a b)^{-1} + (a + b^{-1})^{-1} = a^{-1} \tag{5}$$

(Jacobson [2, p. 2] or [6, p. 1.24]).

2. Separable Jordan Algebras

Definition 1. A finite-dimensional Jordan algebra is called *separable* if its trace bilinear form is nondegenerate.

The structure of separable Jordan algebras of char $\neq 2$ is known. These are direct sums of simple Jordan algebras with separable fields as centers (Jacobson [3, p. 239]). A similar result can be proved for char $\Phi = 2$. Here certain simple algebras have to be dropped, namely, certain algebras of quadratic forms and the algebras $\mathscr{H}(\mathscr{A}, J)$ of symmetric elements of a simple associative algebra $\mathscr{A}$ with involution J of first kind. On the other hand, one has to add the algebra $\mathscr{H}(\mathscr{A}, J)'$, the outer ideal generated by 1 in $\mathscr{H}(\mathscr{A}, J)$, where $\mathscr{A}$ is simple with involution J of first kind and symplectic type. We shall not give the details of the general structure theorem for the case char $\Phi = 2$. Instead we shall indicate some of the most important instances of these Jordan algebras in the algebraically closed case of any characteristic.

If Φ is algebraically closed the following is the list of separable simple Jordan algebras over Φ together with their generic norms.

(I) $M_m(\Phi)^+$ the m^2-dimensional vector space of $m \times m$ matrices over Φ with the Jordan operator U_a defined by $U_a x = axa$. Here $N(a) = \det a$.

(II) $\mathscr{H}(M_n(\Phi), t)$, the $(m(m+1)/2)$-dimensional vector space of ordinary symmetric matrices as subalgebra of $M_m(\Phi)^+$. Here $N(a) = \det a$.

(III) $\mathscr{H}(M_{2m}(\Phi), t_s)$, the $(m(2m-1)/2$-dimensional vector space of $2m \times 2m$ matrices a such that as is alternate for $s = \operatorname{diag}\{q, q, \ldots, q\}$,

$$q = \begin{pmatrix} 0 & 1 \\ -1 & 0 \end{pmatrix},$$

as subalgebra of $M_{2m}(\Phi)^+$. Here $N(a) = Pf(as)$.

(IV) Let V be a finite-dimensional vector space over Φ, 1 an element of V, Q a quadratic form on V such that $Q(1) = 1$. Define $Q(a, b) = Q(a+b) - Q(a) - Q(b)$, $T(a) = Q(a, 1)$, $\bar{a} = T(a)\,1 - \bar{a}$ and

$$U_a b = Q(a, \bar{b}) - Q(a)\,\bar{b}.$$

This gives a Jordan structure on V so that 1 is the unit. We denote this Jordan algebra as Jord$(Q, 1)$. The generic norm in Jord$(Q, 1)$ is $N(a) = Q(a)$. Then Jord$(Q, 1)$ is separable if and only if the bilinear form $Q(a, b)(= T(a, b))$ is nondegenerate.

(V) Let $\mathbb{O}$ be the octonion algebra over Φ and let $\mathscr{H}(M_3(\mathbb{O}))$ denote the set of 3×3 hermitian octonion matrices with diagonal entries in Φ

$$A = \begin{pmatrix} \alpha_1 & a_3 & \bar{a}_2 \\ \bar{a}_3 & \alpha_2 & a_1 \\ a_2 & \bar{a}_1 & \alpha_3 \end{pmatrix}. \tag{6}$$

Following Freudenthal one defines a determinant in $\mathscr{H}(M_3(\mathbb{O}))$ by

$$N(A) = \det A = \alpha_1\alpha_2\alpha_3 + t(a_1 a_2 a_3) - \sum \alpha_i n(a_i), \tag{7}$$

where $t(a_1 a_2 a_3) = a_1(a_2 a_3) + (\bar{a}_3 \bar{a}_2)\,\bar{a}_1$, $n(a_i) = a_i \bar{a}_i$. We can calculate the trace bilinear form of N to be

$$T(A, B) = \sum_1^3 \alpha_i \beta_i + \sum_1^3 t(\bar{a}_i b_i), \tag{8}$$

if B has the form (6) with β_i and b_i replacing α_i and a_i, respectively. It follows that $T(A, B)$ is nondegenerate and hence we can define $A^\#$ by $T(A^\#, B) = \Delta_A^B N$ for all B. It can be verified that the *adjoint* $A^\#$ satisfies the adjoint identity

$$A^{\#\#} = N(A)\,A \tag{9}$$

and this can be used to show that if we define 1 to be the identity matrix and

$$U_A B = T(A, B)\,A - A^\# \times B, \tag{10}$$

where $A \times B = (A+B)^{\#} - A^{\#} - B^{\#}$, then we obtain a Jordan algebra (McCrimmon [11, p. 503]). The generic norm in the algebra is given by (7). Since the trace form is nondegenerate this Jordan algebra is separable.

3. General Results

Definition 2. Let $\bar{\mathcal{J}}$ be a finite-dimensional Jordan algebra over a field $\bar{\Phi}$, $\bar{N}$ the generic norm of $\bar{\mathcal{J}}$, 1 the unit of $\bar{\mathcal{J}}$. Let Φ be a subfield of $\bar{\Phi}$. Then

(i) A Φ-form $\mathcal{J}$ of $(\bar{\mathcal{J}}, \bar{N})$ is Φ-subspace of $\bar{\mathcal{J}}$ such that $\bar{\mathcal{J}} = \bar{\Phi}\mathcal{J} \cong \bar{\Phi} \otimes_{\Phi} \mathcal{J}$ (canonically) and $\bar{N}(\mathcal{J}) \subset \Phi$.

(ii) A Φ-form $\mathcal{J}$ of $(\bar{\mathcal{J}}, \bar{N}, 1)$ is a Φ-form of $(\bar{\mathcal{J}}, \bar{N})$ containing 1.

(iii) An algebra Φ-form of $\bar{\mathcal{J}}$ is a Φ-subalgebra of $\bar{\mathcal{J}}$ such that $\bar{\mathcal{J}} = \Phi\mathcal{J} \cong \bar{\Phi} \otimes_{\Phi} \mathcal{J}$.

The following lemma will be needed for the proof of our first main theorem:

Lemma. *Let $\mathcal{J}_0$ be a subspace of a finite-dimensional Jordan algebra $\mathcal{J}/\Phi$ containing* 1 *and closed under inversion (if $a \in \mathcal{J}_0$ is invertible in $\mathcal{J}$, then $a^{-1} \in \mathcal{J}_0$). Then $\mathcal{J}_0$ is a subalgebra of $\mathcal{J}$.*

Proof. If N denotes the generic norm in $\mathcal{J}$, then the conditions $N(a) \neq 0$, $N(b) \neq 0$, $N(a + b^{-1}) \neq 0$ (equivalently, $N(U_b a + b) \neq 0$) define a nonvacuous Zariski open subset of $\mathcal{J}_0 \times \mathcal{J}_0$. The hypothesis and Hua's identity imply that for (a, b) in the subset, $U_a b \in \mathcal{J}_0$. Now let $(u_1, \ldots, u_n)$ be a base for $\mathcal{J}$ such that $(u_1, \ldots, u_r)$ is a base for $\mathcal{J}_0$. Then for $a, b \in \mathcal{J}_0$, $U_a b = \sum_1^n f_i(a, b)\, u_i$, where f_i are polynomial functions. Our result shows that $f_i(a, b) = 0$ if $i > r$ and (a, b) is in a Zariski open subset of $\mathcal{J}_0 \times \mathcal{J}_0$. Hence $f_i(a, b) = 0$ and $U_a b \in \mathcal{J}_0$ for all $a, b \in \mathcal{J}_0$. Thus $\mathcal{J}_0$ is a subalgebra. ∎

Now let $\bar{\Phi}$ be the algebraic closure of Φ, $\bar{\mathcal{J}}$ a finite-dimensional separable Jordan algebra over $\bar{\Phi}$, $\bar{N}$ and $\bar{T}$ the generic norm and trace bilinear form, respectively of, $\bar{\mathcal{J}}$. Then we have

Theorem 1. (i) *Any Φ-form $\mathcal{J}$ of $(\bar{\mathcal{J}}, \bar{N}, 1)$ is an algebra Φ-form,*

(ii) *Any Φ-form of $(\bar{\mathcal{J}}, \bar{N})$ is an algebra Φ-form of an isotope of $\bar{\mathcal{J}}$.*

Proof. (i) Let $N = \bar{N} \mid \mathcal{J}$. A Vandermonde determinant argument shows that if $a, b \in \mathcal{J}$, then $\Delta_b^a N \in \Phi$. Then, by (3), $T(a, b) \in \Phi$ for

$a, b \in \mathcal{J}$. Let $(u_1, \ldots, u_n)$ be a base for $\mathcal{J}/\Phi$, hence for $\bar{\mathcal{J}}/\Phi$. For given a the system of equations

$$T(a^{\#}, u_i) = \Delta_a^{u_i} N, \qquad i \leqslant i \leqslant n \tag{11}$$

is a linear system with coefficients in Φ. Since T is nondegenerate the determinant of the coefficients of these equations is $\neq 0$. Hence they have a unique solution $a^{\#} \in \mathcal{J}$. Then $a^{-1} = N(a)^{-1} a^{\#} \in \mathcal{J}$ for every $a \in \mathcal{J}$ such that $N(a) \neq 0$.

Let

$$U_{u_i} u_j = \sum \gamma_{ijk} u_k, \qquad U_{u_i, u_j} u_k = \sum \delta_{ijkl} u_l,$$

where the γ_{ijk}, $\delta_{ijkl} \in \bar{\Phi}$. Since the $\gamma_{ijk}, \delta_{ijkl}$ are algebraic over Φ they generate a finite-dimensional extension field E/Φ. Then $\mathcal{J}' = E\mathcal{J}$ is an E-subalgebra of $\bar{\mathcal{J}}$ which is finite dimensional over Φ and contains $\mathcal{J}$. It now follows from the lemma that $\mathcal{J}$ is a Φ-subalgebra of $\mathcal{J}'$ and hence of $\bar{\mathcal{J}}$. Hence $\mathcal{J}$ is an algebra Φ-form of $\bar{\mathcal{J}}$.

(ii) Let $\{\rho_i\}$ be a base for $\bar{\Phi}/\Phi$ and write $N(x) = \sum N_i(x) \rho_i$, where $N_i(x)$ is a polynomial in indeterminates with coefficients in Φ. We may assume $N_1(x) \neq 0$ and choose values of the indeterminates in Φ to obtain an element $u \in \mathcal{J}$ such that $N(u) \neq 0$. Now consider the u^{-1}-isotope of $\bar{\mathcal{J}}^{(u^{-1})}$ of $\bar{\mathcal{J}}$ with unit u. Its generic norm is $\bar{N}^{(u^{-1})} = \bar{N}(u^{-1}) N = N(u)^{-1} \bar{N}$. Since $N^{(u^{-1})}(\mathcal{J}) \subset \Phi$ we can apply (i) to conclude that $\mathcal{J}$ is an algebra Φ-form of the isotope $\bar{\mathcal{J}}^{(u^{-1})}$ of $\bar{\mathcal{J}}$. ∎

We also have

THEOREM 2. *Let $\mathcal{J}_1$ and $\mathcal{J}_2$ be separable and let η be a norm similarity of $\mathcal{J}_1$ onto $\mathcal{J}_2$. Then η is an isotopy. Moreover, if $\eta(1_1) = 1_2$, then η is an isomorphism.*

The proof given for char $\Phi \neq 2$ in [3, p. 244] carries over with minor modifications.

4. SPECIAL CASES

The algebra Φ-forms of the algebras (I)–(II) of Section 1 (with Φ replaced by its algebraic closure $\bar{\Phi}$) were determined by Jacobson and Jacobson [1, pp. 157–160]. For (I) the Φ-forms are either the Jordan algebras $\mathcal{A}^+$, $\mathcal{A}$ finite-dimensional central simple associative or the Jordan algebras

$\mathscr{H}(\mathscr{A}, J)$ of symmetric elements of a finite-dimensional simple associative algebra with center a quadratic extension of Φ and involution J of second kind. Isomorphism of $\mathscr{A}_1^+$ and $\mathscr{A}_2^+$, where $\mathscr{A}_i$ is finite-dimensional central simple associative holds if and only if either $\mathscr{A}_1$ is isomorphic or anti-isomorphic to $\mathscr{A}_2$. If $(\mathscr{A}_i, J_i)$ is finite-dimensional simple associative with involution of second kind then $\mathscr{H}(\mathscr{A}_1, J_1) \cong \mathscr{H}(\mathscr{A}_2, J_2) \Leftrightarrow (\mathscr{A}_1, J_1) \cong (\mathscr{A}_2, J_2)$ as algebras with involution.

The algebras in (III) with $m = 2$ are somewhat exceptional. Hence, we assume $m \neq 2$ in this case. With this restriction the algebra Φ-forms of the Jordan algebras in (II) and (III) are of the form $\mathscr{H}(\mathscr{A}, J)$, where $(\mathscr{A}, J)$ is central simple with involution (of first kind). Two such Jordan algebras are isomorphic if and only if the associative algebras with involution are isomorphic.

The Jordan algebras $\bar{\mathscr{J}} = \mathrm{Jord}(\bar{V}, \bar{Q}, 1)$ over Φ defined in (IV) Section 2 are generically algebraic of degree 2: If $\bar{T}(a) = \bar{Q}(a, 1)$, then $a^2 - \bar{T}(a)\, a + \bar{Q}(a)\, 1 = 0$. If follows that if $\mathscr{J}$ is an algebra Φ-form of $\bar{\mathscr{J}}$, then $\mathscr{J}$ is generically algebraic of degree 2 over Φ and hence $\mathscr{J} = \mathrm{Jord}(V, Q, 1)$ for V, a vector space over Φ, 1 an element of V and Q a quadratic form on V such that $Q(1) = 1$ (Jacobson and McCrimmon [9, p. 9]).

We consider next the algebra Φ-forms of the algebra $\mathscr{H}(M_3(\mathbb{O}))$ in V. These are given in two constructions due to Tits (see Jacobson [6, pp. 2.26–2.30]). For the first of these we begin with a central simple associative algebra $\mathscr{A}$ of degree and form $\mathscr{J} = \mathscr{A}^{(3)}$ whose elements we denote as

$$a = (a_0, a_1, a_2), \qquad a_i \in \mathscr{A}. \tag{12}$$

Let μ be a nonzero element of Φ and define

$$N(a) = n(a_0) + \mu n(a_1) + \mu^{-1} n(a_2) - t(a_0 a_1 a_2), \tag{13}$$

where n and t are the reduced (or generic) norm and trace in $\mathscr{A}$. We define a Jordan algebra structure $\mathrm{Jord}(\mathscr{A}, \mu)$ on $\mathscr{J}$ by taking $1 = (1, 0, 0)$ and defining the U-map by

$$U_a b = T(a, b)\, a - a^{\#} \times b, \tag{14}$$

where $T(a, b)$ is the trace bilinear form determined by N (which is nondegenerate), $a^{\#}$ is defined by $T(a^{\#}, b) = \Delta_a^b N$ for all b and $a \times b = (a + b)^{\#} - a^{\#} - b^{\#}$.

For Tits' second construction we begin with a simple associative algebra $\mathscr{A}$ of degree three over its center P a quadratic extension of Φ, and assume $\mathscr{A}$ has an involution $J\colon a \rightsquigarrow a^*$ of second kind. We assume also that

$\mathscr{H}(\mathscr{A}, J)$ contains an element u such that $n(u) = \mu\mu^*$, where $\mu \neq 0$ in P. Let $\mathscr{J} = \mathscr{H} \oplus \mathscr{A}$, $\mathscr{H} = \mathscr{H}(\mathscr{A}, J)$ and define for $a = (h, a_1)$, $h \in \mathscr{H}$, $a_1 \in \mathscr{A}$,

$$N(a) = n(h) + \mu n(a_1) + \mu^* n(a_1^*) - t(ha_1 u a_1^*), \tag{15}$$

where n and t are as in the first construction. We also define $1 = (1, 0)$ and $U_a b$ as in the first construction. This gives a Jordan algebra Jord$(\mathscr{A}, J, \mu, u)$.

The two constructions give forms of the algebra $\mathscr{H}(M_3(\mathbb{O}))$ and every algebra Φ-form can be obtained in this way (McCrimmon [12, pp. 306–313]). Conditions for isomorphism of these algebras that are not division algebras are known (Racine [16]). Results in the division algebra case are not complete (see Petersson and Racine [14] and a forthcoming paper [15] by these authors).

We conclude by considering one case of an algebra that is not simple. Let $\bar{\Phi}^{(n)}$ be the direct sum of n copies of the base field $\bar{\Phi}$, $\Phi^{(n)+}$ the corresponding Jordan algebra. Since $\bar{\Phi}^{(n)+}$ has degree n (= degree of the generic minimum polynomial), any algebra Φ-form of $\bar{\Phi}^{(n)+}$ has degree n. If $a \in \mathscr{A}^+$, where $\mathscr{A}$ is associative, then the minimum polynomial $\mu_a(\lambda)$ of a in the Jordan algebra $\mathscr{A}^+$ is the same as the minimum polynomial of a in $\mathscr{A}$. We know also that for any finite-dimensional Jordan algebra $\mathscr{J}/\Phi$ the subset of elements a whose minimum polynomials $\mu_a(\lambda)$ have degree equal to the degree of $\mathscr{J}$ is nonvacuous and open [8, p. 222]. It follows that if $\mathscr{J}$ is an algebra Φ-form of $\bar{\Phi}^{(n)+}$, then $\mathscr{J} = \Phi[a]^+$, where $\mu_a(\lambda)$ is a separable polynomial. Moreover, $\Phi[a]$ is a direct sum of separable fields and it is readily seen that any direct sum of separable extension fields is isomorphic to an algebra Φ-form of $\bar{\Phi}^{(n)}$ and the corresponding Jordan algebra is isomorphic to an algebra Φ-form of $\bar{\Phi}^{(n)+}$.

Now let $\mathscr{A}$ be any finite-dimensional separable commutative associative algebra over an infinite field Φ. Then $\mathscr{A} = \oplus_1^r E_i$, where E_i is a separable field extension of Φ. If $a = (a_1, a_2, \ldots, a_r)$, where $a_i \in E_i$, then we define

$$N_{\mathscr{A}/\Phi}(a) = \prod N_{E_i/\Phi}(a_i). \tag{16}$$

It is readily seen that $N_{\mathscr{A}/\Phi}(a) = N(a)$ the generic norm of a in the Jordan algebra $\mathscr{A}+$. We can use Theorem 2 to prove

THEOREM 3. *Let $\mathscr{A}$ and $\mathscr{B}$ be finite-dimensional separable commutative associative algebras over (an infinite) field Φ and let η be a bijective linear map of $\mathscr{A}$ onto $\mathscr{B}$ such that $1 \rightsquigarrow 1$. Then η is an isomorphism if and only if*

$$N_{\mathscr{B}/\Phi}(\eta a) = N_{\mathscr{A}/\Phi}(a), \qquad a \in \mathscr{A}. \tag{17}$$

Proof. We have $\mathscr{A} = \Phi[a]$ and η is an isomorphism if and only if the minimum polynomial $\mu_{\eta(a)}(\lambda)=\mu_a(\lambda)$. Similarly η is an isomorphism of the Jordan algebra $\mathscr{A}^+$ onto $\mathscr{B}^+$ if and only if $\mu_{\eta(a)}=\mu_a(\lambda)$. The result now follows from Theorem 2. ∎

References

1. F. D. Jacobson and N. Jacobson, Classification and representation of semi-simple Jordan algebras, *Trans. Amer. Math. Soc.* **65** (1949), 141–169.
2. N. Jacobson, Forms of algebras, *in* "Some Recent Advances in Basic Sciences," I, Academic Press, New York, 1966.
3. N. Jacobson, "Structure and Representations of Jordan Algebras," *Amer. Math. Soc.* Colloq. Publ. XXXIX, 1968.
4. N. Jacobson, Lectures on quadratic Jordan algebras, *Tata Inst. Fund. Res. Studies in Math.* (1969).
5. N. Jacobson, Structure groups and Lie algebras of Jordan algebras of symmetric elements of associatiave algebras with involution, *Advan. in Math.* **20** (1976), 106–150.
6. N. Jacobson, "Structure Theory of Jordan Algebras," Univ. of Arkansas Lecture Notes, No. 5, 1981.
7. N. Jacobson, Some applications of Jordan norms to involutorial simple associative algebras, *Advan. in Math.* **48** (1983), 149–165.
8. N. Jacobson and J. Katz, Generically algebraic quadratic Jordan algebras, *Scripta Math.* **XXIX** (1971), 215–227.
9. N. Jacobson and K. McCrommon, Quadratic Jordan algebras of quadratic frms with base points, *J. Indian Math. Soc.* **35** (1971), 1–45.
10. K. McCrimmon, A general theory of Jordan rings, *Proc. Nat. Acad. Sci. U.S.A.* **56** (1966), 1072–1079.
11. K. McCrimmon, The Freudenthal–Springer–Tits construction of exceptional Jordan algebras, *Trans. Amer. Math. Soc.* **130** (1969), 495–510.
12. K. McCrimmon, The Fruedenthal–Springer–Tits constructions revisited, *Trans. Amer. Math. Soc.* **145** (1970), 293–314.
13. K. McCrimmon, The generic norm of an isotope as a Jordan algebra, *Scripta Math.* **XXIX** (1971), 229–235.
14. H. Petersson and M. Racine, Exceptional Jordan division algebras, *in* "Algebraists Homage," pp. 307–316, Contemporary Mathematics, Vol. 13, Amer. Math. Soc., Providence, R.I., 1981.
15. H. Petersson and M. Racine, Simple Jordan algebras of degree three and the Tits process, *J. Algebra*, to appear.
16. M. Racine, A note on quadratic Jordan algebras of degree 3, Trans. *Amer Math. Soc.* **164** (1972), 93–103.

Printed by the St. Catherine Press Ltd., Tempelhof 41, Bruges, Belgium

Reprinted from JOURNAL OF ALGEBRA
All Rights Reserved by Academic Press, New York and London

Vol. 97, No. 2, December 1985
Printed in Belgium

Some Projective Varieties Defined by Jordan Algebras*

N. JACOBSON[†]

Department of Mathematics, Yale University, New Haven, Connecticut 06520

Communicated by the Editors

Received September 5, 1984

DEDICATED TO IRVING KAPLANSKY

Two classes of projective varieties are considered in this paper: norm hypersurfaces and varieties of reduced elements defined by finite-dimensional central simple Jordan algebras. The former has been considered previously in [5, 23, 13]. The concept of a reduced Jordan algebra over a field of characteristic $\neq 2$ was introduced in [8]. Such an algebra was defined to be a Jordan algebra with unit $1 = \sum_1^m e_i$ where the e_i are orthogonal idempotents such that the Peirce space $\mathcal{J}_{ii} = Fe_i$, F the base field, $1 \leqslant i \leqslant m$. We determined the structure of the reduced simple Jordan algebra and showed that up to isotopy these are the same simple algebras which we have determined in Section 2. The present definition of reduced algebra is based on the notion of reduced element of a Jordan algebra defined to be an element c such that $U_c \mathcal{J} \subset Fc$ ($U_c = 2L_c^2 - L_{c^2}$ where L_c is $x \rightsquigarrow c \cdot x$ if char $F \neq 2$). An algebra is called reduced if it is spanned by reduced elements.

These concepts have been used recently by John Faulkner to define derivation invariant symmetric bilinear forms on Jordan triple systems over a field [4]. The theory has been extended in the preceding paper by McCrimmon to Jordan triple systems over arbitrary commutative rings. In this paper we restrict our attention to finite-dimensional Jordan algebras over a field F and we take as point of departure some of the basic results noted by McCrimmon in [21]. In particular, Propositions 2.2 and 2.3 below are due to him. We note also that to encompass base fields of

* A preliminary version of this paper entitled "Generic reducing fields for Jordan algebras" was presented at a conference honoring Irving Kaplansky at the University of Chicago, June 12–14, 1984.

[†] This research was partially supported by National Science Foundation Grant MCS 79-05018.

0021-8693/85 $3.00

characteristic two we have adopted the point of view of quadratic Jordan algebras.

Let $\mathscr{J}$ be a finite-dimensional (unital quadratic) Jordan algebra over a field F (definition in Section 1) and let $\bar{F}$ be an algebraically closed field containing F. We define the variety $\mathscr{V}_{\mathscr{J}}$ of reduced elements as the set of points $\bar{F}v$ in the projective space of one-dimensional subspaces of $\mathscr{J}_{\bar{F}}$ such that v is a reduced element ($\neq 0$) of $\mathscr{J}_{\bar{F}}$. This is a variety defined by homogeneous polynomials with coefficients in F. We show that if $\mathscr{J}$ is outer central simple as defined in Section 1 then $\mathscr{V}_{\mathscr{J}}$ is absolutely irreducible and, in fact, $\mathscr{V}_{\mathscr{J}_{F_s}}$ for F_s the separable algebraic closure of F is birationally equivalent to a projective space (see Section 5).

The irreducibility of $\mathscr{V}_{\mathscr{J}}$ for $\mathscr{J}$ outer central simple permits us to define the field $PF(\mathscr{V}_{\mathscr{J}})$ of rational functions on $\mathscr{V}_{\mathscr{J}}$. We show that this field is a generic reducing field for $\mathscr{J}$ in a sense analogous to the definition of generic splitting field for a central simple associative algebra: If F' is an extension field of the base field F then $\mathscr{J}_{F'}$ is reduced if and only if there exists an F-place on $PF(\mathscr{V}_{\mathscr{J}})$ to F'. Moreover, $\mathscr{J}$ outer central simple is reduced if and only if $PF(\mathscr{V}_{\mathscr{J}})$ is rational (= purely transcendental) over F.

The first five sections of the paper are devoted to the general theory of the varieties $\mathscr{V}_{\mathscr{J}}$ and the fields $PF(\mathscr{V}_{\mathscr{J}})$ for $\mathscr{J}$ finite-dimensional central simple. Section 6 deals with norm hypersurfaces for these algebras and Section 7 deals with applications to central simple associative algebras with involutions. In the last section we give an important set of defining equations for $\mathscr{V}_{\mathscr{J}}$ for outer central simple $\mathscr{J}$ due to Faulkner.

1. Definitions

In this paper "Jordan algebra" will mean finite-dimensional unital quadratic Jordan algebra over a field F. This is a triple $(\mathscr{J}, U, 1)$ where $\mathscr{J}$ is a finite-dimensional vector space over F, U is a quadratic map of $\mathscr{J}$ into $\mathrm{End}_F\,\mathscr{J}$, and 1 is distinguished element of $\mathscr{J}$ such that the following axioms (due to McCrimmon) are verified:

QJ1: $U_1 = 1_{\mathscr{J}}$.

QJ2: $U_{U_a b} = U_a U_b U_a$, $a, b \in \mathscr{J}$.

QJ3: If $U_{a,b} = U_{a+b} - U_a - U_b$ and $V_{a,b}$ is defined by $V_{a,b}x = U_{a,x}b$ then

$$U_a V_{b,a} = V_{a,b} U_a = U_{a,U_a b}.$$

QJ4: QJ2, and QJ3 are maintained on extensions of the base field: If F' is an extension field of F and U is the (unique) extension of U on $\mathscr{J}$ to a quadratic map from $\mathscr{J}_{F'}$ to $\mathrm{End}_{F'}\,\mathscr{J}_{F'}$ then QJ2 and QJ3 hold for U on $\mathscr{J}_{F'}$.

QJ4 is superfluous if $|F| \geqslant 4$. If char $F \neq 2$ then the concept of quadratic Jordan algebra is equivalent to the usual one of linear Jordan algebra with unit. Given $(\mathscr{J}, U, 1)$ we obtain the linear Jordan algebra in which the Jordan product is $a \cdot b = \frac{1}{2} a \circ b$ where $a \circ b = U_{a,b} 1$ and 1 is unit for this multiplication. On the other hand, if $\mathscr{J}$ is linear with unit then $(\mathscr{J}, U, 1)$ is quadratic if 1 is the given unit and $U_a = 2L_a^2 - L_{a^2}$, where L_a is $x \rightsquigarrow a \cdot x$.

The following are the standard constructions of quadratic Jordan algebras: Let $\mathscr{A}$ be an associative algebra (with 1). Then $\mathscr{A}$ defines a Jordan algebra $\mathscr{A}^{(q)} = (\mathscr{A}, U, 1)$ where $U_a b = aba$. The algebras $\mathscr{A}^{(q)}$ and their subalgebras are called *special*. If $\mathscr{A}$ has an involution $J: a \rightsquigarrow a^*$ then the subset $H(\mathscr{A}, J)$ of J-symmetric elements is a subalgebra of $\mathscr{A}^{(q)}$.

Another class of Jordan algebras is the algebras $\mathrm{Jord}(Q, 1)$ defined by a quadratic form Q with a base point 1. Here we have a vector space $\mathscr{J}$ equipped with a quadratic form Q and a base point 1 defined by the condition $Q(1) = 1$. If $Q(x, y) = Q(x+y) - Q(x) - Q(y)$, $T(x) = Q(x, 1)$, and $\bar{x} = T(x) 1 - x$ then one defines $\mathrm{Jord}(Q, 1) = (\mathscr{J}, U, 1)$ where $U: a \rightsquigarrow U_a$ and

$$U_a b = Q(a, \bar{b})\, a - Q(a)\, \bar{b}. \tag{1.1}$$

Using a modified definition of Clifford algebras one can show that $\mathrm{Jord}(Q, 1)$ is special (see [16] or [13, Sect. 2.2]).

Next we define the standard exceptional (= non-special) Jordan algebra. Here we begin with an *octonion* or *Cayley algebra* $\mathbb{O}$ and take $\mathscr{J}$ to be the vector space of 3×3 hermitian octonion matrices with diagonal entries in F. These have the form

$$h = \begin{pmatrix} \alpha_1 & a_3 & \bar{a}_2 \\ \bar{a}_3 & \alpha_2 & a_1 \\ a_2 & \bar{a}_1 & \alpha_3 \end{pmatrix}, \qquad \alpha_i \in F1,\ a_i \in \mathbb{O} \tag{1.2}$$

which we denote also as (α, a). We define a cubic form N on $\mathscr{J}$ by

$$N(h) = \alpha_1 \alpha_2 \alpha_3 + t(a_1 a_2 a_3) - \sum \alpha_i n(a_i) \tag{1.3}$$

$t(a) 1 = a + \bar{a}$, $t(a_1 a_2 a_3) = t((a_1 a_2) a_3) = t(a_1 (a_2 a_3))$, and $n(a) 1 = a\bar{a}$, $\bar{a}$ the conjugate of the octonion a. Next we define

$$T(h, k) = (\Delta_1^h N)(\Delta_1^k N) - \Delta_1^h \Delta^k N \tag{1.4}$$

where $\Delta_b^a f$ for a polynomial function f is defined to be the coefficient of λ in $f(b + a\lambda)$. Calculation shows that

$$T(h, k) = \sum_1^3 (\alpha_i \beta_i + t(\bar{a}_i b_i)) \tag{1.5}$$

if $k = (\beta, b)$. T is a non-degenerate symmetric bilinear form on $\mathcal{J}$ so we can use this to define a map $h \rightsquigarrow h^{\#}$ by the condition

$$T(h^{\#}, k) = \Delta_h^k N \tag{1.6}$$

for all k. A calculation shows that $h^{\#} = (\alpha', a')$ where

$$\alpha_i' = \alpha_j \alpha_k - n(a_i), \qquad a_i' = \overline{a_j a_k} - \alpha_i a_i \tag{1.7}$$

and (ijk) is a cyclic permutation of (123). We can define a Jordan algebra $\mathrm{Jord}(N, 1) = (\mathcal{J}, U, 1)$ where 1 is the unit matrix and

$$U_h k = T(h, k)\, h - h^{\#} \times k \tag{1.8}$$

where $h \times k = (h+k)^{\#} - h^{\#} - k^{\#}$. This is a Jordan algebra that is exceptional (see, Sects. 2.3 and 2.4]). It can be shown that if the entries of h, k, l are in an associative subalgebra of $\mathbb{O}$ then

$$U_h k = hkh, \qquad U_{h,l} k = hkl + lkh. \tag{1.9}$$

Moreover, for any h, $h^2 = U_h 1$ coincides with h^2 as defined in $M_3(\mathbb{O})$.

Powers in a Jordan algebra are defined inductively by $a^0 = 1$, $a^1 = a$, $a^{n+2} = U_a a^n$. An element u is *invertible* in $\mathcal{J}$ if U_u is bijective. In this case we define $u^{-1} = U_u^{-1} u$. We refer to [13] for properties of powers and inverses. We recall here only that u and v are invertible if and only if $U_u v$ is invertible in which case $(U_u v)^{-1} = U_{u^{-1}} v^{-1}$. $\mathcal{J}$ is a *division algebra* if every $a \neq 0$ in $\mathcal{J}$ is invertible.

If u is invertible then u defines another Jordan algebra structure $\mathcal{J}^{(u)} = (\mathcal{J}, U^{(u)}, u^{-1})$ on $\mathcal{J}$ where $U^{(u)} = UU_u$. This is called the *u-isotope* of $\mathcal{J}$ and $\mathcal{J}$ and $\mathcal{J}'$ are called *isotopic* if there exists an isomorphism of $\mathcal{J}$ on some u'-isotope of $\mathcal{J}'$. Such an isomorphism is called an *isotopy*. It is readily seen that a map η of $\mathcal{J}$ into $\mathcal{J}'$ is an isotopy if and only if η is bijective linear and there exists a bijective linear map $\eta^*\colon \mathcal{J}' \to \mathcal{J}$ such that

$$U'_{\eta(a)} = \eta U_a \eta^* \tag{1.10}$$

for all $a \in \mathcal{J}$ (U' the U-operator in $\mathcal{J}'$). The group of self-isotopies of $\mathcal{J}$ is called the *structure group* Str $\mathcal{J}$. This contains the automorphism group Aut $\mathcal{J}$ as the subgroup of η such that $\eta(1) = 1$. By QJ2 and (1.10), Str $\mathcal{J}$ contains every $\eta = U_u$, u invertible (with $\eta^* = U_u$). These U_u generate a normal subgroup Instr $\mathcal{J}$ of Str $\mathcal{J}$ called the *inner structure group*. Isotopy is an equivalence relation which is broader than isomorphism. For example, if $(\mathcal{A}, J)$ is an associative algebra with involution and c is an invertible element of $H(\mathcal{A}, J)$ then the map $c_L^{-1} (x \rightsquigarrow c^{-1} x)$ is an isomorphism of

$H(\mathscr{A}, J_c)$ onto $H(\mathscr{A}, J)^{(c)}$ where J_c is $x \rightsquigarrow cJ(x)c^{-1}$. It is easy to give examples in which $H(\mathscr{A}, J)$ and $H(\mathscr{A}, J_c)$ are not isomorphic.

A subspace $\mathscr{B}$ of $\mathscr{J}$ is called an *inner* (*outer*) ideal if $U_{\mathscr{B}}\mathscr{J} \subset \mathscr{B}$ ($U_{\mathscr{J}}\mathscr{B} \subset \mathscr{B}$). If $\mathscr{B}$ is both an inner and an outer ideal then $\mathscr{B}$ is called an *ideal* in $\mathscr{J}$ (notation: $\mathscr{B} \lhd \mathscr{J}$). If char $F \neq 2$ then any outer ideal is an ideal and this is also an ideal in the linear Jordan algebra in which $a \cdot b = \frac{1}{2}U_{a,b}1$. Inner ideals are analogous to one-sided ideals in the associative theory. If $\mathscr{B}$ is a one-sided ideal in the associative algebra $\mathscr{A}$ then $\mathscr{B}$ is an inner ideal of $\mathscr{A}^{(q)}$. Since the intersection of inner ideals is inner the intersection of a left and a right ideal of $\mathscr{A}$ is an inner ideal of $\mathscr{A}^{(q)}$. For any $b \in \mathscr{J}$, $U_b\mathscr{J}$ is an inner ideal, called the *principal inner ideal* determined by b.

A Jordan algebra $\mathscr{J}$ is *simple* (*outer simple*) if $\mathscr{J} \neq 0$ and $\mathscr{J}$ contains no ideals (outer ideals) $\neq 0$, $\mathscr{J}$. The *centroid* of $\mathscr{J}$ is

$$\Gamma(\mathscr{J}) = \{\mathscr{J} \in \operatorname{End}_F \mathscr{J} \mid \gamma U_a = U_a\gamma,\ U_{\gamma a} = \gamma^2 U_a,\ a \in \mathscr{J}\} \tag{1.11}$$

and $\mathscr{J}$ is *central* if $\Gamma(\mathscr{J}) = F1$. If $\mathscr{J}$ is simple then $\mathscr{J}_E$ is simple for every extension field E/F if and only if $\mathscr{J}$ is central. We define the *outer centroid* of $\mathscr{J}$ by

$$\Gamma_0(\mathscr{J}) = \{\gamma \in \operatorname{End}_F \mathscr{J} \mid \gamma U_a = U_a\gamma,\ a \in \mathscr{J}\} \tag{1.12}$$

and $\mathscr{J}$ is *outer central* if $\Gamma_0(\mathscr{J}) = F1$. If $\mathscr{J}$ is outer simple then $\mathscr{J}_E$ is outer simple for every E/F if and only if $\mathscr{J}$ is outer central. If char $F \neq 2$ any simple algebra is outer simple and if $\mathscr{J}$ is central then $\mathscr{J}$ is outer central. (See [13, Sect. 7.6] for these results.)

2. Reduced Elements of a Jordan Algebra

2.1. Definition. An element c of a Jordan algebra $\mathscr{J}/F$ is called *reduced* if $U_c\mathscr{J} \subset Fc$.

We recall that z is an *absolute zero divisor* in $\mathscr{J}$ if $U_z = 0$. If $\mathscr{J}$ is simple or more generally is semiprimitive then the only absolute zero divisor in $\mathscr{J}$ is $z = 0$. Hence in these cases if c is reduced then $U_c\mathscr{J} = Fc$.

2.2. Proposition. *Let $\rho(\mathscr{J})$ denote the set of reduced elements of $\mathscr{J}$. Then*

(1) *If $c \in \rho(\mathscr{J})$ then $\alpha c \in \rho(\mathscr{J})$ for any $\alpha \in F$.*
(2) *If $c \in \rho(\mathscr{J})$ and $a \in \mathscr{J}$ then $U_a c \in \rho(\mathscr{J})$.*
(3) *$F\rho(\mathscr{J})$ is an ideal in $\mathscr{J}$.*
(4) *$\rho(\mathscr{J}) = \rho(\mathscr{J}^{(u)})$ for any isotope $\mathscr{J}^{(u)}$ of $\mathscr{J}$.*

Proof. (1) is clear.

(2) If $c \in \rho(\mathcal{J})$ then for any $x \in \mathcal{J}$ we have a $\gamma(x) \in F$ such that $U_c x = \gamma(x)\, c$. Then

$$U_{U_a c} x = U_a U_c U_a x = U_a(\gamma(U_a x))\, c = \gamma(U_a x)\, U_a c.$$

Hence $U_a c \in \rho(\mathcal{J})$.

(3) By (2), $F\rho(\mathcal{J})$ is an outer ideal of $\mathcal{J}$. We have the following lemma (due to McCrimmon, see [13, Lemma 4.1.9 on p. 4.5]): If $\mathcal{B}$ is an outer ideal in $\mathcal{J}$ and $\mathcal{B} = \sum F b_i$ where $U_{b_i} \mathcal{J} \subset \mathcal{B}$ for every b_i then $\mathcal{B}$ is an ideal. Now $F\rho(\mathcal{J}) = \sum F c_i$ where $c_i \in \rho(\mathcal{J})$. Then $U_{c_i} \mathcal{J} \subset F c_i \subset F\rho(\mathcal{J})$. Hence $F\rho(\mathcal{J})$ is an ideal.

(4) Let $c \in \rho(\mathcal{J})$ and let u be an invertible element of $\mathcal{J}$. Then $U_c^{(u)} \mathcal{J} = U_c U_u J = U_c \mathcal{J} \subset Fc$. Hence $c \in \rho(\mathcal{J}^{(u)})$ and $\rho(\mathcal{J}) = \rho(\mathcal{J}^{(u)})$ by symmetry. ∎

We shall call $\mathcal{J}$ *reduced* if $F\rho(\mathcal{J}) = \mathcal{J}$. By Proposition 2.2(3) if $\mathcal{J}$ is simple then either $\mathcal{J}$ is reduced or $\rho(\mathcal{J}) = 0$. We recall that an idempotent e is called a *division idempotent* if the Jordan algebra $(U_e \mathcal{J}, U, e)$ is a division algebra. We have

2.3. PROPOSITION. *If e is a division idempotent in a reduced Jordan algebra $\mathcal{J}$ then $e \in \rho(\mathcal{J})$.*

Proof. Any element of $U_e(\mathcal{J})$ has the form $\sum U_e c_i$ where $c_i \in \rho(\mathcal{J})$. By 2.2(2), every $d_i = U_e c_i \in \rho(\mathcal{J})$ and since d_i is in the division algebra $U_e \mathcal{J}$, $U_{d_i}(U_e \mathcal{J}) = U_e \mathcal{J}$ if $d_i \neq 0$. On the other hand, $U_{d_i}(U_e \mathcal{J}) \subset F d_i$. Hence $U_e \mathcal{J} = F d_i$ if $d_i \neq 0$ so $U_e \mathcal{J}$ is one dimensional. Then $U_e \mathcal{J} = Fe$ and $e \in \rho(\mathcal{J})$. ∎

We shall now determine the reduced simple Jordan algebras. We note first that a simple algebra $\mathcal{J}$ contains no absolute zero divisors $z \neq 0$ since the existence of such elements implies that $\mathcal{J}$ has a non-zero Jacobson radical or, equivalently, $\mathcal{J}$ contains a non-zero nil ideal. We recall that a Jordan algebra is said to *have a capacity* if $1 = \sum_1^m e_i$ where the e_i are division idempotents that are orthogonal ($U_{e_i} = 0 = e_i \circ e_j$ if $i \neq j$). The minimum m for which such a decomposition of 1 exists is called *the capacity* of $\mathcal{J}$. The structure of (finite-dimensional) semiprimitive (that is, algebras with 0 Jacobson radical) Jordan algebras is known. Such an algebra is isotopic to an algebra having a capacity [13, p. 6.37] and is a finite direct sum of simple Jordan algebras having capacities and the latter can be listed [13, Theorem 6.4.1]. From this result we can deduce the structure of reduced simple Jordan algebras. In the reduced case we can short-cut the usual approach, especially a rather difficult theorem due to

Osborn on simple algebras of capacity two, as we have done earlier in the characteristic $\neq 2$ case [9, p. 203]. We distinguish the three cases: capacity $m=1$, $m=2$, and $m \geqslant 3$. If $m=1$ then $\mathscr{J}=F1$. Now assume $m>1$. The decomposition $1=\sum e_i$ into orthogonal idempotents gives rise to the *Peirce decomposition.*

$$\mathscr{J}=\bigoplus_{i<j} \mathscr{J}_{ij} \tag{2.3}$$

where $\mathscr{J}_{ii}=U_{e_i}\mathscr{J}$ and $\mathscr{J}_{ij}=U_{e_i,e_j}\mathscr{J}$ if $i \neq j$. The Peirce components satisfy many important multiplicative relations. Among these we have $\mathscr{J}_{ij}^2 \subset \mathscr{J}_{ii}+\mathscr{J}_{jj}$ if $i \neq j$. The following result gives the structure of reduced simple Jordan algebras of capacity two.

2.4. THEOREM. *Let $|F|>2$. Then $\mathscr{J}$ is reduced simple of capacity two if and only if $\mathscr{J}=\mathrm{Jord}(Q, 1)$ where Q is a quadratic form with base point 1 such that (1) Q is non-degenerate ($Q(z)=Q(z, a)=0$ for all $a \Rightarrow z \Rightarrow 0$) and (2) Q is isotropic, that is, there exists a $c \neq 0$ such that $Q(c)=0$.*

Proof. We have $\mathscr{J}=\mathscr{J}_{11} \oplus \mathscr{J}_{12} \oplus \mathscr{J}_{22}$ and $\mathscr{J}_{ii}=Fe_i$. Now $\mathscr{J}_{12} \neq 0$ since $\mathscr{J}$ is simple, and if $a_{12} \in \mathscr{J}_{12}$ then $a_{12}^2=\alpha_1 e_2+\alpha_2 e_2$, $\alpha_i \in F$. We have $e_i \circ a_{ij}=a_{ij}$ (PD5 on p. 5.6 of [13]). Hence $\alpha_i a_{ij}=U_{\alpha_i e_i, a_{ij}} 1=U_{\alpha_i e_i, a_{ij}} e_i$ (since $U_{\alpha_i e_i, a_{ij}} e_j=0$ by PD3) $=U_{a_{ij}^2} e_i$ (since $U_{\alpha_j e_j, a_{ij}} e_i=0$) $=V_{a_{ij}, 1} U_{a_{ij}} e_i=V_{a_{ij}, 1} U_{e_j} a_{ij}^2$ (by QJ3 and PD6) $=V_{a_{ij}, 1} \alpha_j e_j=\alpha_j a_{ij}$ (since $V_{a, 1} b=a \circ b$). Hence $\alpha_i=\alpha_j$ and $a_{12}^2=-Q(a_{12}) 1$ where Q is a quadratic form on $\mathscr{J}_{12}$. We extend this to a quadratic form on $\mathscr{J}$ by $Q(x)=\beta_1 \beta_2+Q(a_{12})$ for $x=\beta_1 e_1+\beta_2 e_2+a_{12}$. Then $T(x) \equiv Q(x, 1)=\beta_1+\beta_2$ and direct calculation shows that

$$x^2-T(x)\, x+Q(x)\, 1=0. \tag{2.5}$$

Also since $U_{x^3-T(x) x^2+Q(x) x}=U_x U_{x^2-T(x) x+1}=0$ (by the Weak Macdonald Theorem, p. 3.12 of 13]) and since $\mathscr{J}$ contains no absolute zero divisors we have

$$x^3-T(x)\, x^2+Q(x)\, x=0. \tag{2.6}$$

If we assume $|F|>2$ we can deduce from (2.5) and (2.6) by linearization that

$$U_a b=Q(a, \bar{b})\, a-Q(a)\, \bar{b}, \qquad \bar{b}=T(b)\, 1-b. \tag{2.7}$$

Thus $\mathscr{J}=\mathrm{Jord}(Q, 1)$. If $z \in \mathscr{J}$ satisfies $Q(z)=0=Q(z, b)$ for all b then $U_z=0$ by (2.7). Hence $z=0$ as noted earlier and Q is non-degenerate. Also $Q(e_i)=0$ so Q is isotropic. Thus (1) and (2) hold.

It is straightforward to verify that conversely if $\mathscr{J}=\mathrm{Jord}(Q, 1)$ with Q satisfying (1) and (2) then $\mathscr{J}$ is reduced (see [16, pp. 17–21]). ∎

To describe the structure of reduced simple Jordan algebras of capacity $\geqslant 3$ we need to recall the definitions of composition algebras and standard Jordan matrix algebras $H(M_m(\mathscr{D}), \mathscr{D}_0)$.

2.8. DEFINITION. An alternative algebra with involution $(\mathscr{D}, j)$ over a field F is called a *composition algebra* if (i) $\mathscr{D}$ has no absolute zero divisors $\neq 0$, that is, elements z such that $zaz = 0$ for all $a \in \mathscr{D}$, and (ii) for any $a \in \mathscr{D}$, $a\bar{a} \in F1$, for $\bar{a} = j(a)$.

The list of these algebras is given in

2.9. THEOREM. *Let $(\mathscr{D}, j)$ be a composition algebra over a field F. Then $(\mathscr{D}, j)$ is one of the following:*

(1) *A purely inseparable extension field E/F of characteristic two and exponent* 1 *with $j = 1_{\mathscr{D}}$.*

(2) $\mathscr{D} = F$.

(3) *A quadratic composition algebra (either a separable quadratic extension field E/F or a direct sum of two copies of F), j the automorphism $\neq 1_E$ of E.*

(4) *$\mathscr{D}$ a quaternion algebra, j the standard involution in $\mathscr{D}$.*

(5) *$\mathscr{D}$ an octonion algebra, j the standard involution. (See* [13, p. 6.5f].)

Let $(\mathscr{D}, j)$ be an associative algebra with involution $a \rightsquigarrow \bar{a}$. For any integer m we can form the matrix algebra $M_m(\mathscr{D})$ in which we have the *standard involution* J: $A = (a_{ij}) \rightsquigarrow {}^t(\bar{a}_{ij})$. Then we have the subalgebra $H(M_m(\mathscr{D}))$ of $M_m(\mathscr{D})^{(q)}$ of J-symmetric elements. More generally, if $\mathscr{D}_0$ is a subspace of $H(\mathscr{D}_0)$ containing all the norms $a\bar{a}$, $a \in \mathscr{D}$, then we have the subspace $H(M_m(\mathscr{D}), \mathscr{D}_0)$ of $H(M_n(\mathscr{D}))$ consisting of the J-symmetric matrices with diagonal entries in $\mathscr{D}_0$ and in char $= 2$ this may be a proper subalgebra of $H(M_m(\mathscr{D}))$. We have $H(M_m(\mathscr{D})) = H(M_m(\mathscr{D}), \mathscr{D}_0)$ if and only if $\mathscr{D}_0 = H(\mathscr{D}, j)$. On the other hand, if $\mathscr{D}$ is a quaternion algebra of characteristic two then $H(\mathscr{D}, j)$ is three dimensional and $H(M_n(\mathscr{D}), F)$ is a proper subalgebra of $H(M_m(\mathscr{D}))$.

Now let $(\mathscr{D}, j)$ be an alternative algebra with involution, $\mathscr{D}_0$ a subspace of $H(\mathscr{D}, j)$ containing all $a\bar{a}$. As in the associative case, we let $H(M_3(\mathscr{D}), \mathscr{D}_0)$ be the subspace of $H(M_3(\mathscr{D}))$ of J-symmetric matrices (J as above) with diagonal entries in $\mathscr{D}_0$. As in the special case in which $(\mathscr{D}, j)$ is an octonion algebra with standard involution, we may have a Jordan structure on $\mathscr{J} = H(M_3(\mathscr{D}), \mathscr{D}_0)$. It can be shown that this is unique if we require that (1) $A^2 = U_A 1$ coincides with A^2 as defined in $M_3(\mathscr{D})$, (2) the

e_{ii}, $i=1, 2, 3$, are orthogonal idempotents with $1=\sum e_{ii}$, and (3) the subspaces $H_{ij}=\{ae_{ij}+ae_{ji} \mid a\in \mathscr{D}\}$, $i\neq j$, and $H_{ii}=\{a_0e_{ii} \mid a_0\in \mathscr{D}_0\}$ are the Peirce spaces relative to the e_{ii} [13, p. 5.11]. We call these $H(M_3(\mathscr{D}), \mathscr{D}_0)$ as well as the $H(M_m(\mathscr{D}), \mathscr{D}_0)$ defined for associative $\mathscr{D}$ above *standard Jordan matrix algebras.*

We can now state

2.10. THEOREM. *Let $\mathscr{J}$ be a reduced simple Jordan algebra of capacity $m\geqslant 3$. Then $\mathscr{J}$ is isotopic to a standard Jordan matrix algebra $H(M_m(\mathscr{D}), \mathscr{D}_0)$ where $(\mathscr{D}, j)$ is a composition algebra that is associative if $n\geqslant 4$ and $\mathscr{D}_0=F1$.*

Proof. We have $1=\sum_1^m e_i$ where the e_i are orthogonal division idempotents. Then we have the Peirce decomposition $\mathscr{J}=\bigoplus_{i\leqslant j}\mathscr{J}_{ij}$ where $\mathscr{J}_{ii}=Fe_i$. Since $\mathscr{J}$ is simple the e_i are connected in the sense that for $i\neq j$ there exists an $a_{ij}\in \mathscr{J}_{ij}$ that is invertible in $\mathscr{J}_{ii}+\mathscr{J}_{ij}+\mathscr{J}_{jj}=U_{e_i+e_j}\mathscr{J}$. By passing to an isotope we may assume that $a_{ij}^2=e_i+e_j$. Then we can apply the Strong Coordinatization Theorem to identify $\mathscr{J}$ with a standard Jordan matrix algebra $H(M_m(\mathscr{D}), \mathscr{D}_0)$ so that $e_i=e_{ii}$, $1\leqslant i\leqslant n$, the Peirce spaces $\mathscr{J}_{ii}=\{a_0e_{ii} \,\partial\, a_0\in \mathscr{D}_0\}$, $\mathscr{J}_{ij}=\{ae_{ij}+\bar{a}e_{ji} \mid a\in \mathscr{D}\}$ if $i\neq j$, and $A^2=U_A 1$ is the square of A in $M_m(\mathscr{D})$ [13, Theorem 5.4.2 on p. 5.22]. Since $\mathscr{J}_{ii}=Fe_i$ it is clear that $\mathscr{D}_0=F1$. If $a\in\mathscr{D}$ and $i\neq j$ then $(ae_{ij}+\bar{a}e_{ji})^2=a\bar{a}e_{ii}+\bar{a}ae_{jj}$. Hence $a\bar{a}\in F1$ and if $z\in\mathscr{D}$ satisfies $zaz=0$ for all $a\in\mathscr{D}$ then $U_{ze_{ij}+\bar{z}e_{ji}}(ae_{ij}+\bar{a}e_{ji})=0$ and $U_{ze_{ij}+\bar{z}e_{ji}}(a_ie_{ii})=0$ for all $a\in\mathscr{D}$, $a_0\in\mathscr{D}_0$ (see p. 5.12 of [13]). This and the multiplicative formulas for the Peirce spaces (the PD formulas) imply that $U_{ze_{ij}+\bar{z}e_{ji}}=0$. Hence $z=0$. Thus $(\mathscr{D}, j)$ is a composition algebra. ∎

The converse of 2.10 is also valid. In fact, we have the following result which in characteristic 2 is slightly stronger than the converse of 2.10.

2.11. THEOREM. *Let $\mathscr{J}$ be isotopic to an algebra $H(M_m(\mathscr{D}), F1)$ where $m\geqslant 3$ and $(\mathscr{D}, j)$ is a composition algebra with standard involution. Then $\mathscr{J}$ is simple, and is outer simple except in the case in which* char $F=2$ *and $\mathscr{D}$ is purely inseparable of exponent one.*

Proof. Since an isotope of a simple (outer simple) algebra is simple (outer simple) it suffices to prove the theorem for $\mathscr{J}=H(M_n(\mathscr{D}), F1)$. Let $\mathscr{B}$ be an outer ideal $\neq 0$ in $\mathscr{J}$ and $\mathscr{B}=\sum \beta_i e_{ii}+\sum_{i<j} b_{ij}e_{ij}+\bar{b}e_{ji}$ be a nonzero element of $\mathscr{B}$. Applying the $U_{e_{ii}}$ and the $U_{e_{ii},e_{jj}}$ to B we see that $\mathscr{B}$ contains either one of the e_{ii} or a $be_{ij}+\bar{b}e_{ji}$, $b\neq 0$, in $\mathscr{D}$. In either case it is readily seen that $\mathscr{B}=\mathscr{J}$ unless char $F=2$ and $\mathscr{D}$ is purely inseparable of exponent one (see the QM formulas on p. 5.12 of [1]). In the last case $\mathscr{B}$ contains a $be_{ij}+\bar{b}e_{ji}\neq 0$ and if $\mathscr{B}$ is an ideal it also contains $(be_{ij}+\bar{b}e_{ji})^2=$

$b\bar{b}e_{ii}+\bar{b}be_{jj}$. It follows that $\mathscr{B}=\mathscr{J}$. Finally, it is easy to verify in the case in which char $F=2$ and $\mathscr{D}$ is purely inseparable that $\mathscr{B}=\{\sum b_{ij}e_{ij}+\bar{b}_{ij}e_{ji} \mid b_{ij}\in\mathscr{D},\ i<j\}$ is an outer ideal in $\mathscr{J}$. ∎

3. The Generic Norm and Generic Minimum Polynomial

The theory of the generic norm and minimum polynomial of a Jordan algebra of characteristic $\neq 2$ has been developed in [9, Chap. VI]. The extension of this theory to char $F=2$ was given in [15, 20]. We shall now sketch the definitions and relevant facts which we shall require in this paper.

Let a be an element of a Jordan algebra $\mathscr{J}/F$ and λ an indeterminate. Then we have a homomorphism v_a of $F[\lambda]^{(q)}$ into $\mathscr{J}$ such that $\lambda^r \rightsquigarrow a^r$, $r=0, 1, 2,\ldots$. The image $v_a(F[\lambda])$ is the subalgebra of $\mathscr{J}$ generated by a and is denoted as $F[a]$. The kernel K_a of v_a is an ideal of $F[\lambda]^{(q)}$ which in the case char $F=2$ need not be an ideal of $F[\lambda]$. However, K_a contains a non-zero ideal K_a^* of $F[\lambda]$ that contains all such ideals and $K_a^*=(\mu_a(\lambda))$ where $\mu_a(\lambda)$ is monic. The polynomial $\mu_a(\lambda)$ is called the *minimum polynomial* of a. It can be characterized as the monic polynomial of least degree such that $v_a\mu_a(\lambda)=0=v_a\lambda\mu_a(\lambda)$. Moreover, the first of these conditions suffices in the following cases: (1) char $F\neq 2$, (2) $\mathscr{J}$ is special, (3) $\mathscr{J}$ contains no absolute zero divisor $\neq 0$.

Let $(u_1,\ldots, u_n)$ be a base for $\mathscr{J}/F$, $\xi_1,\ldots, \xi_n$ indeterminates, and let $x\equiv\sum \xi_i u_i \in \mathscr{J}_{F(\xi)}$, $F(\xi)=F(\xi_1,\ldots, \xi_n)$. Then it can be shown that the minimum polynomial $m_x(\lambda)$ of x in $\mathscr{J}_{F(\xi)}$ has the form

$$\lambda^m - \tau_1(\xi)\,\lambda^{m-1} + \tau_2(\xi)\,\lambda^{m-2} - \cdots + (-1)^m\,\tau_m(\xi) \tag{3.1}$$

where $\tau_i(\xi)=\tau_i(\xi_1,\ldots, \xi_n)$ is a homogeneous polynomial of degree i in the ξ's. Then if E is any extension field of F and $a=\sum \alpha_i u_i$, $\alpha_i\in E$, we put

$$m_a(\lambda)=\lambda^m-\tau_1(\alpha_1,\ldots, \alpha_n)\,\lambda^{m-1}+\cdots+(-1)^m\,\tau_m(\alpha_1,\ldots, \alpha_m). \tag{3.2}$$

It follows that $m_a(a)\equiv v_a m_a(\lambda)=0$ and $am_a(a)\equiv v_a(\lambda\mu_a(\lambda))=0$. The polynomial $m_a(\lambda)$ is independent of the choice of the base and is called the *generic minimum polynomial* of a. We put $T(a)=\tau_1(\alpha_1,\ldots, \alpha_n)$, $N(a)=\tau_m(\alpha_1,\ldots, \alpha_n)$ and cell these the *generic trace* and *generic norm*, respectively, of a. The integer m is called the *degree* of $\mathscr{J}$. The maps $a\rightsquigarrow\tau_i(a)\equiv\tau_i(\alpha_1,\ldots, \alpha_n)$ are polynomial functions defined over F. For any infinite extension field E/F there exist $a=\sum \alpha_i u_i\in\mathscr{J}_E$ such that the minimum polynomial $\mu_a(\lambda)=m_a(\lambda)$. In fact, the set of these elements is a non-vacuous Zariski open subset of $\mathscr{J}_E$.

There are many important properties of the generic norm and generic minimum polynomial. We now list some of these:

(i) T is a linear function on $\mathcal{J}$, N is homogeneous of degree m.

(ii) $N(\lambda 1 - a) = m_a(\lambda)$. Here $N(\lambda 1 - a)$ is the generic norm of $\lambda 1 - a \in \mathcal{J}_{F(\lambda)}$, λ an indeterminate.

(iii) a is invertible if and only if $N(a) \neq 0$. Then $a^{-1} = N(a)^{-1} a^{\#}$ where $a^{\#} = (-1)^{m-1}[a^{m-1} - \tau_1(a)\, a^{m-2} + \cdots + (-1)^{m-1} \tau_{m-1}(a)\, 1]$.

(iv) $N(a^{\#}) = N(a)^{m-1}$, $a^{\#\#} = N(a)^{m-2}\, a$.

(v) $m_a(\lambda)$ and the minimum polynomial $\mu_a(\lambda)$ have the same irreducible factors in $F[\lambda]$ and the same roots in the algebraic closure $\bar{F}$ of F.

(vi) $N(U_a b) = N(a)^2\, N(b)$.

(vii) If u is invertible and $N^{(u)}$ denotes the generic norm in the u-isotope $\mathcal{J}^{(u)}$ then

$$N^{(u)}(a) = N(u)\, N(a).$$

A consequence of (v) is that if a is nilpotent then $m_a(\lambda) = \lambda^m$ and if a is idempotent then $m_a(\lambda) = (\lambda - 1)^r \lambda^{m-r}$, $0 \leqslant r \leqslant m$.

We shall now determine the generic norm for the reduced simple algebras. In view of (vii) and (ii) it suffices to consider isotopes of any of these algebras.

We note first that $m_a(\lambda)$ is of degree 1 if and only if $\mathcal{J}$ has capacity 1 or, equivalently, $J = F1$. Next we consider the algebras Jord$(Q, 1)$ where Q is a quadratic form with base point 1 on $\mathcal{J}$. Our results show that for such an algebra, $N = Q$ and

$$m_a(\lambda) = \lambda^2 - T(a)\, \lambda + Q(a) \tag{3.3}$$

where $T(a) = Q(a, 1)$. Next let $\mathcal{J} = H(M_3(\mathbb{O}), F)$ where $\mathbb{O}$ is an octonion algebra and the Jordan structure is determined by the form N defined by (1.3) and the unit matrix 1. In this case N is the generic norm and

$$m_h(\lambda) = \lambda^3 - T(h)\, \lambda^2 + T(h^{\#})\, \lambda - N(h). \tag{3.4}$$

This follows from the fact that $m_h(h) = 0$ (see [13, p. 2.22]) and the fact that we have three non-zero orthogonal idempotents $e_{ii} \neq 0$ in $\mathcal{J}$ with $1 = \sum e_{ii}$ (see [9, p. 229]). The same argument will apply in the remaining cases, namely, for the algebras $H(M_m(\mathcal{D}), F)$ where $m \geqslant 3$ and $(\mathcal{D}, j)$ is an associative composition algebra. We state the result for these as

3.5. THEOREM. *Let* $\mathcal{J} = H(M_m(\mathcal{D}), F)$ *where* $m \geqslant 3$ *and* $(\mathcal{D}, j)$ *is an associative composition algebra with standard involution. Then* $N(h) = \det h$ *and* $m_h(\lambda) = \det(\lambda 1 - h)$ *if* $\mathscr{D}$ *is commutative. On the other hand, if* $\mathscr{D}$ *is a quaternion algebra and* $\bar{F}$ *is the algebraic closure of* F *then* $\mathscr{D}_{\bar{F}} = M_2(\bar{F})$, $\mathscr{J}_{\bar{F}}$ *is the set of matrices* $h \in M_{2m}(\bar{F})$ *such that* hs *is alternate for*

$$s = \operatorname{diag}\left\{\begin{pmatrix} 0 & 1 \\ -1 & 0 \end{pmatrix}, \ldots, \begin{pmatrix} 0 & 1 \\ -1 & 0 \end{pmatrix}\right\}.$$

Then $N(h)$ is the Pfaffian $Pf(hs)$, $m_h(\lambda) = Pf(s\lambda - hs)$.

Proof. We show first that if $\mathscr{D}$ is commutative then $\det h \in F$. This is clear if $\mathscr{D} = F$. Next let $\mathscr{D}$ be a quadratic composition algebra and write $\bar{a} = j(a)$. Let $h \in M_m(\mathscr{D})$. Then $\det\ ({}^t\bar{h}) = \overline{\det h}$ so $\det h = \overline{\det h}$ if $h \in H(M_m(D), F)$. Since F is the set of fixed elements under j we obtain $\det h \in F$. Now let char $F = 2$ and let $\mathscr{D}$ be a purely inseparable extension of F with $j = 1_{\mathscr{D}}$. Let $h \in H(M_m(\mathscr{D}), F)$ so h is a symmetric matrix with entries in $\mathscr{D}$ and diagonal entries in F. We now allow $m \geqslant 1$ and prove by induction on m that $\det h \in F$. We have $\det(\lambda 1 - h) = \lambda^m - \tau_1 \lambda^{m-1} + \cdots + (-1)^{m-1} \tau_{m-1} \lambda + (-1)^m \det h$ and τ_i is the sum of the diagonal i-rowed minors of h. Hence we may assume these are in F. Now $h^m - \tau_1 h^{m-1} + \cdots + (-1)^{m-1} \tau_{m-1} h + (-1)^m (\det h)\, 1 = 0$ and every $h^i \in H(M_m(\mathscr{D}), F)$. If we take the (1–1) entry of this matrix relation we see that $\det h \in F$. In all the foregoing cases the Hamilton–Cayley theorem and the fact that $H(M_m(\mathscr{D}), F)$ contains m orthogonal idempotents e_{ii} imply as in [13, pp. 2.29–2.33] that $N(h) = \det h$ and $m_h(\lambda) = \det(\lambda 1 - h)$. In the same way one sees, as in this reference, that if $\mathscr{D}$ is a quaternion algebra then the result stated holds for $H(M_m(\mathscr{D}), F)$. ∎

Remark. The result in the case char $F = 2$, D purely inseparable, can also be deduced from a general formula for the determinant of a matrix of the form $D + S$ where D is diagonal and S is skew (see [22, p. 410] or [27]).

If e is an idempotent in a Jordan algebra then so is $1 - e$ and $e_1 = e$, $e_2 = 1 - e$ are orthogonal with $e_1 + e_2 = 1$. Then we have the Peirce decomposition $\mathscr{J} = \mathscr{J}_{11} \oplus \mathscr{J}_{12} \oplus \mathscr{J}_{22}$ relative to the e_i. It is customary to denote these also as $\mathscr{J}_1 = \mathscr{J}_1(e)$, $\mathscr{J}_{1/2} = \mathscr{J}_{1/2}(e)$, and $\mathscr{J}_0 = \mathscr{J}_0(e)$, respectively. As we have noted, $m_e(\lambda) = (\lambda - 1)^r \lambda^{m-r}$. We can now obtain a determination of the integer r if $\mathscr{J}$ is reduced and simple.

3.6. THEOREM. *If* e *is an idempotent in a reduced simple Jordan algebra* $\mathscr{J}$ *of degree* m *then* $m_e(\lambda) = (\lambda - 1)^r \lambda^{m-r}$ *where* r *is the degree of* $\mathscr{J}_1(e)$. *In particular* $e \in \rho(\mathscr{J})$ *if and only if* $r = 1$.

Proof. The determination of the structure up to isotropy of reduced simple Jordan algebras in the last section and of their degrees in this one shows that if $1=\sum_{i=1}^{m'} e_i$ where the e_i are division idempotents in $\mathscr{J}$ then m' is the degree m of $\mathscr{J}$. We note next that if the e_i are contained in a subalgebra $\mathscr{K}$ of $\mathscr{J}$ then the generic minimum polynomial in $\mathscr{K}$ is the restriction to $\mathscr{K}$ of the generic minimum polynomial in $\mathscr{J}$. For, the existence of the e_i implies that $\mathscr{K}_{\bar{F}}$ contains an element whose minimum polynomial has degree m. Then $\deg \mathscr{K} \geqslant m$ and hence $\deg \mathscr{K}=m$ and the generic minimum polynomial in $\mathscr{K}$ is obtained by restriction from $\mathscr{J}$. Now let e be an idempotent in $\mathscr{J}$. Then $\mathscr{J}_1=\mathscr{J}_1(e)$ and $\mathscr{J}_0=\mathscr{J}_0(e)$ are semiprimitive [13, p. 4.21] and reduced (2.2, (21)). In passing to an isotope if necessary we have $e=e_1+\cdots+e_r$, $1-e=e_{r+1}+\cdots+e_m$ where the e_i are orthogonal division idempotents such that $e_i\in\mathscr{J}_1$ if $i\leqslant r$ and $e_i\in\mathscr{J}_0$ if $i>r$. Then r and $m-r$ are the degrees of $\mathscr{J}_1$ and $\mathscr{J}_0$, respectively. Moreover, $\mathscr{J}_1+\mathscr{J}_0$ is a subalgebra of $\mathscr{J}$ containing the $\mathscr{J}_i$ as ideals and containing all the e_i. Hence the generic minimum polynomial in $\mathscr{J}_1+\mathscr{J}_0$ is obtained by restriction from $\mathscr{J}$. On the other hand, if $a=a_1+a_0$, $a_i\in\mathscr{J}_i$, then $m_a(\lambda)=m_{1a_1}(\lambda)\,m_{2a_2}(\lambda)$ where m_i is the generic minimum polynomial in $\mathscr{J}_i$. Since e is the unit of $\mathscr{J}_1$ this implies that $m_e(\lambda)=(\lambda-1)^r\lambda^{m-r}$. The second statement is a consequence of the first since an algebra is of degree 1 if and only if it is one dimensional. ∎

4. The Projective Variety of Reduced Elements of a Jordan Algebra

Let $\mathscr{J}$ be a Jordan algebra over F, $(u_1,\ldots,u_n)$ a base for $\mathscr{J}$, $\xi_1,\ldots,\xi_n$ indeterminates. We shall now derive a system $S=\{f\}$ of homogeneous polynomials $f\in F[\xi_1,\ldots,\xi_n]$ such that if E is any extension field of F and $\alpha_i\in E$, $1\leqslant i\leqslant n$, then $a=\sum\alpha_i u_i$ is a reduced element of $\mathscr{J}_E$ if and only if $f(a)=f(\alpha_1,\ldots,\alpha_n)=0$ for all $f\in S$. Consider the element $x=\sum\xi_i u_i$ of $\mathscr{J}_{F(\xi)}$, $F(\xi)=F(\xi_1,\ldots,\xi_n)$. Suppose

$$U_{u_i}u_k=\sum_{l=1}^{n}\gamma_{ikl}u_l,\qquad 1\leqslant i,\,k\leqslant n \tag{4.1}$$

$$U_{u_i,u_j}u_k=\sum_{l=1}^{n}\delta_{ijkl}u_l,\qquad 1\leqslant i,\,j,\,k\leqslant n. \tag{4.2}$$

Since $U_{u_i,u_j}=U_{u_j,u_i}$ and $U_{u_i,u_i}=2U_{u_i}$ we have

$$\delta_{ijkl}=\delta_{jikl},\qquad \delta_{iikl}=2\gamma_{ikl}. \tag{4.3}$$

Now

$$U_x u_k = U_{\sum \xi_i u_i} u_k = \sum_i \xi_i^2 U_{u_i} u_k + \sum_{i<j} \xi_i \xi_j U_{u_i, u_j} u_k = \sum_{l=1}^{n} \zeta_{kl} u_l \tag{4.4}$$

where

$$\zeta_{kl} = \sum_{i=1}^{n} \gamma_{ikl} \xi_i^2 + \sum_{i<j} \delta_{ijkl} \xi_i \xi_j. \tag{4.5}$$

Then $U_x u_k \in F(\xi)\, x$ for all u_k if and only if

$$\xi_j \zeta_{kl} = \xi_l \zeta_{kj}, \qquad 1 \leqslant j, k, l \leqslant n. \tag{4.6}$$

Then if E is any extension field of F and $a = \sum \alpha_i u_i \in \mathscr{J}_E$, $U_a u_k \in Ea$, $1 \leqslant k \leqslant n$, if and only if

$$f_{jkl}(a) = f_{jkl}(\alpha_1, ..., \alpha_n) = 0 \tag{4.7}$$

for all

$$f_{jkl} = \xi_j \zeta_{kl} - \xi_l \zeta_{kj}, \qquad 1 \leqslant j, k, l \leqslant n. \tag{4.8}$$

Since $U_a u_k \in Ea$, $1 \leqslant k \leqslant n$, implies $U_a \mathscr{J}_E \subset Ea$ it is clear that the equations $f_{jkl}(a) = 0$ hold if and only if $a \in \rho(\mathscr{J}_E)$.

The set of homogeneous polynomials $S = \{f_{jkl}\} \subset F[\xi]$ defines a projective F-variety $\mathscr{V}_{\mathscr{J}}$ consisting of the points in the projective space $\mathbb{P}\mathscr{J}_{\bar{F}}$ that are simultaneous zeros of the f_{jkl} where $\bar{F}$ is any algebraically closed field containing F. In general, let S be any set of homogeneous polynomials in $F[\xi]$ then S defines a projective F-variety $\mathscr{V}$ consisting of the points $\bar{F}v$ where $0 \neq v = (\alpha_1, ..., \alpha_n) \in \bar{F}^{(n)}$ that are simultaneous zeros of the set S. The ideal $\mathscr{I}(\mathscr{V})$ of $\mathscr{V}$ is the F-space spanned by the set of homogeneous polynomials in $F[\xi]$ that vanish for every $v \in \mathscr{V}$. $\mathscr{I}(\mathscr{V})$ is the nil radical of any ideal generated by a set of homogeneous polynomials defining $\mathscr{V}$. Thus $\mathscr{I}(\mathscr{V})$ is independent of the choice of the algebraically closed field $\bar{F} \supset F$. Since $\mathscr{I}(\mathscr{V})$ is homogeneous, the *coordinate algebra* $F[\mathscr{V}] \equiv F[\xi]/\mathscr{I}(\mathscr{V})$ is a graded algebra. We recall also that $\mathscr{V}$ is *irreducible* in the sense that it is not the union of two proper F-subvarieties if and only if $\mathscr{I}(\mathscr{V})$ is prime. Then the coordinate algebra $F[\mathscr{V}]$ has a field of fractions $F(\mathscr{V})$. Of greater geometric significance is the field $PF(\mathscr{V})$ of *homogeneous rational functions of degree* 0 on $\mathscr{V}$ defined to be the set of fractions fg^{-1} where f and $g \neq 0$ are homogeneous elements of $F[\mathscr{V}]$ of the same degree. It is readily seen that if V is not vacuous then $F(\mathscr{V})$ is a simple transcendental extension of $PF(V)$.

We shall call $\mathscr{V}_{\mathscr{J}}$ the (*projective*) *variety of reduced elements of* $\mathscr{J}$. Now suppose F is algebraically closed (so we may take $\bar{F} = F$). The generic norm condition implies that if $[\mathcal{J}: F] > 1$ then there exists an $a \neq 0$ in $\mathcal{J}$ that is not invertible. Hence if $\mathcal{J}$ is a division algebra then $\mathcal{J} = F1$. We can use this result to prove.

4.9. THEOREM. *Any simple Jordan algebra over an algebraically closed field is reduced.*

Proof. As we have noted above, if $\mathcal{J}$ is smple then $\mathcal{J}$ is isotopic to an algebra $\mathcal{J}'$ with a capacity. It suffices to show that $\mathcal{J}'$ is reduced. Hence we may suppose that in $\mathcal{J}$ we have $1 = \sum e_i$ where the e_i are orthogonal division idempotents. Since F is algebraically closed every Peirce space $U_{e_i} = Fe_i$. Hence the e_i are reduced elements and $\mathcal{J}$ is reduced. ∎

We can now see that for any $\mathcal{J}$ the projective variety $\mathcal{V}_{\mathcal{J}}$ is not vacuous. We have to show that $\mathcal{J}_{\bar{F}}$ contains reduced elements $\neq 0$. This is clear if $\mathcal{J}_{\bar{F}}$ contains an absolute zero divisor $z \neq 0$ $(U_z = 0)$. Otherwise, $\mathcal{J}$ is semiprimitive and hence, by the structure theory, $\mathcal{J}$ is a direct sum of simple algebras. By 4.9, these contain reduced elements $\neq 0$. Hence $\mathcal{J}$ contains such elements.

If $\mathcal{V}$ is a projective F-variety and F' is an extension field of F then if we assume (as we may) that $\bar{F} \supset F'$, we may regard $\mathcal{V}$ as the F'-variety defined by $\mathcal{I}(\mathcal{V})$. When this is done we shall write $\mathcal{V}_{F'}$ for $\mathcal{V}$. The ideal $I(\mathcal{V}_{F'})$ in $F'[\xi]$ is the nil radical of $F'\mathcal{I}(\mathcal{V})$. We have $F'[\xi] = F' \otimes_F F[\xi]$ and $F'\mathcal{I}(\mathcal{V}) = F' \otimes_F \mathcal{I}(\mathcal{V})$. Hence

$$\mathcal{I}(\mathcal{V}) = F[\xi] \cap F'\mathcal{I}(\mathcal{V}). \tag{4.10}$$

Also it is clear that

$$\mathcal{I}(\mathcal{V}) = F[\xi] \cap \mathcal{I}(\mathcal{V}_{F'}). \tag{4.11}$$

Equation (4.11) implies that the F-subalgebra

$$(F[\xi] + \mathcal{I}(\mathcal{V}_{F'}))/\mathcal{I}(\mathcal{V}_{F'}) \tag{4.12}$$

of the coordinate algebra $F'[\xi]/\mathcal{I}(\mathcal{V}_{F'})$ is isomorphic to the coordinate algebra $F[\xi]/\mathcal{I}(\mathcal{V})$ of $\mathcal{V}$. Thus we have a canonical imbedding of the coordinate algebra $F[\mathcal{V}]$ of $\mathcal{V}$ in the coordinate algebra $F'[\mathcal{V}_{F'}]$ of $\mathcal{V}_{F'}$ and $F[\mathcal{V}_{F'}] = F'(F[\mathcal{V}])$. It is clear that if $\mathcal{V}_{F'}$ is irreducible then $\mathcal{V}$ is irreducible. In particular, this holds if F' is the algebraic closure of F. In this case we say that $\mathcal{V}$ is *absolutely irreducible.*

The F-variety defined by the ideal 0 of $F[\xi] = F[\xi_1, \ldots, \xi_n]$ is the projective space $\mathbb{P}\bar{F}^{(n-1)}$. This is absolutely irreducible with field of rational functions the subfield $PF(\xi)$ of $F(\xi)$ of fractions $f(\xi)\,g(\xi)^{-1}$ where $f(\xi)$,

$g(\xi)$ are homogeneous polynomials of the same degree. We shall show that if $\mathscr{J}$ is reduced and outer central simple then $\mathscr{V}_{\mathscr{J}}$ is birationally equivalent to a projective space. For this purpose we shall use some results on a class of operators that have appeared in other parts of the Jordan theory. If a and b are elements of a Jordan algebra $\mathscr{J}$, we define

$$B_{a,b} = 1 - V_{a,b} + U_a U_b. \tag{4.13}$$

Then we have the identities

$$B_{a,b} B_{-a,b} = B_{a,b} B_{a,-b} = B_{a,U_b a} = B_{U_a b,b} \tag{4.14}$$

$$B_{a,b} U_c B_{b,a} = U_{B_{a,b}c} \tag{4.15}$$

$$B_{a,b} B_{b,a} = 1 - V_c + U_c \qquad \text{where } c = a \circ b - U_a b^2 (V_c = U_{c,1}). \tag{4.16}$$

(See [13, Sect. 3.7] where $B_{a,b}$ is denoted as $T_{a,b}$.) It follows from (4.15) that if c is reduced then so is $B_{a,b}c$ for any a, b.

Let e be an idempotent, $\mathscr{J} = \mathscr{J}_1 \oplus \mathscr{J}_{1/2} \oplus \mathscr{J}_0$ the Peirce decomposition of $\mathscr{J}$ relative to e. If $a \in \mathscr{J}_{1/2}$ then

$$B_{a,e} B_{-a,e} = B_{a,U_e a} = B_{a,0} = 1_{\mathscr{J}}. \tag{4.17}$$

Hence $B_{a,e}$ is invertible with $B_{a,e}^{-1} = B_{-a,e}$.

Moreover, $V_{a,e}e = U_{a,e}e = V_{e,e}a = V_e a$ (Weak Macdonald Theorem, p. 3.10 of [13]) $= a$ (PD5, p. 5.5 of [13]). Hence $B_{a,e}e = e - a + U_a e$ and $U_a e \in \mathscr{J}_0$ by the PD theorem [13, Sect. 5.1]. Hence

$$B_{a,e}e = e - a + U_a e, \qquad U_a e \in \mathscr{J}_0. \tag{4.18}$$

If $a, b \in \mathscr{J}_{1/2}$ then $B_{a,e}b = b - V_{a,e}b + U_a U_e b = b - V_{a,e}b = b - U_{a,b}e$ and $U_{a,b}e \in \mathscr{J}_0$. Thus

$$B_{a,e}b = b - U_{a,b}e, \qquad U_{a,b}e \in \mathscr{J}_0. \tag{4.19}$$

Since $V_{a,e}c = \{aec\} = 0$

$$B_{a,e}c = c, \qquad a \in \mathscr{J}_{1/2},\ c \in \mathscr{J}_0. \tag{4.20}$$

Finally, we note that $U_{e+c}e = U_e e + U_{e,c}e + U_c e = e$. Hence if e is reduced and $c \in \mathscr{J}_0$ then

$$e + c \in \rho(\mathscr{J}) \Leftrightarrow c = 0. \tag{4.21}$$

We can now prove

4.22. Lemma (Faulkner). *If e is a reduced idempotent in a Jordan* *algebra and* $\mathscr{J} = \mathscr{J}_1 \oplus \mathscr{J}_{1/2} \oplus \mathscr{J}_0$ *the corresponding Peirce decomposition then* $u = e + a + c$, *where* $a \in \mathscr{J}_{1/2}$, $c \in \mathscr{J}_0$, *is in* $\rho(\mathscr{J})$ *if and only if* $c = U_a e$, *in which case* $u = B_{-a,e} e$.

Proof. Suppose first that $u \in \rho(\mathscr{J}) \ni B_{a,e} u = B_{a,e}(e + a + c) = e - a + U_a e + a - U_{a,a} e + c$ (by (4.18)–(4.20)) $= e + c'$, $c' = U_a e - U_{a,a} e + c \in \mathscr{J}_0$. Then $c' = U_a e - 2U_a e + c = 0$ by (4.21). Thus $c = U_a e$ and by (4.17) $u = B_{-a,e} e$. Conversely suppose $c = U_a e$. Then $B_{-a,e} e = e + a + U_{-a} e = e + a + U_a e = u$ and since $e \in \rho(\mathscr{J})$, $u \in \rho(\mathscr{J})$. ∎

Let $G = \operatorname{Str} \mathscr{J}$. By (1.10), $\eta \in G$ if and only if η is bijective linear and there exists a bijective linear transformation η^* of $\mathscr{J}$ such that $U_{\eta(a)} = \eta U_a \eta^*$ for all $a \in \mathscr{J}$. Putting $a = 1$ we obtain $U_{\eta(1)} = \eta\eta^*$ so $\eta^* = \eta^{-1} U_{\eta(1)}$. Since $\eta^{-1} = (\det \eta)^{-1} \eta^{\#}$ where $\eta^{\#}$ is the adjoint of the linear transformation the conditions on η become $(\det \eta) U_{\eta(a)} = \eta U_a \eta^{\#} U_{\eta(1)}$. These are equivalent to a set of polynomial equations on the coordinates of the matrix of η relative to a base for $\mathscr{J}/F$. It follows that G is a linear algebraic group. Let G_0 be the component of 1 of G. The map $a \rightsquigarrow U_a$ is a continuous (in the Zariski topology) map of the open subset of $\mathscr{J}$ defined by $N(a) \neq 0$ into $\operatorname{Str} \mathscr{J}$. Since the image contains $1_{\mathscr{J}}$ it follows that every U_a for a invertible is contained in G_0. If e is an idempotent and $a \in \mathscr{J}_{1/2}(e)$ then we have seen that $B_{a,e}$ is invertible. Then $B_{a,e} \in \operatorname{Str} \mathscr{J}$ by (4.15). The argument used for the U_a implies that these $B_{a,e} \in G_0$. We can apply these results to prove

4.23. THEOREM. *Let* $\mathscr{J}$ *be simple over an algebraically closed field and let* G_0 *be the component of* 1 *of* $G = \operatorname{Str} \mathscr{J}$. *Then* G *stabilizes* $\rho(\mathscr{J})^* = \rho(\mathscr{J}) \setminus \{0\}$ *and* G_0 *acts transitively on* $\rho(\mathscr{J})^*$.

Proof. It is clear from the condition $U_{\eta(a)} = \eta U_a \eta^*$ for $\eta \in \operatorname{Str} \mathscr{J}$ that $\rho(\mathscr{J})^*$ is stabilized by $\operatorname{Str} \mathscr{J}$. Now $\mathscr{J}$ contains a reduced idempotent $e \neq 0$. To prove the second assertion we have to show that $G_0 e = \rho(\mathscr{J})^*$. Since G_0 contains the set F^* of non-zero scalar multiplications and every $B_{a,e} a \in \mathscr{J}_{1/2}(e)$ it follows from Faulkner's lemma that $G_0 e$ contains every $u = \alpha e + a + c \in \rho(\mathscr{J})^*$ such that $\alpha \in F^*$, $a \in \mathscr{J}_{1/2}$, $c \in \mathscr{J}_0$. Now let v be any non-zero element of $\rho(\mathscr{J})$. We claim that there exists a k and invertible $a_1, \ldots, a_k$ such that $U_{a_1} \cdots U_{a_k} v \notin \mathscr{J}_{1/2} + \mathscr{J}_0$. Otherwise, by Zariski density, for all k and $a_1, \ldots, a_k \in \mathscr{J}$, $U_{a_1} \cdots U_{a_k} v \in \mathscr{J}_{1/2} + \mathscr{J}_0$. Then v generates a proper outer ideal in $\mathscr{J}$, contrary to outer simplicity. Hence we have invertible a_i such that $U_{a_1} \cdots U_{a_k} v \in G_0 e$. Since $U_{a_1} \cdots U_{a_k} \in G_0$, $v \in G_0 e$ and $\rho(\mathscr{J})^* = G_0 e$. ∎

This result implies that if F is algebraically closed then we have an induced transitive action of G_0 on $\mathscr{V}_{\mathscr{J}}$. Now it is well known that if we have a transitive action of an irreducible algebraic group on an algebraic variety $\mathscr{V}$ and the action is defined by polynomial maps then $\mathscr{V}$ is irreducible (see, e.g., [8, p. 35]). Hence (4.23) implies

4.24. THEOREM. *Let $\mathscr{J}$ be outer central simple (= outer simple and outer central) over a field F then $\mathscr{V}_{\mathscr{J}}$ is absolutely irreducible.*

Again, as in Faulkner's lemma, let $\mathscr{J}$ be a Jordan algebra containing a reduced idempotent e and let $\mathscr{J} = \mathscr{J}_1 \oplus \mathscr{J}_{1/2} \oplus \mathscr{J}_0$ be the corresponding Peirce decomposition. Then if $\bar{F}$ is an algebraically closed field containing F we have the Peirce decomposition $\mathscr{J}_{\bar{F}} = \mathscr{J}_{1\bar{F}} \oplus \mathscr{J}_{1/2\bar{F}} \oplus \mathscr{J}_{0\bar{F}}$. It follows from the lemma that for $\alpha \in \bar{F}^*$ and $a \in \mathscr{J}_{1/2\bar{F}}$, $c \in \mathscr{J}_{0\bar{F}}$, $\alpha e + a + c \in \rho(\mathscr{J}_{\bar{F}})$ if and only if $c = \alpha^{-1} U_a e$. Now consider the projective space of one-dimensional subspaces of the vector space $\mathscr{J}_{1\bar{F}} \oplus \mathscr{J}_{1/2\bar{F}}$. We can identify this with $\mathbb{P}\bar{F}^{(r)}$ where $r = [\mathscr{J}_{1/2}: F]$. We have the open subset O of $\mathbb{P}\bar{F}^{(r)}$ of points $\bar{F}(\alpha e + a)$, $\alpha \neq 0$, and the open subset O' of points $\bar{F}(\alpha e + a + \alpha^{-1} U_a e)$. Now we have

4.25. THEOREM. *The map $\bar{F}(\alpha e + a) \rightsquigarrow \bar{F}(\alpha^2 e + \alpha a + U_a e)$ is bijective regular defined over F of O onto O′ with regular inverse.*

Proof. It is clear that η is bijective. In terms of a suitable choice of homogeneous coordinates the map is given by homogeneous quadratic polynomials with coefficients in F. Hence it is regular and defined over F. Its inverse is a projection map so this is also regular and defined over F. ∎

We can now prove the main theorem of this section.

4.26. THEOREM. *Let $\mathscr{J}$ be reduced outer simple over F. Then $\mathscr{V}_{\mathscr{J}}$ is birationally equivalent to the projective space $\mathbb{P}\bar{F}^{(r)}$ where r is determined in the following way: Let $\mathscr{J}'$ be an isotope of $\mathscr{J}$ containing a reduced idempotent $e \neq 0$, $\mathscr{J}' = \mathscr{J}'_1 \oplus \mathscr{J}'_{1/2} \oplus \mathscr{J}'_0$ the corresponding Peirce decomposition. Then $r = [\mathscr{J}'_{1/2}: F]$.*

Proof. By 4.25 we have an F-isomorphism of an open subset of $\mathscr{V}_{\mathscr{J}}$ $(= \mathscr{V}_{\mathscr{J}'})$ onto an open subset of $\mathbb{P}\bar{F}^{(r)}$. Since $\mathscr{V}_{\mathscr{J}}$ and $\mathbb{P}\bar{F}^{(r)}$ are irreducible, they are birationally equivalent (see, e.g., [5, p. 26]). ∎

The irreducibility of $\mathscr{V}_{\mathscr{J}}$ for $\mathscr{J}$ outer central simple implies that we can define the field $\mathbb{P}F(\mathscr{V}_{\mathscr{J}})$ which is the field of rational functions on $\mathscr{V}_{\mathscr{J}}$. Since birationally equivalent irreducible projective varieties have isomorphic fields of rational functions it follows that if $\mathscr{J}$ is reduced and simple then $\mathbb{P}F(\mathscr{V}_{\mathscr{J}})$ is rational (= purely transcendental) over F of transcendency degree r given by Theorem 4.26.

We shall give next another isomorphism between open subsets of $\mathscr{V}_{\mathscr{J}}$ and of $\mathbb{P}\bar{F}^{(r)}$ for $\mathscr{J}$ of degree $\geqslant 3$ (the one we used originally in the proof of 4.26) which will give additional information on the relation between these varieties. Suppose $\mathscr{J}=H(M_m(D), F)$ where $m\geqslant 3$ and $(\mathscr{D}, j)$ is a composition algebra (including the case in which char $F=2$ and $\mathscr{D}$ is a purely inseparable extension of exponent 1 of F). Suppose first that $\mathscr{D}$ is associative. Consider the row vector $d=(d_1,..., d_m)$, $d_i\in\mathscr{D}$, which we can identify with the matrix $\sum_{i=1}^m d_ie_{1i}$. Then $d^*=\sum_1^m \bar{d}_ie_{i1}$. If $c=(c_1,..., c_m)=\sum c_ie_{1i}$, $c_i\in\mathscr{D}$, we define

$$\langle d, c\rangle=\sum d_i\bar{c}_i. \tag{4.27}$$

Then $dc^*=\langle d, c\rangle e_{11}$ and

$$d^*d=\sum_{i,j=1}^{m} \bar{d}_id_je_{ij}. \tag{4.28}$$

In particular,

$$\begin{aligned}(\bar{d}e_{i1})(de_{1i})&=n(d)\,e_{ii}\\ (e_{i1}+\bar{d}e_{j1})(e_{1i}+de_{1j})&=e_{ii}+n(d)\,e_{ji}+de_{ij}+\bar{d}e_{ji}.\end{aligned} \tag{4.29}$$

It follows that the F-span of the matrices d^*d is $\mathscr{J}$ and since

$$U_{d^*d}c^*c=d^*dc^*cd^*d=n(\langle d, c\rangle)\,d^*d, \tag{4.30}$$

every $d^*d\in\rho(\mathscr{J})$.

The same calculations apply to $\mathscr{J}_{\bar{F}}=H(M_m(\mathscr{D}), \bar{F})$ for any algebraically closed extension field $\bar{F}$ of F (even when $\mathscr{D}_{\bar{F}}$ is not a composition algebra). Consider the projective space $\mathbb{P}F^{(m-1)\delta}$, $\delta=[\mathscr{D}: F]$. We can identify this with the set of one-dimensional subspaces $\bar{F}d$ where $d=(\beta, d_2,..., d_m)\neq 0$, $\beta\in\bar{F}$, $d_i\in\mathscr{D}_{\bar{F}}$. Let O be the open subset of $\mathbb{P}\bar{F}^{(m-1)\delta}$ defined by $\beta\neq 0$ and let O_1 be the open subset of $\mathscr{V}_{\mathscr{J}}$ consisting of the $\bar{F}h$ where $h=\sum h_{ij}e_{ij}\in\rho(\mathscr{J}_{\bar{F}})$ and $h_{11}\neq 0$. If $\bar{F}d\in O$ then $\bar{F}d^*d\in O_1$ and the map

$$\eta\colon \bar{F}d\rightsquigarrow\bar{F}d^*d \tag{4.31}$$

is injective. Also η is surjective on O_1. For, if $h=(h_{ij})\in\rho(\mathscr{J})$ and $h_{11}\neq 0$ then $U_he_{11}=he_{11}h=\sum h_{i1}h_{1j}e_{ij}\in\bar{F}h$ and since $h_{11}\neq 0$, $U_he_{11}=h_{11}h$. Thus $h_{11}h=d^*d$ where $d=\sum h_{1i}e_{1i}$. Hence η is surjective on O_1. Moreover, η is regular since it is given coordinatewise by homogeneous polynomials of the same degree (see, e.g., [5, p. 15]). Also η^{-1} is regular since it is defined by the map of a matrix into its first row. Hence η is an F-isomorphism of O onto O_1.

Next let $\mathcal{J} = H(M_3(\mathbb{O}), F)$. We now denote an element $h \in \mathcal{J}$ as

$$h = \sum_1^3 \alpha_i e_{ii} + a_1[23] + a_2[31] + a_3[12] \tag{4.32}$$

where $a_i[jk] = a_i e_{jk} + \bar{a}_i e_{kj}$. Then we have the formula (1.3) for the generic norm and the adjoint of h is

$$h^\# = \sum_{(123)} (\alpha_i \alpha_j - n(a_k))\, e_{kk} + (\overline{a_i a_j} - \alpha_k a_k)[ij] \tag{4.33}$$

where $\sum_{(123)}$ denotes summation on the cyclic permutation of (1, 2, 3). The U operator is defined by (1.8): $U_h k = T(h, k)\, h - h^\# \times k$ where

$$(h \times k = (h+k)^\# - h^\# - k^\#) \quad \text{and} \quad T(h, k) = (\Delta_1^h N)(\Delta_1^k N) - \Delta_1^h(\Delta^k N).$$

Let $\rho'(\mathcal{J}) = \{h \in \mathcal{J} \mid h^\# = 0\}$, $\rho''(\mathcal{J}) = \{a^*a \mid a = \sum_{i=1}^3 a_i e_{1i}, a_i \in \mathbb{O}\}$. Then $a^*a = \sum n(a_i)\, e_{ii} + \sum_{(123)} b_i[jk]$, $b_i = \bar{a}_j a_k$. Then $a^*a \in \mathcal{J}$ and $(a^*a)^\# = 0$ if the a_i are in an associative subalgebra of $\mathbb{O}$. Also it is clear from (1.8) that $\rho'(\mathcal{J}) \subset \rho(\mathcal{J})$. Let O_1 be the open subset of $\mathcal{V}_{\mathcal{J}}$ defined by $\alpha_1 \neq 0$. Let $h \in O_1$ and consider $U_h e_{11}$. Since $h = \sum \alpha_i e_{ii} + \sum_{(123)} a_i[jk]$, $U_h e_{11}$ is a linear combination of terms of the form $U_{e_{ii}} e_{11}$, $U_{e_{ii}, e_{jj}} e_{11}$, $U_{a_i[jk]} e_{11}$, $U_{e_{pp}, a_i[jk]} e_{11}$, $U_{a_i[jk], a_p[qr]} e_{11}$. Since O is alternative the entries in each of these Jordan products are contained in associative subalgebras. Hence for each of these products we can apply the formulas $U_y e_{11} = y e_{11} y$, $U_{y,z} e_{11} = y e_{11} z + z e_{11} y$ to show that $U_h e_{11} = \sum \bar{h}_{1i} h_{1j} e_{ij}$ where $(h_{11}, h_{12}, h_{13}) = (\alpha_1, a_3, \bar{a}_2)$. Then $h_{11} h = d^*d$ for $d = (h_{11}, h_{12}, h_{13})$. Hence $h \in \rho''(\mathcal{J})$. It now follows as in the proof in the associative case that if we identify $\mathbb{P}\bar{F}^{(16)}$ with the set of one-dimensional subspaces $\bar{F}d$ where $d = (\beta, d_2, d_3)$, $\beta \in \bar{F}$, $d_i \in \mathbb{O}_{\bar{F}}$, and define the open subset O of $\mathbb{P}\bar{F}^{(16)}$ by $\beta \neq 0$ then (4.31) defines an F-isomorphism of O onto O_1.

We prove next

4.34. Theorem. *Let $\mathcal{J} = H(M_m(\mathcal{D}), F)$ where $\mathcal{D}$ is an associative composition algebra and $m \geqslant 3$. Let $d = (d_1, \ldots, d_m) = \sum d_i e_{1i}$ where $d_i \in \mathcal{D}$. Then every element of $\rho(\mathcal{J})$ has the form αd^*d.*

Proof. We note first that the set $\mathcal{V} = \{\alpha d^*d \mid \alpha \in F,\ d = (d_1, \ldots, d_m)\}$ is stabilized by Str. $\mathcal{J}$. If $\mathcal{D} = F$ or a quaternion algebra then it is known that Str $\mathcal{J}$ is the set of maps $x \rightsquigarrow \alpha a^*xa$ where $\alpha \in F^*$ and a is an invertible element of $M_m(\mathcal{D})$ [14, pp. 133–134]. Then $\alpha a^*d^*\, da = \alpha(da)^*\, da \in \mathcal{V}$. If $\mathcal{D}$ is a quadratic composition algebra then Str $\mathcal{J}$ is the set of maps $x \rightsquigarrow \alpha a^*xa$ and the set of maps $x \rightsquigarrow \alpha a^*\bar{x}a$ where $\alpha \in F^*$ and a is invertible in $M_m(\mathcal{D})$. Then $\alpha a^*\bar{d}^*\, \bar{d}a = \alpha(\bar{d}a)^*\, \bar{d}a \in \mathcal{V}$ so the result is clear in this case also.

We have seen that Str $\mathcal{J}$ acts transitively on $\rho(\mathcal{J})$. Since $\mathcal{V} \subset \rho(\mathcal{J})$ and $\mathcal{V}$ is stabilized by Str $\mathcal{J}$ it follows that $\mathcal{V} = \rho(\mathcal{J})$. ∎

We can now strengthen Theorem 4.26 in the special case in which $\mathscr{J} = H(M_m(F))$ by replacing birational equivalence by isomorphism.

4.35. THEOREM. *If $\mathscr{J} = H(M_m(F))$ then $\mathscr{V}_{\mathscr{J}}$ is F-isomorphic to the projective space $\mathbb{P}\bar{F}^{(m-1)}$.*

Proof. We shall show that $\eta: \bar{F}d \rightsquigarrow \bar{F}d^*d$ is an F-isomorphism of $\mathbb{P}\bar{F}^{(m-1)}$ onto $\mathscr{V}_{\mathscr{J}}$. The surjectivity of η follows from Theorem 4.34. Now suppose $d^*d = c^*c \neq 0$. Then $d_i \neq 0$ for some i, $1 \leqslant i \leqslant m$. Then the (i, i)-entry of d^*d is $d_i^2 \neq 0$. Hence $c_i \neq 0$ also and 4.28 implies that $\bar{F}d = \bar{F}c$. Hence η is injective so η is bijective. Also it is clear that η and η^{-1} are regular and defined over F. Hence η is an F-isomorphism. ∎

We shall establish next another isomorphism theorem for projective varieties which in a different form is well known.

4.36. THEOREM. *Let $\mathscr{J} = (\mathrm{End}_F V)^{(q)}$ where V is any m-dimensional vector space over F. Then $\mathscr{V}_{\mathscr{J}} \cong \mathbb{P}\bar{F}^{(m-1)} \times \mathbb{P}\bar{F}^{(m-1)}$.*

We remark that $\mathscr{J} \cong M_m(F)^{(q)} \cong H(M_m(\mathscr{D}), F)$ where $\mathscr{D} = F1 \oplus F1$ and j is the exchange involution.

Proof. Let $V^* = \mathrm{Hom}_F(V, F)$. We have the standard isomorphism of $V^* \otimes V$ onto $\mathrm{End}_F V$ ($\cong M_m(F)$) sending $u^* \otimes v$, $u^* \in V^*$, $v \in V$, into

$$l_{u^*,v}: x \rightsquigarrow \langle u^*, x\rangle v \tag{4.37}$$

where $\langle u^*, x\rangle = u^*(x)$. Then we have

$$(u_1^* \otimes v_1)(u_2^* \otimes v_2) = \langle u_1^*, v_2\rangle (u_2^* \otimes v_1). \tag{4.38}$$

Hence

$$(u^* \otimes v)(x^* \otimes y)(u^* \otimes v) \in F(u^* \otimes v) \tag{4.39}$$

so $u^* \otimes v \in \rho(\mathscr{J})$. Conversely, suppose $l = \sum_1^r u_i^* \otimes v_i \in \rho(\mathscr{J})$. We may suppose the u_i^* are linearly independent and the v_i are independent. Then $r \leqslant m$ and there exist $u_i \in V$, $v_i^* \in V^*$ such that

$$\langle u_i^*, u_j\rangle = \delta_{ij} = \langle v_i^*, v_j\rangle.$$

Hence

$$\left(\sum u_i^* \otimes v_i\right)(v_k^* \otimes u_k)\left(\sum u_i^* \otimes v_i\right) = u_k^* \otimes v_k.$$

This implies that $r = 1$. Hence $\rho(\mathscr{J}) = \{u^* \otimes v \mid u^* \in V^*, v \in V\}$.

Similarly if $\bar{F}$ is an algebraically closed field containing F then $\mathscr{J}_{\bar{F}} = \mathrm{End}_{\bar{F}} V_{\bar{F}}$ and $\rho(\mathscr{J}_{\bar{F}}) = \{u^* \otimes v \mid u^* \in V_{\bar{F}}^*, v \in V_{\bar{F}}\}$. We can identity $\mathbb{P}\bar{F}^{(m-1)}$ with the set of one-dimensional subspaces $\bar{F}u^*$ of $V_{\bar{F}}^*$ and with the set of one-dimensional subspaces $\bar{F}v$ of V. Then we have the surjective map of $\mathbb{P}\bar{F}^{(m-1)} \times \mathbb{P}\bar{F}^{(m-1)}$ onto $\mathscr{V}_{\mathscr{J}}$ such that

$$\eta\colon (\bar{F}u^*, \bar{F}v) \rightsquigarrow \bar{F}(u^* \otimes v). \tag{4.40}$$

Since $u_1^* \otimes v_1$ and $v_2^* \otimes v_2$ are linearly independent if u_1^* and u_2^* are linearly independent and v_1 and v_2 are linearly independent it follows that η is injective. It is also clear that η and η^{-1} are regular and defined over F. Hence η is an isomorphism of the projective varieties. ∎

5. Reducing Fields and Generic Reducing Fields

We recall that if E and E' are fields then a place $\mathscr{P}$ on E to E' is a homomorphism of a subring R of E into E' such that if $a \in E$ and $a \notin R$ then $a^{-1} \in R$ and $\mathscr{P}(a^{-1}) = 0$. R is a valuation ring in E in the sense that $E = R \cup (R^*)^{-1}$ where $(R^*)^{-1} = \{r^{-1} \mid r \neq 0 \text{ in } R\}$. R is a local ring and if P is its maximal ideal and U is the multiplicative group of invertible elements of R then $\Gamma = E^*/U$ can be regarded as an ordered group in which the subset P^*U/U is the set of elements <1. Moreover, if we take $V = \Gamma \cup \{0\}$ then we obtain a V-valuation φ of E by defining

$$\varphi(0) = 0, \qquad \varphi(a) = aU \in \Gamma \qquad \text{if} \quad a \neq 0. \tag{5.1}$$

(See, e.g., [12, Sect. 9.7].) If E and E' are extensions of the same field F then $\mathscr{P}$ is called an F-place if $F \subset R$ and $\mathscr{P}$ is an F-algebra homomorphism.

We can now give the following

5.2. Definition. Let $\mathscr{J}$ be outer central simple over F. Then an extension field E/F is called a *reducing field* for $\mathscr{J}$ if $\mathscr{J}_E$ is reduced. E is called a *generic reducing field* if it has the property that an extension E'/F is a reducing field for $\mathscr{J}$ if and only if there exists an F-place on E to E'.

If $\mathscr{J}$ is outer central simple then $\mathscr{J}_E$ is simple for any extension E/F. Hence $\mathscr{J}_E$ is reduced if and only if it contains a reduced element $v \neq 0$. It is clear from the definition of $\mathscr{V}_{\mathscr{J}}$ that this is the case if and only if $\mathscr{V}_{\mathscr{J}}$ has an E-rational point. Hence we have

5.3. Theorem. *E/F is a reducing field for an outer central simple Jordan algebra $\mathscr{J}$ if and only if $\mathscr{V}_{\mathscr{J}}$ has an E-rational point.*

We can use this criterion to prove

5.4. THEOREM. *If E/F is a reducing field for the outer central simple $\mathcal{J}$ and there exists an F-place on E to F'/F then F' is a reducing field for $\mathcal{J}$.*

Evidently this follows from 5.3 and the following

5.5. LEMMA. *Let $\mathcal{V}$ be a projective F-variety. Suppose:* (1) *$\mathcal{V}$ has an E-rational point for the field E/F, and* (2) *there is an F-Place $\mathcal{P}$ on E to E'/F. Then $\mathcal{V}$ has an E'-rational point.*

Proof. Let $(\alpha_1, \ldots, \alpha_n)$ be the homogeneous coordinates of an E-rational point of $\mathcal{V}$ and let R be the valuation ring of $\mathcal{P}$. We can divide by an α_i with the biggest valuation to obtain one $\alpha_i = 1$ and the rest in R. Then $(\mathcal{P}\alpha_1, \ldots, \mathcal{P}\alpha_n)$ is an E'-rational point. ∎

An immediate consequence of 5.4 is

5.6. COROLLARY. *If $F(\xi_1, \ldots, \xi_r)$, ξ_i indeterminates, is a reducing field for $\mathcal{J}$ then $\mathcal{J}$ is reduced.*

Proof. This follows from 5.4 since we have an F-place on $F(\xi_1, \ldots, \xi_r)$ to F (see Exercises 2–4 on p. 561 of [12]). ∎

We can now prove

5.7. THEOREM. *If $\mathcal{J}$ is outer central simple the field $PF(\mathcal{V}_{\mathcal{J}})$ of rational functions on $\mathcal{V}_{\mathcal{J}}$ is a generic reducing field for $\mathcal{J}$. Moreover, $\mathcal{J}$ is reduced if and only if $PF(\mathcal{V}_{\mathcal{J}})$ is rational over F.*

Proof. Consider the coordinate algebra $F[\mathcal{V}_{\mathcal{J}}] = F[\xi]/\mathcal{I}(\mathcal{V}_{\mathcal{J}})$, and the field $PF(\mathcal{V}_{\mathcal{J}})$. It is clear that $(\xi_1 + \mathcal{I}(\mathcal{V}_{\mathcal{J}}), \ldots, \xi_n + \mathcal{I}(\mathcal{V}_{\mathcal{J}})$, is an $\mathcal{F}(\mathcal{V}_{\mathcal{J}})$-rational point of $\mathcal{V}_{\mathcal{J}}$. Hence $PF(\mathcal{V}_{\mathcal{J}})$ is a reducing field for $\mathcal{J}$. Thus $\mathcal{J}_{F(\mathcal{V}_{\mathcal{J}})}$ is reduced. On the other hand, $PF(\mathcal{V}_{\mathcal{J}})$ is the subfield of $F(\mathcal{V}_{\mathcal{J}})$ of fractions fg^{-1} where $f, g \in F[\mathcal{V}_{\mathcal{J}}]$ and are homogeneous of the same degree. Now $\mathcal{J}_{F(\mathcal{V}_{\mathcal{J}})} = (\mathcal{J}_{PF(\mathcal{V}_{\mathcal{J}})})_{F(\mathcal{V}_{\mathcal{J}})}$ and since $F(\mathcal{V}_{\mathcal{J}})$ is a simple transcendental extension of $PF(\mathcal{V}_{\mathcal{J}})$ it follows from Corollary 5.6 that $\mathcal{J}_{PF(\mathcal{V}_{\mathcal{J}})}$ is reduced. Hence $PF(\mathcal{V}_{\mathcal{J}})$ is a reducing field and hence, by 5.4, if we have an F-place on $PF(\mathcal{V}_{\mathcal{J}})$ to a field E'/F then E' is a reducing field. Conversely, suppose E'/F is a reducing field. Then we have a canonical imbedding of the coordinate algebra of $\mathcal{V}_{\mathcal{J}}$ into the coordinate algebra of $\mathcal{V}_{\mathcal{J}F'}$ and this can be extended to an imbedding of the field $PF(\mathcal{V}_{\mathcal{J}})$ into the field $PE'(\mathcal{V}_{\mathcal{J}_{E'}})$. On the other hand, since $\mathcal{J}_{F'}$ is reduced $PF'(\mathcal{V}_{\mathcal{J}_{F'}})$ is rational over F'. Then we have an E'-place on $PE'(\mathcal{V}_{\mathcal{J}_{E'}})$ to F'. The restriction of this place to $PF(\mathcal{V}_{\mathcal{J}})$ $(\subset PE(\mathcal{V}_{\mathcal{J}_{E'}}))$ is an F-place on $PF(\mathcal{V}_{\mathcal{J}})$ to E'. This concludes the proof of the first statement.

We have seen before that if $\mathscr{J}$ is reduced then $PF(\mathscr{V}_{\mathscr{J}})$ is rational over F. Conversely, suppose $PF(\mathscr{V}_{\mathscr{J}})$ is rational. Since $PF(\mathscr{V}_{\mathscr{J}})$ is a reducing field it follows from 5.6 that $\mathscr{J}$ is reduced. ∎

We shall prove next an analogue for Jordan algebras of the classical existence theorem for separable splitting fields of central simple associative algebras.

5.8. THEOREM. *Any finite-dimensional outer central simple Jordan algebra $\mathscr{J}$ has a finite-dimensional separable reducing field.*

Proof. As we shall indicate in Section 7, it follows from the structure theory that if the base field F is finite then $\mathscr{J}$ is reduced. Hence we may assume F is infinite. Let m be the degree of $\mathscr{J}$. Let $(u_1,..., u_n)$ be a base for $\mathscr{J}/F$, $x=\sum_1^n \xi_i u_i$ where the ξ_i are indeterminates, and let $\delta(x)=\delta(\xi_i,..., \xi_n)$ be the discriminant of $m_x(\lambda)$, the minimum polynomial of $x\in \mathscr{J}_{F(\xi)}$. We claim that $\delta(x)\neq 0$. To see this let $\bar{F}$ be an algebraically closed extension of F. Then $\mathscr{J}_{\bar{F}}$ is reduced and hence by the structure theorems of Section 2, $\mathscr{J}_{\bar{F}}$ contains m orthogonal idempotents $e_i\neq 0$. Let $a=\sum \alpha_i e_i$ where the α_i are distinct elements of $\bar{F}$. Then the minimum polynomial $\mu_a(\lambda)$ of a is $f(\lambda)=\prod(\lambda-\alpha_i)$ so $m_a(\lambda)=\mu_a(\lambda)$. Hence if $a=\sum \beta_i u_i$, $\beta_i\in\bar{F}$, then $\delta(\beta_1,..., \beta_n)\neq 0$ so $\delta(x)\neq 0$. Since F is infinite we can choose $\gamma_i\in F$ so that if $c=\sum \gamma_i u_i\in\mathscr{J}$ then $\delta(c)\neq 0$. Then $m_c(\lambda)=\mu_c(\lambda)$ has distinct roots in $\bar{F}$. Now let F_s be the separable algebraic closure of F and consider c as element of $\mathscr{J}_{F_s}$. Since $\mu_c(\lambda)$ has distinct roots in F_s, the subalgebra $F_s[c]$ of $\mathscr{J}_{F_s}$ and hence $\mathscr{J}_{F_s}$ contain m non-zero orthogonal idempotents. This implies that $\mathscr{J}_{F_s}$ is reduced. Let $d=\sum \delta_i u_i$ be a reduced element of $\mathscr{J}_{F_s}$ and let $E=F(\delta_1,..., \delta_n)$. Then E is finite dimensional separable over F and $\mathscr{J}_E$ contains the reduced element d. Hence $\mathscr{J}_E$ is reduced. ∎

A projective variety $\mathscr{V}$ defined over F is called a *Brauer–Severi variety* if there exists a separable algebraic extension E/F such that $\mathscr{V}_E$ is isomorphic to a projective space. Hence Theorem 5.8 and Theorem 4.35 imply that if $\mathscr{J}=H(\mathscr{A}, J)$ where $\mathscr{A}$ is central simple with involution J of orthogonal type then $\mathscr{V}_{\mathscr{J}}$ is a Brauer–Severi variety.

It is well-known result of the associative theory that if $\mathscr{A}$ is central simple over F and E is a subfield such that $[E\colon F]=\deg\mathscr{A}$ then E is a splitting field. We shall now prove an analogue of this result for separable subfields that will be needed in the next section.

5.9. THEOREM. *Let $\mathscr{J}$ be an outer central simple Jordan algebra over F. Suppose E is a separable extension field of F such that $[E\colon F]=m=\deg\mathscr{J}$ and $E^{(q)}$ is isomorphic to a subalgebra of $\mathscr{J}$. Then E is a reducing field for $\mathscr{J}$.*

Proof. We have $E=F(\rho)$ where ρ has minimum polynomial $f(\lambda)$ of degree m with distinct roots. Also $\mathscr{J}$ contains an element a with minimum polynomial $f(\lambda)$. Let $f_i(\lambda)=f(\lambda)/(\lambda-\rho)\in E[\lambda]$ and put

$$e_1=f(a)/f'(\rho) \tag{5.10}$$

where $f'(\lambda)$ is the formal derivative of $f(\lambda)$. We claim that e_1 is a reduced idempotent in $\mathscr{J}_E$. To see this we consider the separable algebraic closure F_s which we assume contains E. Then $f(\lambda)=\prod(\lambda-\rho_i)=\prod_{j\neq i}(\lambda\neg\rho_j)$ in $F_s[\lambda]$ with distinct ρ_i and $\rho_1=\rho$. By elementary algebra, if we put $f_i(\lambda)=f(\lambda)/(\lambda-\rho_i)$ and $e_i=f_i(a)/f'(\rho_i)$ then the e_i are non-zero orthogonal idempotents in $\mathscr{J}_{F_s}$ with sum 1. Hence the e_i are reduced idempotents in $\mathscr{J}_{F_s}$ and since $e_1\in\mathscr{J}_E$ this is a reduced idempotent in $\mathscr{J}_E$. Hence $\mathscr{J}_E$ is reduced. ∎

Remark. It seems likely that 5.9 holds without the hypothesis of separability.

6. Norm Hypersurfaces

If $\mathscr{J}$ is a Jordan algebra over F we define the *norm hypersurface* $\mathscr{N}_{\mathscr{J}}$ of $\mathscr{J}$ as the set of zeros in $\mathbb{P}\mathscr{J}_{\bar{F}}$ of the generic norm, $\bar{F}$ an algebraically closed field containing F. If $(u_1,\ldots,u_n)$ is a base for $\mathscr{J}/F$ and $x=\sum\xi_iu_i$ in $\mathscr{J}_{F(\xi)}$ where the ξ_i are indeterminates then $\mathscr{N}_{\mathscr{J}}$ is defined by the polynomial $N(x)\in F[\xi]$. We have shown in [8] that $N(x)$ is irreducible if $\mathscr{J}$ is simple of characteristic $\neq 2$. The preliminary Lemma 1 on p. 38 of [8] carries over to arbitrary characteristic and this implies that $N(x)$ is absolutely irreducible if $\mathscr{J}$ is outer central simple. Then $(N(x))$ is the ideal of $\mathscr{N}_{\mathscr{J}}$ and the coordinate algebra $F[\mathscr{N}_{\mathscr{J}}]=F[\xi]/(N(x))$ is a domain. Hence we can define the field of rational functions (homogeneous of degree 0) of $\mathscr{N}_{\mathscr{J}}$. We shall call this the *projective norm field* of $\mathscr{J}$ and denote it as $PNF(\mathscr{J})$.

It can be shown by case considerations that if $\mathscr{J}$ is reduced and outer simple then $\mathscr{N}_{\mathscr{J}}$ is birationally equivalent over F to $\mathbb{P}\bar{F}^{(n-2)}$. We shall give a uniform proof of this result due to John Faulkner which gives some additional information on equations satisfied by reduced elements. (We shall show in Section 8 that these characterize $\mathscr{V}_{\mathscr{J}}$.)

If a and b are elements of a Jordan algebra $\mathscr{J}$, N the generic norm and m the degree, then we write

$$N(a+\lambda b)=\sum_{i=0}^{m}N_i(a,b)\,\lambda^i. \tag{6.1}$$

Then $N_0(a,b)=N(a)$, $N_m(a,b)=N(b)$, and, in general, $N_i(a,b)=N_{m-i}(b,a)$ is a homogeneous polynomial function of degree $m-i$ in a and of degree i in b. The generic minimum polynomial $m_a(\lambda) = \lambda^m - \tau_1(a)\lambda^{m-1} + \cdots + (-1)^n \tau_m(a) = N(\lambda 1 - a)$. Hence

$$\tau_i(a) = N_i(1, a). \tag{6.2}$$

Now let $\mathscr{J}$ be reduced outer simple. We have seen in Theorem 3.6 that if e is a redued idempotent in $\mathscr{J}$ then $m_e(\lambda) = (\lambda - 1)\lambda^{m-1}$ so $N_i(1, e) = 0$ for $i \geqslant 2$. If a is any reduced element in $\mathscr{J}$ then $a^2 = U_a 1 = \alpha a$ so if $\alpha \neq 0$ then $e = \alpha^{-1} a$ is a reduced idempotent. Otherwise $a^2 = 0$ in which case $m_a(\lambda) = \lambda^m$. Thus $N_i(1, a) = 0$ for $i \geqslant 2$ if a is reduced. If c is an invertible element of $\mathscr{J}$ then a is also reduced in $\mathscr{J}^{(c)}$ whose unit is c^{-1}. Since the generic norm $N^{(c)}(x) = N(c) N(x)$ we have $N(c) N_i(c^{-1}, a) = 0$ so $N_i(c^{-1}, a) = 0$ if a is reduced. By extending the base field to an infinite one if necessary and using Zariski density we obtain

$$N_i(b, a) = 0, \qquad i \geqslant 2 \tag{6.3}$$

for all b if a is reduced.

We can now prove

6.4. Theorem. *Let $\mathscr{J}$ be reduced and outer simple. Then the norm hypersurface $\mathscr{N}_{\mathscr{J}}$ of $\mathscr{J}$ is birationally equivalent over F to $\mathbb{P}\bar{F}^{(n-2)}$* $(n = [\mathscr{J} : F])$.

Proof. We may assume that $\mathscr{J}$ contains a reduced idempotent e. If $a \in \mathscr{J}$ we can write $a = \alpha e + b$ where $b \in V \equiv \mathscr{J}_0 + \mathscr{J}_{1/2}$. Since e is reduced we have, by (6.3) and (6.1), that

$$N(a) = \alpha N_1(b, e) + N(b). \tag{6.5}$$

Let O be the open subset of $\mathbb{P}V_{\bar{F}}$ defined by $N(b) N_1(b, e) \neq 0$ and O_1 the open subset of $\mathscr{N}_{\mathscr{J}}$ defined by $\alpha \neq 0$. Then we have the bijective map $\eta: \bar{F}b \rightsquigarrow \bar{F}a$, $a = -N(b) e + N_1(b, e) b$, of O onto O_1. This is regular with regular inverse. Since both $\mathscr{N}_{\mathscr{J}}$ and $\mathbb{P}V_{\bar{F}}$ are irreducible it follows that these varieties are birationally equivalent. Since $[V: F] = [\mathscr{J} : F] - 1$, $\mathbb{P}V_{\bar{F}}$ can be identified with $\mathbb{P}\bar{F}^{(n-2)}$. ∎

We also have the immediate

6.6. Corollary. *If $\mathscr{J}$ is reduced and outer simple, the projective norm field of $\mathscr{J}$ is a rational function field over F transcendency degree $n-2$.*

We can now prove the following analogue of Theorem 5.7 for the projective norm field.

6.7. Theorem. *If $\mathscr{J}$ is outer central simple then the projective norm field*

$PNF(\mathscr{J})$ is a generic reducing field for $\mathscr{J}$ and $\mathscr{J}$ is reduced if and only if $PNF(\mathscr{J})$ is rational over F.

Proof. Let $x = \sum_1^n \xi_i u_i \in \mathscr{J}_{F(\xi)}$ where the ξ_i are indeterminates and the u_i constitute a base for $\mathscr{J}/F$. The minimum polynomial of x is $m_x(\lambda) = N(\lambda 1 - x)$. This is irreducible in $F[\xi_1, \ldots, \xi_n, \lambda]$ since $N(x)$ is irreducible in $F[\xi]$. Hence $m_x(\lambda)$ is irreducible in $F(\xi)[\lambda]$. This implies that $F(\xi)[x]$ is isomorphic to a field over $F(\xi)$. Moreover, since the discriminant $\delta(x)$ of $m_x(\lambda) \neq 0$, $F(\xi)[x]$ is separable over $F(\xi)$. It follows from Theorem 5.9 that $F(\xi)[x]$ is a reducing field for $\mathscr{J}_{F(\xi)}$ over $F(\xi)$. Then $F(\xi)[x]$ is also a reducing field for $\mathscr{J}/F$. By a theorem of Heuser's, $F(\xi)[x]$ is a simple transcendental extension of the field $F(\mathscr{N}_{\mathscr{J}})$ of fractions of $F[\mathscr{N}_{\mathscr{J}}] = F[\xi]/(N(x))$ ([6, p. 107] or [14, p. 158]). Hence, by 5.6, $F(\mathscr{N}_{\mathscr{J}})$ is a reducing field. Since $F(\mathscr{N}_{\mathscr{J}})$ is a simple transcendental extension of $PNF(\mathscr{J})$ it follows that $PNF(\mathscr{J})$ is a reducing field for $\mathscr{J}$. The rest of the argument is similar to the proof of Theorem 5.7 and is left to the reader. ∎

7. Applications to Involutorial Simple Associative Algebras

We recall that an associative algebra with involution $(\mathscr{A}, \mathscr{J})$ is called *simple* (or $\mathscr{A}$ is *-simple) if $\mathscr{A}$ has no ideals $\neq 0$, $\neq \mathscr{A}$ stabilized by J. The *center* $\mathscr{C}_0$ of $(\mathscr{A}, J)$ is $\mathscr{C} \cap H(\mathscr{A}, J)$ where $\mathscr{C}$ is the center of $\mathscr{A}$ and $(\mathscr{A}, J)$ is *central* if $\mathscr{C}_0 = F1$. Now $(\mathscr{A}, J)$ is central simple if and only if it is of one of the following types:

A_I. $\mathscr{A} = \mathscr{B} \oplus \mathscr{B}^{op}$, $\mathscr{B}$ central simple, J the exchange involution: $(a_1, a_2) \rightsquigarrow (a_2, a_1)$.

A_{II}. $\mathscr{A}$ simple with center $\mathscr{C}$ a separable quadratic field over F, J an involution of second kind.

B. $\mathscr{A}$ central simple, J of orthogonal type.

C. $\mathscr{A}$ central simple, J of symplectic type.

The corresponding Jordan algebra $H(\mathscr{A}, J)$ are outer central simple except those defined by algebras with involution of type $\mathscr{C}$ in characteristic two. In this case the outer ideal $H(\mathscr{A}, J)'$ generated by 1 is outer central simple. The following is the list of the degrees and dimensionalities of the $\mathscr{J} = H(\mathscr{A}, J)$ or $H(\mathscr{A}, J)'$ that are outer central simple:

A_I. $\deg \mathscr{J} = \deg \mathscr{B} = m$, $[\mathscr{J} : F] = m^2$.

A_{II}. $\deg \mathscr{J} = \deg \mathscr{A}/\mathscr{C} = m$, $[\mathscr{J} : F] = m^2$.

B. $\deg \mathscr{J} = \deg \mathscr{A} = m$, $[\mathscr{J} : F] = m(m+1)/2$.

C. $\deg \mathscr{J} = \frac{1}{2} \deg \mathscr{A} = m$, $[\mathscr{J} : F] = m(2m-1)$.

If the degree m of any of the absolutely simple $\mathscr{J} = H(\mathscr{A}, J)$ or $H(\mathscr{A}, J)'$ is $\geqslant 3$ then any homomorphism of $\mathscr{J}$ into a Jordan algebra $\mathscr{B}^{(q)}$, $\mathscr{B}$ associative, has a unique extension to a homomorphism (of associative algebras) of $\mathscr{A}$ into $\mathscr{B}$ [13, Sects. 5 and 6]. It follows from functorial considerations (loc. cit., p. 118) that if $(\mathscr{A}_i, J_i)$, $i = 1, 2$, are central simple of degree $\geqslant 3$ then $\mathscr{J}_1 \cong \mathscr{J}_2$ if and only if $(\mathscr{A}_1, J_1) \cong (\mathscr{A}_2, J_2)$ (that is, there is an isomorphism η of $\mathscr{A}_1$ onto $\mathscr{A}_2$ such that $\eta J_1 = J_2 \eta$). Also if $\mathscr{J}_1$ and $\mathscr{J}_2$ are isotopic then $\mathscr{A}_1 \cong \mathscr{A}_2$ (loc. cit., p. 119). Finally, we note that it follows from the structure theory that the $\mathscr{J}$'s listed comprise all the outer central simple Jordan algebras of degree $\geqslant 3$. One can deduce from this result and the determination of the outer simple Jordan algebras of degree 2 which we consider below that any outer central simple $\mathscr{J}$ over a finite field is reduced, a result which we used in the proof of 5.8.

Now let $\mathscr{D}$ be a composition algebra, j the standard involution in $\mathscr{D}$, and let J be the standard involution in $M_m(\mathscr{D})$ extending j. If $\mathscr{D} = F1$ then $(M_m(\mathscr{D}), J)$ is of type B. If $\mathscr{D}$ is a quadratic algebra then $(M_m(\mathscr{D}), J)$ is of type $\mathrm{A_I}$ or $\mathrm{A_{II}}$ according as $\mathscr{D}$ is a direct sum of two copies of F or is a field. If $\mathscr{D}$ is a quaternion algebra then $(M_m(\mathscr{D}), J)$ is of type C. If $\operatorname{char} F = 2$ and $\mathscr{D}$ is quaternion then $H(M_m(\mathscr{D}), J)' = H(M_m(\mathscr{D}), F)$. In all other cases $H(M_m(\mathscr{D}), J) = H(M_m(\mathscr{D}), F)$.

We now treat the various cases separately.

$\mathrm{A_I}$. $(\mathscr{A}, J) = (\mathscr{B} \oplus \mathscr{B}^{\mathrm{op}}, J)$, $\mathscr{B}$ central simple associative, J the exchange involution. The map $a \rightsquigarrow (a, a)$ is an isomorphism of $H(\mathscr{A}, J)$ onto $\mathscr{B}^{(q)}$ so we can state our results in terms of the algebras $\mathscr{B}$ and $\mathscr{B}^{(q)}$. First, we have

7.1. Theorem. *Let $\mathscr{B}$ be central simple associative of degree $m \geqslant 3$ over F. Then an extension field E/F is a reducing field (generic reducing field) for $\mathscr{B}^{(q)}$ if and only if E is a splitting field (generic splitting field) for $\mathscr{B}$.*

Proof. Suppose E is a reducing field for $\mathscr{B}^{(q)}$. Then $H(\mathscr{B}_E \oplus \mathscr{B}_E^{\mathrm{op}}, J)$ is isotopic to $H(M_m(\mathscr{D}), E)$ for a composition algebra $\mathscr{D}$ over E. Hence $\mathscr{B}_E \oplus \mathscr{B}_E^{\mathrm{op}} \cong M_m(\mathscr{D})$. Since the only $M_m(\mathscr{D})$, $\mathscr{D}$ a composition algebra over E, that is not simple is $\mathscr{D} = E \oplus E$ it follows that $\mathscr{B}_E \oplus \mathscr{B}_E^{\mathrm{op}} \cong M_m(E) \oplus M_m(E)$. Then $\mathscr{B}_E \cong M_m(E)$ so E is a splitting field for $\mathscr{B}$. Conversely suppose $\mathscr{B}_E \cong M_m(E)$. Then $\mathscr{B}_E^{(q)} \cong M_m(E)^{(q)} \cong H(M_m(E) \oplus M_m(E), J) \cong H(M_m(E \oplus E), E)$. Hence E is a reducing field. This proves the statement on reducing fields and splitting fields. The statement on generic reducing fields and generic splitting fields follows from this and the definitions of the generic objects. ∎

An immediate consequence of this result and our results on generic reducing fields is

7.2. COROLLARY. *Let $\mathscr{B}$ be central simple associative and let $\mathscr{J} = \mathscr{B}^{(q)}$. Then the fields $PF(\mathscr{V}_{\mathscr{J}})$ and $PF(\mathscr{N}_{\mathscr{J}})$ are generic splitting fields for $\mathscr{B}$.*

We recall that if $\mathscr{B}$ is a central simple associative algebra of degree m the Brauer–Severi variety of $\mathscr{B}$ is the subvariety of the Grassmannian $\mathrm{Gr}(\mathscr{B}, m)$ of m-dimensional subspaces of $\mathscr{B}_{\bar{F}}$ whose points correspond to the m-dimensional right ideals of $\mathscr{B}_{\bar{F}}$ [1, 2]. This is an irreducible variety defined over F of $m-1$ dimensions. It is known that the field of rational functions on this variety is a generic splitting field of $\mathscr{B}$. Hence we have the following result which reverses the roles of $\mathscr{B}$ and $\mathscr{B}^{(q)}$.

7.3. COROLLARY. *Let $\mathscr{B}$ be as in 7.2. Then the field of rational functions on the Brauer–Severi variety of $\mathscr{B}$ is a generic reducing field for $\mathscr{B}^{(q)}$.*

We recall that the transcendency degrees over F of $PF(\mathscr{V}_{\mathscr{J}})$ and $PF(\mathscr{N}_{\mathscr{J}})$ are respectively $2m-2$ and m^2-2 (4.26 and 6.4) whereas that of the field of the Brauer–Severi variety is $m-1$.

$\mathrm{A_{II}}$. An extension field E/F will be called a *splitting field* for $(\mathscr{A}, J)$ of type $\mathrm{A_{II}}$ if $\mathscr{A}_E = M_m(\mathscr{C}_E)$ where $\mathscr{C}$ is the center of $\mathscr{A}$. Since $\mathscr{C}$ is separable quadratic, $\mathscr{C}_E$ is either a direct sum of two copies of F or a field according as E contains a subfield isomorphic to $\mathscr{C}$ or not. In the first case we say that E is a *type* I *splitting field* and in the second a *type* II *splitting field* for $\mathscr{A}$. Now suppose that E is a reducing field for $\mathscr{J} = H(\mathscr{A}, J)$. Then comparison of the degrees and dimensionalities shows that $\mathscr{J}_E$ is isotopic to a Jordan algebra $H(M_m(\mathscr{D}), E)$ where $\mathscr{D}$ is a quadratic composition algebra and we shall say that E is a *type* I *reducing field* or a *type* II *reducing field* according as $\mathscr{D}$ is a direct sum of two copies of F or is a field. Finally, we define *generic type* I and *type* II *splitting fields* for $(\mathscr{A}, J)$ and *generic type* I and *type* II *reducing fields* in the obvious way. Then we have

7.4. THEOREM. *Let $(\mathscr{A}, J)$ be simple associative with involution of second kind and assume the degree of $\mathscr{J} = H(\mathscr{A}, J) \geqslant 3$. Then an extension field E/F is a type* I *or type* II *splitting (generic splitting field) for $(\mathscr{A}, J)$ if and only if it is respectively a type* I *or type* II *reducing field for $\mathscr{J}$.*

We omit the proof which is similar to that of 7.1.

Continuing with the same notations we have

7.5. THEOREM. *The fields $PF(\mathscr{V}_{\mathscr{J}})$ and $PF(\mathscr{N}_{\mathscr{J}})$ are generic type* II *reducing fields for $\mathscr{J}$ and hence by 7.4, if the degree $\geqslant 3$, these are generic type* II *splitting fields for $(\mathscr{A}, J)$.*

Proof. Let $\mathscr{V}$ be an absolutely irreducible projective variety defined over F, $PF(\mathscr{V})$ its field of rational functions, and let E be a finite-dimensional separable extension field of F. If $\mathscr{I}(\mathscr{V})$ is the ideal in $F[\xi] =$

$F[\xi_1, \ldots, \xi_n]$ of $\mathscr{V}$ then $E \otimes_F \mathscr{I}(\mathscr{V})$ is the ideal in $E[\xi] = E \otimes_F F[\xi]$ of $\mathscr{V}_E$. For, $E \otimes_F \mathscr{I}(\mathscr{V})$ is a defining ideal for $\mathscr{V}_E$ and this ideal coincides with its nil radical [28, Vol. II, p. 226]. Hence $\mathscr{I}(\mathscr{V}_E) = E \otimes_F \mathscr{I}(\mathscr{V})$. Since $\mathscr{V}_E$ is irreducible, $E \otimes_F \mathscr{I}(\mathscr{V})$ is a prime ideal and the coordinate algebra $E[\mathscr{V}_E] \cong E \otimes_F F[\mathscr{V}]$. This is a domain. Hence $E \otimes_F F(\mathscr{V})$ is a domain and so is its subalgebra $E \otimes_F PF(\mathscr{V})$. Since $[E: F] < \infty$, $E \otimes PF(\mathscr{V})$ is finite dimensional over $PF(\mathscr{V})$. Hence $E \otimes PF(\mathscr{V})$ is a field. Since $E \otimes_F E$ is not a field of $E \supsetneq F$ it follows that $PF(\mathscr{V})$ contains no finite-dimensional separable $E \supsetneq F$. This applies to $PF(\mathscr{V}_{\mathscr{J}})$ and $PF(\mathscr{N}_{\mathscr{J}})$ and shows that these fields do not contain copies of the center $\mathscr{C}$ of $\mathscr{A}$. Since $PF(\mathscr{V}_{\mathscr{J}})$ and $PF(\mathscr{N}_{\mathscr{J}})$ are generic reducing fields for $\mathscr{J} = H(\mathscr{A}, J)$ it follows that they are type II reducing fields. ∎

Now $(\mathscr{A}, J)_{\mathscr{C}}$ is of type $\mathrm{A_I}$ as algebra over $\mathscr{C}$ and if $E/\mathscr{C}$ is a splitting field for $\mathscr{A}$ then E/F is a type I splitting field for $(\mathscr{A}, J)$ and a type I reducing field for $\mathscr{J}$. It is clear also that if E is a type I splitting field for $(\mathscr{A}, J)$ then E contains a copy of $\mathscr{C}$ and E is a type I reducing field for $\mathscr{J}_{\mathscr{C}}$. Thus we have a reduction of the theory to the $\mathrm{A_I}$-type algebras we considered first.

B. Let $(\mathscr{A}, J)$ be of type B, $\mathscr{J} = H(\mathscr{A}, J)$, and let the degree $m \geqslant 3$. If E is a reducing field for $\mathscr{J}$ then $\mathscr{J}_E$ is isotopic to $H(M_m(E), E)$ and $\mathscr{A}_E \cong M_m(E)$ so E is a splitting field for $\mathscr{A}$. Conversely, any splitting field for $\mathscr{A}$ is a reducing field for $\mathscr{J}$. It follows as in the case $\mathrm{A_I}$ that $PF(\mathscr{V}_{\mathscr{J}})$ and $PF(\mathscr{N}_{\mathscr{J}})$ are generic splitting fields for $\mathscr{A}$ and generic reducing fields for $\mathscr{J}$. Moreover, the field of rational functions on the Brauer–Severi variety of $\mathscr{A}$ is a generic reducing field for $\mathscr{J}$.

C. An extension field E/F is called a $1/s$-splitting field for a central simple algebra $\mathscr{A}$ if $\mathscr{A}_E$ has index $\leqslant s$ [14, p. 161]. This is equivalent to: $\mathscr{A}_E$ has a splitting field of degree $\leqslant s$. We shall now say that E/F is a *strictly* $1/s$*-splitting field* for $\mathscr{A}$ in $\mathscr{A}_E$ has index s. In a similar manner we define *generic* $1/s$ and *generic strictly* $1/s$*-splitting fields* for $\mathscr{A}$.

Now suppose $(\mathscr{A}, J)$ is central simple with involution of type C. Then $\mathscr{A} = M_n(\Delta)$ where Δ is a central division algebra with involution j and J is a canonical involution, that is, there exist invertible $c_i \in H(\Delta, j)$ such that J has the form

$$A = (a_{ij}) \rightsquigarrow C'(\bar{a}_{ij})\bar{C}^{-1} \tag{7.6}$$

where $\bar{a}_{ij} = j(a_{ij})$ and $C = \operatorname{diag}\{c_1, \ldots, c_r\}$ [10, p. 0.11]. The conditions that $A \in \mathscr{J} = H(\mathscr{A}, J)$ if $\operatorname{char} F \neq 2$ and $= H(\mathscr{A}, J)'$ if $\operatorname{char} F = 2$ are that $a_{ji} = c_j \bar{a}_{ij} c_i^{-1}$ if $i \neq j$ and $a_{ii} c_i \in H(\Delta, j)$ if $\operatorname{char} F \neq 2$. Comparison of dimensionalities shows that (Δ, j) is a division algebra with involution of type C. We can now prove

7.7. THEOREM. *Let (Δ, j) be a central division algebra with involution of type* C *such that $\Delta \neq F$ and let $\Gamma = H(\Delta, j)$ if* char $F \neq 2$ *and* $= H(\Delta, j)'$ *if* char $F = 2$. *Let d be the degree of Δ so $d = 2\delta$, $d \in \mathbb{Z}$. Then Γ contains a separable subfield of dimension δ over F. Moreover, any δ-dimensional subfield of Δ is a strictly $\frac{1}{2}$-splitting field for Δ.*

Proof. Since $\Delta \neq F$ it follows from Wedderburn's theorem on the commutativity of finite division algebras that F is infinite. The degree of Γ is δ and, as in the proof of 5.8, the discriminant of $m_a(\lambda)$ for generic x is $\neq 0$. Then we can specialize to obtain an a such that $m_a(\lambda)$ has distinct roots. Then $F[a]$ is a separable subfield of Γ such that $[F[a]: F] = \delta$. Now let E be any subfield of Δ such that $[E: F] = \delta$. Then E is not a splitting field for Δ since $[E: F] < \deg \Delta$. On the other hand, $[\Delta^E: F][E: F] = [\Delta: F] = 4\delta^2$ so $[\Delta^E: F] = 4\delta$ and $[\Delta^E: E] = 4$. Thus Δ^E is a quaternion algebra over E and so contains a quadratic subfield E'/E. Then $E' \supset E$ and $[E': F] = 2\delta$. Then E' is a splitting field for Δ. It follows that the index of Δ_E is two. Hence E is a strictly $\frac{1}{2}$-splitting field for Δ. ∎

7.8. COROLLARY. *Let $(\mathcal{A}, J)$ be central simple with involution of type* C *such that $\mathcal{A} \not\sim 1$ ($\mathcal{A}$ is not split). Then there exist strictly $\frac{1}{2}$-splitting fields for $\mathcal{A}$.*

Our main result on the type C algebra is

7.9. THEOREM. *Let $(\mathcal{A}, J)$ be a central simple algebra with involution of type* C, $\mathcal{J} = H(\mathcal{A}, J)$ *if* char $F \neq 2$ *and* $= H(\mathcal{A}, J)'$ *if* char $F = 2$. *Suppose* $\deg \mathcal{J} \geqslant 3$. *Then an extension field E/F is a reducing field (generic reducing field) for $\mathcal{J}$ if and only if it is a $\frac{1}{2}$-splitting field(generic $\frac{1}{2}$-splitting field) for $\mathcal{A}$. Moreover, if $\mathcal{A} \not\sim 1$ then any generic $\frac{1}{2}$-splitting field for $\mathcal{A}$ is a strictly $\frac{1}{2}$-splitting field.*

Proof. The first statement follows as in the proof of Theorem 7.1. Now suppose $\mathcal{A} \not\sim 1$ and let E/F be a generic $\frac{1}{2}$-splitting field. Suppose E is not a strictly $\frac{1}{2}$-splitting field. Then E is a splitting field. By 7.8 there exists a strictly $\frac{1}{2}$-splitting field E' for $\mathcal{A}$ so $\mathcal{A}_{E'} = M_m(\mathcal{D})$ where $\mathcal{D}$ is a quaternion division algebra over E'. Since E/F is a generic $\frac{1}{2}$-splitting field and E'/F is a $\frac{1}{2}$-splitting field there is an F-place on E to E'. Since E/F is a splitting field and we have an F-place on E to F', E' is a splitting field contrary to $\mathcal{A}_{E'} = M_m(\mathcal{D})$ where $\mathcal{D}$ is a division algebra. ∎

Since $PF(\mathcal{V}_{\mathcal{J}})$ and $PF(\mathcal{N}_{\mathcal{J}})$ are generic reducing fields for $\mathcal{J}$ we have

7.10. COROLLARY. *The fields $PF(\mathcal{V}_{\mathcal{J}})$ and $PF(\mathcal{N}_{\mathcal{J}})$ are generic $\frac{1}{2}$-splitting fields for $\mathcal{A}$ and if $\mathcal{A} \not\sim 1$ then these are strictly $\frac{1}{2}$-splitting fields.*

8. Norm Equations Defining $\mathscr{V}_{\mathscr{J}}$

We shall now derive an important system of equations defining the variety of reduced elements of an outer central simple Jordan algebra $\mathscr{J}$. These were communicated to the author by John Faulkner.

8.1. Theorem. *Let $\mathscr{J}$ be outer central simple over an infinite field F, $\bar{F}$ an algebraically closed field containing F. Then $b \in \mathscr{J}_{\bar{F}}$ is reduced if and only if*

$$N_i(a, b) = 0, \qquad i \geqslant 2 \tag{8.2}$$

holds for all $a \in \mathscr{J}$.

Proof. For each $a \in \mathscr{J}$ Eq. (8.2) is a homogeneous polynomial equation of degree i with coefficients in F for b. These equations define a projective variety $\mathscr{V}$. We have seen in Section 6 that if $b \in \rho(\mathscr{J}_{\bar{F}})$ then (8.2) holds for all $a \in \mathscr{J}_{\bar{F}}$ (which is reduced). Hence (8.2) holds for all $a \in \mathscr{J}$. Thus $\mathscr{V}_{\mathscr{J}} \subset \mathscr{V}$. Also it is clear that $\mathscr{V} \subset \mathscr{N}_{\mathscr{J}}$. Now let $b \in \mathscr{V}$. Then $N_i(a, b) = 0$ is a homogeneous polynomial equation of degree $m - i$ in a with coefficients in $\bar{F}$. By equating to zero the coefficients of a base for $\bar{F}/F$ this equation is equivalent to a system of homogeneous polynomial equations of degree $m - i$ with coefficients in F. Since these hold for all $a \in \mathscr{J}$ and F is infinite they hold for all $a \in \mathscr{J}_{\bar{F}}$. Thus we may assume $\bar{F} = F$ or, equivalently, F is algebraically closed. Then $\mathscr{J}$ is reduced and we may assume $\mathscr{J}$ contains a reduced idempotent e.

Let $\mathscr{J} = \mathscr{J}_1 \oplus \mathscr{J}_{1/2} \oplus \mathscr{J}_0$ be the Peirce decomposition of $\mathscr{J}$ relative to e. Then $b = \alpha e + c + d$, $\alpha \in F$, $c \in \mathscr{J}_{1/2}$, $d \in \mathscr{J}_0$. We wish to show that there exists an $\eta \in \operatorname{Str} \mathscr{J}$ such that η stabilizes $\mathscr{V}$ and $\eta b = e$. Now any invertible U_u and any invertible $B_{u,v}$ stabilizes $\mathscr{V}$. The first is clear since $N(U_a x) = N(u)^2 N(x)$ and hence $N_i(U_u x, U_u y) = N(u)^2 N(x, y)$. To see the second we use the fact that if u is invertible then $B_{u,v} = U_u U_{u^{-1} - v}$ [13, p. 3.41]. Hence $N(B_{u,v} x) = N(u)^2 N(u^{-1} - v)^2 N(x) = N(B_{u,v} 1) N(x)$. Then $N_i(B_{u,v} x, B_{u,v} y) = N(B_{u,v} 1) N_i(x, y)$ which implies that $B_{u,v}$ stabilizes $\mathscr{V}$ if $B_{u,v}$ is invertible. It now follows as in the proof of Theorem 4.23 that there exists an η that is a product of a non-zero scalar, of invertible U_u, and of $B_{e,v}$ such that $\eta b = e + c'$, $c' \in \mathscr{J}_0$. Such an $\eta \in \operatorname{Str} \mathscr{J}$ and stabilizes $\mathscr{V}$. Hence $e + c' \in \mathscr{V}$.

We claim that $c' = 0$. Let N' denote the generic norm in $\mathscr{J}_0$. Then $N(\beta e + d') = \beta N'(d')$ for $d' \in \mathscr{J}_0$ and hence $N_i(\beta e + d', e + c') = \beta N'_i(d', c') + N'_{i-1}(d', c')$. The condition that $e' + c' \in \mathscr{V}$ implies that $N'_i(d', c') = 0$ for all $i \geqslant 1$ and all $d' \in \mathscr{J}_0$. Then $N'(d' + i') = N'(d')$ for $d' \in \mathscr{J}_0$. By a result of McCrimmon's [20, p. 235] this implies that $c' \in \operatorname{rad} \mathscr{J}$ and hence, since $\mathscr{J}_0$ is semiprimitive, $c' = 0$.

We now have $\eta^{-1}e = h$ and since $\eta^{-1} \in \operatorname{Str} \mathscr{J}$ and so stabilizes $\mathscr{V}_{\mathscr{J}}$, $\eta^{-1}e = b \in \mathscr{V}_{\mathscr{J}}$. Thus $\mathscr{V} = \mathscr{V}_{\mathscr{J}}$. ∎

ACKNOWLEDGMENTS

Some of the research reported on in this paper was done while the author was a visiting scholar at the mathematical research institute of the University of Virginia from mid February to mid April of 1984. We benefitted greatly from a number of stimulating conversations on the subject of the paper with John Faulkner, with Kevin McCrimmon, and with Craig Huneke, who was also a visiting scholar at the University of Virginia. Finally, it is a pleasure to dedicate this paper to my old friend, Irving Kaplansky, with whom I have shared many common mathematical interests over the years.

Note added in proof. After this paper was finished I learned by letter and preprint that it overlaps substantially with a paper by Holger Petersson, "Generic Reducing Fields for Jordan Pairs," which has now appeared in *Trans. Amer. Math. Soc.* **285** (1984), 825–843. However, in the parts where the overlap occurs the points of view and methods are different.

REFERENCES

1. S. A. AMITSUR, Generic splitting fields of central simple associative algebras, *Ann. of Math.* **62** (1955), 8–43.
2. M. ARTIN, Brauer Severi varieties, *in* "Lecture Notes in Mathematics," pp. 194–211, Springer-Verlag, Berlin/Heidelberg/New York, 1981.
3. J. FAULKNER, "Octonion Planes Defined by Quadratic Jordan Algebras," Amer. Math. Soc. Memoirs, Vol. 104, Amer. Math. Soc., Providence, R. I., 1970.
4. J. FAULKNER, Derivation invariant bilinear forms without traces, *J. Algebra*, (1983).
5. R. HARSHORNE, "Algebraic Geometry," Graduate Texts in Mathematics 52, Springer-Verlag, Berlin/Heidelberg/New York, 1977.
6. A. HEUSER, Über den Funktionenkorper der Norm-fläche..., *J. Math.* **301** (1978), 105–113.
7. J. HUMPHREYS, "Linear Algebraic Groups," Springer-Verlag, New York/Heidelberg/Berlin, 1975.
8. N. JACOBSON, Generic norm of an algebra, *Osaka J. Math.* **15** (1963), 25–50.
9. N. JACOBSON, "Structure and Representations of Jordan Algebras," Colloq. Publ., Amer. Math. Soc., Providence, R. I., 1968.
10. N. JACOBSON, "Lectures on Quadratic Jordan Algebras," Tata Inst. of Fundamental Research, Bombay, 1969.
11. N. JACOBSON, Structure groups and Lie algebras of Jordan algebras..., *J. Algebra* **20** (1976), 106–150.
12. N. JACOBSON, "Basic Algebra," Vol. II, Freeman, San Francisco/New York, 1980.
13. N. JACOBSON, "Structure theory of Jordan algebras," Univ. of Arkansas lecture notes, teville, Art., 1981.
14. N. JACOBSON, Some applications of Jordan norms to involutorial simple associative algebras, *Adv. in Math.* **48** (1983), 149–165.
15. N. JACOBSON, AND J. KATZ, Generically algebraic quadratic Jordan algebras, *Scripta Math.* **29** (1975), 215–227.
16. N. JACOBSON, AND K. MCCRIMMON, Quadratic Jordan of quadratic forms with base points, *J. Indian Math. Soc.* **35** (1971), 1–45.

17. O. Loos, "Jordan Pairs," Lecture Notes in Mathematics, Vol. 460, Springer-Verlag, Berlin/Heidelberg/New York, 1975.
18. K. McCrimmon, A general theory of Jordan rings, *Proc. Nat. Acad. Sci. U.S.A.* **56** (1956), 1072–1079.
19. K. McCrimmon, A note on reduced Jordan algebras, *Proc. Amer. Math. Soc.* **19** (1968), 964–970.
20. K. McCrimmon, The generic norm of an isotope of a Jordan algebra, *Scripta Math.* **29** (1975), 229–236.
21. K. McCrimmon, Reduced elements in Jordan triple systems, *J. Algebra* **97** (1985), 540–564.
22. T. Muir, "A Treatise on the Theory of Determinants," Vol. 4, Dover, New York, 1960.
23. H. Petersson, Conjugacy of idempotents in Jordan pairs, *Comm. Algebra* **6** (1978), 673–715.
24. P. Roquette, On the Galois theory of the projective linear group, *Math. Ann.* **150** (1963), 411–439.
25. D. Saltman, Norm polynomials and algebras, *J. Algebra* **62** (1980), 333–345.
26. I. R. Shafarevich, "Basic Algebraic Geometry," Springer-Verlag, Berlin/Heidelberg/New York, 1977.
27. W. C. Waterhouse, Symmetric determinants and Jordan norm similarities in characteristic 2, *Proc. Amer. Math. Soc.*, in press.
28. O. Zariski and P. Samuel, "Commutative Algebra," Springer-Verlag, Berlin/Heidelberg/New York, 1975 (first published by Van Nostrand, Princeton, N. J., 1960).

Printed by the St. Catherine Press Ltd., Tempelhof 41, Bruges, Belgium

JORDAN ALGEBRAS OF REAL SYMMETRIC MATRICES

To Yitz

Nathan Jacobson[1]
Department of Mathematics
Yale University
Box 2155 Yale Station
New Haven, CT 06520

Received March 27, 1987
Revised June 22, 1987

Abstract

This paper gives a determination of the orbits under the orthogonal group of Jordan algebras of real symmetric matrices.

[1] *This article was presented at a Conference honoring I.N. Herstein at the University of Chicago, March 25–27, 1987. This work has been partially supported by National Science Foundation Grant No. DMS–8512904.*

AMS 1980 Subject Classification: 17C15, 17C25

In this paper we shall determine the orbits under the orthogonal group $O(n)$ of the subalgebras of the Jordan algebra $H(M_n(\mathbb{R}),t)$ of $n \times n$ real symmetric matrices. Here the action of $O(n)$ on an algebra of real symmetric matrices is given by similarity. Our results are an outgrowth of an affirmative answer we gave to the following question that had been proposed to us by J. D. Malley: Can every Jordan algebra $\mathbb{R}1 \oplus V$ of a positive definite symmetric bilinear be realized as a Jordan algebra of real symmetric matrices? Malley was interested in this question for applications to statistics (see Malley [4]). In [5] he also showed by a method different from ours that the Jordan algebra $\mathbb{R}1 \oplus V$ can be imbedded in a Jordan algebra of real symmetric matrices.

We observe first that $H(M_n(\mathbb{R}),t)$ is formally real in the sense that $\sum_1^k a_i^2 = 0 \Rightarrow$ every $a_i = 0$. This is clear since if $a = (\alpha_{rs})$ then $\operatorname{tr} a^2 = \sum_{r,s} \alpha_{rs}^2$. Hence if $a_i = (\alpha_{irs})$ then $\operatorname{tr} \sum_1^k a_i^2 = \sum \alpha_{irs}^2 = 0$ implies every $\alpha_{irs} = 0$.

Evidently if J is a subalgebra of $H(M_n(\mathbb{R}),t)$ then J is formally real. Hence we can apply the following fundamental structure theorem proved by Jordan, v. Neumann and Wigner in 1934.

JNW <u>Theorem</u>. If J is a finite dimensional formally real Jordan algebra then J is a direct sum of ideals of one of the following types:

I. $H(M_n(C),*)$ the Jordan algebra of $n \times n$ $*$-symmetric matrices where C is one of the following:

(1) $C = \mathbb{R}$

(2) $C = \mathbb{C}$

(3) $C = \mathbb{H}$, Hamilton's quaternion algebra

(4) $C = \mathbb{O}$, the algebra of octonions of Cayley-Graves.

In all cases $*$ is the hermitian involution $a \mapsto a^* = {}^t\bar{a}$ where if $a = (\alpha_{rs})$ then ${}^t\bar{a}$ is the transposed of $\bar{a} = (\bar{\alpha}_{rs})$. Moreover, in all cases the multiplication in J is given by

$$a \cdot b = \frac{1}{2}(ab + ba) \tag{1}$$

and we must take $n \leq 3$ if $C = \mathbb{O}$.

II. Jordan algebras of positive definite symmetric bilinear forms. Here we take V to be a real n dimensional vector space over $\mathbb{R}$, f a positive definite symmetric bilinear form on V and we form the vector space $J = \mathbb{R}1 \oplus V$ and introduce the product

$$(\alpha 1 + u) \cdot (\beta 1 + v) = (\alpha\beta + f(u,v)1) + (\alpha v + \beta u). \tag{2}$$

In all of this Jordan algebras are defined axiomatically by the identities

$$a \cdot b = b \cdot a \tag{3}$$

$$(a^2 \cdot b) \cdot a = a^2 \cdot (b \cdot a), \quad a^2 = a \cdot a \tag{4}$$

and we always assume the existence of a unit 1 for associative and for Jordan algebras. If A is an associative algebra, A becomes a Jordan algebra A^+ by replacing the given associative multiplication by the Jordan multiplication (1). The algebras A^+ and their subalgebras are called <u>special</u>. Otherwise, the algebras are <u>exceptional</u>. In the foregoing list all the algebras are special except $H(M_3(\mathbb{O}),*)$. It was shown by Albert that this algebra is exceptional. We note also that all the Jordan algebras listed in the JNW Theorem are formally real and if we assume $n \geq 3$ in the class I and $n \geq 0$ in II then none of the algebras listed are isomorphic (see Jacobson [3]).

In view of the JNW Theorem, for the determination of the Jordan algebras of real symmetric matrices, we should consider first the following question.

Which Jordan algebras in the above list can be imbedded in an algebra of real symmetric matrices?

Evidently the exceptional algebra $H(M_3(\mathbb{O}),*)$ can not. What about the others? We show first that every $H(M_n(C),*)$ where $C = \mathbb{R}$, $\mathbb{C}$ or $\mathbb{H}$ can be imbedded in an $H(M_n(\mathbb{R}),t)$. This is clear if $C = \mathbb{R}$. Next suppose $C = \mathbb{C}$. Let $a = (a_{rs}) \in M_n(\mathbb{C})$ where $a_{rs} = \alpha_{rs1} + \alpha_{rs2}i$, $\alpha_{rsj} \in \mathbb{R}$. Then we can replace each entry a_{rs} by the matrix

$$(5) \qquad \begin{pmatrix} \alpha_{rs1} & -\alpha_{rs2} \\ \alpha_{rs2} & \alpha_{rs1} \end{pmatrix}$$

in $M_2(\mathbb{R})$. Let η denote the map of $M_n(\mathbb{C})$ into $M_{2n}(\mathbb{R})$ that sends a into ηa which is obtained by replacing the entry a_{rs} by the 2×2 matrix (5). Then η is a monomorphism of $M_n(\mathbb{C})$ as associative algebra over $\mathbb{R}$ into $M_{2n}(\mathbb{R})$ as associative algebra over $\mathbb{R}$. Also, as is readily checked,

$$(6) \qquad \eta(a^*) = {}^t(\eta a)$$

where $a^* = {}^t(\bar{a}_{rs})$. It follows that if $a \in H(M_n(\mathbb{C}),*)$ then $\eta a \in H(M_{2n}(\mathbb{R}),t)$ the Jordan algebra of real symmetric matrices and η is a monomorphism of Jordan algebras.

Now consider $H(M_n(\mathbb{H}),*)$. We have the monomorphism of $\mathbb{H}$ into $M_4(\mathbb{R})$ such that

$$(7) \qquad \alpha_1 + \alpha_2 i + \alpha_3 j + \alpha_4 k \rightsquigarrow \begin{pmatrix} \alpha_1 & -\alpha_2 & -\alpha_3 & -\alpha_4 \\ \alpha_2 & \alpha_1 & -\alpha_4 & \alpha_3 \\ \alpha_3 & \alpha_4 & \alpha_1 & -\alpha_2 \\ \alpha_4 & -\alpha_3 & \alpha_2 & \alpha_1 \end{pmatrix} .$$

This gives the monomorphism η of $M_n(\mathbb{H})$ mapping $a = (a_{rs})$, into $M_{4n}(\mathbb{R})$ obtained by replacing the entry a_{rs} by the 4×4 real matrix given in (7). η is a monomorphism of associative algebras such that $\eta(a^*) = {}^t(\eta a)$. It follows that the restriction of η to $H(M_n(\mathbb{H}),*)$ is a monomorphism into the Jordan algebra of $4n \times 4n$ real symmetric matrices.

It remains to consider the Jordan algebras of positive definite symmetric bilinear forms. Let f be such a form on a vector space V and let $Q(x) = f(x,x)$. Let $C(V,Q)$ be the Clifford algebra of Q. Then we have the canonical imbedding of $\mathbb{R}1 \oplus V$ into $C(V,Q)$ and we have (2) where $x \cdot y = \frac{1}{2}(xy + yx)$. Hence we have an imbedding of $\mathbb{R}1 \oplus V$ into $C(V,Q)^+$. Now $C(V,Q)$ has a unique involution π called its <u>main involution</u> such that $\pi(\alpha 1+u) = \alpha 1+u$ for $\alpha \in \mathbb{R}$, $u \in V$. Thus $J = \mathbb{R}1 \oplus V \subset H(C(V,Q),\pi)$ the subalgebra of $C(V,Q)^+$ of π-symmetric elements and we shall have an isomorphism of J with a Jordan algebra of real symmetric matrices if we have a monomorphism of the associative algebra with involution $(C(V,Q),\pi)$ into some $(M_m(\mathbb{R}),t)$.

We now consider $(C(V,Q),\pi)$. Let $(u_1,\ldots,u_n)$ be an orthonormal base for V. Then in $C(V,Q)$ we have the relations

$$(8) \qquad u_i^2 = 1, \quad u_i u_j = -u_j u_i, \quad i \neq j$$

and the elements $u_{i_1} \ldots u_{i_r}$, $1 \leq i_1 < \ldots < i_r \leq n$, constitutes a base for $C(V,Q)$. Then $[C(V,Q):\mathbb{R}] = 2^n$.

Lemma 1. We have the following isomorphisms for algebras with involution

(i) $(C(V,Q),\pi) \cong (\mathbb{R} \oplus \mathbb{R}, 1_{\mathbb{R} \oplus \mathbb{R}})$ if $[V:\mathbb{R}] = 1$

(ii) $(C(V,Q),\pi) \cong (M_2(\mathbb{R}),t)$ if $[V:\mathbb{R}] = 2$

(iii) $(C(V,Q),\pi) \cong (M_2(\mathbb{C}),*)$ if $[V:\mathbb{R}] = 3$

(iv) $(C(V,Q),\pi) \cong (M_2(\mathbb{H}),*)$ if $[V:\mathbb{R}] = 4$.

Proof. (i) The linear map η of $C(V,Q)$ into $\mathbb{R} \oplus \mathbb{R}$ such that $1 \rightsquigarrow (1,1)$, $u_1 \rightsquigarrow (1,-1)$ is an isomorphism of $(C(V,Q),\pi)$ onto $(\mathbb{R} \oplus \mathbb{R}, 1_{\mathbb{R} + \mathbb{R}})$.

(ii) Let

$$v_1 = \begin{pmatrix} 1 & 0 \\ 0 & -1 \end{pmatrix} \quad , \quad v_2 = \begin{pmatrix} 0 & 1 \\ 1 & 0 \end{pmatrix} .$$

Then $v_i^2 = 1$, $v_1v_2 = -v_2v_1$. It follows that we have an isomorphism of $C(V,Q)$ onto $M_2(\mathbb{R})$ such that $u_i \rightsquigarrow v_i$, $i = 1,2$. Since ${}^tv_i = v_i$ this is an isomorphism of $(C(V,Q),\pi)$ onto $(M_2(\mathbb{R}),\pi)$. (Note in general, that an algebra homomorphism $\eta: A \to A'$ is a homomorphism of algebras with involution $(A,*) \to (A',*')$ if $\eta(v_i^*) = \eta(v_i)^*$ for a generating set $\{v_i\}$ of A, in particular, if $\eta(x_i)$ are symmetric in A' for a symmetric generating set $\{v_i\}$ of A.)

(iii) Let v_1 and v_2 be as in (ii) and let

$$v_3 = \begin{pmatrix} 0 & -i \\ i & 0 \end{pmatrix} .$$

Then $v_i^2 = 1$ and $v_iv_j = -v_jv_i$, $i \neq j$. Hence the linear map η such that $1 \to 1$, $u_i \to v_i$, $1 \le i \le 3$, is an isomorphism of $C(V,Q)$ onto $M_2(\mathbb{C})$. Since $v_i^* = v_i$ this is an isomorphism of $(C(V,Q),\pi)$ onto $(M_2(\mathbb{C}),*)$.

(iv) Let v_1, v_2, v_3 be as in (iii) and let

$$v_4 = \begin{pmatrix} 0 & -j \\ j & 0 \end{pmatrix}.$$

Then $v_i^2 = 1$, $v_i v_j = -v_j v_i$, $i \neq j$, and $v_i^* = v_i$. It follows that the linear map η of $C(V,Q)$ into $M_2(\mathbb{H})$ such that $1 \rightsquigarrow 1$, $u_i \rightsquigarrow v_i$ is an isomorphism of $(C(V,Q),\pi)$ onto $(M_2(\mathbb{H}),*)$. □

<u>Lemma 2</u>. Let $n > 4r$ and $U = \sum_1^{4r} \mathbb{R}u_i$, $W = \sum_{4r+1}^{n} \mathbb{R}u_j$. Then $(C(V,Q),\pi) \cong (C(U,Q) \otimes C(W,Q),\ \pi \otimes \pi)$.

<u>Proof</u>. Let $d = u_1u_2u_3 \ldots u_{4r}$. Then $d^2 = 1$, $\pi d = d$, $du_i = -u_i d$, $1 \leq i \leq 4r$ and $u_j' = du_j = u_j d$, $j \geq 4r+1$, satisfy $u_j'^2 = 1$, $u_j' u_k' = -u_k' u_j'$, $u_j' u_i = u_i u_j'$, $1 \leq i < 4r+1 \leq j < k \leq n$ and $\pi u_j' = u_j'$. It follows that the subalgebra of $C(V,Q)$ generated by the u_j' is isomorphic to $C(W,Q)$ and together with the restriction of π is isomorphic as algebra with involution to $(C(W,Q),\pi)$. Hence we can identify $(C(W,Q),\pi)$ with the algebra generated by the u_j' together with the restriction of π to this algebra. Since the u_i, $1 \leq i \leq 4r$, generate $C(U,Q)$ and the u_j' generate $C(W,Q)$, $C(U,Q)$ and $C(W,Q)$ commute element-wise. Hence $C(V,Q) \cong C(U,Q) \otimes C(W,Q)$. Since $\pi u_i = u_i$ and $\pi u_j' = u_j'$ and the u_i and u_j' are generators of $C(U,Q)$ and $C(W,Q)$ we have $(C(V,Q),\pi) \cong (C(U,Q) \otimes C(W,Q),\ \pi \otimes \pi)$.

<u>Lemma 3</u>. Let c and s denote the standard involution $a \rightsquigarrow \bar{a}$ in $\mathbb{C}$ and $\mathbb{H}$ respectively. Then $(\mathbb{C} \otimes \mathbb{H},\ c \otimes s) \cong (M_2(\mathbb{C}),*)$ and $(\mathbb{H} \otimes \mathbb{H},\ s \otimes s) \cong (M_4(\mathbb{R}),t)$.

<u>Proof</u>. 1. We have $\mathbb{C} \otimes \mathbb{H} \cong M_2(\mathbb{C})$ and the involution $c \otimes s$ is of second kind. Hence either $(\mathbb{C} \otimes \mathbb{H},\ c \otimes s) \cong (M_2(\mathbb{C}),*)$ or $(\mathbb{C} \otimes \mathbb{H},\ c \otimes s) \cong (M_2(\mathbb{C}),K)$ where

$$K: \begin{pmatrix} a & b \\ c & d \end{pmatrix} \rightsquigarrow \begin{pmatrix} 1 & 0 \\ 0 & -1 \end{pmatrix} \begin{pmatrix} \bar{a} & \bar{c} \\ \bar{b} & \bar{d} \end{pmatrix} \begin{pmatrix} 1 & 0 \\ 0 & -1 \end{pmatrix} = \begin{pmatrix} \bar{a} & -\bar{c} \\ -\bar{b} & \bar{d} \end{pmatrix}.$$

Now $H(M_2(\mathbb{C}),K)$ contains the matrix $\begin{pmatrix} -1 & 1 \\ -1 & 1 \end{pmatrix}$ whose square is 0. On the other hand, $H(\mathbb{C} \otimes \mathbb{H}, c \otimes s)$ has the base $1, \ell \otimes i, \ell \otimes j, \ell \otimes k$ where $\ell = \sqrt{-1}$ in $\mathbb{C}$. If $a = \alpha 1 + \beta\ell \otimes i + \gamma\ell \otimes j + \delta\ell \otimes k$, $\alpha, \beta, \gamma, \delta \in \mathbb{R}$ then $a^2 = (\alpha^2 + \beta^2 + \gamma^2 + \delta^2)1 + b$ where b is a linear combination of $\ell \otimes i$, $\ell \otimes j$, and $\ell \otimes k$. Hence $a^2 = 0 \Rightarrow a = 0$. Hence $(\mathbb{C} \otimes \mathbb{H}, c \otimes s) \not\cong (M_2(\mathbb{C}),K)$. Then $(C \otimes \mathbb{H}, c \otimes s) \cong (M_2(\mathbb{C}),*)$.

2. Consider the left and right regular matrix representations of $\mathbb{H}$ determined by the base $(1,i,j,k)$. These are respectively

$$\lambda: \alpha + \beta i + \gamma j + \delta k \rightsquigarrow \begin{pmatrix} \alpha & -\beta & -\gamma & -\delta \\ \beta & \alpha & -\delta & \gamma \\ \gamma & \delta & \alpha & -\beta \\ \delta & -\gamma & \beta & \alpha \end{pmatrix}$$

$$\rho: \alpha + \beta i + \gamma j + \delta k \rightsquigarrow \begin{pmatrix} \alpha & -\beta & -\gamma & -\delta \\ \beta & \alpha & \delta & -\gamma \\ \gamma & -\delta & \alpha & \beta \\ \delta & \gamma & -\beta & \alpha \end{pmatrix}.$$

We have $\mathbb{H} \cong \lambda(\mathbb{H})$ and since $\mathbb{H}$ has an involution we also have $\mathbb{H} \cong \rho(\mathbb{H})$. Any matrix in $\lambda(\mathbb{H})$ commutes with any in $\rho(\mathbb{H})$. It follows that $M_4(\mathbb{R}) = \lambda(\mathbb{H})\,\rho(\mathbb{H}) \cong \mathbb{H} \otimes \mathbb{H}$. The matrices $\lambda(i)$, $\lambda(j)$, $\lambda(k)$, $\rho(i)$, $\rho(j)$, $\rho(k)$ are skew symmetric. Hence the matrices 1, $\lambda(i)\,\rho(i)$, $\lambda(i)\,\rho(j)$, $\lambda(i)\,\rho(k)$, $\lambda(j)\,\rho(i)$, $\lambda(j)\,\rho(j)$, $\lambda(j)\,\rho(k)$, $\lambda(k)\,\rho(i)$, $\lambda(k)\,\rho(j)$, $\lambda(k)\,\rho(k)$ are symmetric. These form a base for $H(M_4(\mathbb{R}),t)$. It follows that $(\mathbb{H} \otimes \mathbb{H}, s \otimes s) \cong (M_4(\mathbb{R}),t)$. □

Theorem 1. Let $n = 2\nu$: Then

(9) $(C(V,Q),\pi) \cong (M_{2^{\nu}}(\mathbb{R}),t)$ if $\nu \equiv 0$ or $1 \pmod 4$ ($n \equiv 0$ or $2 \pmod 8$)

(10) $(C(V,Q),\pi) \cong (M_{2^{\nu-1}}(\mathbb{H}),*)$ if $\nu \equiv 2$ or $3 \pmod 4$ ($n \equiv 4$ or $6 \pmod 8$).

In the first case the Jordan algebra $J = \mathbb{R}1 \oplus V$ of the positive definite symmetric bilinear form for V can be imbedded in $\mathbb{H}(M_{2^{\nu}}(\mathbb{R}),t)$ and in the second case J can be imbedded in $\mathbb{H}(M_{2^{\nu+1}}(\mathbb{R}),t)$.

Proof. The isomorphisms (9) and (10) hold if $\nu = 1$ and 2 by Lemma 1. If $\nu = 3$ we can apply Lemma 2 to obtain $(C(V,Q),\pi) \cong (C(U,Q) \otimes C(W,Q),\ \pi \otimes \pi)$ where $[U:\mathbb{R}] = 4$ and $[W:\mathbb{R}] = 2$. Then $(C(U,Q),\pi) \cong (M_2(\mathbb{H}),*)$ and $(C(W,Q),\pi) \cong (M_2(\mathbb{R}),t)$ by Lemma 1. Then $(C(V,Q),\pi) \cong (C(U,Q) \otimes C(W,Q),\ \pi \otimes \pi) \cong (M_2(\mathbb{H}) \otimes M_2(\mathbb{R}),\ * \otimes t) \cong (M_4(\mathbb{H}),*)$ as in (10). Next let $\nu = 4$. Then $(C(V,Q),\pi) \cong (C(U,Q) \otimes C(W,Q),\ \pi \otimes \pi)$ where $[U:\mathbb{R}] = 4 = [W:\mathbb{R}]$. Then $(C(U,Q),\pi) \cong (M_2(\mathbb{H}),*) \cong (C(W,Q),\pi)$. Now by Lemma 2 $(M_2(\mathbb{H}) \otimes M_2(\mathbb{H}),\ * \otimes *) \cong (M_4(\mathbb{H} \otimes \mathbb{H}),J)$ where J is the extension of the involution $s \otimes s$ on $\mathbb{H} \otimes \mathbb{H}$ to $M_4(\mathbb{H} \otimes \mathbb{H})$. Since $(\mathbb{H} \otimes \mathbb{H},\ s \otimes s) \cong (M_4(\mathbb{R}),t)$ by Lemma 3, $(M_2(\mathbb{H}) \otimes M_2(\mathbb{H}),\ * \otimes *) \cong (M_{16}(\mathbb{R}),t)$. Hence $(C(V,Q),\pi) \cong (M_{16}(\mathbb{R}),t)$ as in (9).

Now suppose $\nu > 4$. Then $(C(V,Q),\pi) \cong (C(U,Q) \otimes C(W,Q),\ \pi \otimes \pi)$ where $[U:\mathbb{R}] = 8$ and $[W:\mathbb{R}] = 2\nu - 8 = 2(\nu-4)$. Then $(C(U,Q),\pi) \cong M_{16}(\mathbb{R})$ and we can apply induction on dimensionality to $(C(W,Q),\pi)$ to obtain the result for $(C(V,Q),\pi)$. □

Theorem 2. Let $n = 2\nu - 1$. Then we have the following cases:

(11) $$(C(V,Q),\pi) \cong (M_{2^{\nu-1}}(\mathbb{R}),t) \oplus (M_{2^{\nu-1}}(\mathbb{R}),t)$$

if $\nu \equiv 1 \pmod 4$ ($n \equiv 1 \pmod 8$).

(12) $$(C(V,Q),\pi) \cong (M_{2^{\nu-2}}(\mathbb{H}),*) \oplus (M_{2^{\nu-2}}(\mathbb{H}),*)$$

if $\nu \equiv 3 \pmod 4$ ($n \equiv 5 \pmod 8$).

(13) $$(C(V,Q),\pi) \cong (M_{2^{\nu-1}}(\mathbb{C}),*)$$

if $\nu \equiv 0,2 \pmod 4$ ($n \equiv 7$ or $3 \pmod 8$). In the first case $J = \mathbb{R}1 \oplus V$ can be imbedded in $H(M_{2^{\nu-1}}(\mathbb{R}),t)$, in the second case $H(M_{2^\nu}(\mathbb{R}),t)$ and in the third case in $H(M_{2^\nu}(\mathbb{R}),t)$.

<u>Proof</u>. Let $d = u_1 \ldots u_{2\nu-1}$. Then $du_i = u_i d$ so d is in the center. Also $d^2 = (-1)^{\nu-1}\,1$ and $\pi d = (-1)^{\nu-1}d$. For the proof we need to distinguish only the cases ν odd and ν even.

Case I. ν odd. Then $d^2 = 1$ and $\pi d = d$. Let $e_1 = (1+d)/2$, $e_2 = (1-d)/2$. Then the e_i are orthogonal central idempotents such that $e_1 + e_2 = 1$. These give the decomposition $C(V,Q) = C_1 \oplus C_2$ when $C_i = C(V,Q)e_i$. Since C_i is generated by $u_1e_i,\ldots,u_{2\nu-2}e_i$ and $(u_je_i)^2 = e_i$, $(u_je_i)(u_ke_i) = -(u_ke_i)(u_je_i)$ if $j \neq k$, and $u_ke_i = e_iu_k$, it follows that $(C_i,\pi) \cong (C(U,Q),\pi)$ where U is $2\nu-2$ dimensional. Then, by Theorem 1, $(C(U,Q),\pi) \cong (M_{2^{\nu-1}}(\mathbb{R}),t)$ if $\nu \equiv 1 \pmod 4$ and $(C(U,Q),\pi) \cong (M_{2^{\nu-2}}(\mathbb{H}),*)$ if $\nu \equiv 3 \pmod 4$. We note also that since J is simple, the homomorphism $x \rightsquigarrow xe_i$ of J into C_i^+ is injective. Hence we have an imbedding of J in $H(M_{2^{\nu-1}}(\mathbb{R}),t)$ if $\nu \equiv 1 \pmod 4$ and an imbedding of J in $H(M_{2^{\nu-2}}(\mathbb{H}),*)$ and hence also in $H(M_{2^\nu}(\mathbb{R}),t)$ if $\nu \equiv 3 \pmod 4$. This proves the result in Case I.

Case II. ν even. Here $(\mathbb{R}[d],\pi) \cong (\mathbb{C},c)$ and $(C(V,Q),\pi) \cong (C(U,Q) \otimes \mathbb{C}, \pi \otimes c)$ where U is $(2\nu-2)$-dimensional. Again the result follows from Theorem 1 applied to $(C(U,Q),\pi)$ using Lemma 3. □

Now let J be any subalgebra of the Jordan algebra $H(M_m(\mathbb{R}),t)$ of $m \times m$ real symmetric matrices. Since the matrices of J can be identified with the matrices of self-adjoint linear transformations of a Euclidean vector space relative to an orthonormal base, it follows from the orthogonal complete reducibility of sets of self-adjoint linear transformations of a Euclidean space that J is orthogonally similar to a Jordan algebra of matrices in completely reduced form. We shall use this to prove

<u>Theorem 3</u>. Let J be a Jordan algebra over $\mathbb{R}$, ρ_1 and ρ_2 representations of J by real symmetric matrices. Suppose ρ_1 and ρ_2 are similar. Then ρ_1 and ρ_2 are orthogonally similar.

<u>Proof</u>. Suppose first that ρ_1 and ρ_2 are irreducible. We have a real invertible matrix s such that $\rho_2(a) = s\rho_1(a)s^{-1}$, $a \in J$. Then $\rho_2(a) = ({}^ts)^{-1}\,\rho_1(a)({}^ts)$ and $({}^ts)s\,\rho_1(a) = \rho_1(a)({}^ts)s$. It follows from Schur's lemma that the minimum polynomial of $({}^ts)s$ over $\mathbb{R}$ is irreducible. On the other hand, $({}^ts)s$ is a positive definite symmetric matrix. Hence its minimum polynomial has distinct positive roots. Then $({}^ts)s = \beta 1$, $\beta > 0$. We can replace s by the orthogonal matrix $\mathcal{O} = \beta^{-\frac{1}{2}}s$ to obtain $\rho_2(a) = \mathcal{O}\rho_1(a)\mathcal{O}^{-1}$.

Now let ρ_1 and ρ_2 be arbitrary. We can replace the ρ_i by orthogonally similar representations so we may assume

$$\rho_i(a) = \mathrm{diag}\{\rho_{i1}(a), \rho_{i2}(a), \ldots, \rho_{ir_i}(a)\}$$

where the ρ_{ij} are irreducible. By the Krull-Schmidt theorem we have $r_1 = r_2$ and we may assume there exist invertible matrices s_j such that $\rho_{2j}(a) = s_j\rho_{1j}(a)s_j^{-1}$, $1 \leq j \leq r_1$. By the case considered first we may assume s_j is orthogonal. Then $\rho_2(a) = \mathcal{O}\rho_1(a)\mathcal{O}^{-1}$, $\mathcal{O}$ orthogonal. □

We shall now show that the imbeddings in real symmetric matrices we determined are irreducible. We recall that we are assuming $n \geq 3$ for the Jordan algebras $H(M_n(C),*)$. It is a known result of F.D. Jacobson and N. Jacobson ([2]) that if $n \geq 3$ the special universal envelope of $H(M_n(C),*)$ is $(M_n(C),\iota)$, where ι is the injection map. The special universal envelope for $\mathbb{R}1 \oplus V$ is $C(V,Q)$. By Theorems 1 and 2 we have the following information:

$$
\begin{aligned}
(C(V,Q) &\cong M_{2^\nu}(\mathbb{R}) \quad \text{if} \quad [V:\mathbb{R}] = 2\nu \quad \text{and} \quad \nu \equiv 0,\ 1 \pmod 4)\\
&\cong M_{2^{\nu-1}}(\mathbb{H}) \quad \text{if} \quad [V:\mathbb{R}] = 2\nu \quad \text{and} \quad \nu \equiv 2,\ 3 \pmod 4\\
&\cong M_{2^{\nu-1}}(\mathbb{R}) \oplus M_{2^{\nu-1}}(\mathbb{R}) \quad \text{if} \quad [V:\mathbb{R}] = 2\nu-1 \text{ and } \nu \equiv 1 \pmod 4\\
&\cong M_{2^{\nu-2}}(\mathbb{H}) \oplus M_{2^{\nu-2}}(\mathbb{H}) \quad \text{if} \quad [V:\mathbb{R}] = 2\nu-1 \text{ and } \nu \equiv 3 \pmod 4\\
&\cong M_{2^{\nu-1}}(\mathbb{C}) \quad \text{if} \quad [V:\mathbb{R}] = 2\nu-1 \quad \text{and} \quad \nu \equiv 0,\ 2 \pmod 4.
\end{aligned}
$$

Since all the algebras we are considering except $C(V,Q)$ with $[V:\mathbb{R}] = 2\nu-1$, $\nu \equiv 1,\ 3 \pmod 4$ are simple, they have only one irreducible representation the degree of which is $n[C:\mathbb{R}]$ ($= n$, $2n$ or $4n$ for $M_n(C)$) and is 2^ν, $2^{\nu+1}$, 2^ν in the first, second and fifth case listed for $C(V,Q)$. In the third case we have two inequivalent irreducible representations of degree $2^{\nu-1}$ and in the fourth case we have two inequivalent irreducible representations of degree 2^ν. Now any representation of a Jordan algebra has a unique extension to its special universal algebra. In particular, this is true of the imbeddings in algebras of real symmetric matrices we gave. Since the degrees of these imbeddings coincide with the degrees of the irreducible representations of the enveloping algebras the imbeddings we gave are all irreducible.

Our results imply the following theorem which gives a determination up to orthogonal similarity of the Jordan algebras of real symmetric matrices.

Main Theorem. Let J be a Jordan algebra of real symmetric matrices. Then

1. J is orthogonally similar to an algebra in completely reduced form.

2. Any irreducible component of J is isomorphic to one and only one of the following: (i) $\mathbb{R}1$, (ii) $H(M_n(C),*)$ where $n > 2$ and $C = \mathbb{R}$, $\mathbb{C}$ or $\mathbb{H}$ and $*$ is the usual hermitian involution, (iii) $\mathbb{R}1 \oplus V$ the Jordan algebra of a positive definite symmetric bilinear form on an n-dimensional real vector space with $n \geq 1$. Moreover, there exist irreducible Jordan algebras of real symmetric matrices isomorphic to any of the algebras we have listed.

3. For (i) the degree of the irreducible component is 1 and for (ii) it is $n[C:\mathbb{R}]$ and any two irreducible Jordan algebras of real symmetric matrices isomorphic to $H(M_n(C),*)$ are orthogonally similar. For (iii) the degree of an irreducible algebra of real symmetric matrices isomorphic to $\mathbb{R}1 \oplus V$ is 2^ν if $n = 2\nu$ and $\nu \equiv 0,1 \pmod 4$, $2^{\nu+1}$ if $n = 2\nu$ and $\nu \equiv 2, 3 \pmod 4$, $2^{\nu-1}$ if $n = 2\nu-1$ and $\nu \equiv 1 \pmod 4$, 2^ν if $n = 2\nu-1$ and $\nu \equiv 3 \pmod 4$, 2^ν if $n = 2\nu-1$ and $\nu \equiv 0, 2 \pmod 4$. Any two irreducible algebras of real symmetric matrices isomorphic to a Jordan algebra $\mathbb{R}1 \oplus V$ are orthogonally similar except in the case in which $n = 2\nu-1$, $\nu \equiv 1, 3 \pmod 4$. In this case there are two orbits of irreducible algebras of real symmetric matrices isomorphic to $\mathbb{R}1 \oplus V$ under the action of the orthogonal group.

4. Two Jordan algebras of real symmetric matrices are orthogonally similar if and only if they have similar irreducible components each occurring with the same multiplicity.

Remarks. 1. For a given m there are only a finite number of orbits of subalgebras of $H(M_m(\mathbb{R}),t)$ under the action of the orthogonal group.

2. The methods we have used give explicit "canonical forms", that is, representatives of the orbits of subalgebras of $H(M_m(\mathbb{R}),t)$.

I am indebted to Kevin McCrimmon for a critical reading of an earlier version of this paper.

References

1. P. Jordan, J. v. Neumann and E. Wigner, "On an algebraic generalization of the quantum mechanical formalism", Annals of Math. v. 35 (1937), 29-14

2. F. D. Jacobson and N. Jacobson, "Classification and representation of semi-simple Jordan algebras", Trans. Amer. Math. Soc. v. 65 (1949), 141-169.

3. N. Jacobson, Structure and Representations of Jordan Algebras, Amer. Math. Soc. Colloq. v. 39.

4. J. D. Malley, Optimal Unbiased Estimation of Variance Components, Lecture Notes in Statistics, v. 39, Springer-Verlag.

5. J. D. Malley, "Finitely generated algebras of real symmetric matrices", the preceding paper in this journal.

Expository Papers

Reprinted from the
BULLETIN OF THE AMERICAN MATHEMATICAL SOCIETY
Vol. 46, No. 7, pp. 592-595

The Classical Groups. By Hermann Weyl. Princeton, University Press, 1939. 11+302 pp.

It is a curious fact that while almost all the textbooks on higher algebra written prior to 1930 devote considerable space to the subject of invariants, the recent ones written from the axiomatic point of view disregard it completely. Because of this neglect the phrase "invariant theory" is apt to suggest a subject that was once of great interest but one that has little bearing on modern algebraic developments. For this reason it is an important and original accomplishment that Professor Weyl has made here in connecting the theory of invariants with the main stream of algebra and in indicating that the subject has a future as well as a distinguished past.

In his treatment the theory of invariants becomes a part of the theory of representations. The most natural way to begin the study of invariants of a particular group is therefore to determine its representations. A large part of the book is concerned with this problem as it applies to the "classical" groups $GL(n)$, the full linear group; $O(n)$, the orthogonal, and $S(n)$, the symplectic (complex or abelian) group.

In discussing the representations of an abstract group $\mathfrak{g}$ one finds it convenient to adjoin to the set of representing matrices their linear combinations. The resulting set is an algebra, the enveloping algebra of the original set, and defines a representation of a certain abstract algebra, the group algebra, that is completely determined by $\mathfrak{g}$ and the field of the coefficients. In this way the theory of algebra is applicable. The author has gone somewhat beyond his immediate needs in discussing this domain. Consequently his book may serve also as an excellent introduction to this theory.

The book begins with a brief review of the basic concepts of field, abstract group, vector space, linear transformation (matrix) and representation of a group. These are used to give the following definition of an invariant: Let $x, y, \cdots$ be variable vectors varying in different representation spaces of an abstract group $\mathfrak{g}$ and let $A(s), B(s), \cdots$ be the corresponding matrices. A polynomial $f(x, y, \cdots)$ in the coordinates of $x, y, \cdots$ is an invariant of $\mathfrak{g}$ if $f(A(s)x, B(s)y, \cdots) = f(x, y, \cdots)$ for all s in $\mathfrak{g}$. Relative invariants and covariants are defined along similar lines. The central problems to which this definition gives rise are: (1) Given a class of invariants, to determine, if possible, a basis, that is, a finite number $I_1, I_2, \cdots, I_r$ of these invariants such that every invariant in the class is expressible as a polynomial in the I's. (2) To determine a fundamental set of relations obtaining between the basic invariants. The solution of these problems for vector invariants[1] of the unimodular group $SL(n)$ and of the orthogonal group is the subject of Chapter II. There are two essential steps in the proofs, the formal part by which the theorems are reduced by certain identities (Capelli's) to the cases where the number of vector variables of each kind does not exceed $(n-1)$ and the numerical argument which disposes of the reduced cases. The proofs are less direct for $O(n)$ than for $SL(n)$. For here it is necessary to suppose first that the underlying field is real and then to extend the results to arbitrary fields of characteristic 0 by using Cayley's parametrization of the "general" orthogonal matrix of determinant 1 as $(E-S)(E+S)^{-1}$ where E is the identity and S is an arbitrary skew-symmetric matrix.

The third chapter introduces the theory of algebras. The treatment given is based on the concept of a completely reducible matrix algebra. It is shown that any representation of an algebra $\mathfrak{A}$ of this type is completely reducible and there is a reciprocity between $\mathfrak{A}$ and the algebra $\mathfrak{B}$ consisting of the matrices commutative with those of $\mathfrak{A}$: $\mathfrak{B}$ is completely reducible and $\mathfrak{A}$ in turn may be characterized as the set of matrices commutative with those of $\mathfrak{B}$. Together with Schur's lemma this result yields Wedderburn's fundamental theorems. The methods used here are well suited for handling problems of the type discussed in this book, where algebras occur in concrete form as sets of matrices. More abstract methods would be required if one wished to obtain these theorems in their most general form. In the second half of the chapter an independent treatment of group algebras is given. Their complete reducibility and reciprocity with the commutator algebras is derived in a very elementary way. As a consequence the results, particularly the latter, are given a much more explicit form than in the general situation.

After this machinery has been set up, it is possible to determine the (integral) representations of $GL(n)$. One must establish first the complete reducibility of the tensor representation $A\rightarrow A\times A\times\cdots\times A$ (Kronecker product of f factors) and to split this representation into its irreducible parts. This is accomplished by showing that the enveloping algebra $\mathfrak{A}_f$ is the commutator of a certain representation of σ_f, the group algebra of the symmetric group on f letters. Explicit formulas for the idempotent elements of σ_f (Young symmetrizers)

[1] That is, functions of an arbitrary number of covariant and contravariant vectors where by a covariant vector we mean one belonging to the original representation $x\rightarrow Ax$ and contravariant refers to vectors in $\xi\rightarrow(A')^{-1}\xi$, A' the transposed matrix. There is no distinction between covariant and contravariant vectors if A is orthogonal.

yield the irreducible subspaces of tensor space. These results may be extended to the representation obtained by stringing along the diagonal the tensor representations of ranks 1, 2, $\cdots$, f. The enveloping algebra is here denoted as $\mathfrak{A}^{(f)}$. Now, if $A \rightarrow T(A)$ is any representation of $GL(n)$ whose coordinates are polynomials of degree not greater than f in the coordinates of A, then the enveloping algebra of $T(A)$ represents $\mathfrak{A}^{(f)}$. The general theory shows that $T(A)$ is completely reducible into parts similar to some of those contained in the tensor representations.

To obtain these results for $O(n)$ the author follows a procedure due to Brauer (used first by Weyl for $S(n)$). The complete reducibility of the tensor representation is established for the case of real fields. Its commutator algebra is determined by using the first main theorem for orthogonal invariants. Finally one obtains the enveloping algebra of the tensor representation by the reciprocity theorem. The tool by which the reality restriction is removed is again Cayley's parametrization. The real significance of the parametrization is brought out in the following beautiful theorem: The ideal generated by the relations defining an orthogonal matrix of determinant 1 is prime and has the generic zero $(E-S)(E+S)^{-1}$. To obtain the irreducible parts of tensor space relative to $O(n)$ a process of contraction is used. By this means the problem is reduced to one of decomposing certain spaces relative to the symmetric group. These results are then extended to the proper orthogonal group. The discussion of the symplectic group is reserved for a separate chapter. It parallels that of $O(n)$ and enables the reader to get a bird's-eye view of the entire structure as it appears up to this point.

In cases such as those under discussion here, in which the representations decompose into absolutely irreducible parts, a knowledge of these parts, though not their explicit construction, may be obtained by expressing the character of the given representation as a sum of primitive (irreducible) characters. This is done in Chapter VII for $GL(n)$, $O(n)$ and $S(n)$. The method is that of Weyl's "unitarian trick." One replaces these groups by certain compact Lie groups (thus, in place of $GL(n)$ one uses the group of unitary matrices). The characters of the latter may be obtained by using an invariant integration defined on the group manifolds. Since any algebraic relation with rational coefficients in the elements of the compact group $\mathfrak{u}$ holds also for the original group $\mathfrak{g}$, the formulas obtained for $\mathfrak{u}$ are valid for $\mathfrak{g}$ after being expressed as polynomials in the coefficients of the matrices. It is shown incidentally that the representations, determined by the earlier constructions form a complete set of continuous irreducible representations for the compact groups. As an illustration of the use of characters, the author determines the irreducible components of the Kronecker product of irreducible representations.

In the first part of Chapter VIII the general definition of an invariant is specialized to the classical case. Here the representation space is determined by the coefficients of a homogeneous polynomial. The symbolic method as well as certain irrational methods are discussed as tools to obtain the first fundamental theorem. Hilbert's basis theorem for polynomial ideals settles the question of the finiteness of relations among the basic invariants. The symbolic method is then used to extend these results to general orthogonal and symplectic invariants. The ideas in the last half of the chapter cluster around the concept of a Lie group and its Lie algebra. The local properties of these groups are all expressible in terms of the algebras and if the group is simply connected one may readily pass to its properties in the large. On the other hand Lie algebras may be studied independently of the group theory and questions analogous to those regarding representations, invariants, and so on, may be raised. For compact Lie groups the powerful integration method finds another application here in a very simple derivation of the first fundamental theorem for general invariants. The last chapter resumes the discussion of algebras. A number of interesting results concerning automorphisms and direct products are obtained by matrix methods.

We have confined our description to the broad outlines of the subject. Space does not permit a detailed account of the interesting bypaths indicated by the author. We may mention, for example, a new formulation of Klein's Erlanger program, the close connection between representation theory and the theory of almost periodic functions and Fourier series, the topology of the classical groups. There are ample indications in the notes and bibliography to enable the reader to pursue these questions further.

In short, the book is heartily recommended to the present generation of algebraists as an introduction to a rich and rather neglected field and to those educated in the classical tradition for an insight into important recent algebraic ideas and their applicability to familiar problems.

N. JACOBSON

REPRESENTATION THEORY FOR JORDAN RINGS[1]

N. Jacobson

It is well known that the theory of Lie algebras is equivalent to the study of subspaces of associative algebras which are closed relative to the composition $[ab] \equiv ab - ba$. Similarly, the theory of Jordan algebras has arisen in the attempt to study subspaces of an associative algebra which are closed relative to the composition $\{ab\} = ab + ba$. The characterization of the Lie composition by identities is well known ([5] and [21]). (Besides the bilinearity, the axioms which characterize $[ab]$ are $[aa] = 0$ and the Jacobi identity.) On the other hand, we do not possess a set of identities for $\{ab\}$ which characterize this composition. It is easy to see that $\{ab\}$ satisfies

$$(1) \qquad \{ab\} = \{ba\}, \qquad \{\{aa\}\{ba\}\} = \{\{\{aa\}b\}a\},$$

and this observation has led to the definition of an *(abstract) Jordan algebra* as a (nonassociative) algebra whose composition ab satisfies the properties (1) of $\{ab\}$, that is,

$$(1') \qquad ab = ba, \qquad a^2(ba) = (a^2b)a.$$

The algebras obtained from subspaces of associative algebras closed relative to $\{ab\}$ are called *special Jordan algebras*. It is known that there exist Jordan algebras which are not special [1] so that the properties (1) do not give an exact characterization of the special systems. However, this may be the best that one can do in the way of identities; for it is conceivable that every Jordan algebra is a homomorphic image of a special one. At any rate, an extensive theory can be built on the axioms (1′) (see particularly [16] and [3]). In this note we shall be concerned primarily with an extension of this theory of abstract Jordan algebras.

1. Definition and elementary properties of representations of Jordan algebras. The second identity $(a^2b)a = a^2(ba)$ for Jordan algebras is cubic in a. If the characteristic is not two or three (and we assume this unless otherwise stated), then this identity is equivalent to the multilinear identity:

$$(2) \qquad xybz + yzbx + zxby = (xy)(bz) + (yz)(bx) + (zx)(by)$$

in which we have abbreviated $((xy)b)z$ to $xybz$, etc. If we denote the right multiplication mapping $x \to xa$ by R_a, then we obtain from (2) the following relations:

[1] The main results stated in the first five sections of this paper can be found in [14]. The results in the last section are contained in the joint paper [15] by Rickart and the present author.

$$(3)\qquad \begin{gathered}[R_aR_{bc}] + [R_bR_{ac}] + [R_cR_{ab}] = 0\\ R_aR_bR_c + R_cR_bR_a + R_{(ac)b} = R_aR_{bc} + R_bR_{ac} + R_cR_{ab}\,.\end{gathered}$$

We define a (general) representation of a Jordan algebra in the following way.

DEFINITION 1. A mapping $a \to S_a$ of a Jordan algebra into a set of linear transformations of a vector space $\mathfrak{M}$ is called a *representation* if (1) S_a is linear in a and (2) S_a satisfies

$$(4)\qquad \begin{gathered}[S_aS_{bc}] + [S_bS_{ac}] + [S_cS_{ab}] = 0\\ S_aS_bS_c + S_cS_bS_a + S_{(ac)b} = S_aS_{bc} + S_bS_{ac} + S_cS_{ab}\,.\end{gathered}$$

This concept is equivalent to that of a Jordan module which has been given by Eilenberg [7]. We shall not require the alternative formulation here.

Evidently the mapping $a \to R_a$ is a representation. As usual we call this mapping *the regular representation.* This representation plays a fundamental role in the structure theory. On the other hand, if one recalls the origin of the Jordan theory, one is led to consider also mappings $a \to U_a$ of a Jordan algebra into linear transformations such that U_a is linear in a and

$$(5)\qquad U_{ab} = U_aU_b + U_bU_a\,.$$

Such mappings, which are just the homomorphisms of Jordan rings into the special Jordan rings of linear transformations, have been considered before ([6] and [11]). It is noteworthy that these mappings are representations in the sense defined here. We call these representations *special.* Thus the general concept of representations can be used to unify to some extent the structure theory and the theory of special representations.

There is another interesting connection between special and general representations. Let $a \to U_a$, $a \to V_a$ be two special representations acting in the same vector space. Assume that these *commute* in the sense that $[U_aV_b] = 0$ for all a, b. Then it can be verified that $a \to S_a \equiv U_a + V_a$ is a representation. In particular, if $a \to U_a$ and $a \to V_a$ are arbitrary special representations, then $a \to S_a = U_a \times 1 + 1 \times V_a$, where the $\times$ denotes Kronecker multiplication, is a representation.

2. Universal associative algebras. Let $\mathfrak{A}$ be a Jordan algebra and let $\mathfrak{F} = \mathfrak{A} \oplus \mathfrak{A} \times \mathfrak{A} \oplus \mathfrak{A} \times \mathfrak{A} \times \mathfrak{A} \oplus \cdots$ be the free associative algebra determined by the vector space $\mathfrak{A}$. Multiplication in $\mathfrak{F}$ is defined by the distributive laws and the rule

$$(x_1 \times x_2 \times \cdots \times x_r) \times (x_{r+1} \times \cdots \times x_s) = x_1 \times x_2 \times \cdots \times x_s\,.$$

Let $\mathfrak{K}$ be the ideal in $\mathfrak{F}$ generated by the elements

$$(6)\qquad \begin{gathered}a \times bc - bc \times a + b \times ac - ac \times b + c \times ab - ab \times c,\\ a \times b \times c + c \times b \times a + (ac)b - a \times bc - b \times ca - c \times ab,\end{gathered}$$

and let $\mathfrak{U} = \mathfrak{F}/\mathfrak{K}$. We shall call $\mathfrak{U}$ the *universal associative algebra* (*of the representations*) of $\mathfrak{A}$. The algebra $\mathfrak{U}$ has the following properties: (1) Any representation of $\mathfrak{A}$ can be extended to one of $\mathfrak{U}$ (in the associative sense). (2) Any (associative) representation of $\mathfrak{U}$ defines a representation of $\mathfrak{A}$. A basic result in this connection is that $\mathfrak{U}$ *is finite-dimensional if* $\mathfrak{A}$ *is finite-dimensional*. This has the consequence that a finite-dimensional Jordan algebra has only a finite number of inequivalent irreducible representations. A fundamental problem which is as yet unsolved is the determination of these representations.

We can define also a *special universal associative algebra* $\mathfrak{U}_s$ for the special representations of $\mathfrak{A}$. This is obtained by replacing the ideal $\mathfrak{K}$ by the ideal $\mathfrak{K}_s$ generated by the elements $a \times b + b \times a - ab$. The algebra $\mathfrak{U}_s$ bears the same relation to special representations that $\mathfrak{U}$ does to all the representations. It is of interest to introduce also another universal associative algebra which can be used to study Kronecker sums $(U_a \times 1 + 1 \times V_a)$ of special representations. We obtain this algebra by adjoining a (new) identity 1 to $\mathfrak{U}_s$ to obtain $\mathfrak{U}_s^*$. Let $\mathfrak{U}_s^{(2)}$ be the subalgebra of $\mathfrak{U}_s^* \times \mathfrak{U}_s^*$ generated by the elements $a_s \times 1 + 1 \times a_s$ where a_s is the coset of $a + \mathfrak{K}_s$ in $\mathfrak{U}_s$. Any Kronecker sum of special representations defines a representation of $\mathfrak{U}_s^{(2)}$.

The algebras $\mathfrak{U}_s$ and $\mathfrak{U}_s^{(2)}$ are homomorphic images of $\mathfrak{U}$. Also $\mathfrak{U}_s$ is a homomorphic image of $\mathfrak{U}_s^{(2)}$. It would be interesting to have more precise results on the relation between $\mathfrak{U}$ and $\mathfrak{U}_s^{(2)}$ for special Jordan algebras $\mathfrak{A}$. At the present time we know of no example of a special Jordan algebra for which $\mathfrak{U} \not\cong \mathfrak{U}_s^{(2)}$. Such examples could be used to construct extensions of $\mathfrak{A}$ which are not special.

3. Associator structure Lie triple systems. If $\mathfrak{A}$ is a special Jordan algebra with composition $\{ab\}$, a direct verification shows that the Jordan associator

$$A(b, c, a) \equiv \{\{bc\}a\} - \{b\{ca\}\} = [[ab]c]. \tag{7}$$

Hence $\mathfrak{A}$ is closed also relative to the Lie ternary composition $[abc] \equiv [[ab]c]$. A subspace of an associative algebra having this property is called a (*special*) *Lie triple system*.

It is possible to give a characterization by identities of Lie triple systems. Thus let $\mathfrak{T}$ be a vector space in which a ternary composition $[abc]$ is defined. Assume that this composition is trilinear and that it satisfies the following relations:

$$\begin{gathered}
[aab] = 0 \\
[abc] + [bca] + [cab] = 0 \\
[[abc]de] + [[bad]ce] + [ba[cde]] + [cd[abe]] = 0 \\
[[abc]de] + [[bad]ce] + [[dcb]ae] + [[cda]be] = 0 \\
[[[abc]de]fg] + [[[bac]df]eg] + [[[bad]ce]fg] \\
+ [[[abd]cf]eg] + Q + R = 0
\end{gathered} \tag{8}$$

where Q and R are obtained from the displayed terms by cyclic permutation of the pairs (a, b), (c, d), (e, f). Then it can be shown that $\mathfrak{T}$ can be imbedded in a Lie algebra $\mathfrak{L}$ in such a way that $\mathfrak{T}$ becomes a subspace of $\mathfrak{L}$ closed relative to $[[ab]c]$ and that $[[ab]c] = [abc]$. Since every Lie algebra can be identified with a subspace of an associative algebra closed under $[ab] \equiv ab - ba$, $\mathfrak{T}$ can be identified with a subspace of an associative algebra closed relative to $[[ab]c]$.

One can also associate a Lie triple system with every abstract Jordan algebra. If $\mathfrak{A}$ is such an algebra, we define $[abc] = A(b, c, a) = (bc)a - b(ca)$. Then it can be proved that this composition satisfies (8). This can be established easily by noting the following consequence of the second relation in (3):

$$[[R_a R_b]R_c] = R_{[abc]} \,. \tag{9}$$

The system consisting of the vector space $\mathfrak{A}$ and the composition $[abc]$ is called the *associator* (*Lie triple*) *system* of the Jordan algebra. As a generalization of (9) we have

$$[[S_a S_b]S_c] = S_{[abc]} \,, \tag{10}$$

and this shows that if $a \to S_a$ is a representation of $\mathfrak{A}$, then it is also a representation of the associator system, that is, it is a Lie triple system homomorphism of $\mathfrak{A}$ into the Lie triple system of linear transformations.

It is possible to define a universal Lie algebra $\mathfrak{L}_u$ for any Lie triple system $\mathfrak{T}$ in a manner similar to that indicated in the preceding section. If $\mathfrak{T}$ is a Lie triple system contained in a Lie algebra, the Lie algebra generated by $\mathfrak{T}$ is the set $\mathfrak{T} + [\mathfrak{T}\mathfrak{T}]$ of sums $a + \sum [b_i c_i]$. It is clear from this that if $\mathfrak{T}$ is finite-dimensional, then so is $\mathfrak{T} + [\mathfrak{T}\mathfrak{T}]$. In particular the universal Lie algebra of a finite-dimensional Lie triple system is finite-dimensional. Thus we can associate with every finite-dimensional Jordan algebra a finite-dimensional Lie algebra, the universal Lie algebra of its associator system. This association is very useful in the representation theory.

4. Representation theory for finite-dimensional algebras. At the present time we are in possession of some of the basic facts on representations of finite-dimensional Jordan algebras. These have been obtained by making extensive use of Lie algebra theory. Consequently we have had to assume that the base fields of our algebras are of characteristic 0.

We state now two of the main results of the representation theory of finite-dimensional algebras.

THEOREM A. *Every representation of a finite-dimensional semi-simple Jordan algebra* $\mathfrak{A}$ *of characteristic* 0 *is completely reducible.*

The analogous result for Lie algebras is required for the proof of this theorem. We recall that the Lie algebra result is best proved by establishing first a certain cohomology lemma due to Whitehead [9]. In the Jordan case we have found it more convenient to reverse this procedure and prove Theorem A first. Using this result we can establish the following analogue of Whitehead's (first) lemma.

THEOREM B. *Let $\mathfrak{A}$ be as in the preceding theorem and let $a \to S_a$ be a representation of $\mathfrak{A}$ acting in the vector space $\mathfrak{M}$. Let $a \to f(a)$ be a linear mapping of $\mathfrak{A}$ into $\mathfrak{M}$ such that*

$$f(ab) = f(a)S_b + f(b)S_a .$$

Then there exist elements $w_i \in \mathfrak{M}$ and $b_i \in \mathfrak{A}$ such that

$$f(a) = \sum w_i(S_a S_{b_i} - S_{ab_i}).$$

The analogue for Jordan algebras of Levi's theorem on Lie algebras has been proved recently by Penico [20]: If $\mathfrak{A}$ is a finite-dimensional Jordan algebra of characteristic 0, then $\mathfrak{A} = \mathfrak{S} + \mathfrak{N}$ when $\mathfrak{N}$ is the radical (maximal solvable ideal) and $\mathfrak{S}$ is semi-simple. The analogues of the supplementary results on the Levi decomposition which are due to Malcev and Harish-Chandra ([8] and [18]) can be obtained for Jordan algebras by using Theorem B.

5. Jordan homomorphisms of rings. At the present time little is known on the structure of Jordan algebras of infinite dimensions or of Jordan rings (see, however, [19]). It is therefore somewhat premature to consider the representation theory for such systems. However, it does seem to be of interest to study certain special Jordan rings obtained from associative rings. For example, the problem of semi-automorphisms of rings which was introduced by Ancochea [4] can be regarded as a problem on automorphisms of Jordan rings. If $\mathfrak{A}$ is an associative ring, a *semi-automorphism* S of $\mathfrak{A}$ is a 1–1 mapping of $\mathfrak{A}$ onto itself satisfying

$$(a + b)^S = a^S + b^S, \qquad (ab + ba)^S = a^S b^S + b^S a^S. \tag{11}$$

Evidently this is just an automorphism of the special Jordan ring $\mathfrak{A}_J$ obtained from $\mathfrak{A}$ by replacing the associative composition ab by $\{ab\}$. Because of this it seems to be more appropriate to call S a *Jordan automorphism* of $\mathfrak{A}$. Also we can generalize this notion and consider Jordan homomorphisms of one associative ring into a second one. The second condition in (11) loses a good deal of its force if the rings have elements of additive order 2 and in order to be able to treat this case, too, it is necessary to replace this condition by

$$(a^2)^S = (a^S)^2, \qquad (aba)^S = a^S b^S a^S. \tag{12}$$

Mappings of this type have been studied by a number of writers ([4], [10], [12], and [17]). Recently C. E. Rickart and the present author have undertaken a systematic study of Jordan homomorphisms in the sense of (11) [15]. Some rather surprising results have been obtained. Thus, for example, we have shown that if $\mathfrak{A} = \mathfrak{B}_n$ is a ring of n by n, $n \geqq 2$, matrices over *any* ring $\mathfrak{B}$ with an identity, then any Jordan homomorphism of $\mathfrak{A}$ is obtained by combining in an obvious way an associative homomorphism and an associative anti-homomorphism. A similar result holds also for rings which are *locally matrix* in the sense that any finite subset of elements can be embedded in a subring $\mathfrak{B}_n$, $n \geqq 2$, $\mathfrak{B}$ with an identity. A noteworthy class of locally matrix rings is the class of simple rings possessing minimal one-sided ideals. Hence the Jordan homomorphisms of these rings have been completely determined. We have also obtained the Jordan automorphisms of primitive rings having minimal one-sided ideals. We recall that a ring is *primitive* if it has a 1–1 irreducible module. The Jordan automorphisms (and more generally the Jordan homomorphisms of one such ring onto another) are either automorphisms (homomorphisms) or anti-automorphisms (anti-homomorphisms).

The methods which we have used are based in part on matrix calculations, in part, on the identity

$$[(ab)^s - a^s b^s][(ab)^s - b^s a^s] = 0, \tag{13}$$

and in part on Lie ring techniques.

In addition to the Jordan rings obtained from associative rings by using $\{ab\}$, another "classical" type of Jordan ring is the following: Let $\mathfrak{A}$ be an associative ring which has an anti-automorphism I of period two. Then the subset $\mathfrak{H}(\mathfrak{A}, I)$ of self-adjoint elements of $\mathfrak{A}$ is a special Jordan ring. The set of ordinary symmetric matrices and the set of hermitian matrices are obvious examples of this type. Rickart and I have recently considered the problem of the homomorphisms of Jordan rings of this type and we have found that a substantial portion of the results for the rings $\mathfrak{A}_J$ have analogues for the rings $\mathfrak{H}(\mathfrak{A}, I)$. We hope to publish these results shortly.

Bibliography

1. A. A. Albert, *On a certain algebra of quantum mechanics*, Ann. of Math. vol. 35 (1934) pp. 65-73.

2. ———, *On Jordan algebras of linear transformations*, Trans. Amer. Math. Soc. vol. 59 (1946) pp. 524–555.

3. ———, *A structure theory for Jordan algebras*, Ann. of Math. vol. 48 (1947) pp. 446–467.

4. G. Ancochea, *On semi-automorphisms of division algebras*, Ann. of Math. vol. 48 (1947) pp. 147–154.

5. G. Birkhoff, *Representability of Lie algebras and Lie groups by matrices*, Ann. of Math. vol. 38 (1937) pp. 526–532.

6. G. Birkhoff and P. Whitman, *Representation of Jordan and Lie algebras*, Trans. Amer. Math. Soc. vol. 61 (1942) pp. 116–136.

7. S. Eilenberg, *Extensions of general algebras*, Annales de la Société Polonaise de Mathématique vol. 21 (1948) pp. 125–134.

8. Harish-Chandra, *On the radical of a Lie algebra*, Proc. Amer. Math. Soc. vol. 1 (1950) pp. 14–17.

9. G. Hochschild, *Semi-simple algebras and generalized derivations*, Amer. J. Math. vol. 64 (1941) pp. 677–694.

10. L. K. Hua, *On the automorphisms of a field*, Proc. Nat. Acad. Sci. U. S. A. vol. 35 (1949) pp. 386–389.

11. F. D. Jacobson and N. Jacobson, *Classification and representation of semi-simple Jordan algebras*, Trans. Amer. Math. Soc. vol. 65 (1949) pp. 141–169.

12. N. Jacobson, *Isomorphisms of Jordan rings*, Amer. J. Math. vol. 70 (1948) pp. 317-328.

13. ———, *Lie and Jordan triple systems*, Amer. J. Math. vol. 71 (1949) pp. 149–170.

14. ———, *General representation theory of Jordan algebras*, Trans. Amer. Math. Soc. vol. 70 (1951) pp. 510–530.

15. N. Jacobson and C. E. Rickart, *Jordan homomorphisms of rings*, Trans. Amer. Math. Soc. vol. 69 (1950) pp. 479–502.

16. P. Jordan, E. Wigner, and J. von Neumann, *On an algebraic generalization of the quantum mechanical formalism*, Ann. of Math. vol. 35 (1934) pp. 29–64.

17. I. Kaplansky, *Semi-automorphisms of rings*, Duke Math. J. vol. 14 (1947) pp. 521-527.

18. A. Malcev, *On the representation of an algebra as direct sum of the radical and a semi-simple algebra*, C. R. (Doklady) Acad. Sci. URSS. N.S. vol. 36 (1942) pp. 42–45.

19. J. von Neumann, *On an algebraic generalization of the quantum mechanical formalism (Part 1)*, Rec. Math (Mat. Sbornik) N.S. 4 (43) (1936) pp. 415–484.

20. A. J. Penico, *The Wedderburn principal theorem for Jordan algebras*, to appear.

21. E. Witt, *Treue Darstellung Liescher Rings*, J. Reine Angew. Math. vol. 177 (1937) pp. 152–160.

Yale University,
New Haven, Conn., U. S. A.

Reprinted from
Proceedings of the International Congress of Mathematicians
American Mathematical Society, 1952.

Séminaire BOURBAKI
(Mai 1952)

Le Problème de Kurosch

par N. Jacobson

1. Histoire

Nous considérons dans la suite des algèbres $\mathfrak{a}$ sur un corps commutatif Φ. On dit que $\mathfrak{a}$ est *algébrique* si la sous-algèbre engendrée par chaque élément de $\mathfrak{a}$ est de dimension finie. On dit que $\mathfrak{a}$ est *localement finie* si chaque sous-ensemble fini de $\mathfrak{a}$ engendre une sous-algèbre de dimension finie. En 1941, KUROSCH a posé la question: est-il vrai que toute algèbre algébrique est localement finie? En 1948, KAPLANSKY a introduit la notion d'une algèbre qui satisfait à une identité polynomials (les PI algèbres). En 1950, KAPLANSKY a démontré que toute PI algèbre algébrique est localement finie. La méthode de Kaplansky est basée sur la notion de l'espace des idéaux primitifs dans un anneau. Récemment, LEVITZKI a donné une démonstration du résultat de Kaplansky qui est plus simple que celle de Kaplansky. C'est la démonstration de Levitzki [10] que je donnerai ici.

2. Préliminaires

Rappelons d'abord les notions de la théorie des anneaux arbitraires. On dit qu'un anneau $\mathfrak{a}$ est *primitif* s'il possède une représentation irréductible fidèls. Si $I = \{\mathfrak{M}\}$ est l'ensemble des modules irréductibles pour un anneau quelconque $\mathfrak{a}$, on définit le *radical* $\mathscr{R}$ comme l'ensemble des éléments z tels que $xz = 0$ pour tout x de tout $\mathfrak{M}$ de I. $\mathscr{R}$ est un idéal bilatère. $\mathfrak{a}$ est *semi-simple* si $\mathscr{R} = 0$. On obtient le même radical si on considère des modules à gauche et les anti-représentations au lieu de modules (à droite) et les représentations. Aussi on peut donner la caractérisation suivante: $\mathscr{R}$ est le plus grand idéal à droite (à gauche) *quasi-régulier* dans le sens que pour chaque $z \in \mathscr{R}$ il existe un z' tel que $z \circ z' = z + z' - zz' = 0$. Un idempotent $e \neq 0$ ne peut être quasi-régulier. Alors $\mathscr{R}$ ne contient pas d'idempotents $\neq 0$. Tout élément nilpotent est quasi-régulier et ceci entraîne que $\mathscr{R}$ contient tous les nilidéaux de l'anneau. On dit qu'un idéal bilatèrs $\mathfrak{p}$ est *primitif* (dans $\mathfrak{a}$) si $\mathfrak{a}/\mathfrak{p}$ est un anneau primitif. Par définition (à peu près) $\mathfrak{a}$ est semi-simple si et seulement si $\cap\mathfrak{p} = 0$ pour l'ensemble $\{\mathfrak{p}\}$ des idéaux primitifs. Rappelons aussi que tout idéal bilatère dans un anneau semi-simple est un anneau semi-simple. On a le même résultat pour $e\mathfrak{a}e$, e un idempotent dans l'anneau semi-simple $\mathfrak{a}$.

Ensuite supposons que $\mathfrak{a}$ est primitif et $\mathfrak{M}$ est un $\mathfrak{a}$-module irréductible fidèle. Soit $\mathfrak{c}$ le commutant de $\mathfrak{a}$ dans $\mathfrak{M}$, c'est-à-dire, l'ensemble des endomorphismes qui commutent avec les opérateurs correspondant aux éléments de $\mathfrak{a}$. On a le *lemme de Schur*: $\mathfrak{c}$ est un corps (non-commutatif). Si Γ est un anneau anti-isomorphe à $\mathfrak{c}$ on peut traiter $\mathfrak{M}$ comme un espace vectoriel (à gauche) sur Γ. Alors on a le *théorème de Burnside généralisé ou théorème de densité*: Si $x_1, x_2, \ldots, x_n$ sont des éléments de $\mathfrak{M}$ linéairement indépendants par rapport à Γ et $y_1, y_2, \ldots, y_n$ sont quelconques dans $\mathfrak{M}$, il existe un $a \in \mathfrak{a}$ tel que $x_i a = y_i$, $i = 1, 2, \ldots, n$.

3. Anneaux Matriciels

Je suppose connues les notions de décompositions de Peirce à droite, à gauche et bilatère par rapport à un ensemble fini d'idempotents. Rappelons seulement à ce sujet quelques résultats qui sont fondamentaux pour la suite.

Soient e_1, e_2 deux idempotents dans $\mathfrak{a}$. Alors les idéaux à droite $e_1\mathfrak{a}$, $a_2\mathfrak{a}$ sont $\mathfrak{a}$-isomorphes si et seulement s'il existe des éléments e_{12}, e_{21} tels que $e_{12}e_{21} = e_1$, $e_{21}e_{12} = e_2$. Supposons que $\mathfrak{a}$ possède un élément unité 1 et que $\mathfrak{a} = \mathfrak{Z}_1 \oplus \mathfrak{Z}_2 \oplus \cdots \oplus \mathfrak{Z}_n$ où les $\mathfrak{Z}_j$ sont des idéaux à droite isomorphes deux à deux. On a $1 = \sum_1^n e_j$, $e_j \in \mathfrak{Z}_j$, les e_j sont des idempotents orthogonaux et $\mathfrak{a} = e_j\mathfrak{a}$. On peut démontrer qu'il existe des *unités matricielles* e_{ij}, $i, j = 1, 2, \ldots, n$ telles que $e_{ii} = e_i$ et

$$e_{ij}e_{kl} = \delta_{jk}e_{il}, \qquad \Sigma e_{ii} = 1.$$

Si $\mathfrak{b}$ est le sous-anneau des éléments commutant avec les e_{ij}, tout $a \in \mathfrak{a}$ peut s'écrire dans la forme $a = \Sigma b_{ij}c_{ij}$, $b_{ij} \in \mathfrak{b}$. Alors $\mathfrak{a} \cong \mathfrak{b}_n$, l'anneau des matrices d'ordre n sur $\mathfrak{b}$. On a aussi $\mathfrak{b} \cong e_{11}\mathfrak{a}e_{11}$.

4. Algèbresalgébriques

Tous les résultats de 2 et 3 sont valables pour des algèbres. Désormais nous considèrerons des algèbres sur un corps commutatif Φ.

On dit qu'un élément $a \in \mathfrak{a}$ est algébrique si la sous-algèbre A engendrée par a est de dimension finie. On appelle $\dim A + 1$ le *degré* de a et si $A \supset aA \supset \cdots \supset a^{k-1}A = a^kA = \ldots$, k est *l'indice* de a. On a $\deg a = \operatorname{ind} a$ si et seulement si a est nilpotent.

Proposition 1. *Si a est algébrique d'indice k, il existe un x (dans A) tel que*

$$a^kxa^k = a^k.$$

On a $a^k \in a^{k-1}A = a^kA = a^{k+1}A$. Alors

$$a^k = a^{k+1}b = a^{k+2}b^2 = \cdots = a^{2k}b^k = a^kb^ka^k$$

et $$x = b^k.$$

Remarquons que $e = a^kx$ est idempotent et $ea^k = a^k$. Par conséquent, si a n'est pas nilpotent, $e \neq 0$. Ceci entraîne

Théorème 1. *Si $\mathfrak{a}$ est algébrique, ou bien $\mathfrak{a}$ est une nilalgèbre, ou bien $\mathfrak{a}$ contient un élément idempotent.*

Puisque le radical $\mathscr{R}$ ne contient aucun idempotent, il est clair que $\mathscr{R}$ est une nilalgèbre.

On dit que $\mathfrak{a}$ est de *degré borné* (*indice borné*) s'il existe une borne finie pour les degrés (indices) des éléments de $\mathfrak{a}$. Evidemment toute sous-algèbre et toute image homomorphe d'une algèbre algébrique de degré (indice) borné est de la même sorte.

Proposition 2. *Une algèbre $\mathfrak{a}$ est d'indice borné si et seulement si les indices de tous les éléments nilpotents sont bornés.*

Soit $\varphi(\lambda)$ le polynôme de degré minimal (sans terme constant) tel que $\varphi(a) = 0$, $a \in \mathfrak{a}$. Soit λ^k la plus grande puissance de λ divisant $\varphi(\lambda)$. On voit que k est l'indice de a. Posons $\Psi(\lambda) = \varphi(\lambda)\lambda^{-(k-1)}$, $b = \Psi(a)$. On vérifie que b est nilpotent d'indice k.

Théorème 2. *Soit $\mathfrak{a}$ une algèbre algébrique sans éléments nilpotents* (*sauf* 0) *et soit $a \in \mathfrak{a}$. Alors l'idéal bilatère* (a) *engendré par a est de la forme* (e) *où e est un indempotent contenu dans le centre de $\mathfrak{a}$.*

D'après la proposition 1 on a un idempotent e tel que (a) = (e). Ensuite nous démontrons que si $\mathfrak{a}$ est un anneau sans éléments nilpotents, tout idempotent dans $\mathfrak{a}$ est contenu dans le centre. Pour cela considérons la décomposition bilatère de Peirce $\mathfrak{a} = e\mathfrak{a}e + (1-e)\mathfrak{a}e + e\mathfrak{a}(1-e) + (1-e)\mathfrak{a}(1-e)$. Les éléments de $e\mathfrak{a}(1-e)$ et $(1-e)\mathfrak{a}e$ sont nilpotents. Alors $e\mathfrak{a}(1-e) = 0 = (1-e)\mathfrak{a}e$, $\mathfrak{a} = e\mathfrak{a}e + (1-e)\mathfrak{a}(1-e)$ ce qui entraîne que e est dans le centre.

Corollaire. *Soit $\mathfrak{a}$ comme dans le théorème 2. Alors pour chaque sous-ensemble fini $\{a_1, a_2, \ldots, a_k\}$ d'éléments de $\mathfrak{a}$ il existe un idempotent e dans le centre tel que $ea_i = a_i$, $i = 1, 2, \ldots, k$.*

Soit e_i un idempotent tel que $e_i a_i = a_i$; on prend

$$e = e_1 \circ e_2 \circ \cdots \circ e_k \qquad (a \circ b = a + b - ab).$$

5. PI-Algèbres

On dit qu'une algèbre $\mathfrak{a}$ (pas nécessairement algébrique) satisfait à une *identité polynomiale* ou que $\mathfrak{a}$ est une PI-*algèbre*, s'il existe un élément $p(x_1, \ldots, x_r) \neq 0$ dans l'algèbre libre tel que $p(a_1, \ldots, a_r) = 0$ pour tout choix des a_i dans $\mathfrak{a}$. Nous verrons que la classe des PI-algèbres contient celle des algébriques de degré borné.

Théorème 3 (JACOBSON). *Toute algébre algébrique de degré borné est une PI-algèbre.*

Soit N la borne supérieure des degrés des éléments de $\mathfrak{a}$. Alors on a une équation $a^N + \alpha_1 a^{N-1} + \cdots + \alpha_{N-1} a = 0$ pour chaque $a \in \mathfrak{a}$. Soit b un autre élément de $\mathfrak{a}$. On a $[a^N b] + \alpha_1 [a^{N-1} b] + \cdots + \alpha_{N-1}[ab] = 0$, où généralement $[xy] = xy - yx$. Ensuite on a

$$[[a^N b], [ab]] + \alpha_1 [[a^{N-1} b][ab]] + \cdots + \alpha_{N-2}[[a^2 b], [ab]] = 0.$$

Après ça on prend le commutateur avec $[[a^2 b], [ab]]$, etc. Cela donne une identité

$$[[[[a^N b], [ab]]; [[a^2 b], [ab]]; [[[a^3 b], [ab]]; [a^2 b], [ab]]]] \ldots = 0$$

valable pour tout a, $b \in \mathfrak{a}$. Cette identité n'est pas triviale car le terme $a^N baba^2 baba^3 baba^2 bab \ldots$ ne paraît qu'une fois.

Ensuite nous considèrerons la structure des PI-algèbres primitives.

Proposition 4. *Si $\mathfrak{a}$ satisfait à une identité polynomiale de degré m, $\mathfrak{a}$ satisfait à une identité multilinéaire de degré $\leq m$.*

On entend par un élément *multilinéaire* un élément de la forme

$$\Sigma\rho_{i_1}, \ldots, i_m x_{i_1} x_{i_2} \ldots x_{i_m}$$

où la somme est prise sur toutes les permutations des chiffres 1, 2, ..., n. On obtient la démonstration par le procédé usuel de linéarisation.

Une conséquence immédiate de la proposition 4 est la

Proposition 5. *Si $\mathfrak{a}$ est une PI-algèbre et P est un corps commutatif contenant le corps de base Φ, $\mathfrak{a}_P$ est PI.*

Le théorème 2 montre que l'algèbre Φ_n de toutes les matrices sur Φ est une PI-algèbre. D'autre part le degré des identités pour Φ_n ne peut pas être trop petit. En effet, on a la

Proposition 6. *L'algèbre Φ_n ne satisfait à aucune identité de degré $< 2n$.*

Autrement il existerait une identité de la forme

$$a_1 a_2 \ldots a_m = \Sigma'' \rho_{i_1 \ldots i_m} a_{i_1} \ldots a_{i_m},$$

$m < 2n$, la somme prise sur les permutations $\neq$ l'identité. Posons $a_1 = e_{11}$, $a_2 = e_{12}$, $a_3 = e_{22}$, $a_4 = a_{23}$, ..., ce qui donne $a_1 \ldots a_m \neq 0$, $a_{i_1} \ldots a_{i_m} = 0$ ce qui contredit l'identité.

Proposition 7 (AZUMAYA et NAKAYAMA). *Soit Γ un corps, Σ son centre et P un sous-corps commutatif maximal. Alors $\Gamma \otimes_\Sigma P$ considéré comme une algèbre sur P est isomorphe à une algèbre dense (c'est-à-dire, ayant la propriété donnée dans le théorème de densité) des transformations linéaires d'un espace vectoriel sur P.*

On voit facilement que $\Gamma \otimes_\Sigma P$ est simple. Nous pouvons identifier Γ avec l'ensemble des multiplications à droite $\xi \to \xi\gamma$ dans Γ et P avec l'ensemble des multiplications à gauche par les éléments de P. On a un homomorphisme naturel de $\Gamma \otimes_\Sigma P$ sur l'ensemble d'endormorphismes ΓP dans Γ. Puisque $\Gamma \otimes_\Sigma P$ est simple cela donne un isomorphisme. Puisque Γ est un corps l'ensemble de multiplications à droite est irréductible. Par conséquent ΓP est irréductible, le commutant de cet ensemble est d'après la maximalité de P, P lui-même. Alors on obtient le résultat par utilisation du théorème de densité.

Théorème 4 (KAPLANSKY). *Supposons que $\mathfrak{a}$ soit une algèbre qui satisfait à une identité du degré d. Alors le centre Σ de $\mathfrak{a}$ est un corps et $\mathfrak{a}$ est une algèbre simple de dimension $\leq [d/2]^2$ sur Σ.*

D'après le théorème de densité on peut considérer $\mathfrak{a}$ comme une algèbre dense de transformations linéaires d'un espace vectoriel $\mathfrak{M}$ sur un corps Γ. Soit $\mathfrak{M}_k$ un sous-espace de $\mathfrak{M}$ de dimension finie k et soit $\mathfrak{b}$ la sous-algèbre de $\mathfrak{a}$ appliquant $\mathfrak{M}_k$ dans lui-même. En tenant compte du théorème de densité on voit que toute transformation linéaire de $\mathfrak{M}_k$ dans $\mathfrak{M}_k$ est indutte par un élément convenable appartement à $\mathfrak{b}$. Alors l'algèbre Γ_k de toutes les matrices d'ordre k sur Γ, qui est isomorphe à l'algèbre des transformations linéaires de $\mathfrak{M}_k$, est une image homomorphe de $\mathfrak{b}$. Il suit que Γ_k et par consequent Φ_k satisfait à une identité de degré d. D'après la Proposition 6, $d \geq 2k$. Alors dim $\mathfrak{M} \leq [d/2]$ et $\mathfrak{a} \cong \Gamma_k$, $k \leq [d/2]$ et $\mathfrak{a}$ est simple. Le centre de Γ_k est le centre Σ de Γ. Evidemment Γ satisfait à l'identité donnée pour $\mathfrak{a}$ Cela est le cas aussi pour $\Gamma \otimes_\Sigma P$, P un sous-corps commutatif maximal. Si l'on emploie la proposition 7 et le raisonnement donné dans la première partie de la démonstration on voit que la dimension de $\Gamma \otimes_\Sigma P$ sur P est finie. Par conséquent la dimension de Γ sur Σ est finie et celle de $\mathfrak{a}$ est $\mathfrak{a}$ est finie. On voit maintenant que $\mathfrak{a} \otimes_\Sigma P \cong P_h$ où $h^2 = \dim \mathfrak{a}$ sur Σ. La proposition 6 s'applique encore pour montrer que $h \leq [d/2]$.

6. Idéaux Matriciels

Proposition 8 (LEVITZKI). *Soit a un élément nilpotent d'indice n dans une algèbre algébrique semi-simple. Alors l'idéal* (a) *contient un système de n^2 unités matricielles.*

On a $a^{n-1}\mathfrak{a} \neq 0$ et par conséquent il existe un élément b tel que $a^{n-1}b = e = e^2 \neq 0$. Posons $e_i = a^{n-i}ba^{i-1}$, $i = 1, 2, \ldots, n$. On trouve que $e_i^2 = e_i$ et $e_i \neq 0$ puisque $e_1 = e \neq 0$. De plus $e_ie_j = 0$ si $j < i$ ce qui entraîne que $u = e_1 \circ e_2 \circ \cdots \circ e_n$ est idempotent et $e_iu = e_i = ue_i$. Alors $\mathfrak{b} = u\mathfrak{a}u \subseteq (a)$ et u est un élément unité pour $\mathfrak{b}$. On voit que $\mathfrak{b} = \Sigma_\oplus e_i b$. Puisque $(a^{n-1}b)a^{i-1} = e_i$ et $a^{i-1}(a^{n-i}b) = e_1$, $e_1\mathfrak{b} \cong e_i\mathfrak{b}$. D'après le résultat cité en paragraphe 3, $\mathfrak{c}$ contient un système de n^2 unités matricielles.

Proposition 9. *Soit $\mathfrak{a}$ algébrique et semi-simple, $\{e_{ij}\}$ un système de n^2 unités matricielles $u = \sum_i^n e_{ii}$. Supposons ou bien*

1° *$e_{11}\mathfrak{a}e_{11}$ contient un élément nilpotent $\neq 0$, ou bien*

2° *u n'est pas dans le centre. Alors il existe un système $\{f_{kl}\}$ de au moins $(n+1)^2$ unités matricielles tel que $f_{11} \in e_{11}\mathfrak{a}e_{11}$.*

(1) $$\mathfrak{b} = u\mathfrak{a}u = \mathfrak{c}_n, \qquad \mathfrak{c}_n \cong e_{11}\mathfrak{b}e_{11} = e_{11}\mathfrak{a}e_{11}.$$

D'après la proposition 8, $\mathfrak{c}$ contient un système de $r^2 > 1$ unités matricielles. Cela entraîne que $\mathfrak{a}$ possède un système de n^2r^2 unités matricielles.

(2) On obtient la démonstration facilement par la considération de la décomposition de Peirce $\mathfrak{a} = u\mathfrak{a}u + u\mathfrak{a}(1-u) + (1-u)\mathfrak{a}u + (1-u)\mathfrak{a}(1-u)$.

Proposition 10 (KAPLANSKY). *Soit $e_1, e_2, \ldots$ une suite d'idempotents $\neq 0$ tels que $e_{i+1} \in e_i\mathfrak{a}e_i$, $i = 1, 2, \ldots$ Alors il existe un idéal primitif $\mathfrak{p}$ tel que $e_i \in \mathfrak{p}$, $i = 1, 2, \ldots$*

On voit facilement que l'idéal à droite $\mathfrak{J} = \Sigma(1 - e_i)\mathfrak{a}$ ne contient pas e_1. Soit $\mathfrak{J}'$ un idéal à droite maximal contenant $\mathfrak{J}$ et soit $\mathfrak{p}$ l'idéal de tous les éléments b tels que $\mathfrak{a}b \subseteq \mathfrak{J}'$. On voit que $\mathfrak{p}$ est primitif et $e_i \in \mathfrak{p}$, $i = 1, 2, \ldots$, puisque $\mathfrak{p} \subseteq \mathfrak{J}'$.

Théorème 4 (LEVITZKI). *Soit $\mathfrak{a}$ une algèbre algébrique semi-simple ayant la propriété que pour chaque idéal primitif $\mathfrak{p}$, $\mathfrak{a}/\mathfrak{p}$ est d'indice borné. Alors tout idéal $\mathfrak{b} \neq 0$ de $\mathfrak{a}$ contient un idéal $\mathfrak{c} = \mathfrak{d}_n$ où $\mathfrak{d}$ est une algèbre avec élément unité u appartenant au centre de $\mathfrak{a}$ et $\mathfrak{d}$ est sans éléments nilpotents $\neq 0$.*

La démonstration est assez facile d'après les propositions 8, 10.

7. Le Problème de Kurosch

Proposition 11. *Une algèbre* $\mathfrak{a}$ *est localement finie* (l.f.) *si elle contient un idéal* 6 *tel que* $\mathfrak{b}$ *et* $\mathfrak{a}/\mathfrak{b}$ *sont l.f.*

Soit $\{x_1, x_2, \ldots, x_r\}$ une partie finie de $\mathfrak{a}$ et X la sous-algèbre engendrée. Puisque $\mathfrak{a}/\mathfrak{b}$ est l.f. on peut trouver $x_{r+1}, \ldots, x_n \in X$ tel que $\sum_1^n \Phi\bar{x}_i$, $\bar{x}_i = x_i + \mathfrak{b}$, est une sous-algèbre de $\mathfrak{a}/\mathfrak{b}$. On a $x_i x_j = \Sigma\gamma_{ijk}x_k + z_{ij}$, $\gamma_{ijk} \in \Phi$, $z_{ij} \in X \cap \mathfrak{b}$. Soit M la sous-algèbre engendrée par les éléments $z_{ij}, x_k z_{ij}, z_{ij}x_k, x_k z_{ij} x_l$. Puisque $M \subseteq \mathfrak{b}$, $\dim M < \infty$. On vérifie que M est un idéal dans X. Puisque $x_i x_j - \Sigma\gamma_{ijk}x_k \in M$ X/M est de dimension finie. Par conséquent $\dim X < \infty$.

Soient $\mathfrak{b}$ et $\mathfrak{c}$ des idéaux l.f. de $\mathfrak{a}$. Puisque $(\mathfrak{b} + \mathfrak{c})/\mathfrak{b} \cong \mathfrak{b}/(\mathfrak{b} \cap \mathfrak{c})$ est l.f., $\mathfrak{b} + \mathfrak{c}$ est l.f. Il suit que la somme $\mathfrak{F}$ de tous les idéaux l.f. de $\mathfrak{a}$ est l.f. on appelle $\mathfrak{F}$ le *noyau localement fini* de $\mathfrak{a}$. D'après la proposition 11, il est clair que le noyau l.f. de $\mathfrak{a}/\mathfrak{F}$ est nul. Nous avons besoin aussi de la proposition suivante:

Proposition 12. *Le noyau l.f. contient tout idéal à droite* (*à gauche*) *l.f.*

Soit $\mathfrak{I}$ un idéal à droite l.f. et posons $\mathfrak{b} = \mathfrak{a}\mathfrak{I} + \mathfrak{I}$, l'idéal engendré par $\mathfrak{I}$. Soit $\{x_i\}$ une partie finie de $\mathfrak{b}$. Alors

$$x_i = \Sigma a_{ij}u_{ij} + u_i, \qquad a_{ij} \in \mathfrak{a}, \qquad u_i, u_{ij} \in \mathfrak{I}.$$

Posons $u_{ijk} = u_i a_{jk}$, $u_{ijkl} = u_{ij}a_{kl} \in \mathfrak{I}$. Les éléments u_i, u_{ij}, u_{ijk}, u_{ijkl} engendrent une sous-algèbre avec une base finie $y_1, y_2, \ldots, y_m$. On voit que la sous-algèbre X engendrée par les x_i est contenue dans l'espace engendré par less y_l et les $a_{ij}y_l$. Alors $\dim X < \infty$.

Théorème 5 (KAPLANSKY). *Toute PI-algèbre qui est algébrique est localement finie.*

On voit facilement qu'il suffit de démontrer que le noyau l.f., $\mathfrak{F}$, est $\neq 0$. Considérons d'abord le cas où $\mathfrak{a}$ est une nilalgèbre. Soit a un élément de $\mathfrak{a}$ tel que $a \neq 0$, $a^2 = 0$ et considérons l'idéal a $\mathfrak{a}$. Si $a\mathfrak{a} = 0$, $\mathfrak{a}$ contient un idéal nilpotent $\neq 0$ et $\mathfrak{F}$ est $\neq 0$. Supposons ensuite que a $\mathfrak{a} \neq 0$. On a une identité de la forme $f(x_2, \ldots, x_d)x_1 + g(x_1, x_2, \ldots, x_d) = 0$ pour $\mathfrak{a}$ tel que tout terme de g contient x_1 avant la dernière place. Alors si on pose $x_1 = a_1 = a$, $x_i = a_i \in a\mathfrak{a}$, $i > 1$ on obtient $g(a_1, \ldots, a_d) = 0$ et par conséquent $f(a_2, \ldots, a_d)a = 0$. Soit $\mathfrak{d}$ l'annulateur à gauche de $a\mathfrak{a}$. Alors on voit que $a\mathfrak{a}/\mathfrak{d}$ satisfait à une identité de degré $\leq d - 1$. Par induction sur d, on voit que $a\mathfrak{a}/\mathfrak{d}$ est l.f. Puisque $\mathfrak{d}^2 = 0$, $a\mathfrak{a}$ est l.f. et le noyau l.f. de $\mathfrak{a}$ est $\neq 0$. Considérons ensuite le cas où $\mathfrak{a}$ ne contient aucun élément nilpotent $\neq 0$. Nous pouvons supposer aussi que $\mathfrak{a}$ possède un nombre fini de générateurs $x_1, x_2, \ldots, x_r$. Soit $\mathfrak{p}$ un idéal primitif dans $\mathfrak{a}$. L'algèbre $\mathfrak{a}/\mathfrak{p}$ est de dimension finie sur le centre Σ et puisque elle *a* un nombre fini de générateurs $\dim \mathfrak{a}/\mathfrak{p}$ (sur Φ) $< \infty$. On peut trouver alors des éléments $y_1, y_2, \ldots y_m$ tels que,

$$x_i = \Sigma\alpha_{ij}y_j + z_i, \quad \alpha_{ij} \in \Phi, \quad z_i \in \mathfrak{p}$$

et

$$y_i y_j = \Sigma\gamma_{ijk}y_k + z_{ij}, \quad \gamma_{ijk} \in \Phi, \quad z_{ij} \in \mathfrak{p}.$$

On voit facilement que l'idéal $(z_i, z_{ij}) = \mathfrak{p}$. D'après le corollaire au théorème 2 il existe un élément unité u dans $\mathfrak{p}$. Par conséquent $\mathfrak{a} = \mathfrak{p} \oplus \mathfrak{d}$ et $\mathfrak{d} \cong \mathfrak{a}/\mathfrak{p}$ est de dimension finie.

Ainsi on voit dans ce cas aussi que le noyau l.f. est $\neq 0$. Supposons finalement que $\mathfrak{a}$ est arbitraire. Le premier résultat donne une réduction au cas où $\mathfrak{a}$ est semi-simple. D'après le théorème 4, $\mathfrak{a}$ contient un idéal matriciel $\mathfrak{b} = \mathfrak{d}_n$ où $\mathfrak{d}$ est sans éléments nilpotents. Alors $\mathfrak{d}$ est l.f. Il suit que $\mathfrak{b}$ est l.f. Le noyau l.f. est donc encore $\neq 0$.

Le même raisonnement donne le résultat suivant:

Théorème 6. *Soit* $\mathfrak{a}$ *une algèbre algébrique d'indice fini tel que* $\mathfrak{a}/\mathfrak{p}$ *soit une PI-algèbre pour tout idéal primitif de* $\mathfrak{a}$. *Alors* $\mathfrak{a}$ *est l.f.*

Bibliographie

[1] AMITSUR (A. S.) and LEVITZKI (J.). Minimal identities for algebras, Proc. Amer. math. Soc., t. 1, 1950, p. 449–463.

[2] JACOBSON (Nathan). Structure theory for algebraic algebras of bounded degree, Annals of Math., Series 2, t. 46, 1945, p. 695–707.

[3] KAPLANSKY (Irving). On a problem of Kurosch and Jacobson, Bull. Amer. math. Soc., t. 52, 1946, p. 496–500.

[4] KAPLANSKY (Irving). Rings with a polynomial identity, Bull. Amer. math. Soc., t. 54, 1948, p. 575–580.

[5] KAPLANSKY (Irving). Groups with representations of bounded degree, Canadian J. of Math., t. 1, 1949, p. 105–112.

[6] KAPLANSKY (Irving). Topological representation of algebras, II., Trans. Amer. math. Soc., t. 68, 1950, p. 62–75.

[7] KAPLANSKY (Irving). The structure of certain operator algebras, Trans. Amer. math. Soc., t. 70, 1951, p. 219–255.

[8] KUROSCH (A.). Ringtheoretische Probleme, die mit dem Burnsideschen Problem über periodische Gruppen in Zusammenhang stehen, Bull. Acad. Sc. URSS, Série mathématique, t. 5, 1941, p. 233–240.

[9] LEVITZKI (Jakob). On a problem of A. Kurosch, Bull. Amer. math. Soc., t. 52, 1946, p. 1033–1035.

[10] LEVITZKI (Jakob). On the structure of algebraic algebras and related rings, Trans. Amer. math. Soc., t. 74, 1953, p. 384–409.

[11] MALČEV (A. I.). O predstavlenijakh beskonečnykh algebr, Recueil math. Acad. Sc. Moscou (Mat. Sbornik), N.S., t. 13, 1943, p. 263–286.

[Juin 1958]

Reprinted from
The Seminaire Bourbaki, May 1952.

SOME ASPECTS OF THE THEORY OF REPRESENTATIONS OF JORDAN ALGEBRAS[1])

N. Jacobson

A *Jordan algebra* is a non-associative algebra whose multiplication composition satisfies

$$ab = ba \tag{1}$$

$$(a^2b)a = a^2(ba). \tag{2}$$

If $\mathfrak{A}$ is an associative algebra, $\mathfrak{A}$ defines a Jordan algebra $\mathfrak{A}_J$ relative to the composition $\{ab\} = ab + ba$. Such algebras and their subalgebras are called *special*. There exist also *exceptional* (non-special) algebras, the best known example being the algebra M_3^8 of three-rowed hermitian Cayley matrices relative to $\{ab\} = ab + ba$.

If the characteristic is $\neq 2, 3$, and we shall assume this from now on, then (1), (2) are equivalent to (1) and (2') $abcd + adcb + a(bdc) = (ab)(cd) + (ac)(bd) + (ad)(bc)$ where, in general, $abc \ldots u = (((ab)c) \ldots u)$. Let R_a denote the mapping $z \to xa$ $(= ax)$. Then (2') is equivalent to either

$$R_aR_bR_c + R_cR_bR_a + R_{acb} = R_aR_{bc} + R_bR_{ac} + R_cR_{ab} \tag{3}$$

or

$$[R_aR_{bc}] + [R_bR_{ac}] + [R_cR_{ab}] = 0, \quad ([AB] = AB - BA). \tag{4}$$

A *representation* of a Jordan algebra $\mathfrak{J}$ is a linear mapping $a \to R_a$ of $\mathfrak{J}$ into the algebra of linear transformations of a vector space $\mathfrak{M}$ such that (3) and (4) hold. If R is such a representation then we can define two compositions of $\mathfrak{J}$ and $\mathfrak{M}$ into $\mathfrak{M}$ by setting $xa = xR_a$, x in $\mathfrak{M}$, a in $\mathfrak{J}$ and $ax = xR_a$. Then (3) and (4) guarantee that all the relations obtained from the fundamental identity (2) by taking one of the arguments in $\mathfrak{M}$ the other three in $\mathfrak{J}$ are satisfied. A space $\mathfrak{M}$ with multiplications $xa = ax$, a in $\mathfrak{J}$, satisfying these conditions is called a *Jordan bimodule*. If we drop the superfluous composition ax then we obtain a *Jordan module*. The concepts of representation, Jordan bimodule,

---

[1]) The representation theory discussed here has been developed in two papers by the author: (1) *General representation theory of Jordan algebras*, Trans. Amer. Math. Soc. vol 70 (1951), pp. 509—530. (2) *Structure of alternative and Jordan bimodules*, Osaka Math. Jour., vol. 6 (1954), pp. 1—71.

28

Jordan module are fully equivalent. The concepts of equivalence, etc. for representations or modules are defined as usual.

If $\mathfrak{M}$ is a Jordan bimodule and $\mathfrak{E}$ is the space direct sum $\mathfrak{J} \oplus \mathfrak{M}$ then we can define a composition in $\mathfrak{E}$ by $(a + x)\ (b + y) = ab + ay + xb$, a, $b \in \mathfrak{J}$, x, $y \in \mathfrak{M}$. The conditions on $\mathfrak{M}$ are just the ones needed to insure that $\mathfrak{E}$ is a Jordan algebra. This Jordan algebra is called the *split null extension* of $\mathfrak{J}$ by $\mathfrak{M}$.

Let $a \to S_a$ be a linear mapping of $\mathfrak{J}$ into the space of linear transformations of $\mathfrak{M}$ such that

$$S_{ab} = \{S_a S_b\}. \tag{5}$$

Clearly such a mapping is a homomorphism of $\mathfrak{J}$ into the special Jordan algebra $\mathfrak{L}_J$ of linear transformations in $\mathfrak{M}$. It can be verified that $R_a = S_a$ satisfies (3) and (4). We call such a representation *special*. Next let $a \to S_a^{(1)}$, $a \to S_a^{(2)}$ be two special representations which *commute*: $[S_a^{(1)} S_b^{(2)}] = 0$. Then one can verify that $R_a = S_a^{(1)} + S_a^{(2)}$ defines a representation. We call this the *sum* of the two commuting special representations. In a sense one can form the sum of any two special representations. Thus suppose $S^{(i)}$, $i = 1, 2$, acts in $\mathfrak{M}_i$ and form the Kronecker product $\mathfrak{M}_1 \otimes \mathfrak{M}_2$. Then $a \to S_a^{(1)} \otimes 1_2$ and $a \to 1_1 \otimes S_a^{(2)}$ are commuting special representations. Hence $a \to S_a^{(1)} \otimes 1_2 + 1_1 \otimes S_a^{(2)}$ is a representation, the *Kronecker sum* of $S^{(1)}$ and $S^{(2)}$.

In order to study the various types of representations one introduces corresponding universal associative algebras. Thus one defines the *universal associative algebra* $\mathfrak{U}(\mathfrak{J})$ of all representations as $\mathfrak{F}/\mathfrak{K}$ where $\mathfrak{F}$ is the free algebra $\mathfrak{J} \oplus (\mathfrak{J} \otimes \mathfrak{J}) \oplus (\mathfrak{J} \otimes \mathfrak{J} \otimes \mathfrak{J}) \oplus \ldots$ determined by $\mathfrak{J}$, and $\mathfrak{K}$ is the ideal generated by the elements $a \otimes b \otimes c + c \otimes b \otimes a + acb - a \otimes bc - b \otimes ac - c \otimes ab$ and $a \otimes bc - bc \otimes a + b \otimes ac - ac \otimes b + c \otimes ab - ab \otimes c$. Every representation determines a representation of this associative algebra. Similarly we define the *universal associative algebra* $\mathfrak{U}_s(\mathfrak{J})$ *of special representations* as $\mathfrak{F}/\mathfrak{K}_s$ where $\mathfrak{K}_s$ is the ideal generated by the elements $a \otimes b + b \otimes a - ab$ Let a_s denote the coset of a in $\mathfrak{U}_s$. If $\mathfrak{J}$ has an identity element 1 then $u = 2.1_s$ is the identity in $\mathfrak{U}_s$, Consider now the algebra $\mathfrak{U}_s \otimes \mathfrak{U}_s$ and let $\mathfrak{U}_s^{(2)}(\mathfrak{J})$ be the subalgebra generated by the elements $a_s \otimes u + u \otimes a_s$. We call $\mathfrak{U}_s^{(2)}(\mathfrak{J})$ the *universal associative algebra* of sums of *commuting special representation*, for it can be shown that every such representation of $\mathfrak{J}$ determines one for the associative algebra $\mathfrak{U}_s^{(2)}(\mathfrak{J})$. If $\mathfrak{J}$ has an identity we can also define a *universal algebra* $\mathfrak{U}_1(\mathfrak{J})$ for *unital representations*, that is, for representations S such that $S_1 = 1$.

There are two fundamental relations connecting the various universal associative algebras for a Jordan algebra with an identity. The first of these is $\mathfrak{U}(\mathfrak{J}) \cong \mathfrak{U}_1(\mathfrak{J}) \oplus \mathfrak{U}_s(\mathfrak{J})$. Also we have a natural homomorphism of $\mathfrak{U}_1(\mathfrak{J})$ onto

$\mathfrak{U}^{(2)}(\mathfrak{J})$. In important cases this is an isomorphism. The tool for establishing this result is the following criterion:

Theorem 1. Let $\mathfrak{J}$ be a special Jordan algebra, $\mathfrak{M}$ a module for $\mathfrak{J}$ and $\mathfrak{E} = \mathfrak{J} \oplus \mathfrak{M}$ the corresponding split null extension. Then $\mathfrak{M}$ is equivalent to a submodule of a sum of two commuting special modules (corresponding to special representations) if and only if $\mathfrak{E}$ is a special Jordan algebra.

In a number of important cases the relation between $\mathfrak{U}_s(\mathfrak{J})$ and $\mathfrak{U}_s^{(2)}(\mathfrak{J})$ can be made more explicit as follows. Thus we have the automorphism $P: \Sigma a_i \otimes b_i \rightarrow \Sigma b_i \otimes a_i$ in $\mathfrak{U}_s \otimes \mathfrak{U}_s$. The elements $a_s \otimes u + u \otimes a_s$ which generate $\mathfrak{U}_s^{(2)}$ are evidently contained in the subalgebra $\mathfrak{B}$ of elements invariant under P. In a number of cases it turns out that $\mathfrak{U}_s^{(2)}$ coincides with the subalgebra $\mathfrak{B}$.

Our main aim in this paper is to sketch the theory of representations for finite dimensional semi-simple Jordan algebras. For the sake of simplicity we shall also assume that the base field is algebraically closed. It can be shown that if $\mathfrak{J}$ is finite dimensional then $\mathfrak{U}(\mathfrak{J})$ is finite dimensional. Our problem can be defined as that of determining the structure of this associative algebra. However, it does not seem easy to attack this directly. Instead it appears to be necessary to devise a number of other methods to obtain the representations. The one we shall indicate here leans heavily on Albert's structure theory. [2]) According to this every semi-simple finite dimensional algebra has an identity and is a direct sum of simple ones. One can classify the simple ones according to their degree: the maximum number of pairwise orthogonal idempotent elements contained in such an algebra. At the present time the structure of simple algebras of degree 1 over an algebraically closed field is not known. One might conjecture that it is trivial, namely, that every such algebra is one dimensional. This is known to be the case for special algebras and for algebras of characteristic 0. The structure of simple algebras of degree > 1 has been determined by Albert, as follows.

I. If $\mathfrak{J}$ is simple of degree 2 over an algebraically closed field then $\mathfrak{J}$ is *the Jordan algebra of a non-degenerate scalar product.* Such an algebra is obtained by starting with a vector space $\mathfrak{J}_0$ in which a non-degenerate scalar product (symmetric bilinear form) (x, y) is defined. One forms $\mathfrak{J} = \mathfrak{J}_0 + \Phi 1$ and defines a multiplication in $\mathfrak{J}$ by

$$(x + \alpha 1)(y + \beta 1) = (\beta x + \alpha y) + ((x, y) + \alpha\beta)1 \tag{6}$$

for x, y in $\mathfrak{J}_0$, α, β in Φ. One verifies that $\mathfrak{J}$ is a Jordan algebra and $\mathfrak{J}$ is simple if dim $\mathfrak{J}_0 > 1$.

[2]) See Albert's papers: *A structure theory for Jordan algebras*, Ann. of Math., vol. 48 (1947), pp. 446—467 and *A theory of power associative commutative algebras*, Trans. Amer. Math. Soc. vol 69 (1950), pp. 503—527.

30

II. Let $\mathfrak{J}$ be simple of degree $\geqq 3$ over an algebraically closed field. Then there exists an alternative algebra [3]) $\mathfrak{D}$ with an involution whose self-adjoint elements are the multiples of 1 and whose norm form $x\bar{x}$ is non-degenerate, such that $\mathfrak{J}$ is isomorphic to the algebra of hermitian matrices of the n-rowed matrix algebra $\mathfrak{D}_n$ relative to the composition $\{ab\} = ab + ba$. Moreover, if $n \geqq 4$, then $\mathfrak{D}$ is necessarily associative.

To complete the determination of the algebras $\mathfrak{D}$ one has to determine the algebras $\mathfrak{D}$ and the involution. The following are the only possibilities:

(i) $\mathfrak{D} = \Phi 1$.

(ii) $\mathfrak{D} = \Phi(q)$, $q^2 = \mu 1 \neq 0$, $\bar{q} = -q$.

(iii) $\mathfrak{D}$ is a quaternion algebra, usual involution.

(iv) $\mathfrak{D}$ is the algebra of Cayley numbers, usual involution.

The structure theorem II for algebras of degree $\geqq 3$ is an easy consequence of the following result which plays a fundamental role also in the representation theory for these algebras.

Theorem 2. Let $\mathfrak{J}$ be a Jordan algebra containing an identity 1 and suppose $\mathfrak{J}$ contains $n \geqq 3$ orthogonal idempotent elements e_i such the $\Sigma e_i = 1$ and $n - 1$ elements u_{i1}, $i = 2, \ldots, n$, such that $e_i u_{i1} = \frac{1}{2} u_{i1} = u_{i1} e_1$, $u_{i1}^2 = 4(e_1 + e_i)$. Then $\mathfrak{J}$ is isomorphic to the subalgebra $\mathfrak{H}$ of all hermitian matrices of an algebra $\mathfrak{D}_n$ where $\mathfrak{D}$ is an algebra with an involution. Moreover, $\mathfrak{D}$ is associative if $n > 3$ and alternative if $n \geqq 3$ and its self-adjoint elements are in the mucleus.

We can now outline the main results of the representation of finite dimensional semi-simple Jordan algebras. Such an algebra $\mathfrak{J}$ has an identity. Hence any $\mathfrak{J}$-module $\mathfrak{M}$ can be decomposed as $\mathfrak{M} = \mathfrak{M}_0 + \mathfrak{M}_1 + \mathfrak{M}_{1/2}$ where $R_1 = 0$ in $\mathfrak{M}_0$, $= 1$ in $\mathfrak{M}_1$ and $= \frac{1}{2}$ in $\mathfrak{M}_{1/2}$. The submodule $\mathfrak{M}_0$ is trivial since $R_a = 0$ in it. Also it can be shown easily that $\mathfrak{M}_{1/2}$ is special. If $\mathfrak{J} = \mathfrak{J}_1 \oplus \ldots \oplus \mathfrak{J}_r$ is the direct decomposition of $\mathfrak{J}$ into simple ideals then one sees easily that $\mathfrak{M}_{1/2}$ is a direct sum of modules which are special for one of the $\mathfrak{J}_i$ and trivial for the others. Next one can decompose the unital module $\mathfrak{M}_1$ as a direct sum of two types of submodules denoted as $\mathfrak{M}_{ij}$, $i < j$, and $\mathfrak{M}^{(i)}$. The module $\mathfrak{M}_{ij}$ is a sum of two commuting special modules one of which is special for $\mathfrak{J}_i$, trivial for the other $\mathfrak{J}_i$, and the second of which is special for $\mathfrak{J}_j$ and trivial for the $\mathfrak{J}_k$, $k \neq j$. The module $\mathfrak{M}^{(i)}$ is unital for $\mathfrak{J}_i$ and trivial for the other $\mathfrak{J}_j$. In this way one reduces the considerations to those of special and unital modules for the simple components $\mathfrak{J}_i$.

---

[3]) An algebra $\mathfrak{M}$ is called *alternative* if the associator $(ab)c - a(bc)$ is a skew symmetric function of the arguments a, b, c. A vector space $\mathfrak{M}$ is an alternative *bimodule* for $\mathfrak{A}$ if bilinear compositions am, ma $\mathfrak{M}$ for a $\mathfrak{A}$, m $\mathfrak{M}$ are defined so that all associators with two arguments in A and one in M are skew-symmetric functions.

We assume now that $\mathfrak{J}$ is simple and the base field is algebraically closed. If $\mathfrak{J}$ is of degree 1 then we can proceed no further unless $\mathfrak{J}$ is 1 dimensional. In this case the representation theory is trivial. Next suppose $\mathfrak{J}$ is of degree $= 2$. Then $\mathfrak{J}$ is the Jordan algebra $\mathfrak{J}_0 + 1\Phi$ where for x, y in $\mathfrak{J}_0$, $xy = (x, y)\,1$ and (x, y) is a nondegenerate scalar product. To obtain the representations for this algebra one can proceed directly to determine the universal associative algebras. One considers first the universal associative algebra $\mathfrak{U}_s(\mathfrak{J})$ of special representations. It is clear from the definition that $\mathfrak{U}_s(\mathfrak{J})$ is the *Clifford algebra* of the scalar product (x, y). Its structure is well known: Let $\dim \mathfrak{J}_0 = n$. Then $\dim \mathfrak{U}_s(\mathfrak{J}) = 2^n$ and $\mathfrak{U}_s(\mathfrak{J})$ is the full matrix algebra Φ_{2^ν} if $n = 2\nu$ while $\mathfrak{U}_s(\mathfrak{J}) = \Phi_{2^\nu} \oplus \Phi_{2^\nu}$ if $n = 2\nu + 1$. In any case $\mathfrak{U}_s(\mathfrak{J})$ is semi-simple so that all special representations are completely reducible. There are either one or two distinct irreducible representations according as n is even or odd. We consider next the universal associative algebra $\mathfrak{U}_1(\mathfrak{J})$ of unital representations. One can show that this algebra can be obtained from the free algebra $\mathfrak{F}^*(\mathfrak{J}_0) = \Phi \oplus \mathfrak{J}_0 \oplus (\mathfrak{J}_0 \otimes \mathfrak{J}_0) \oplus (\mathfrak{J}_0 \otimes \mathfrak{J}_0 \otimes \mathfrak{J}_0) \oplus \ldots$ by factoring out the ideal generated by the elements of the form

$$xyz + zyx - (x, y)z - (y, z)x. \tag{7}$$

This algebra has been considered in the theory of mesons and hence we shall call it the *meson algebra* of $\mathfrak{J}$ (or of the scalar product (x, y)). The principal facts on this algebra are contained in the following

Theorem 3. Let $\mathfrak{U}_1(\mathfrak{J})$ be the meson algebra of $\mathfrak{J}$ (of dim $n + 1$) then (1) $\dim \mathfrak{U}_1(\mathfrak{J}) = \binom{2n+1}{n}$, (2) $\mathfrak{U}_1(\mathfrak{J}) \cong \mathfrak{U}_s^{(2)}(\mathfrak{J})$, (3) $\mathfrak{U}_1(\mathfrak{J}) = \Phi_{\binom{n+1}{0}} +$

$$\Phi_{\binom{n+1}{1}} + \ldots + \Phi_{\binom{n+1}{\nu}} \quad \text{if} \quad n = 2\nu, \quad \text{while} \quad \mathfrak{U}_1(\mathfrak{J}) =$$

$$\Phi_{\binom{n+1}{0}} = \Phi_{\binom{n+1}{1}} + \ldots + \Phi_{\binom{n+1}{\nu-1}} + \Phi_{\frac{1}{2}\binom{n+1}{\nu}} + \Phi_{\frac{1}{2}\binom{n+1}{\nu}} \quad \text{if } n = 2\nu - 1.$$

This result shows that $\mathfrak{U}_1(\mathfrak{J})$ is semi-simple. It follows that the unital representations of $\mathfrak{J}$ are completely reducible. Moreover, the result gives all the irreducible representations.

We consider next the algebras of degree $\geqq 3$. Let $\mathfrak{J}$ be such an algebra and suppose $\mathfrak{J}$ is isomorphic to the Jordan algebra of hermitian matrices of $\mathfrak{D}_n$ where $\mathfrak{D}$ is an alternative algebra with 1, having the structure indicated above. We recall that $\mathfrak{D}$ is either the Cayley algebra or $\mathfrak{D}$ is associative and the first alternative can occur only when $n = 3$. We consider first the special representations. The result here is

Theorem 4. If $\mathfrak{D}$ is associative and $\mathfrak{J}$ is the Jordan algebra of hermitian matrices of $\mathfrak{D}_n$, $n \geqq 3$, then $\mathfrak{D}_n$ is the universal associative algebra $\mathfrak{U}_s(\mathfrak{J})$ of the special representations of $\mathfrak{J}$.

32

This result is a special case of a more general one due to N. Jacobson and C. E. Rickart. It follows also from earlier results of F. D. Jacobson and N. Jacobson. [4])

If $\mathfrak{D}$ is the Cayley algebra then there are no special representations. Hence in this case $\mathfrak{U}_s(\mathfrak{J}) = 0$.

The determination of $\mathfrak{U}_1(\mathfrak{J})$ is fairly complicated and hence we can only give some general indications here. Let $\mathfrak{M}$ be a unital bimodule for $\mathfrak{J}$ and let $\mathfrak{E}$ be the corresponding split null extension $\mathfrak{J} \oplus \mathfrak{M}$. Then $\mathfrak{E}$ has an identity 1 and $\mathfrak{E}$ contains the e_i, $i = 1, \ldots, n$, u_{j1}, $j = 2, \ldots, n$ given in the key structure theorem. Hence there exists an alternative algebra $\mathfrak{F}$ with an involution such that $\mathfrak{E}$ is isomorphic to the algebra of hermitian matrices of $\mathfrak{F}_n$. If $n > 3$ $\mathfrak{F}$ is associative. In this case we can conclude by Theorem 1 that $\mathfrak{M}$ is a submodule of a sum of commuting special modules. This implies that $\mathfrak{U}_1(\mathfrak{J}) \cong \mathfrak{U}_s^{(2)}(\mathfrak{J})$ and one can determine the structure of this algebra. If $n = 3$, $\mathfrak{F}$ need not be associative. In this case one has to apply a second method which is also useful for $n > 3$. One shows that $\mathfrak{F} = \mathfrak{D} \oplus \mathfrak{N}$ where $\mathfrak{N}$ is an ideal such that $\mathfrak{N}^2 = 0$. This reduces the consideration to that of representations or bimodules for the alternativs algebra $\mathfrak{D}$. Actually, somewhat more is involved since the involution in $\mathfrak{F}$ has to be taken into account. We should remark also that to obtain the results with a minimum of calculation it seems best to combine the two methods.

The main results that one obtains are that $\mathfrak{U}_1(\mathfrak{J})$ is semi-simple and its structure can be given. The results for $\mathfrak{U}_s(\mathfrak{J})$ and $\mathfrak{U}_1(\mathfrak{J})$ taken together show that the representations are completely reducible and the irreducible representa tion can be given. If we combine these results with the earlier considerations on module decompositions we can see that the representations of known semi-simple algebras are completely determined and the irreducible representations can be deduced from the structure of $\mathfrak{J}$. In concluding we should remark also that similar results can be obtained for arbitrary base fields. For example complete reducibility holds for the representation if $\mathfrak{J}$ is a known separable algebra.

---

[4]) See the references in (2) of footnote 1.

Reprinted from
Proceedings of the International Congress of Mathematicians, Vol. III, 1984.

Jordan Algebras

N. Jacobson
Yale University

The purpose of the present paper is to give a brief account of the recent progress and present status of the following aspects of the theory of Jordan algebras: (1) foundations, (2) structure and representation theory, (3) cohomology theory.

1. Foundations. Throughout this paper the term Jordan algebra will mean Jordan algebra over a field Φ of characteristic not two. The dimensionality over the base field may be infinite. An algebra is said to be Jordan if it satisfies the following two identities

$$ab = ba, \tag{1}$$

$$(a^2b)a = a^2(ba). \tag{2}$$

The most important examples of such algebras are the subspaces of associative algebras which are closed relative to the Jordan multiplication $ab = \frac{1}{2}(a \times b + b \times a)$ (or $a \times b + b \times a$) where $a \times b$ is the associative product. Algebras of this type are called *special* Jordan algebras. It is known that there exist *exceptional* (non-special) Jordan algebras. The classical example is the algebra M_3^8 of 3×3 hermitian matrices with entries taken in a Cayley algebra and with the multiplication $AB = \frac{1}{2}(A \times B + B \times A)$, where $\times$ is the usual matrix multiplication. It has also been shown by P. Cohn [4] that there exist special Jordan algebras possessing homomorphic images which are not special.

The class of homomorphic images of special Jordan algebras is closed relative to the processes of taking subalgebras, homomorphic images, forming (unrestricted) direct sums. It follows from a result of G. Birkhoff that this class can be defined by identities. Are these identities just the identities (1) and (2)? This is equivalent to the following: Is every Jordan algebra a homomorphic image of a special one? An affirmative answer to this question would follow from an affirmative answer to the following question: Is the free Jordan algebra generated by any number of elements special? In this connection the following important result has been proved recently by Širšov [12].

Theorem 1. The free Jordan algebra generated by two elements is special.

Previously it had been proved by Cohn that any homomorphic image of a special Jordan algebra generated by two elements is special. We therefore have the following consequences:

(1) Every Jordan algebra generated by two elements is special.

(2) Every two variable identity which holds for all special Jordan algebras holds for all Jordan algebras.

(3) The exceptional Jordan algebra M_3^8 cannot be generated by two elements. (It can be generated by three.)

Little is known at the present time on Jordan algebras generated by more than two elements or about identities in more than two variables. It is easy to formulate identities for special Jordan algebras involving the ternary composition $\{abc\} = (ab)c + (bc)a - (ac)b$. In a special Jordan algebra this reduces to $\frac{1}{2}(a \times b \times c + c \times b \times a)$ in terms of the associative multiplication $\times$. Hence it is easy to check that

$$\{\{axa\}\ x\ \{axb\} = \{\{\{axa\}xb\}xa\}, \tag{3}$$

$$\{\{xbx\}\ a\ \{xbx\}\} = \{x\{b\{xax\}b\}x\} \tag{4}$$

are valid in special Jordan algebras. These identities have some interesting applications. At the present time it is not known whether or not they are generally valid for Jordan algebras. We proceed to prove that these identities hold for the exceptional Jordan algebra M_3^8 and we shall obtain some general results on identities in this algebra.

Thus let $J = M_3^8$. Any element of M_3^8 has the form

$$X = \begin{bmatrix} \xi_1 & x_3 & \bar{x}_2 \\ \bar{x}_3 & \xi_2 & x_1 \\ x_2 & \bar{x}_1 & \xi_3 \end{bmatrix} \tag{5}$$

where the x_i are Cayley numbers, $\bar{x}_i$ their conjugates and the ξ_i are in the base field. Following Freudenthal [5] we define

$$\det X = t(x_1 x_2 x_3) + \xi_1\xi_2\xi_3 - \xi_1\, n(x_1) - \xi_2\, n(x_2) - \xi_3\, n(x_3) \tag{6}$$

where(1) $t(x) = x + \bar{x}$ and $n(x) = x\bar{x} = \bar{x}x$. Then it can be verified (Freudenthal) that X is a root of the *characteristic polynomial.*

$$f(\lambda,X) = \det(\lambda - X) \tag{7}$$

The coefficients of this polynomial are polynomials in the ξ_i and the coordinates of the Cayley numbers x_i and the same is true for the discriminant Δ of $f(\lambda,X)$. An element X which does not satisfy $\Delta \det X = 0$ has a characteristic polynomial with three distinct non-zero roots. If the base field is algebraically closed then the algebra generated by X contains three orthogonal idempotent elements E_i. Then we may choose the representation so that

$$E_1 = \mathrm{diag}\,\{1,0,0\}, \quad E_2 = \mathrm{diag}\,\{0,1,0\}, \quad E_3 = \mathrm{diag}\,\{0,0,1\}.$$

Then

$$X = \mathrm{diag}\,\{\rho_1, \rho_2, \rho_3\} \tag{8}$$

where the ρ_i are the roots of the characteristic polynomials. Let

$$Y = \begin{bmatrix} \eta_1 & y_3 & \bar{y}_2 \\ \bar{y}_3 & \eta_2 & y_1 \\ y_2 & \bar{y}_1 & \eta_3 \end{bmatrix} \tag{9}$$

be a second element of M_3^8 The subalgebra $\mathfrak{L}$ generated by X and Y is the same as the subalgebra generated by the idempotents E_i and the elements

$$(10) \qquad Y_1 = \begin{bmatrix} 0 & 0 & 0 \\ 0 & 0 & y_1 \\ 0 & \bar{y} & 0 \end{bmatrix}, \quad Y_2 = \begin{bmatrix} 0 & 0 & \bar{y}_2 \\ 0 & 0 & 0 \\ y_2 & 0 & 0 \end{bmatrix}, \quad Y_3 = \begin{bmatrix} 0 & y_3 & 0 \\ \bar{y}_3 & 0 & 0 \\ 0 & 0 & 0 \end{bmatrix}.$$

Suppose now that $n(y_i) \neq 0$, $i = 1,2,3$. Then we set $U_{12} = n(y_3)^{-\frac{1}{2}} Y_3$, $U_{13} = n(y_2)^{-\frac{1}{2}} Y_2$ and we have $U_{12}^2 = E_1 + E_2$, $U_{13}^2 = E_1 + E_3$. It is known (Jacobson ([9], p. 35) that we may choose the representation so that

$$U_{12} = \begin{bmatrix} 0 & 1 & 0 \\ 1 & 0 & 0 \\ 0 & 0 & 0 \end{bmatrix}, \quad U_{13} = \begin{bmatrix} 0 & 0 & 1 \\ 0 & 0 & 0 \\ 1 & 0 & 0 \end{bmatrix}.$$

It is then easy to verify that the subalgebra $\mathfrak{L} \cong \Phi_{3J}$ the special Jordan algebra of three-rowed matrices with entries in Φ. Moreover, in this representation the elements y_2 and y_3 are in the base field.

To prove that an identity $P(X,Y, \ldots ,T) = 0$ holds in M_3^8 it suffices to show that this holds in the algebra obtained by extending the base field to any extension field. Hence we may assume that the base field is algebraically closed. The algebra M_3^8 is 27 dimensional. Let $(B_1, \ldots ,B_{27})$ be a basis. Then $P(X,Y, \ldots ,T) = \sum_1^{27} P_i B_i$ where the P_i are polynomials in the coordinates of the matrices $X, \ldots ,T$ in terms of the basis (B_i). We have $P(X,Y, \ldots ,T) = 0$ if and only if the $P_i = 0$ for all choices of the coordinates in Φ. Since Φ is infinite it suffices to know that the $P_i = 0$ for all choices of the coordinates which do not satisfy a finite set of algebraic equations. We can now prove the following

Theorem 2. Let Φ be algebraically closed.
(1) A two variable identity $P(X,Y) = 0$ holds in M_3^8 if and only if it holds in the Jordan algebra Φ_{3J} of three-rowed matrices.
(2) An identity $P(X,Y, \ldots ,T) = 0$ holds in M_3^8 if it holds for every X,Y of the form

$$(11) \qquad X = \begin{bmatrix} \xi_1 & & 0 \\ & \xi_2 & \\ 0 & & \xi_3 \end{bmatrix}, \quad Y = \begin{bmatrix} \eta_1 & \alpha & \beta \\ \alpha & \eta_2 & y \\ \beta & \bar{y} & \eta_3 \end{bmatrix}.$$

where the ξ_i, η_i, α,β are in Φ (the other arguments arbitrary).

Proof. It suffices to prove $P(X,Y, \ldots ,T) = 0$ for all X such that Δ det $X \neq 0$, Δ the discriminant of the characteristic polynomial. If X is a fixed element satisfying this condition we may take a basis so that X has diagonal form. Similarly $P(X,Y, \ldots ,T) = 0$ holds for this fixed X if it holds for all Y of the form (9) with $n(y_1)\ n(y_2)\ n(y_3) \neq 0$. For such a Y we may choose a basis so that X is unchanged while Y is as in (11). Thus the identity holds if and only if it holds for all X,Y as in (11). This proves (2). The subalgebra $\mathfrak{C}$ generated by X and Y is isomorphic to Φ_{3J}. Hence $P(X,Y) = 0$ holds in M_3^8 if and only if it holds in Φ_{3J}. This proves (1).

We shall now use this result to prove that the identities (3) and (4) hold in M_3^8. Let $X = \operatorname{diag}\{\xi_1, \xi_2, \xi_3\}$. Then X is in the nucleus of the algebra C_3 of three-rowed Cayley matrices. An easy direct calculation then shows that $\{AXB\} = \frac{1}{2}(A \times X \times B + B \times X \times A)$. Now assume that the $\xi \neq 0$. Then if we set $X^{-1} = \operatorname{diag}\{\xi_1^{-1}, \xi_2^{-1}, \xi_3^{-1}\}$, the mapping $\theta: C \longrightarrow C \times X$ is $1-1$ of C_3 onto itself and θ^{-1} is $C \longrightarrow C \times X^{-1}$. We use this to define a new composition in C_3 by $A \circ B = (A^\theta \times B^\theta)^{\theta^{-1}} = A \times X \times B$ (either association). Evidently $(C_3,\circ)$ is an algebra isomorphic to $(C_3,\times)$. Consider the involution $J: C \longrightarrow M^{-1} \times \bar{C}' \times M$ in $(C_3,\times)$ where $M = \operatorname{diag}\{\mu_1, \mu_2, \mu_3\}$. The corresponding involution in $(C_3,\circ)$ is $\theta^{-1}J\theta$ and this has the form $C \longrightarrow N^{-1} \times \bar{C}' \times N$ where $N = M \times X$. Hence if we take $M = X^{-1}$ then the corresponding involution is $C \longrightarrow \bar{C}'$. The set of self-adjoint elements relative to this is M_3^8 but the new composition in M_3^8 is $\{AXB\}$ where X is the given fixed element. Our argument shows that this algebra is isomorphic to the Jordan algebra of self-adjoint elements of $(C_3,\times)$ relative to the involution defined by $M = X^{-1}$. Hence M_3^8 is a Jordan algebra relative to the product composition $\{AXB\}$. The identity (3) is just the Jordan identity $(a^2b)a = a^2(ba)$ for this composition. Hence (3) holds for the $x = X$ of the form $\operatorname{diag}\{\xi_1, \xi_2, \xi_3\}$, $\xi_1\xi_2\xi_3 \neq 0$. This implies that (3) holds for all A,B,C in M_3^8. Our argument shows also that we may state this result in the following way.

Theorem 3. Let X be a fixed element in a Jordan algebra M_3^8 and define a new multiplication composition $(A,B) \longrightarrow \{AXB\} = (AX)B + (XB)A - (AB)X$ in M_3^8. Then M_3^8 relative to this composition is a Jordan algebra.

To prove (4) for M_3^8 we note first that if $X = \operatorname{diag}\{\xi_1, \xi_2, \xi_3\}$ and Y is as in (9) then, since $\{XYX\} = X \times Y \times X$,

$$\{XYX\} = \begin{bmatrix} \xi_1^2\eta_1 & \xi_1\xi_2 y_3 & \xi_1\xi_3\bar{y}_2 \\ \xi_2\xi_1\bar{y}_3 & \xi_2^2\eta_2 & \xi_2\xi_3 y_1 \\ \xi_3\xi_1 y_2 & \xi_3\xi_2\bar{y}_1 & \xi_3^2\eta_3 \end{bmatrix} \tag{12}$$

We use the notation $Y = (\eta, y)$ for the matrix in (9) and similarly write $A = (\alpha, a)$, $B = (\beta, b)$ where α_i and a_i are used for η_i, y_i for A, β_i, b_i for η_i, y_i for B. Also we write abc for $(ab)c$, etc. A direct calculation shows that $\{BYB\} = (y,c)$ where

$$y_1 = \beta_1^2\eta_1 + \beta_1 t(b_3\bar{y}_3 + \bar{b}_2 y_2) + \eta_2 n(b_3) + \eta_3 n(b_2) + t(b_3 y_1 b_2) \tag{13}$$

$$\begin{aligned} c_3 = {}& \beta_1\beta_2 y_3 + \beta_1\bar{y}_2\bar{b}_1 + b_3\bar{y}_3 b_3 + \beta_2\eta_2 b_3 + b_3 y_1\bar{b}_1 + \bar{b}_2 y_2 b_3 \\ & + \eta_3\bar{b}_2\bar{b}_1 + \beta_1\eta_1 b_3 + \beta_2\bar{b}_2\bar{y}_1 \end{aligned} \tag{14}$$

and the other y_i, c_i are obtained by cyclic permutation of the indices 1,2,3. It follows from (12), (13), and (14) that $\{\{XBX\}A\{XBX\}\} = (\delta, d)$ where

$$\delta_1 = \xi_1^4\beta_1^2\alpha_1 + \xi_1^2\beta_1 t(\xi_1\xi_2 b_3\bar{a}_3 + \xi_1\,\xi_3\bar{b}_2 a_2) + \alpha_2\xi_1^2\xi_2^2\, n(b_3) + \alpha_3\xi_1^2\xi_3^2\, n(b_3) +$$
$$t(\xi_1^2\xi_2\xi_3 b_3 a_1 b_2)$$

$$d_3 = \xi_1^2\xi_2^2\beta_1\beta_2 a_3 + \xi_1^2\xi_2\xi_3\bar{a}_2\bar{b}_1 + \xi_1^2\xi_2^2 b_3\bar{a}_3 b_3 + \xi_1\xi_2^3\alpha_2 b_3 + \xi_2^2\xi_1\xi_3 b_3 a_1\bar{b}_1 +$$
$$\xi_1^2\xi_2\xi_3\bar{b}_2 a_2 b_3 + \xi_3^2\xi_1\xi_2\alpha_3\bar{b}_2\bar{b}_1 + \xi_1^3\xi_2\alpha_1 b_3 + \xi_1\xi_2^2\xi_3\bar{b}_2\bar{a}_1.$$

Similarly, $\{X\{B\{XAX\}B\}X\} = (\delta',d')$ where

$$\delta'_1 = \xi_1^2[\beta_1^2\xi_1^2\alpha_1 + \beta_1 t(b_3\xi_1\xi_2\bar{a}_3 + b_2\xi_1\xi_3 a_2) + \xi_2^2\alpha_2 n(b_3) + \xi_3^2\alpha_3 n(b_3) +$$
$$t(\xi_2\xi_3 b_3 a_1 b_2)]$$

$$d'_3 = \xi_1\xi_2[\beta_1\beta_2\xi_1\xi_2 a_3 + \beta_1\xi_1\xi_3\bar{a}_2\bar{b}_1 + \xi_1\xi_2 b_3\bar{a}_3 b_3 + \beta_2\xi_2^2\alpha_2 b_3 + \xi_2\xi_3 b_3 a_1\bar{b}_1 +$$
$$\xi_1\xi_3\bar{b}_2 a_2 b_3 + \xi_3^2\alpha_3\bar{b}_2\bar{b}_1 + \beta_1\xi_1^2\alpha_1 b_3 + \xi_2\xi_3\bar{b}_2\bar{a}_1]$$

Comparison of these formulas shows that $\delta_1 = \delta'_1$, $d_3 = d'_3$. Similarly $\delta_i = \delta'_i$, $d_i = d'_i$. Hence we have verified (4) for X diagonal. Consequently we have the following result.

Theorem 4. The identity (4) holds in M_3^8.

2. Structure and Representations. Until recently there was a gap in the structure and representation theory of finite dimensional semi-simple Jordan algebras. This was due to the fact that we did not know the structure of a simple algebra of this type having only one idempotent, the identity, and with base field algebraically closed. It was known that any special Jordan algebra or any Jordan algebra of characteristic zero satisfying these conditions is necessarily one dimensional. The author has recently established this result in general (Jacobson [10]). This enables one to prove the theorem on complete reducibility of representations of finite dimensional separable algebras without any restriction and to complete the determination of the irreducible representations (cf. Jacobson [9]).

The proof of the structure theorem makes use of properties of the ternary composition $\{abc\}$ defined before and of a new concept of regularity and inverses. In particular, the following identity

$$\{aba\}^2 = \{a\{ba^2b\}a\} \tag{15}$$

is essential in the proof. This identity was formulated by the present author and proved independently by Hall [6] and by Harper [7]. It is also a consequence of Sirsov's theorem.

If J is a Jordan algebra with an identity u we call an element a of J *regular* in J if there exists an element b in J such that

$$ab = u \qquad a^2b = a.^{(2)}$$

It can be shown that these conditions are symmetric in a and b and that b is uniquely determined. We call b the *inverse* of a. An important property of inverses is contained in the following

Theorem 5. If a and b are inverses and U_a is the operator $x \longrightarrow \{axa\}$ then $U_aU_b = 1 = U_bU_a$.

This is proved under the restriction that the characteristic is not three in [10]. This restriction has recently been removed by D. Sasser.

3. Cohomology. One defines a *representation* S of a Jordan algebra J to be a linear mapping $a \longrightarrow S_a$ of J into the space of linear transformations of a vector space $\mathfrak{m}$ such that

$$S_{a^2}S_a = S_aS_{a^2} \tag{17}$$

$$S_{a^2b} + 2S_aS_bS_a = S_bS_{a^2} + 2S_aS_{ab}{}^{(3)} \tag{18}$$

Let $xa = xS_a$ for x in $\mathfrak{m}$, a in J. Then the product xa is bilinear and

$$(xa^2)a = (xa)a^2 \tag{19}$$

$$x(a^2b) + 2((xa)b)a = (xb)a^2 + 2(xa)(ab). \tag{20}$$

These conditions define a *module* $\mathfrak{m}$ for J. Let $\mathfrak{I}$ be the free associative algebra $\Phi \oplus J \oplus (J \times J) \oplus \ldots$ and let $\mathfrak{k}$ be the ideal in $\mathfrak{I}$ generated by all the elements of the form

$$a^2 \times a - a \times a^2 \tag{21}$$

$$a^2b + 2a \times b \times a - b \times a^2 - 2a \times ab. \tag{22}$$

Then the algebra $\mathfrak{u} = \mathfrak{I}/\mathfrak{k}$ is called the *universal associative algebra* of J. Any representation (module) for J defines a representation (module) for $\mathfrak{u}$ in the usual associative sense.

It is known that $\mathfrak{u}$ is finite dimensional if J is finite dimensional and $\mathfrak{u}$ is semi-simple if J is finite dimensional and separable (Jacobson [8] and [9]). The latter result is equivalent to the complete reducibility of modules for finite dimensional separable Jordan algebras. In the characteristic zero case this result can be proved by using Lie algebra methods. In the general case the known proof depends on the structure theory and involves a detailed analysis of the representations.

The following two results are analogous to Whitehead's first and second lemmas on Lie and associative algebras.

Theorem 6 (Jacobson [8]). Let J be a finite dimensional Jordan algebra over a field of characteristic 0 and let $\mathfrak{m}$ be a J-module. Let $a \longrightarrow f(a)$ be a linear mapping of J into $\mathfrak{m}$ such that

$$f(ab) = f(a)b + f(b)a \tag{23}$$

Then there exist $v_i \in \mathfrak{m}$, $b_i \in J$ such that

$$f(a) = \sum_i ((v_ia)b_i - v_i(ab_i)). \tag{24}$$

Theorem 7 (Penico [11]). Let J be a finite dimensional separable Jordan algebra, $\mathfrak{m}$ a module for J. Let $f(a,b)$ be a linear mapping of $J \times J$ into $\mathfrak{m}$ such that

$$f(a,b) = f(b,a) \tag{25}$$

$$f(a^2,ab) + f(a,b)a^2 + f(a,a)ab = f(a^2b,b) + f(a^2,b)a + (f(a,a)b)a. \tag{26}$$

Then there exists a linear mapping $g(a)$ of J into $\mathfrak{m}$ such that

$$f(a,b) = g(ab) - g(b)a - g(a)b. \tag{27}$$

Theorem 7 is the usual factor set formulation of a special case of the theorem that if $\mathfrak{k}$ is a finite dimensional Jordan algebra with radical and $J = \mathfrak{k}/\mathfrak{R}$ is separable then $\mathfrak{k} = \mathfrak{R} + J_1$ where J_1 is a subalgebra. A simplified proof of this result has been given recently by E. Taft. Theorem 6 is the main step in the proof of a Malcev type uniqueness theorem for the decomposition $\mathfrak{k} = \mathfrak{R} + J_1$. It seems likely that this result is valid for separable algebras and that it could be proved by using the analysis of the modules for separable algebras. However, the problem of extending the Malcev uniqueness theorem to the characteristic p case appears to involve an additional difficulty of finding an adequate substitute for the exponential mapping of derivations into automorphisms. Perhaps the recent work of Dieudonné's on formal Lie groups may be pertinent to this question.

The associative and Lie theorems corresponding to Theorems 6 and 7 are special cases of results on cohomology groups. The various important cohomology theories have been recently unified by Cartan and Eilenberg [3]. Can their theory be applied to obtain the foregoing results? From the Cartan-Eilenberg point of view it appears natural to investigate the functors $\mathrm{Ext}_{\mathfrak{u}}(J,\mathfrak{m})$, $\mathrm{Tor}^{\mathfrak{u}}(\mathfrak{m},J)$ where $\mathfrak{u}$ is the universal associative algebra of J and is a module. A basic question in this connection is construction of useful projective resolutions for J as $\mathfrak{u}$-module.

(1)It is known that $t((x_1x_2)x_3) = t(x_1(x_2x_3))$. We have written $t(x_1x_2x_3)$ for this element of Φ.

(2)This and Theorem 5 below appear in slightly different forms in [10]. The change is necessitated by the change from the composition $a \times b + b \times a$ in [10] to the composition $\frac{1}{2}(a \times b + b \times a)$ in special algebras used in the present paper.

(3)These conditions are slightly different from those in [9]. The present ones take care of the characteristic 3 case.

BIBLIOGRAPHY

[1] ALBERT, A. A., A structure theory for Jordan algebras. Annals of Math. 48 (1947), 446–467.

[2] ______, A theory of power associative commutative algebras. Trans. A. M. S., vol. 36 (1950), 503–527.

[3] CARTAN, H. and EILENBERG, S., *Homological Algebra*, Princeton University, 1956.

[4] COHN, P. M., On homomorphic images of special Jordan algebras. Canadian Math. Journal, vol. 6 (1954), 253–264.

[5] FREUDENTHAL, H., *Octaven, Ausnamengruppen und Oktavengeometrie*, Utrecht 1951.

[6] HALL, M., An identity in Jordan rings. Proc. A. M. S., vol. 7 (1956), 990–998.

[7] HARPER, L. R., Proof of an identity on Jordan algebras. Proc. Nat. Acad. Sci., vol. 42 (1956), 137–140.

[8] JACOBSON, N., General representation theory of Jordan algebras. Trans. A. M. S., vol. 70 (1951), 509–530

[9] ______, Structure of Alternative and Jordan bimodules. Osaka Math. Journal, vol. 6 (1954), 1–70.

[10] ______, A theorem on the structure of Jordan algebras. Proc. Nat. Acad. Sci. vol. 42 (1956), 140–147.

[11] PENICO, A. J., The Wedderburn principal theorem for Jordan algebras. Trans. A. M. S., vol. 70 (1951) 404–421.

[12] SIRSOV, A. I., On special J-rings. Math. Sbornik, vol. 38 (80) (1956) 149–160.

Reprinted from
A Conference On Linear Algebras, June 1956.

Representation Theory of Jordan Algebras

by

N. Jacobson
(University-Yale)

Chapter I

Multiplication Representations in Classes of Algebras Defined by Identities.

In this chapter we develop the basic concepts of representation theory for an arbitrary class of algebras defined by identities. If f is an element of a free non-associative algebra over a field Φ then we say that an algebra $\mathfrak{A}/\Phi$ satisfies the identity $f = 0$ if f is mapped into 0 by every homomorphism of the free algebra into $\mathfrak{A}$. If S is a subset of a free non-associative algebra then we denote by $C(S)$ the class of algebras satisfying every identity $f = 0$, $f \in S$. The representation theory for $C(S)$ has as its starting point the notion of an S-bimodule for an $\mathfrak{A}$ in the class $C(S)$. This is a vector space $\mathfrak{M}/\Phi$ with bilinear compositions $(a, u) \to au$, $(a, u) \to ua$ of $(\mathfrak{A}, \mathfrak{M})$ into $\mathfrak{M}$ such that the algebra $\mathscr{E} = \mathfrak{A} \oplus \mathfrak{M}$ with multiplication $(a_1 + u_1)(a_2 + u_2) = a_1 a_2 + a_1 u_2 + u_1 a_2$, $a_i \in \mathfrak{A}$, $u_i \in \mathfrak{M}$ is in the class $C(S)$. We can derive the explicit conditions on au and ua for an S-bimodule of $\mathfrak{A}$ from the set S of defining identities. Moreover, these conditions can be expressed as conditions on the linear transformations $u \to au$, $u \to ua$ in $\mathfrak{M}$ and this leads to the notion of an S-multiplication representation (S-birepresentation) of $\mathfrak{A}$ in the associative algebra $\mathrm{Hom}_\Phi(\mathfrak{M}, \mathfrak{M})$. It is convenient to generalize this concept to that of an S-multiplication specialization in which $\mathrm{Hom}(\mathfrak{M}, \mathfrak{M})$ is replaced by an arbitrary associative algebra with an identity element 1. This leads to the notion of a universal S-multiplication envelope for $\mathfrak{A}$ in $C(S)$. The determination of such envelopes is one of the basic problems of the representation theory since the S-bimodules and S-multiplication representations for $\mathfrak{A}$ can be identified with right modules and representations of the associative universal envelope.

We consider also the problem of extension of an algebra. For given $\mathfrak{A}$ and $\mathfrak{M}$ in $C(S)$ such an extension is a short exact sequence $0 \to \mathfrak{M} \to \mathscr{E} \to \mathfrak{A} \to 0$ where $\mathscr{E}$ is in $C(S)$. The special case of extensions of $\mathfrak{A}$ by null algebras ($\mathfrak{M}^2 = 0$) leads to the consideration of factor sets associated with an S-bimodule $\mathfrak{M}$. We shall derive the explicit conditions for these from the given f in S. As in the known special cases, one has a bijection of certain equivalence classes of extensions and elements of a cohomology group.

The most significant classes of algebras $C(S)$ are associative, Lie, alternative and Jordan algebras. We shall consider the explicit form of the results for these.

1. *Free associative and non-associative algebras.* Let X be an arbitrary non-vacuous set. We shall usually write $X = \{x_1, x_2, \ldots, x_n\}$ if X is finite, $X = \{x_i\}$, $i = 1, 2, 3, \ldots$, if X is countably infinite and $X = \{x_\alpha\}$ where α is in some set of indices if there is no restriction on the cardinality $|X|$ of X. Let $M(X)$ denote the set of (associative) words $x_{\alpha_1} x_{\alpha_2} \ldots x_{\alpha_k}$ formed of the $x_\alpha \in X$. In this set we include the vacuous set which we denote as 1, and we define a product in $M(X)$ by juxtaposition: $(x_{\alpha_1} \ldots x_{\alpha_k})(x_{\beta_1} \ldots x_{\beta_l}) = x_{\alpha_1} \ldots x_{\alpha_k} x_{\beta_1} \ldots x_{\beta_l}$. In this way $M(X)$ becomes an associative multiplicative system with

1 as (two-sided) identity element. The set $M(X)$ with its multiplication has the following universal mapping property: If S is any set with an associative binary composition and with an identity element then any mapping $x \to y$ of X into S has a unique extension to a homomorphism of $M(X)$ into S mapping $1 \to 1$.

Let Φ be a field and let $\Phi\{X\}$ denote the algebra over Φ with $M(X)$ as basis and with the multiplication of the base elements as given in $M(X)$. Thus the elements of $\Phi\{X\}$ have the form $\Sigma\varphi_{\alpha_1}\ldots{}_{\alpha_k}x_{\alpha_1}\ldots x_{\alpha_k}$, $\varphi_{\alpha_1}\ldots{}_{\alpha_k} \in \Phi$ and the product in $\Phi\{X\}$ is given by

$$(\Sigma\varphi_{\alpha_1\ldots\alpha_k}x_{\alpha_1}\ldots x_{\alpha_k})(\Sigma\psi_{\beta_1\ldots\beta_l}x_{\beta_1}\ldots x_{\beta_l}) = \Sigma\varphi_{\alpha_1\ldots\alpha_k}\psi_{\beta_1\ldots\beta_l}x_{\alpha_1}\ldots x_{\alpha_k}x_{\beta_1}\ldots x_{\beta_l}. \tag{1}$$

$\Phi\{X\}$ is an associative algebra with 1 as identity element. We shall call this *the free associative algebra* (freely) generated by the set X. The universal mapping property of $M(X)$ implies the following universal mapping property of $\Phi\{X\}$: if $x \to y$ is any mapping of X into an associative algebra $\mathfrak{A}$ with identity 1 then this mapping has a unique extension to a homomorphism of $\Phi\{X\}$ into $\mathfrak{A}$ sending 1 into 1.

We call k the (total) *degree* of the element $x_{\alpha_1}x_{\alpha_2}\ldots x_{\alpha_k} \in M(X)$. More generally, if Y is a subset of X then the Y-degree of $m = x_{\alpha_1}x_{\alpha_2}\ldots x_{\alpha_k}$ is the number of $x_{\alpha_j} \in Y$. If $Y = \{x\}$ then we speak of the x-degree of m rather than the Y-degree. An element of $\Phi\{X\}$ which is a Φ-linear combination of elements of $M(X)$ of degree (Y-degree) k is said to be *homogeneous of degree* (*Y-degree*) *k*. The subset of these elements is a subspace of $\Phi\{X\}$ and $\Phi\{X\}$ becomes a graded algebra relative to these subspaces.

We proceed next to define free non-associative multiplicative systems and algebras. We note first that if S is any set with an associative binary multiplication and p is a fixed element of S then we can define three p-products in S by the formulas

$$x \cdot y = pxy, \qquad x * y = xpy, \qquad x \circ y = xyp. \tag{2}$$

The middle one of these is associative but not the other two. Either of the extreme ones can be used in the construction we shall now give of a free non-associative algebra. For this purpose let X be an arbitrary set as before, and let $\{p\}$ be a single element set disjoint to X. Write $M(X, p)$ for $M(X \cup \{p\})$ the free associative multiplicative system with 1 generated by $X \cup \{p\}$. We introduce the right p multiplication $x \circ y = xyp$ in $M(X, p)$ and we let $N(X)'$ be the subsystem of $(M(X, p), \circ)$ generated by the set X. One sees easily that $N(X)'$ can be defined inductively on degree by the rules that $X \subseteq N(X)'$, and if a, $b \in N(X)'$ then $abp \in N(X)'$. It is clear from this that the X-degree of any $m \in N(X)'$ exceeds its p-degree by one. The elements of $N(X)'$ of degree >1 have the form uvp where u, $v \in N(X)'$. It follows by induction on degree that if $w \in N(X)'$ then no right factor of w, $\neq w$ is of X-degree exceeding its degree in p. We use this remark to prove the following

Lemma. *If $w \in N(X)'$ has degree ≥ 2 then the representation of w as $w = uvp$, u, $v \in N(X)'$ is unique.*

Proof. Let $w = uvp$, u, $v \in N(X)'$ and let v' be the right hand factor of least degree of $w' = uv$ with the property that its X-degree exceeds is p-degree. Since v has this property, v' is a right factor of v. Hence by the result noted we have $v' = v$. Hence v and consequently u is uniquely determined by w.

We consider $N(X)'$ as a set with the binary composition $a \circ b = abp$ and now define *the degree* (or *Y-degree* for a subset Y of X) of $w \in (N(X)', \circ)$ to be the X-degree (Y-degree) of w as element of $M(X)$. Let S be any set with an arbitrary binary composition cd and let $x \to y$ be a mapping α of X into S. Then we can extend this inductively on the degree to a mapping α on $N(X)'$ by specifying that if $w = uvp = u \circ v$, $u, v \in N(X)'$ then $w^\alpha = u^\alpha v^\alpha$. Since the representation of w as $u \circ v$ is unique it is clear that α is single-valued. Also it is clear that α is a homomorphism of $(N(X)', \circ)$ into S and this is the only homomorphism of $(N(X)', \circ)$ which coincides with the mapping $x \to y$ on the set X.

Now let $\Phi\{\{X\}\}'$ be the algebra over the field Φ with basis $N(X)'$ and product $\circ$ given by the multiplication table of the base elements as defined in $(N(X)', \circ)$. It is immediate from the universal property just noted that if $\mathfrak{A}$ is any (non-associative) algebra and $x \to y$ is a mapping of X into $\mathfrak{A}$ then this has a unique extension to a homomorphism of $\Phi\{\{X\}\}'$ into $\mathfrak{A}$. We shall call $\Phi\{\{X\}\}'$ the *free non-associative algebra (freely) generated by the set X*.

The algebra $\Phi\{\{X\}\}'$ does not have an identity element. However, we can adjoin one to obtain the algebra $\Phi\{\{X\}\} = \Phi\{\{X\}\}' \oplus \Phi 1$ which we shall call the *free non-associative algebra with* 1 *generated (freely) by the set X*. This has the universal mapping property that any mapping of X into an algebra with 1 has a unique extension to a homogeneous of $\Phi\{\{X\}\}$ mapping $1 \to 1$.

We define homogeneous element of $\Phi\{\{X\}\}'$ and $\Phi\{\{X\}\}$ as for the free associative algebra $\Phi\{X\}$. In this way these algebras are graded.

2. *Classes of algebras defined by identities.* Let $\mathfrak{A}$ be a non-associative algebra over a field Φ and let f be an element of a free non-associative algebra $\Phi\{\{X\}\}'$. Then we say that $\mathfrak{A}$ *satisfies the identity* $f = 0$ (or $f = 0$ *is an identity for* $\mathfrak{A}$) if the element f is mapped into 0 under every homomorphism of $\Phi\{\{X\}\}'$ into $\mathfrak{A}$. There is no loss in generality in assuming $X = \{X_i\}$, $i = 1, 2, 3, \ldots$ and we do this from now on. Then f is contained in a subalgebra generated by a finite set $\{x_1, x_2, \ldots, x_n\}$, and accordingly, we write $f = f(x_1, x_2, \ldots, x_n)$. Also we denote the image of f under the homomorphism sending $x_i \to a_i$ by $f(a_1, \ldots, a_n)$. Hence $\mathfrak{A}$ satisfies the identity $f = 0$ if and only if $f(a_1, \ldots, a_n) = 0$ for all $a_i \in \mathfrak{A}$. If S is a subset of $\Phi\{\{X\}\}'$ then we denote the class of algebras (over Φ) satisfying all the identities $f = 0$, $f \in S$ by $C(S)$. If $f = (x_1x_2)x_3 - x_1(x_2x_3)$ then the class $C(S)$ is the class of associative algebras. (Note: we now use the standard notation ab for the product in a free algebra.) If S consists of $f = x_1^2$ and $g = (x_1x_2)x_3 + (x_2x_3)x_1 + (x_3x_1)x_2$ then $C(S)$ is the class of Lie algebras. If S consists of $f = x_1^2x_2 - x_1(x_1x_2)$, $g = x_2x_1^2 - (x_2x_1)x_1$ then $C(S)$ is the class of *alternative algebras*. If S consists of $f = x_1x_2 - x_2x_1$, $g = (x_1^2x_2)x_1 - x_1^2(x_2x_1)$ and the characteristic of Φ is $\neq 2$ then $C(S)$ is the class of *Jordan algebras*.

If $\mathfrak{A}$ is an algebras then the subset $I(\mathfrak{A})$ of identities f (more exactly $f = 0$) satisfied by $\mathfrak{A}$ is an ideal in $\Phi\{\{X\}\}'$. It is clear that $I(\mathfrak{A})^\eta \subseteq I(\mathfrak{A})$ for every endomorphism η of the free non-associative algebra $\Phi\{\{X\}\}'$. Clearly the subsets S and S' of $\Phi\{\{X\}\}'$ define the same class $C(S)$ and $C(S')$ if the T-ideals generated by S and S' are the same.

Let $f \in I$ a T-ideal and write $f = f_0 + f_1 + \cdots + f_r$ where f_j is homogeneous of degree j in x_i. Let $\alpha \in \Phi$ and consider the endomorphism of $\Phi\{\{X\}\}'$ such that $x_i \to \alpha x_i$, $x_k \to x_k$ if $k \neq i$. This maps f into $f_0 + \alpha f_1 + \alpha^2 f_2 + \cdots + \alpha^r f_r$. If follows that if Φ contains $r + 1$ distinct elements then every $f_j \in I$. For this reason it is natural to make the following hypothesis on the sets S of defining identities for classes of algebras:

(α) The elements of S are homogeneous in every x_i.

We now introduce another set $Y = \{y_i\}$, $i = 1, 2, \ldots$ such that $X \cap Y = \phi$ and put $Z = X \cup Y$. For each $i = 1, 2, \ldots$ we have the homomorphism λ_i of $\Phi\{\{X\}\}'$ into $\Phi\{\{Z\}\}'$ such that $x_i \to x_i + y_i$, $x_k \to x_k$ if $k \neq i$ and if $f \in \Phi\{\{X\}\}'$ has degree m_i in x_i then we can write

$$f(x_1, x_2, \ldots, x_{i-1}, x_i + y_i, x_{i+1}, \ldots, x_n) = \sum_{j=0}^{m_i} f_{ij} \tag{3}$$

where f_{ij} is homogeneous of degree j in y_i. The f_{ij} are uniquely determined and we have the mappings D_{ij}: $f \to f_{ij}$, $i = 1, 2, \ldots,$ $j = 0, 1, 2, \ldots$ of $\Phi\{\{X\}\}'$ into $\Phi\{\{Z\}\}'$. The D_{ij} are linear and

$$(fg)D_{ij} = \sum_{k=0}^{j} (fD_{ik})(gD_{i,j-k}). \tag{4}$$

It is clear also from (3) that $f_{io} = fD_{io} = f(x_1, \ldots, x_n)$ so $D_{io} = 1$. Hence $\{D_{io} = 1, D_{i1}, D_{i2}, \ldots\}$ is a higher derivation of infinite rank of $\Phi\{\{X\}\}'$ into $\Phi\{\{Z\}\}'$ (cf. Jacobson, *Lectures in Abstract Algebra*, vol. III, p. 191). In particular, $D_i \equiv D_{i1}$ is a derivation of $\Phi\{\{X\}\}'$ into $\Phi\{\{Z\}\}'$. This is characterized by the conditions

$$x_k D_i = \delta_{ik} y_i, \qquad i, k = 1, 2, \ldots. \tag{5}$$

If f is homogeneous of degree m_k in $x_k \neq x_i$ then the same holds for fD_{ij}, and if f is homogeneous of degree m_i in x_i and $j \leq m_i$ then fD_{ij} is homogeneous of degree $m_i - j$ in x_i. Also in this case $fD_{\mathrm{im}_i} = f(x_1, \ldots, x_{i-1}, y_i, x_{i+1}, \ldots, x_n)$ and $fD_{ij} = 0$ if $j > m_i$. All of these assertions are easily proved for elements of $N(X)'$ by induction on the degree. Then they follow for arbitrary f by the linearity of the operators D_{ij}. In the same way, one establishes

$$f(x_1 + y_1, x_2 + y_2, \ldots, x_n + y_n) = f(x_1, \ldots, x_n) + \sum_{i=1}^{n} fD_i \pmod{\mathscr{Y}^2} \tag{6}$$

where $\mathscr{Y}$ is the ideal in $\Phi\{\{Z\}\}'$ generated by all the y's.

As an example, let $g = (x_1^2 x_2)x_1 - x_1^2(x_2 x_1)$ one of the polynomials defining the class of Jordan algebras. Then

$$\begin{aligned} gD_{11} &= ((x_1y_1 + y_1x_1)x_2)x_1 + (x_1^2x_2)y_1 - (x_1y_1 + y_1x_1)(x_2x_1) - x_1^2(x_2y_1) \\ gD_{12} &= (y_1^2x_2)x_1 + ((x_1y_1 + y_1x_1)x_2)y_1 - (x_1y_1 + y_1x_1)(x_2y_1) - y_1^2(x_2x_1). \end{aligned} \tag{7}$$

We now apply a 1-1 mapping of the set Z onto X. The corresponding homomorphism of $\Phi\{\{Z\}\}'$ into $\Phi\{\{X\}\}'$ maps the elements fD_{ij} into elements of $\Phi\{\{X\}\}'$. The elements obtained in this way and by iterating the process will be called *derived elements* of f. For example, the elements

$$g_1 = ((x_1x_3 + x_3x_1)x_2)x_1 + (x_1^2x_2)x_3 - (x_1x_3 + x_3x_1)(x_2x_1) - x_1^2(x_2x_3)$$

$$g_2 = (x_3^2x_2)x_1 + ((x_1x_3 + x_3x_1)x_2)x_3 - (x_1x_3 + x_3x_1)(x_2x_3) - x_3^2(x_2x_1)$$

$$\begin{aligned} g_3 = {} & ((x_1x_3 + x_3x_1)x_2)x_4 + ((x_4x_3 + x_3x_4)x_2)x_1 + ((x_1x_4 + x_4x_1)x_2)x_3 \\ & - (x_1x_3 + x_3x_1)(x_2x_4) - (x_4x_3 + x_3x_4)(x_2x_1) - (x_1x_4 + x_4x_1)(x_2x_3) \end{aligned} \tag{8}$$

$$\begin{aligned} g_4 = {} & ((x_3x_4 + x_4x_3)x_2)x_1 + ((x_1x_3 + x_3x_1)x_2)x_4 + ((x_1x_4 + x_4x_1)x_2)x_3 \\ & - (x_1x_3 + x_3x_1)(x_2x_4) - (x_1x_4 + x_4x_1)(x_2x_3) - (x_3x_4 + x_4x_3)(x_2x_1) \end{aligned} \tag{9}$$

are derived elements of $g = (x_1^2x_2)x_1 - x_1^2(x_2x_1)$. Here g_1 and g_2 are obtained from gD_{11}, gD_{12} by replacing y_1 by x_3, g_3 is obtained by replacing y_3 in g_2D_{31} by x_4.

Let $f = f(x_1, x_2, \ldots, x_n)$ be homogeneous of degree m_i in x_i, $i = 1, \ldots, n$. Then we cal $\sum_1^n (m_i - 1)$ the *height* of f. Thus the height is 0 if and only if f^1 is a *multilinear element*, that is, f is homogeneous of degree ≤ 1 in all the $x'w$. It is clear from the results noted before that the derived elements of f are homogeneous and have heights $\leq$ the height h of f. Moreover, the height of a derived element of $f = f(x_{i_1}, x_{i_2}, \ldots, x_{i_n})$ is h only if the element has the form $f(x_{i_1}, x_{i_2}, \ldots, x_{i_n})$ where the i_j are distinct. Now suppose that Φ contais r distinct elements where $r \geq m_i$, the degree of f in x_i, $i = 1, 2, 3, \ldots,$ and let I be a T-ideal containing f. Let η be a homomorphism $\Phi\{\{Z\}\}'$ into $\Phi\{\{X\}\}'$ mapping Z in a 1-1 way onto X. Then I contains

$$f(x_1, \ldots, x_{i-1}, x_i + y_i, x_{i+1}, \ldots, x_n)^\eta - f(x_1, \ldots, x_n)^\eta - fD^\eta_{\mathrm{im}_i} = \sum_{j=i}^{m_i - 1} fD^\eta_{ij}$$

and fD^η_{ij} is homogeneous of degree j in y_i^η. Since there exist $m_i - 1$ distinct non-zero elements of Φ it follows that every derived element $fD^\eta_{ij} \in I$. Also it is clear that Φ contains as many elements as the degree of these derived elements in any of x's. Hence it is clear that I contains every derived element of f. We shall now impose the following condition on the sets S of defining identities of classes of algebras $C(S)$:

(β). If $f \in S$ then every derived element of f is contained in the T-ideal I generated by S. The result we have just shown is that (β) is automatically satisfied if the elements of S are homogeneous and Φ contains as many distinct elements as the degree of every $f \in S$ in every x_i. It is clear for the specification of sets S defining associative, Lie and alternative algebras that these satisfy the conditions (α) and (β). If we recall that for Jordan algebras the characteristic of the base field Φ is $\neq 2$ so that Φ contains at least three distinct elements we see also that the set S defining Jordan algebras also satisfies the conditions.

Theorem 1. *Let S be a subset of $\Phi\{\{X\}\}'$ satisfying (α) and (β) and let $\mathfrak{A}$ be an algebra over Φ' with basis $\{u_i\}$. Then $\mathfrak{A} \in C(S)$ if and only if* $f(u_{i_1}, u_{i_2}, \ldots, u_{i_n}) = 0$ for all $f \in S$ and all $u_{i_j} \in \{u_i\}$.

Proof. We have to show that the conditions stated imply that $f(a_1, \ldots, a_n) = 0$ for all $a_i \in \mathfrak{A}$ and $f \in S$. The condition (α) and (β) on S imply that we may assume that if $f \in S$ then all the derived elements of f are contained in S. If $f \in S$ is multilinear then it is clear that the condition $f(a_1, a_2, \ldots, a_n) = 0$, $a_j \in \mathfrak{A}$, is a consequence of the conditions $f(u_{i_1}, u_{i_2}, \ldots, u_{i_r})$; $u_{i_j} \in \{u_i\}$. Hence we assume that the height h of f is positive and we may assume that $g(a_1, a_2, \ldots, a_m) = 0$ for all $a_j \in \mathfrak{A}$ and $g \in S$ of height less than h. We assume also that $f(a_1, \ldots, a_{r-1}, u_{i_r}, \ldots, u_{i_n}) = 0$ for all $a_j \in \mathfrak{A}$, $u_{i_k} \in \{u_i\}$. Let $\mathfrak{B}$ be the subset of element $b \in \mathfrak{A}$ such that $f(a_1, \ldots, a_{r-1}, b, u_{i_{r+1}}, \ldots, u_{i_n}) = 0$, $a_j \in \mathfrak{A}$, $u_{i_k} \in \{u_i\}$. Then $\mathfrak{B}$ contains the basis $\{u_i\}$. Also since f is homogeneous in x_r, $\mathfrak{B}$ is closed under multiplication by elements of Φ. Now let $a_j \in \mathfrak{A}$, $u_{i_k} \in \{u_i\}$, $b \in \mathfrak{B}$ and consider $f(a_1, \ldots, a_{r-1}, b + u_{i_r}, u_{i_{r+1}}, \ldots, u_{i_n}) = f(a_1, \ldots, a_{r-1}, b, u_{i_{r+1}}, \ldots, u_{i_n}) + f(a_1, \ldots, a_{r-1}, u_{i_r}, u_{i_{r+1}}, \ldots, u_{i_n}) + \Sigma f_j(a_1, \ldots, a_{r-1}, b, u_{i_r}, \ldots, u_{i_n})$ where the f_j are derived elements of f and have lower height than f. Our induction assumptions imply that $f(a_1, \ldots, a_{r-1}, b + u_{i_r}, u_{i_{r+1}}, \ldots, u_i) = 0$. Hence $b + u_{i_r} \in \mathfrak{B}$. This now implies that $\mathfrak{B} = \mathfrak{A}$. Thus $f(a_1, \ldots, a_r, u_{i_{r+1}}, \ldots, u_{i_n}) = 0$ and $f(a_1, \ldots, a_n) = 0$ by induction on r.

Let Π be a extension field of Φ. Then the elements $f \in S$ can be regarded as elements of $\Pi\{\{X\}\}'$. Hence these determine the class of algebras over Π which satisfy the identities of S. We denote this as $C(S,\Pi)$ and the class determined by S and Φ as $C(S,\Phi)$ $(=C(S))$. Then we have the

Corollary 1. *Let S be a subset of $\Phi\{\{X\}\}'$ satisfying* (α) *and* (β) *and let $\mathfrak{A} \in C(S,\Phi)$. Then the extension algebra $\mathfrak{A}_\Pi = \Pi \otimes_\Phi \mathfrak{A} \in C(S,\Pi)$.*

Proof. This is clear from the criterion of Theorem 1 since any basis $\{u_i\}$ for $\mathfrak{A}/\Phi$ is a basis for $\mathfrak{A}_\Pi/\Pi$.

In particular, the corollary shows that if $\mathfrak{A}$ is associative, Lie, alternative or Jordan then the same is true of $\mathfrak{A}_\Pi$.

If $\mathfrak{A}$ is an algebra over Φ than we can adjoin an identity element 1 to $\mathfrak{A}$ by forming $\mathfrak{A}^* = \mathfrak{A} \oplus \Phi_1$ and extending the multiplication in $\mathfrak{A}$ to one in $\mathfrak{A}^*$ so that $la^* = a^* = a^*$ 1 for $a^* \in \mathfrak{A}^*$. We shall now give a condition on a subset S of $\Phi\{\{X\}\}'$ that along with (α) and (β) will insure that if $\mathfrak{A} \in C(S)$ then $\mathfrak{A}^* \in C(S)$. This is

(γ) If $f(x_1,\ldots,x_n) \in S$ then for every $i = 1,\ldots,n$ $f(x_1,\ldots,x_{i-1},1,\ldots,x_n)$ which is an element of $\Phi\{\{X\}\}$ is necessarily in $\Phi\{\{X\}\}'$ and is in the T-ideal generated by the elements of S.

The following is an immediate consequence of Theorem 1.

Corollary 2. *Let S be a subset of $\Phi\{\{X\}\}'$ satisfying* (α), (β), *and* (γ) *and let $\mathfrak{A}$ be an algebra in the class $C(S)$. Then the algebra $\mathfrak{A}^* = \mathfrak{A} \oplus \Phi\, 1$ obtained by adjoining an identity element 1 to $\mathfrak{A}$ is in $C(S)$.*

We omit the proof.

The specializations $x_i = 1$ in the identities f, g defining the classes of alternative and Jordan algebras give 0. Hence (γ) and Corollary 2 holds for these classes. On the other hand, if we put $x_1 = 1$ in the identity x_1^2 we obtain 1 so (γ) does not hold for the identities defining Lie algebras.

We remark finally that if $\mathfrak{A}$ is in a class $C(S)$ then every subalgebra of $\mathfrak{A}$ and every homomorphic image of $\mathfrak{A}$ is in $C(S)$. Also if $\{\mathfrak{A}_\alpha\}$ is a set of algebras in $C(S)$ then the complete direct sum of the $\mathfrak{A}_\alpha$ is in $C(S)$. Conversely, by a theorem of G. Birkhoff, any class of algebras having these closure properties has the form $C(S)$. (Cohn, *Universal Algebra*, p. 169).

3. *Bimodules for algebras in a class defined by identities.* Let $\mathfrak{A}/\Phi$ be an algebra, $\mathfrak{M}/\Phi$ a vector space and suppose we have a pair of bilinear mappings $(a,u) \to au$, $(a,u) \to ua$ of the pair $(\mathfrak{A},\mathfrak{M})$ into $\mathfrak{M}$. We can use these to define an algebra extension $\mathscr{E} = \mathfrak{A} \oplus \mathfrak{M}$ with multiplication given by

$$(10) \qquad (a_1 + u_1)(a_2 + u_2) = a_1 u_2 + a_1 u_2 + u_1 a_2,$$

$a_i \in \mathfrak{A}$, $u_i \in \mathfrak{M}$. It is clear that this product is bilinear so $\mathscr{E}$ is an algebra. Moreover, $\mathfrak{A}$ is a subalgebra and $\mathfrak{M}$ is an ideal such that $\mathfrak{M}^2 = 0$. We shall call $\mathscr{E}$ the *split null extension* of $\mathfrak{A}$ determined by the given bilinear mappings of $(\mathfrak{A},\mathfrak{M})$ into $\mathfrak{M}$. Now suppose S is a subset of the free non-associative algebra $\Phi\{\{X\}\}'$ and $C(S)$ is the class of algebras defined by the identities $f \in S$. Let $\mathfrak{A} \in C(S)$. Then we shall say that $\mathfrak{M}$ (with the compositions au, ua) is an *S-bimodule* for $\mathfrak{A}$ if $\mathscr{E} \in C(S)$. In particular, if S is the set defining associative, Lie, alternative or Jordan algebras then we say that $\mathfrak{M}$ is an *associative, Lie, alternative or Jordan* bimodule for .

As in § 2 let $f = f(x_1, x_2, \ldots, x_n) \in \Phi\{\{X\}\}'$ and let D_{ij} be defined as before. Then it is clear that fD_{ij} is in the subalgebra of $\Phi\{\{Z\}\}'$ generated by $x_1, \ldots, x_n, y_i$. Accordingly, we write $fD_{ij}(x_1, \ldots, x_n; y_i)$. The image of this element under a homomorphism of $\Phi\{\{Z\}\}$ such that $x_k \to a_k$, $k = 1, \ldots, n$ and $y_i \to u$ is denoted as $fD_{ij}(a_1, \ldots, a_n; u)$. Also we abbreviate $D_i = D_{i1}$. Then we have the following criterion.

Theorem 2. *$\mathfrak{M}$ is an S-bimodule for $\mathfrak{A} \in C(S)$ if and only if $fD_i(a_1, \ldots, a_n; u) = 0$ for all $a_k \in \mathfrak{A}$, $u \in \mathfrak{M}$, $f \in S$.*

Proof. If $\mathfrak{M}$ is an S-bimodule then $\mathscr{E} = \mathfrak{A} \oplus \mathfrak{M}$ is in $C(S)$. Hence if $f \in S$ then $f(a_1, \ldots, a_{i-1}, a_i + u, a_{i+1}, \ldots, a_n) = 0$ for $a_k \in \mathfrak{A}$, $u \in \mathfrak{M}$. By (3) and $\mathfrak{M}^2 = 0$ we have $f(a_1, \ldots, a_{i-1}, a_i + u, a_{i+1}, \ldots, a_n) = f(a_1, \ldots, a_n; u) = 0$. Conversevely, assume these conditions hold and consider $f(a_1 + u_1, a_2 + u_2, \ldots, a_n + u_n)$, $a_i \in \mathfrak{A}$ $u_i \in \mathfrak{M}$. By (6) and $\mathfrak{M}^2 = 0$ we have $f(a_1 + u_1, \ldots, a_n + u_n) = f(a_1, \ldots, a_n) + \Sigma fD_i - (a_1, \ldots, a_n; u_i) = 0$. Hence $\mathscr{E} \in C(S)$ and $\mathfrak{M}$ is an S-bimodule for $\mathfrak{A}$.

Let $\mathfrak{M} = \mathfrak{A}$ and let ua, au, $a \in \mathfrak{A}$, $u \in \mathfrak{M}$, be the products of these elements in $\mathfrak{A}$. If $f \in S$, $fD_i(a_1, \ldots, a_n; u)$ is the image of a derived element of f under a homomorphism of $\Phi\{\{X\}\}'$ into $\mathfrak{A}$. Hence if the conditions (α) and (β) on S are satisfied then $fD_i(a_1, \ldots, a_n; u) = 0$ for $a_k \in \mathfrak{A}$, $u \in \mathfrak{M}$ $f \in S$. Thus $\mathfrak{A}$ is an S-bimodule. Hence we have the

Corollary. *If S satisfies conditions (α) and (β) then $\mathfrak{A}$ is an S-bimodule relative to the module compositions given by the product in $\mathfrak{A}$.*

We shall call $\mathfrak{M} = \mathfrak{A}$ the *regular* bimodule for $\mathfrak{A}$.

Examples 1) $C(S)$ the class of associative algebras. Here $S = \{f\}$ where $f = (x_1x_2)x_3 - x_1(x_2x_3)$. We have

$$
\begin{aligned}
fD_1 &= (y_1x_2)x_3 - y_1(x_2x_3) \\
fD_2 &= (x_1y_2)x_3 - x_1(y_2x_3) \\
fD_3 &= (x_1x_2)y_1 - x_1(x_2y_1).
\end{aligned}
\tag{11}
$$

Hence $\mathfrak{M}$ is a bimodule for the associative algebra $\mathfrak{A}$ if and only if $(ua)b = u(ab)$, $(au)b = a(ub)$, $(ab)u = a(bu)$ for a, $b \in \mathfrak{A}$, $u \in \mathfrak{M}$. This is the usual definition of a bimodule (or two-sided module) for an associative algebra.

2) $C(S)$ the class of Lie algebras. Here $S = \{f, g\}$ where $f = x_1^2$, $g = (x_1x_2)x_3 + (x_2x_3)x_1 + (x_3x_1)x_2$. Then

$$
\begin{aligned}
fD_1 &= x_1y_1 + y_1x_1 \\
gD_1 &= (y_1x_2)x_3 + (x_2x_3)y_1 + (x_3y_1)x_2 \\
gD_2 &= (x_1y_2)x_3 + (y_2x_3)x_1 + (x_3x_1)y_2 \\
gD_3 &= (x_1x_2)y_3 + (x_2y_3)x_1 + (y_3x_1)x_2.
\end{aligned}
\tag{12}
$$

The corresponding conditions for a Lie bimodule $\mathfrak{M}$ for the Lie algebra $\mathfrak{A}$ are

$au = -ua$, $(ua)b + (ab)u + (bu)a = 0$, $(au)b + (ub)a + (ba)u = 0$, $(ab)u + (bu)a + (ua)b = 0$, $a, b \in \mathfrak{A}$, $u \in \mathfrak{M}$.

3) $C(S)$ the class of alternative algebras. We have $S = \{f, g\}$, $f = x_1^2x_2 - x_1(x_1x_2)$, $g = x_2x_1^2 - (x_2x_1)x_1$. Then

$$
(13) \qquad \begin{aligned}
fD_1 &= (x_1y_1 + y_1x_1)x_2 - x_1(y_1x_2) - y_1(x_1x_2) \\
fD_2 &= x_1^2y_2 - x_1(x_1y_2) \\
gD_1 &= x_2(x_1y_1 + y_1x_1) - (x_2x_1)y_1 - (x_2y_1)x_1 \\
gD_2 &= y_2x_1^2 - (y_2x_1)x_1.
\end{aligned}
$$

The conditions for an alternative bimodule are; $(au + ua)b = a(ub) + u(ab)$, $a^2u = a(au)$, $b(au + ua) = (ba)u + (bu)a$, $ua^2 = (ua)a$.

4) $C(S)$ the class of Jordan algebras. Here $S = \{f, g\}$ where $f = x_1x_2 - x_2x_1$, $g = (x_1^2x_2)x_1 - x_1^2(x_2x_1)$. Then

$$
(14) \qquad \begin{aligned}
&fD_1 = y_1x_2 - x_2y_1, fD_2 = x_1y_2 - y_2x_1 \\
&gD_1 = ((x_1y_1 + y_1x_1)x_2)x_1 - (x_1y_1 + y_1x_1)(x_2x_1) \\
&gD_2 = (x_1^2y_2)x_1 - x_1^2(y_2x_1).
\end{aligned}
$$

Then the Jordan bimodule conditions are $au = ua$, $((au + ua)b)a = (au + ua)(ba)$, $(a^2u)a = a^2(ua)$.

4. *Multiplication specializations and representations.* We shall now seek to formulate the S-bimodule conditions given in Theorem 2 as conditions on the linear mappings $a_L: u \to au$, $a_R: u \to au$ in $\mathfrak{M}$. Accordingly, we consider the free non-associative algebra $\Phi\{\{X, y\}\}'$ generated by $X = \{x_i\}$ and an element $y \notin X$. The subalgebra of this algebra generated by the x's can be identified with the free non-associative algebra $\Phi\{\{X\}\}'$. Now let $\mathscr{V}$ be the subalgebra (with 1) of $\mathrm{Hom}_\Phi(\Phi\{\{X, y\}\}', \Phi\{\{X, y\}\}')$ generated by 1 and the multiplications A_L, A_R in $\Phi\{\{X, y\}\}'$ determined by the elements $A \in \Phi\{\{X\}\}'$. Clearly $\mathscr{V}$ is generated by 1 and the transformations m_L, m_R, $m \in N(X)'$. We now call the elements m $N(X)'$ *monomials* in the x_i and we introduce a set P which is a disjoint union of two copies of $N(X)'$. Thus we have two injective mappings $m \to l(m)$, $m \to r(m)$ of $N(X)'$ into P such that $P = l(N(X)') \cup r(N(X)')$ and $l(N(X)') \cap r(N(X)') = \phi$. Let $\Phi\{P\}$ be the free associative algebra (with 1) generated by the set P. Then we have the homomorphism ν of $\Phi\{P\}$ onto ν such that $l(m)^\nu = m_L$, $r(m)^\nu = m_R$. More generally, let η be a homomorphism of $\Phi\{\{X\}\}'$ into an algebra $\mathfrak{A}$. Then we have a homomorphism η' of $\Phi\{P\}$ into $\mathrm{Hom}_\Phi(\mathfrak{A}, \mathfrak{A})$ such that $1 \to 1$, $l(m)^{\eta'} = (m^\eta)_L$, $r(m)^{\eta'} = (m^\eta)_R$ where these now denote multiplications in $\mathfrak{A}$. Thus we have the commutative diagrams

$$
(15) \qquad
\begin{array}{ccc}
N(X)' & \xrightarrow{\eta} & \mathfrak{A} \\
{\scriptstyle l}\downarrow & & \downarrow{\scriptstyle L} \\
\Phi\{P\} & \xrightarrow{\eta'} & \mathrm{Hom}_\Phi(\mathfrak{A}, \mathfrak{A})
\end{array}
\qquad
\begin{array}{ccc}
N(X)' & \xrightarrow{\eta} & \mathfrak{A} \\
{\scriptstyle r}\downarrow & & \downarrow{\scriptstyle R} \\
\Phi\{P\} & \xrightarrow[\eta']{} & \mathrm{Hom}_\Phi(\mathfrak{A}, \mathfrak{A})
\end{array}
$$

Lemma 1. *There exist a homomorphism η'' of $\mathscr{V}$ into* $\mathrm{Hom}_\Phi(\mathfrak{A}, \mathfrak{A})$ *such that*

$$
(16) \qquad m_L^{\eta''} = (m^\eta)_L, \qquad m_R^{\eta''} = (m^\eta)_R.
$$

Proof. We show first that the kernel of ν, $\ker \nu \subseteq \ker \eta'$.

Let $u \in \mathfrak{A}$ and let (η, u) be the extension of η to a homomorphism of $\Phi\{\{X, y\}\}'$ into $\mathfrak{A}$ such that $y \to u$. We claim that

$$(17) \qquad (yQ^{\nu})^{(\eta,u)} = uQ^{\eta'}$$

for every $Q \in \Phi\{P\}$. Since 1, $l(m)$, $r(m)$, $m \in N(X)'$, generate $\Phi\{P\}$ this will follow if we can show that if (17) holds for an element Q then it holds for $Ql(m)$ and for $Qr(m)$. We have $y(Ql(m))^{\nu} = y(Q^{\nu}l(m)^{\nu}) = yQ^{\nu}m_L = m(yQ^{\nu})$. Hence $(y(Ql(m)^{\nu}))^{(\eta,u)} = (m(yQ^{\nu}))^{(\eta,u)} = m^{\eta}(yQ^{\nu})^{(\eta,u)} = m^{\eta}(uQ^{\eta'}) = (uQ^{\eta'})(m^{\eta})_L = (uQ^{\eta'})l(m)^{\eta'} = u(Ql(m))^{\eta'}$. Hence (17) holds for $Ql(m)$. Similarly, it holds for $Qr(m)$. Hence it holds for all Q in $\Phi\{P\}$. Now suppose $Q^{\nu} = 0$. Then (17) implies that $uQ^{\eta'} = 0$. Since this holds for all $u \in \mathfrak{A}$, we have $Q^{\eta'} = 0$. This proves the assertion on the kernels of ν and η'. It now follows that we have a homomorphism η'' of $\mathscr{V}$ into $\mathrm{Hom}_{\Phi}(\mathfrak{A}, \mathfrak{A})$ such that

$$(18) \qquad \begin{array}{ccc} P & \xrightarrow{\nu} & \mathscr{V} \\ & \searrow^{\eta'} & \downarrow \eta'' \\ & & \mathrm{Hom}_{\Phi}(\mathfrak{A}, \mathfrak{A}) \end{array}$$

is commutative. Then we have $m_L^{\eta''} = l(m)^{\nu\eta'} = l(m)^{\eta'} = (m^{\eta})_L$, by (15). Similarly, $m_R^{\eta''} = (m^{\eta})_R$.

Lemma 2. *ν is an isomorphism of $\Phi\{P\}$ onto $\mathscr{V}$.*

Proof. We take a $1-1$ representation σ of $\Phi\{P\}$ by linear transformations in a vector space $\mathfrak{N}/\Phi$ and we use this to define bilinear compositions of $(\Phi\{\{X\}\}', \mathfrak{N})$ into $\mathfrak{N}$ such that for $m \in N(X)'$ and $n \in \mathfrak{N}$ we have $(m, n) \to nl(m)^{\sigma}$, $(m, n) \to nr(m)^{\sigma}$ respectively. These bilinear compositions define the split null extension $\mathscr{E} = \Phi\{\{X\}\}' \oplus \mathfrak{N}$ of $\Phi\{\{X\}\}'$. Let η be the injection of $\Phi\{\{X\}\}'$ into $\mathscr{E}$, η'' the homomorphism of $\mathscr{V}$ into $\mathrm{Hom}_{\Phi}(\mathscr{E}, \mathscr{E})$ such that $m_L^{\eta''} = (m^{\eta})_L$, $m_R^{\eta''} = (m^{\eta})_R$, as in Lemma 1. By the definition of $\mathscr{E}$, we have $n(m^{\eta})_L = nl(m)^{\sigma}$, $n(m^{\eta})_R = nr(m)^{\sigma}$ if $n \in \mathfrak{N}$. Hence $m_L^{\eta''}$ and $m_R^{\eta''}$ map $\mathfrak{N}$ into itself and consequently the elements of $\mathscr{V}^{\eta''}$ map $\mathfrak{N}$ into itself. If $v^{\eta'''}$ denotes the restriction to $\mathfrak{N}$ of $v^{\eta''}$, $v \in \mathscr{V}$, then η''' is a homomorphism of $\mathscr{V}$ into $\mathrm{Hom}_{\Phi}(\mathfrak{N}, \mathfrak{N})$ such that $m_L^{\eta} = l(m)^{\sigma}$, $m_R^{\eta'''} = r(m)^{\sigma}$. Since σ is an isomorphism we have the homomorphism of $\mathscr{V}$ into $\Phi\{P\}$ such that $m_L \to l(m)$, $m_R \to r(m)$. It is clear that this is the inverse of the mapping ν, so ν is an isomorphism.

It is clear from the definition of $\Phi\{\{X, y\}\}'$ that any element q of this algebra which is homogeneous of degree 1 in y has the form yQ^{ν} where $Q \in \Phi\{P\}$. It is an easy consequence of Lemma 2 that Q is unique. We can now give the following

Definition 1. Let $C(S)$ be a class of algebras defined by a set of identities $S \subseteq \Phi\{\{X\}\}'$, and let $\mathfrak{A} \in C(S)$. Then a pair of linear mappings (λ, ρ) of $\mathfrak{A}$ into an associative algebra $\mathfrak{G}$ with 1 is called an *S-multiplication specialization of* $\mathfrak{A}$ *in* $\mathfrak{G}$ if the following conditions hold: Let $f \in S$ and write $fD_i = y_i F_i^{\nu}$ where $F_i \in \Phi\{P\}$. Let η be any homomorphism of $\Phi\{\{X\}\}'$ into $\mathfrak{A}$ and let τ be the homomorphism of $\Phi\{P\}$ into $\mathfrak{G}$ such that $1^{\tau} = 1$, $l(m)^{\tau} = m^{\eta\lambda}$, $r(m)^{\tau} = m^{\eta\rho}$. Then $F_i^{\tau} = 0$. This is required to hold for all $f \in S$, all $i = 1, 2, \ldots,$ and every homomorphism η of $\Phi\{X\}'$ into $\mathfrak{A}$. If $\mathfrak{A} = \mathrm{Hom}_{\Phi}(\mathfrak{M}, \mathfrak{M})$ then (λ, ρ) is called an *S-multiplication representation of* $\mathfrak{A}$.

If S is the set defining associative, Lie, alternative or Jordan algebras then we call (λ, ρ) an associative, Lie etc. multiplication specialization or representation of $\mathfrak{A}$. We now consider the explicit form of the conditions on λ and ρ in these cases.

1) *Associative algebras.* If we refer to (11) we obtain $fD_1 = y_1F_1$, $fD_2 = y_2F_2$, $fD_3 = y_3F_3$ where

$$
\begin{aligned}
F_1 &= r(x_2)r(x_3) - r(x_2x_3) \\
F_2 &= l(x_1)r(x_3) - r(x_3)l(x_1) \\
F_3 &= l(x_1x_2) - l(x_2)l(x_1).
\end{aligned} \tag{19}
$$

Let $\mathfrak{A}$ be an associative algebra, λ, ρ linear mapping of $\mathfrak{A}$ into an associative algebra $\mathfrak{G}$ with 1. Let η be the homomorphism of $\Phi\{\{X\}\}'$ into $\mathfrak{A}$ such that $x_1^\eta = a$, $x_2^\eta = b$, $x_3^\eta = c$ and let τ be the homomorphism of $\Phi\{P\}$ into $\mathfrak{G}$ such that $1 \to 1$, $l(m)^\tau = m^{\eta\lambda}$, $r(m)^\tau = m^{\eta\rho}$, $m \in N(X)'$. Then $r(x_2)^\tau = x_2^{\eta\rho} = b^\rho$, $r(x_3)^\tau = c^\rho$, $r(x_2x_3)^\tau = (x_2x_3)^{\eta\rho} = (bc)^\rho$, $l(x_1)^\tau = a^\lambda$, $l(x_1x_2)^\tau = (x_1x_2)^{\eta\lambda} = (ab)^\lambda$, $l(x_2)^\tau = b^\lambda$. Hence (λ, ρ) is an associative multiplication specialization of $\mathfrak{A}$ in $\mathfrak{G}_\lambda$ if and only if $F_1^\tau = b^\rho c^\rho - (bc)^\rho = 0$, $F_2^\tau = a^\lambda c^\rho - c^\rho a^\lambda = 0$, $F_3^\tau = (ab)^\lambda - b^\lambda a^\lambda = 0$. In a slightly different form the conditions are

$$
(ab)^\rho = a^\rho b^\rho, \qquad a^\rho b^\lambda = b^\lambda a^\rho, \qquad (ab)^\lambda = b^\lambda a^\lambda \tag{20}
$$

for all $a, b \in \mathfrak{A}$.

2) *Lie algebras.* By (12) we have $fD_1 = y_1F^\vee, gD_1 = y_1G_1^\vee, gD_2 = y_2G_2^\vee, gD_3 = y_3G_3^\vee$ where

$$
\begin{aligned}
F &= \ell(x_1) + r(x_1) \\
G_1 &= r(x_2)r(x_3) + \ell(x_2x_3) + \ell(x_3)r(x_2) \\
G_2 &= \ell(x_1)r(x_3) + r(x_3)r(x_1) + \ell(x_3x_1) \\
G_3 &= \ell(x_1x_2) + \ell(x_2)r(x_1) + r(x_1)r(x_2).
\end{aligned} \tag{21}
$$

These lead to the conditions

$$
\begin{aligned}
\lambda &= -\rho \\
(ab)^\rho &= a^\rho b^\rho - b^\rho a^\rho
\end{aligned} \tag{22}
$$

for a *Le* multiplication specialization

3) *Alternative algebras.* By (13) we have $fD_1 = y_1F_1^\vee$, $fD_2 = y_2F_2^\vee$, $gD_1 = y_1G_1^\vee$, $gD_2 = y_2G_2^\vee$ where

$$
\begin{aligned}
F_1 &= (\ell(x_1) + r(x_1))r(x_2) - r(x_1x_2)\ell(x_1) \\
F_2 &= \ell(x_1^2) - \ell(x_1)^2 \\
G_1 &= (\ell(x_1) + r(x_1))\ell(x_2) - \ell(x_2x_1) - \ell(x_2)r(x_1) \\
G_2 &= r(x_1^2) - r(x_1)^2.
\end{aligned} \tag{23}
$$

These give the conditons

$$(24)\qquad \begin{aligned} a^{\lambda+\rho}b^{\rho} &= (ab)^{\rho} + b^{\rho}a^{\lambda} \\ (a^2)^{\lambda} &= (a^{\lambda})^2 \\ a^{\lambda+\rho}b^{\lambda} &= (ba)^{\lambda} + b^{\lambda}a^{\rho} \\ (a^2)^{\rho} &= (a^{\rho})^2 \end{aligned}$$

for an alternative multiplication specialization.

4) *Jordan algebras.* By (14) we have $fD_1 = y_1F_1^{\nu}$, $fD_2 = y_2F_2^{\nu}$, $gD_1 = y_1G_1^{\nu}$, $gD_2 = y_2G_2^{\nu}$ where

$$(25)\qquad \begin{aligned} F_1 &= r(x_2) - \ell(x_2) \\ F_2 &= \ell(x_1) - r(x_1) \\ G_1 &= (\ell(x_1)+r(x_1))r(x_2)r(x_1)+\ell(x_1^2x_2)-(\ell(x_1)+r(x_1))r(x_2x_1)-\ell(x_2)\ell(x_1^2) \\ G_2 &= \ell(x_1^2)r(x_1) - r(x_1)\ell(x_1^2). \end{aligned}$$

These give the following conditions for a Jordan multiplications specialization:

$$(26)\qquad \begin{aligned} \lambda &= \rho \\ (a^2)^{\rho}a^{\rho} &= a^{\rho}(a^2)^{\rho} \\ 2a^{\rho}b^{\rho}a^{\rho} + (a^2b)^{\rho} &= 2a^{\rho}(ba)^{\rho} + b^{\rho}(a^2)^{\rho}. \end{aligned}$$

The following result establishes the equivalence of the concepts of S-bimodule and S-multiplication representation.

Theorem 3. *Let $\mathfrak{A}$ be an algebra in a class $C(S)$, S a subset of $\Phi\{[X]\}'$, $\mathfrak{M}$ a vector space, $(a,u) \to > au$, $(a,u) \to ua$ bilinear mappings of $(\mathfrak{A},\mathfrak{M})$ into $\mathfrak{M}$. Then $\mathfrak{M}$ is an S-bimodule for $\mathfrak{A}$ if and only if the linear mappings λ: $u \to au$, ρ: $u \to ua$* constitute an *S*-multiplication representation of $\mathfrak{A}$ in $\mathrm{Hom}_{\Phi}(\mathfrak{M},\mathfrak{M})$.

Proof. Let $\mathscr{E} = \mathfrak{A} \oplus \mathfrak{M}$ the split null extension defined by the given bilinear mappings. Let η be a homomorphism of $\Phi\{[X]\}'$ into $\mathfrak{A}$, η' the homomorphism of $\Phi\{P\}$ into $\mathrm{Hom}_{\Phi}(\mathscr{E},\mathscr{E})$ such that $\ell(m)^{\eta'} = (m^{\eta})_L$, $r(m)^{\eta'} = (m^{\eta})_R$ where $m \in N(X)'$ and the L and R denote multiplications in $\mathscr{E}$. Also we have the homomorphism τ of $\Phi\{P\}$ into $\mathrm{Hom}_{\Phi}(\mathfrak{M},\mathfrak{M})$ such that $\ell(m)^{\tau} = m^{\eta\lambda}$, $r(m)^{\tau} = m^{\eta\rho}$. Since a^{λ} is the restriction to $\mathfrak{M}$ of a_L in $\mathscr{E}$ and a^{ρ} is the restriction of a_R, we have for $u \in \mathfrak{M}$, $u\ell(m)^{\tau} = um^{\eta\lambda} = u(m^{\eta})_L = u\ell(m)^{\eta'}$ and $ur(m)^{\tau} = ur(m)^{\eta'}$. It follows that we have $uF^{\tau} = uF^{\eta'}$ for all $F \in \Phi\{P\}$. Now let $f \in S$ and write $fD_i = y_iF_i^{\nu}$ where $F_i \in \Phi\{P\}$. Consider the homomorphism (η,u) of $\Phi\{\{X,y_i\}\}$ into $\mathscr{E}$ which extends η and maps y_i into the element $u \in \mathfrak{M}$. Then, by (17), $(y_iF_i^{\nu})^{(\eta,u)} = uF_i^{\eta'} = uF_i^{\tau}$. Hence we have $(fD_i)^{(\eta,u)} = uF_i^{\tau}$. By Theorem 2, $\mathfrak{M}$ is an S-bimodule for $\mathfrak{A}$ if and only if $(fD_i)^{(\eta,u)} = 0$ for all $f \in S$, $i = 1, 2, \ldots$, $u \in \mathfrak{M}$. Evidently this is equivalent to $F_i^{\tau} = 0$ for all $f \in S$, $i = 1, 2, \ldots$ and all homomorphism τ of the type indicated.

5. *Universal multiplication envelopes.* From now on we assume that the set S of identities satisfies the conditions (α) and (β) of § 2. We recall that these imply that if $\mathfrak{A}$ is in $C(S,\Phi)$ then $\mathfrak{A}_{\Gamma}$ is in $C(S,\Gamma)$ for Γ an extension field of Φ and that $\mathfrak{A}$ with the multiplications as given in this algebra is an S-bimodule for $\mathfrak{A}$. The corresponding representation (L,R) is called the *regular representation* of $\mathfrak{A}$ (in $C(S)$).

Definition 2. Let $\mathfrak{A}$ be an algebra in a class $C(S)$ defined by a set of identities S. Then an associative algebra $\mathscr{U}$ with 1 and an S-multiplication specialization (λ_u, ρ_u) of $\mathfrak{A}$ in $\mathscr{U}$ is called a *universal S-multiplication envelope of* $\mathfrak{A}$ if for any S-multiplication specialization (λ, ρ) of $\mathfrak{A}$ in an associative algebra $\mathfrak{G}$ with 1 there exists a unique homomorphism μ of $\mathscr{U}$ into $\mathfrak{G}$ such that $1 \to 1$ and $\lambda_u \mu = \lambda$, $\rho_u \mu = \rho$, that is, the following diagrams are commutative:

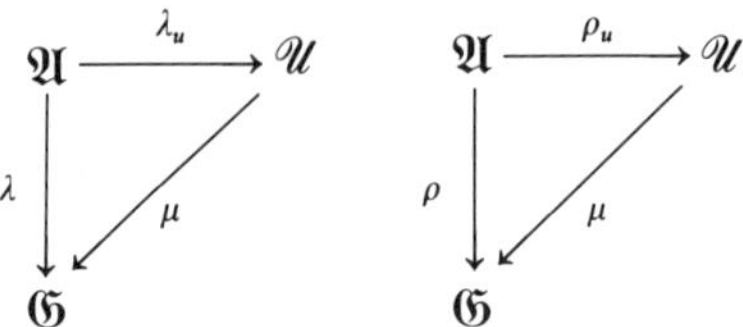

We can now state the following basic result.

Theorem 4. *There exists a universal S-multiplication envelope for any algebra $\mathfrak{A}$ in the class $C(S)$. Universal multiplication envelopes have the following properties:*

(1) *If $(\mathscr{U}, \lambda_u, \rho_u)$ and $(\mathscr{U}', \lambda_u', \rho_u')$ are universal S-multiplication envelopes for $\mathfrak{A}$ then there exists a unique isomorphism j of $\mathscr{U}$ onto $\mathscr{U}'$ such that $\lambda_u' = \lambda_u j$, $\rho_u' = \rho_u j$.*

(2) *$\mathscr{U}$ is generated by $\mathfrak{A}^{\lambda_u} \cup \mathfrak{A}^{\rho_u} \cup \{1\}$.*

(3) *If ξ is a homomorphism of $\mathfrak{A}$ into another algebra $\mathfrak{A}'$ in $C(S)$ and $(\mathscr{U}', \lambda_u', \rho_u')$ is a universal S-multiplication envelope for $\mathfrak{A}'$ then there exists a unique homomorphism ξ_u of $\mathscr{U}$ onto $\mathscr{U}'$ such that $1 \to 1$ and the following diagrams are commutative:*

$$\begin{array}{ccc} \mathfrak{A} & \longrightarrow & \mathfrak{A}' \\ \downarrow{\scriptstyle \lambda_u} & & \downarrow{\scriptstyle \lambda_u'} \\ \mathscr{U} & \underset{u}{\longrightarrow} & \mathscr{U}' \end{array} \qquad \begin{array}{ccc} \mathfrak{A} & \longrightarrow & \mathfrak{A}' \\ \downarrow{\scriptstyle u} & & \downarrow{\scriptstyle u} \\ \mathscr{U} & \underset{u}{\longrightarrow} & \mathscr{U}' \end{array}$$

(4) *If $0 \longrightarrow \mathfrak{A}'' \underset{\eta}{\longrightarrow} \mathfrak{A} \underset{\xi}{\longrightarrow} \mathfrak{A}' \longrightarrow 0$ is exact and $\mathscr{U}$, $\mathscr{U}'$ and ξ_u are as in* (3) *then $\mathscr{U} \underset{\xi_u}{\longrightarrow} \mathscr{U}' \longrightarrow 0$ is exact and $\ker \xi_u$ is the ideal $\mathfrak{K}$ in $\mathscr{U}$ generated by $\mathfrak{A}''^{\eta\lambda_u} \cup \mathfrak{A}''^{\eta\rho_u}$.*

(5) *Let Γ be an extension of the base field Φ of $\mathfrak{A}$ and let λ_u, ρ_u* be the *Γ-linear extensions of λ_u, ρ_u to $\mathfrak{A}_\Gamma$ into $\mathscr{U}_\Gamma$. Then $(\mathscr{U}_\Gamma, \lambda_u, \rho_u)$ is a universal S-multiplication envelope for $\mathfrak{A}_\Gamma$.*

Proof. To give a construction of $(\mathscr{U}, \lambda_u, \rho_u)$ for $\mathfrak{A}$ in $C(S)$ we form $\mathfrak{B} = \mathfrak{A}_1 \oplus \mathfrak{A}_2$ where $\mathfrak{A}_i$ is a vector space isomorphic to $\mathfrak{A}$ under an isomorphism $a \to a_i$, $i = 1, 2$. Let $\mathfrak{J}(\mathfrak{B})$ be the tensor algebra $\Phi 1 \oplus \mathfrak{B} \oplus (\mathfrak{B} \oplus \mathfrak{B}) \oplus (\mathfrak{B} \oplus \mathfrak{B} \oplus \mathfrak{B}) \oplus \dots$ based on the vector space $\mathfrak{B}$. Let $f \in S$ and write $fD_i = y_i F_i^v$ where $F_i \in \Phi\{P\}$ (cf. Definition 1). Let η be any homomorphism of $\Phi\{\{X\}\}'$ into $\mathfrak{A}$ and let η_i be the resultant of η and the mapping $a \to a_i$ considered as a mapping of $\mathfrak{A}$ into $\mathfrak{B} = \mathfrak{A}_1 \oplus \mathfrak{A}_2$. Then there exists a unique homomorphism ψ of $\Phi\{P\}$ into $\mathfrak{J}(\mathfrak{B})$ such that $1 \to 1$ and for any monomial m (element of $N(X)'$) we have $\ell(m)^\psi = m^{\eta_1}$, $r(m)^\psi = m^{\eta_2}$.

Let $\mathfrak{N}$ be the ideal in $\mathfrak{J}(\mathfrak{B})$ generated by the elements F_i^ψ obtained by taking all the $f \in S$, $i = 1, 2, \dots$ and all the homomorphism η of $\Phi\{\{X\}\}'$ into $\mathfrak{A}$. Set $\mathscr{U} = \mathfrak{J}(\mathfrak{B})/\mathfrak{N}$ and let λ_u and ρ_u be the linear mappings of $\mathfrak{A}$ into $\mathscr{U}$ such that $a^{\lambda_u} = a_1 + \mathfrak{N}$, $a^{\rho_u} = a_2 + \mathfrak{N}$. We claim that $(\mathscr{U}, \lambda_u, \rho_u)$ is a universal S-multiplication envelope for

$\mathfrak{A}$. Thus let $f \in S$ and write $fD_i = y_i F_i^\nu$ as before. Let η be a homomorphism of $\Phi\{\{X\}\}'$ into $\mathfrak{A}$ and let τ be the homomorphism of $\Phi\{P\}$ into $\mathscr{U}$ such that $1 \to 1$, $\ell(m)^\tau = m^{\eta\lambda_u} = m^{\eta_1} + \mathfrak{N}$, $r(m)^\tau = m^{\eta\rho_u} = m^{\eta_2} + \mathfrak{N}$, $m \in N(X)'$. Since $\ell(m)^\psi = m^{\eta_1}$, $r(m)^\psi = m^{\eta_2}$ we have $\ell(m)^\tau = \ell(m)^\psi + \mathfrak{N}$, $r(m)^\tau = r(m)^\psi + \mathfrak{N}$. Since ψ and τ are homomorphisms of $\Phi\{P\}$ into $\mathfrak{J}(\mathfrak{B})$ and $\mathscr{U} = \mathfrak{J}(\mathfrak{B})/\mathfrak{N}$ respectively and 1 and the $\ell(m)$ and $r(m)$ generate $\Phi\{P\}$, we have $F^\tau = F^\psi + \mathfrak{N}$ for any $F \in \Phi\{P\}$. In particular $F_i^\tau = F_i^\psi + \mathfrak{N} = 0$ since $F_i^\psi \in \mathfrak{N}$. This implies that λ_u, ρ_u is an S-multiplication specialization of $\mathfrak{A}$ in $\mathscr{U}$. Next let (λ, ρ) be any S-multiplication specialization of $\mathfrak{A}$ in an associative algebra $\mathfrak{G}$ with 1. We have a homomorphism μ of $\mathfrak{J}(\mathfrak{B})$ into $\mathfrak{G}$ such that $1 \to 1$, $a_1^{\mu'} = a^\lambda$, $a_2^{\mu'} = a^\rho$, $a \in \mathfrak{A}$. Let η be a homomorphism of $\Phi\{\{X\}\}'$ into $\mathfrak{A}$, τ the corresponding homomorphism of $\Phi\{P\}$ into $\mathfrak{G}$ given in Definition 1, ψ the homomorphism of $\Phi\{P\}$ into $\mathfrak{J}(\mathfrak{B})$ defined before. We have $\ell(m)^{\psi\mu'} = m^{\eta\mu'} = (m^\eta)_1^{\mu'} = m^{\eta\lambda}$ and $\ell(m)^\tau = m^{\eta\lambda}$. Hence $\ell(m)^{\psi\mu'} = \ell(m)^\tau$ and, similarly, $r(m)^{\psi\mu'} = r(m)^\tau$. It follows that $\psi\mu' = \tau$. Hence if $f \in S$ and $fD_i = y_i F_i^\nu$ then $F_i^{\psi\mu'} = F_i^\tau = 0$, since (λ, ρ) is an S-multiplication specialization. This implies that the ideal $\mathfrak{N}$ in $\mathfrak{J}(\mathfrak{B})$ which is generated by the elements F_i^ψ is contained in the kernel of μ'. Hence we have a homomorphism μ of $\mathscr{U} = \mathfrak{J}(\mathfrak{B})/\mathfrak{N}$ into $\mathfrak{G}$ such that $1 \to 1$ and $a^{\lambda_u} = (a_1 + \mathfrak{N})^\mu = a_1^{\mu'} = a^\lambda$ and $a^{\rho_u\mu} = a^\rho$. Thus we have $\lambda_u\mu = \lambda$, $\rho_u\mu = \rho$. Since the elements $1, a_1, a_2$ for $a \in \mathfrak{A}$ generate $\mathfrak{J}(\mathfrak{B})$, the elements $1, a^{\lambda_u}a^{\rho_u}$ generate $\mathscr{U}$. Consequently, the homomorphism μ is unique and we have proved that $(\mathscr{U}, \lambda_u, \rho_u)$ is a universal S-multiplication envelope for $\mathfrak{A}$.

The properties (1)–(4) for the special case of Lie algebras are proved in Jacobson's *Lie Algebras*, pp. 152–155. The proofs go over without change for the general case (see also Cohn, *Universal Algebra*, pp. 108–111). Hence we omit these. To prove (5) we suppose we have an S-multiplication specialization (λ, ρ) of $\mathfrak{A}_\Gamma$ in an associative algebra $\mathfrak{G}/\Gamma$ with 1. We can consider $\mathfrak{A}$ as a Φ-subalgebra of $\mathfrak{A}_\Gamma$ and $\mathfrak{G}$ as algebra over Φ. Then the restrictions $\lambda/\mathfrak{A}$, $\rho/\mathfrak{A}$ of λ and ρ to $\mathfrak{A}$ constitute an S-multiplication specialization of $\mathfrak{A}$ in $\mathfrak{G}/\Phi$. Hence there exists a homomorphism μ of $\mathscr{U}$ into $\mathfrak{G}/\Phi$ such that $1 \to 1$, $\lambda_u\mu = \lambda/\mathfrak{A}$, $\rho_u\mu = \rho/\mathfrak{A}$. Now let λ_u, ρ_u be the Γ-linear extensions of λ_u, ρ_u to $\mathfrak{A}_\Gamma$ into $\mathscr{U}_\Gamma$ and let μ be the linear extension of $\mathscr{U}_\Gamma$ to $\mathfrak{G}/\Gamma$. Then we have $\lambda_u\mu = \lambda$, $\rho_u\mu = \rho$ and μ is a homomorphism of $\mathscr{U}_\Gamma$ into $\mathfrak{G}/\Gamma$ sending $1 \to 1$. Since $\mathfrak{A}^{\lambda_u} \cup \mathfrak{A}^{\rho_u} \cup \{1\}$ generates $\mathscr{U}$, $\mathfrak{A}_\Gamma^{\lambda_u} \cup \mathfrak{A}_\Gamma^{\rho_u} \cup \{1\}$ generates $\mathscr{U}_\Gamma$. Hence μ is unique and $(\mathscr{U}_\Gamma, \lambda_u, \rho_u)$ is a universal S-multiplication envelope for $\mathfrak{A}_\Gamma$.

It is convenient to make a particular choice of the universal envelope and we do this by taking the one constructed in the foregoing proof. We write $\mathscr{U}(\mathfrak{A})$ for the algebra in the construction. We now consider $C(S)$ as a category with homomorphisms as morphism and we consider also the category of associative algebras with identity elements with the morphisms as homomorphisms mapping 1 into 1. Then we have the mapping $\mathfrak{A} \to \mathscr{U}(\mathfrak{A})$ of the first category into the second. Also if ζ is a homomorphism of $\mathfrak{A}$ into $\mathfrak{A}' \in C(S)$ then we have the corresponding homomorphism ζ_u of $\mathscr{U}(\mathfrak{A})$ into $\mathscr{U}(\mathfrak{A})$ given in (3) of the theorem. Next let θ be a homomorphism $\mathfrak{A}' \to \mathfrak{A}'' \in C(S)$. Then it is clear from the commutativity: $\zeta\lambda_u' = \lambda_u\zeta_u$, $\theta\lambda_u'' = \lambda_u'\theta_u$, $\zeta\rho_u' = \rho_u\zeta_u$, $\theta\rho_u'' = \rho_u'\theta_u$ of the squares in

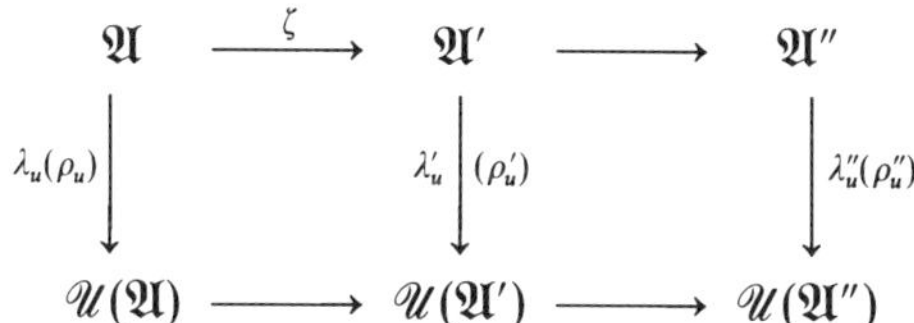

that the rectangle is commutative: $(\zeta\theta)\lambda''_u = \lambda_u(\zeta_u\theta_u)$, $(\zeta\theta)\rho''_u = \rho_u(\zeta_u\theta_u)$. Hence, by (3), $(\zeta\theta)_u = \zeta_u\theta_u$. Also it is clear from (3) that if $\mathfrak{A}' = \mathfrak{A}$ and $\zeta = 1$ then $\zeta_u = 1$. Hence the mappings $\mathfrak{A} \to \mathfrak{A}'$, $\zeta \to \zeta_u$ define a functor for *universal multiplication functor* from the category $C(S)$ to the category of associative algebras with 1.

6. *Extensions of algebras and factor sets.* If $\mathfrak{A}$ and $\mathfrak{M} \in C(S)$ then we define an *S-extension* of $\mathfrak{A}$ by $\mathfrak{M}$ as a short exact sequence

$$0 \longrightarrow \mathfrak{M} \underset{\alpha}{\longrightarrow} \mathscr{E} \underset{\beta}{\longrightarrow} \mathfrak{A} \longrightarrow 0 \tag{27}$$

where $\mathscr{E}$ is in $C(S)$. Two S-extensions are called *equivalent* if there can be imbedded in a commutative diagramm

(28)

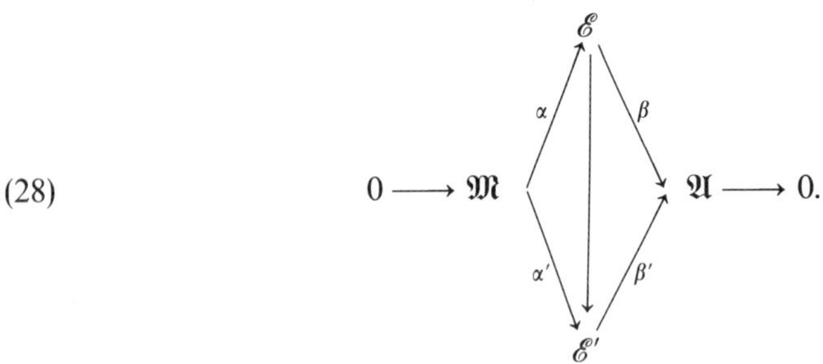

It follows from this that γ is an isomorphism of $\mathscr{E}$ onto $\mathscr{E}'$. The extension (27) is called *inessential* (or *split*) if there exists a homomorphism $\delta\colon \mathfrak{A} \to \mathscr{E}$ such that $\delta\beta = 1_{\mathfrak{A}}$ the identity mapping on $\mathfrak{A}$. If this is the case we have the vector space decomposition $\mathscr{E} = \mathfrak{A}^\delta \oplus \mathfrak{M}^\alpha$ and $\mathfrak{A}^\delta$ is a subalgebra of $\mathscr{E}$ isomorphic to $\mathfrak{A}$. If $\mathfrak{M}$ is a trivial algebra in the sense that $\mathfrak{M}^2 = 0$ (necessarily contained in $C(S)$) then the extension of $\mathfrak{A}$ by $\mathfrak{M}$ is called a *null* extension. We shall now analyze these extensions.

We note first (27) implies that α is injective. Hence we may identify $\mathfrak{M}$ and $\mathfrak{M}^\alpha$ and so take α to be the injection of $\mathfrak{M}(\leq \mathscr{E})$ into $\mathscr{E}$. Since $\mathfrak{M}$ has a complementary vector space in $\mathscr{E}$ there exists a linear mapping δ of $\mathfrak{A}$ into $\mathscr{E}$ such that $\delta\beta = 1_{\mathfrak{A}}$. Then we have the vector space decomposition $\mathscr{E} = \mathfrak{M} \oplus \mathfrak{A}^\delta$. If $a, b \in \mathfrak{A}$ then $(a^\delta b^\delta - (ab)^\delta)^\beta = ab - ab = 0$. Since (27) is exact this implies that $a^\delta b^\delta - (ab)^\delta \in \mathfrak{M}$. Hence we have

$$a^\delta b^\delta = (ab)^\delta + h(a,b) \tag{29}$$

where $h(a,b) \in \mathfrak{M}$. It is clear that the mapping $(a,b) \to h(a,b)$ is a bilinear mapping of $(\mathfrak{A}, \mathfrak{A})$ into $\mathfrak{M}$.

We shall show next that $\mathfrak{M}$ is an S-bimodule for $\mathfrak{A}$ relative to the bilinear products

$$au = a^\delta u, \qquad ua = ua^\delta, \qquad a \in \mathfrak{A}, \qquad u \in \mathfrak{M}, \tag{30}$$

where the right hand sides indicate products in $\mathscr{E}$. To see this we require some simple remarks on bimodules which are well known for associative and Lie algebras. First, let $\mathfrak{A}$ and $\mathfrak{A}' \in C(S)$, let θ be a homomorphism of $\mathfrak{A}$ into $\mathfrak{A}'$ and let $\mathfrak{M}$ be an S-bimodule for $\mathfrak{A}'$. Then $\mathfrak{M}$ becomes an S-bimodule for $\mathfrak{A}$ if we define $au = a^\theta u$, $ua = ua^\theta$, $a \in \mathfrak{A}$, $u \in \mathfrak{M}$. For the proof we form the split null extensions $\mathfrak{A} \oplus \mathfrak{M}$ and $\mathfrak{A}' \oplus \mathfrak{M}$ for the given bilinear compositions and we note that if $F = F(x_1, \ldots, x_n; y) \in \Phi\{\{X, y\}\}'_\theta$ is homogeneous of degree 1 in y then $F(a_1, \ldots, a_n; u) = F(a_1^\theta, \ldots, a_n^\theta, u)$ for $a_i \in \mathfrak{A}$, $u \in \mathfrak{M}$. This is readily proved for monomials by (30) using induction on the degree. It then follows for all F by linearity. The result just established and the criterion for an

S-bimodule given in Theorem 2 make it clear that $\mathfrak{M}$ is an S-bimodule for $\mathfrak{A}$ as indicated. Next we observe that if $\mathfrak{M}$ is an S-bimodule for $\mathfrak{A} \in C(S)$ and $\mathfrak{B}$ is an ideal in $\mathfrak{A}$ such that $\mathfrak{B}\mathfrak{M} = 0 = \mathfrak{M}\mathfrak{B}$ then $\mathfrak{M}$ is an S-bimodule for $\bar{\mathfrak{A}} = \mathfrak{A}/\mathfrak{B}$ relative to $\bar{a}u = au$, $u\bar{a} = ua$, $a \in \mathfrak{A}$, $\bar{a} = a + \mathfrak{B}$. It is clear that $(\bar{a}, u) \to au$, $(\bar{a}, u) \to ua$ are bilinear. Also we have the homomorphism $a + u \to \bar{a} + u$ of the split null extension $\mathfrak{A} + \mathfrak{M}$ into $\bar{\mathfrak{A}} + \mathfrak{M}$. Since $\mathfrak{A} + \mathfrak{M} \in C(S)$, $\bar{\mathfrak{A}} + \mathfrak{M} \in C(S)$ and $\mathfrak{M}$ is an S-bimodule for $\bar{\mathfrak{A}}$. We now consider $\mathscr{E}$ as before and we consider the regular S-bimodule $\mathscr{E}$. Since $\mathfrak{M}$ is an ideal in $\mathscr{E}$, $\mathfrak{M}$ is a sub-bimodule of $\mathscr{E}$. Since $\mathfrak{M}^2 = 0$, $\mathfrak{M}$ is an S-bimodule for $\bar{\mathscr{E}} = \mathscr{E}/\mathfrak{M}$ relative to $\bar{e}u = eu$, $u\bar{e}, = ue$, $e \in \mathscr{E}$, $\bar{e} = e + \mathfrak{M}$, $u \in \mathfrak{M}$. Since (27) is exact, $\bar{e} \to e^{\beta}$ is an isomorphism of $\bar{\mathscr{E}} = \mathscr{E}/\mathfrak{M}$ onto $\mathfrak{A}$. Hence $\mathfrak{M}$ is an S-bimodule for $\mathfrak{A}$ relative to $e^{\beta}u = eu$, $ue^{\beta} = ue$. If we take $e = a^{\delta}$ in this we obtain $e^{\beta}u = a^{\delta\beta}u = au = eu = a^{\delta}u$ and $ua = ua^{\delta}$. Hence $\mathfrak{M}$ is an S-bimodule for $\mathfrak{A}$ relative to (30).

We shall now derive the conditions on $h(a, b)$ which are imposed by the idintities in the set S. Let η be a homomorphism of $\Phi\{\{X\}\}'$ into $\mathfrak{A}$. Then we define a mapping (η, h) of $\Phi\{\{X\}\}'$ into the $\mathfrak{A}$-bimodule $\mathfrak{M}$ as follows: (η, h) is Φ-linear, $x_i^{(\eta,h)} = 0$, $(x_i x_j)^{(\eta,h)} = h(x_i^{\eta}, x_j^{\eta})$, and if M is a monomial of degree ≥ 3 and $M = M_1 M_2$ is the factorization of M as a product of two monomials then

$$M^{(\eta,h)} = M_1^{\eta} M_2^{(\eta,h)} + M_1^{(\eta,h)} M_2 + h(M_1^{\eta}, M_2^{\eta}). \tag{31}$$

Let η be the homomorphism of $\Phi\{\{X\}\}'$ into $\mathfrak{A}$ such that $x_i \to a_i$ and let η' be the homomorphism of $\Phi\{\{X\}\}'$ into $\mathscr{E}$ such that $x_i \to a_i^{\delta}$. Then we claim that

$$f(a_1^{\delta}, \ldots, a_n^{\delta}) = f(a_1, \ldots, a_n)^{\delta} + f^{(\eta,h)} \tag{32}$$

for all $f = f(x_1, \ldots, x_n) \in \Phi\{\{X\}\}'$. This is clear if f is of degree 1 since in this case $f^{(\eta,h)} = 0$. Also, it holds if f is a monomial of degree two, since $f^{(\eta,h)} = h(x_i^{\eta}, x_j^{\eta}) = h(a_i, a_j)$ and (29) holds. Noa assume (32) holds for the monomials M_1 and M_2 and let $M = M_1 M_2$. Then

$$\begin{aligned} M(a_1^{\delta}, \ldots, a_n^{\delta}) &= M_1(a_1^{\delta}, \ldots, a_n^{\delta}) M_2(a_1^{\delta}, \ldots, a_n^{\delta}) \\ &= (M_1(a_1, \ldots, a_n)^{\delta} + M_1^{(\eta,h)})(M_2(a_1, \ldots, a_n)^{\delta} + M_2^{(\eta,h)}) \\ &= M(a_1, \ldots, a_n)^{\delta} + h(M_1(a_1, \ldots, a_n), M_2(a_1, \ldots, a_n)) \\ &\quad + M_1(a_1, \ldots, a_n) M_2^{(\eta,h)} + M_1^{(\eta,h)} M_2(a_1, \ldots, a_n) \end{aligned}$$

by (29), (30) and $\mathfrak{M}^2 = 0$. Hence, by (31),

$$M(a_1^{\delta}, \ldots, a_n^{\delta}) = M(a_1, \ldots, a_n)^{\delta} + M^{(\eta,h)},$$

which implies (32) for all $f \in \Phi\{\{X\}\}'$.

Now suppose $f \in S$. Then since $\mathscr{E}$ and $\mathfrak{A} \in C(S)$, $f(a_1, \ldots, a_n)^{\delta} = 0$ and $f(a_1, \ldots, a_n) = 0$. Hence (31) implies that

$$f^{(\eta,h)} = 0 \tag{33}$$

for every homomorphism η of $\Phi\{\{X\}\}'$ into $\mathfrak{A}$. This leads us to formulate the following

Definition 3. Let $\mathfrak{A} \in C(S)$ and let $\mathfrak{M}$ be an S-bimodule for $\mathfrak{A}$. Then a bilinear mapping h of $(\mathfrak{A}, \mathfrak{A})$ into $\mathfrak{M}$ is called an *S-factor set* for $\mathfrak{A}$ in $\mathfrak{M}$ if (33) holds for all $f \in S$ and all homomorphisms η of $\Phi\{\{X\}\}'$ into $\mathfrak{A}$.

Our result is that if we have an extension (27) of $\mathfrak{A} \in C(S)$ by $\mathfrak{M}$ such that $\mathfrak{M}^2 = 0$ and α is the injection mapping then if δ is a linear mapping of $\mathfrak{A}$ into $\mathscr{E}$ such that

$\delta\beta = 1_{\mathfrak{A}}$, then $\mathfrak{M}$ is an S-bimodule for $\mathfrak{A}$ via (30) and h defined by $h(a,b) = a^\delta b^\delta - (ab)^\delta$ is an S-factor set for $\mathfrak{A}$ in $\mathfrak{M}$. Conversely, suppose $\mathfrak{A} \in C(S)$, $\mathfrak{M}$ is an S-bimodule for $\mathfrak{A}$ and h is an S-factor set of $\mathfrak{A}$ in $\mathfrak{M}$.

Let $\mathscr{E}$ be a vector space containing $\mathfrak{M}$ such that we have a linear isomorphism δ of $\mathfrak{A}$ into $\mathscr{E}$ such that $\mathscr{E} = \mathfrak{M} \oplus \mathfrak{A}^\delta$. Then we define an algebra structure on $\mathscr{E}$ by defining

$$(a^\delta + u)(b^\delta + v) = (ab)^\delta + h(a,b) + av + bu, \tag{34}$$

$a, b \in \mathfrak{A}$, $u, v \in \mathfrak{M}$. Clearly $\mathfrak{M}$ is an ideal in $\mathscr{E}$ such that $\mathfrak{M}^2 = 0$. Hence, by (6),

$$f(a_1^\delta + u_1, \ldots, a_n^\delta + u_n) = f(a_1^\delta, \ldots, a_n^\delta) + \sum_{i=1}^{n} fD_i(a_1^\delta, \ldots, a_n^\delta; u_i)$$

for $f \in \Phi\{\{X\}\}'$, $a_i \in \mathfrak{A}$, $u_i \in \mathfrak{M}$. Also (32) holds as before for η the homomorphism such that $x_i^\eta = a_i$. Hence

$$f(a_1^\delta + u_1, \ldots, a_n^\delta + u_n) = f(a_1, \ldots, a_n)^\delta + f^{(\eta,h)} + \Sigma fD_i(a_1^\delta, \ldots, a_n^\delta; u_i).$$

If $f \in S$, $f(a_1, \ldots, a_n) = 0$ and $\Sigma fD_i(a_i^\delta, \ldots, a_n^\delta; u_i) = 0$ since (33) gives $a^\delta u = au$, $ua^\delta = ua$ and $\mathfrak{M}$ is an S-bimodule for $\mathfrak{A}$. Also $f^{(\eta,h)} = 0$ by hypothesis. Hence $f(a_1^\delta + u_1, \ldots, a_n^\delta + u_n) = 0$ and $\mathscr{E} \in C(S)$.

We now replace δ by another linear mapping δ' of $\mathfrak{A}$ into $\mathscr{E}$ such that $\delta'_\beta = 1$. Such a δ' has the form $\delta + \lambda$ where λ is a linear mapping of $\mathfrak{A}$ into $\mathfrak{M}$, then $a^{\delta'}u = a^\delta u$, $ua^{\delta'} = ua^\delta$ so the bimodule structure on $\mathfrak{M}$ is unchanged. The bilinear mapping h is replaced by h' where

$$\begin{aligned} h'(a,b) &= (ab)^{\delta'} - a^{\delta'}b^{\delta'} = (ab)^\delta + (ab)^\lambda - (a^\delta + a^\lambda)(b^\delta + b^\lambda) \\ &= h(a,b) + (ab)^\lambda + ab^\lambda + a^\lambda b. \end{aligned}$$

Accordingly, we call the S-factor sets h and h' of $\mathfrak{A}$ in $\mathfrak{M}$ *equivalent* if there exists a linear mapping λ of $\mathfrak{A}$ into $\mathfrak{M}$ such that

$$h'(a,b) = h(a,b) + (ab)^\lambda - ab^\lambda - a^\lambda b, \tag{35}$$

$a, b \in \mathfrak{A}$. This is an equivalence relation in the set of factor sets of $\mathfrak{A}$ in $\mathfrak{M}$. The set of equivalence classes thus determined is denoted as $H^2(\mathscr{S}, \mathfrak{A}, \mathfrak{M})$. It is now easy to complete the proof of the following

Theorem 5. *If $\mathfrak{A} \in C(S)$ and $\mathfrak{M}$ is an S-bimodule then we have a bijection of $H^2(\mathscr{S}, \mathfrak{A}, \mathfrak{M})$ with the set of equivalence classes of null S-extensions of $\mathfrak{A}$ by $\mathfrak{M}$ such that the associated bimodule structure on $\mathfrak{M}$ (by (30)) is the given one. In this correspondence the equivalence class of 0 in $H^2(S, \mathfrak{A}, \mathfrak{M})$ corresponds to the isomorphism class of inessential extensions.*

We leave the rest of the details to the reader (cf. Cartan-Eilenberg's *Homological Algebra*, p. 295). We now list the factor set conditions in the cases which we have singled out throughout the discussion.

1) *Associative algebras.* Let η be a homomorphism of $\Phi\{\{X\}\}'$ such that $x_1^\eta = a$, $x_2^\eta = b$, $x_3^\eta = c$, h a bilinear mapping of $(\mathfrak{A}, \mathfrak{A})$ into $\mathfrak{M}$. Then $(x_1x_2)^{(\eta,h)} = h(a,b)$, $x_3^{(\eta,h)} = 0$, $((x_1x_2)x_3)^{(\eta,h)} = h(a,b)c + h(ab,c)$, $x_1^{(\eta,h)} = 0$, $(x_2x_3)^{(\eta,h)} = h(b,c)$,

$$(x_1(x_2x_3))^{(\eta,h)} = ah(b,c) + h(a,bc).$$

Hence the factor set condition $f^{(\eta,h)} = 0$ for $f = (x_1x_2)x_3 - x_1(x_2x_3)$ is

(36) $$h(a,b)c + h(ab,c) = ah(b,c) + h(a,bc).$$

2) *Lie algebras.* Let η be as in 1). Then $x_1^{2(\eta,h)} = h(a,a)$, $((x_1x_2)x_3^{(\eta,h)} = h(a,b)c + h(ab,c)$. Hence the factor set conditions $f^{(\eta,h)} = 0$ for $f = x_1^2$ and $g^{(\eta,h)} = 0$ for $g = (x_1x_2)x_3 + (x_2x_3) + (x_3x_1)x_2$ are

(37) $$\begin{gathered} h(a,a) = 0 \\ h(a,b)c + h(ab,c) + h(b,c)a + h(bc,a) + h(c,a)b + h(ca,b) = 0. \end{gathered}$$

3) *Alternative algebras.* Let η be as before. Then $(x_1^2x_2)^{(\eta,h)} = h(a,a)b + h(a^2,b)$, $(x_1(x_1x_2))^{(\eta,h)} = ah(a,b) + h(a,ab)$, $(x_2x_1^2)^{(\eta,h)} = bh(a,a) + h(b,a^2)$, $((x_2x_1)x_1)^{(\eta,h)} = h(b,a)a + h(ba,a)$. Hence the factor set conditions given by $f = x_1^2x_2 - x_1(x_1x_2)$ and $g = x_2x_1^2 - (x_2x_1)x_1$ are

(38) $$\begin{gathered} h(a,a)b + h(a^2,b) = ah(a,b) + h(a,ab) \\ bh(a,a) + h(b,a^2) = h(b,a)a + h(ba,a). \end{gathered}$$

4) *Jordan algebras.* Let η be as before. Then $(x_1x_2)^{(\eta,h)} = h(a,b)$, $(x_2x_1)^{(\eta,h)} = h(b,a)$, $((x_1^2x_2)x_1)^{(\eta,h)} = (h(a,a)b)a + h(a^2b,a)$, $(x_1^2(x_2x_1))^{(\eta,h)} = a^2h(b,a) + h(a,a)(ba) + h(a^2,ba)$. Hence the conditions given by $f = x_1x_2 - x_2x_1$ and $g = (x_1^2x_2)x_1 - x_1^2(x_2x_1)$ are

(39) $$\begin{gathered} h(a,b) = h(b,a) \\ (h(a,a)b)a + h(a^2,b)a + h(a^2b,a) = a^2h(b,a) + h(a,a)(ba) + h(a^2,ba). \end{gathered}$$

Chapter II
Foundations of Representation Theory for Jordan Algebras

In this chapter we shall develop the basic notions and general results of representation theory for the class of Jordan algebras. If $\mathfrak{M}$ is a bimodule for a Jordan algebra $\mathfrak{J}$ then we have $a \cdot u = u \cdot a$, $a \in \mathfrak{J} \cdot u \in \mathfrak{M}$. Hence we may drop the bilinear composition $(a \cdot u) \to a \cdot u$ in our considerations and confine our attention to $(a, u) \to u \cdot a$. Similarly, if (λ, ρ) is a multiplication specialization then $\lambda = \rho$ so λ can be dropped and the conditions can be formulated in terms of ρ alone. Accordingly, we define a multiplication specialization in the Jordan case as a single linear mapping ρ of $\mathfrak{J}$ in an associative algebra $\mathfrak{G}$ with 1 satisfying the conditions (26) of Chapter I.

If $\mathfrak{U}$ is an associative algebra over a field of characteristic $\neq 2$ then $\mathfrak{U}$ gives rise to a Jordan algebra $\mathfrak{U}^+$ which has the same vector space structure as $\mathfrak{U}$ and has the product composition $a \cdot b = \frac{1}{2}(ab + ba)$. Such Jordan algebras and their subalgebras are called special. If $\mathfrak{J}$ is an arbitrary Jordan algebra then we define an associative specialization of $\mathfrak{J}$ in an associative algebra $\mathfrak{U}$ with 1 to be a homomorphism of $\mathfrak{J}$ into $\mathfrak{U}^+$. These mappings for fixed $\mathfrak{J}$ give rise to a special universal envelope $(\mathscr{U}_s(\mathfrak{J}), s)$. Here $\mathscr{U}_s(\mathfrak{J})$ is an associative algebra 1, s is an associative specialization of $\mathfrak{J}$ in $\mathscr{U}_s$ and if σ is any associative specialization of $\mathfrak{J}$ in $\mathscr{U}_s$ then there exists a unique homomorphism μ of $\mathscr{U}_s$ into $\mathfrak{G}$ such that $1 \to 1$ and $a^\sigma = a^{s\mu}$, $a \in \mathfrak{J}$.

There is a fundamental relation between associative specializations and multiplication specializations, namely, if σ_1 and σ_2 are two associative specializations of $\mathfrak{J}$ in $\mathfrak{G}$ and these commute in the sense that $[a^{\sigma_1}b^{\sigma_2}] = 0$, a, $b \in \mathfrak{J}$ then $\rho = \frac{1}{2}(\sigma_1 + \sigma_2)$ is a multiplication specialization. It follows that the mapping $a \to a'' = \frac{1}{2}(a^s \otimes 1 + 1 \otimes a^s)$ in $\mathscr{U}_s(\mathfrak{J}) \otimes \mathscr{U}_s(\mathfrak{J})$ is a multiplication specialization. This and the subalgebra $\mathscr{U}''(\mathfrak{J})$ of $\mathscr{U}_s(\mathfrak{J}) \otimes \mathscr{U}_s(\mathfrak{J})$ generated by 1 and the a'', $a \in \mathfrak{J}$, constitute a universal envelope for multiplication specializations of the form $\rho = \frac{1}{2}(\sigma_1 + \sigma_2)$.

If $\mathfrak{J}$ is a Jordan algebra with identity element 1 then a unital multiplication (associative) specialization $\rho(\sigma)$ is a multiplication (associative) specialization satisfying $1^\rho = 1$ $(1^\sigma = 1)$. We can define universal envelopes for these mappings. The associative algebras thus defined are direct components of $\mathscr{U}(\mathfrak{J})(\mathscr{U}_s(\mathfrak{J}))$. We consider this decompositions. Also we define a canonical homomorphism of $\mathscr{U}(\mathfrak{J})$ onto $\mathscr{U}''(\mathfrak{J})$. The nice situation is that in which this is an isomorphism. Instances of this will be encountered in Chapter III. In this chapter we derive some simple criteria for this and for other "nice" situations involving the basic concepts. We shall also develop a few of the basic properties of Jordan algebras which we shall need.

1. *Special Jordan algebras. Some basic identities.* Let $\mathfrak{U}$ be an algebra over a field of characteristic $\neq 2$ and let $a \cdot b = \frac{1}{2}(ab + ba)$. We call this the *Jordan product* in $\mathfrak{U}$. It is clearly bilinear and so it endows the vector space $\mathfrak{U}$ with another algebra structure, the *Jordan structure* of $\mathfrak{U}$. We denote the resulting algebra as $\mathfrak{U}^+$. Clearly $\mathfrak{U}^+$ is a commutative algebra. Moreover, a direct verification shows that if $\mathfrak{U}$ is associative then we have $(a^{\cdot 2} \cdot b) \cdot a = a^{\cdot 2} \cdot (b \cdot a)$ where $a^{\cdot 2} = a \cdot a$. Hence $\mathfrak{U}^+$ is a Jordan algebra, Jordan algebras of this type and their subalgebras are called special Jordan algebras. More generally, if $\mathfrak{J}$ is an algebra then $\mathfrak{J}$ is called a *special Jordan algebra* if there exists an isomorphism σ of $\mathfrak{J}$ into an algebra $\mathfrak{U}^+$ where $\mathfrak{U}$ is an associative algebra. Since we shall be interested primarily in special Jordan algebras we shall use the notation $a \cdot b$

for the given product in any Jordan algebra (just as $[ab]$ is usually used for the product in a Lie algebra). We consider some

Examples. 1) Let $\mathfrak{A}$ be an associative algebra with an involution J. Thus J is a linear mapping in $\mathfrak{A}$ such that $J^2 = 1$ and $(ab)^J = b^J a^J$, $a, b \in \mathfrak{A}$. Let $\mathscr{H}(\mathfrak{A}, J)$ denote the subest of $\mathfrak{A}$ of symmetric elements relative to J $(a^J = a)$. It is immediate that $\mathscr{H}(\mathfrak{A}, J)$ is a subalgebra of $\mathfrak{A}^+$. Hence this is a special Jordan algebra. 2) Let $\mathfrak{M}$ be a vector space equipped with a symmetric bilinear form $(\ ,\ )$ into the base field Φ. Let $\mathfrak{J} = \Phi 1 \oplus \mathfrak{M}$ where $\Phi 1$ is a one dimensional space with 1 as basis. Define a product in $\mathfrak{J}$ by

$$(1) \qquad (\alpha 1 + u)\cdot(\beta 1 + v) = (\alpha\beta + (u, v))1 + \alpha v + \beta u$$

for $\alpha, \beta \in \Phi$, $u, v \in \mathfrak{M}$. One checks directly that $\mathfrak{J}$ is a Jordan algebra. We shall see later that this is special. We shall call $\mathfrak{J}$ the *Jordan algebra of the symmetric bilinear form* $(\ ,\ )$ on $\mathfrak{M}$. 3) Let $\mathfrak{C}$ be a Cayley algebra over Φ (definition in Schafer's *An Introduction to Nonassociative Algebras*, p. 29) and let $\mathfrak{C}_3$ be the algebra of 3×3 matrices with elements in $\mathfrak{C}$ with the usual matrix multiplication. Then $\mathfrak{C}_3$ has the involution $X \to \bar{X}'$ where the denotes transpose and if $X = (x_{ij})$ then $\bar{X} = (\bar{x}_{ij})$, $\bar{x}$, the image of x under the standard involution in $\mathfrak{C}$. Then the set $\mathscr{H}(\mathfrak{C}_3)$ of symmetric elements under the involution $X \to \bar{X}'$ is a sub-algebra of $\mathfrak{C}_3^+$. It can be proved by a fairly lengthy calculation that this is a Jordan algebra. Moreover, it can be shown that this is not special.

Let $\mathfrak{J}$ be a Jordan algebra, so that we have the defining relations $a\cdot b = b\cdot a$, $(a^{\cdot 2}\cdot b)\cdot a = a^{\cdot 2}\cdot(b\cdot a)$ for all $a, b \in \mathfrak{J}$. Also since that set S defining the class of Jordan algebras satisfies conditions (α), (β) of § 1.2, every Jordan algebra satisfies the derived identities of the given ones. In particular, the multilinear identity g_3 (or equivalently g_4) in (8) of Chapter I is an identity for Jordan algebras. If we take into account commutativity and the fact that the characteristic is $\neq 2$, this implies that

$$(2) \qquad \begin{aligned} &((a\cdot c)\cdot b)d + ((c\cdot d)\cdot b)\cdot a + ((a\cdot d)\cdot b)\cdot c \\ &\quad = (a\cdot c)\cdot(b\cdot d) + (c\cdot d)\cdot(a\cdot b) + (a\cdot d) + (b\cdot c) \end{aligned}$$

holds for all a, b, c, d in any Jordan algebra. In any Jordan algebra we denote the mappings $x \to x\cdot a$ and $x \to a\cdot x$ by R_a and L_a respectively. Then the defining conditions are equivalent to $L_a = R_a$ and $[R_a, R_a\cdot 2] = R_a R_{a\cdot 2} - R_{a\cdot 2}R_a = 0$. If we interchange b and d in (2) then the resulting relation can be written as $d(R_{a\cdot c}R_b + R_{b\cdot c}R_a + R_{a\cdot b}R_c) = d(R_b R_{a\cdot c} + R_c R_{a\cdot c} + R_c R_{a\cdot b})$. Hence

$$(3) \qquad [R_a R_{b\cdot c}] + [R_b R_{a\cdot c}] + [R_c R_{a\cdot b}] = 0$$

for a, b, c in $\mathfrak{J}$. If we interchange c and d in (2) then the result can be written as $d(R_a R_b R_c + R_c R_b R_a + R_{(a\cdot c)\cdot b}) = d(R_a R_{b\cdot c} + R_c R_{a\cdot b} + R_b R_{a\cdot c})$. Hence we have

$$(4) \qquad R_a R_b R_c + R_c R_b R_a + R_{(a\cdot c)\cdot b} = R_a R_{b\cdot c} + R_b R_{a\cdot c} + R_c R_{a\cdot b}.$$

In any algebra we define a^n inductively by $a^1 = a$, $a^k = a^{k-1}a$. For Jordan algebras we write $a^{\cdot n}$ for a^n. If we take $c = a^{\cdot k}$, $b = a$ in (4) we obtain

$$(5) \qquad R_{a^{\cdot k+2}} = 2R_a R_{a^{\cdot k+1}} + R_{a^{\cdot k}} R_{a^{\cdot 2}} - R_a^2 R_{a^{\cdot k}} - R_{a^{\cdot k}} R_a^2.$$

This recursion formula implies that all the $R_{a^{\cdot 2}}$ are contained in the subalgebra of $\mathrm{Hom}_\Phi(\mathfrak{J}, \mathfrak{J})$ generated by R_a and $R_{a^{\cdot 2}}$. Since $[R_a R_{a^{\cdot 2}}] = 0$ we have $[R_{a^{\cdot m}} R_{a^{\cdot n}}] = 0$ for all $m, n = 1, 2, 3, \ldots$.

A non-associative algebra is called *power associative* if the subalgebras generated by single elements of the algebra are associative. This is equivalent to $a^m a^n = a^{m+n}$. We have

Theorem 1. *Any Jordan algebra is power associative.*

Proof. We have $a^{\cdot m} \cdot a = a^{\cdot m+1}$. Assuming that $a^{\cdot m} a^{\cdot k} = a^{\cdot m+k}$ we have $a^{\cdot m} \cdot a^{\cdot m} \cdot a^{\cdot k+1} = a^{\cdot m} \cdot (a^{\cdot k} \cdot a) = a^{\cdot k} R_a R_{a^{\cdot m}} = a^{\cdot k} R_{a^{\cdot m}} R_a = (a^{\cdot k} \cdot a^{\cdot m}) \cdot a = a^{\cdot k+m} \cdot a = a^{\cdot k+m+1}$. Hence $a^{\cdot m+n}$ for all m, n.

2. *Jordan bimodules and universal multiplication envelopes.* If $\mathfrak{J}$ is a Jordan algebra and $\mathfrak{M}$ is a Jordan bimodule for $\mathfrak{J}$ (or $\mathfrak{J}$-bimodule) then we shall denote the bilinear products of $u \in \mathfrak{M}$ and $a \in \mathfrak{J}$ by $a \cdot u$ and $u \cdot a$. The bimodule conditions (after (14) of Chapter I) are $a \cdot u = u \cdot a$, $((a \cdot u + u \cdot a) \cdot b) \cdot a = (a \cdot u + u \cdot a)(b \cdot a)$, $(a^{\cdot 2} \cdot u) \cdot a = a^{\cdot 2} \cdot (u \cdot a)$. Because of the first of these it is convenient to drop the composition $a \cdot u$ and to formulate everyting in terms of $u \cdot a$ alone. Then the conditions on this product are

$$(6) \qquad \begin{aligned} (u \cdot a) \cdot a^{\cdot 2} &= (u \cdot a^{\cdot 2}) \cdot a \\ 2((u \cdot a) \cdot b) \cdot a + u \cdot (a^{\cdot 2} \cdot b) &= 2(u \cdot a) \cdot (a \cdot b) + (u \cdot b) \cdot a^{\cdot 2}. \end{aligned}$$

Accordingly, we redefine a bimodule in the Jordan case to be a vector space $\mathfrak{M}$ with a bilinear product $(a, u) \to u \cdot a$ of $(\mathfrak{J}, \mathfrak{M})$ into $\mathfrak{M}$ satisfying (6). Similarly, we have $\lambda = \rho$ for a multiplication specialization (λ, ρ) of $\mathfrak{J}$ in an associative algebra $\mathfrak{G}$. Hence we drop λ and we have the following conditions on ρ given in (26) of Chapter I:

$$(7) \qquad \begin{aligned} &[a^\rho, (a^{\cdot 2})^\rho] = 0 \\ &2a^\rho b^\rho a^\rho + (a^{\cdot 2} \cdot b)^\rho = 2a^\rho (a \cdot b)^\rho + b^\rho (a^{\cdot 2})^\rho. \end{aligned}$$

Hence we now define a multiplication specialization of $\mathfrak{J}$ in $\mathfrak{G}$ to be a single linear mapping ρ of $\mathfrak{J}$ in $\mathfrak{G}$ such that (7) holds for all $a, b \in \mathfrak{J}$. Obviously this is equivalent to the earlier definition. We make the corresponding alteration in the definition of a universal multiplication envelope $(\mathscr{U}(\mathscr{J}), \rho_u)$. We remark that this permits us to give a somewhat simpler construction of such an object. Here we let $J(\mathfrak{J}) = \Phi \oplus \mathfrak{J} \oplus (\mathfrak{J} \otimes \mathfrak{J}) \oplus (\mathfrak{J} \otimes \mathfrak{J} \otimes \mathfrak{J}) + \cdots$ the tensor algebra based on the vector space $\mathfrak{J}$ and we let $\mathfrak{N}$ be the idea in $J(\mathfrak{J})$ generated by the elements $a \otimes a^{\cdot 2} - a^{\cdot 2} \otimes a$, $2a \otimes b \otimes a + a^{\cdot 2} \cdot b - 2a \otimes a \cdot b - b \otimes a^{\cdot 2}$, a, $b \in \mathfrak{J}$. Set $\mathscr{U}(\mathfrak{J}) = J(\mathfrak{J})/\mathfrak{N}$ and $a^{\rho u} = a + \mathfrak{N}$, $a \in \mathfrak{J}$. Then one sees exactly as in the proof of Theorem 1.4 that $(\mathscr{U}(\mathfrak{J}), \rho_u)$ is a universal multiplication envelope for $\mathfrak{J}$. We shall take this to be our standard construction of $\mathscr{U}(\mathfrak{J})$ henceforth and call this *the* universal multiplication envelope of $\mathfrak{J}$. Also we write xy for the product in $\mathscr{U}(\mathfrak{J})$. Besides the general properties of universal multiplication envelopes give in Theorem 1.4 we have the following ones in the Jordan case.

Theorem 2. *Let $\mathfrak{J}$ be a Jordan algebra, $(\mathscr{U}(\mathfrak{J}), \rho)$ the universal multiplication envelope of $\mathfrak{J}$. Then we have the following properties*:

(1) *ρ is injective.*

(2) *There exists a unique involution π in $\mathscr{U}(\mathfrak{J})$ such that $1^\pi = 1$ and $a^{\rho\pi} = a^\rho$, $a \in \mathfrak{J}$.*

(3) *If* $\dim \mathfrak{J}/\Phi = n < \infty$ *then* $\dim \mathscr{U}(\mathfrak{J}) < \binom{2n+1}{n}$.

Proof. (1) If $\mathfrak{J}$ has an identity 1, the regular multiplication representation $a \to R_a$ (in $\mathrm{Hom}_\Phi(\mathfrak{J}, \mathfrak{J})$) is injective. There exists a homomorphism μ of $\mathscr{U}(\mathfrak{J})$ into $\mathrm{Hom}_\Phi(\mathfrak{J}, \mathfrak{J})$

such that $a^{\rho\mu} = R_a$. Hence ρ is injective. If $\mathfrak{J}$ does not have an identity element 1, then we adjoin one and obtain the Jordan algebra $\mathfrak{J}^* = \Phi 1 \oplus \mathfrak{J}$. Then we apply the foregoing argument to the representation $a \to R_a$, $a \in \mathfrak{J}$, R_a acting in $\mathfrak{J}^*$.

In view of (1) we identify henceforth a with a^ρ, $\mathfrak{J}$ with the subset $\mathfrak{J}^\rho$ of $\mathscr{U}(\mathfrak{J})$. Then we avete associative product ab in $\mathscr{U}(\mathfrak{J})$ and the Jordan product a, b in the subspace $\mathfrak{J}$ of $\mathscr{U}(\mathfrak{J})$. The latter should *not* be interpreted as the Jordan product in $\mathscr{U}(\mathfrak{J})$. Instead we have the following basic relations:

$$\begin{aligned} &[a, a^{\cdot 2}] = 0 \\ (7') \qquad &2aba + a^{\cdot 2}b = 2a(a \cdot b) + ba^{\cdot 2}, \qquad a, b \in \mathfrak{J}. \end{aligned}$$

Also the relations (3), (4) and (5) for the R_a give the following relations in $\mathscr{U}(\mathfrak{J})$.

$$(8) \qquad [a, b \cdot c] + [b, a \cdot c] + [c, a \cdot b] = 0$$

$$(9) \qquad abc + cba + (a \cdot c) \cdot b = a(b \cdot c) + b(a \cdot c) + c(a \cdot b)$$

$$(10) \qquad a^{\cdot k+2} = 2aa^{\cdot k+1} + (a^{\cdot 2} - 2a^2)a^{\cdot k}, \qquad k \geq 1.$$

To prove these we let $x \to x^\mu$ be a $1 - 1$ representation of the associative algebra $\mathscr{U}(\mathfrak{J})$ in an algebra $\mathrm{Hom}_\Phi(\mathfrak{M}, \mathfrak{M})$ (e.g. the regular representation of $\mathscr{U}(\mathfrak{J})$). The restriction of μ to $\mathfrak{J}$ is a Jordan multiplication representation of $\mathfrak{J}$. Hence $\mathfrak{M}$ with the compositions $a \cdot u = ua^\mu = u \cdot a$ is a Jordan bimodule for $\mathfrak{J}$. Let $\mathscr{E}$ be the corresponding split null extension $\mathfrak{J} \oplus \mathfrak{M}$. Then a^μ is the restriction to $\mathfrak{M}$ of R_a, $a \in \mathfrak{J}$, R_a acting in $\mathscr{E}$. Hence (3), (4) and (5) imply respectively $[a^\mu, (b \cdot c)^\mu] + [b^\mu, (a \cdot c)^\mu] + [c^\mu, (a \cdot b)^\mu] = 0$, $a^\mu b^\mu c^\mu + c^\mu b^\mu a^\mu + ((a \cdot c) \cdot b)^\mu = a^\mu (b \cdot c)^\mu + b^\mu (a \cdot c)^\mu + c^\mu (a \cdot b)^\mu$, $(a^{\cdot k+2})^\mu = 2a^\mu (a^{\cdot k+1})^\mu + (a^{\cdot k})^\mu (a^{\cdot 2})^\mu - (a^\mu)^2 (a^{\cdot k})^\mu - (a^{\cdot k})^\mu (a^\mu)^2$. Also since $[R_{a^{\cdot k}} R_{a^{\cdot \ell}}] = 0$, $(a^{\cdot k})^\mu$, $(a^{\cdot \ell})^\mu = 0$ so the last relation becomes $(a^{\cdot k+2})^\mu = 2a^\mu (a^{\cdot k+1})^\mu + (a^{\cdot 2} - 2a^2)^\mu (a^{\cdot k})^\mu$. Since μ is an isomorphism of $\mathscr{U}(\mathfrak{J})$ these relations give (8), (9) and (10) respectively. We remark that the same procedure can be applied to any relation for the multiplications R_a which is valid for all Jordan algebras. More generally, this can be applied to the universal multiplication envelopes in any class $C(S)$. We can now complete the proof of the theorem.

(2) Let $\mathscr{U}(\mathfrak{J})'$ be an algebra anti-isomorphic to $\mathscr{U}(\mathfrak{J})$, $x > x'$ an anti-isomorphism of $\mathscr{U}(\mathfrak{J})$ onto $\mathscr{U}(\mathfrak{J})'$. By (7′) and (8) we have the relation $2aba + a^{\cdot 2}b = 2(a \cdot b)a + a^{\cdot 2}b$, $a, b \in \mathfrak{J}$, in $\mathscr{U}(\mathfrak{J})$. Hence $2a'b'a' + (a^{\cdot 2}b)' = 2a'(a \cdot b)' + b'(a^{\cdot 2})'$ in $\mathscr{U}(\mathfrak{J})'$. Also, by the first equation in (7′) we have $[a', (a^{\cdot 2})'] = 0$. Hence we have the multiplication specialization $a \to a'$ of $\mathfrak{J}$ in $\mathscr{U}(\mathfrak{J})'$. Consequently, we have a unique homomorphism μ of $\mathscr{U}(\mathfrak{J})$ into $\mathscr{U}(\mathfrak{J})'$ such that $1 \to 1$, $a^\mu = a'$, $a \in \mathfrak{J}$. This implies that we have an anti-homomorphism π of $\mathscr{U}(\mathfrak{J})$ such that $1^\pi = 1$, $a^\pi = a$. Since $\mathscr{U}(\mathfrak{J})$ is generated by 1 and $\mathfrak{J}$ it is immediate that $\pi^2 = 1_{\mathscr{U}(\mathfrak{J})}$ and π is unique.

(3) We prove first that if $\{e_k\}$, $k \in I$ an ordered set, is a basis for $\mathfrak{J}$ then every element of $\mathscr{U}(\mathfrak{J})$ is a linear combination of 1 and the "standard" monomials

$$(11) \qquad e_{k_1}^2 e_{k_2}^2 \cdots e_{k_r}^2 e_{k_{r+1}} e_{k_{r+2}} \cdots e_{k_s}$$

where the k_i are distinct and satisfy

$$\begin{aligned} &k_1 < k_2 < \cdots < k_r \\ (12) \qquad &k_{r+1} < k_{r+3} < k_{r+5} < \cdots \\ &k_{r+2} < k_{r+4} < k_{r+6} < \cdots. \end{aligned}$$

By (7′), (8) and (9) we have the relations $2aba = 2a(a \cdot b) + ba^{\cdot 2} - a^{\cdot 2} \cdot b$, $abc = -cba + a(b \cdot c) + b(a \cdot c) - (a \cdot c) \cdot b$, $ab^2 = -b^2a + ab^{\cdot 2} + 2b(a \cdot b) - (a \cdot b) \cdot b$ for a, b, $c \in \mathfrak{J}$. Now consider a monomial in the e's. The middle relation just noted permits us to move a $c = u_k$ in such a monomial two places to the left at the expense of a linear combination of monomials in the e's of lower formal degree. If a monomial contains an e_k in three places then by moving two of these to the left we can replace the monomial by one containing the factor $e_k e_\lambda e_k$. Then the first relation noted shows that such a monomial is a linear combination of monomials in the e's of lower formal degree. Hence we may assume our monomial contains an e_k in at most two positions and if it contains two then these occur next to each other. Then the factor e_k^2 can be permuted with any e_λ preceding it at the expense of monomials of lower formal degree, by the last relations we noted. We may therefore suppose that the e_k^2 occur at the beginning and that the subscripts for these are in increasing order. Then the remaining e_k in the monomial can be arranged by our straightening process of moving an e_k two units to the left so that the conditions (12) are achieved. Now suppose the basis is finite: $(e_1, e_2, \ldots, e_n)$. Then the number of standard monomials in these e's together with 1 does not exceed

$$(13) \qquad N = \sum_{s=0}^{n} \binom{n}{s} \sum_{r=0}^{s} \binom{s}{r} \binom{s-r}{[\frac{1}{2}(s-r)]}$$

where $[i]$ denotes the largest integer in i. A simple calculation with generating functions shows that $N = \binom{2n+1}{n}$. This proves (3).

Let e be an idempotent element of $\mathfrak{J}$ ($e^{\cdot 2} = e$) and put $a = b = c = e$ in (9). This gives $2e^3 + e = 3e^2$ in $\mathscr{U}(\mathfrak{J})$. Hence we have

$$(14) \qquad e(e-1)(2e-1) = 0.$$

Set $E_0 = (e-1)(2e-1)$, $E_1 = e(2e-1)$, $E_{\frac{1}{2}} = -4e(e-1)$. Then the E_i are orthogonal idempotent elements of $\mathscr{U}(\mathfrak{J})$ and $E_0 + E_1 + E_{\frac{1}{2}} = 1$. Now suppose e is an identity element for $\mathfrak{J}$. Then if we put $b = c = e$ in (8) we obtain $[a, c] = 0$, which implies that e is in the center of $\mathscr{U}(\mathfrak{J})$. Hence the E_i are in the center and we have the decomposition of $\mathscr{U} = \mathscr{U}(\mathfrak{J})$ as $\mathscr{U} = \mathscr{U}_0 \oplus \mathscr{U}_1 \oplus \mathscr{U}_{\frac{1}{2}}$ where $\mathscr{U}_i = \mathscr{U}E_i$ is an ideal in $\mathscr{U}$. If $x \in \mathscr{U}$ we set $x_i = xE_i$. Since $eE_0 = 0$, $(e-1)E_1 = 0$, $(e - \frac{1}{2})E_{\frac{1}{2}} = 0$ we have $e_0 = 0$, $e_1 = E_1$, $e_{\frac{1}{2}} = \frac{1}{2}E_{\frac{1}{2}}$. Since $x \to x_i$ is a homomorphism of $\mathscr{U}$ it is clear that $a \to a_i$, $a \in \mathfrak{J}$, is a multiplication specialization of $\mathfrak{J}$. By (9), we have $a_i b_i c_i + c_i b_i a_i + ((a, c) \cdot b)_i = a_i(b \cdot c)_i + b_i(a \cdot c)_i + c_i(a \cdot b)_i$. If we take $b = c = e$, $i = 0$ in this we obtain $a_0 = 0$ for all a. If we take $c = e$, $i = \frac{1}{2}$ in this we obtain $(ab)_{\frac{1}{2}} = a_{\frac{1}{2}} b_{\frac{1}{2}} + b_{\frac{1}{2}} a_{\frac{1}{2}}$.

We shall now call a multiplication specialization ρ of a Jordan algebra with identity element e in an associative algebra $\mathfrak{G}$ (with 1) *unital* if $e^\rho = 1$. We call $(\mathfrak{G}, \rho)$ a *universal unital multiplication envelope* for $\mathfrak{J}$ if given any unital multiplication specialization ρ' of $\mathfrak{G}$ in $\mathfrak{G}'$ there exists a unique homomorphism μ_1 of $\mathfrak{G}$ into $\mathfrak{G}'$ such that $a^{\rho\mu_1} = a^{\rho'}$, $a \in \mathfrak{J}$. We can now prove

Theorem 3. *Let $\mathfrak{J}$ be a Jordan algebra with identity element e. Then $\mathscr{U} = \mathscr{U}(\mathfrak{J})$ is a direct sum $\mathscr{U} = \mathscr{U}_0 \oplus \mathscr{U}_1 \oplus \mathscr{U}_{\frac{1}{2}}$ where $\mathscr{U}_i$ is an ideal with identity element E_i such that if $a_i = aE_i$, $a \in \mathfrak{J}$, then* 1) $a_0 = 0$, $a \in \mathfrak{J}$, 2) $(a \cdot b)_{\frac{1}{2}} = a_{\frac{1}{2}} b_{\frac{1}{2}} + b_{\frac{1}{2}} a_{\frac{1}{2}}$, 3) *$\mathscr{U}_1$ and the mapping $a \to a_1$ is a universal unital multiplication envelope of $\mathfrak{J}$.*

Proof. Everything except 3) has been proved above for the E_i we defined. Also we have $e_1 = E_1$ so $a \to a_1$ is a unital multiplication specialization of $\mathfrak{J}$ in $\mathscr{U}_1$. Suppose ρ

is a unital multiplication specialization of $\mathfrak{J}$ in the associative algebra $\mathfrak{G}$. We have the homomorphism μ of $\mathscr{U}$ into $\mathfrak{G}$ such that $1 \to 1$ and $a \to a^\rho$, $a \in \mathfrak{J}$. Then $E_0^\mu = (e-1)^\mu(2e-1)^\mu = (1-1)(2e-1)^\mu = 0$ and $E_{\frac{1}{2}}^\mu = -4e^\mu(e-1)^\mu = 0$. Hence $\mathscr{U}_0^\mu = 0 = \mathscr{U}_0^\mu$ and since $a = a_0 + a_{\frac{1}{2}} + a_1$, $a^\mu = a_1^\mu$. Then $a_1^\mu = a^\rho$ so the restriction μ_1 of μ to $\mathscr{U}_1$ is a homomorphism of $\mathscr{U}_1$ into $\mathfrak{G}$ such that $a_1^{\mu_1} = a^\rho$. Clearly, $\mathfrak{J}_1 = \{a_1\}$ generates $\mathscr{U}_1$. Hence μ_1 is unique and so $\mathscr{U}_1$ and $a \to a_1$ satisfy the definition of a universal unital multiplication envelope.

The properties (1)–(5) of Theorem 1. hold also for universal unital multiplication envelopes of Jordan algebras. Also properties (1) and (2) of Theorem 2 above are valid for the universal unital multiplication envelopes. The proofs carry over without change. Property (3) of Theorem 3 has the analogue that if $\dim \mathfrak{J} = n < \infty$ then $\dim \mathscr{U}_1 < \binom{2n-1}{n}$. The proof is similar to the one given for (3).

(3) *Associative specializations. Special universal envelopes.*

Definition 1. Let $\mathfrak{J}$ be a Jordan algebra, $\mathfrak{G}$ an associative algebra with identity element 1. Then an *associative specialization* σ of $\mathfrak{J}$ in $\mathfrak{G}$ is a homomorphism of $\mathfrak{J}$ into the Jordan algebra $\mathfrak{G}^+$. A pair $(\mathscr{U}_s, s)$ where $\mathscr{U}_s$ is an associative algebra with 1 and s is an associative specialization of $\mathfrak{J}$ in $\mathfrak{G}$ is called a *special universal envelope* if given any associative specialization σ of $\mathfrak{J}$ in $\mathfrak{G}$ there exists a unique homomorphism μ of $\mathscr{U}_s$ into $\mathfrak{G}$ such that $1 \to 1$ and $\sigma = s\mu$.

The following theorem gives the main properties of special universal envelopes.

Theorem 4. *There exists a special universal envelope for any Jordan algebra $\mathfrak{J}$. Moreover, special universal envelopes have the following properties*:

(1) *If $(\mathscr{U}_s, s)$ and $(\mathscr{U}_s', s')$ are special universal envelopes for $\mathfrak{J}$ then there exists a unique isomorphism j of $\mathscr{U}_s$ onto $\mathscr{U}_s'$ such that $s' = sj$.*

(2) *$\mathscr{U}_s$ is generated by 1 and $\mathfrak{J}^s$.*

(3) *There exists a unique involution π, called the main involution of $\mathscr{U}_s$ such that $1^\pi = 1$ and $a^{s\pi} = a^s$, $a \in \mathfrak{J}$.*

(4) *If ζ is a homomorphism of $\mathfrak{J}$ into a second Jordan algebra $\mathfrak{J}'$ and $(\mathscr{U}_s, s)$, $(\mathscr{U}_s', s')$ are special universal envelopes for $\mathfrak{J}$ and $\mathfrak{J}'$ respectively then there exists a unique homomorphism ζ_u of $\mathscr{U}_s$ into $\mathscr{U}_s'$ such that $1 \to 1$ and the following diagram is commutative*

$$\begin{array}{ccc} \mathfrak{J} & \longrightarrow & \mathfrak{J}' \\ {\scriptstyle s}\downarrow & & \downarrow{\scriptstyle s'} \\ \mathscr{U}_s & \xrightarrow[\zeta_u]{} & \mathscr{U}_{s'} \end{array} \tag{15}$$

(5) *If $0 \to \mathfrak{J}'' \xrightarrow[\eta]{} \mathfrak{J} \xrightarrow[\zeta]{} \mathfrak{J}' \to 0$ is exact and $\mathscr{U}_s$ and $\mathscr{U}_s'$ are as in* (4) *then $\mathscr{U}_s \xrightarrow[\zeta_u]{} \mathscr{U}_s' \to 0$ is exact and $\ker \zeta_u$ is the ideal $\mathfrak{K}$ in $\mathscr{U}_s$ generated by $\mathfrak{J}''^{\eta s}$.*

(6) *If Γ is an extension field of the base field Φ of $\mathfrak{J}$ and s is the linear extension of s to $\mathfrak{J}_\Gamma$* into *$\mathscr{U}_{s\Gamma}$ then $(\mathscr{U}_{s\Gamma}, s)$ is a special universal envelope for $\mathfrak{J}_\Gamma$.*

(7) *If D is a derivation in $\mathfrak{J}$ then there exists a unique derivation D_u in $\mathscr{U}_s$ such that $Ds = sD_u$.*

(8) *If* $\dim \mathfrak{J} = n$ *then* $\dim \mathscr{U}_s(\mathfrak{J}) \leq 2^n$.

Proof. To construct $\mathscr{U}_s(\mathfrak{J})$ we form the tensor algebra $J(\mathfrak{J}) = \Phi 1 \oplus \mathfrak{J} \oplus (\mathfrak{J} \oplus \mathfrak{J}) \oplus \cdots$, let $\mathfrak{R}$ be the ideal in $J(\mathfrak{J})$ generated by all the elements of the form $\frac{1}{2}(a \otimes b + b \otimes a) - a \cdot b$, $a, b \in \mathfrak{J}$. Let $\mathscr{U}_s = J(\mathfrak{J})/\mathfrak{R}$ and put $a^s = a + \mathfrak{R}$, $a \in \mathfrak{J}$. Then one sees directly, as in the corresponding construction given in Theorem 1.4, that $(\mathscr{U}_s, s)$ is a special universal envelope for $\mathfrak{J}$. The proofs of (1)–(5) and (7) are the same as for the corresponding properties of the Birkhoff-Witt algebra given in Jacobson's *Lie Algebras* pp. 152–155. The proof of (6) is the same as that of (5) of Theorem 1.4. To prove (8) we use a straightening argument based on the relations $e_k e_\lambda = -e_\lambda e_k + 2e_k e_\lambda$ to show that if $\{e_k\}$ is an ordered bases for $\mathfrak{J}$ then every element of $\mathscr{U}_s(\mathfrak{J})$ is a linear combination of 1 and monomials of the form $e_{k_1} e_{k_2} \ldots e_{k_r}$ with $k_1 < k_2 < \cdots < k_r$.

The property (4) permits us to define a functor from the category of Jordan algebras to the category of associative algebras with 1. If $\mathfrak{J}$ is in the first category then we let $\mathscr{U}_s(\mathfrak{J})$ be associative algebra with 1 constructed in the proof and if ζ is a homomorphism of $\mathfrak{J}$ into the Jordan algebra $\mathfrak{J}'$ then we let ζ_u be the homomorphism in (4) of $\mathscr{U}_s(\mathfrak{J})$ into $\mathscr{U}_s(\mathfrak{J}')$. If λ is a homomorphism of $\mathfrak{J}'$ into another Jordan algebra $\mathfrak{J}''$ then (4) implies that $(\zeta\lambda)_u = \zeta_u \lambda_u$. Also $1_u = 1$. Hence it is clear that the mappings $\mathfrak{J} > \mathscr{U}_s(\mathfrak{J})$, $\zeta > \zeta_u$ define a functor from the category of Jordan algebras to the category of associative algebras with identity. We remark also that ζ_u respects the involutions in $\mathscr{U}_s(\mathfrak{J})$ and $\mathscr{U}_s(\mathfrak{J}')$ since $a^{s\pi\zeta_u} = a^{s\zeta_u} = a^{\zeta s} = a^{\zeta s\pi} = a^{s\zeta_u\pi}$ and $1^{\pi\zeta_u} = 1 = 1^{\pi\zeta_u}$, which imply that $x^{\zeta_u\pi} = x^{\pi\zeta_u}$, $x \in \mathscr{U}_s(\mathfrak{J})$. Hence we can also regard the functor we have defined as one into the category of associative algebras with 1 and involution. In this category the morphisms are the algebra homomorphisms mapping 1 into 1 and respecting the given involutions.

If s is injective then clearly $\mathfrak{J}$ is special. On the other hand, if $\mathfrak{J}$ is special then we have an isomorphism σ of $\mathfrak{J}$ in an associative algebra $\mathfrak{G}$. Since we may adjoin an identity to $\mathfrak{G}$ we may suppose $\mathfrak{G}$ already has one. Then $\sigma = s\mu$ where μ is a homomorphism of $\mathfrak{G}$. It follows that s is injective. Hence $\mathfrak{J}$ is special if and only if s is injective. More generally, let $\mathfrak{K}_s$ be the kernel of s. Then $\mathfrak{J}_s \equiv \mathfrak{J}/\mathfrak{K}_s$ is special and (5) shows that $(\mathscr{U}_s, \bar{s})$ where $(a + \mathfrak{K}_s)^{\bar{s}} = a^s$ is a special universal envelope for $\mathfrak{J}_s$. It is clear also from the universal property of s that $\mathfrak{K}_s$ is contained in the kernel of every homomorphism of $\mathfrak{J}$ into an algebra $\mathfrak{G}^+$, $\mathfrak{G}$ associative. Hence every special homomorphic image of $\mathfrak{J}$ is a homomorphic image of $\mathfrak{J}_s$.

An associative specialization in an algebra $\mathrm{Hom}_\Phi(\mathfrak{M}, \mathfrak{M})$ will be called a *special representation* of $\mathfrak{J}$. If σ is such a special representation then we have a unique homomorphism μ of $\mathscr{U}_s(\mathfrak{J})$ into $\mathrm{Hom}_\Phi(\mathfrak{M}, \mathfrak{M})$ such that $1^\mu = 1$ and $a^{s\mu} = a^\sigma$. It follows that we can define a unique unital (right) module structure on $\mathfrak{M}$ relative to the associative algebra $\mathscr{U}_s(\mathfrak{J})$ so that

$$ua^s = ua^\sigma, \qquad a \in \mathfrak{J}. \tag{16}$$

Also, since we have the main involution π in $\mathscr{U}_s(\mathfrak{J})$ we have a unique anti-homomorphism μ' of $\mathscr{U}_s(\mathfrak{J})$ into $\mathrm{Hom}_\Phi(\mathfrak{M}, \mathfrak{M})$ such that $a^{s\mu'} = a^s$. Accordingly, we have a unital left module structure on $\mathfrak{M}$ relative to $\mathscr{U}_s(\mathfrak{J})$ such that

$$a^s u = ua^\sigma, \qquad a \in \mathfrak{J}. \tag{17}$$

If a is in an associative algebra $\mathfrak{G}$, then the Jordan square $a^{\cdot 2} = \frac{1}{2}(a^2 + a^2) = a^2$ and one proves by induction that $a^{\cdot n} = a^n$, $n = 1, 2, 3, \ldots$. An important trilinear composition in a Jordan algebra is given by

$$\{abc\} = (a \cdot b) \cdot c + (b \cdot c) \cdot a - (a \cdot c) \cdot b. \tag{18}$$

Direct evaluation shows that in $\mathfrak{G}^+$ we have

$$\{abc\} = \tfrac{1}{2}(abc + cba). \tag{19}$$

In particular, we have

$$\{aba\} = aba. \tag{20}$$

Now let σ be an associative specialization of the Jordan algebra $\mathfrak{J}$ in the associative algebra $\mathfrak{G}$. Let a, b, $c \in \mathfrak{J}$. Then the homomorphism property of σ implies that $(a^{\cdot n})^\sigma = (a^\sigma)^{\cdot n}\{abc\}^\sigma = \{a^\sigma b^\sigma c^\sigma\}, \{aba\}^\sigma = \{a^\sigma b^\sigma a^\sigma\}$. Hence by the above formulas we have the following important relations for associative specializations

$$\begin{aligned} (a^{\cdot n})^\sigma &= (a^\sigma)^n \\ \{abc\}^\sigma &= \tfrac{1}{2}(a^\sigma b^\sigma c^\sigma + c^\sigma b^\sigma a^\sigma) \\ \{abc\}^\sigma &= a^\sigma b^\sigma a^\sigma. \end{aligned} \tag{21}$$

Let e be an idempotent element of $\mathfrak{J}$. Then $e^{\cdot 2} = e$ and (21) gives $(e^\sigma)^2 = e^\sigma$ so e^σ is an idempotent element of $\mathfrak{J}$. Next let $a \in \mathfrak{J}$ satisfy $a \cdot e = 0$. Then the definition of $\{aba\}$ and (21) imply that $e^\sigma a^\sigma e^\sigma = 0$. Also we have $a^\sigma e^\sigma + e^\sigma a^\sigma = 2a^\sigma \cdot e^\sigma = 2(a \cdot e)^\sigma = 0$. Since $(e^\sigma)^2 = e^\sigma$ this gives $a^\sigma e^\sigma = 0 = e^\sigma a^\sigma$. An important consequence of these two results is that if $e_1, e_2, \ldots, e_r$ are orthogonal idempotent elements of $\mathfrak{J}$ ($e_i^{\cdot 2} = e_i$, $e_i \cdot e_j = 0$ if $i \neq j$) then $e_1^\sigma, e_2^\sigma, \ldots, e_n^\sigma$ are orthogonal idempotents in $\mathfrak{G}$. Next let e be idempotent in $\mathfrak{J}$ and let a be an element such that $e \cdot a = a$. Then we have $\{eae\} = a$ by the definition of $\{abc\}$. Consequently, by (21) $e^\sigma a^\sigma e^\sigma = a^\sigma$. Since $(e^\sigma)^2 = e^\sigma$ this implies that $e^\sigma a^\sigma = a^\sigma = a^\sigma e^\sigma$.

Now let e be an identity element for $\mathfrak{J}$. Then e^s acts as identity for the elements of $\mathfrak{J}^s$ in the associative algebra $\mathscr{U}_s = \mathscr{U}_s(\mathfrak{J})$. Since $\mathfrak{J}^s$ and 1 generate $\mathscr{U}_s$ it is clear that e^s is a central idempotent element and $\mathscr{U}_s = \mathscr{U}_{s1} \oplus \Phi z$ where $\mathscr{U}_{s1}$ is the subalgebra (without 1) generated by $\mathfrak{J}^s$ and $z = 1 - e^s$. $\mathscr{U}_{s1}$ and Φz are ideals. Moreover $\Phi z \neq 0$ since we have an associative specialization σ of $\mathfrak{J}$ in a the one dimensional algebra with identity $\Phi 1$ such that $a^\sigma = 0$, $a \in \mathfrak{J}$. Hence if μ is the homomorphism of $\mathscr{U}_s$ into Φ such that $a^\sigma = a^{s\mu}$, then $\mathscr{U}_{s1}^\mu = 0$ and $\mathscr{U}_s^\mu = \Phi 1 \neq 0$. Hence $\Phi z \neq 0$. The associative specialization s can be regarded as one in $\mathscr{U}_{s1}$. When this is done $s_1 = s$ is *unital* in the sense that the identity e of $\mathfrak{J}$ is mapped into the identity of the associative algebra. Moreover, one sees as in the corresponding consideration of the universal multiplication envelope that $(\mathscr{U}_{s1}, s_1)$ is universal for central associative specializations in the obvious sense.

Again suppose e is the identity for $\mathfrak{J}$ and assume $\mathfrak{J} = \mathfrak{J}_1 \oplus \mathfrak{J}_2$ where the $\mathfrak{J}_i$ are ideals. Then $e = e_1 + e_2$, $e_i \in \mathfrak{J}_i$, and e_i is the identity of $\mathfrak{J}_i$ and $e_i \cdot \mathfrak{J}_j = 0$, $i \neq j$. The foregoing results imply that the e_i^s are central orthogonal idempotent elements of $\mathscr{U}_s$. Hence we have $\mathscr{U}_s = \mathscr{U}_{s1} e_1^s \oplus \mathscr{U}_{s1} e_2^s$ and one sees easily using property (5) that $\mathscr{U}_{si}$ and the mapping $a \to a^s e_i^s$ is a special unital universal envelope for $\mathfrak{J}_i$. We leave the proof to the reader.

4. *Special multiplication specialization and envelopes.* Let $\mathfrak{J}$ be Jordan and let σ_1 and σ_2 be associative specializations of $\mathfrak{J}$ in the same associative algebra $\mathfrak{G}$ with 1. Then we shall say that σ_1 and σ_2 *commute* if $[a^{\sigma_1}, b^{\sigma_2}] = 0$, $a, b \in \mathfrak{J}$. Assume this is the case and assume also that $\mathfrak{G} = \mathrm{Hom}_\Phi(\mathfrak{M}, \mathfrak{M})$ where $\mathfrak{M}$ is a vector space. Then we have seen in § 3 that $\mathfrak{M}$ can be regarded as a unital right $\mathscr{U}_s = \mathscr{U}_s(\mathfrak{J})$-module so that $ua^s = ua^1$,

$u \in \mathfrak{M}$, $a \in \mathfrak{J}$. Also $\mathfrak{M}$ is a unital left $\mathscr{U}_s$-module so that $a^s u = ua^{\sigma_2}$. We have $(a^s u)b^s = a^s(ub^s)$, a, $b \in \mathfrak{J}$, since σ_1 and σ_2 commute. Since $\mathfrak{J}^s$ and 1 generate $\mathscr{U}_s$ this implies $(cu)d = c(ud)$, $c, b \in \mathscr{U}_s$. Thus $\mathfrak{M}$ is an associative bimodule for $\mathscr{U}_s$. Hence this gives rise to the split null extension $\mathscr{E} = \mathscr{U}_s \oplus \mathfrak{M}$ which is an associative algebra. We now consider the Jordan algebra $\mathscr{E}^+$. This contains $\mathfrak{M}$ as ideal and $\mathscr{U}_s^+$ as a subalgebra. Also $\mathfrak{J}^s$ is a subalgebra of $\mathscr{U}_s^+$; hence of $\mathscr{E}^+$. We have the multiplication representation $a^s \to R_{a^s}|\mathfrak{M}$ of $\mathfrak{J}^s$ where $R_{a^s}|\mathfrak{M}$ is the restriction of the right multiplication R_{a^s} to $\mathfrak{M}$. Since $a \to a^s$ is a homomorphism this gives the multiplication representation $a \to R_s$ of $\mathfrak{J}$ in $\mathrm{Hom}_\Phi(\mathfrak{M}, \mathfrak{M})$. Now $uR_{a^s} = \frac{1}{2}(ua^s + a^s u)$. Hence if $u \in \mathfrak{M}$ then

(22) $$uR_{a^s} = \tfrac{1}{2}(ua^s + a^s u) = \tfrac{1}{2}u(a^{\sigma_1} + a^{\sigma_2})$$

We have therefore shown that the mapping $a \to \frac{1}{2}(a^{\sigma_1} + a^{\sigma_2})$ of $\mathfrak{J}$ into $\mathrm{Hom}_\Phi(\mathfrak{M}, \mathfrak{M})$ is a multiplication representation of $\mathfrak{J}$.

More generally, suppose σ_1 and σ_2 are associative specializations of $\mathfrak{J}$ in the associative algebra $\mathfrak{G}$ with 1 and σ_1 and σ_2 commute. Let τ be a $1-1$ representation of $\mathfrak{G}$ in $\mathrm{Hom}_\Phi(\mathfrak{M}, \mathfrak{M})$ and put $\tau_i = \sigma_i \tau$. Then τ_i is a special representation of $\mathfrak{J}$ in $\mathrm{Hom}_\Phi(\mathfrak{M}, \mathfrak{M})$ and τ_1 and τ_2 commute. Then $\frac{1}{2}(\tau_1 + \tau_2)$ is a multiplication representation of $\mathfrak{J}$. Since τ is $1-1$ and $\frac{1}{2}(\sigma_1 + \sigma_2)\tau = \frac{1}{2}(\tau_1 + \tau_2)$ it follows that $\frac{1}{2}(\sigma_1 + \sigma_2)$ is a multiplication specialization of $\mathfrak{J}$ in $\mathfrak{G}$. We have now proved the following

Theorem 5. *Let $\mathfrak{J}$ be a Jordan algebra, σ_1 and σ_2 two associative specializations of $\mathfrak{J}$ in the same associative algebra $\mathfrak{G}$ with 1 and assume σ_1 and σ_2 commute. Then $\rho = \frac{1}{2}(\sigma_1 + \sigma_2)$ is a multiplication specialization of $\mathfrak{J}$ in $\mathfrak{G}$.*

We shall call $\rho = \frac{1}{2}(\sigma_1 + \sigma_2)$ the *average* of the two commutling associative specializations of $\mathfrak{J}$. A multiplication specialization which has this form will be called *special.* We shall call a bimodule $\mathfrak{M}$ for $\mathfrak{J}$ *special* if there exists an isomorphism μ of $\mathfrak{M}$ into a bimodule $\mathfrak{N}$ for $\mathfrak{J}$ which is obtained from an associative unital bimodule for $\mathscr{U}_s(\mathfrak{J})$ so that for $a \in \mathfrak{J}$, $u \in \mathfrak{N}$ we have $u \cdot a = \frac{1}{2}(a^s u + ua^s)$. The bimodule we constructed above from σ_1 and σ_2 is of this type.

We now form $\mathscr{U}_s(\mathfrak{J}) \otimes \mathscr{U}_s(\mathfrak{J})$ and set

(23) $$a'' = \tfrac{1}{2}(a^s \otimes 1 + 1 \otimes a^s)$$

$a \in \mathfrak{J}$. Let $\mathscr{U}''(\mathfrak{J})$ be the subalgebra of the tensor product generated by 1 and $\mathfrak{J}'' = \{a'' | a \in \mathfrak{J}\}$. We shall call $\mathscr{U}''(\mathfrak{J})$ and the mapping $a \to a''$ the *universal special multiplication envelope* for $\mathfrak{J}$. It is clear that $a \to a^s \otimes 1$ and $a \to 1 \otimes a^s$ are associative specializations of $\mathfrak{J}$ in $\mathscr{U}_s \otimes \mathscr{U}_s$ and these commute. Moreover, $a \to a''$ is the average of the two associative specializations. Now let σ_1 and σ_2 be commuting associative specializations of $\mathfrak{J}$ in $\mathfrak{G}$. We have homomorphisms of $\mathscr{U}_s$ into $\mathfrak{G}$ such that $1 \to 1$ and $a^s > a^{\sigma_i}$, $a \in \mathfrak{J}$. Since $[a^{\sigma_1}, b^{\sigma_2}] = 0$, a, $b \in \mathfrak{J}$ it follows from the basic property of the tensor product of algebras that we have a homomorphism of $\mathscr{U}''$ into $\mathfrak{G}$ such that $1 \to 1$, $a'' \to \frac{1}{2}(a^{\sigma_1} + a^{\sigma_2})$, $a \in \mathfrak{J}$. Since $\mathscr{U}''$ is generated by 1 and $\mathfrak{J}''$ this establishes the universal property of $\mathscr{U}''$ and $a \to a''$ in the class of special multiplication specializations of $\mathfrak{J}$. An important consequence of the universal property just noted is the following: Let μ be a homomorphism of $\mathfrak{J}$ in a special Jordan algebra $\mathfrak{K}$. Then we have the multiplication specialization $a \to R_{a^\mu_a}$ of $\mathfrak{J}$ in $\mathrm{Hom}_\Phi(\mathfrak{K}, \mathfrak{K})$. It follows directly from the definitions and the universal property of $\mathscr{U}''$ that there exists a unique homomorphism of $\mathscr{U}''$ into $\mathrm{Hom}_\Phi(\mathfrak{K}, \mathfrak{K})$ such that $1 \to 1$ and $a'' \to R_{a^s}$, $a \in \mathfrak{J}$.

The universal property of the universal multiplication envelope $\mathscr{U} = \mathscr{U}(\mathfrak{J})$ implies that we have an exact sequence

$$(24) \qquad \mathscr{U} \underset{\zeta}{\rightarrow} \mathscr{U}'' \rightarrow 0$$

where $1^\xi = 1$ and $a^\xi = a''$ for $a \in \mathfrak{J}$ $(\subseteq \mathscr{U})$. Of particular interest is the situation in which ξ is an isomorphism since in this case the consideration of $\mathscr{U}$ is reduced to that of $\mathscr{U}''$ which is determined by $\mathscr{U}_s$. Moreover, the relations defining $\mathscr{U}_s$ are simpler that those defining $\mathscr{U}$. If ξ is an isomorphism, clearly $a \rightarrow a'' = a^s \otimes 1 + 1 \otimes a^s$, $a \in \mathfrak{J}$, is injective. This implies that $a \rightarrow a^s$ is injective and hence that $\mathfrak{J}$ is special. Accordingly we shall say that $\mathfrak{J}$ is *strongly special* if ξ is an isomorphism. There exist Jordan algebras which are special but not strongly special.

The tensor product $\mathscr{U}_s \otimes \mathscr{U}_s$ has an automorphism $\mathscr{E}$ of period two such that $u \otimes v \rightarrow v \otimes u$, $u, v \in \mathscr{U}_s$. We shall call $\mathscr{E}$ the *exchange automorphism* in $\mathscr{U}_s \otimes \mathscr{U}_s$. The generators $1\,(= 1 \otimes 1)$ and $a'' = a^s \otimes 1 + 1 \otimes a^s$ of $\mathscr{U}''$ are clearly fixed under $\mathscr{E}$. Hence $\mathscr{U}'' \subseteq \mathscr{V}$ the subalgebra of $\mathscr{U}''$ of $\mathscr{E}$-fixed elements. Accordingly, we have the exact sequence

$$(25) \qquad 0 \rightarrow \mathscr{U}'' \underset{i}{\rightarrow} \mathscr{V}$$

where i is the injection of $\mathscr{U}''$ into $\mathscr{V}$.

If σ is an associative specialization of $\mathfrak{J}$ then we can average σ with the commuting associative specialization 0: $a \rightarrow 0$ in the given associative algebra $\mathfrak{G}$. It follows that $\frac{1}{2}\sigma$ is a multiplication specialization of $\mathfrak{J}$. In particular, we have the multiplication specialization $\frac{1}{2}s$ of $\mathfrak{J}$ in $\mathscr{U}_s$. Clearly this is special so we have the homomorphism of $\mathscr{U}''$ onto $\mathscr{U}_s$ such that $1 \rightarrow 1$, $a'' = a^s \otimes 1 + 1 \otimes a^s \rightarrow a^s$, $a \in \mathfrak{J}$. We denote this as η and we have the exact sequence

$$(26) \qquad \mathscr{U}'' \underset{\eta}{\rightarrow} \mathscr{U}_s \rightarrow 0.$$

We now assume that $\mathfrak{J}$ has an identity element e. We have seen that $\mathscr{U} = \mathscr{U}_1 \oplus \mathscr{U}_{\frac{1}{2}} \oplus \mathscr{U}_0$ where the $\mathscr{U}_i$ are the ideals defined in § 2. If we write $a = a_1 + a_{\frac{1}{2}} + a_0$, $a_i \in \mathscr{U}_i$, then $\mathscr{U}_1$ and the mapping $a \rightarrow a_i$, $a \in \mathfrak{J}$ is the unital universal multiplication envelope. Also $a_0 = 0$ and $(ab)_{\frac{1}{2}} = a_{\frac{1}{2}}b_{\frac{1}{2}} + b_{\frac{1}{2}}a_{\frac{1}{2}}$, $e_{\frac{1}{2}} = \frac{1}{2}E_{\frac{1}{2}}$ where $E_{\frac{1}{2}}$ is the identity element of $\mathscr{U}_{\frac{1}{2}}$. The first relation implies that $a \rightarrow 2a_{\frac{1}{2}}$ is an associative specialization. The same is true of $a \rightarrow 2a_0 = 0$ and of $a \rightarrow 2(a_0 + a_{\frac{1}{2}})$. Hence we have the homomorphism of $\mathscr{U}_s$ into $\mathscr{U}_0 + \mathscr{U}_{\frac{1}{2}}$ such that $1 \rightarrow E_0 + E_{\frac{1}{2}}$ the identity element of $\mathscr{U}_0 + \mathscr{U}_{\frac{1}{2}}$ and $a_s \rightarrow 2(a_0 + a_{\frac{1}{2}})$. On the other hand, $a \rightarrow \frac{1}{2}a^s$ is a multiplication specialization so we have the homomorphism of $\mathscr{U}$ into $\mathscr{U}_s$ such that $1 \rightarrow 1$ and $a \rightarrow \frac{1}{2}a_s$. The image of $e(e - 1) = e^2 - e$ under this homomorphism is $(e^s)^2 - e^s = 0$. Hence $\mathscr{U}_1 \rightarrow 0$ in the homomorphism and we have a homomorphism of $\mathscr{U}_0 + \mathscr{U}_{\frac{1}{2}}$ such that $E_0 + E_{\frac{1}{2}} \rightarrow 1$, $2(a_0 + a_{\frac{1}{2}}) \rightarrow a^s$. If follows that we have an isomorphism of $\mathscr{U}_s$ onto $\mathscr{U}_0 + \mathscr{U}_{\frac{1}{2}}$ such that $a^s > 2(a_0 + a_{\frac{1}{2}})$. Hence ${}_0 + {}_{\frac{1}{2}}$ and the mapping $a \rightarrow 2(a_0 + a_{\frac{1}{2}})$ is a special universal envelope for $\mathfrak{J}$.

In a similar manner, one sees that $\mathscr{U}_{\frac{1}{2}}$ and the mapping $a \rightarrow 2a_{\frac{1}{2}}$ is a unital special envelope for $\mathfrak{J}$. Let $\mathscr{U}_1''$ denote the subalgebra of $\mathscr{U}_{s1} \otimes \mathscr{U}_{s1}$ generated by the elements $a_1'' = \frac{1}{2}(a^{s_1} \otimes 1 + 1 \otimes a^{s_1})$. We shall call this the *unital special multiplication envelope* of $\mathfrak{J}$. If σ_1 and σ_2 are unital associative specializations of $\mathfrak{J}$ in the same associative algebra $\mathfrak{G}$ with 1 and σ_1 and σ_2 commute then we have a homomorphism of $\mathscr{U}_1''$ into $\mathfrak{G}$ such that $1 \rightarrow 1$ and $\frac{1}{2}(a^s \otimes 1 + 1 \otimes a^s)$. Also let $\mathscr{V}_1$ be the subalgebra of $\mathscr{U}_{s1} \otimes \mathscr{U}_{s1}$ of elements fixed under the exchange automorphism ε_1 of this algebra. Then $\mathscr{U}_1''$ is a subalgebra of $\mathscr{V}_1$. It is clear that analogous to (24), (25) and (26) we have the following exact sequences:

(27) $$\mathscr{U}_1 \underset{\xi_1}{\rightarrow} \mathscr{U}_1'' \rightarrow 0$$

(28) $$0 \rightarrow \mathscr{U}_1'' \underset{i_1}{\rightarrow} \mathscr{V}$$

(29) $$\mathscr{U}_1'' \underset{\eta_1}{\rightarrow} \mathscr{U}_{s1} \rightarrow 0$$

where $a_1^{\xi_1} = a^{s_1} \otimes 1 + 1 \otimes a^{s_1}$, $a_1''^{i_1} = a_1''$, $a_1''^{\eta_1} = a^{s_1}$, $a \in \mathfrak{J}$.

We remark finally that if $\mathfrak{J}$ is n $(< \infty)$ dimensional with eidentity element then $\dim \mathscr{U}_{s1} \leq 2^{n-1}$ and $\dim \mathscr{U}_1 \leq \binom{2n-1}{n}$. Since $\mathscr{U}_{s1} \cong \mathscr{U}_{\frac{1}{2}}$ and $\mathscr{U} = \mathscr{U}_1 \oplus \mathscr{U}_{\frac{1}{2}} \oplus \mathscr{U}_1$ we obtain the upperbound $N_1 = \binom{2n-1}{n} + 2^{n-1} + 1$ for $\dim \mathscr{U}$. This is smaller than $N = \binom{2n+1}{n}$ the upper bound for $\dim \mathscr{U}$ given in Theorem 2, if $n > 1$.

5. *Some basic criteria.* In this section we shall derive some useful sufficient conditions that the homomorphisms ξ, i of (24) and (25) be isomorphisms and that bimodules be special. We prove first

Theorem 6. *Let $\mathfrak{J}$ be a special Jordan algebra, $\mathfrak{M}$ a bimodule for $\mathfrak{J}$. Then $\mathfrak{M}$ is special if and only if the split null extension $\mathscr{E} = \mathfrak{J} \oplus \mathfrak{M}$ is a special Jordan algebra.*

Proof. Suppose first that $\mathfrak{M}$ is special so that we have an isomorphism μ of $\mathfrak{M}$ into a bimodule $\mathfrak{N}$ for $\mathfrak{J}$ such that $v \cdot a = \frac{1}{2}(va^{\sigma_1} + va^{\sigma_2})$ where σ_1 and σ_2 are commuting associative specializations of $\mathfrak{J}$ in $\mathrm{Hom}_\Phi(\mathfrak{N}, \mathfrak{N})$. We have seen that $\mathfrak{N}$ is a unital associative bimodule for $\mathscr{U}_s(\mathfrak{J})$ so that $va^s = va^{\sigma_1}$, $a^s v = va^{\sigma_2}$, $a \in \mathfrak{J}$. Then the split null extension $\mathfrak{F} = \mathscr{U}_s \oplus \mathfrak{N}$ is an associative algebra. Consider the mapping $a + u \rightarrow > a^s + u^\mu$, $a \in \mathfrak{J}$, $u \in \mathfrak{M}$, of $\mathscr{E}$ into $\mathfrak{F}$. If $b \in \mathfrak{J}$, $v \in \mathfrak{M}$ then

$$\begin{aligned}(a + u)\cdot(b + v) &= a \cdot b + v \cdot a + u \cdot b > (a \cdot b)^s + (v \cdot a)^\mu + (u \cdot b)^\mu \\ &= a^s \cdot b^s + v^\mu \cdot a + u^\mu \cdot b \\ &= a^s \cdot b^s + \tfrac{1}{2}(v^\mu a^{\sigma_1} + v^\mu a^{\sigma_2}) + \tfrac{1}{2}(u^\mu \cdot b^{\sigma_1} + u^\mu b^{\sigma_2}) \\ &= a^s \cdot b^s + \tfrac{1}{2}(v^\mu a^s + a^s v^\mu) + \tfrac{1}{2}(u^\mu b^s + b^s u^\mu) \\ &= a^s \cdot b^s + v^\mu \cdot a^s + u^\mu \cdot b^s\end{aligned}$$

where $v^\mu \cdot a^s$ and $u^\mu \cdot b^s$ are the Jordan products evaluated in $\mathfrak{F}^+$. Also $(a^s + u^\mu)\cdot(b^s + v^\mu) = a^s \cdot b^s + v^\mu \cdot a^s + u^\mu \cdot b^s$. Hence $a + u \rightarrow a^s + u^\mu$ is an algebra homomorphism. Since $\mathfrak{J}$ is special, s is injective. Also μ is injective. Hence we have an isomorphism of $\mathscr{E}$ in $\mathfrak{F}^+$ and $\mathscr{E}$ is special. Conversely, assume $\mathscr{E}$ is special so we have an isomorphism σ of $\mathscr{E}$ into $\mathfrak{G}^+$, $\mathfrak{G}$ associative. If c_L and c_R denote the left and right multiplications in $\mathfrak{G}$ then we have the associative specializations $a \rightarrow a^{\sigma_1} = (a^\sigma)_L$, $a \rightarrow a^{\sigma_2}(a^\sigma)_R$ of $\mathfrak{J}$ in $\mathrm{Hom}_\Phi(\mathfrak{G}, \mathfrak{G})$. These commute and their average is $a \rightarrow R_{a^\sigma}$ where $R_c = \frac{1}{2}(c_L + c_R)$ is the right multiplication in the Jordan algebra $\mathfrak{G}^+$. Hence $a \rightarrow R_{a^\sigma}$ is a special multiplication representation of $\mathfrak{J}$ in $\mathrm{Hom}_\Phi(\mathfrak{G}, \mathfrak{G})$. If $u \in \mathfrak{M}$, $a \in \mathfrak{J}$ then $(u \cdot a)^\sigma = u^\sigma \cdot a^\sigma = uR_{a^\sigma}$. Hence the restriction of σ to $\mathfrak{M}$ is an isomorphism of $\mathfrak{M}$ as $\mathfrak{J}$-bimodule into $\mathfrak{G}$. It follows that $\mathfrak{M}$ is a special bimodule for $\mathfrak{J}$.

We shall derive next a criterion that the fact that the canonical homomorphism ξ of $\mathscr{U}(\mathfrak{J})$ onto $\mathscr{U}''(\mathfrak{J})$ is an isomorphism carries over to a homomorphic image $\bar{\mathfrak{J}}$ of $\mathfrak{J}$.

Suppose we have an exact sequence

$$\mathfrak{J} \underset{\zeta}{\to} \bar{\mathfrak{J}} \to 0. \tag{30}$$

Then we have seen that

$$\begin{gathered} \mathscr{U}(\mathfrak{J}) \underset{\zeta_u}{\to} \mathscr{U}(\bar{\mathfrak{J}}) \to 0 \\ \mathscr{U}_s(\mathfrak{J}) \underset{\zeta_s}{\to} \mathscr{U}_s(\bar{\mathfrak{J}}) \to 0 \end{gathered} \tag{31}$$

are exact. The homomorphism ζ_s gives the homomorphism $\zeta_s \otimes \zeta_s$ of $\mathscr{U}_s(\mathfrak{J}) \otimes \mathscr{U}_s(\mathfrak{J})$ into $\mathscr{U}_s(\bar{\mathfrak{J}}) \otimes \mathscr{U}_s(\bar{\mathfrak{J}})$ and by restriction we obtain the homomorphism ζ'' of $\mathscr{U}''(\mathfrak{J})$ into $\mathscr{U}''(\bar{\mathfrak{J}})$. Then

$$\mathscr{U}''(\mathfrak{J}) \underset{\zeta''}{\to} \mathscr{U}''(\bar{\mathfrak{J}}) \to 0 \tag{32}$$

is exact. Also we have the homomorphism ξ of $\mathscr{U}(\mathfrak{J})$ onto $\mathscr{U}''(\mathfrak{J})$ as in (24), and a similar homomorphism $\bar{\xi}$ of $\mathscr{U}(\bar{\mathfrak{J}})$ onto $\mathscr{U}''(\bar{\mathfrak{J}})$. It follows directly from the definitions that

$$\begin{array}{ccccc} \mathscr{U}(\mathfrak{J}) & \xrightarrow{\zeta_u} & \mathscr{U}(\bar{\mathfrak{J}}) & \longrightarrow & 0 \\ \Big\downarrow \xi & & & & \\ \mathscr{U}''(\mathfrak{J}) & \xrightarrow[\zeta'']{} & \mathscr{U}''(\bar{\mathfrak{J}}) & \longrightarrow & 0 \end{array} \tag{33}$$

is commutative. Let $\mathfrak{K}_s = \ker \zeta_s$. Then, as is well known, $\ker \zeta_s \otimes \zeta_s = \mathscr{U}_s(\mathfrak{J}) \otimes \mathfrak{K}_s + \mathfrak{K}_s \otimes \mathscr{U}_s(\mathfrak{J})$. Hence

$$\ker \zeta'' = \mathscr{U}''(\mathfrak{J}) \cap (\mathscr{U}_s(\mathfrak{J}) \otimes \mathfrak{K}_s + \mathfrak{K}_s \otimes \mathscr{U}_s(\mathfrak{J})). \tag{34}$$

Let $\mathfrak{B} = \ker \zeta$. If $b \in \mathfrak{B}$ then $b^{\zeta} = 0$ and $b^{\zeta_u} = 0$. Hence by (33), $b^{\xi\zeta''} = 0$. Since $b^{\xi} = b''$ we see that $b'' \in \ker \zeta''$. Hence the ideal $\mathfrak{K}''$ in $\mathscr{U}''$ generated by $\mathfrak{B}'' = \{b'' | b \in \mathfrak{B}\}$ is contained in $\ker \zeta''$:

$$\mathfrak{K}'' \subseteq \ker \zeta'' \tag{35}$$

We can now prove

Theorem 7. *Let $\mathfrak{J}$ be a Jordan algebra, ζ a homomorphism of $\mathfrak{J}$ onto $\bar{\mathfrak{J}}$. Suppose the canonical homomorphism ξ of $\mathscr{U}(\mathfrak{J})$ onto $\mathscr{U}''(\mathfrak{J})$ is an isomorphism. Then the same thing is true of the canonical homomorphism $\bar{\xi}$ of $\mathscr{U}(\bar{\mathfrak{J}})$ onto $\mathscr{U}''(\bar{\mathfrak{J}})$ if and only if the ideal $\mathfrak{K}''$ in $\mathscr{U}''(\mathfrak{J})$ generated by the elements b'', $b \in \ker \zeta$, coincides with $\ker \zeta''$, ζ'' the canonical homomorphism of $\mathscr{U}''(\mathfrak{J})$ onto $\mathscr{U}''(\bar{\mathfrak{J}})$.*

Proof. Suppose first that $\mathfrak{K}'' = \ker \zeta''$ and let $\bar{v} \in \ker \bar{\zeta}$. Let u be an element of $\mathscr{U}(\mathfrak{J})$ such that $v^{\zeta}u = \bar{v}$. Then $v^{\zeta_u \xi} = 0$ and the commutativity of (33) gives $v^{\xi\zeta''} = 0$ so $v^{\xi} \in \ker \zeta''$. Hence $v^{\xi} \in \mathfrak{K}''$. Since ξ is an isomorphism it follows that v is in the ideal in $\mathscr{U}(\mathfrak{J})$ generated by $\mathfrak{B}$. By Theorem 1.4 (4) the ideal in $\mathscr{U}(\mathfrak{J})$ generated by $\mathfrak{B}$ is $\ker \zeta_u$. Hence $\bar{v} = v^{\zeta_u} = 0$ and $\ker \bar{\xi} = 0$. Hence $\bar{\xi}$ is an isomorphism. In the same way we see that if $\mathfrak{K}'' \neq \ker \zeta''$ then, by (35), we can choose a $v'' \in \ker \zeta''$, $v'' \notin \mathfrak{K}''$. We have $v'' = v^{\xi}$ for a $v \in \mathscr{U}(\mathfrak{J})$. Since $v^{\xi} \notin \mathfrak{K}''$ and ξ is an isomorphism v is not contained in the ideal in $\mathscr{U}(\mathfrak{J})$ generated by $\mathfrak{B}$. Hence $v^{\zeta_u} \neq 0$. On the other hand, $v^{\zeta_u \bar{\xi}} = v^{\xi\zeta''} = 0$. Hence $\bar{\xi}$ is not an isomorphism.

We remark that the condition $\mathfrak{K}'' = \ker \zeta''$ is equivalent, by (34) and (35), to

$$\mathfrak{K}'' \supseteq \mathscr{U}''(\mathfrak{J}) \cap (\mathscr{U}_s(\mathfrak{J}) \otimes \mathfrak{K}_s + \mathfrak{K}_s \otimes \mathscr{U}_s(\mathfrak{J})). \tag{36}$$

We remark also that Theorem 7 carries over word for word to the unital envelopes $\mathscr{U}_1(\mathfrak{J})$ and $\mathscr{U}_1''(\mathfrak{J})$.

Let $\mathfrak{J}$ be Jordan algebra with identity element e, $(\mathscr{U}_{s1}(\mathfrak{J}), s_1)$ the unital special envelope of $\mathfrak{J}$, π_1 the main involution in $\mathscr{U}_{s1}$ and $\mathscr{H}(\mathscr{U}_{s1}, \pi_1)$ the set of π_1-symmetric elements of $\mathscr{U}_{s1}$. Then $\mathfrak{J}^{s1}$ is a subalgebra of $\mathscr{H}(\mathscr{U}_{s1}, \pi_1)$ and we shall call $\mathfrak{J}$ *reflexive* if $\mathfrak{J}^s 1 = \mathscr{H}(\mathscr{U}_{s1}, \pi_1)$. Then we have the following sufficient conditions, that $\mathscr{U}_1''(\mathfrak{J})$ coincide with the subalgebra $\mathscr{V}_1$ of $\mathscr{U}_{s1}(\mathfrak{J}) \otimes \mathscr{U}_{s1}(\mathfrak{J})$ of elements fixed under the exchange automorphism.

Theorem 8. *Let $\mathfrak{J}$ be a Jordan algebra with identity element e. Assume* (1) *$\mathfrak{J}$ is reflexive and* (2) *the subspace $\gamma(\mathscr{U}_{s1}, \pi_1)$ of π_1-skew elements satisfies $\gamma(\mathscr{U}_{s1}, \pi_1) = [\mathscr{H}(\mathscr{U}_{s1}, \pi_1), \mathscr{H}(\mathscr{U}_{s1}, \pi_1)]$. Then $\mathscr{U}_1''$ coincides with the subalgebra $\mathscr{V}_1$ of $\mathscr{U}_s \otimes \mathscr{U}_s$ of fixed points under the exchange automorphism.*

Proof. We show first that for any $x \in \mathscr{U}_{s1}$, $x \otimes 1 + 1 \otimes x \in \mathscr{U}_1''$. Since x is reflexive this is clear if $x \in \mathscr{H}(\mathscr{U}_{s1}, \pi_1)$. Next let y, $z \in \mathscr{H}(\mathscr{U}_{s1}, \pi_1)$. Then $[y, z] \otimes 1 + 1 \otimes [y, z] = [y \otimes 1 + 1 \otimes y, z \otimes 1 + 1 \otimes z] \in \mathscr{U}_1''$. Then (2) implies that every element of the form $x \otimes 1 + 1 \otimes x$ with x in $\gamma(\mathscr{U}_{s1}, \pi_1)$ is in $\mathscr{U}_1''$. Since $\mathscr{U}_{s1} = \mathscr{H} + \gamma$ every $x \otimes 1 + 1 \otimes x$, $x \in \mathscr{U}_{s1}$ is in $\mathscr{U}_1''$. Then if x, $y \in \mathscr{U}_{s1}$,

$$x \otimes y + y \otimes x = (x \otimes 1 + 1 \otimes x)(y \otimes 1 + 1 \otimes y) - (xy \otimes 1 + 1 \otimes xy)$$

$\varepsilon \mathscr{U}_1''$. It follows that $\mathscr{U}_1'' = \mathscr{V}_1$.

Chapter III
Representation Theory for Some Important Classes of Jordan Algebras

In this chapter we shall sketch the representation theory of two important classes of Jordan algebras: 1) Jordan algebra defined by symmetric bilinear forms and 2) Jordan matrix algebras. In both of these cases we are dealing with Jordan algebras with identity elements so we have the reduction of the problem of determining the universal multiplication envelope $\mathscr{U}(\mathfrak{J})$ to that of determining the unital universal multiplication envelope $\mathscr{U}_1(\mathfrak{J})$ and the unital special universal envelope $\mathscr{U}_{s_1}(J)$. If J is the Jordan algebra $\Phi_1 \oplus \mathfrak{M}$ defined by the symmetric bilinear form (,) on $\mathfrak{M}$ then $\mathscr{U}_{s_1}(\mathfrak{J})$ is the Clifford algebra $C(\mathfrak{M}, Q)$ where $Q(x) = (x, x)$. The structure of these algebras is classical. A full account of this can be found in Chevalley's work *Algebraic Theory of Spinors*. The algebra $\mathscr{U}_1(\mathfrak{J})$ for $\mathfrak{J} = \Phi_1 \oplus \mathfrak{M}$ has arisen in the theory of mesons. For this reason we have called this algebra the meson algebra of (,). We shall sketch the structure of these algebras.

Let $\mathscr{V}$ be an arbitrary non associative algebra with 1, $\mathscr{V}_n$ the algebra of $n \times n$ matrices with entries in $\mathscr{V}$ and usual matrix multiplication. Now assume that $\mathscr{V}$ has an involution $d \to \bar{d}$. Then $X \to {}^t\bar{X}$, where $\bar{X} = (\bar{x}_{ij})$ if $X = (x_{ij})$ and t denotes the transpose, is an involution in $\mathscr{V}_n$. More generally, let a_i, $i = 1, \ldots, n$, be an element of the nucleus $N(\mathscr{V})$ of $\mathscr{V}$ such that $\bar{a}_i = a_i$ and a_i^{-1} exists in $N(\mathscr{V})$. Put $a = \operatorname{diag}(a_1, a_2, \ldots, a_n)$ the diagonal matrix with entries a_i. Then $X \to X^+ = a^{-1}({}^t\bar{X})a$ is an involution in $\mathscr{V}_n$ and $\mathscr{H}(\mathscr{V}_n, \mathfrak{J}_n) =$ the set of matrices satisfying $X^+ = X$ is a subalgebra of $\mathscr{V}_n$ (for characteristic $\neq 2$). If $n \geq 3$ and $\mathscr{H}(\mathscr{V}_n, \mathfrak{J}_a)$ is Jordan then we call $\mathscr{H}(\mathscr{V}_n, \mathfrak{J}_a)$ a Jordan matrix algebra. It is known that $\mathscr{H}(\mathscr{V}_n, \mathfrak{J}_a)$, $n \geq 3$, is Jordan if and only if one of the following sets of conditions holds: (2) $n \geq 4$ and $\mathscr{V}$ is associative, (21) $n = 3$ and $\mathscr{V}$ is alternative with symmetric elements relative to $d \to \bar{d}$ in the nucleus. The special Jordan matrix algebras are precisely those such that $\mathscr{V}$ is associative.

If $\mathfrak{J} = \mathscr{H}(\mathscr{V}_n, S_a)$ is a special Jordan matrix algebra then $\mathscr{V}_n$ and the injection mapping of $\mathscr{H}(\mathscr{V}_n, \mathfrak{J}_a)$ in $\mathscr{V}_n$ constitute a special universal envelope for $\mathfrak{J}$. This result, due to Jacobson and Rickart, has been given an extensive generalisation by Martindale. We shall sketch the proof of the latter theorem.

The key result for determining $\mathscr{U}_1(\mathfrak{J})$ if $n \geq 4$ is that the canonical homomorphism of $\mathscr{U}_1(\mathfrak{J})$ into $\mathscr{U}_1''(\mathfrak{J})$ is an isomorphism—that is, the algebras $\mathfrak{J}$ are strongly special. The key result for proving this is a characterization of Jordan matrix algebras by the existence of certain types of elements in the algebra. This characterization is called the Coordinatization Theorem since it is somewhat reminiscent of the coordinatization of projective spaces. Another basic result for the determination of the bimodules for Jordan matrix algebras (n and $\mathscr{V}$ arbitrary) are certain category isomorphism between the category of unital bimodules for $\mathfrak{J}$ and certain categories of bimodules with involution for the coefficient algebra. Complete results can be obtained in the important case in which $\mathscr{V}$ is a composition algebra. Improved proofs of this have been given by Mc Crimmon (to appear).

1. *Jordan algebras of symmetric bilinear forms and Clifford algebras.* Let $\mathfrak{J} = \Phi 1 \oplus \mathfrak{M}$ the Jordan algebra of a symmetric bilinear form (,) on $\mathfrak{M}$. We recall that the product in $\mathfrak{J}$ is

$$(\alpha 1 + x)\cdot(\beta 1 + y) = (\alpha\beta + (x, y))1 + \alpha y + \beta x$$

if $\alpha, \beta \in \Phi$ and $x, y \in \mathfrak{M}$. Let $Q(x) = (x, x)$. Then the Clifford algebra $C(\mathfrak{M}, Q)$ is defined as $\mathfrak{T}(\mathfrak{M})/\mathfrak{I}$ where $\mathfrak{T}(\mathfrak{M}) = \Phi 1 \oplus \mathfrak{M} \oplus (\mathfrak{M} \otimes \mathfrak{M}) \oplus \cdots$ the tensor algebra over $\mathfrak{M}$ and $\mathfrak{I}$ is the ideal in $\mathfrak{T}(\mathfrak{M})$ generated by all elements of the form

$$x \otimes x - Q(x)1, \qquad x \in M \tag{1}$$

(Chevalley's Algebraic Theory of Spinors, p. 37). We recall that $C(\mathfrak{M}, Q)$ can be characterized by the following universal mapping property: If σ is a linear mapping of $\mathfrak{M}$ into an associative algebra $\mathfrak{G}$ with 1 and $(x^\sigma)^2 = Q(x)1$, then there exists a unique homomorphism η of $C(\mathfrak{M}, Q)$ into $\mathfrak{G}$ such that $1^\eta = 1$ $(= i + \mathfrak{I})$ and $(x + \mathfrak{I})^\eta = x^\sigma$, $x \in \mathfrak{M}$ (loc. cit. p. 39). We now claim that if we set $(\alpha 1 + x)^{s_1} = \alpha 1 + x + \mathfrak{I}$, $\alpha \in \Phi$, $x \in \mathfrak{M}$, then $C(\mathfrak{M}, Q)$ and this s_1 constitute a unital special universal envelope for $\mathfrak{J} = \Phi 1 \oplus \mathfrak{M}$. First, we note that $1^{s_1} = 1$ in $C(\mathfrak{M}, Q)$ and $(x^{s_1})^s = Q(x)1$, $x \in \mathfrak{M}$. The latter can be linearized to give $x^{s_1} \cdot y^{s_1} = (x, y)1$. Then

$$(\alpha 1 + x)^{s_1}\cdot(\beta 1 + y)^{s_1} = \alpha\beta 1 + x^{s_1}\cdot y^{s_1} + \alpha y^{s_1} + \beta x^{s_1} = (\alpha\beta + (x, y))1 + (\alpha y + \beta x)^{s_1}$$
$$= ((\alpha 1 + x)\cdot(\beta 1 + y))^{s_1}.$$

Hence s_1 is a unital associative specialization of $\mathfrak{J}$ in $C(\mathfrak{M}, Q)$. Next let σ be any unital associative specialization of $\mathfrak{J}$ in an associative algebra $\mathfrak{G}$ with 1. Then $(x^\sigma)^2 = (x^{\cdot 2})^\sigma = Q(x)1$, $x \in \mathfrak{M}$. Hence we have a unique homomorphism μ of $C(\mathfrak{M}, Q)$ into $\mathfrak{G}$ such that $1^\mu = 1$ and $x^{s_1\mu} = x^\sigma$, $x \in \mathfrak{M}$. Also $(\alpha 1 + x)^{s_1\mu} = \alpha 1 + x^\sigma = (\alpha 1 + x)^\sigma$. Hence $(C(\mathfrak{M}, Q), s_1)$ is a unital special universal envelope for $\mathfrak{J}$.

The properties of Clifford algebras are well known and will not be derived here. The readers may consult Chevalley's book for these. We recall the following facts: the mapping s of $\mathfrak{J}$ into $C(\mathfrak{M}, Q)$ is injective.[(o)] Hence $\mathfrak{J} = \Phi 1 \oplus \mathfrak{M}$ is a special Jordan algebra. We recall also that if $\dim \mathfrak{M} = 2r$, r a non-negative integer, and $(\ , \ )$ is non-degenerate then $C(\mathfrak{M}, Q)$ is central simple and if $\dim \mathfrak{M} = 2r + 1$ then $C(\mathfrak{M}, Q)$ is simple or a direct sum of two isomorphic central simple components according as $(-1)^{r-1}\delta$, δ the discriminant of $(\ , \ )$, is not or is a square in Φ.

2. *Meson algebras.* We consider next the unital universal multiplication envelope of $\mathfrak{J} = \Phi 1 \oplus \mathfrak{M}$ the Jordan algebra of the symmetric bilinear form $(\ , \ )$ on $\mathfrak{M}$. As in § 2,2 we identify $\mathfrak{J}$ with its image in $\mathscr{U}(\mathfrak{J})$. Then by (1′) of Chapter II we have

$$2xyx = -x^{\cdot 2}\cdot y + yx^{\cdot 2} + 2x(x\cdot y)$$
$$= -Q(x)y + Q(x)y + 2(x, y)x,$$

for any x in $\mathfrak{M}$. Hence

$$xyx = (x, y)x \tag{2}$$

in $\mathscr{U}(\mathfrak{J})$. If we denote the image of $a - \mathfrak{J}$ in $\mathscr{U}_1(\mathfrak{J})$ by a_1 then we have

$$x_1 y_1 x_1 = (x, y)x_1. \tag{3}$$

We now give the following

(o) This is proved in Chevalley's book in the finite dimensional case. The extension to the infinite dimensional case can be made by a direct limit argument.

Definition 1. Let $\mathfrak{M}$ be a vector space equipped with symmetric bilinear form $(\ ,\ )$. Let $\mathfrak{T}(\mathfrak{M})$ be the tensor algebra $\Phi 1 \oplus \mathfrak{M} \oplus (\mathfrak{M} \otimes \mathfrak{M}) \oplus \cdots$, $\mathfrak{K}$ the ideal in $\mathfrak{T}(\mathfrak{M})$ generated by the elements $x \otimes y \otimes x - (x, y)x$, $x \in \mathfrak{M}$. Then $\mathfrak{T}(\mathfrak{M})/\mathfrak{K}$ is called the *meson algebra* $D(\mathfrak{M}, (\ ,\ ))$ of the symmetric bilinear form $(\ ,\ )$.[(○)]

Consider the homomorphism of $\mathfrak{T}(\mathfrak{M})$ into $\mathscr{U}_1(\mathfrak{J})$ such that $1 \to 1$ and $x \to x_1$, $x \in \mathfrak{M}$. Since (3) holds it is clear that $\mathfrak{K}$ is in the kernel of this homomorphism. Hence we have the homomorphism of $D = D(\mathfrak{M}, (\ ,\ ))$ into $\mathscr{U}_1 = \mathscr{U}_1(\mathfrak{J})$ sending $(\alpha 1 + x)^\rho \equiv \alpha 1 + x + \mathfrak{K}$ into $\alpha 1 + x_1$. We show next that $\alpha 1 + x \to (\alpha 1 + x)^\rho$ is a unital multiplication specialization of $\mathfrak{J}$. We have $(\alpha 1 + x)^{\cdot 2} = (\alpha^2 + Q(x))1 + 2\alpha x =$ so

$$((\alpha 1 + x)^{\cdot 2})^\rho = (\alpha^2 + Q(x))1 + 2\alpha x^\rho.$$

This commutes with $(\alpha 1 + x)^\rho = \alpha 1 + x^\rho$. Hence the condition $[a^\rho, a^{2\rho}]$, $a \in \mathfrak{J}$, for multiplication specialization is satisfield. If $x, y \in \mathfrak{M}$ we have

$$(\alpha 1 + x) \otimes (\beta 1 + y) \otimes (\alpha 1 + x) = \alpha^2\beta 1 + 2\alpha\beta x + \alpha^2 y + \beta x \otimes x + \alpha(x \otimes y + y \otimes x) + x \otimes y \otimes x.$$

Hence if $a = \alpha 1 + x$, $b = \beta 1 + y$ then

$$a^\rho \otimes b^\rho \otimes a^\rho = \alpha^2\beta 1 + 2\alpha\beta x^\rho + \alpha^2 y^\rho + \beta x^\rho \otimes x^\rho + \alpha(x^\rho \otimes y^\rho + y^\rho \otimes x^\rho) + (x, y)x^\rho. \tag{3}$$

Also,

$$(\alpha 1 + x)^{\cdot 2} = (\alpha^2 + Q(x))1 + 2\alpha x,$$

$$(\alpha 1 + x)\cdot(\beta 1 + y) = (\alpha\beta + (x, y))1 + \alpha y + \beta x;$$

$$(\alpha 1 + x)^{\cdot 2}\cdot(\beta 1 + y) = (\alpha^2\beta + Q(x)\beta + 2\alpha(x, y))1 + (\alpha^2 y + Q(x)y + 2\alpha\beta x).$$

Hence

$$(a^{\cdot 2}\cdot b^\cdot)^\rho = (\alpha^2\beta + Q(x)\beta + 2\alpha(x, y))1 + (\alpha^2 + Q(x))y^\rho + 2\alpha\beta x^\rho$$

$$a^\rho \otimes (a, b)^\rho = (\alpha^2\beta + \alpha(x, y))1 + \alpha^2 y^\rho + 2\alpha\beta x^\rho + (x, y)x^\rho + \alpha x^\rho \otimes y^\rho + \beta x^\rho \otimes x^\rho$$

$$b^\rho \otimes (a^{\cdot 2})^\rho = (\alpha^2\beta + \beta Q(x))1 + 2\alpha\beta x^\rho + (\alpha^2 + Q(x))y^\rho + 2\alpha y^\rho \otimes x^\rho.$$

These relations and (3) imply that

$$2a^\rho \otimes b^\rho \otimes a^\rho + (a^{\cdot 2}\cdot b)^\rho = 2a^\rho \otimes (a\cdot b)^\rho + b^\rho \otimes a^{\cdot 2\rho}.$$

Since $1^\rho = 1$ it is now clear that ρ is a unital multiplication specialization of $\mathscr{J}$. Hence we have the homomorphism of $\mathscr{U}$ into $\mathscr{D}$ such that $\alpha 1 + x_1 \to \alpha 1 + x^\rho$. It follows that the two homomorphisms are isomorphisms. Consequently, we have

Theorem 1. *Let $\mathfrak{M}$ be a vector space with a symmetric bilinear form $(\ ,\ )$, $\mathscr{J}$ the associated Jordan algebra $\Phi 1 \oplus \mathfrak{M}$, $D(\mathfrak{M}, (\ ,\ ))$ the associated meson algebra. If $a = \alpha 1 + x$, $x \in M$ then we write a^ρ for the image $a + \mathfrak{K}$ of a in D. Then (D, ρ) is a unital universal multiplication envelope for $\mathfrak{J}$.*

[(○)] These algebras have been used in physics in the theory of mesons. I believe they were introduced in this connection by Duppin and by Kenner.

We consider next the canonical homomorphism $a_1 \to a_1''$ of $\mathscr{U}_1(\mathscr{J})$ onto $\mathscr{U}_1''(\mathscr{J})$. Owing to the equivalence we have noted we can shift our attention to the meson algebra $D(\mathfrak{M},(\ ,\ ))$ and the algebra $C(\mathfrak{M},Q) \otimes C(\mathfrak{M},Q)$ and the homomorphism of $D = D(\mathfrak{M},(\ ,\ ))$ into $C \otimes C$, $C = C(\mathfrak{M},Q)$, such that $1 \to 1$ and $a^\rho \to a'' = 1/2(a^{s_1} \otimes 1 + 1 \otimes a^{s_1})$, $a \in \mathfrak{J}$. For the sake of simplicity we now assume $\dim \mathfrak{M} = n < \infty$ and we let $(u_1, u_2, \ldots, u_n)$ be an orthogonal basis for $\mathfrak{M}$ (relative to $(\ ,\ )$). As in the proof of Theorem 2.2(3), one sees that every element of D is a linear combination of 1 and the standard monomials

(4) $$(u_{k_1}^\rho)^2(u_{k_2}^\rho)^2\cdots(u_{k_r}^\rho)^2 u_{k_{r+1}}^\rho \cdots u_{k_r}^\rho$$

where the k_i are distinct, $k_1 < k_2 < \cdots < k_r$; $k_{r+1} < k_{r+3} < \cdots$; $k_{r+2} < k_{r+4} < \cdots$. Hence, $\dim D \leq \binom{2n+1}{n}$. We now consider the corresponding elements in $C \otimes C$: $1, (u_{k_1}'')^2(u_{k_2}'')^2 \ldots (u_{k_r}'')^2 u_{k_{r+1}}'' \ldots u_{k_s}''$. One can show by a fairly simple combinatorial argument based on the linear independence of the elements $u_{i_1}^{s_1}u_{i_2}^{s_1}\ldots u_{i_r}^{s_1} \otimes u_{j_1}^{s_1}u_{j_2}^{s_1}\ldots u_{j_t}^{s_1}$, $i_1 < i_2 < \cdots; j_1 < j_2 < \cdots$ in $C \otimes C$ that the indicated elements are linearly independent. This implies

Theorem 2. *The dimensionality of the meson algebra D defined by a symmetric bilinear form in an n dimensional vector space $\mathfrak{M}$ is $\binom{2n+1}{n}$ and the canonical homomorphism of D into $C \otimes C$, C the Clifford algebra of the given form, is a monomorphism.*

We shall now indicate how one can determine the structure of D for non-degenerate $(\ ,\ )$ and we assume first that $n = 2\nu$ is even. For the sake of simplicity we now identify $\mathscr{J} = \Phi 1 \oplus \mathfrak{M}$ with the corresponding subspace of the Clifford algebra C. Then $\mathscr{J}$ is a subalgebra of C^+ containing 1 and we have the unital multiplication representation $a \to R_a$ of $\mathscr{J}$ where R_a acts in C^+. We have $R_a = 1/2(a_L + a_R)$ where a_L, a_R denote multiplications in C. Since C is central simple we know that if $a_1, \ldots, a_r$ and $b_1, \ldots, b_s$ are linearly independent elements of C then the rs operators $a_{iL}b_{jR}$ are linearly independent. This implies that the representation of $\mathscr{U}_1''(\mathfrak{J})$ determined by $a \to R$ is a monomorphism. Hence by the last theorem this gives a monomorphism of D. Thus D is isomorphic to the algebra D' of linear transformations in C generated by 1 and the $R_a = 1/2(a_L + a_R)$, $a \in \mathfrak{J}$.

We now define some invariants subspaces of C relative to D'. For the moment we drop the restriction on $\dim \mathfrak{M}$ and the non-degeneracy of $(\ ,\ )$. If $x_1, x_2, \ldots, x_r \in \mathfrak{M}$ we define $[x_1x_2\ldots x_r]$ inductively by $[x_1] = x_1$, $[x_1\ldots x_{2k-1}x_{2k}] = [[x_1\ldots x_{2k-1}], x_{2k}]$, $[x_1\ldots x_{2k}x_{2k+1}] = x_1\ldots x_{2k}x_{2k+1}$. Let $\mathfrak{M}^{[0]} = \Phi 1$, $\mathfrak{M}^{[i]}$ for $i > 0$ be the subspace of C spanned by the elements $[x_1x_2\ldots x_i]$, $x_i \in \mathfrak{M}$. It is easily seen that $C = \mathfrak{M}^{[0]} \oplus \mathfrak{M}^{[1]} \oplus \mathfrak{M}^{[2]} \oplus \cdots$. If $\dim \mathfrak{M} = n < \infty$ then $\mathfrak{M}^{[i]}$ is the subspace spanned by all products $e_{i_1}e_{i_2}\cdots e_{i_r}$, $i_1 < i_2 < \cdots < i_r$ where $(e_1, e_2, \ldots, e_n)$ is an orthogonal basis for $\mathfrak{M}$ and $C = \mathfrak{M}_0 \oplus \mathfrak{M}_1 \oplus \cdots$, $\dim \mathfrak{M}_i = \binom{n}{i}$. It is easy to see that $\mathfrak{M}^{[2k+1]}R_y \subseteq \mathfrak{M}^{[2k]}$ if $x \in \mathfrak{M}$. This implies that the spaces $\mathfrak{M}^{[0]} + \mathfrak{M}^{[1]}, \mathfrak{M}^{[2]} + \mathfrak{M}^{[3]}, \ldots$ are invariant under every R_a, $a \in \mathfrak{J}$. Now let $n = 2\nu$ and $(\ ,\ )$ non-degenerate. Then we obtain the invariant subspaces $\mathfrak{M}^{[0]} + \mathfrak{M}^{[1]}, \mathfrak{M}^{[2]} + \mathfrak{M}^{[3]}, \ldots, \mathfrak{M}^{[2\nu-2]} + \mathfrak{M}^{[2\nu-1]}, \mathfrak{M}^{[2\nu]}$ of respective dimensions $\binom{n+1}{1}, \binom{n+1}{3}, \ldots, \binom{n+1}{n+1}$. The sum of the dimensionality of the algebras of linear transformations in these spaces is

(5) $$\binom{n+1}{1}^2 + \binom{n+1}{3}^2 + \cdots + \binom{n+1}{n+1}^2 = \binom{2n+1}{n}$$

if $n = 2\nu$. Since this is the same as $\dim D' = \dim D$ we see that the mappings induced by D' in the employed spaces in the complete algebra of linear transformations in these and D' hence D is isomorphic to a direct sum of these algebras. Hence we have

Theorem 3. *Let* $\dim \mathfrak{M} = n$ *be even and* $(\ ,\ )$ *non-degenerate. Then the meson algebra* $D = D(\mathfrak{M}, (\ ,\ ))$ *is isomorphic to a direct sum of ideals which are isomorphic to matrix algebras* Φ_k, $k = \binom{n+1}{1}, \binom{n+1}{3}, \ldots, \binom{n+1}{n+1}$.

The analysisi in the odd dimensional case is more complicated but can be made along the same lines after one imbeds $\mathfrak{M}$ in an $n + 1$ dimensional vector space $\mathfrak{M}'$ with symmetric form so that $(\mathfrak{M}, e_{n+1}) = 0$, $(e_{n+1}, e_{n+1}) = 1$ for an $e_{n+1} \notin \mathfrak{M}$. The result one obtains in this case is the following

Theorem 4. *Let* $\dim \mathfrak{M} = n = 2\nu - 1$, $(\ ,\ )$ *non-degenerate. Then the meson algebra* $D = D(\mathfrak{M}, (\ ,\ ))$ *is isomorphic to a direct sum of ideals which are matrix algebras* Φ_k, $k = \binom{n+1}{0}, \binom{n+1}{1}, \ldots, \binom{n+1}{\nu-1}$ *and* Γ_k, $k = 1/2\binom{n+1}{\nu}$ *where* Γ *is the center of the Clifford algebra of* $\mathfrak{M}$ *and* $(\ ,\ )$.

3. *Jordan matrix algebras.* Let $\mathscr{V}$ be a non-associative algebra with 1 and involution $d \to \bar{d}$ and let $a = \operatorname{diag}(a_1, a_2 \ldots a_n)$ where $n \geq 3$ and $\bar{a}_i = a_i$ is in the nucleus $N(\ )$ and has an inverse in this associative algebra. Then we have the involution $\mathfrak{J}_a$: $X \to a^{-1t}\bar{X}a$ in ϑ_n and the subalgebra $\mathscr{H}(\mathscr{V}_n, \mathfrak{J}_a)$ of $\mathscr{V}_n^+$. As we indcated in the introduction $\mathscr{H}(\mathscr{V}_n, \mathfrak{J}_a)$ is a Jordan algebra if and only if either $n \geq 4$ and $\mathscr{V}$ is associative or $n = 3$ and $\mathscr{V}$ is alternative with symmetric elements $(\bar{d} = d)$ in the nucleus. We assume $\mathfrak{J} = \mathscr{H}(\mathscr{V}_n, a)$ is Jordan and call this a *Jordan matrix algebra* $(n \geq 3)$. Since $\mathfrak{J}$ has a 1 the analysis of $\mathscr{U}(\mathfrak{J})$ is reduced to that of $\mathscr{U}_1(\mathfrak{J})$ and $\mathscr{U}_{s1}(\mathfrak{J})$ the unital universal multiplication envelope and unital special universal envelope of $\mathfrak{J}$. We consider $\mathscr{U}_{s1}$ first assuming $\mathscr{V}$ is associative. Thus we have the following

Theorem 5 (jacobson-Rickart). *Let* $\mathscr{V}$ *be an associative algebra with* 1 *and involution* $d \to \bar{d}$, $\mathscr{H}(\mathscr{V}_n, \mathfrak{J}_a)$ *the Jordan algebra of* $\mathfrak{J}_a$*-symmetric ellements of* $\mathscr{V}_n$, $n \geq 3$, *where* $\mathfrak{J}_a$ *is the involution* $X \to a^{-1t}\bar{X}a = \operatorname{diag}\{a_1, a_2, \ldots, a_n\}$, $\bar{a}_i = a_i$ *in* $\mathscr{V}$. *Then* $\mathscr{V}_n$ *and the injection mapping is a unital special universal envelope for* $\mathscr{H}(\mathscr{V}_n, \mathfrak{J}_a)$.

This result has been generalized by Martindale to the following result

Theorem 6 (martindale, to appear). *Let* $\mathfrak{U}$ *be an associative algebra with* 1 *and involution* J *such that the following conditions hold*: $\mathscr{H}(\mathfrak{U}, J)$ *contains three orthogonal idempotents* e_i *such that* $e_1 + e_2 + e_3 = 1$ *and* $\mathfrak{U}e_i\mathfrak{U} = \mathfrak{U}$, $i = 1, 2, 3$. *Then* $\mathfrak{U}$ *and the injection mapping are a unital special universal envelope for* $\mathscr{H}(\mathfrak{U}, J)$.

It is trivial to see that this implies Theorem 5. We now sketch the proof of Martindale's theorem. The assertion of the theorem is that any homomorphism σ of $\mathscr{H}(\mathfrak{U}, \mathfrak{J})$ in a Jordan algebra $\mathfrak{B}^+$ where $\mathfrak{B}$ is associative with 1 has a unique extension to a homomorphism η of $\mathfrak{U}$ into $\mathfrak{B}$. Let $\mathfrak{U} = \sum \mathfrak{U}_{ij}$, $\mathfrak{U}_{ij} = e_i\mathfrak{U}e_j$ be the two-sided Peirce decomposition of $\mathfrak{U}$ relative to the e_i. The space $\mathfrak{U}_{ii}$ is characterized by the Jordan condition $\mathfrak{U}_{ii} = \{a_{ii} | \{e_i a_{ii} e_i\} = a_{ii}\}$ and if $i \neq j$ then $\mathfrak{U}_{ij} + \mathfrak{U}_{ji}$ is characterized by $\mathfrak{U}_{ij} + \mathfrak{U}_{ji} = \{a_{ij} | \{e_i a_{ij} e_j\} = 1/2 a_{ij}\}$. Let $f_i = e_i^\sigma$. Then we have seen that the f_i are orthogonal idempotents and $f_i + f_2 + f_3 = 1$. Let $i \neq j$ and let $e_iae_j \in \mathfrak{U}_{ij}$. Then $e_iae_j + e_ja^Je_i \in \mathscr{H}(\mathfrak{U}, J) \cap (\mathfrak{U}_{ij} + \mathfrak{U}_{ji})$. Hence $(e_iae_j + e_ja^Je_i)^\sigma$ is defined and is contained in $\mathfrak{B}_{ij} + \mathfrak{B}_{ji}$, $\mathfrak{B}_{ij} = f_i\mathfrak{B}f_j$. Let $(e_iae_j)^{\eta_{ij}} = e_i(e_iae_j + a^Ja_j)^\sigma e_j$ be the part of $(e_iae_j + e_ja^Je_i)^\sigma$ in $\mathfrak{B}_{ij}$. Then η_{ij} is a linear mapping of $\mathfrak{U}_{ij}$ into $\mathfrak{B}_{ij}$. One establishes easily the following basic formula: if $a_{ij} \in \mathfrak{U}_{ij}$ and $a_{jk} \in \mathfrak{U}_{jk}$, $i, j, k \neq$, then $(a_{ij}a_{jk})^{\eta_{ik}} = a_{ij}^{\eta_{ij}}a_{jk}^{\eta_{jk}}$. This can be used to show that if $\sum a_{ij}^{(b)}a_{ji}^{(b)} = \sum a_{ik}^{(g)}a_{ki}^{(g)}$, $a_{ij}^{(b)} \in \mathfrak{U}_{ij}$, $a_{ji}^{(b)} \in \mathfrak{U}_{ji}$ etc. then $\sum a_{ij}^{(b)\eta_{ij}}a_{ji}^{(b)\eta_{ji}} = \sum a_{ik}^{(g)\eta_{ik}}a_{ki}^{(g)\eta_{ki}}$. Now let $a_{ii} \in \mathfrak{U}_{ii}$. Then the hypotheses imply that we can write $a_{ii} = \sum a_{ij}^{(b)}a_{ji}^{(b)}$, $a_{ij}^{(b)} \in \mathfrak{U}_{ij}$, $a_{ji}^{(b)} \in \mathfrak{U}_{ji}$ and the result noted implies that if we put $a_{ii}^{\eta_{ii}} = \sum a_{ij}^{(b)\eta_{ij}}a_{ji}^{(b)\eta_{ji}}$ then we obtain a linear mapping η_{ii} of $\mathfrak{U}_{ii}$ into $\mathfrak{B}_{ii}$. Let η be the linear mapping of $\mathfrak{U}$ into $\mathfrak{B}$ which coincides with η_{ij} ($i = j$ or $i \neq j$) on $\mathfrak{U}_{ij}$. Then we can show that η is a homomorphism of $\mathfrak{U}$ into $\mathfrak{B}$ which coincides with σ on $\mathscr{H}(\mathfrak{U}, \mathfrak{J})$.

We indicate next a characterization of Jordan matrix algebras. We need first the notion of inversion in Jordan algebras. If $\mathfrak{J}$ is any Jordan algebra with identity element 1 then b is said to be an *inverse* of a if $a \cdot b = 1$ and $a^{\cdot 2} \cdot b = a$. If $\mathfrak{J} = \mathfrak{U}^+$, $\mathfrak{U}$ associative, then it is easy to see that these Jordan conditions are equivalent to the associative conditions $ab = 1 = ba$. In any case it can be shown that the relation between a and b is a reciprocal one and each element determines the other uniquely. If e is an idempotent in a Jordan algebra $\mathfrak{J}$ then the subspace $\mathfrak{J}_1(e) = \{x | x \cdot e = x\}$ is a subalgebra. If e and f are orthogonal idempotent elements of $\mathfrak{J}$ then we say that *eabdf* are *connected* if there exists an element a_{12} in $\mathfrak{J}_1(e + f)$ satisfying $a_{12} \cdot e = \frac{1}{2}a_{12} = a_{12} \cdot f$ which has an inverse in $\mathfrak{J}_1(e + f)$. We can now give the characterization of Jordan matrix rings in the following

Coordinatization Theorem. Let $\mathfrak{J}$ be a Jordan algebra satisfying the following conditions: has an identity $1 = \sum_1^n e_i$ where the e_i are non-zero orthogonal idempotents such that e_i, $i > 1$, is connected with e_i and $n \geq 3$. Then $\mathfrak{J}$ is isomorphic to a Jordan matrix algebra $\mathscr{H}(\mathscr{V}_n, \mathfrak{J}_a)$ and conversely.

Now let $\mathfrak{J} = \mathscr{H}(\mathscr{V}_n, \mathfrak{J}_a)$ with $n \geq 4$. Then we have stated before that $\mathscr{V}$ must be associative and so $\mathfrak{J}$ is special. Let $\mathfrak{M}$ be any unital bimodule for $\mathfrak{J}$ and $\mathscr{E} = \mathfrak{J} \oplus \mathfrak{M}$. Then the e_i satisfying the hypotheses of the above theorem for $\mathfrak{J}$ also satisfy these for $\mathscr{E}$. Hence we see that $\mathscr{E} = \mathscr{H}(\mathfrak{J}_n, \mathfrak{J}_b)$ where $n \geq 4$. Hence $\mathscr{E}$ is special and so we have a homomorphism of $\mathscr{U}_1''(\mathfrak{J})$ into $\mathrm{Hom}_\Phi(\mathfrak{M}, \mathfrak{M})$ such that $a_1'' \to R_a|\mathfrak{M}$ the restriction of R_a, $a \in \mathfrak{J}$ acting in $\mathscr{E}$ to $\mathfrak{M}$. Now consider $\mathscr{U}_1(\mathfrak{J})$ and take a unital right module $\mathfrak{M}$ for this such that the compouding representation is faithful. Then this gives a unital bimodule for $\mathfrak{J}$ and if we apply the result just noted we see that we have a homomorphism of $\mathscr{U}_1''(\mathfrak{J})$ into $\mathscr{U}_1(\mathfrak{J})$ such that $a_1'' \to a_1$, $a \in \mathfrak{J}$. This implies that $\mathfrak{J}$ is strongly special if $n \geq 4$.

The method just indicated of passing from $\mathfrak{J}\mathscr{H}(\mathscr{V}_n, \mathfrak{J}_a)$ to $\mathscr{E} = \mathfrak{J} \oplus \mathfrak{M}$ also gives a category isomorphism of the category of unital Jordan bimodules with $n \geq 4$ to the category unital of associative bimodules with involution for the associative algebra $\mathscr{V}$ with its involution. Here if $\mathscr{V}$ is an algebra with involution $d \to \bar{d}$, a bimodule for $(\mathscr{V}, -)$, that is, $\mathscr{V}$ with its involution $d \to \bar{d}$, is a bimodule $\mathfrak{M}$ for $\mathscr{V}$ in the usual sense together with a linear mapping $u \to \bar{u}$ of period two in $\mathfrak{M}$ such that

$$\overline{a\,u} = \bar{u}\,\bar{a}, \qquad \overline{u\,a} = \bar{a}\,\bar{u}, \tag{6}$$

$a \in \mathscr{V}$, $u \in \mathfrak{M}$. The case $n = 3$ be handled also. In this case one obtains a category isomorphism of the category of unital Jordan bimodules for $\mathfrak{J}$ to a certain category of unital alternative bimodules with involution for the coefficient algebra $\mathscr{V}$. Both of these results can be stated in terms of appropriate associative algebras.

We note formally that if $\mathscr{V}$ is a composition algebra so that $\mathscr{V}$ is either Φ, a two dimensional semi-simple commutative associative algebra with involution $\neq 1$, a quaternion algebra or a Cayley algebra with standard involution, then one can determine completely the alternative bimodules with involution for $\mathscr{V}$. This gives complete results for the Jordan problem in the case of interest for the finite dimensional theory.

Bibliography

1. G. Birkhoff and P. Whitman, Representation of Jordan and Lie algebras, Trans. Amer. Math. Soc., 65 (1949), 116–136.
2. H. Cartan and S. Eilenberg, Homological Algebra, Princeton.
3. C. Chevalley, The Algebraic Theory of Spinors, New York, 1954.
4. P. Cohn, Universal Algebra, New York 1965.
5. S. Eilenberg, Extensions of general algebras, Annales de la Soc. Polonaise de Math. *21* (1948), 125–134.
6. F. D. Jacobson and N. Jacobson, Classification and representation of semi-simple Jordan algebras, Trans. Amer. Math. Soc., *65* (1949), 141–169.
7. N. Jacobson, General representation theory of Jordan algebras, Trans. Amer. Math. Soc., *70* (1951), 509–530.
8. N. Jacobson, Structure of alternative and Jordan bimodules, Osaka Math. Jour., *6* (1954); 1–71.
9. N. Jacobson, A coordinatization theorem for Jordan algebras, Proc. Nat. Acad. Sci. U.S.A. *48* (1962), 1154–1160.
10. N. Jacobson and C. E. Rickart, Homomorphisms of Jordan rings of self adjoint elements. Trans. Amer. Math. Soc., *72* (1952), 310–322.
11. J. Knopfmacher, Universal envelopes for non-associative algebras, Quart J. Math. Oxford Sci. (2) *13* (1962), 264–282.
12. W. S. Martindale, Jordan homomorphisms of the symmetric elements of a ring with involution, to appear.
13. K. Mc Crimmon, Bimodules for composition algebras, to appear.
14. R. D. Schafer, Representations of alternative algebras, Trans. Amer. Math. Soc., *72* (1952), 1–17.
15. R. D. Schafer, Structure and representation of non-associative algebras, Bull. Amer. Math. Soc., *61* (1955), 469–484.
16. R. D. Schafer, An introduction to non-associative algebras, Oklahoma State University, Stillwater, Okla., 1961.

Reprinted from
Corso tenuto a Varenna (Como) dal 23al 31 agost, 1965.

REPRINTED FROM

SOME RECENT ADVAN. BASIC SCIENCES, VOLUME 1

ACADEMIC PRESS INC., NEW YORK

FORMS OF ALGEBRAS[†]

N. Jacobson

DEPARTMENT OF MATHEMATICS

YALE UNIVERSITY, NEW HAVEN, CONNECTICUT

INTRODUCTION

It is well known that the problem of classifying simple Lie algebras of types A, B, C, and D, except D_4, is equivalent to that of classifying finite dimensional simple associative algebras with involution (Landherr [1, 2], Jacobson [3–6], Chapter X of "Lie Algebras," and Weil [7]). Also, the problem of classifying finite dimensional simple Jordan algebras other than those defined by quadratic forms and the exceptional ones is equivalent to the same problem for simple associative algebras with involution (Kalisch [8] and Jacobson and Jacobson [9]). The problems of classifying Lie algebras of types G_2 and F_4 are equivalent, respectively, to the same problems for finite dimensional exceptional alternative and Jordan algebras (Jacobson [10], Tomber [11, 12], and Barnes [13]). In this paper, we shall give a uniform way of deriving these results from a general theorem on classes of forms of a vector space.

Our starting point is the consideration of an arbitrary vector space $\tilde{\mathfrak{M}}$ over P without any further structure. If Φ is a subfield of P, then a Φ-form $\mathfrak{M}$ of $\tilde{\mathfrak{M}}$ is a Φ-subspace such that the canonical mapping of $P \otimes_\Phi \mathfrak{M}$ into $\tilde{\mathfrak{M}}$ (sending $\rho \otimes x \rightarrow \rho x$, $\rho \in P$, $x \in \mathfrak{M}$) is an isomorphism. Let $[P : \Phi] < \infty$, $\mathfrak{E} = \mathrm{Hom}_\Phi(P, P)$, and identify P with the set of multiplications $\xi \rightarrow \rho\xi$, $\xi, \rho \in P$, in P. Also consider $\tilde{\mathfrak{M}}$ as right P-module by defining $\tilde{x}\rho = \rho\tilde{x}$, $\rho \in P$, $\tilde{x} \in \tilde{\mathfrak{M}}$. Then any Φ-form $\mathfrak{M}$ of $\tilde{\mathfrak{M}}$ determines an extension ϵ of the P-module structure on $\tilde{\mathfrak{M}}$ to an $\mathfrak{E}$-module structure

† This research has been supported by the Air Force Office of Scientific Research under grant AF-AFOSR-402-64.

on $\tilde{\mathfrak{M}}$, and we have a bijection of the set of Φ-forms of $\tilde{\mathfrak{M}}$ with the set of such extensions. We use this general result to obtain known special ones for the cases P/Φ Galois and P/Φ purely inseparable of exponent one. Also, we extend the correspondence for P/Φ finite dimensional Galois to the infinite Galois case in the case $[\tilde{\mathfrak{M}} : P] < \infty$. The correspondence in the Galois case can be modified slightly to give a bijection of the set of Φ-forms of $\tilde{\mathfrak{M}}$ into the set $Z^1(G, \text{Aut } \tilde{\mathfrak{M}}/P)$ of 1-cocycles of the Galois group G of P/Φ into the group Aut $\tilde{\mathfrak{M}}/P$ of 1-1 linear transformations of $\tilde{\mathfrak{M}}$ into itself, where G acts on Aut $\tilde{\mathfrak{M}}/P$ in a manner determined by choosing a particular Φ-form $\mathfrak{M}_0$ of $\tilde{\mathfrak{M}}$. An analogous result holds for P/Φ finite dimensional purely inseparable of exponent one. Here G is replaced by $\mathfrak{D} = \text{Der } P/\Phi$ and the set of derivatives of P/Φ and Aut $\tilde{\mathfrak{M}}/P$ by $\text{Hom}_P(\tilde{\mathfrak{M}}, \tilde{\mathfrak{M}})$.

To obtain the applications to algebras, we introduce the notion of a descent class of forms on a finite dimensional vector spece $\tilde{\mathfrak{M}}/P$. This consists of a set $\mathscr{C}$ of forms of $\tilde{\mathfrak{M}}$, a set $\mathscr{S}$ of 1-1 semilinear mappings between these, and a set $\mathscr{D}$ of semiderivations of the forms satisfying certain axioms. If $\tilde{\mathfrak{A}}/P$ is an algebra in a very general sense (cf. Sec. III), then the set $\mathscr{C}$ of algebra forms, $\mathscr{S}$ the set of semiisomorphisms between these, and $\mathscr{D}$ the set of semiderivations of the algebra forms satisfy our axioms. Also, we can define a descent class by means of a homogeneous polynomial f on a vector space $\tilde{\mathfrak{M}}/P$ (Sec. IV). If $(\tilde{\mathfrak{M}}, \mathscr{C}, \mathscr{S}, \mathscr{D})$ is a descent class and $\overline{\mathfrak{M}}/\Lambda \in \mathscr{C}$, then the correspondence indicated before between the set of Φ-forms of the vector space $\overline{\mathfrak{M}}/\Lambda$ and the set of 1-cocycles of the Galois group (derivation set) of Λ/Φ if Λ/Φ is Galois (finite dimensional purely inseparable of exponent one) gives a similar correspondence between the set of Φ-forms contained in $\mathscr{C}$ and the 1-cocycles whose values belong to $\mathscr{S}(\mathscr{D})$. We define a notion of a derived class $(\tilde{\mathfrak{M}}', \mathscr{C}', \mathscr{S}', \mathscr{D}')$ of a descent class $(\tilde{\mathfrak{M}}, \mathscr{C}, \mathscr{S}, \mathscr{D})$. Here we have canonical mappings of $\mathscr{C}$ into $\mathscr{C}'$, $\mathscr{S}$ into $\mathscr{S}'$, and $\mathscr{D}$ into $\mathscr{D}'$ satisfying certain conditions of a functorial character. For the applications to classification of algebras, one is interested in the situation in which P is algebraically closed and Φ is a subfield such that P/Φ is algebraic, and one seeks conditions that the canonical mapping of the set $\mathscr{C}_\Phi$ of Φ-forms of $\tilde{\mathfrak{M}}$ into the set $\mathscr{C}_\Phi'$ of Φ-forms of $\tilde{\mathfrak{M}}'$ is bijective, and the same thing holds for the sets of isomorphisms between these forms. Our main result (Theorem 6) gives some simple conditions for this. These can be easily applied to obtain the type of results which we have indicated at the beginning.

I. FORMS OF A VECTOR SPACE

Let $\tilde{\mathfrak{M}}$ be a (not necessarily finite dimensional) vector space over a field P, and let Φ be a subfield of P. By a *form* or, more precisely, a *Φ-form* of $\tilde{\mathfrak{M}}$, we shall mean a Φ-subspace $\mathfrak{M}$ of $\tilde{\mathfrak{M}}$ such that 1) the P-subspace $P\mathfrak{M}$ spanned by $\mathfrak{M}$ is $\tilde{\mathfrak{M}}$, and 2) $\mathfrak{M}$ and P are linearly disjoint over Φ in the sense that, if $x_1, x_2, \ldots, x_r \in \mathfrak{M}$ are Φ-independent, then they are necessarily P-independent. These conditions are equivalent to the single condition that the canonical homomorphism $\Sigma\rho_j \otimes x_j \to \Sigma\rho_j x_j$ of $P \otimes_\Phi \mathfrak{M}$ into $\tilde{\mathfrak{M}}$ is bijective. Another equivalent condition is that $\mathfrak{M}$ contains a Φ-basis which is also a P-basis for $\tilde{\mathfrak{M}}$. If this holds, then every Φ-basis of $\mathfrak{M}$ is a P-basis. If $\mathfrak{M}$ is a Φ-form of $\tilde{\mathfrak{M}}$ and Σ is a subfield of P containing Φ, then the Σ-subspace $\Sigma\mathfrak{M}$ is a Σ-form of $\tilde{\mathfrak{M}}$. Conversely, if $\bar{\mathfrak{M}}$ is a Σ-form of $\tilde{\mathfrak{M}}$ and $\mathfrak{M}$ is a Φ-form of $\bar{\mathfrak{M}}$, then $\mathfrak{M}$ is a Φ-form of $\tilde{\mathfrak{M}}$.

Now assume $[P : \Phi] < \infty$ and let $\mathfrak{E} = \mathrm{Hom}_\Phi(P, P)$, the ring of linear transformations of P as vector space over Φ. This contains the set of multiplications $\xi \to \xi\rho$, $\xi, \rho \in P$, which is a subring of $\mathfrak{E}$ which can be identified with P. Moreover, if we consider $\mathfrak{E}$ as right vector space over P relative to the product $E\rho$ that coincides with the ring product of $E \in \mathfrak{E}$ with $\rho \in P$, then the right dimensionality $[\mathfrak{E} : P]_R = [P : \Phi]$. In fact, if $(\rho_1, \rho_2, \ldots, \rho_n)$ is a basis for P/Φ and we define $E_i \in \mathfrak{E}$ by $\rho_j E_i = \delta_{ij}$, then $(E_1, E_2, \ldots, E_n)$ is a right basis for $\mathfrak{E}/P$ (Jacobson, "Lectures in Abstract Algebra," Vol. III, p. 20).

In dealing with Φ-forms of $\tilde{\mathfrak{M}}$, we shall find it convenient to consider $\tilde{\mathfrak{M}}$ as right vector space over P by defining $\tilde{x}\rho = \rho\tilde{x}$, $\rho \in P$, $\tilde{x} \in \tilde{\mathfrak{M}}$. Let $\mathfrak{M}$ be a Φ-form of $\tilde{\mathfrak{M}}$. Then we can use $\mathfrak{M}$ to define a right $\mathfrak{E}$-module structure on $\tilde{\mathfrak{M}}$ in the following way. If $E \in \mathfrak{E}$ and $\tilde{x} \in \tilde{\mathfrak{M}}$, we write $\tilde{x} = \Sigma\rho_i x_i$, $x_i \in \mathfrak{M}$, $\rho_i \in P$, and we define

$$(\Sigma\rho_i x_i)E = \Sigma(\rho_i E)x_i \,. \tag{1}$$

By properties of tensor products, this is well defined and gives a module structure on $\tilde{\mathfrak{M}}$. Moreover, if we take $E = \rho \in P$, then $(\Sigma\rho_i x_i)E = \Sigma(\rho_i\rho)x_i = \rho(\Sigma\rho_i x_i) = \rho\tilde{x}$. Hence the $\mathfrak{E}$-module structure is an extension of the given (right) P-module structure on $\tilde{\mathfrak{M}}$. It is clear from the definition that, if $x \in \mathfrak{M}$, then $(\rho x)E = (\rho E)x$, $\rho \in P$, $E \in \mathfrak{E}$. Conversely, let x be any element of $\tilde{\mathfrak{M}}$ satisfying this condition. Write $x = \Sigma\rho_i x_i$, where the $x_i \in \mathfrak{M}$ are Φ-independent and, hence, also P-independent. Then $(\rho x)E = (\rho E)x$ gives $\Sigma(\rho\rho_i)Ex_i = \Sigma(\rho E)\rho_i x_i$. Hence $(\rho\rho_i)E = (\rho E)\rho_i$, and so the multiplication by ρ_i commutes with every $E \in \mathrm{Hom}_\Phi(P, P)$. It follows that $\rho_i = \varphi_i \in \Phi$ and, hence,

$x = \Sigma\varphi_i x_i \in \mathfrak{M}$. We have therefore shown that we can recover the form $\mathfrak{M}$ from the $\mathfrak{E}$-module structure on $\tilde{\mathfrak{M}}$ as

$$\mathfrak{M} = \{x \in \tilde{\mathfrak{M}} \mid (\rho x)E = (\rho E)x,\ \rho \in P,\ E \in \mathfrak{E}\}. \tag{2}$$

Our result gives half of a correspondence between Φ-forms of $\tilde{\mathfrak{M}}$ and extensions to $\mathfrak{E}$ of the right P-module structure on $\tilde{\mathfrak{M}}$ which we establish in the following.

Theorem 1. *Let $\tilde{\mathfrak{M}}$ be an arbitrary vector space over a field P, Φ a subfield of P such that $[P : \Phi]$ is finite, and let $\mathfrak{E} = \mathrm{Hom}_\Phi(P, P)$. Also consider $\tilde{\mathfrak{M}}$ as right P-module by means of $\tilde{x}\rho = \rho\tilde{x}$. Then if $\mathfrak{M}$ is a Φ-form of $\tilde{\mathfrak{M}}$, formula* (1) *gives an extension of the right P-module structure to a right $\mathfrak{E}$-module structure on $\tilde{\mathfrak{M}}$. Conversely, let ϵ denote any extension of the right P-module structure on $\tilde{\mathfrak{M}}$ to a right $\mathfrak{E}$-module structure on $\tilde{\mathfrak{M}}$, and let $\mathfrak{M}$ be the subset of $\tilde{\mathfrak{M}}$ defined by* (2); *then $\mathfrak{M}$ is a Φ-form of $\tilde{\mathfrak{M}}$. The correspondences between Φ-forms and extensions ϵ are inverses of each other.*

Proof. It remains to show that, if ϵ is a given extension of module structure and $\mathfrak{M}$ is given by (2), then $\mathfrak{M}$ is a Φ-form, and the extension of module structure defined by $\mathfrak{M}$ is ϵ. For this we observe that $\mathfrak{E}$ is a simple Artinian ring. Hence $\tilde{\mathfrak{M}}$ as right $\mathfrak{E}$-module is a direct sum $\Sigma \oplus \mathfrak{M}_\iota$ of irreducible submodules isomorphic to the irreducible right module P for $\mathfrak{E}$. Let x_ι be the image of 1 under an $\mathfrak{E}$-isomorphism of P onto $\mathfrak{M}_\iota$. Since $1P = P$, $Px_\iota = x_\iota P = \mathfrak{M}_\iota$. Since Φ is the subset of elements of P such that $(\rho\varphi)E = (\rho E)\varphi$, $\rho \in P$, $E \in \mathfrak{E}$, and the image of $\Phi = 1\Phi$ under our $\mathfrak{E}$-isomorphism is $x_\iota\Phi = \Phi x_\iota$, it is clear that Φx_ι is the subset of $\mathfrak{M}_\iota$ of elements x such that $(\rho x)E = (\rho E)x$, $\rho \in P$, $E \in \mathfrak{E}$. Hence $\mathfrak{M} \equiv \Sigma\Phi x_\iota = \{x \in \tilde{\mathfrak{M}} \mid (\rho x)E = (\rho E)x,\ \rho \in P,\ E \in \mathfrak{E}\}$. The set $\{x_\iota\}$ is a P-basis for $\tilde{\mathfrak{M}}$ and a Φ-basis for $\mathfrak{M}$. Hence $\mathfrak{M}$ is a Φ-form of $\tilde{\mathfrak{M}}$. Since $(\Sigma\rho_\iota x_\iota)E = \Sigma(\rho_\iota E)x_\iota$, it is clear that the $\mathfrak{E}$-module structure defined by the form $\mathfrak{M}$ is the given one. Hence the proof is complete.

Now an $\mathfrak{E}$-module structure on $\tilde{\mathfrak{M}}$ determines a homomorphism α of $\mathfrak{E}$ into Hom $(\tilde{\mathfrak{M}}, \tilde{\mathfrak{M}})$, the ring of endomorphisms of the additive group $(\tilde{\mathfrak{M}}, +)$, and conversely. The relation is that, if $\tilde{x} \in \tilde{\mathfrak{M}}$ and $E \in \mathfrak{E}$, then $\tilde{x}E = \tilde{x}\alpha(E)$. If the given $\mathfrak{E}$-module structure is an extension of the P-module structure, then we have $\alpha(\rho) = \rho$, where the right-hand side denotes the mapping $\tilde{x} \to \tilde{x}\rho = \rho\tilde{x}$ in $\tilde{\mathfrak{M}}$. Conversely, this condition is sufficient. Hence Theorem 1 provides a bijection of the set $\{\mathfrak{M}\}$ of Φ-forms of $\tilde{\mathfrak{M}}$ with the set of homomorphisms α of $\mathfrak{E}$ into Hom $(\tilde{\mathfrak{M}}, \tilde{\mathfrak{M}})$ such that $\alpha(\rho) = \rho$ for $\rho \in P \subseteq \mathfrak{E}$. The determination of $\mathfrak{M}$ from α and α from $\mathfrak{M}$ is immediate from Theorem 1.

We now look at the homomorphisms α in terms of a right P-basis of $\mathfrak{E}/P$. Let $(L_1, L_2, \ldots, L_n)$ be such a basis. Then if $\rho \in P\ (\subseteq \mathfrak{E})$, we have

$$\rho L_i = \sum_{j=1}^{n} L_j \tau_{ji}(\rho) \qquad (i = 1, \ldots, n), \tag{3}$$

where $\tau_{ji}(\rho) \in P$. Also,

$$L_i L_j = \sum_k L_k \mu_{kij}, \qquad \mu_{kij} \in P, \tag{4}$$

and we can write

$$1 = \sum_i L_i \delta_i \tag{5}$$

for suitable $\delta_i \in P$. Now let α be a homomorphism of $\mathfrak{E}$ into Hom$(\mathfrak{M}, \mathfrak{M})$ such that $\alpha(\rho) = \rho$, $\rho \in P$. Then (3–5) give the following conditions on α:

$$\rho\alpha(L_i) = \sum_j \alpha(L_j)\tau_{ji}(\rho) \tag{6}$$

$$\alpha(L_i)\alpha(L_j) = \sum_k \alpha(L_k)\mu_{kij} \tag{7}$$

$$\sum_i \alpha L(_i)\delta_i = 1. \tag{8}$$

If $L_1 = 1$, then this is just the condition $\alpha(1) = 1$. Conversely, if we have a mapping $L_i \to \alpha(L_i)$ of the basis $(L_1, \ldots, L_n)$ into Hom $(\mathfrak{M}, \mathfrak{M})$ satisfying (6–8), then we can extend this to $\mathfrak{E}$ by defining

$$\alpha(L) = \sum_i \alpha(L_i)\rho_i, \qquad \rho_i \in P, \tag{9}$$

and it is easy to check that α is a homomorphism of $\mathfrak{E}$ such that $\alpha(\rho) = \rho$, $\rho \in P$.

We shall now apply this to obtain two special results on Φ-forms which will be basic in the sequel. First, we assume P/Φ is Galois with Galois group $G = \{1, s, t, \ldots, u\}$, and we write the image of $\rho \in P$ under $s \in G$ as ρ^s (rather than ρs). Since s is an automorphism of P, condition (3) in the present case is $\rho s = s\rho^s$. Also, (4) is just the multiplication table for G. Hence (6) and (7) become $\rho\alpha(s) = \alpha(s)\rho^s$ and $\alpha(st) = \alpha(s)\alpha(t)$. Corresponding to (8), we have $\alpha(1) = 1$. Hence we see that $\alpha(s)$ is an s-semilinear mapping in $\mathfrak{M}/P$, and $s \to \alpha(s)$ is a homomorphism of G into the group of 1-1 semilinear transformations of $\mathfrak{M}/P$ onto itself. We denote the latter group as s-Aut $\mathfrak{M}/P$. Also, we shall call a homomorphism of the Galois group of P/Φ into s-Aut $\tilde{\mathfrak{M}}/P$ such that, for every $s \in G$, $\alpha(s)$ is s-semilinear, a *precocycle* of G into s-Aut $\tilde{\mathfrak{M}}/P$. We note also that, if $\tilde{\mathfrak{M}}$ is a Φ-form of $\tilde{\mathfrak{M}}$ and $s \in G$, then we have an endomorphism $\alpha(s)$ of the additive group $\tilde{\mathfrak{M}}$ such that $(\Sigma\rho_i x_i)\alpha(s) = \Sigma\rho_i{}^s x_i$, $\rho_i \in P$, $x_i \in \mathfrak{M}$. It is clear that $\alpha(s)$ is s-semilinear and is the identity on $\mathfrak{M}$. Moreover, these properties characterize $\alpha(s)$ and $\alpha(s) \in s$-Aut $\tilde{\mathfrak{M}}/P$. We can now deduce from Theorem 1 the following.

Corollary 1. *Let $\tilde{\mathfrak{M}}$ be an arbitrary vector space over P, and Φ a subfield of P such that P/Φ is finite dimensional Galois. Then we have a* 1-1 *correspondence between the set of Φ-forms $\mathfrak{M}$ of $\tilde{\mathfrak{M}}$ and the set of precocycles α of the Galois group G of P/Φ into* s-Aut $\tilde{\mathfrak{M}}/P$. *For given $\mathfrak{M}$, $\alpha(s)$ is the s-semilinear transformation of $\tilde{\mathfrak{M}}/P$ which is the identity on $\mathfrak{M}$, and for given α the corresponding Φ-form is $\mathfrak{M} = \{x \mid x\alpha(s) = x,\ s \in G\}$.*

Proof. Everything except the last statement is clear from Theorem 1 and the foregoing discussion. Now let $\mathfrak{M}$ be a given Φ-form and let $E \in \mathfrak{E} = \mathrm{Hom}_\Phi(P, P)$. Then by Theorem 1 the homomorphism α of $\mathfrak{E}$ into Hom $(\tilde{\mathfrak{M}}, \tilde{\mathfrak{M}})$ defined by $\mathfrak{M}$ satisfies $(\Sigma\rho_i x_i)\alpha(E) = \Sigma(\rho_i E)x_i$, $\rho_i \in P$, $x_i \in \mathfrak{M}$. In particular, if $s \in G$, then $(\Sigma\rho_i x_i)\alpha(s) = \Sigma\rho_i{}^s x_i$, which shows that $\alpha(s)$ is the s-semilinear transformation of $\tilde{\mathfrak{M}}/P$, which is the identity on $\mathfrak{M}$. Conversely, let α be a precocycle of G into s-Aut $\tilde{\mathfrak{M}}/P$, and extend α to a homomorphism of $\mathfrak{E}$ into Hom $(\tilde{\mathfrak{M}}, \tilde{\mathfrak{M}})$ so that $\alpha(\rho) = \rho$. Then the corresponding Φ-form is $\mathfrak{M} = \{x \in \tilde{\mathfrak{M}} \mid (\rho x)\alpha(E) = (\rho E)x,\ \rho \in P,\ x \in \mathfrak{M}\}$. One sees directly that this coincides with the set of fixed points under the $\alpha(s)$, $s \in G$. Hence $\mathfrak{M}$ is as given in the statement of the corollary.

We remark that Corollary 1 is well known. A somewhat more elementary proof of this result is given in the author's "Lie Algebras," p. 300.

We consider next the situation in which P/Φ is purely inseparable of exponent one. Here the role of the Galois group in the Galois case is taken by the set $\mathfrak{D} = \mathrm{Der}\, P/\Phi$ of derivations d of P/Φ. As for automorphisms, we denote the image of ρ under d as ρ^d. The conditions on d are that d is an endomorphism of the additive group of P and $\rho d = d\rho + \rho^d$, $\rho \in P$, where ρ and ρ^d in this equation denote the multiplications $\xi \to \rho\xi = \xi\rho$ and $\xi \to \rho^d\xi = \xi\rho^d$ in P. Also $d \in \mathfrak{E} = \mathrm{Hom}_\Phi(P, P)$. Hence, if α is a homomorphism of $\mathfrak{E}$ into Hom $(\tilde{\mathfrak{M}}, \tilde{\mathfrak{M}})$ such that $\alpha(\rho) = \rho$, $\rho \in P$, then we have $\rho\alpha(d) = \alpha(d)\rho + \rho^d$. This is equivalent to $(\rho\tilde{x})\alpha(d) = \rho(\tilde{x}\alpha(d)) + \rho^d\tilde{x}$, $\rho \in P$, $x \in \tilde{\mathfrak{M}}$. We shall call an endomorphism of $\tilde{\mathfrak{M}}$ having this property a d-semiderivation of $\tilde{\mathfrak{M}}/P$.[1]

[1] These are usually called d-differential transformations. However, we propose to use the term d-semiderivation to stress the analogy with s-semilinear transformation.

We recall that $\mathfrak{D} = \text{Der } P/\Phi$ is a right vector space over P (subspace of $\mathfrak{E}/P$), and $\mathfrak{D}$ is a restricted Lie ring (Jacobson, "Lectures in Abstract Algebra," Vol. III, p. 186). Moreover, $\Phi = \{\varphi \in P \mid \varphi^d = 0, d \in \mathfrak{D}\}$. We shall now denote the set of semiderivations of $\tilde{\mathfrak{M}}/P$ by s-Der $\tilde{\mathfrak{M}}/P$. One sees easily that, if D is a d-semiderivation of $\tilde{\mathfrak{M}}/P$ and E is an e-semiderivation of $\tilde{\mathfrak{M}}/P$, then $D + E$ is a $d + e$-semiderivation, $[D, E]$ is a $[d, e]$-semiderivation, and D^p is a d^p-semiderivation for p the characteristic of P. Also, if $\rho \in P$, then $D\rho$ is a $d\rho$-semiderivation. It follows that s-Der $\tilde{\mathfrak{M}}/P$ is a subspace of Hom $(\tilde{\mathfrak{M}}, \tilde{\mathfrak{M}})$ as right vector space over P closed under Lie commutation and pth powers. We shall now define a *precocycle* of $\mathfrak{D} = \text{Der } P/\Phi$ into s-Der $\tilde{\mathfrak{M}}/P$ as a mapping α of $\mathfrak{D}$ into s-Der $\tilde{\mathfrak{M}}/P$ such that 1) $\alpha(d)$ is a d-semiderivation of $\tilde{\mathfrak{M}}/P$, 2) α is P-linear for the two right vector spaces, 3) $\alpha(d^p) = \alpha(d)^p$. We shall see that these conditions imply 4) $\alpha[de] = [\alpha(d), \alpha(e)]$, so that α is also a Lie homomorphism.

Let $\mathfrak{M}$ be a Φ-form, and let $d \in \mathfrak{D} \subseteq \mathfrak{E}$. Then we have a unique endomorphism $\alpha(d)$ of $\tilde{\mathfrak{M}}$ such that $(\Sigma\rho_i x_i)\alpha(d) = \Sigma\rho_i{}^d x_i$, $\rho_i \in P$, $x_i \in \mathfrak{M}$. It is immediate that $\alpha(d)$ is a d-semiderivation of $\tilde{\mathfrak{M}}/P$, which is characterized by the property that $x\alpha(d) = 0$ for all $x \in \mathfrak{M}$.

We can now state the following.

Corollary 2. *Let $\tilde{\mathfrak{M}}$ be an arbitrary vector space over a field P and assume P finite dimensional purely inseparable of exponent one over Φ (characteristic $p \neq 0$). Then we have a* 1-1 *correspondence between the set of Φ-forms $\mathfrak{M}$ of $\tilde{\mathfrak{M}}$ and the set of precocycles α of $\mathfrak{D} =$* Der P/Φ *into* s-Der $\tilde{\mathfrak{M}}/P$. *For a given $\mathfrak{M}$, $\alpha(d)$ is the d-semiderivation of $\tilde{\mathfrak{M}}/P$ which is* 0 *on $\mathfrak{M}$ and for given α, $\mathfrak{M} = \{x \in \tilde{\mathfrak{M}} \mid x\alpha(d) = 0, d \in \mathfrak{D}\}$.*

Proof. Suppose first that the Φ-form $\mathfrak{M}$ is given. For $d \in \mathfrak{D}$, let $\alpha(d)$ be the d-semiderivation of $\tilde{\mathfrak{M}}$ which annihilates $\mathfrak{M}$. Then it is straightforward to verify that $\alpha : d \rightarrow \alpha(d)$ is a precocycle of $\mathfrak{D}$ into s-Der $\tilde{\mathfrak{M}}/P$. Also, since Φ is the set of elements that are d-constants ($\varphi^d = 0$) for all $d \in \mathfrak{D}$, it follows that $\mathfrak{M}$ is the set of $\alpha(d)$-constants for all $d \in \mathfrak{D}$. Next let α be any precocycle of $\mathfrak{D}$ into s-Der $\tilde{\mathfrak{M}}/P$. We proceed to show that α can be extended to a homomorphism of $\mathfrak{E}$ into Hom $(\tilde{\mathfrak{M}}, \tilde{\mathfrak{M}})$ such that $\alpha(\rho) = \rho$, $\rho \in P$. We recall that there exists a derivation e in P/Φ such that $(e, e^p, e^{p^2}, \ldots, e^{p^{r-1}})$ is a right basis for $\mathfrak{D}/P$. Then $(1, e, e^2, \ldots, e^{p^r-1})$ will be a right basis for $\mathfrak{E}/P$, and the minimum equation for e as linear transformation in P/Φ has the form

$$e^{p^r} + e^{p^{r-1}}\beta_1 + \cdots + e\beta_r = 0 \tag{10}$$

where $\beta_i \in \Phi$ (Jacobson, "Lectures," Vol. III, Ex. 3, p. 185, Ex. 3, p. 190). The operator form $\rho e = e\rho + \rho^e$ of the derivation condition gives, by induction,

$$\rho e^i = \sum_{k=0}^{i} \binom{i}{k} e^k \rho^{e^{i-k}} \qquad (i = 1, 2, \ldots). \tag{11}$$

This corresponds to (3) for the general basis $(L_1, L_2, \ldots)$ for $\mathfrak{E}/P$. The multiplication table for $e^i e^j$ corresponding to (4) can be deduced from (10). Now let α be the P-linear mapping of $\mathfrak{E}/P$ into Hom $(\tilde{\mathfrak{M}}, \tilde{\mathfrak{M}})$ as right vector space over P such that $e^i \to \alpha(e)^i$, $i = 0, \ldots, p^r - 1$. Since $\alpha(e^{p^k}) = \alpha(e)^{p^k}$, $k = 0, 1, \ldots, r-1$, it is clear that α is an extension of the given mapping of $\mathfrak{I}$. We have $\alpha(1) = 1$ and $\rho\alpha(e) = \alpha(e)\rho + \rho^e$, since $\alpha(e)$ is an e-semiderivation. Also, we can iterate this to obtain by induction that

$$\rho\alpha(e^i) = \sum_{k=0}^{i} \binom{i}{k} \alpha(e)^k \rho^{e^{i-k}} \qquad (i = 1, 2, \ldots). \tag{12}$$

Since α is P-linear and commutes with pth powers, we can obtain from (10) that

$$\alpha(e)^{p^r} + \alpha(e)^{p^{r-1}}\beta_1 + \cdots + \alpha(e)\beta_r = 0. \tag{13}$$

Now (12) is (6) for the basis $(1, e, e^2, \ldots, e^{p^{r-1}})$, and we can deduce from (13) and $\alpha(e^i) = \alpha(e)^i$ that (7) holds for our basis. It follows that α is a homomorphism of $\mathfrak{E}$ into Hom $(\tilde{\mathfrak{M}}, \tilde{\mathfrak{M}})$ satisfying $\alpha(\rho) = \rho$. Hence, by Theorem 1, $\mathfrak{M} = \{x \in \tilde{\mathfrak{M}} \mid (\rho x)\alpha(E) = (\rho E)x,\ \rho \in P,\ E \in \mathfrak{E}\}$ is a Φ-form. One sees directly that $\mathfrak{M}$ can be specified also as the set of constants of every $\alpha(d)$, $d \in \mathfrak{D}$. Moreover, one sees also directly that $\alpha(d)$, is the d-semiderivation of $\tilde{\mathfrak{M}}/P$ which annihilates $\mathfrak{M}$. This completes the proof.

It is clear that, if $\mathfrak{M}$ is a Φ-form in the situation of Corollary 2 and $\alpha(d)$, $\alpha(d')$ are d and d' semiderivations in $\tilde{\mathfrak{M}}/P$ which are 0 in $\mathfrak{M}$, then $[\alpha(d), \alpha(d')]$ is the $[d, d']$-semiderivation in $\tilde{\mathfrak{M}}/P$ which is 0 in $\mathfrak{M}$. Hence we have $\alpha([d, d']) = [\alpha(d), \alpha(d')]$ for every $d, d' \in \mathfrak{I}$. This shows that any precocycle of $\mathfrak{D}$ into s-Der $\tilde{\mathfrak{M}}/P$ is necessarily a Lie homomorphism. This result is similar to a recent theorem of Gerstenhaber's [14] to the effect that the Lie structure is superfluous for the Galois theory of purely inseparable extensions of exponent one. We remark also that Corollary 2 in a less invariant form was proved by the author in [6] II, p. 118.

For the applications we have in mind, it will be useful to extend Corollary 1 to the case in which P/Φ is an arbitrary (possibly infinite) Galois extension. To compensate for this, we shall assume $\tilde{\mathfrak{M}}/P$ finite dimensional. We topologize the Galois group G and the ring of endomorphisms of $\tilde{\mathfrak{M}}$ by the so-called finite topology (Jacobson, "Lectures," Vol. II, p. 248). An equivalent topology for G is given by considering this group as an inverse limit of finite groups (Serre, "Cohomologie Galoisienne," I-1). At any rate, G and the group of units of the ring of endomorphisms of $\tilde{\mathfrak{M}}$ are topological groups relative to their topologies. Now let $\mathfrak{M}$ be a Φ-form of the space $\tilde{\mathfrak{M}}$. If $s \in G$ and $x = \Sigma \rho_j x_j$, $x_j \in \mathfrak{M}$, $\rho_j \in P$, then we define $\alpha(s)$ to be the s-semilinear transformation of $\tilde{\mathfrak{M}}$ which is 1 on $\mathfrak{M}$. As before, $\alpha(1) = 1$, $\alpha(st) = \alpha(s)\alpha(t)$, and $\mathfrak{M}$ is the set of common fixed points of the $\alpha(s)$. Also, $s \rightarrow \alpha(s)$ is a continuous mapping of G into the group of units of the endomorphism ring. For, let N be a neighborhood of 1 in the latter, given by a finite subset $\{x_1, x_2, \ldots, x_r\}$ of $\tilde{\mathfrak{M}}$ as $N = \{E \mid x_i E = x_i, i = 1, \ldots, r\}$. Let $(u_1, u_2, \ldots, u_m)$ be a basis for $\mathfrak{M}/\Phi$ and hence for $\tilde{\mathfrak{M}}/P$, and write $x_i = \Sigma \xi_{ij} u_j$, $\xi_{ij} \in P$. Then $x_i \alpha(s) = \Sigma \xi_{ij}{}^s u_j$, and the finite set of conditions $\xi_{ij}{}^s = \xi_{ij}$ defines a neighborhood M of G such that $\alpha(G) \subseteq N$. We shall now define a *precocycle* α *of* G *into* s-Aut $\tilde{\mathfrak{M}}/P$ as a continuous homomorphism of G into the group of semilinear transformations of $\tilde{\mathfrak{M}}$ such that $\alpha(s)$ is s-semilinear for $s \in G$. Then we have the following.

Corollary 3. *Let $\tilde{\mathfrak{M}}$ be a finite dimensional vector space over a field P which is Galois (not necessarily finite dimensional) over Φ. Then we have a* 1-1 *correspondence between the set of Φ-forms of $\tilde{\mathfrak{M}}$ and the set of precocycles of the Galois group G into* s-Aut $\tilde{\mathfrak{M}}/P$, *as in Corollary* 1.

Proof. The half of the correspondence beginning with $\mathfrak{M}$ is clear. Now let α be a precocycle of G into s-Aut $\tilde{\mathfrak{M}}/P$. Then $\alpha(G)$ is compact, since G is compact. If $\tilde{x}$ is a fixed element of $\tilde{\mathfrak{M}}$, we define $C_{\tilde{y}} = \{\alpha(s) \mid \tilde{x}\alpha(s) = \tilde{y}\}$. Then $C_{\tilde{y}}$ is an open (and closed) subset of $\alpha(G)$, and $\alpha(G) = \bigcup C_{\tilde{y}}$. Hence we have a finite subset $\{\tilde{y}_1, \tilde{y}_2, \ldots, \tilde{y}_r\}$ of $\tilde{\mathfrak{M}}$ such that $\alpha(G) = \bigcup C_{\tilde{y}_i}$. This means that $\tilde{x}\alpha(G)$ is finite for every $\tilde{x} \in \tilde{\mathfrak{M}}$. Now let $(u_1, u_2, \ldots, u_m)$ be a basis of $\tilde{\mathfrak{M}}/P$. If $s \in G$, we have $u_i\alpha(s) = \Sigma \mu_{ij}(s) u_j$, and the set $\{\mu_{ij}(s) \mid s \in G,\ i, j = 1, \ldots, m\}$ is finite. Hence this subset is contained in a finite dimensional Galois subfield Γ/Φ of P/Φ. We may assume also that Γ contains a finite subset F of P which will be specified later. It is clear that the Γ-subspace $\overline{\mathfrak{M}} = \Sigma \Gamma u_i$ is invariant under all of the $\alpha(s)$, and the set of restrictions $\overline{\alpha(s)}$ of the $\alpha(s)$ to $\overline{\mathfrak{M}}$ is finite. The mapping $s \rightarrow \overline{\alpha(s)}$ is a homomorphism of G whose kernel H is a closed invariant subgroup of finite index in G. If $\gamma \in \Gamma$ and $x \neq 0$ is in $\overline{\mathfrak{M}}$, then $(\gamma x)\alpha(t) = \gamma x$ for $t \in H$ gives $\gamma^t = \gamma$. Hence Γ is contained in the subfield $\Delta = I(H)$ of fixed points under the elements of H. Also, Δ/Φ is finite dimensional Galois, and $\mathfrak{M}' = \Sigma \Delta u_i$ is invariant under all of the $\alpha(s)$. We have the homomorphism $s \rightarrow \alpha'(s)$ of G, where $\alpha'(s)$ is the restriction of $\alpha(s)$ to $\mathfrak{M}'$. Let s' denote the restriction of s to Δ. Then $\alpha'(s)$ is s'-semilinear in $\mathfrak{M}'$. If $t' = 1$, $t \in H$, and so the restriction of $\alpha(t)$ to $\overline{\mathfrak{M}}$ is the identity. Since $\mathfrak{M}' = \Delta\overline{\mathfrak{M}}$, we have $\alpha'(t) = 1$. It follows that $s' \rightarrow \alpha'(s)$ is a homomorphism of the Galois group G' of Δ/Φ into Hom$(\mathfrak{M}', \mathfrak{M}')$ $\alpha'(1) = 1$, and $\alpha'(s')$ is s'-semilinear. It follows that α' is a precocycle of G' into s-Aut $\mathfrak{M}'/\Delta$. Hence, by Corollary 1, $\mathfrak{M}' = \Delta(\mathfrak{M}' \cap \mathfrak{M})$, where $\mathfrak{M} = \{x \in \tilde{\mathfrak{M}} \mid x\alpha(s) = x;\, s \in G\}$. Then $\mathfrak{M}'$ contains the u_i, and consequently, $P\mathfrak{M} = \tilde{\mathfrak{M}}$. Now let $x_1, x_2, \ldots, x_r$ be a subset of $\mathfrak{M}$ which is Φ-independent, and suppose that these elements are P-dependent. Let F now be specified as the set of coefficients for the expressions for the x's in terms of the basis (u_i), together with a set of γ_i such that $\Sigma\gamma_i x_i = 0$, γ_i not all 0. Then the $x_i \in \mathfrak{M}' \cap \mathfrak{M}$, and these are Δ-dependent. This contradicts Corollary 1. Hence the result is proved.

II. FORMS OF A VECTOR SPACE AND COHOMOLOGY

We shall now modify the correspondences given in Corollaries 1–3 of Sec. I to make them fit into the setting of homological algebra. We observe first that, if Φ is any subfield of the base field P of the given vector space $\tilde{\mathfrak{M}}$, then there exist Φ-forms $\mathfrak{M}$ of $\tilde{\mathfrak{M}}$. For, let $\{e_\alpha\}$ be a basis for $\tilde{\mathfrak{M}}/P$, and let $\mathfrak{M}$ be the set of Φ-linear combinations of the e_α. Then it is clear that $\mathfrak{M}$ is a Φ-form of $\tilde{\mathfrak{M}}$. Now let P/Φ be finite dimensional Galois, and let $\mathfrak{M}_0$ be any Φ-form. Consider the precocycle α_0 of the Galois group given by $\mathfrak{M}_0$ as in Corollary 1. Let Aut $\tilde{\mathfrak{M}}/P$ denote the group of 1-1 linear transformations of $\tilde{\mathfrak{M}}/P$ onto itself. If $\tau \in$ Aut $\tilde{\mathfrak{M}}/P$, then we define

$$\tau^s = \alpha_0(s)^{-1}\tau\alpha_0(s) \qquad (s \in G). \tag{14}$$

Since $\alpha_0(s)$ is s-semilinear, it is clear the $\tau^s \in$ Aut $\tilde{\mathfrak{M}}/P$. Moreover, we have

$$(\tau_1\tau_2)^s = \tau_1{}^s\tau_2{}^s, \qquad \tau^{st} = (\tau^s)^t \tag{15}$$

for $\tau_i, \tau \in$ Aut $\tilde{\mathfrak{M}}/P$, $s, t \in G$. Thus Aut $\tilde{\mathfrak{M}}/P$ is a G-operator group relative to the composition τ^s. Next let $\mathfrak{M}$ be any Φ-form of $\tilde{\mathfrak{M}}$, and let α be the precocycle of G determined by $\mathfrak{M}$. Then it is clear that

$$\eta(s) \equiv \alpha_0(s)^{-1}\alpha(s) \tag{16}$$

$\in$ Aut $\tilde{\mathfrak{M}}/P$. We have $\eta(st) = \alpha_0(st)^{-1}\alpha(st) = \alpha_0(t)^{-1}\alpha_0(s)^{-1}\alpha(s)\alpha(t) = \alpha_0(t)^{-1}\eta(s)\alpha(t) = \alpha_0(t)^{-1}\eta(s)\alpha_0(t)\alpha_0(t)^{-1}\alpha(t) = \eta(s)^t\eta(t)$. Thus

$$\eta(st) = \eta(s)^t\eta(t), \tag{17}$$

and so the mapping $\eta : s \to \eta(s)$ is a crossed homomorphism of G into Aut $\mathfrak{M}/P$ as G-operator group. We shall now call such a crossed homomorphism of G into Aut $\tilde{\mathfrak{M}}/P$ a 1-*cocycle* of G into Aut $\tilde{\mathfrak{M}}/P$ as G-group. Conversely, let η be any 1-cocycle of G into Aut $\tilde{\mathfrak{M}}/P$, and let $\alpha(s) = \alpha_0(s)\eta(s)$. Then $\alpha : s \to \alpha(s)$ satisfies the conditions of Corollary 1. For, $\alpha(s)$ is an s-semilinear automorphism of $\tilde{\mathfrak{M}}/P$, and $\alpha(st) = \alpha(s)\alpha(t)$ follows from the cocycle property and the definition of τ^s. Hence, if $\mathfrak{M} = \{x \mid x\alpha(s) = x\}$, then $\mathfrak{M}$ is a Φ-form. It is now clear that Corollary 1 has the following translation into cohomology language.

Theorem 2. *Let $\tilde{\mathfrak{M}}$ be an arbitrary vector space over P,* Aut $\tilde{\mathfrak{M}}/P$ *the group of* 1-1 *linear transformations of $\tilde{\mathfrak{M}}$ onto itself, Φ a subfield of P such that P/Φ is finite dimensional Galois, G the Galois group of P/Φ. Let $\mathfrak{M}_0$ be a particular Φ-form of $\mathfrak{M}$, α_0 the corresponding precocycle of G defined as in Corollary* 1 *to Theorem* 1, *and make* Aut $\tilde{\mathfrak{M}}/P$ *into a G-group by defining $\tau^s = \alpha_0(s)^{-1}\tau\alpha_0(s)$. Then we have a* 1-1 *correspondence between the set of Φ-forms of $\tilde{\mathfrak{M}}$ and the set of* 1-*cocycles of G into* Aut $\tilde{\mathfrak{M}}/P$. *If $\mathfrak{M}$ is a given Φ-form and $\alpha(s)$ is the s-semilinear mapping of $\tilde{\mathfrak{M}}$ which is the identity on $\mathfrak{M}$, then the corresponding* 1-*cocycle is η, where $\eta(s) = \alpha_0(s)^{-1}\alpha(s)$. On the other hand, if η is a given* 1-*cocycle, then the corresponding Φ-forms is $\mathfrak{M} = \{x \in \tilde{\mathfrak{M}} \mid x\alpha_0(s)\eta(s) = x\}$.*

A similar translation can be made for Corollary 3. Here $\mathfrak{M}/P$ is finite dimensional, and P/Φ is Galois but not necessarily finite dimensional. We make Aut $\mathfrak{M}/P$ a G-group by (14), and, if $s \to \alpha(s)$ is a continuous homomorphism of G into Aut $\tilde{\mathfrak{M}}/\Phi$, then we get the continuous crossed homomorphism $\eta : s \to \eta(s) = \alpha_0(s)^{-1}\alpha(s)$ of G into Aut $\tilde{\mathfrak{M}}/\Phi$. We shall call such a continuous crossed homomorphism a 1-*cocycle* of G into Aut $\tilde{\mathfrak{M}}/\Phi$. Then we have the following.

Theorem 3. *Theorem* 2 *holds also if $\tilde{\mathfrak{M}}/P$ is finite dimensional, P/Φ arbitrary Galois, provided it is understood that* 1-*cocycles are continuous.*

We consider next the homological form of Corollary 2 to Theorem 1. Let Φ be a subfield of P such that P/Φ is finite dimensional purely inseparable of exponent one, $\mathfrak{D} =$ Der P/Φ the set of derivations of P/Φ. Let $\mathfrak{M}_0$ be a particular Φ-form of $\tilde{\mathfrak{M}}$, α_0 the corresponding precocycle of $\mathfrak{D}$ into s-Der $\tilde{\mathfrak{M}}/P$ given by Corollary 2. Let $E \in \mathrm{Hom}_P(\tilde{\mathfrak{M}}, \tilde{\mathfrak{M}})$. Then it is clear that

$$E^d \equiv [E, \alpha_0(d)] \in \mathrm{Hom}_P(\tilde{\mathfrak{M}}, \tilde{\mathfrak{M}}). \tag{18}$$

The mapping $E \to E^d$ is a derivation of $\mathrm{Hom}_P(\tilde{\mathfrak{M}}, \tilde{\mathfrak{M}})$ as associative ring, that is, $E \to E^d$ is additive, and $(E_1E_2)^d = E_1E_2{}^d + E_1{}^dE_2$. If ρ is the mapping $\tilde{x} \to \rho\tilde{x}$ in $\tilde{\mathfrak{M}}$, then $\rho \in \mathrm{Hom}_P(\tilde{\mathfrak{M}}, \tilde{\mathfrak{M}})$, and

$[\rho, \alpha_0(d)]$ is the multiplication $\tilde{x} \to \rho^d \tilde{x}$. One checks directly that $E^{d_1+d_2} = E^{d_1} + E^{d_2}$, $E^d \rho = \rho E^d = E^d \rho$ and $E^{d^p} = [\cdots (E^d)^d \cdots]^d$, p-times. In view of this there is no inconsistency in defining $E^{d^k} = (E^{d^{k-1}})^d$ for all $k = 1, 2, \ldots$.

Now let $\mathfrak{M}$ be any Φ-form of $\tilde{\mathfrak{M}}$, α the corresponding precocycle of $\mathfrak{D}$ into s-Der $\tilde{\mathfrak{M}}/P$. Set

$$\lambda(d) = \alpha(d) - \alpha_0(d). \tag{19}$$

Then $\lambda(d) \in \mathrm{Hom}_P(\tilde{\mathfrak{M}}, \tilde{\mathfrak{M}})$, and the mapping $\lambda : d \to \lambda(d)$ is a P-linear mapping considering $\mathfrak{D}$ and $\mathrm{Hom}_P(\tilde{\mathfrak{M}}, \tilde{\mathfrak{M}})$ as right vector spaces over P ($E\rho = \rho E$ for the latter). In order to obtain the condition on λ corresponding to $\alpha(d^p) = \alpha(d)^p$, we require the following.

Lemma. *Let $\mathfrak{A}$ be an associative ring of characteristic $p \neq 0$, $\mathfrak{A}^-$ the Lie ring obtained by replacing the associative multiplication by $[ab] = ab - ba$, D a derivation in $\mathfrak{A}$. If $a \in \mathfrak{A}$, define*

$$a^{[1]} = a, \qquad a^{[k]} = a^{[k-1]}a + a^{[k-1]}D. \tag{20}$$

Then $a^{[p]} - a^p$ is in the subring of $\mathfrak{A}^-$ generated by a, aD, $aD^2, \ldots$. Moreover, if D is the inner derivation $x \to [xu]$, then

$$a^{[p]} = (a + u)^p - u^p. \tag{21}$$

Proof. Assume first that D is the inner derivation $x \to [xu]$. We have $a^{[1]} = a = (a + u) - u$, and one establishes, by induction, that

$$a^{[k]} = \sum_{i=0}^{k} (-1)^i \binom{k}{i} u^i (a + u)^{k-i}. \tag{22}$$

Since the characteristic is p, this gives (21) for $k = p$. Now, it is well known that

$$(a + u)^p = a^p + u^p + \sum_{i=1}^{p-1} s_i(a, u) \tag{23}$$

where $is_i(a, u)$ is the coefficient of t^{i-1} in

$$[\cdots[a, \overbrace{tu + a]tu + a] \cdots tu}^{p-1} + a] \tag{24}$$

(Jacobson, "Lie Algebras," p. 187). Now consider the subring $\mathfrak{M}$ of the Lie ring $\mathfrak{A}^-$ generated by u, a, $aD^2, \ldots$. Since $[aD^{k-1}, u] = aD^k$

it is clear that the subring $\mathfrak{L}$ of $\mathfrak{A}^-$ generated by a, aD, aD^2, ... is an ideal in $\mathfrak{M}$. It is clear from (24) that $s_i(a, b) \in \mathfrak{L}$. By (21) and (23),

$$a^{[p]} - a^p = (a + u)^p - u^p - a^p = \sum_{i=1}^{p-1} s_i(a, b) \in \mathfrak{L}.$$

This proves both statements of the lemma if D is inner. Now, if D is any derivation in $\mathfrak{A}$, then we can construct the ring $\mathfrak{A}[y, D]$ of differential polynomials in an indeterminate y with coefficients in $\mathfrak{A}$ such that $ay = ya + aD$. Then D extends to the inner derivation determined by y in $\mathfrak{A}[y, D]$. Then the result just proved shows that $a^{[p]} - a^p$ is in the Lie ring generated by a, aD, aD^2, ... This completes the proof.

Now suppose that P/Φ is finite dimensional purely inseparable of exponent one, $\mathfrak{D} = \text{Der } P/\Phi$, as before. Let $\mathfrak{A}$ be an associative algebra with an identity element 1 over P. Then $\mathfrak{A}$ contains $P = P1$ in its center. We shall say that $\mathfrak{A}$ is a *$\mathfrak{D}$-module* if a product $a^d \in \mathfrak{A}$ for $a \in \mathfrak{A}$, $d \in \mathfrak{D}$ is defined such that

(i) $(a + b)^d = a^d + b^d$
(ii) $(ab)^d = ab^d + a^d b$
(iii) ρ^d for $\rho \in P \subseteq \mathfrak{A}$ is the same as ρ^d defined originally in P.
(iv) $a^{d_1 + d_2} = a^{d_1} + a^{d_2}$
(v) $a^{d\rho} = \rho a^d = a^d \rho$
(vi) $a^{d^p} = (\cdots (a^d)^d) \cdots)^d$, p-times.

By i and ii, $a \to a^d$ is a derivation in $\mathfrak{A}$ as ring. We now define a *1-cocycle of $\mathfrak{D}$ in $\mathfrak{A}$* as a linear mapping λ of $\mathfrak{D}$ as right vector space over P into $\mathfrak{A}$ as right P-space such that $a\rho = \rho a$ such that if $\lambda^{[k]}$ is defined by

$$\lambda^{[1]}(d) = \lambda(d), \qquad \lambda^{[k]}(d) = \lambda^{[k-1]}(d)\lambda(d) + \lambda^{[k-1]}(d)^d \tag{25}$$

[cf. (20)], then

$$\lambda^{[p]}(d) = \lambda(d^p). \tag{26}$$

It is clear from the lemma that $\lambda^{[p]}(d) - \lambda(d)^p$ is in the Lie subring of $\mathfrak{A}^-$ generated by $\lambda(d)$, $\lambda(d)^d$, $\lambda(d)^{d^2}$, ...

We can now return to the consideration of $\lambda(d) = \alpha(d) - \alpha_0(d)$, where α and α_0 are precocycles of $\mathfrak{D}$ into s-Der $\tilde{\mathfrak{M}}/P$. As before, we define $E^d = [E, \alpha_0(d)]$, $E \in \text{Hom}_P(\tilde{\mathfrak{M}}, \tilde{\mathfrak{M}})$, $d \in P$. Then $\text{Hom}_P(\tilde{\mathfrak{M}}, \tilde{\mathfrak{M}})$ becomes a $\mathfrak{D}$-module relative to this product. Let $\mathfrak{A}$ be the ring of endomorphisms of $\tilde{\mathfrak{M}}$, and let D denote the inner derivation

$A \to [A, \alpha_0(d)]$ in $\mathfrak{A}$. Then, applying the lemma to $\mathfrak{A}$ and D, we see that, if $\lambda^{[p]}(d)$ is defined by (25), then

$$\begin{aligned}\lambda^{[p]}(d) &= [\lambda(d) + \alpha_0(d)]^p - \alpha_0(d)^p \\ &= \alpha(d)^p - \alpha_0(d)^p \\ &= \alpha(d^p) - \alpha_0(d^p) \\ &= \lambda(d^p).\end{aligned}$$

Hence λ is a 1-cocycle of $\mathfrak{D}$ in $\mathrm{Hom}_P(\tilde{\mathfrak{M}}, \tilde{\mathfrak{M}})$. The converse follows by retracing the steps. Thus $\lambda = \alpha - \alpha_0$ is a 1-cocycle if and only if α is a precocycle.

The foregoing considerations evidently yield the following cocycle version of Corollary 2 to Theorem 1.

Theorem 4. *Let $\tilde{\mathfrak{M}}$ be an arbitrary vector space over P, Φ a subfield of P such that P/Φ is finite dimensional purely inseparable of exponent one,* $\mathfrak{D} =$ Der P/Φ. *Let $\tilde{\mathfrak{M}}_0$ be a particular Φ-form of $\tilde{\mathfrak{M}}$, α_0 the corresponding precocycle of $\mathfrak{D}$ into s-*Der $\tilde{\mathfrak{M}}/P$ *given by Corollary 2, and define $E^d = [E, \alpha_0(d)]$ for $E \in \mathrm{Hom}_P(\tilde{\mathfrak{M}}, \tilde{\mathfrak{M}})$. Then this makes $\mathrm{Hom}_P(\tilde{\mathfrak{M}}, \tilde{\mathfrak{M}})$ a $\mathfrak{D}$-module, and we have a 1-1 correspondence between the set of Φ-forms of $\tilde{\mathfrak{M}}$ and the set of 1-cocycles of $\mathfrak{D}$ into $\mathrm{Hom}_P(\tilde{\mathfrak{M}}, \tilde{\mathfrak{M}})$. If $\tilde{\mathfrak{M}}$ is a Φ-form and $\alpha(d)$ is the d-semiderivation of $\tilde{\mathfrak{M}}/P$ which is 0 on $\mathfrak{M}$, then the corresponding 1-cocycle* is $\lambda = \alpha - \alpha_0$, *and, if λ is given, then $\mathfrak{M} = \{x \in \tilde{\mathfrak{M}} \mid x\lambda(d) = -x\alpha_0(d),\ d \in \mathfrak{D}\}$.*

Now let $\mathfrak{M}_1$ and $\mathfrak{M}_2$ be Φ-forms of $\tilde{\mathfrak{M}}/P$, Φ any subfield of P, and let τ be a 1-1 linear mapping of $\mathfrak{M}_1$ onto $\mathfrak{M}_2$. Then τ has a unique extension to an element $\tau \in \mathrm{Aut}\ \tilde{\mathfrak{M}}/P$. Suppose we are in the situation of Theorem 2 or 3, and let η_1 and η_2 be the 1-cocycles corresponding to $\mathfrak{M}_1$ and $\mathfrak{M}_2$ respectively. Then, if $\alpha_i(s) = \alpha_0(s)\eta_i(s)$, $\alpha_i(s)$ is s-semilinear, and $\mathfrak{M}_i = \{x_i \mid x_i\alpha_i(s) = x_i\ , s \in G\}$. On the other hand, $\tau^{-1}\alpha_1(s)\tau$ is s-semilinear, and this transformation is the identity on $\mathfrak{M}_2$. Hence we must have $\tau^{-1}\alpha_1(s)\tau = \alpha_2(s)$, $s \in G$. Then

$$\tau^{-1}\alpha_0(s)\eta_1(s)\tau = \alpha_0(s)\eta_2 \qquad \text{and} \qquad (s)\alpha_0(s)^{-1}\tau^{-1}\alpha_0(s)\eta_1(s)\tau = \eta_2(s),$$

and so

$$\eta_2(s) = (\tau^s)^{-1}\eta_1(s)\tau \qquad (s \in G). \tag{27}$$

Conversely, if τ is a 1-1 linear mapping of $\tilde{\mathfrak{M}}$ into itself satisfying (27), then, retracing the steps, we see that τ maps $\mathfrak{M}_1$ onto $\mathfrak{M}_2$.

If we are in the situation of Theorem 4 P/Φ finite dimensional of exponent one, then a similar calculation shows that a 1-1 linear mapping τ of $\tilde{\mathfrak{M}}$ onto itself maps the form $\mathfrak{M}_1$ onto $\mathfrak{M}_2$ if and only if

$$\lambda_2(d) = \tau^{-1}\lambda_1(d)\tau - \tau^{-1}\tau^d \qquad (d \in \mathfrak{D}). \tag{28}$$

Evidently, if $\mathfrak{M}_1$ and $\mathfrak{M}_2$ are any two Φ-forms of $\tilde{\mathfrak{M}}$, then there exists a 1-1 linear transformation τ of $\mathfrak{M}_1$ onto $\mathfrak{M}_2$. For P/Φ Galois, we let $\mathfrak{M}_1 = \mathfrak{M}_0$, the form whose precocycle α_1 is used to make Aut $\tilde{\mathfrak{M}}/P$ a G-group. The corresponding 1-cocycle of G into Aut $\tilde{\mathfrak{M}}/P$ is $\eta_1(s) = \alpha_1(s)^{-1}\alpha_1(s) = 1$. Hence, if η is the 1-cocycle corresponding to $\mathfrak{M}_2 = \mathfrak{M}$, an arbitrary Φ-form of $\tilde{\mathfrak{M}}$, and τ is the extension to $\tilde{\mathfrak{M}}$ of a linear isomorphism of $\mathfrak{M}_1$ onto $\mathfrak{M}_2$, then (27) gives $\eta(s) = (\tau^s)^{-1}\tau$. Thus we have shown that, in the Galois case, if η is any 1-cocycle of G into Aut $\tilde{\mathfrak{M}}/P$, then there exists a $\tau \in$ Aut $\tilde{\mathfrak{M}}/P$ such that $\eta(s) = (\tau^s)^{-1}\tau$. In a similar fashion, one sees that, if P/Φ is finite dimensional purely inseparable of exponent one and λ is a 1-cocycle of $\mathfrak{I}$ in $\mathrm{Hom}_P(\tilde{\mathfrak{M}}, \tilde{\mathfrak{M}})$, then there exists a $\tau \in$ Aut $\tilde{\mathfrak{M}}/P$ such that $\lambda(d) = -\tau^{-1}\tau^d$. We have therefore proved the following.

Theorem 5. (1) *Let P/Φ be Galois and $\tilde{\mathfrak{M}}$ a vector space over P which is finite dimensional if P/Φ is infinite dimensional. Make* Aut $\tilde{\mathfrak{M}}/P$ *a G-group for G the Galois group of P/Φ as in Theorems* 2 *and* 3, *and let η be an arbitrary* 1*-cocycle of G in* Aut $\tilde{\mathfrak{M}}/P$. *Then there exists a $\tau \in$* Aut $\tilde{\mathfrak{M}}/P$ *such that $\eta(s) = (\tau^s)^{-1}\tau$.* (2) *Let P/Φ be finite dimensional purely inseparable of exponent one, $\tilde{\mathfrak{M}}$ an arbitrary vector space over P. Make* $\mathrm{Hom}_P(\tilde{\mathfrak{M}}, \tilde{\mathfrak{M}})$ *a $\mathfrak{D}$-module for $\mathfrak{D} =$* Der P/Φ *as in Theorem* 4, *and let λ be any* 1*-cocycle of $\mathfrak{D}$ in* $\mathrm{Hom}_P(\tilde{\mathfrak{M}}, \tilde{\mathfrak{M}})$. *Then there exists a $\tau \in$* Aut $\tilde{\mathfrak{M}}/P$ *such that $\lambda(d) = -\tau^{-1}\tau^d$, $d \in \mathfrak{D}$.*

We remark that part 1 is a generalization of a well-known result of Speiser's, and part 2 is an analog of this and is a generalization of an earlier result of the author's.

III. ALGEBRAS AND FORMS OF ALGEBRAS

From now on we shall assume, for the sake of simplicity, that $\tilde{\mathfrak{M}}/P$ is finite dimensional. We shall specify certain classes of forms of $\tilde{\mathfrak{M}}$ by additional structures on $\tilde{\mathfrak{M}}$ together with appropriate notions of isomorphisms and derivations. One instance is an algebra structure that we define in a very general sense. Another that we consider in the next section is a vector space $\tilde{\mathfrak{M}}$ equipped with a homogeneous polynomial mapping into the base field. Later we shall give a set of axioms for classes of forms which will be applicable to these two cases and to many others.

Let $\mathfrak{M}$ and $\mathfrak{N}$ be (finite dimensional) vector spaces over an infinite field P. We recall that a polynomial mapping of $\mathfrak{M}$ into $\mathfrak{N}$ is a mapping of the form

$$\sum_1^m \xi_i u_i \to \sum_1^n f_j(\xi_1, ..., \xi_m) v_j ,$$

where $(u_1, u_2, ..., u_m)$ is a basis for $\mathfrak{M}/P$, and $(v_1, v_2, ..., v_n)$ is a basis for $\mathfrak{N}/P$, and the f_j are fixed polynomials in indeterminates $X_1, ..., X_m$ with coefficients in P. A similar definition can be given for a polynomial mapping in r variables in $\mathfrak{M}$ into $\mathfrak{N}$. If the arguments are $x_1, x_2, ..., x_r \in \mathfrak{M}$ the value $f(x_1, ..., x_r) = \Sigma f_j(\xi_{11}, ..., \xi_{1m}, ..., \xi_{rl}, ..., \xi_{rm})v_j$, $x_j = \Sigma \xi_{ji} u_i$, f_j a polynomial. Since P is infinite, the f_j are uniquely determined. We shall say that f is homogeneous of degree l in x_i if $f(..., \lambda x_i, ...) = \lambda^l f(..., x_i, ...)$, $\lambda \in P$. The condition for this is that all the f_j are homogeneous of degree l in the indeterminates $X_{il}, ..., X_{im}$. The notion of degree in one x_i and of (total) degree of f is clear.

If Σ is an extension field of P, then a polynomial mapping of r arguments in $\mathfrak{M}$ to $\mathfrak{N}$ has a unique extension to a polynomial mapping of r arguments in the extension space $\mathfrak{M}_\Sigma$ to $\mathfrak{N}_\Sigma$. In particular, let $\Sigma = P(t)$, t an indeterminate. If f is of degree l in x_1, then we have

$$f(x_1 + ty_1, x_2, ..., x_r) = \sum_{i=0}^{l} t^i \frac{\partial^{(i)} f}{\partial y_1}(x_1, ..., x_r) \tag{29}$$

where $(\partial^{(i)} f)_1$ is a polynomial function in $r + 1$ variables $y_1, x_1, ..., x_r$ whose value at $(y_1, x_1, ..., x_r)$ is denoted as $(\partial^{(i)} f/\partial y_1)(x_1, ..., x_r)$. If f is homogeneous of degree l in x_1, then $(\partial^{(i)} f)_1$ is homogeneous of degree i in y_1 and of degree $l - i$ in x_1, and, if f is homogeneous of degree k in x_j, $j > 1$, then the same is true for $(\partial^{(i)} f)_1$. In a similar manner, replacement of x_j by $x_j + ty_j$ leads to a definition of the polynomial mappings $(\partial^{(i)} f)_j$ in $(y_j, x_1, ..., x_r)$. We can repeat this process also on the new mappings. The mappings thus obtained including f itself will be called the *derived* mappings of the given mapping f. It is clear that $(\partial^{(0)} f/\partial y_1)(x_1, ..., x_r) = f(x_1, x_2, ..., x_r)$ and $(\partial^{(l)} f/\partial y_1)(x_1, ..., x_r) = f(y_1, x_2, ..., x_r)$ if f is homogeneous of degree l in x_1. In this sense f and $(\partial^{(0)} f)_1$ and $(\partial^{(l)} f)_1$ define the same mapping. In particular, if f is multilinear, that is, f is homogeneous of degree one in all of its variables, then the set of derived mappings of f consists of f alone.

It is clear by induction on k that

$$f(x_1 + ty_1, ..., x_k + ty_k, x_{k+1}, ..., x_r) = \sum_j t^j F_j(y_1, ..., y_k, x_1, ..., x_r),$$

where the F_j are sums of derived mappings of f. In particular, we have

$$f(x_1 + ty_1, \ldots, x_r + ty_r) = \sum_{i=0}^{h} t^i \Delta_y^{(i)} f(x) \tag{30}$$

where h is the degree of f and $\Delta^{(i)} f$ is a sum of derived mappings whose value at $(y_1, \ldots, y_r, x_1, \ldots, x_r)$ is denoted as $\Delta_y^{(1)} f(x)$.

We now define an *algebra* $\tilde{\mathfrak{A}}/P$ to be a finite dimensional vector space over an infinite field P equipped with a set π of polynomial mappings in any (finite) number of arguments in $\tilde{\mathfrak{A}}$ to $\tilde{\mathfrak{A}}$ such that every $f \in \pi$ is homogeneous in all of its arguments with positive degree of homogeneity. Let π' denote the set of derived mappings of the $f \in \pi$. If Φ is a subfield of P, we define a *Φ-form of the algebra* $\tilde{\mathfrak{A}}/P$ to be a Φ-form of the vector space $\tilde{\mathfrak{A}}/P$, which is closed under every $f \in \pi'$ in the sense that $f(x_1, \ldots, x_r) \in \mathfrak{A}$ for all $x_i \in \mathfrak{A}$. It is easy to see that, if the number of elements of $\Phi >$ the degree of every $f \in \pi$ in all of its arguments, then the closure condition of the definition can be replaced by the customary one that $\mathfrak{A}$ is closed under every $f \in \pi$. In particular, this is the case if Φ is infinite, and this shows that $\tilde{\mathfrak{A}}/P$ is a form of itself.

We shall now define isomorphisms and derivations and more generally semi-isomorphisms and semiderivations of forms of an algebra. If $\mathfrak{A}$ and $\mathfrak{B}$ are forms of the algebra $\mathfrak{A}$, and s is an isomorphism of the base field Φ of $\mathfrak{A}$ onto the base field Φ^s of $\mathfrak{B}$, then an s-semilinear transformation of $\mathfrak{A}/\Phi$ onto $\mathfrak{B}/\Phi^s$ is called an *s-semi-isomorphism* of the algebra form $\mathfrak{A}$ onto $\mathfrak{B}$ if $f(xS) = f(x)S$ for every $f \in \pi'$. It is clear that the resultant of two such mappings is another one, and the set of semiautomorphisms is a group of transformations of the form $\mathfrak{A}/\Phi$.

Next let $\mathfrak{A}/\Phi$ be a Φ-form of the algebra $\tilde{\mathfrak{A}}/P$, and let d be a derivation of Φ into Φ. Then we define a d-semiderivation D of the algebra form $\mathfrak{A}/\Phi$ to be a d-derivation of the vector space $\mathfrak{A}/\Phi$ such that for every $f \in \pi'$ we have

$$f[x + t(xD)] \equiv f(x) + tf(x)D \pmod{t^2}. \tag{31}$$

In view of (30), this is equivalent to

$$f(x)D = \Delta_{xD}^{(1)} f(x) \qquad (f \in \pi'). \tag{32}$$

If D and E are d- and e-semiderivations, respectively, of the algebra form $\mathfrak{A}/\Phi$, then the linearity of $\Delta_y^{(1)} f(x)$ in y and (32) imply that $D + E$ is a $d + e$-semiderivation and $D\varphi$ is a $d\varphi$-semiderivation of $\mathfrak{A}$ as form of the algebra $\tilde{\mathfrak{A}}/P$. We shall show next that $[DE]$ is a $[de]$-algebra form semiderivation. We note that it follows directly from the definition of $\Delta^{(1)}$ that

$$\Delta^{(1)}_{y'x'}[\Delta^{(1)}_{y}f(x)] = \Delta^{(1)}_{x'}[\Delta^{(1)}_{y}f(x)] + \Delta^{(1)}_{y'}f(x) \tag{33}$$

where $\Delta^{(1)}_{x'y'}$ is applied to $\Delta^{(1)}_{y}f(x)$ regarded as a function of the y_i and x_i, whereas on the right-hand side $\Delta^{(1)}_{x'}$ is applied to $\Delta^{(1)}_{y}f(x)$ considered as a function of the x_i alone with the y_i fixed. If we put $x' = xE$, $y' = yE$, then (32) and (33) give

$$[\Delta^{(1)}_{y}f(x)]E = \Delta^{(1)}_{xE}[\Delta^{(1)}_{y}f(x)] + \Delta^{(1)}_{yE}f(x). \tag{34}$$

Setting $y = xD$, we obtain

$$f(x)DE = [\Delta^{(1)}_{xD}f(x)]E = \Delta^{(1)}_{xE}\Delta^{(1)}_{xD}f(x) + \Delta^{(1)}_{xDE}f(x). \tag{35}$$

Since $\Delta^{(1)}_{xE}\Delta^{(1)}_{xD}f(x) = \Delta^{(1)}_{xD}\Delta^{(1)}_{xE}f(x)$ by the commutativity of partial derivatives, (35) implies that

$$f(x)[DE] = \Delta^{(1)}_{x[DE]}f(x) \qquad (f \in \pi').$$

Hence $[DE]$ is a $[de]$-semiderivation of the form $\mathfrak{A}$. We remark that it is not true in general that, if the characteristic is $p \neq 0$, then the pth power of a d-semiderivation D is a semiderivation. This is the case if every f is bilinear and in other cases, but it may not be the case if an $f \in \pi$ is trilinear.

Now let d be a derivation in Φ and s be an isomorphism of Φ onto the subfield Φ' of P. Then one verifies directly that $s^{-1}ds$ is a derivation in Φ'. Let $\mathfrak{A}/\Phi$ and $\mathfrak{A}'/\Phi'$ be forms of $\tilde{\mathfrak{A}}/P$, and let D be a d-semiderivation of the form $\mathfrak{A}/\Phi$, S an s-semi-isomorphism of $\mathfrak{A}/\Phi$ onto $\mathfrak{A}'/\Phi'$. If $f \in \pi'$ and $x' \in \mathfrak{A}'$, then

$$\begin{aligned} f(x' + tx'S^{-1}DS) &= f(x'S^{-1}S + tx'S^{-1}DS) \\ &= f(x'S^{-1} + tx'S^{-1}D)S \\ &\equiv f(x'S^{-1})S + tf(x'S^{-1})DS (\text{mod } t^2) \\ &\equiv f(x') + tf(x')S^{-1}DS (\text{mod } t^2) \end{aligned}$$

Also, if $\alpha' \in \Phi'$, then $(\alpha'x')S^{-1}DS = (\alpha'^{s^{-1}}x'S^{-1})DS = [\alpha'^{s^{-1}d}(x'S^{-1})]S + [\alpha'^{s^{-1}}(x'S^{-1}D)]S = \alpha'^{s^{-1}ds}x' + \alpha'(x'S^{-1}DS)$. Hence $S^{-1}DS$ is an $s^{-1}ds$-semiderivation of $\mathfrak{A}'$.

An s-semi-isomorphism for which $s = 1$ is an *isomorphism* of $\mathfrak{A}$, and a d-semiderivation for which $d = 0$ is a *derivation* of $\mathfrak{A}$. The set of automorphisms of $\mathfrak{A}/\Phi$ is a group that we denote as Aut $\mathfrak{A}/\Phi$, and the set of derivations of $\mathfrak{A}/\Phi$ is a Lie algebra that we shall denote as Der $\mathfrak{A}/\Phi$.

Examples. 1) Let $\tilde{\mathfrak{A}}/P$ be a restricted Lie algebra of characteristic $p \neq 0$ (Jacobson, "Lie Algebras," p. 186). This is defined by a set π constisting of a bilinear Lie multiplication $[xy]$ and a p-power mapping $x \to x^{[p]}$ satisfying certain conditions given in the indicated reference. It is easily seen that D is a derivation of a form $\mathfrak{A}/\Phi$ of $\tilde{\mathfrak{A}}$ if and only if D is a derivation of $\mathfrak{A}/\Phi$ as a Lie algebra and $x^{[p]}D = (xD)(a\ dx)^{p-1}$. This is the usual definition of a restricted derivation. The set Der $\mathfrak{A}/\Phi$ of these derivations is a restricted Lie algebra of linear transformations.

2) Let $\tilde{\mathfrak{A}}/P$ be an associative algebra with an involution J. Here π consists of the given associative product and the linear mapping J. A Φ-form of $\tilde{\mathfrak{A}}/P$ is a Φ-subalgebra closed under J which is a Φ-form of the vector space $\tilde{\mathfrak{A}}/P$. Isomorphisms of two such forms are algebra isomorphisms that commute with J. A derivation of the form $\mathfrak{A}/\Phi$ is a derivation in the usual sense commuting with J.

Let $\tilde{\mathfrak{A}}/P$ be a finite dimensional algebra defined by a set π of polynomial mappings (homogeneous in all the arguments), and let π' be the set of derived mappings. Let $\mathfrak{A}$ be a Φ-form of the vector space $\tilde{\mathfrak{A}}/P$, and let $(u_1, u_2, \ldots, u_m)$ be a basis for $\mathfrak{A}/\Phi$. We claim that $\mathfrak{A}$ is an algebra form of $\tilde{\mathfrak{A}}/P$ if and only if $f(u_{i_1}, u_{i_2}, \ldots, u_{i_r}) \in \mathfrak{A}$ for all $f(x_1, \ldots, x_r) \in \pi'$ and $i_j \in \{1, 2, \ldots, m\}$. This is clearly a necessary condition. To prove sufficiency, we use an inductive argument based on the *grade* of the mapping f which is defined as

$$l = \sum_{i=1}^{r} (l_i - 1),$$

where l_i ($\geqslant 1$) is the degree of homogeneity of f in x_i. If $l = 0$, then f is multilinear, and the result is clear in this case. Assume now that $g(x_1, \ldots, x_s) \in \mathfrak{A}$ for all $x_i \in \mathfrak{A}$ and all $g \in \pi'$ of grade $< l$. Let f be a mapping contained in π' of grade l. We have $f(u_{i_1}, \ldots, u_{i_r}) \in \mathfrak{A}$ for all $i_j = 1, \ldots, m$. Suppose that we already know also that $f(x_1 \ldots, x_{k-1}, u_{i_k}, \ldots, u_{i_r}) \in \mathfrak{A}$ for all x_i in $\mathfrak{A}$ and all u_{i_j} in the given basis. We now use an additional induction on k and we let $\mathfrak{R}$ be the set of element $x \in \mathfrak{A}$ such that $f(x_1, \ldots, x_{k-1}, x, u_{i_{k+1}}, \ldots, u_{i_r}) \in \mathfrak{A}$ for all $x_i \in \mathfrak{A}$ and all u_{i_j} in the basis. It is clear from the homogeneity that, if $x \in \mathfrak{R}$ and $\varphi \in \Phi$, then $\varphi x \in \mathfrak{R}$. We now observe that the derived function $(\partial^{(i)} f)_k$, $1 \leqslant i \leqslant l_k - 1$, where l_k is the degree of homogeneity of f in x_k, are of grade less than l. Hence, by the formula analogous to (29) for x_k ($t = 1$), we see that $f(x_1, \ldots, x_{k-1}, x + u_{i_k}, u_{i_{k+1}}, \ldots) \equiv f(x_1, \ldots, x_{k-1}, x, u_{i_{k+1}}, \ldots, u_{i_m}) + f(x_1, \ldots, x_{k-1}, u_{i_k}, \ldots, u_{i_m})$ (mod $\mathfrak{A}$). The first term belongs to $\mathfrak{A}$, since $x \in \mathfrak{R}$, and the second belongs to $\mathfrak{A}$ by the induction hypothesis. Hence we see that $x + u_{i_k} \in \mathfrak{R}$. This and

$\varphi x \in \mathfrak{N}$ for x in $\mathfrak{N}$ imply that $\mathfrak{N} = \mathfrak{A}$. We have therefore established the inductive step, and so we can conclude that $f(x_1, x_2, \ldots, x_r) \in \mathfrak{A}$ for all $x_i \in \mathfrak{A}$. Hence the conditions that we assumed on the basis are sufficient for $\mathfrak{A}$ to be a Φ-form of the algebra $\tilde{\mathfrak{A}}$.

Next let S be a 1-1 s-semilinear transformation of the form $\mathfrak{A}/\Phi$ of the algebra $\tilde{\mathfrak{A}}$ onto the form $\mathfrak{L}/\Phi^s$. The inductive argument that we have just given can be repeated to show that S is a semi-isomorphism for the algebra forms if and only if $f(u_{i_1}S, u_{i_2}S, \ldots, u_{i_r}S) = f(u_{i_1}, u_{i_2}, \ldots, u_{i_r})S$ for all i_j in $\{1, 2, \ldots, m\}$ and all $f \in \pi'$. Now let Λ/Φ be a subfield of P/Φ and $\bar{s}$ an extension of s to an isomorphism of Λ into P. It is immediate from the properties of tensor products that there is a unique $\bar{s}$-semilinear transformation of $\bar{S}\Lambda\mathfrak{A} \cong \Lambda \otimes_\Phi \mathfrak{A}$ into $\Lambda^{\bar{s}}\mathfrak{B}$ such that, if $x_i \in \mathfrak{A}$, $\lambda_i \in \Lambda$, then $\Sigma\lambda_i x_i)\bar{S} = \Sigma\lambda_i^{\bar{s}}(x_i S)$. Moreover, $\bar{S}$ is bijective. Since $(u_1, u_2, \ldots, u_m)$ is a basis for $\Lambda\mathfrak{A}/\Lambda$, the criterion for semi-isomorphism shows that $\bar{S}$ is an $\bar{s}$-semi-isomorphism of the Λ form $\Lambda\mathfrak{A}$ of the algebra $\tilde{\mathfrak{A}}/P$ onto the $\Lambda^{\bar{s}}$-form $\Lambda^{\bar{s}}\mathfrak{B}$. It is clear also that, if $\bar{S}$ is any $\bar{s}$-semi-isomorphism of a Λ-form $\bar{\mathfrak{A}}$ of the algebra $\tilde{\mathfrak{A}}/P$ into a $\Lambda^{\bar{s}}$-form $\bar{\mathfrak{B}}$ of this algebra, and $\mathfrak{A}$ is an algebra Φ-form of $\bar{\mathfrak{A}}$, then the restriction of $\bar{S}$ to $\mathfrak{A}$ is a semi-isomorphism of $\mathfrak{A}$ as algebra form.

Next let D be a d-semiderivation of $\mathfrak{A}$ as vector space. Then the inductive argument that we have used before can be repeated for congruences modulo t^2 to show that D is a d-semiderivation of $\mathfrak{A}$ as form of the algebra $\tilde{\mathfrak{A}}/P$ if and only if $f(u_{i_1}, u_{i_2}, \ldots, u_{i_r})D = \Delta_{uD}^{(1)} f(u_{i_1}, u_{i_2}, \ldots, u_{i_r})$, where $(u_1, u_2, \ldots, u_m)$ is a basis for $\mathfrak{A}/\Phi$, $uD \equiv (u_{i_1}D, u_{i_2}D, \ldots, u_{i_r}D)$, $i_j \in \{1, 2, \ldots, m\}$, and $f \in \pi'$. Now let Λ/Φ be a subfield of P/Φ, and let $\bar{d}$ be an extension of d to a derivation in Λ. Then there exists a unique $\bar{d}$-semiderivation $\bar{D}$ of $\Lambda\mathfrak{A}$ as vector space which coincides with D on $\mathfrak{A}$. If $\bar{D}$ exists, then $(\Sigma\lambda_i x_i)\bar{D} = \Sigma\lambda_i^{\bar{d}} x_i + \Sigma\lambda_i(x_i D)$, $\lambda_i \in \Lambda$, $x_i \in \mathfrak{A}$. Hence uniqueness is clear. To prove the existence of $\bar{D}$, we define for $\lambda \in \Lambda$, $x \in \mathfrak{A}$,

$$\lambda \otimes' x = \lambda^d x + \lambda(xD).$$

This is additive in both factors and, if $\varphi \in \Phi$, then $\varphi\lambda \otimes' x = \lambda \otimes' \varphi x$. It follows that we have an endomorphism $\bar{D}$ in $\Lambda\mathfrak{A}$ such that $(\Sigma\lambda_i x_i)\bar{D} = \Sigma\lambda_i^{\bar{d}} x_i + \Sigma\lambda_i(x_i D)$, $\lambda_i \in \Lambda$, $x_i \in \mathfrak{A}$. Clearly this coincides with D on $\mathfrak{A}$, and direct verification shows that $\bar{D}$ is a $\bar{d}$-semiderivation of the vector space $\Lambda\mathfrak{A}$. It is now clear from the criterion in terms of the basis (u_i) that $\bar{D}$ is a $\bar{d}$-derivation of $\Lambda\mathfrak{A}$ as form of the algebra $\tilde{\mathfrak{A}}/P$. Also, if $\bar{D}$ is any semiderivation of $\Lambda\mathfrak{A}$ which maps $\mathfrak{A}$ into itself, then the restriction D of $\bar{D}$ to $\mathfrak{A}$ is a semiderivation of $\mathfrak{A}$ as form of the algebra $\tilde{\mathfrak{A}}/P$.

Again let $\mathfrak{A}$ be a Φ-form of the algebra $\tilde{\mathfrak{A}}/P$, (u_i) a basis for $\mathfrak{A}/\Phi$ and, hence, for $\tilde{\mathfrak{A}}/P$. If D is a linear transformation in $\mathfrak{A}/\Phi$ or $\tilde{\mathfrak{A}}/P$, the conditions that D be a derivation are that $f(u_{i_1}, \ldots, u_{i_r})D = \Delta_{uD}^{(1)} f(u_{i_1}, \ldots, u_{i_r})$ for all $f \in \pi'$ and u_{i_j} in the given basis. If we write $u_i D = \Sigma d_{ij} u_j$, then each of these conditions is equivalent to a set of homogeneous linear equations in the d_{ij} with coefficients in Φ. It follows that, if we identify a linear transformation in $\mathfrak{A}/\Phi$ with its extension to $\tilde{\mathfrak{A}}/P$, then we can obtain a basis for the derivation Lie algebra Der $\mathfrak{A}/\Phi$ which is also a basis for Der $\tilde{\mathfrak{A}}/P$. Hence Der $\mathfrak{A}/\Phi$ is a Φ-form of Der $\tilde{\mathfrak{A}}/P$.

IV. POLYNOMIAL FUNCTIONS ON A VECTOR SPACE

We consider next another type of structure on a vector space $\tilde{\mathfrak{M}}/P$, namely, that given by a polynomial mapping of $\tilde{\mathfrak{M}}$ into P which is homogeneous in all of its arguments. Again we assume $\tilde{\mathfrak{M}}$ infinite and we call f a *homogeneous polynomial function in r arguments* in $\tilde{\mathfrak{M}}$. We define a *Φ-form of the pair* $(\tilde{\mathfrak{M}}, f)$ to be a Φ-form $\mathfrak{M}$ of the vector space $\tilde{\mathfrak{M}}/P$ such that $g(x_1, \ldots, x_r) \in \Phi$ for all $x_i \in \mathfrak{M}$ and every derived function g of f. If $(\mathfrak{M}, f)$ and $(\mathfrak{N}, f)$ are forms of $(\tilde{\mathfrak{M}}, f)$, and S is a 1-1 s-semilinear transformation of $\mathfrak{M}$ onto $\mathfrak{N}$, then we say that S is an *s-semi-isomorphism* of $(\mathfrak{M}, f)$ onto $(\mathfrak{N}, f)$ if $g(x_1 S, x_2 S, \ldots, x_r S) = g(x_1, x_2, \ldots, x_r)^s$ for every derived function g of f. If D is a d-semiderivation of $\mathfrak{M}$, then D is called a *d-semiderivation* of $(\mathfrak{M}, f)$ if $g[x_1 + t(x_1 D), x_2 + t(x_2 D), \ldots, x_r + t(x_r D)] \equiv g(x_1, x_2, \ldots, x_r)(1 + td)(\text{mod } t^2)$. This is equivalent to $\Delta_{xD}^{(1)} g(x_1, x_2, \ldots, x_r) = g(x_1, \ldots, x_r)^d$, $x = (x_1, x_2, \ldots, x_r)$, and the indicated condition is required to hold for all $x_i \in \mathfrak{A}$ and every derived function g of f.

It is easy to carry over the discussion of algebras of the last section to polynomial functions. We shall state the results without repeating the proofs. We note first that the resultant of semi-isomorphisms is a semi-isomorphism. If D and E are semiderivations of a Φ-form $(\mathfrak{M}, f)$ of $(\tilde{\mathfrak{M}}, f)$, then $D + E$, $[DE]$, $D\varphi$ for $\varphi \in \Phi$ are semiderivations of $(\mathfrak{M}, f)$. If S is a semi-isomorphism of $(\mathfrak{M}, f)$ onto the form $(\mathfrak{N}, f)$ and D is a semiderivation of $(\mathfrak{M}, f)$, then $S^{-1}DS$ is a semiderivation of $(\mathfrak{N}, f)$.

If $(u_1, u_2, \ldots, u_m)$ is a basis for the Φ-form $\mathfrak{M}$ of the vector space $\tilde{\mathfrak{M}}/P$, then $\mathfrak{M}$ defined a Φ-form $(\mathfrak{M}, f)$ of $(\overline{\mathfrak{M}}, f)$ if and only if $g(u_{i_1}, u_{i_2}, \ldots, u_{i_r}) \in \Phi$ for all $i_j \in \{1, 2, \ldots, m\}$ and g a derived function of f. This implies that, if Λ/Φ is a subfield of P/Φ and $(\mathfrak{M}, f)$ is a Φ-form of $(\tilde{\mathfrak{M}}, f)$, then $(\Lambda\mathfrak{M}, f)$ is a Λ-form of $(\tilde{\mathfrak{M}}, f)$. Also if S is a semilinear isomorphism of the form $(\mathfrak{M}, f)$ onto the form $(\mathfrak{N}, f)$ and $\bar{S}$ is a semilinear transformation of $\Lambda\mathfrak{M}$ which coincides with S on $\mathfrak{M}$, then $\bar{S}$ is a semi-isomorphism of the Λ-form $(\Lambda\mathfrak{M}, f)$ onto another form of $(\tilde{\mathfrak{M}}, f)$. If $\bar{S}$ is a semi-isomorphism of a form $(\overline{\mathfrak{M}}, f)$ and $(\mathfrak{M}, f)$ is a form of $(\overline{\mathfrak{M}}, f)$, then the restriction of $\bar{S}$ to $\mathfrak{M}$ is a semi-isomorphism of $(\mathfrak{M}, f)$

onto another form of $(\tilde{\mathfrak{M}}, f)$. Analogous results hold also for semiderivations. Finally, it is easy to see that if $(\mathfrak{M}, f)$ is a form of $(\mathfrak{M}, f)$, then the Lie algebra of derivations of $(\mathfrak{M}, f)$ is a form of that of $(\tilde{\mathfrak{M}}, f)$ provided that we identify, as usual, a linear mapping of $\mathfrak{M}$ with its linear extension to $\tilde{\mathfrak{M}}$.

V. AXIOMS FOR A CLASS OF FORMS OF A VECTOR SPACE

We shall now give a set of axioms for classes of forms of a vector space which are satisfied by the forms of an algebra in the general sense of Sec. III and also by the forms of a vector space relative to a homogeneous polynomial function.

Let $\tilde{\mathfrak{M}}$ be a finite dimensional vector space over a field P. Our axioms concern three sets $\mathscr{C}$, $\mathscr{S}$, $\mathscr{D}$, where $\mathscr{C}$ is a set of forms of $\tilde{\mathfrak{M}}$, $\mathscr{S}$ a set of 1-1 semilinear mappings of elements $\mathfrak{M} \in \mathscr{C}$ into elements $\mathfrak{N} \in \mathscr{C}$, $\mathscr{D}$ a set of semiderivations of elements $\mathfrak{M} \in \mathscr{C}$ into $\mathfrak{M}$. We make the following assumptions on the set $\mathscr{C}$, $\mathscr{S}$, $\mathscr{D}$:

1) $\tilde{\mathfrak{M}} \in \mathscr{C}$.

2) If $\mathfrak{M}/\Phi \in \mathscr{C}$ and Λ/Φ is a subfield of P/Φ, then $\Lambda\mathfrak{M} \in \mathscr{C}$.

3) If S, $T \in \mathscr{S}$ and the range of S is the same as the domain of T, then $ST \in \mathscr{S}$. If $\mathfrak{M} \in \mathscr{C}$, then the identity mapping $1_{\mathfrak{M}}$ on $\mathfrak{M}$ belongs to $\mathscr{S}$. If $S \in \mathscr{S}$, then $S^{-1} \in \mathscr{S}$.

4) If $S \in \mathscr{S}$ with domain $\mathfrak{M}/\Phi$ and $\bar{S}$ is a semilinear extension of S to $\Lambda\mathfrak{M}$, where Λ/Φ is a subfield of P/Φ, then $\bar{S} \in \mathscr{S}$. If $\mathfrak{M}$, $\mathfrak{N} \in \mathscr{C}$, $\mathfrak{N}$ a form of $\mathfrak{M}$ and $S \in \mathscr{S}$ has domain $\mathfrak{M}$, then the restriction of S to $\mathfrak{N}$ belongs to $\mathscr{S}$.

5) If D, $E \in \mathscr{D}$ and have the same domain $\mathfrak{M}/\Phi$, then $D + E$, $[D, E]$ and $D\varphi$, $\varphi \in \Phi$ are contained in $\mathscr{D}$. If $S \in \mathscr{S}$ has domain $\mathfrak{M}$ and range $\mathfrak{N}$, then $S^{-1}DS \in \mathscr{D}$ (with domain $\mathfrak{N}$).

6) If $D \in \mathscr{D}$ with domain $\mathfrak{M}/\Phi$ and $\bar{D}$ is a semiderivation extension of D to $\Delta\mathfrak{M}$, Λ/Φ a subfield of P/Φ, then $\bar{D} \in \mathscr{D}$. Also, if $\mathfrak{N} \in \mathscr{C}$ is a form of $\mathfrak{M}$ and D maps $\mathfrak{N}$ into itself, then the restriction of D to $\mathfrak{N}$ belongs to $\mathscr{D}$.

7) If $\mathfrak{M}/\Phi \in \mathscr{C}$ and Der $(\mathfrak{M}/\Phi, \mathscr{D})$ denotes the Lie algebra of linear transformations of $\mathfrak{M}$ which belong to $\mathscr{D}$, then Der $(\mathfrak{M}/\Phi, \mathscr{D})$ is a Φ-form of Der $(\tilde{\mathfrak{M}}/P, \mathscr{D})$ (if we identify the elements of Der $(\mathfrak{M}/\Phi, \mathscr{D})$ with their linear extensions to $\tilde{\mathfrak{M}}/P$).

8) Let $\overline{\mathfrak{M}}/\Lambda \in \mathscr{C}$, Φ a subfield of Λ such that Λ/Φ is Galois (finite dimensional purely inseparable of exponent one). Let $\mathfrak{M}$ be a Φ-form

of the vector space $\bar{\mathfrak{M}}/\Lambda$, α the corresponding precocycle of the Galois group H of Λ/Φ (of $\mathfrak{F} = \mathrm{Der}\, \Lambda/\Phi$). Then $\mathfrak{M} \in \mathscr{C}$ if every $\alpha(t) \in \mathscr{S}$, $t \in H$ $[\alpha(f) \in \mathscr{D}, f \in \mathrm{Der}\, \Lambda/\Phi]$.

We remark that the converse of assumption 8 is a consequence of assumptions 3–6. Suppose we have the situation of assumption 8 and Λ/Φ is Galois and assume that $\mathfrak{M}$ is a Φ-form of $\bar{\mathfrak{M}}/\Lambda$ which belongs to $\mathscr{C}$. If α is the precocycle of H corresponding to α, then $\alpha(t)$, $t \in H$, is the t-semilinear extension to $\bar{\mathfrak{M}}$ of the identity $1_{\mathfrak{M}}$ on $\mathfrak{M}$. Hence $\alpha(t) \in \mathscr{S}$ by assumptions 3 and 4. A similar argument applies in the purely inseparable case using assumptions 5 and 6.

We shall call a vector space $\tilde{\mathfrak{M}}$ with sets $\mathscr{C}$, $\mathscr{S}$, $\mathscr{D}$ satisfying conditions 1–8 a *descent class* on $\tilde{\mathfrak{M}}/P$.

Let $\tilde{\mathfrak{A}}/P$ be an algebra in the general sense of Sec. III, $\mathscr{C}$ the set of algebra forms of $\tilde{\mathfrak{A}}/P$, $\mathscr{S}$ the set of semi-isomorphisms of algebra forms of $\tilde{\mathfrak{A}}$ onto algebra forms, $\mathscr{D}$ the set of semiderivations of the algebra forms as defined in Sec. III. It is immediate from the results of Sec. III that axioms 1–7 hold for $\mathscr{C}$, $\mathscr{S}$, $\mathscr{D}$. Now let $\bar{\mathfrak{A}}/\Lambda$ be a Λ-form of the algebra $\tilde{\mathfrak{A}}/P$, Φ a subfield of Λ such that Λ/Φ is Galois, and $\mathfrak{A}/\Phi$ a Φ-form of the vector space $\bar{\mathfrak{A}}/\Lambda$, α the corresponding precocycle of the Galois group H of Λ/Φ. Suppose $\alpha(t)$, $t \in H$, is a semi-isomorphism of $\bar{\mathfrak{A}}/\Lambda$ as algebra form. If $f \in \pi'$, the derived set of the set π of polynomial mappings defining $\tilde{\mathfrak{A}}$, then, for $x_i \in \mathfrak{A}$, $t \in H$, we have $f(x_1, \ldots, x_r)\alpha(t) = f(x_1\alpha(t), \ldots, x_1\alpha(t)) = f(x_1, \ldots, x_r)$. Since $\mathfrak{A}$ is the subset of $\mathfrak{A}$ of elements fixed under every $\alpha(t)$, this implies that $\mathfrak{A}$ is an algebra form of $\tilde{\mathfrak{A}}$. A similar argument applies in the situation in which Λ/Φ is finite dimensional purely inseparable of exponent one. Hence assumption 8 is valid for the $\mathscr{C}$, $\mathscr{S}$, $\mathscr{D}$ defined by an algebra $\tilde{\mathfrak{A}}/P$. It is easy to verify also that, if $\tilde{\mathfrak{M}}/P$ is a vector space equipped with a homogeneous polynomial mapping f into the base field P, $\mathscr{C}$ the set of forms $(\mathfrak{M}, f)$, $\mathscr{S}$ the set of semi-isomorphisms of these forms, $\mathscr{D}$ the set of semiderivations of the $(\mathfrak{M}, f)$, then $\mathscr{C}$, $\mathscr{S}$, $\mathscr{D}$ satisfy the axioms.

Now let $(\tilde{\mathfrak{M}}, \mathscr{C}, \mathscr{S}, \mathscr{D})$ be any descent class on the vector space $\tilde{\mathfrak{M}}/P$. If $\mathfrak{M}/\Phi \in \mathscr{C}$, then we denote the group of 1-1 linear mappings of $\mathfrak{M}/\Phi$ onto itself which belong to $\mathscr{S}$ by $\mathrm{Aut}(\mathfrak{M}/\Phi, \mathscr{S})$. It is clear from assumption 4 that the linear extensions to $\tilde{\mathfrak{M}}$ of the elements of $\mathrm{Aut}(\mathfrak{M}/\Phi, \mathscr{S})$ are contained in $\mathrm{Aut}(\tilde{\mathfrak{M}}/P, \mathscr{S})$, and $\mathrm{Aut}(\mathfrak{M}/\Phi, \mathscr{S})$ can be identified with the subgroup of $\mathrm{Aut}(\tilde{\mathfrak{M}}/P, \mathscr{S})$ of elements mapping $\mathfrak{M}$ into itself.

Let $\bar{\mathfrak{M}}/\Lambda \in \mathscr{C}$, Φ a subfield of Λ such that Λ/Φ is Galois, H the Galois group of Λ/Φ, and let $\mathfrak{M}_0/\Phi \in \mathscr{C}$ be a Φ-form of $\bar{\mathfrak{M}}$, α_0 the corresponding precocycle of H. If we make the group $\mathrm{Aut}\, \bar{\mathfrak{M}}/\Lambda$ of 1-1 linear transformations of $\bar{\mathfrak{M}}$ into $\bar{\mathfrak{M}}$ an H-group by means of α_0 so that

$\tau^t = \alpha_0(t)^{-1}\tau\alpha_0(t)$, $t \in H$, then $\mathrm{Aut}(\bar{\mathfrak{M}}/\Lambda, \mathscr{S})$ is an H-subgroup. Moreover, it is clear that, if $\mathfrak{M}$ is a Φ-form of the vector space $\bar{\mathfrak{M}}/\Lambda$, then $\mathfrak{M} \in \mathscr{C}$ if and only if $\eta(t) \in \bar{A} = \mathrm{Aut}(\bar{\mathfrak{M}}/\Lambda, \mathscr{S})$, $t \in H$, for the 1-cocycle of H corresponding to $\mathfrak{M}$. A 1-cocycle having this property will be called a *1-cocycle of H into the H-group $\bar{A}$*. We have a bijection of the set $Z^1(H, A)$ of these 1-cocycles with the subset of $\mathscr{C}$ of elements that are Φ-forms of $\bar{\mathfrak{M}}/\Lambda$. If $\mathfrak{M}_1/\Phi$ and $\mathfrak{M}_2/\Phi \in \mathscr{C}$, then an element $\tau \in \mathscr{S}$ which is linear and has domain $\mathfrak{M}_1$ and range $\mathfrak{M}_2$ will be called an *$\mathscr{S}$-isomorphism* of $\mathfrak{M}_1$ onto $\mathfrak{M}_2$. It is clear that the linear extension of τ to $\tilde{\mathfrak{M}}/P$ is an element of $\mathrm{Aut}(\tilde{\mathfrak{M}}/P, \mathscr{S})$, and, conversely, if $\tau \in \mathrm{Aut}(\tilde{\mathfrak{M}}/P, \mathscr{S})$, then the restriction of τ to $\mathfrak{M}_1/\Phi \in \mathscr{C}$ is an $\mathscr{S}$-isomorphism of $\mathfrak{M}_1$ onto an element $\mathfrak{M}_2/\Phi \in \mathscr{C}$. If η_1 and η_2 are two elements of $Z^1(H, \bar{A})$, $\mathfrak{M}_1/\Phi$, $\mathfrak{M}_2/\Phi$ the corresponding forms of $\bar{\mathfrak{M}}/\Lambda$, then, as in Sec. III, it is clear that a linear mapping $\tau \in \mathscr{S}$ mapping $\mathfrak{M}_1$ into $\bar{\mathfrak{M}}$ is an $\mathscr{S}$-isomorphism of $\mathfrak{M}_1$ onto $\mathfrak{M}_2$ if and only if $\eta_2(t) = (\tau^t)^{-1}\eta(t)\tau$, $t \in H$. If such a relation holds, then we say that η_2 is *cohomologous* to η_1 (relative to $\bar{A}$). The set $H^1(H, \bar{A})$ of cohomology classes defined in this way is bijective with the set of $\mathscr{S}$-isomorphism classes of Φ-forms of $\bar{\mathfrak{M}}/\Lambda$ contained in $\mathscr{C}$.

Similar considerations apply in the situation Λ/Φ finite dimensional purely inseparable of exponent one. Again let $\bar{\mathfrak{M}}/\Lambda \in \mathscr{C}$, and let $\mathfrak{M}_1/\Phi$ be a Φ-form of $\bar{\mathfrak{M}}$ contained in $\mathscr{C}$, α_0 the corresponding precocycle of $\mathfrak{F} = \mathrm{Der}\,\Lambda/\Phi$. If we make $\mathrm{Hom}_\Lambda(\bar{\mathfrak{M}}, \bar{\mathfrak{M}})$ an $\mathfrak{F}$-module as in Sec. II by defining $E^f = [E, \alpha_0(f)]$, $E \in \mathrm{Hom}_\Lambda(\bar{\mathfrak{M}}, \bar{\mathfrak{M}})$, $f \in \mathfrak{F}$, then $\bar{\mathfrak{D}} = \mathrm{Der}(\bar{\mathfrak{M}}/P, \mathscr{D})$ is a submodule, and we define $Z^1(\mathfrak{F}, \bar{\mathfrak{I}})$ to be the set of 1-cocycles of $\mathfrak{F}$ whose values $\lambda(f) \in \bar{\mathfrak{D}}$. We have a bijection of $Z^1(\mathfrak{F}, \bar{\mathfrak{D}})$ with the set of Φ-forms of $\bar{\mathfrak{M}}/\Lambda$ which belong to $\mathscr{C}$. If $\mathfrak{M}_1/\Phi$, $\mathfrak{M}_2/\Phi$ are two such Φ-forms with corresponding 1-cocycles λ_1 and λ_2, then the $\mathscr{S}$-isomorphisms of $\mathfrak{M}_1$ onto $\mathfrak{M}_2$ are restrictions to $\mathfrak{M}_1$ of elements $\tau \in \mathrm{Aut}(\bar{\mathfrak{M}}/\Lambda, \mathscr{S})$ such that $\lambda_2(f) = \tau^{-1}\lambda_1(f)\tau - \tau^{-1}\tau^f$, $f \in \mathfrak{F}$. If such a τ exists, then λ_2 and λ_1 are said to be cohomologous. This is an equivalence relation. The set $H^1(\mathfrak{F}, \bar{\mathfrak{I}})$ of equivalence classes in $Z^1(\mathfrak{F}, \bar{\mathfrak{I}})$ defined by this relation is bijective with the set of $\mathscr{S}$-isomorphism classes of Φ-forms of $\bar{\mathfrak{M}}/\Lambda$ contained in $\mathscr{C}$.

We remark that the situation considered in Secs. I and II is the special case of the present one in which $\mathscr{C}$ is the complete set of forms of $\tilde{\mathfrak{M}}/P$, $\mathscr{S}$ the complete set of semi-isomorphisms of these forms, and $\mathscr{D}$ the complete set of semiderivations of the forms. The results established in Theorem 5 can now be stated as 1) $H^1(G, \mathrm{Aut}\,\tilde{\mathfrak{M}}/P) = 1$ if P/Φ is Galois and G is the Galois group, and 2) $H^1(\mathfrak{I}, \mathrm{Hom}_P(\tilde{\mathfrak{M}}, \tilde{\mathfrak{M}})) = 0$, where P/Φ is finite dimensional purely inseparable of exponent one. Also, in 1, 1 is the cohomology class of the 1-cocycle such that $\lambda(s) = 1$

and, in 2 it is the 1-cocycle of the cohomology class 0 such that $0(d) = 0$, $d \in \mathfrak{Z}$.

VI. DERIVED CLASSES

Let $(\tilde{\mathfrak{M}}/P, \mathscr{C}, \mathscr{S}, \mathscr{D})$ and $(\tilde{\mathfrak{M}}'/P, \mathscr{C}', \mathscr{S}', \mathscr{D}')$ be two descent classes based on the vector spaces $\tilde{\mathfrak{M}}/P$ and $\tilde{\mathfrak{M}}'/P$, respectively. We shall say that the second is a *derived* class of the first if the following conditions hold:

1) For every subfield Φ of P we have a mapping $\mathfrak{M} \to \mathfrak{M}'$ of the set of Φ-forms contained in $\mathscr{C}$ into the set of Φ-forms contained in $\mathscr{C}'$. This commutes with extension of the base field in the sense that, if Λ/Φ is a subfield of P/Φ, then $(\Lambda\mathfrak{M})' = \Lambda\mathfrak{M}'$.

2) To each $\sigma \in \mathscr{S}$ which is an s-semilinear mapping of the Φ-form $\mathfrak{M} \in \mathscr{C}$ onto the Φ^s-form $\mathfrak{N} \in \mathscr{C}$, there corresponds a $\sigma' \in \mathscr{S}'$ which is an s-semilinear mapping of $\mathfrak{M}'$ onto $\mathfrak{N}'$. If $\mathfrak{P} \in \mathscr{C}$ and $\tau \in \mathscr{S}$ has domain $\mathfrak{N}$ and range $\mathfrak{P}$, then $(\sigma\tau)' = \sigma'\tau'$. Also, if Λ/Φ is a subfield of P/Φ and $\bar{\sigma} \in \mathscr{S}$ is an extension of σ to $\Lambda\mathfrak{M}$, then $\bar{\sigma}'$ is an extension of σ' to $\Lambda\mathfrak{M}'$.

3) To each $D \in \mathscr{D}$ which is a d-semiderivation of a Φ-form $\mathfrak{M} \in \mathscr{C}$ there corresponds a d-semiderivation D' of $\mathfrak{M}'$. We have $(D_1 + D_2)' = D_1' + D_2'$, $(D\varphi)' = D'\varphi$ for $\varphi \in \Phi$, $[D_1D_2]' = [D_1'D_2']$. If Λ/Φ is a subfield of P/Φ and $\bar{D} \in \mathscr{D}$ is an extension of D to $\Lambda\mathfrak{M}$, then $\bar{D}'$ is an extension of D' to $\Lambda\mathfrak{M}'$.

4) If σ and D are as in 1 and 2 then $(\sigma^{-1}D\sigma)' = (\sigma')^{-1}D'\sigma'$.

We shall call the mappings $\mathfrak{M} \to \mathfrak{M}'$, $\sigma \to \sigma'$, $D \to D'$ *canonical.*

Examples. 1) Let $\tilde{\mathfrak{A}}/P$ be an associative algebra with an involution J (Ex. 2 of Sec. III) and let $(\tilde{\mathfrak{A}}, \mathscr{C}, \mathscr{S}, \mathscr{D})$ be the descent class on $\tilde{\mathfrak{A}}$ determined by the algebra structure (including J) as indicated in Sec. V. Let $\tilde{\mathfrak{A}}' = \mathfrak{K}(\tilde{\mathfrak{A}}, J)$, the Lie algebra of J-skew elements of $\tilde{\mathfrak{A}}$ (multiplication $[ab] = ab - ba$), and let $(\tilde{\mathfrak{A}}', \mathscr{C}', \mathscr{S}', \mathscr{D}')$ be the descent class determined by the algebra structure on $\tilde{\mathfrak{A}}'$. If $\mathfrak{A}$ is a Φ-form of $\tilde{\mathfrak{A}}$ contained in $\mathscr{C}$, then $\mathfrak{A}$ is a Φ-subalgebra of $\tilde{\mathfrak{A}}$ in the usual sense mapped into itself by J, and $\mathfrak{A}' = \mathfrak{K}(\mathfrak{A}, J)$ is a form of $\tilde{\mathfrak{A}}'$. If σ is an s-semi-isomorphism of a form $\mathfrak{A}$ of the algebra $\tilde{\mathfrak{A}}/P$ onto the form $\mathfrak{B}$, then $J\sigma = \sigma J$, which implies that σ maps $\mathfrak{A}'$ onto $\mathfrak{B}'$. Then the restriction σ' of σ to $\mathfrak{A}'$ is an s-semi-isomorphism of $\mathfrak{A}'$ onto $\mathfrak{B}'$. If D is a d-semiderivation of $\mathfrak{A}$, then D maps $\mathfrak{A}'$ into itself, and the restriction D' is a d-semiderivation of $\mathfrak{A}'$. It is easy to check that the mappings $\mathfrak{A} \to \mathfrak{A}'$, $\sigma \to \sigma'$, $D \to D'$ satisfy the foregoing conditions. Hence the descent class defined by the algebra structure on $\tilde{\mathfrak{A}}'$ is a derived class of that defined by $\tilde{\mathfrak{A}}$ with J. We shall say also that $\tilde{\mathfrak{A}}'$ is a *derived algebra* of $\tilde{\mathfrak{A}}$.

In place of the $\tilde{\mathfrak{A}}'$ just used, we can also take $\tilde{\mathfrak{A}}' = [\mathfrak{K}(\mathfrak{A}, J), \mathfrak{K}(\mathfrak{A}, J)]$ the derived algebra in the usual sense of the Lie algebra $\mathfrak{K}(\mathfrak{A}, J)$. Then we let $\mathfrak{A}' = [\mathfrak{K}(\mathfrak{A}, J), \mathfrak{K}(\mathfrak{A}, J)]$, s' the restriction of s to $\mathfrak{A}'$, D' the restriction of D to $\mathfrak{A}'$.

2) Let $\tilde{\mathfrak{A}}$ be as in 1, and let $\tilde{\mathfrak{A}}' = \mathfrak{H}(\tilde{\mathfrak{A}}, J)$ the space of J-symmetric elements of $\tilde{\mathfrak{A}}$. If the characteristic is $\neq 2$, we consider $\tilde{\mathfrak{A}}'$ as Jordan algebra relative to the product $a \cdot b = \frac{1}{2}(ab + ba)$. If the characteristic is two, then we use the polynomial mappings $a \rightarrow a^2$, $(a, b) \rightarrow aba$ to make $\tilde{\mathfrak{A}}'$ an algebra. As in example 1, we let $\mathfrak{A}' = \mathfrak{A} \cap \mathfrak{H}(\tilde{\mathfrak{A}}, J)$ and let σ' and D' for the semi-isomorphism σ and the semiderivation D be the restriction mappings. Then $\mathfrak{A} \rightarrow \mathfrak{A}'$, $\sigma \rightarrow \sigma'$, $D \rightarrow D'$ define a derived algebra of $\tilde{\mathfrak{A}}/P$.

3) Let $(\tilde{\mathfrak{M}}, \mathscr{C}, \mathscr{S}, \mathscr{D})$ be an arbitrary descent class, and let $\tilde{\mathfrak{M}}' = \mathrm{Der}(\tilde{\mathfrak{M}}/P, \mathscr{D})$ considered as a Lie algebra over P. By axiom 7 for descent classes, $\mathfrak{M}' \equiv \mathrm{Der}(\mathfrak{M}/\Phi, \mathscr{D})$ is a Φ-form of $\tilde{\mathfrak{M}}'$ if $\mathfrak{M}/\Phi \in \mathscr{C}$. Let $\sigma \in \mathscr{S}$ be an s-semilinear mapping of $\mathfrak{M}/\Phi \in \mathscr{C}$ onto $\mathfrak{N}/\Phi^s \in \mathscr{C}$. If $D \in \mathfrak{M}'$, then, by axiom 5, $\sigma^{-1}D\sigma \in \mathfrak{N}'$. Also, it is clear that σ': $D \rightarrow \sigma^{-1}D\sigma$ is an s-semi-isomorphism of the Lie algebra $\mathfrak{M}'/\Phi$ onto the Lie algebra $\mathfrak{N}'/\Phi$. Next let E be a d-semiderivation of $\mathfrak{M}/\Phi$ contained in $\mathscr{D}$. Then $[D, E] \in \mathrm{Der}(\mathfrak{M}/\Phi, \mathscr{D})$, and $E' : D \rightarrow [D, E]$ is a d-semiderivation of the Lie algebra $\mathfrak{M}'/\Phi$. Let D, E, and σ be as indicated. Then

$$\begin{aligned} D(\sigma^{-1}E\sigma)' &= [D, \sigma^{-1}E\sigma] \\ &= D\sigma^{-1}E\sigma - \sigma^{-1}E\sigma D \\ &= \sigma^{-1}\sigma D\sigma^{-1}E\sigma - \sigma^{-1}E\sigma D\sigma^{-1}\sigma \\ &= \sigma^{-1}(\sigma D\sigma^{-1}E - E\sigma D\sigma^{-1})\sigma \\ &= D(\sigma')^{-1}E'\sigma'. \end{aligned}$$

Hence condition 4 for derived classes holds. It is easy to check that the other conditions hold for the mappings $\mathfrak{M} \rightarrow \mathfrak{M}'$, $\sigma \rightarrow \sigma'$, $D \rightarrow D'$. Hence $\tilde{\mathfrak{M}}'$ with the descent class determined by the Lie algebra structure is a derived class of $(\tilde{\mathfrak{M}}, \mathscr{C}, \mathscr{S}, \mathscr{D})$.

VII. COMPLETENESS THEOREMS

Let $(\tilde{\mathfrak{M}}'/P, \mathscr{C}', \mathscr{S}', \mathscr{D}')$ be a derived class of $(\tilde{\mathfrak{M}}, \mathscr{C}, \mathscr{S}, \mathscr{D})$, $\mathfrak{M} \rightarrow \mathfrak{M}'$, $\sigma \rightarrow \sigma'$, $D \rightarrow D'$ the canonical mappings. Let $\overline{\mathfrak{M}}/\Lambda \in \mathscr{C}$, and consider the groups $\bar{A} = \mathrm{Aut}(\overline{\mathfrak{M}}/\Lambda, \mathscr{S})$, $\bar{A}' = \mathrm{Aut}(\overline{\mathfrak{M}}'/\Lambda, \mathscr{S}')$. It is clear from condition 2 on derived classes that $\tau \to \tau'$ is a homomorphism of $\bar{A}$ into $\bar{A}'$. Next let Λ/Φ be Galois, and let $\mathfrak{M}/\Phi \in \mathscr{C}$ be a Φ-form of $\bar{\mathfrak{M}}$ with corresponding precocycle α of the Galois group H of Λ/Φ. Since the canonical mapping of $\mathrm{Aut}(\mathfrak{M}/\Phi, \mathscr{S})$ into $\mathrm{Aut}(\mathfrak{M}'/\Phi, \mathscr{S}')$ is a homomorphism, the identity mapping on $\mathfrak{M}$ is mapped into the identity mapping on $\mathfrak{M}'$ by the canonical mapping. If $t \in H$, then $\alpha(t)$ is the t-semilinear extension to $\bar{\mathfrak{M}}$ of the identity mapping on $\mathfrak{M}$. Hence by 2, $\alpha(t)'$ is the t-semilinear extension to $\bar{\mathfrak{M}}'$ of the identity on $\mathfrak{M}'$. Hence, if α' denotes the precocycle of $\mathfrak{M}'$ as form of $\bar{\mathfrak{M}}'$, then we have $\alpha'(t) = \alpha(t)'$. Now let $\mathfrak{M}_0/\Phi \in \mathscr{C}$ be a Φ-form of $\bar{\mathfrak{M}}$, α_0 the corresponding precocycle. We use this to make $\bar{A}$ an H-group in the usual way. Similarly, we use the precocycle α_0' of $\mathfrak{M}'$ to make $\bar{A}'$ an H-group. Then if $\tau \in \bar{A}$ and $t \in H$, we have $(\tau^t)' = [\alpha_0(t)^{-1}\tau\alpha_0(t)]' = \alpha_0'(t)^{-1}\tau'\alpha_0(t)' = \tau'^t$. Hence, the canonical homomorphism $\tau \to \tau'$ is an H-homomorphism of $\bar{A}$ into $\bar{A}'$. If η is the 1-cocycle of H into $\bar{A}$ corresponding to $\mathfrak{M}$, then we have $\eta(t) = \alpha_0(t)^{-1}\alpha(t)$. The 1-cocycle of H into $\bar{A}'$ corresponding to $\mathfrak{M}'$ is η', where $\eta'(t) = \alpha_0'(t)^{-1}\alpha'(t) = \eta(t)'$.

Similar results hold in the case Λ/Φ finite dimensional purely inseparable of exponent one. Let $\mathfrak{F} = \mathrm{Der}\,\Lambda/\Phi$, $\bar{\mathfrak{D}} = \mathrm{Der}(\bar{\mathfrak{M}}/\Lambda, \mathscr{D})$, $\bar{\mathfrak{D}}' = \mathrm{Der}(\bar{\mathfrak{M}}'/\Lambda, \mathscr{D}')$. The canonical mapping gives a Lie algebra homomorphism of $\bar{\mathfrak{D}}/\Lambda$ into $\bar{\mathfrak{D}}'/\Lambda$. Moreover, if we make $\bar{\mathfrak{D}}$ and $\bar{\mathfrak{D}}'$ into $\mathfrak{F}$-modules by means of the precocycles α_0 and α_0' of $\mathfrak{M}_0/\Phi \in \mathscr{C}$, a Φ-form of $\bar{\mathfrak{M}}/\Lambda$ and of its image $\mathfrak{M}_0'$, then the canonical homomorphism is an $\mathfrak{F}$-homomorphism. If λ is the 1-cocycle of the Φ-form $\mathfrak{M} \in \mathscr{C}$ of $\bar{\mathfrak{M}}$, then the 1-cocycle λ' of $\mathfrak{M}'$ satisfies $\lambda'(f) = \lambda(f)'$, $f \in \mathfrak{F}$.

We now assume P algebraically closed, Φ a subfield such that P/Φ is algebraic. We shall give conditions that the mapping $\mathfrak{M} \to \mathfrak{M}'$ of the set of Φ-forms of $\tilde{\mathfrak{M}}/P$ contained in $\mathscr{C}$ is bijective with the set of Φ-forms of $\tilde{\mathfrak{M}}'/P$ contained in $\mathscr{C}'$ and that the same thing holds for the mapping $\tau \to \tau'$ of the $\mathscr{S}$-isomorphisms of these forms. These results are directly applicable to classification problems for algebras. However, we shall not indicate these applications explicitly but refer the reader to the references given in the introduction.

We make the following two hypotheses:

(α) The canonical homomorphism $\tau \to \tau'$ of $\mathrm{Aut}(\tilde{\mathfrak{M}}/P, \mathscr{S})$ into $\mathrm{Aut}(\tilde{\mathfrak{M}}'/P, \mathscr{S}')$ is bijective.

(β) The canonical homomorphism $D \to D'$ of $\mathrm{Der}(\bar{\mathfrak{M}}/P, \mathscr{D})$ into $\mathrm{Der}(\tilde{\mathfrak{M}}'/P, \mathscr{D}')$ is bijective.

We note first that (β) carries over to forms, that is, if $\mathfrak{M}/\Phi \in \mathscr{C}$, then the canonical homomorphism of $\mathrm{Der}(\mathfrak{M}/\Phi, \mathscr{D})$ into $\mathrm{Der}(\mathfrak{M}'/\Phi, \mathscr{D}')$ is bijective. This is clear since $\mathrm{Der}(\mathfrak{M}/\Phi, \mathscr{D})$ and $\mathrm{Der}(\mathfrak{M}'/\Phi, \mathscr{D}')$ are Φ-forms of $\operatorname{Der}(\tilde{\mathfrak{M}}/P, \mathscr{D})$ and $\operatorname{Der}(\tilde{\mathfrak{M}}'/P, \mathscr{D}')$, respectively, and the canonical homomorphism between the latter is the linear extension of that between the former Lie algebras (cf. condition 3 of the definition of derived classes). We note next that (α) implies that, if ρ_1 and ρ_2 are $\mathscr{S}$-isomorphisms of $\mathfrak{M}/\Phi \in \mathscr{C}$ such that $\rho_1' = \rho_2'$, then $\rho_1 = \rho_2$. Let $\bar{\rho}_1$ and $\bar{\rho}_2$ be the linear extensions of ρ_1 and ρ_2 to $\tilde{\mathfrak{M}}$. Then $\bar{\rho}_i \in \operatorname{Aut}(\tilde{\mathfrak{M}}/P, \mathscr{S})$, and $\bar{\rho}_i'$ is the linear extension of ρ_i' to $\tilde{\mathfrak{M}}'$. Hence $\bar{\rho}_i' \in \operatorname{Aut}(\tilde{\mathfrak{M}}'/P, \mathscr{S}')$, and $\bar{\rho}_1' = \bar{\rho}_2'$. Then ($\alpha$) implies that $\bar{\rho}_1 = \bar{\rho}_2$, which gives $\rho_1 = \rho_2$. We now prove the following.

Theorem 6. *Let P be algebraically closed, $(\tilde{\mathfrak{M}}/P, \mathscr{C}, \mathscr{S}, \mathscr{D})$ a descent class on $\tilde{\mathfrak{M}}/P$, $(\tilde{\mathfrak{M}}'/P, \mathscr{C}', \mathscr{S}', \mathscr{D}')$ a derived class satisfying the conditions (α) and (β). Let Φ be a subfield of P such that P/Φ is algebraic, $\mathfrak{M}/\Phi$, $\mathfrak{N}/\Phi \in \mathscr{C}$, σ an $\mathscr{S}'$-isomorphism of $\mathfrak{M}'$ onto $\mathfrak{N}'$. Then there exists a unique $\mathscr{S}$-isomorphism ρ of $\mathfrak{M}$ onto $\mathfrak{N}$ such that $\rho' = \sigma$. If $\mathfrak{M} = \mathfrak{N}$, then $\rho \to \rho'$ is an isomorphism of* $\operatorname{Aut}(\mathfrak{M}/\Phi, \mathscr{S})$ *onto* $\operatorname{Aut}(\mathfrak{M}'/\Phi, \mathscr{S}')$.

Proof. The last statement is an immediate consequence of the first one. Also, uniqueness of ρ has been established above. Hence only the existence of ρ has to be proved. Now the linear extension of σ to $\tilde{\mathfrak{M}}'$ is contained in $\operatorname{Aut}(\tilde{\mathfrak{M}}'/P, \mathscr{S}')$. Hence, by ($\alpha$), this has the form τ', where $\tau \in \operatorname{Aut}(\tilde{\mathfrak{M}}/P, \mathscr{S})$. We wish to prove that $\mathfrak{M}^\tau = \mathfrak{N}$. Let Δ/Φ be the perfect closure of Φ in P, so that $\Delta = \Phi$ if the characteristic is 0 and Δ is the set of p^eth roots of the elements of Φ, $e = 1, 2, 3, \ldots$, if the characteristic is $p \neq 0$. In any case, P/Δ is Galois. Let G be the Galois group of P/Δ, and make $\operatorname{Aut}(\tilde{\mathfrak{M}}/P, \mathscr{S})$ and $\operatorname{Aut}(\tilde{\mathfrak{M}}'/P, \mathscr{S}')$ G-groups by means of the pre-cocycles corresponding to a Δ-form $\overline{\mathfrak{M}}_0 \in \mathscr{C}$ (e.g., $\Delta\mathfrak{N}$) and its corresponding Δ-form $\overline{\mathfrak{M}}_0' \in \mathscr{C}'$. Thus the canonical mapping of $\operatorname{Aut}(\tilde{\mathfrak{M}}/P, \mathscr{S})$ into $\operatorname{Aut}(\tilde{\mathfrak{M}}'/P, \mathscr{S}')$, which is a G-homomorphism, is a G-isomorphism of the first of these groups onto the second one, by (α). Also, we have seen that, if η is the 1-cocycle of G into $\operatorname{Aut}(\tilde{\mathfrak{M}}/P, \mathscr{S})$ corresponding to a Δ-form $\overline{\mathfrak{M}} \in \mathscr{C}$, then the 1-cocycle η' of G into $\operatorname{Aut}(\tilde{\mathfrak{M}}'/P, \mathscr{S}')$ corresponding to $\overline{\mathfrak{M}}'$ satisfies $\eta'(s) = \eta(s)'$, $s \in G$. It follows from this that the canonical mapping $\overline{\mathfrak{M}} \to \overline{\mathfrak{M}}'$ of the set $\mathscr{C}_\Delta$ of Δ-forms of $\tilde{\mathfrak{M}}$ contained in $\mathscr{C}$ into the set $\mathscr{C}_\Delta'$ of Δ-forms of $\tilde{\mathfrak{M}}'$ contained in $\mathscr{C}'$ is 1–1. Now $\Delta\mathfrak{M}^\tau = (\Delta\mathfrak{M})^\tau$, and $\Delta\mathfrak{N} \in \mathscr{C}_\Delta$. Since τ maps $\Delta\mathfrak{M}$ onto $\Delta\mathfrak{M}^\tau$, τ' maps $\Delta\mathfrak{M}'$ onto $(\Delta\mathfrak{M}^\tau)'$. Hence $(\Delta\mathfrak{M}^\tau)' = (\Delta\mathfrak{M}')^{\tau'} = \Delta\mathfrak{M}'^{\tau'} = \Delta\mathfrak{N}' = (\Delta\mathfrak{N})'$. It follows that $\Delta\mathfrak{M}^\tau = \Delta\mathfrak{N}$. If the characteristic is 0, we have $\Delta = \Phi$, and the required relation $\mathfrak{M}^\tau = \mathfrak{N}$ holds. Otherwise, let the characteristic be $p \neq 0$, and let Λ/Φ be any subfield of Δ/Φ such that $\Lambda\mathfrak{M}^\tau = \Lambda\mathfrak{N}$. Let Γ/Φ be a subfield of Λ/Φ such that Λ/Γ is finite dimensional purely inseparable of exponent one, and let

$\mathfrak{F} = \text{Der}\,\Lambda/\Gamma$. Let $\overline{\mathfrak{M}} = \Lambda\mathfrak{M}^\tau = \Lambda\mathfrak{N}$, so that. $\overline{\mathfrak{M}}$ is a Λ-form of $\tilde{\mathfrak{M}}$ contained in $\mathscr{C}$, and let $\overline{\mathscr{C}}_\Gamma$ be the set of Γ-forms of $\overline{\mathfrak{M}}$ contained in $\mathscr{C}$. Then $\Gamma\mathfrak{M}^\tau, \Gamma\mathfrak{N} \in \overline{\mathscr{C}}_\Gamma$. Let $\overline{\mathscr{C}}_\Gamma{}'$ be the set of Γ-forms of $\overline{\mathfrak{M}}'$ contained in $\mathscr{C}$. We now make $\text{Der}(\overline{\mathfrak{M}}/\Lambda, \mathscr{D})$ and $\text{Der}(\overline{\mathfrak{M}}'/\Lambda, \mathscr{D}')$ $\mathfrak{F}$-modules by means of a precocycle corresponding to $\mathfrak{M}_0 \in \overline{\mathscr{C}}_\Gamma$ and $\mathfrak{M}_0{}' \in \overline{\mathscr{C}}_\Gamma{}'$. Then the canonical mapping $\bar{D} \to \bar{D}'$ of $\text{Der}(\overline{\mathfrak{M}}/\Lambda, \mathscr{D})$ into $\text{Der}(\overline{\mathfrak{M}}'/\Lambda, \mathscr{D}')$ is bijective. Hence the argument used in the Galois case implies that $\Gamma\mathfrak{M}^\tau = \Gamma\mathfrak{N}$. We now consider the collection of subfield Λ/Φ of Δ/Φ such that $\Lambda\mathfrak{M}^\tau = \Lambda\mathfrak{N}$, and we order this set by reverse inclusion. The result just established and Zorn's lemma imply that Φ is in our collection. Hence, we have $\mathfrak{M}^\tau = \mathfrak{N}$. The restriction ρ of τ to $\mathfrak{M}$ is an $\mathscr{S}$-isomorphism of $\mathfrak{M}$ onto $\mathfrak{N}$, and the linear extension of ρ' to $\tilde{\mathfrak{M}}'$ is τ'. Since the linear extension of σ to $\tilde{\mathfrak{M}}'$ is also τ', we have $\rho' = \sigma$, as required.

Let $(\tilde{\mathfrak{M}}/P, \mathscr{C}, \mathscr{S}, \mathscr{D})$ be a descent class, Φ a subfield of P, $\overline{\mathfrak{M}} \in \mathscr{C}$ a Λ-form of $\tilde{\mathfrak{M}}$, where Λ/Φ is a subfield of P/Φ. Then we shall say that the subfield Σ/Λ of P/Λ is a Φ-splitting field of $\overline{\mathfrak{M}}$ (relative to the class) if there exists a Φ-form $\mathfrak{M} \in \mathscr{C}$ such that $\Sigma\mathfrak{M} = \Sigma\overline{\mathfrak{M}}$. It is clear on considering bases that, if Σ has this property, then a suitable finitely generated subfield also has the property. Hence, if P/Φ is algebraic, then the existence of a *splitting field* Σ implies the existence of such a field that is a subfield of Σ and is finite dimensional over Λ. We shall now impose the following further condition which is a condition on the descent class $(\tilde{\mathfrak{M}}/P, \mathscr{C}, \mathscr{S}, \mathscr{D})$ and on the subfield Φ of P:

(γ) Every Λ-form $\overline{\mathfrak{M}} \in \mathscr{C}$ for Λ/Φ a subfield of P/Φ has a Φ-splitting field which is separable over Λ.

Example. Let $\tilde{\mathfrak{A}}/P$ be the associative algebra P_n of $n \times n$ matrices over P, $\mathscr{C}$ the set of algebra forms of $\tilde{\mathfrak{A}}$. These are central simple algebras of n^2 dimensions. Let P be algebraically closed, Φ a subfield such that P/Φ is algebraic, and let $\overline{\mathfrak{A}}$ be a Λ-form of the algebra $\tilde{\mathfrak{A}}$, where Λ/Φ is a subfield of P/Φ. Let Σ/Λ be a separable splitting field for $\mathfrak{A}/\Lambda$ in the classical sense. Since P is the algebraic closure of Φ, we may assume that Σ/Λ is a subfield of P/Λ. Also, the classical definition shows that we can choose a basis $(u_1, u_2, \ldots, u_{n^2})$ for $\Sigma\overline{\mathfrak{A}}/\Sigma$ such that multiplication constants γ_{ijk} $(u_iu_j = \Sigma\gamma_{ijk}u_k)$ are in the prime field and hence are in Φ. Then $\mathfrak{A} = \Phi u_1 + \Phi u_2 + \cdots + \Phi u_{n^2}$ is a Φ-form of the algebra $\tilde{\mathfrak{A}}/P$ such that $\Sigma\mathfrak{A} = \Sigma\overline{\mathfrak{A}}$. Hence condition ($\gamma$) is fulfilled for the descent class determined by $\tilde{\mathfrak{A}}$ and the subfield Φ.

We can now prove our main result.

Theorem 7. *Let P be algebraically closed, $(\tilde{\mathfrak{M}}/P, \mathscr{C}, \mathscr{S}, \mathscr{D})$ a descent class, Φ a subfield of P such that P/Φ is algebraic, and condition (γ) holds.*

Let $(\tilde{\mathfrak{M}}'/P, \mathscr{C}', \mathscr{S}', \mathscr{D}')$ *be a derived class satisfying* (α) *and* (β). *Then the canonical mapping* $\mathfrak{M} \to \mathfrak{M}'$ *is a bijection of the set of* Φ*-forms of* $\tilde{\mathfrak{M}}$ *contained in* $\mathscr{C}$ *into the set of* Φ*-forms of* $\tilde{\mathfrak{M}}'$ *contained in* $\mathscr{C}'$. *Also, the canonical mapping* $\tau \to \tau'$ *is a bijection of the set of* $\mathscr{S}$*-isomorphisms of the* Φ*-forms contained in* $\mathscr{C}$ *into the set of* $\mathscr{S}'$*-isomorphisms of the set of* Φ*-forms contained in* $\mathscr{C}'$.

Proof. We note first that the hypothesis (γ) applied to $\bar{\mathfrak{M}} = \tilde{\mathfrak{M}}$ implies that there exists a Φ-form $\mathfrak{M}_0$ of $\tilde{\mathfrak{M}}$ contained in $\mathscr{C}$. Let $\mathfrak{P}$ be a Φ-form of $\tilde{\mathfrak{M}}'$ contained in $\mathscr{C}'$. We have $P\mathfrak{M}_0' = \tilde{\mathfrak{M}}' = P\mathfrak{P}$. Hence there exists a finite dimensional subfield Λ/Φ of P/Φ such that $\Lambda\mathfrak{M}_0' = \Lambda\mathfrak{P}$. We now consider pairs Λ, $\mathfrak{M}_0$, Λ/Φ finite dimensional, $\mathfrak{M}_0$ a Φ-form of $\tilde{\mathfrak{M}}$ contained in $\mathscr{C}$ such that $\Lambda\mathfrak{M}_0' = \Lambda\mathfrak{P}$, and we shall prove by induction on the degree of inseparability of Λ/Φ that there exists a Φ-form $\mathfrak{M} \in \mathscr{C}$ such that $\mathfrak{M}' = \mathfrak{P}$. If the degree of inseparability is 1, which means that Λ/Φ is separable, then, by replacing Λ by a larger subfield of P, if necessary, we may assume that Λ/Φ is Galois. Let $\bar{\mathfrak{M}} = \Lambda\mathfrak{M}_0$, and make $\bar{A} = \mathrm{Aut}(\bar{\mathfrak{M}}/\Lambda, \mathscr{S})$, and $\bar{A}' = \mathrm{Aut}(\bar{\mathfrak{M}}'/\Lambda, \mathscr{S}')$ into H-groups, H the Galois group of Λ/Φ, by using the precocycles of $\mathfrak{M}_0$ and $\mathfrak{M}_0'$. Then, by Theorem 6, the canonical mapping of $\bar{A}$ into $\bar{A}'$ is an H-isomorphism of $\bar{A}$ onto $\bar{A}'$. It follows as in the proof of Theorem 6 that the canonical mapping $\mathfrak{M} \to \mathfrak{M}'$ of the set $\bar{\mathscr{C}}_\Phi$ of Φ-forms of $\bar{\mathfrak{M}}$ contained in $\mathscr{C}$ into the set $\bar{\mathscr{C}}_\Phi'$ of Φ-forms of $\bar{\mathfrak{M}}'$ contained in $\mathscr{C}'$ is bijective. Now $\mathfrak{P} \in \bar{\mathscr{C}}_\Phi'$, since $\bar{\mathfrak{M}}' = \Lambda\mathfrak{M}_0' = \Lambda\mathfrak{P}$. Hence $\mathfrak{P} = \mathfrak{M}'$ for an $\mathfrak{M} \in \bar{\mathscr{C}}_\Phi$. Now assume Λ/Φ has degree of inseparability > 1. Then there exists a subfield Γ/Φ of Λ/Φ of lower degree of inseparability than Λ/Φ such that Λ/Γ is purely inseparable of exponent one. Let $\mathfrak{F} = \mathrm{Der}\,\Lambda/\Gamma$, $\bar{\mathfrak{M}} = \Lambda\mathfrak{M}_0$, and make $\mathfrak{D} = \mathrm{Der}(\bar{\mathfrak{M}}/\Lambda, \mathscr{D})$ and $\bar{\mathfrak{D}}' = \mathrm{Der}(\bar{\mathfrak{M}}'/\Lambda, \mathscr{D}')$ $\mathfrak{F}$-modules by using the precocycles of the Γ-forms $\Gamma\mathfrak{M}_0$ and $\Gamma\mathfrak{M}_0'$. Then the canonical mapping of $\bar{\mathfrak{D}}$ into $\bar{\mathfrak{D}}'$ is bijective. Hence, as in the proof of Theorem 6, the canonical mapping of the set $\bar{\mathscr{C}}_\Gamma$ of Γ-forms of $\bar{\mathfrak{M}}$ contained in $\mathscr{C}$ into the set $\bar{\mathscr{C}}_\Gamma'$ of Γ-forms of $\bar{\mathfrak{M}}'$ contained in $\mathscr{C}'$ is bijective. Since $\Lambda\mathfrak{P} = \bar{\mathfrak{M}}'$, $\Gamma\mathfrak{P} \in \bar{\mathscr{C}}_\Gamma'$. Hence there exists an $\mathfrak{N} \in \bar{\mathscr{C}}_\Gamma$ such that $\mathfrak{N}' = \Gamma\mathfrak{P}$. By (γ), there exists a separable subfield Σ/Γ of P/Γ such that $\Sigma\mathfrak{N} = \Sigma\mathfrak{N}_0$ for a suitable Φ-form $\mathfrak{N}_0 \in \mathscr{C}$. Also, we may assume Σ/Γ finite dimensional. Then Σ/Φ is finite dimensional and of lower degree of inseparability that Λ/Φ. Moreover, $\Sigma\mathfrak{N}_0' = \Sigma\mathfrak{N}' = \Sigma\mathfrak{P}$. Hence, the induction hypothesis applied to Σ and $\mathfrak{N}_0$ implies that there exists a Φ-form $\mathfrak{M}$ contained in $\mathscr{C}$ such that $\mathfrak{M}' = \mathfrak{P}$. Now assume that $\mathfrak{M}_1$ and $\mathfrak{M}_2$ are Φ-forms of $\tilde{\mathfrak{M}}$ contained in $\mathscr{C}$, and $\mathfrak{M}_1' = \mathfrak{P} = \mathfrak{M}_2'$. Then Theorem 6 implies that there exists an $\mathscr{S}$-isomorphism ρ of $\mathfrak{M}_1$ onto $\mathfrak{M}_2$ such that ρ' is the identity mapping on $\mathfrak{P}$, which can be considered an $\mathscr{S}'$-isomorphism of $\mathfrak{M}_1'$ onto $\mathfrak{M}_2'$. Since the extension of ρ' is the identity mapping in $\tilde{\mathfrak{M}}'$, it follows that the same thing holds for ρ. Then $\mathfrak{M}_1 = \mathfrak{M}_2$. The rest of the theorem follows from Theorem 6.

References

1. W. Landherr, Über einfache Liesche Ringe. *Abh. Math. Sem. Univ. Hamburg* **11**, 41–64 (1935).
2. W. Landherr, Liesche Ringe von Typus *A*. *Abh. Math. Sem. Univ. Hamburg* **12**, 200–241 (1938).
3. N. Jacobson, A class of normal simple Lie Algebras of characteristic zero. *Ann. of Math.* **38**, 508–517 (1937).
4. N. Jacobson, Simple Lie algebras of type A. *Ann. of Math.* **39**, 181–188 (1938).
5. N. Jacobson, Simple Lie Algebras over a field of characteristic zero. *Duke Math. J.* **4**, 534–551 (1938).
6. N. Jacobson, Classes of restricted Lie algebras of characteristic *p*. I, *Amer. J. Math.* **63**, 481–515 (1941); II. *Duke Math. J.* **10**, 107–121 (1943).
7. A. Weil, Algebras with involution and the classical groups. *J. Indian Math. Soc.* **24**, 589–623 (1960).
8. G. K. Kalisch, On special Jordan algebras. *Trans. Amer. Math. Soc.* **61**, 482–494 (1947).
9. F. D. Jacobson and N. Jacobson, Classification and representation of semi-simple Jordan algebras. *Trans. Amer. Math. Soc.* **65**, 141–149 (1949).
10. N. Jacobson, Cayley numbers and simple Lie algebras of type *G*. *Duke Math. J.* **5**, 775–783 (1939).
11. M. L. Tomber, Lie algebras of type F. *Proc. Amer. Math. Soc.* **4**, 759–768 (1953).
12. M. L. Tomber, Lie algebras of type *A*, *B*, *C*, *D*, *F*. *Trans. Amer. Math. Soc.* **88**, 99–106 (1958).
13. R. T. Barnes, On Derivation Algebras and Lie Algebras of Prime Characteristic. Yale Univ. dissertation (1963).
14. M. Gerstenhaber, On the Galois theory of inseparable extensions. *Bull. Amer. Math. Soc.* (1964).
15. T. Springer, On the equivalence of quadratic forms. *Proc. Amsterdam Akad. Sci.* **A62**, 241–253 (1959).

KONINKL. NEDERL. AKADEMIE VAN WETENSCHAPPEN – AMSTERDAM
Reprinted from Proceedings, Series A, 69, No. 2 and Indag. Math., 28, No. 2, 1966

MATHEMATICS

ASSOCIATIVE ALGEBRAS WITH INVOLUTION AND JORDAN ALGEBRAS

BY

N. JACOBSON [1])

Dedicated to Professor Hans Freudenthal on his sixtieth birthday

(Communicated by Prof. H. D. KLOOSTERMAN at the meeting of September 25, 1965)

If $\mathfrak{A}$ is an associative algebra with involution J over a field of characteristic $\neq 2$ then $\mathfrak{A}$ determines the Jordan algebra $\mathfrak{H}(\mathfrak{A}, J)$ of J-symmetric elements relative to the Jordan product $a \cdot b = \frac{1}{2}(ab+ba)$. We obtain in this way a functor from the category of associative algebras with involution into the category of Jordan algebras. On the other hand, if $\mathfrak{J}$ is a Jordan algebra, then $\mathfrak{J}$ determines an associative algebra with involution $(\mathrm{su}(\mathfrak{J}), \pi)$ where $\mathrm{su}(\mathfrak{J})$ is the special universal envelope (=universal associative algebra for special representations) and π is the main involution in $\mathrm{su}(\mathfrak{J})$. This defines a functor S from the category of Jordan algebras into the category of associative algebras with involution.

If $\mathfrak{J}$ is a Jordan algebra then we have a natural homomorphism s of $\mathfrak{J}$ into $\mathfrak{H}(\mathrm{su}(\mathfrak{J}), \pi)$ and if $(\mathfrak{A}, J)$ is an associative algebra with involution then we have a natural homomorphism h of $(\mathrm{su}(\mathfrak{H}(\mathfrak{A}, J), \pi)$ into $(\mathfrak{A}, J)$. It is immediate that s is injective if and only if $\mathfrak{J}$ is special and we call $\mathfrak{J}$ reflexive if s is surjective. The homomorphism h is surjective if and only if the subalgebra of $\mathfrak{A}$ generated by $\mathfrak{H}(\mathfrak{A}, J)$ is $\mathfrak{A}$. We call the associative algebra with involution $(\mathfrak{A}, J)$ perfect if h is an isomorphism of $(\mathrm{su}\,\mathfrak{H}(\mathfrak{A}, J), \pi)$ (onto $(\mathfrak{A}, J)$). In § 2 we shall obtain some sufficient conditions for reflexivity of Jordan algebras and for perfection of associative algebras with involution. In § 3 we shall apply the general notions and one of the general results of § 2 to obtain an improved formulation and derivation of the main theorem on the classification of special finite dimensional simple Jordan algebras (cf. F. D. JACOBSON and N. JACOBSON [1]).

1. *Basic notions.* Throughout this paper "algebra" will mean not necessarily associative algebra with an identity element 1 over a field of characteristic $\neq 2$. The usual conventions for algebras with 1 will be adopted: subalgebras contain 1, homomorphisms map 1 into 1, etc. By an *associative algebra with involution* we mean a pair $(\mathfrak{A}, J)$ where $\mathfrak{A}$ is an associative algebra and J is an involution in A, that is, J is an anti-

[1]) This research has been supported by the Air Force Office of Scientific Research under Grant AF–AFOSR–402–64.

automorphism in $\mathfrak{A}$ such that $J^2 = 1$, the identity mapping. An *ideal* $\mathfrak{K}$ of $(\mathfrak{A}, J)$ is an ideal of $\mathfrak{A}$ such that $\mathfrak{K}^J \subseteq \mathfrak{K}$ and a *homomorphism* η of the associative algebra with involution $(\mathfrak{A}, J)$ into the associative algebra with involution $(\mathfrak{B}, K)$ is a homomorphism of $\mathfrak{A}$ into $\mathfrak{B}$ such that $J\eta = \eta K$. The class of associative algebras with involution over a fixed field Φ together with the homomorphisms as morphisms is a category which we shall denote as cat AI/Φ.

Let $(\mathfrak{A}, J)$ be an associative algebra with involution over the field Φ and let $\mathfrak{H}(\mathfrak{A}, J)$ denote the set of J-symmetric elements $(a^J = a)$ of $\mathfrak{A}$. Then $\mathfrak{H}(\mathfrak{A}, J)$ is a subspace of $\mathfrak{A}$ closed under the composition $a \cdot b = \frac{1}{2}(ab + ba)$. We recall that if $\mathfrak{A}$ is any associative algebra then $\mathfrak{A}$ determines a Jordan algebra $\mathfrak{A}^+$ whose underlying vector space is the same as that of $\mathfrak{A}$ and whose multiplication composition is the Jordan product $a \cdot b = \frac{1}{2}(ab + ba)$. We recall also that an algebra $\mathfrak{J}$ over a field Φ is a Jordan algebra if its multiplication composition $a \cdot b$ satisfies the identities: $a \cdot b = b \cdot a$, $(a^{\cdot 2} \cdot b) \cdot a = a^{\cdot 2} \cdot (b \cdot a)$ where $a^{\cdot 2} = a \cdot a$. Let cat J/Φ denote the category of Jordan algebras over the field Φ with the morphisms as homomorphisms. Then the result we noted before is that any associative algebra with involution $(\mathfrak{A}, J)$ determines a Jordan algebra $\mathfrak{H}(\mathfrak{A}, J)$. Next let η be a homomorphism of $(\mathfrak{A}, J)$ into the associative algebra with involution $(\mathfrak{B}, K)$. Then it is clear that η maps $\mathfrak{H}(\mathfrak{A}, J)$ into $\mathfrak{H}(\mathfrak{B}, K)$ and that the restriction $\eta|\mathfrak{H}$ of η to $\mathfrak{H}(A, J)$ is a homomorphism of the Jordan algebra $\mathfrak{H}(\mathfrak{A}, J)$ into $\mathfrak{H}(\mathfrak{B}, K)$. It is immediate that the mappings $(\mathfrak{A}, J) \to \mathfrak{H}(\mathfrak{A}, J)$, $\eta \to \eta|\mathfrak{H}$ where η is a homomorphism of $(\mathfrak{A}, J)$ into $(\mathfrak{B}, K)$ define a functor H of cat AI/Φ into cat J/Φ.

If $\mathfrak{A}$ is an algebra over Φ and P is an extension field of Φ then $\mathfrak{A}_P = P \otimes_\Phi \mathfrak{A}$ is an algebra over P. If $(\mathfrak{A}, J)$ is an associative algebra with involution then the linear mapping J in $\mathfrak{A}$ has a unique extension to a linear mapping J in $\mathfrak{A}_P$. The pair $(\mathfrak{A}_P, J)$ is an associative algebra with involution over P. We have an isomorphism of the Jordan algebra $\mathfrak{H}(\mathfrak{A}_P, J)$ with $\mathfrak{H}(\mathfrak{A}, J)_P$ and we can identify these two Jordan algebras.

We proceed next to define a functor of the category cat J/Φ of Jordan algebras over Φ into cat AI/Φ. If $\mathfrak{J}$ is a Jordan algebra over Φ and $\mathfrak{A}$ is an associative algebra over Φ then we define an *associative specialization* σ of $\mathfrak{J}$ in $\mathfrak{A}$ to be a homomorphism of $\mathfrak{J}$ into the Jordan algebra $\mathfrak{A}^+$. A *special universal envelope* of $\mathfrak{J}$ is a pair $(\mathfrak{U}, s)$, where $\mathfrak{U}$ is an associative algebra and s is an associative specialization of $\mathfrak{J}$ in $\mathfrak{U}$, such that if σ is any associative specialization of $\mathfrak{J}$ in $\mathfrak{A}$, then there exists a unique homomorphism η of $\mathfrak{U}$ into $\mathfrak{A}$ such that the following diagram is commutative:

(1)

The following theorem gives the main properties of special universal envelopes.

Theorem 1. (1) *If* $(\mathfrak{U}, s)$ *and* $(\mathfrak{U}', s')$ *are special universal envelopes for* $\mathfrak{J}$ *then there exists a unique isomorphism* j *of* $\mathfrak{U}$ *onto* $\mathfrak{U}'$ *such that* $s' = sj$. (2) $\mathfrak{U}$ *is generated by the image* $\mathfrak{J}^s$ *of* $\mathfrak{J}$ *under* s. (3) *There exists a unique involution* π *called the main involution of* $\mathfrak{U}$ *such that* $a^{s\pi} = a^s$, $a \in \mathfrak{J}$. (4) *If* ζ *is homomorphism of* $\mathfrak{J}$ *into a second Jordan algebra* $\mathfrak{J}'$ *and* $(\mathfrak{U}, s)$ $(\mathfrak{U}', s')$ *are special universal envelopes for* $\mathfrak{J}$ *and* $\mathfrak{J}'$ *respectively then there exists a unique homomorphism* ζ_u *of* $\mathfrak{U}$ *into* $\mathfrak{U}'$ *such that the following diagram is commutative:*

(2)

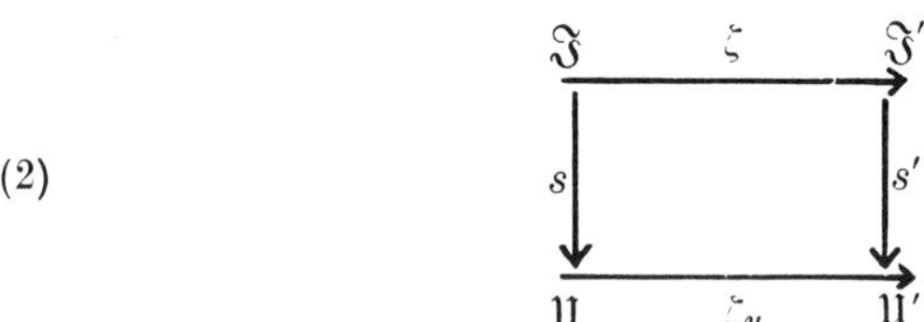

(5) *If* $\mathfrak{K}$ *is an ideal in* $\mathfrak{J}$ *and* $\mathfrak{B}$ *is the ideal in* $\mathfrak{U}$ *generated by* $\mathfrak{K}^s$ *then* $\bar{s}$: $a + \mathfrak{K} \to a^s + \mathfrak{B}$ *is an associative specialization of* $\overline{\mathfrak{J}} = \mathfrak{J}/\mathfrak{K}$ *in* $\overline{\mathfrak{U}} = \mathfrak{U}/\mathfrak{B}$ *and* $(\overline{\mathfrak{U}}, \bar{s})$ *is a special universal envelope for* $\overline{\mathfrak{J}}$. (6) *If* P *is an extension field of the base field* Φ *of* $\mathfrak{J}$ *and* s *is the linear extension of* s *to* $\mathfrak{J}_P$ *then* $(\mathfrak{U}_P, s)$ *is a special universal envelope for* $\mathfrak{J}_P$. (7) *If* D *is a derivation in* $\mathfrak{J}$ *then there exists a unique derivation* D_u *in* $\mathfrak{U}$ *such that* $Ds = sD_u$. (8) *If* $\mathfrak{J} = \mathfrak{J}_1 \oplus \mathfrak{J}_2$ *where* $\mathfrak{J}_i$ *is an ideal in* $\mathfrak{J}$ *and* $(\mathfrak{U}_i, s_i)$ *is a special universal envelope for* $\mathfrak{J}_i$ *then* $(\mathfrak{U} = \mathfrak{U}_1 \oplus \mathfrak{U}_2, s = s_1 + s_2)$ *(that is,* $(a_1 + a_2)^s = a_1^{s_1} + a_2^{s_2}$*) is a special universal envelope for* $\mathfrak{J}$. (9) *Let* $\mathfrak{J}$ *be a Jordan algebra which is a direct limit of the Jordan algebras* $\mathfrak{J}_\alpha$ *defined by the homomorphisms* $\varphi_{\alpha\beta}$ *for* $\alpha < \beta$ *and let* $(\mathfrak{U}_\alpha, s_\alpha)$ *be a special universal envelope for* $\mathfrak{J}_\alpha$. *Let* $\Phi_{\alpha\beta}$ *be the homomorphism of* $\mathfrak{U}_\alpha$ *into* $\mathfrak{U}_\beta$ *such that* $s_\alpha \Phi_{\alpha\beta} = \varphi_{\alpha\beta} s_\beta$. *Then the* $\Phi_{\alpha\beta}$ *define a direct limit* $\mathfrak{U}$ *of the associative algebras* $\mathfrak{U}_\alpha$. *Moreover, if* φ_α *and* Φ_α *denote the canonical homomorphisms of* $\mathfrak{J}_\alpha$ *into* $\mathfrak{J}$ *and of* $\mathfrak{U}_\alpha$ *into* $\mathfrak{U}$ *respectively, then there exists a unique associative specialization* s *of* $\mathfrak{J}$ *in* $\mathfrak{U}$ *such that* $\varphi_\alpha s = s_\alpha \Phi_\alpha$ *and* $(\mathfrak{U}, s)$ *is a special universal envelope for* $\mathfrak{J}$. (10) *If the dimensionality* $\dim \mathfrak{J}/\Phi = n < \infty$ *and* $(\mathfrak{U}, s)$ *is a special universal envelope for* $\mathfrak{J}$ *then* $\dim \mathfrak{U} \leqslant 2^{n-1}$.

The proofs of (1)–(5) and (7) are exactly like the corresponding ones for the Birkhoff–Witt algebra of a Lie algebra given in the author's *Lie Algebras*, pp. 152–157. Hence we omit these. To prove (6) we suppose we have an associative specialization σ of $\mathfrak{J}_P$ in an associative algebra $\mathfrak{A}/P$. Considering $\mathfrak{J}$ as a Φ-subalgebra of $\mathfrak{J}_P$ and $\mathfrak{A}$ as algebra over Φ then the restriction $\sigma|\mathfrak{J}$ of σ to $\mathfrak{J}$ is an associative specialization of $\mathfrak{J}/\Phi$ in $\mathfrak{A}/\Phi$. Hence there exists a homomorphism η of $\mathfrak{U}$ into $\mathfrak{A}/\Phi$ such that $s\eta = \sigma|\mathfrak{J}$. Since $\mathfrak{A}$ is also an algebra over P we can extend η to a homomorphism η of $\mathfrak{U}_P$ into $\mathfrak{A}$. Also the linear extension s of s to $\mathfrak{J}_P$ is an associative specialization of $\mathfrak{J}_P$ in $\mathfrak{U}_P$. Since $s\eta = \sigma|\mathfrak{J}$ holds in $\mathfrak{J}$, $s\eta = \sigma$ holds in $\mathfrak{J}_P$. Hence we have proved the existence of a homomorphism η of $\mathfrak{U}_P$ into $\mathfrak{A}$ such that $s\eta = \sigma$ for the given associative specialization σ of $\mathfrak{J}_P$ in $\mathfrak{A}$. Since $\mathfrak{J}^s$ generates $\mathfrak{U}$, $\mathfrak{J}_P{}^s$ generates $\mathfrak{U}_P$. Hence η is unique. Thus $(\mathfrak{U}_P, s)$ is a special universal envelope for $\mathfrak{J}_P$. The direct sum property stated in (8) is proved in N. Jacobson and F. D. Jacobson's paper [1], p. 147. The property of commutativity with direct limits given in (9) is a standard one for universal mappings (see Milnor and Moore [1], p. 239). Hence we omit the proof of this. Property (10) is proved in F. D. Jacobson and N. Jacobson [1], p. 145 and in Birkhoff and Whitman [1], p. 122. More generally, it is easy to see that if $\{1, u_\alpha\}$, where the indices α take on values in an ordered set I, is a basis for $\mathfrak{J}$ then every element of $\mathfrak{U}$ is a linear combination of 1 and the monomials $u_{\alpha_1} u_{\alpha_2} \dots u_{\alpha_r}$ where $\alpha_1 < \alpha_2 < \dots < \alpha_r$.

Examples. (1) Let $\mathfrak{M}$ be a vector space over Φ, Q a quadratic form on $\mathfrak{M}$ and let $\mathfrak{J} = \Phi 1 \oplus \mathfrak{M}$ a direct sum of $\mathfrak{M}$ and a one-dimensional space with 1 as basis. If $\alpha, \beta \in \Phi$ and $x, y \in \mathfrak{M}$ then we define $(\alpha 1 + x)\cdot(\beta 1 + y) = (\alpha\beta + (x, y))1 + \alpha y + \beta x$ where $(x, y) = \frac{1}{2}[Q(x+y) - Q(x) - Q(y))$. Then $\mathfrak{J}$ is a Jordan algebra called the *Jordan algebra of the quadratic form Q*. Let $C(\mathfrak{M}, Q)$ denote the Clifford algebra of Q and identify $\mathfrak{M}$ as usual with a subspace of $C(\mathfrak{M}, Q)$ (Chevalley, *Algebraic Theory of Spinors*, p. 37). Then it is easily seen that $C(\mathfrak{M}, Q)$ and s defined as $(\alpha 1 + x)^s = \alpha 1 + x$ is a special universal envelope for $\mathfrak{J}$. (2) Let $FJ^{(r)}$ denote the free Jordan algebra with r (free) generators $x_1, x_2, \dots, x_r$ and $FA^{(r)}$ the free associative algebra with r (free) generators $u_1, u_2, \dots, u_r$. Let s be the associative specialization of $FJ^{(r)}$ in $FA^{(r)}$ such that $x_i{}^s = u_i$, $i = 1, 2, \dots, r$. Then it follows directly from the definition and the defining property of $FA^{(r)}$ that $(FA^{(r)}, s)$ is special universal envelope for $FJ^{(r)}$. The main involution in $FA^{(r)}$ is the reversal (linear) operator, which maps a monomial $u_{i_1} u_{i_2} \dots u_{i_k}$ into its reverse $u_{i_k} u_{i_{k-1}} \dots u_{i_1}$. The image $FJ^{(r)s}$ in $FA^{(r)}$ is the subalgebra of $FA^{(r)+}$ generated by the u_i. This is called the free special Jordan algebra with r generators u_i.

The two examples just given indicate that special universal envelopes may be obtained by different constructions. We shall now give one which we shall take as our standard construction. Let $\mathfrak{J}$ be a given Jordan algebra and let $\mathfrak{T}(\mathfrak{J})$ be the tensor algebra based on the vector space $\mathfrak{J}$: $\mathfrak{T}(\mathfrak{J}) = \Phi 1 \oplus \mathfrak{J} \oplus (\mathfrak{J} \otimes \mathfrak{J}) \oplus \dots$. Let $\mathfrak{R}$ be the ideal in $\mathfrak{T}(\mathfrak{J})$ generated by the elements of the form $a\cdot b - \frac{1}{2}(a \otimes b + b \otimes a)$, $1_{\mathfrak{J}} - 1$ where $a, b \in \mathfrak{J}$ and $1_{\mathfrak{J}}$ is the identity element of $\mathfrak{J}$ and set $\mathrm{su}(\mathfrak{J}) = \mathfrak{T}(\mathfrak{J})/\mathfrak{R}$. Then it is straight-forward to check that $(\mathrm{su}(\mathfrak{J}), s)$ where $a^s = a + \mathfrak{R}$, $a \in \mathfrak{J}$, is a special universal envelope for $\mathfrak{J}$. We shall refer to this one as *the* special universal envelope for $\mathfrak{J}$. As before, we let π denote the main involution in $\mathrm{su}(\mathfrak{J})$.

We now have a mapping $\mathfrak{J} \to (\mathrm{su}(\mathfrak{J}), \pi)$ of Jordan algebras over a field Φ into associative algebras with involution over Φ. Let ζ be a homomorphism of $\mathfrak{J}$ into $\mathfrak{J}'$. Then we have a unique homomorphism ζ_u of su $(\mathfrak{J})$ into su $(\mathfrak{J}')$ such that $\zeta s' = s\zeta_u$ where s and s' are the given associative specializations of $\mathfrak{J}$ and $\mathfrak{J}'$ in su $(\mathfrak{J})$ and su $(\mathfrak{J}')$ respectively. If π' denotes the main involution su $(\mathfrak{J}')$ then we have $a^{s\pi\zeta_u} = a^{s\zeta_u} = a^{\zeta s'}$ and $a^{s\zeta_u\pi'} = a^{\zeta s'}$, for $a \in \mathfrak{J}$. Since $\mathfrak{J}^s$ generates su $(\mathfrak{J})$ we have $\zeta_u\pi' = \pi\zeta_u$ so ζ_u is a homomorphism of (su $(\mathfrak{J}), \pi$) into (su $(\mathfrak{J}'), \pi'$). Next let λ be a homomorphism of $\mathfrak{J}'$ into $\mathfrak{J}''$ then it is clear from Theorem 1 (4) that $(\zeta\lambda)_u = \zeta_u\lambda_u$. It follows that the mappings $\mathfrak{J} \to$ (su $(\mathfrak{J}), \pi$), $\zeta \to \zeta_u$ define a functor S from cat J/Φ to cat AI/Φ.

2. *General theorems.* Let $\mathfrak{J}/\Phi$ be a Jordan algebra and let s be the associative specialization of $\mathfrak{J}$ into the special universal envelope su $(\mathfrak{J})$. Since $a^{s\pi} = a^s$, $a \in \mathfrak{J}$, $\mathfrak{J}^s \subseteq \mathfrak{H}(\text{su}\,(\mathfrak{J}), \pi)$. Clearly s is a homomorphism of $\mathfrak{J}$ into the Jordan algebra $\mathfrak{H}(\text{su}\,(\mathfrak{J}), \pi)$. It is clear also from the definition that $\mathfrak{J}$ is a special Jordan algebra, that is, $\mathfrak{J}$ has an injective associative specialization if and only if s is injective. We shall now call $\mathfrak{J}$ a *reflexive* Jordan algebra if s is surjective on $\mathfrak{H}(\text{su}\,(\mathfrak{J}), \pi)$. Then $\mathfrak{J}$ is special and reflexive if and only if s is an isomorphism of $\mathfrak{J}$ onto $\mathfrak{H}(\text{su}\,(\mathfrak{J}), \pi)$.

Next let $(\mathfrak{A}, J)$ be an associative algebra with involution over the field Φ and let $\mathfrak{H}(\mathfrak{A}, J)$ be the Jordan algebra of J-symmetric elements. Then the injection mapping $a \to a$ of $\mathfrak{H}(\mathfrak{A}, J)$ is an associative specialization of $\mathfrak{H}(\mathfrak{A}, J)$ in $\mathfrak{A}$. Hence we have a unique homomorphism h of su $(\mathfrak{H}(\mathfrak{A}, J))$ into $\mathfrak{A}$ such that $a^{sh} = a$, $a \in \mathfrak{H}(\mathfrak{A}, J)$. We have $a^{s\pi h} = a$ and $a^{shJ} = a$. Since the a^s, $a \in \mathfrak{H}(\mathfrak{A}, J)$, generate su (J) it is clear that h is a homomorphism of (su $(J), \pi$) into $(\mathfrak{A}, J)$. It is clear also that h is surjective if and only if $\mathfrak{H}(\mathfrak{A}, J)$ generates $\mathfrak{A}$. We shall now call $(\mathfrak{A}, J)$ a *perfect* associative algebra with involution if h is an isomorphism of (su $(\mathfrak{J}), \pi$) onto $(\mathfrak{A}, J)$. This is the case if and only if $\mathfrak{A}$ together with the injection mapping of $\mathfrak{H}(\mathfrak{A}, J)$ into $\mathfrak{A}$ is a special universal envelope for $\mathfrak{H}(\mathfrak{A}, J)$. More explicitly, $(\mathfrak{A}, J)$ is perfect if and only if any associative specialization of $\mathfrak{H}(\mathfrak{A}, J)$ has a unique extension to a homomorphism of $\mathfrak{A}$. It is clear from the definitions that if $(\mathfrak{A}, J)$ is perfect then $\mathfrak{H}(\mathfrak{A}, J)$ is special and reflexive.

We shall now give some sufficient conditions for perfection and reflexivity.

Theorem 2. *Any Jordan algebra with* $\leqslant 3$ *generators is reflexive.*

Proof. We have seen that if $FJ^{(r)}$ is the free Jordan algebra with r generators $x_1, x_2, \ldots, x_r$ and $FA^{(r)}$ is the free associative algebra with r generators $u_1, u_2, \ldots, u_r$ then $FA^{(r)}$ together with the associative specialization such that $x_i \to u_i$ constitute a special universal envelope for $FJ^{(r)}$. We have seen also that the main involution in $FA^{(r)}$ is the reversal operator. It has been proved by P. Cohn that the space of symmetric elements under the reversal operator coincides with the Jordan subalgebra of $FA^{(r)+}$ generated by the u_i if and only if $r \leqslant 3$ (Cohn [1], pp. 257–259).

It follows that $FJ^{(r)}$ is reflexive if and only if $r \leqslant 3$. The proof of Theorem 2 will now follow by showing that if $\mathfrak{J}$ is reflexive then any homomorphic image of $\mathfrak{J}$ is reflexive. Let ζ be a homomorphism of $\mathfrak{J}$ onto $\overline{\mathfrak{J}}$. Since the images of $\mathfrak{J}$ and $\overline{\mathfrak{J}}$ in su $(\mathfrak{J})$ and su $(\overline{\mathfrak{J}})$ generate these algebras it is clear that the corresponding homomorphism ζ_u of su $(\mathfrak{J})$ is surjective on su $(\overline{\mathfrak{J}})$. Let $\bar{a} \in \mathfrak{H}(\text{su}\,(\overline{\mathfrak{J}}), \bar{\pi})$, $\bar{\pi}$ the main involution in su $(\overline{\mathfrak{J}})$. Then $\bar{a} = \frac{1}{2}(\bar{a} + \bar{a}^{\bar{\pi}})$ and $\bar{a} = a^{\zeta_u}$ for some $a \in \text{su}\,(\mathfrak{J})$. Hence $\bar{a} = \frac{1}{2}(\bar{a} + \bar{a}^{\bar{\pi}}) = \frac{1}{2}(a^{\zeta_u} + a^{\zeta_u \bar{\pi}}) = \frac{1}{2}(a^{\zeta_u} + a^{\pi\zeta_u}) = b^{\zeta_u}$ where $b = \frac{1}{2}(a + a^{\pi}) \in \mathfrak{H}(\text{su}\,(\mathfrak{J}), \pi)$. If $\mathfrak{J}$ is reflexive then $b \in \mathfrak{J}^s$. Then $b = c^s$, $c \in \mathfrak{J}$ and $\bar{a} = c^{s\zeta_u} = c^{\zeta\bar{s}} \in \overline{\mathfrak{J}}^{\bar{s}}$, $\bar{s}$ the associative specialization of $\overline{\mathfrak{J}}$ in su $(\overline{\mathfrak{J}})$. Hence $\overline{\mathfrak{J}}$ is reflexive.

It is a known result due to Shirshov and Cohn (SHIRSHOV [1], p. 159) that every Jordan algebra with $\leqslant 2$ generators is special. Together with Theorem 2 we have that any such algebra is special and reflexive. Thus such an algebra $\mathfrak{J}$ is isomorphic to the Jordan algebra $\mathfrak{H}(\text{su}\,(\mathfrak{J}), \pi)$. Consequently, any Jordan algebra with two generators is isomorphic to an algebra $\mathfrak{H}(\mathfrak{A}, J)$ for some associative algebra with involution $(\mathfrak{A}, J)$. Clearly this is a strengthening of the Theorem of Shirshov–Cohn.

Theorem 3. *For any Jordan algebra* $\mathfrak{J}$ *the associative algebra with involution* (su $(\mathfrak{J}), \pi$) *is perfect.*

Proof. Set $\mathfrak{U} = \text{su}\,(\mathfrak{J})$, $\mathfrak{K} = \mathfrak{H}(\mathfrak{U}, \pi)$, $\mathfrak{V} = \text{su}\,(\mathfrak{K})$ and let s, t denote the given associative specializations of $\mathfrak{J}$ and $\mathfrak{K}$ in $\mathfrak{U}$ and $\mathfrak{V}$ respectively. Then $\mathfrak{J}^s$ generates $\mathfrak{U}$ and $\mathfrak{K}^t$ generates $\mathfrak{V}$. Hence $\mathfrak{J}^{st}$ generates $\mathfrak{V}$. Considering s as a homomorphism of the Jordan algebra $\mathfrak{J}$ into the Jordan algebra $\mathfrak{K}$ we obtain a homomorphism s_u of $\mathfrak{U}$ into $\mathfrak{V}$ such that

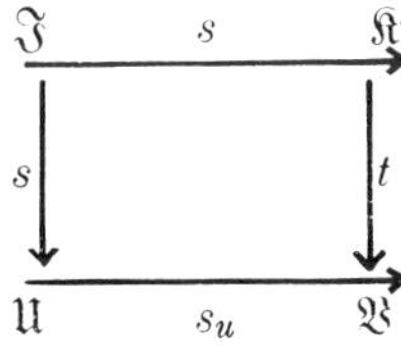

is commutative. Then if $a \in \mathfrak{J}$ we have $a^{st} = a^{ss_u}$. Let k be the homomorphism of $\mathfrak{V}$ into $\mathfrak{U}$ such that $b^{tk} = b$, $b \in \mathfrak{K} = \mathfrak{H}(\mathfrak{U}, \pi)$. We have to show that k is an isomorphism of $\mathfrak{V}$ onto $\mathfrak{U}$. Now if $a \in \mathfrak{J}$ then $a^s \in \mathfrak{J}^s \subseteq \subseteq \mathfrak{H}(\mathfrak{U}, \pi) = \mathfrak{K}$. Hence $a^s = a^{stk} = a^{ss_uk}$. Since the a^s, $a \in \mathfrak{J}$, generate $\mathfrak{U}$ this implies that $s_uk = 1_{\mathfrak{U}}$ the identity mapping on $\mathfrak{U}$. Next, if $a \in \mathfrak{J}$, then $a^{stks_u} = a^{ss_u} = a^{st}$. Since the elements a^{st}, $a \in \mathfrak{J}$, generate $\mathfrak{V}$ this implies that $ks_u = 1_{\mathfrak{V}}$. Hence k is an isomorphism of $\mathfrak{V}$ onto $\mathfrak{U}$.

In the next section we shall give an improved formulation and new derivation of the main theorem on the classification of special finite dimensional central simple Jordan algebras (cf. F. D. JACOBSON and N. JACOBSON [1], pp. 157–167). For this a special case of the following theorem due to JACOBSON and RICKART ([1], p. 315) will be fundamental.

Theorem 4. *Let $\mathfrak{D}$ be an associative algebra with an involution $d \to \bar{d}$, $\mathfrak{D}_n$, $n \geqslant 3$, the $n \times n$ matrix algebra over $\mathfrak{D}$, J_a the involution $X \to a^{-1}\bar{X}'a$ where $\bar{X}'$ is the transpose of $\bar{X} = (\bar{x}_{ij})$, $X = (x_{ij})$, and a is a diagonal matrix* diag $\{a_1, a_2, \ldots, a_n\}$ *where $\bar{a}_i = a_i$ has an inverse in $\mathfrak{D}$. Then $(\mathfrak{D}_n, J_a)$ is perfect.*

We remark that the result of Jacobson–Rickart is more general than the foregoing in that $\mathfrak{D}$ can be an arbitrary associative ring rather than an algebra. We note also that this result has been generalized recently by W. Martindale ([1]) and Martindale's theorem has the following immediate consequence.

Theorem 5. *Let $(\mathfrak{A}, J)$ be an associative algebra with an involution having the following properties: $\mathfrak{A}$ contains three J-symmetric idempotent elements such that $e_ie_j = 0$, $i \neq j$, $e_1 + e_2 + e_3 = 1$ and $\mathfrak{A}e_i\mathfrak{A} = \mathfrak{A}$, $i = 1, 2, 3$. Then $(\mathfrak{A}, J)$ is perfect.*[2])

To obtain Theorem 4 from Theorem 5 we remark first that the latter result has an immediate extension in which the hypothesis on the three idempotents is replaced by one on $n \geqslant 3$ idempotents. For, suppose $e_1, e_2, \ldots, e_n$ satisfy the conditions $e_i^2 = e_i = e_i^J$, $e_ie_j = 0$, $i \neq j$, $\sum e_i = 1$, $\mathfrak{A}e_i\mathfrak{A} = \mathfrak{A}$. Set $f_1 = e_1$, $f_2 = e_2$, $f_3 = \sum_{i>2} e_i$. Then the f_i satisfy the conditions of the theorem. Hence $(\mathfrak{A}, J)$ is perfect. In particular, let $\mathfrak{A} = \mathfrak{D}_n$, $J = J_a$ as in Theorem 4. Then the diagonal idempotents $e_i = \text{diag}\ \{0, \ldots, 0, 1, 0, \ldots, 0\}$ satisfy the indicated conditions. Hence $(\mathfrak{D}_n, J_a)$ is perfect.

3. *Classification of finite dimensional special central simple Jordan algebras.* Let $(u_1, u_2, \ldots, u_n)$ be a basis for a Jordan algebra $\mathfrak{J}/\Phi$ and let $\Xi = \Phi(\xi_1, \xi_2, \ldots, \xi_n)$ be the field of rational expressions in n indeterminates ξ_i. The element $x = \sum_1^n \xi_i u_i$ of $\mathfrak{J}_P$ is called a generic element of $\mathfrak{J}$ and the degree of the minimum polynomial $m_x(\lambda)$ of x is called the degree of $\mathfrak{J}$ (Jacobson [4], p. 179). It is easily seen that the degree is independent of the basis and $\mathfrak{J}$ and $\mathfrak{J}_P$ have the same degree for every extension field P/Φ. Moreover, if $\mathfrak{J}$ is simple over an algebraically closed field then the degree of $\mathfrak{J}$ is the maximum cardinality for sets of non-zero orthogonal idempotent elements of $\mathfrak{J}$.

The finite dimensional simple Jordan algebras over an algebraically closed field Ω have been determined by Albert ([1], p. 252, [2], p. 567, [3], p. 524, also Jacobson [2]). His results can be formulated as follows. If $\mathfrak{J}/\Omega$ is simple and the degree

deg $\mathfrak{J} = 1$, then $\mathfrak{J} = \Omega 1$.

deg $\mathfrak{J} = 2$, then $\mathfrak{J}$ is isomorphic to the Jordan algebra of a non-degenerate quadratic form Q in a vector space $\mathfrak{M}$ with deg $\mathfrak{M} \geqslant 2$,

deg $\mathfrak{J} = n \geqslant 3$, then $\mathfrak{J}$ is isomorphic to a Jordan algebra $\mathfrak{H}(\mathfrak{D}_n) \equiv$

[2]) Martindale's hypotheses are different but equivalent to the foregoing in that the condition $\mathfrak{A}e_i\mathfrak{A} = \mathfrak{A}$ is replaced by $e_i\mathfrak{A}e_j\mathfrak{A}e_i = e_i\mathfrak{A}e_i$, $i \neq j$.

$\equiv \mathfrak{H}(\mathfrak{D}_n, J_1)$ where $\mathfrak{D}$ is a composition algebra and the involution $\mathfrak{J}_1$ is $X \to \bar{X}'$, $d \to \bar{d}$ the standard involution in $\mathfrak{D}$. Moreover, $\mathfrak{D}$ is associative if $n > 3$. We recall that a composition algebra is an algebra equipped with a non-degenerate quadratic form Q such that $Q(xy) = Q(x)Q(y)$. Such an algebra is necessarily alternative and has a uniquely determined involution $d \to \bar{d}$, called the standard involution, such that $d\bar{d} = Q(d)1 = \bar{d}d$ (Jacobson [3]). In the algebraically closed case the possibilities for $\mathfrak{D}$ and its standard involution are the following: I. $\mathfrak{D} = \Omega 1$, involution the identity mapping, II. $\mathfrak{D} = \Omega 1_1 \oplus \Omega 1_2$, involution exchanging the two ideals $\Omega 1_i$, III. $\mathfrak{D} = \Omega_2$, involution $d \to tr\,(d)1 - d$ where $tr\, d$ is the trace of the matrix d, IV. $\mathfrak{D}$ a split Cayley algebra with its standard involution.

Since IV is not associative it can be used only if $n = 3$. The resulting algebra is an exceptional (non-special) Jordan algebra. All the other Jordan algebras in the list are special. Also any Jordan algebra of the type listed above is simple and has the indicated degree.

An associative algebra with involution $(\mathfrak{A}, J)$ is *simple* if it has no ideals other than $\mathfrak{A}$ and 0 and $\mathfrak{A} \neq 0$. If $(\mathfrak{A}, J)$ is simple then either $\mathfrak{A}$ is simple or $\mathfrak{A} = \mathfrak{B} \oplus \mathfrak{B}^J$ where $\mathfrak{B}$ is a simple ideal. It is well known that any finite dimensional simple associative algebra with involution over an algebraically closed field Ω is isomorphic to one of the following algebras with involution: *A*. $\Omega_n \oplus \Omega_n$, involution $(X, Y) \to (Y', X')$, $X, Y \in \Omega_n$, X', Y' the transposes of X, Y. *B*. Ω_n, involution $X \to X'$, *C*. Ω_{2n}, involution $X \to S^{-1}X'S$ where $S = \mathrm{diag}\,\{q, \ldots, q\}$, $q = \begin{pmatrix} 0 & 1 \\ -1 & 0 \end{pmatrix}$. It is easily seen that the algebras listed can be described in a more uniform way as the algebras $\mathfrak{D}_n$, $\mathfrak{D}$ an associative composition algebra with the *standard involution* J_1: $X \to \bar{X}'$ in $\mathfrak{D}_n$. The case A corresponds to II in the list of composition algebras over Ω and B and C correspond to I and III respectively. Conversely, any $(\mathfrak{D}_n, J_1)$, $\mathfrak{D}$ a composition algebra over Ω, J_1 the standard involution, is simple. Moreover, two such algebras are isomorphic if and only if the n's indicated are the same and the composition algebras are isomorphic. We shall call the invariant n the *degree* of $(\mathfrak{D}_n, J_1)$.

Albert's result for special algebras of degree $\geqslant 3$ can now be stated in the following way: Any finite dimensional special simple Jordan algebra of degree $n \geqslant 3$ over an algebraically closed field Ω is isomorphic to an algebra $\mathfrak{H}(\mathfrak{A}, J)$ where $(\mathfrak{A}, J)$ is a finite dimensional simple associative algebra of degree n over Ω and conversely. We shall now consider the extension of this result to arbitrary base fields.

We recall that a Jordan algebra $\mathfrak{J}/\Phi$ is central simple if $\mathfrak{J}$ is simple and the only elements c in $\mathfrak{J}$ such that $[x, y, c] \equiv (x \cdot y) \cdot c - x \cdot (y \cdot c) = 0 = [x, c, y] = [x, y, c]$ are the elements of $\Phi 1$. A simple algebra over an algebraically closed field is necessarily central simple. $\mathfrak{J}/\Phi$ is central simple if and only if $\mathfrak{J}_P$ is simple for every extension field P/Φ. If $\mathfrak{J}$ is central simple of degree 1 then $\mathfrak{J}\Omega = \Omega 1$ if Ω is the algebraic closure of the base field Φ. Hence $\mathfrak{J}=\Phi 1$. If $\mathfrak{J}$ is central simple of degree 2 then $\mathfrak{J}_\Omega$ is isomorphic to the Jordan algebra of a non-degenerate quadratic form on a vector space $\mathfrak{M}$ with dim $\mathfrak{M}>1$. It follows easily that $\mathfrak{J}$ is of the same type and the classification of these algebras is equivalent to the classification of non-degenerate quadratic forms. The central simple Jordan algebras of degree $\geqslant 3$ include exceptional ones. Our considerations play no role in the study of these algebras. (For these the main references are Albert–Jacobson [1], Albert [4], Springer [1].) We shall now consider the problem of classifying the finite dimensional special central simple Jordan algebras of degree $\geqslant 3$.

We define the *center* $\mathfrak{C}_J$ of an associative algebra with involution $(\mathfrak{A}, J)$ to be $\mathfrak{C}_J=\mathfrak{C}\cap\mathfrak{H}(\mathfrak{A}, J)$ where $\mathfrak{C}$ is the center of $\mathfrak{A}$. An associative algebra with involution is *central simple* if it is simple and its center is $\Phi 1$ (Φ the base field). It is easy to see that $(\mathfrak{A}, J)$ is central simple if and only if $(\mathfrak{A}_P, J)$ is simple for every extension field P/Φ. Also if $\mathfrak{A}$ is finite dimensional over an algebraically closed field Ω and $(\mathfrak{A}, J)$ is simple then $(\mathfrak{A}, J)$ is central simple and we have noted that such an algebra with involution is isomorphic to one of the pairs $(\mathfrak{D}_n, J_1)$ where $\mathfrak{D}$ is a composition algebra and J_1 is the standard involution. If $(\mathfrak{A}, J)$ is finite dimensional central simple over Φ and Ω is the algebraic closure of Φ then $(\mathfrak{A}_\Omega, J)$ is isomorphic to one of the algebras $(\mathfrak{D}_n, J_1)$. We shall call the invariant n the *degree* of $(\mathfrak{A}, J)$.

We shall call a property $\mathfrak{A}$ of Jordan algebras (associative algebras with involution) *linear* if the validity of A for $\mathfrak{J}((\mathfrak{A}, J))$ is equivalent to its validity for $\mathfrak{J}_P((\mathfrak{A}_P, J))$, P any extension field of Φ. Clearly the property that the degree is a fixed integer n is linear. Also it is easily seen that the property of central simplicity is linear. We also have the following

Lemma. *The property that Jordan algebras are special or reflexive is linear. The property that associative algebras with involution are perfect is linear.*

Proof. Suppose $\mathfrak{J}$ is special (reflexive). Then the associative specialization s of $\mathfrak{J}$ in su $(\mathfrak{J})$ is injective (surjective). Hence the linear extension s of s to $\mathfrak{J}_P$ into su $(\mathfrak{J})_P$ is injective (surjective). Since (su $(\mathfrak{J})_P, s$) is a special universal envelope for $\mathfrak{J}_P$ it follows that the given associative specialization s of $\mathfrak{J}_P$ in su $(\mathfrak{J}_P)$ is injective (surjective). Hence $\mathfrak{J}_P$ is special (reflexive). Conversely, assume $\mathfrak{J}_P$ is special (reflexive). Then a reversal of the steps of the argument shows that $\mathfrak{J}$ is special (reflexive). Next assume $(\mathfrak{A}, J)$ is perfect. Then $\mathfrak{A}$ and the injection mapping of $\mathfrak{H}(\mathfrak{A}, J)$ into $\mathfrak{A}$ constitute a special universal envelope for $\mathfrak{H}(\mathfrak{A}, J)$. Hence by Theorem 1 (6), $\mathfrak{A}_P$ and the canonical isomorphism of $\mathfrak{H}(\mathfrak{A}, J)_P$ in $\mathfrak{A}_P$ constitute a special universal envelope for $\mathfrak{H}(\mathfrak{A}, J)_P$. Hence $\mathfrak{A}_P$ and the injection mapping form a special universal envelope for $\mathfrak{H}(\mathfrak{A}_P, J)$ and $(\mathfrak{A}_P, J)$ is perfect. Conversely, assume $(\mathfrak{A}_P, J)$ perfect and let σ be an associative specialization of $\mathfrak{H}(\mathfrak{A}, J)$ in an associative algebra $\mathfrak{B}$. Then the linear extension of σ to $\mathfrak{H}(\mathfrak{A}, J)_P = \mathfrak{H}(\mathfrak{A}_P, J)$ can be extended to a homomorphism of $\mathfrak{A}_P$ into $\mathfrak{B}_P$. This implies that σ can be extended to a homomorphism of $\mathfrak{A}$ into $\mathfrak{B}$. Since $\mathfrak{H}(\mathfrak{A}_P, J)$ generates $\mathfrak{A}_P$, $\mathfrak{H}(\mathfrak{A}, J)$ generates $\mathfrak{A}$. Hence the extension of σ to $\mathfrak{A}$ is unique. Hence $\mathfrak{A}$ and the injection mapping form a special universal envelope for $\mathfrak{H}(\mathfrak{A}, J)$ so $(\mathfrak{A}, J)$ is perfect.

We can now prove the following key result for the classification of special central simple Jordan algebras.

Theorem 6. (1) *Let $(\mathfrak{A}, J)$ be a finite dimensional central simple associative algebra of degree $n \geqslant 3$. Then $(\mathfrak{A}, J)$ is perfect and $\mathfrak{H}(\mathfrak{A}, J)$ is a finite dimensional central simple Jordan algebra of degree n.* (2) *Let $\mathfrak{J}$ be a finite dimensional special central simple Jordan algebra of degree $n \geqslant 3$. Then $\mathfrak{J}$ is reflexive and $(\mathrm{su}\,(\mathfrak{J}), \pi)$ is a finite dimensional central simple associative algebra with involution of degree n.*

Proof. The results we have noted show that all the properties stated in the hypotheses and conclusions of the theorem are linear. Hence it suffices to prove the theorem for an algebraically closed base field Ω. Let $(\mathfrak{A}, J)$ be a finite dimensional central simple associative algebra with involution over Ω of degree $n \geqslant 3$. Then $(\mathfrak{A}, J)$ is isomorphic to $(\mathfrak{D}_n, J_1)$, $\mathfrak{D}$ an associative composition algebra. Then $(\mathfrak{D}_n, J_1)$, hence $(\mathfrak{A}, J)$ is perfect by Theorem 4. Also $\mathfrak{H}(\mathfrak{D}_n) \equiv \mathfrak{H}(\mathfrak{D}_n, J_1)$ is central simple of degree n, so the same is true of $\mathfrak{H}(\mathfrak{A}, J)$. Next let $\mathfrak{J}$ be a finite dimensional special central simple Jordan algebra of degree $n \geqslant 3$ over Ω. Then $\mathfrak{J}$ is isomorphic to a Jordan algebra $\mathfrak{H}(\mathfrak{D}_n)$, $\mathfrak{D}$ an associative composition algebra, by Albert's theorem. Since $(\mathfrak{D}_n, J_1)$ is perfect, $\mathfrak{H}(\mathfrak{D}_n)$ is reflexive and $\mathfrak{J}$ is reflexive. Also $(\mathrm{su}\,(\mathfrak{J}), \pi)$ is isomorphic to $(\mathfrak{D}_n, J_1)$ by Theorem 4 and $(\mathfrak{D}_n, J_1)$ is a finite dimensional central simple associative algebra with involution of degree n. Hence the same is true of $(\mathrm{su}\,(\mathfrak{J}), \pi)$.

We remark that the reflexivity of $\mathfrak{J}$ can also be deduced from Theorem 2 since one can show that $\mathfrak{H}(\mathfrak{D}_n)$, $\mathfrak{D}$ a composition algebra, is generated by three elements. One can use this to prove also that any finite dimensional central simple Jordan algebra of degree $\geqslant 3$ can be generated by three elements. The following is the main result on the classification of special central simple Jordan algebras.

Theorem 7. *A Jordan algebra is finite dimensional special central simple of degree $\geqslant 3$ if and only if it is isomorphic to an algebra $\mathfrak{H}(\mathfrak{A}, J)$ where $(\mathfrak{A}, J)$ is a finite dimensional central simple associative algebra with involution of degree $\geqslant 3$. If $(\mathfrak{A}, J)$, $(\mathfrak{B}, K)$ are finite dimensional central simple associative algebras with involutions of degree $\geqslant 3$ then these are isomorphic if and only if the Jordan algebras $\mathfrak{H}(\mathfrak{A}, J)$, $\mathfrak{H}(\mathfrak{B}, K)$ are isomorphic.*

Proof. If $\mathfrak{J}$ is a special finite dimensional central simple Jordan algebra of degree $\geqslant 3$ then $\mathfrak{J}$ is isomorphic to a Jordan algebra $\mathfrak{H}(\mathfrak{A}, J)$

where $(\mathfrak{A}, J)$ is a finite dimensional central simple associative algebra with involution of degree $\geqslant 3$, by Theorem 6 (2). The converse follows the second statement in Theorem 6 (1). If $(\mathfrak{B}, K)$ is a second associative algebra with involution of the kind indicated then an isomorphism of $\mathfrak{H}(\mathfrak{A}, J)$ onto $\mathfrak{H}(\mathfrak{B}, K)$ has a unique extension to an isomorphism of $\mathfrak{A}$ onto $\mathfrak{B}$ since $\mathfrak{A}$ and the injection mapping and $\mathfrak{B}$ and the injection mapping are special universal envelopes for $\mathfrak{H}(\mathfrak{A}, J)$ and $\mathfrak{H}(\mathfrak{B}, K)$ by Theorem 6 (1). Thus $\mathfrak{H}(\mathfrak{A}, J) \cong \mathfrak{H}(\mathfrak{B}, K)$ implies $(\mathfrak{A}, J) \cong (\mathfrak{B}, K)$. The converse is clear.

This result reduces the problem of classification for the Jordan algebras to the corresponding one on associative algebras with involutions. What the latter amounts to explicitly can be seen in JACOBSON [1], pp. 541–551.

Yale University and University of Chicago

REFERENCES

ALBERT, A. A., [1] On Jordan algebras of linear transformations, Trans. Amer. Math. Soc., **59**, 524–555 (1946).

———, [2] A structure theory for Jordan algebras, Ann. Math., **48**, 546–567 (1947).

———, [3] A theory of power associative commutative algebras, Trans. Amer. Math. Soc., **69**, 503–527 (1950).

———, [4] A construction of exceptional Jordan division algebras, Ann. Math. **67**, 1–28 (1958).

ALBERT, A. A. and N. JACOBSON, [1] Reduced exceptional simple Jordan algebras, Ann. Math., **67** 400–417 (1957).

BIRKHOFF, G. and P. WHITMAN, [1] Representation of Jordan and Lie algebras, Trans. Amer. Math. Soc., **65**, 116–136 (1949).

COHN, P. M., [1] Special Jordan algebras, Can. Jour. Math., **6**, 253–264 (1954).

JACOBSON, F. D. and N. JACOBSON, [1] Classification and representation of semi-simple Jordan algebras, Trans. Amer. Math. Soc., **65**, 141–169 (1949).

JACOBSON, N., [1] Simple Lie algebras over a field of characteristic zero, Duke Math. Jour., **4**, 534–551 (1938).

———, [2] A theorem on the structure of Jordan algebras, Proc. Nat. Acad. Sci., **42**, 140–147 (1956).

———, [3] Composition algebras and their automorphisms, Rend. Cir. Mat. di Palermo (II), **7**, 55–80 (1958).

———, [4] Some groups of transformations defined by Jordan algebras I, Jour. für reine und angew. Math., **201**, 178–195 (1959).

JACOBSON, N. and C. E. RICKART, [1] Homomorphisms of Jordan rings of self-adjoint elements, Trans. Amer. Math. Soc., **72**, 310–322 (1952).

MARTINDALE, W. S., [1] Jordan homomorphisms of the symmetric elements of a ring with involution, to appear in Journal of Algebra.

MILNOR, J. W. and J. C. MOORE, [1]. On the structure of Hopf algebras, Ann. Math., **81**, 211–264 (1965).

SHIRSHOV, A. I., [1] On special J-rings, Rec. Math. (Mat. Sbornik) N. S. **38**, (80) 149–166 (1956).

SPRINGER, T., [1] The classification of reduced exceptional Jordan algebras, Proc. Neth. Acad. Sci., Series **A63** (Indag. Math. 22), 414–420 (1960).

Istituto Nazionale di Alta Matematica
Symposia Mathematica
Volume VIII (1972)

CONNECTIONS BETWEEN ASSOCIATIVE AND JORDAN RINGS (*)

N. Jacobson

We propose in this lecture to describe some recent developments in the theory of Jordan algebras which have interesting interpretations and consequences for the theory of associative algebras. Throughout we assume that all algebras are *unital* in the sense that they contain a unit element 1 and we adopt the usual conventions on subalgebras and homomorphisms for unital algebras. Until recently the Jordan theory was restricted to algebras over fields of characteristic $\neq 2$ or more generally to algebras over commutative rings Φ containing $\frac{1}{2}$. Now, thanks to the work of McCrimmon, a good part of the theory has been extended to algebras over an arbitrary commutative ring Φ. This, of course, applies in particular to rings, which can be regarded as algebras over $\mathbb{Z}$. The extension to the general case has been achieved by passing from a linear to a quadratic composition. Suppose first that $\mathfrak{A}$ is an associative algebra over a commutative ring $\Phi \ni \frac{1}{2}$. We obtain the Jordan algebra $\mathfrak{A}^+$ by retaining the given Φ-module structure and replacing the given product ab by the *Jordan product* $a\cdot b = \frac{1}{2}(ab + ba)$. If $\Phi \not\ni \frac{1}{2}$ the composition $a \cdot b$ is not defined. We could replace this by $\{ab\} = ab + ba$ but this has the disadvantage that 1 is not a unit since $\{a1\} = 2a$. It has been known for some time that the natural way of introducing a Jordan structure on any associative algebra is to work with the product aba, which is linear in b and quadratic in a. Thus if $\mathfrak{A}$ is an associative algebra over any commutative ring Φ we define the *quadratic Jordan algebra* $\mathfrak{A}^{(q)}$ as the structure consisting of the given Φ-module, the distinguished element 1 and the binary product $(a, b) \to aba$. It is usually more convenient to work with the Φ-endomorphism U_a defined as $x \to axa$ instead of with the product aba. If Φ contains $\frac{1}{2}$ we can pass from the product

(*) Conferenza tenuta il 26 novembre 1970.

$a \cdot b$ which we write as bR_a to U_a and back. This is done by observing that $U_a = 2R_a^2 - R_{a^2}$ and $U_{a,b} = U_{a+b} - U_a - U_b$ is $x \to axb + bxa$ so $bR_a = \frac{1}{2} 1 U_{a,b}$. It is clear from this that the theory based on $a \cdot b$ for $\Phi \ni \frac{1}{2}$ is equivalent to that based on aba.

We now consider $\mathfrak{A}^{(q)}$, $\mathfrak{A}$ an associative algebra over an arbitrary commutative ring Φ. It is possible to give a simple set of working axioms for the operator U_a. This has been done by McCrimmon. We shall not need this and shall mention only one of the axioms, the so-called *fundamental identity*:

$$U_a U_b U_a = U_{bU_a} . \tag{1}$$

This is clearly satisfied in $\mathfrak{A}^{(q)}$ since

$$x U_a U_b U_a = a(b(axa)b)a ,$$

$$x U_{bU_a} = (aba)x(aba) .$$

The notion of a *subalgebra* $\mathfrak{B}$ of $\mathfrak{A}^{(q)}$ is clear: a Φ-submodule of $\mathfrak{A}$ containing 1 and closed under aba. Such a subalgebra is called a *special* (*unital*) *quadratic* Jordan algebra. In this lecture we abbreviate this simply to: a *Jordan algebra*. The important class of examples we shall consider is defined as follows. Let $(\mathfrak{A}, J)$ be an associative algebra with involution and let $\mathfrak{H}(\mathfrak{A}, J)$ be the set of J-symmetric elements of $\mathfrak{A}$, that is, the h satisfying $h^J = h$. It is clear that $\mathfrak{H}(\mathfrak{A}, J)$ is a subalgebra of $\mathfrak{A}^{(q)}$. Our primary concern will be with Jordan algebras of this type. The algebras $\mathfrak{A}^{(q)}$ are included here by the following device. Let $\mathfrak{A}_0$ denote the opposite algebra of $\mathfrak{A}$ and form $\mathfrak{B} = \mathfrak{A} \oplus \mathfrak{A}_0$. This has the exchange involution $J: (a, b) \to (b, a)$ and $\mathfrak{H}(\mathfrak{B}, J)$ is the subalgebra of elements (a, a). It is immediate that $a \to (a, a)$ is an isomorphism of $\mathfrak{A}^{(q)}$ into $\mathfrak{H}(\mathfrak{B}, J)$. The notion of *homomorphism* of a Jordan algebra $\mathfrak{J}$ into a second one $\mathfrak{J}'$ is a mapping η of $\mathfrak{J}$ into $\mathfrak{J}'$ which is: 1) a Φ-homomorphism, 2) $1\eta = 1'$, 3) $(aba)\eta = (a\eta)(b\eta)(a\eta)$.

If $\mathfrak{J}$ is a Jordan algebra then a homomorphism of $\mathfrak{J}$ into a Jordan algebra $\mathfrak{A}^{(q)}$ will be called an *associative specialization* of $\mathfrak{J}$ in $\mathfrak{A}$. (This seems a bit odd but the terminology comes from the abstract theory.) One can construct a universal object for associative specializations of $\mathfrak{J}$. This consists of an associative algebra $S(\mathfrak{J})$ together with an associative specialization σ_u of $\mathfrak{J}$ in $S(\mathfrak{J})$ such that if σ is any associative specialization of $\mathfrak{J}$ in an associative algebra $\mathfrak{A}$ then there exists a unique homomorphism η of associative algebras making the following diagram commutative:

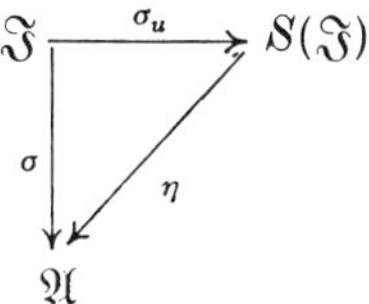

We call $(S(\mathfrak{J}), \sigma_u)$ a *special universal envelope* for $\mathfrak{J}$. To construct this one forms the tensor algebra $T(\mathfrak{J}) = \bigoplus_0^\infty \mathfrak{J}^{(i)}$ where $\mathfrak{J}^{(0)} = \Phi$, $\mathfrak{J}^{(i)}$ for $i > 0$, the i-fold tensor product $\mathfrak{J} \otimes \mathfrak{J} \otimes \dots \otimes \mathfrak{J}$ with the usual associative multiplication. Let $\mathfrak{K}$ denote the ideal in $T(\mathfrak{J})$ generated by $1_\Phi - 1$, 1_Φ the unit of Φ and the elements $aba - a \otimes b \otimes a$, $a, b \in \mathfrak{J}$. Put $S(\mathfrak{J}) = T(\mathfrak{J})/\mathfrak{K}$ and let σ_u be the restriction to $\mathfrak{J}$ of the canonical homomorphism of $T(\mathfrak{J})$ into $S(\mathfrak{J})$. Then it is easy to see, using the universal property of the tensor algebra, that $(S(\mathfrak{J}), \sigma_u)$ is a special universal envelope. Since, by definition, we have an associative specialization σ which is an injection in an associative algebra $\mathfrak{A}$, it follows that σ_u is injective. Hence we may identify $\mathfrak{J}$ with its image $\mathfrak{J}^{\sigma_u}$ and so replace σ_u by the injection mapping in $S(\mathfrak{J})$. In this way $\mathfrak{J}$ becomes a subalgebra of $S(\mathfrak{J})^{(q)}$, $\mathfrak{J}$ generates $S(\mathfrak{J})$ and the universal property is that if σ is any homomorphism of $\mathfrak{J}$ into an $\mathfrak{A}^{(q)}$ then σ has a unique extension to a homomorphism of the associative algebra $S(\mathfrak{J})$ into $\mathfrak{A}$. It follows that we have the functorial property that if ζ is a homomorphism of $\mathfrak{J}$ into a second Jordan algebra $\mathfrak{J}'$ then ζ has a unique extension to a homomorphism ζ_u of $S(\mathfrak{J})$ into $S(\mathfrak{J}')$. It follows that if ζ is an isomorphism then so is ζ_u.

Clearly all of this is analogous to the situation of a Lie algebra and its universal envelope. As in the Lie case which is discussed in Chapter V of the author's *Lie algebras* one can show easily that $S(\mathfrak{J})$ has an involution π characterized by the property that $a^\pi = a$ for $a \in \mathfrak{J}$, or, $\mathfrak{J} \subseteq \mathfrak{H}(S(\mathfrak{J}), \pi)$. π is called the *main involution* of $S(\mathfrak{J})$. We shall call $S(\mathfrak{J})$ *reflexive* if $\mathfrak{J} = \mathfrak{H}(S(\mathfrak{J}), \pi)$. Another important property of $S(\mathfrak{J})$ is that this commutes with extension of the base ring Φ. Let P be a commutative algebra over Φ and let $\mathfrak{A}^P = P \otimes_\Phi \mathfrak{A}$ regarded as algebra over P. Then $S(\mathfrak{J}^P) = S(\mathfrak{J})^P$.

We define a *derivation* D of a Jordan algebra $\mathfrak{J}$ to be a Φ-endomorphism such that $1D = 0$ and $(aba)D = (aD)ba + ab(aD) + a(bD)a$. In operator form the latter condition is

$$[U_a, D] = U_{a,aD} \tag{2}$$

where, as before, $U_{a,b} = U_{a+b} - U_a - U_b$. Let Δ be the algebra of dual numbers over Φ: Δ has the basis $(1, \delta)$ when $\delta^2=0$ and forms the Jordan algebra $\mathfrak{J}^{\Delta}$. Then it is easily seen that D is a derivation in $\mathfrak{J}$ if and only if $1+\delta D$ is an automorphism of $\mathfrak{J}^{\Delta}$. Using this and the commutativity of $S(\mathfrak{J})$ with extension of the base ring one sees easily that every derivation of $\mathfrak{J}$ has a unique extension to a derivation in the usual sense of the associative algebra $S(\mathfrak{J})$.

Let $(\mathfrak{A}, J)$ be an associative algebra with involution, $\mathfrak{H}(\mathfrak{A}, J)$ the corresponding Jordan algebra of J-symmetric elements. We shall say that $(\mathfrak{A}, J)$ is a *perfect algebra with involution* if $\mathfrak{A}$ together with the injection mapping of $\mathfrak{H}(\mathfrak{A}, J)$ in $\mathfrak{A}$ is a special universal envelope for $\mathfrak{H}(\mathfrak{A}, J)$. In other words this means that if σ is a homomorphism of $\mathfrak{H}(\mathfrak{A}, J)$ in a $\mathfrak{L}^{(q)}$, $\mathfrak{L}$ associative, then σ has a unique extension to an (associative) homomorphism of $\mathfrak{A}$ into $\mathfrak{L}$. This implies that $\mathfrak{H}(\mathfrak{A}, J)$ generates $\mathfrak{A}$. In 1950 Rickart and I proved that if $\mathfrak{A}=\mathfrak{D}_n$, $(\mathfrak{D}, j)$ any associative algebra with involution, J a canonical involution in $\mathfrak{D}_n$ then $(\mathfrak{D}_n, J)$ is perfect if $n\geqslant 3$. By a canonical involution we mean one of the form $X\to D^{-1}\bar{X}^t D$ where D is a diagonal matrix with diagonal entries d_i invertible and j-symmetric, $\bar{X}=(x_{ik}^j)$ if $X=(x_{ik})$ and t denotes the transpose. It is well known that any involution in a simple Artinian algebra is canonical. An extensive generalization of the Jacobson-Rickart theorem was given by Martindale in 1967. This is the following

Martindale's Theorem: *Let $(\mathfrak{A}, J)$ be an associative algebra with involution such that $\mathfrak{H}(\mathfrak{A}, J)$ contains $n\geqslant 3$ orthogonal idempotents e_i such that $e_i\in\mathfrak{A}e_j\mathfrak{A}$ for any i, j. Then $(\mathfrak{A}, J)$ is perfect.*

Martindale has also proved perfection in the case $n=2$ assuming some additional conditions which we shall not state here.

We shall now consider some applications of these concepts and results to certain groups and Lie algebras associated with $\mathfrak{J}$. Following Koecher one defines the *structure group* $\operatorname{Str}\mathfrak{J}$ of $\mathfrak{J}$ as the set of module automorphisms η of $\mathfrak{J}/\Phi$ for which there exists a module automorphism η^* of $\mathfrak{J}$ such that

$$U_{a\eta}=\eta^* U_a\eta\,, \qquad a\in\mathfrak{J}\,. \tag{3}$$

It is easily seen that the set of η thus defined is a group of transformations in $\mathfrak{J}$. If η is an automorphism of $\mathfrak{J}$ then $(bU_a)^\eta=(b^\eta)U_{a^\eta}$ so $U_{a^\eta}=\eta^{-1}U_a\eta$. Hence $\operatorname{Aut}\mathfrak{J}\subseteq\operatorname{Str}\mathfrak{J}$. It is easily seen that $\operatorname{Aut}\mathfrak{J}$ is the subgroup of $\operatorname{Str}\mathfrak{J}$ fixing the element 1. The structure group arises in considering a problem on matrices which was originally solved by Frobenius. Let Φ_n be the algebra of $n\times n$ matrices over the field Φ. Determine the bijective linear mappings η of Φ_n into Φ_n such that

$\det A^{\eta} = \varrho \det A$, $A \in \Phi_n$. Frobenius showed that these were either of the form $X \to BXC$ or the form of $X \to BX^{t}C$. By an elementary calculation using the differential calculus of rational mappings one can show that if η satisfies Frobenius' conditions then $\eta \in \operatorname{Str} \mathfrak{J}$. The same result can be obtained for any subalgebra $\mathfrak{J}$ of Φ_n which is a non-isotropic subspace of Φ_n relative to the trace bilinear form $\operatorname{tr}(AB)$. Hence the solution of Frobenius' problem for $\mathfrak{J}$ is reduced to the problem of determining $\operatorname{Str} \mathfrak{J}$. For Koecher the structure group seems to have arisen in determining the automorphism group of a domain of positivity (see Koecher's University of Minnesota lecture notes).

There is an alternative definition of $\operatorname{Str} \mathfrak{J}$ due to the author which leads to its determination for the algebras $\mathfrak{H}(\mathfrak{A}, J)$ where $(\mathfrak{A}, J)$ is a perfect algebra with involution. This is based on the notion of an isotope of a Jordan algebra. First, we consider isotopes of associative algebras. Let c be an invertible element of the associative algebra $\mathfrak{A}$. Then we can define a c product in $\mathfrak{A}$ by $a_c b = acb$. This is associative and c^{-1} is the unit. The algebra thus obtained is called the *c-isotope*. $\mathfrak{A}^{(c)}$ of $\mathfrak{A}$. It is immediate that $c_R : x \to xc$ is an isomorphism of $\mathfrak{A}^{(c)}$ into $\mathfrak{A}$. If $\mathfrak{J}$ is a subalgebra of $\mathfrak{A}^{(q)}$ and $c, c^{-1} \in \mathfrak{J}$ then $\mathfrak{J}$ is also a subalgebra of $\mathfrak{A}^{(c)(q)}$ and the U-operator in this algebra is $U_a^{(c)} = U_c U_a$. We denote the new algebra as $\mathfrak{J}^{(c)}$ and call this the *c-isotope* of $\mathfrak{J}$. It is easily seen that isotopy is an equivalence relation, that is, the isotope of an isotope is an isotope and the relation is symmetric and reflexive. In fact one has that $\mathfrak{J}^{(c)(d)} = \mathfrak{J}^{(cdc)}$ and $\mathfrak{J}^{(c)(c^{-2})} = \mathfrak{J}$, $\mathfrak{J}^{(1)} = \mathfrak{J}$. Now let $\eta \in \operatorname{Str} \mathfrak{J}$, so we have (3). Since $U_1 = 1$ this gives $\eta^* \eta = U_{1^\eta}$ so U_{1^η} is bijective. It follows from this that 1^η is invertible and $c = (1^\eta)^{-1} \in \mathfrak{J}$. Then $U_{1^\eta}^{-1} = U_c$ and we have $\eta^* = U_{1^\eta} \eta^{-1} = U_c^{-1} \eta^{-1}$ and $U_c U_{a^\eta} = \eta^{-1} U_a \eta$ or $U_{a^\eta}^{(c)} = \eta^{-1} U_a \eta$. This is just the condition that η is an isomorphism of $\mathfrak{J}$ into the isotope $\mathfrak{J}^{(c)}$. The converse is easily seen also. Hence $\operatorname{Str} \mathfrak{J}$ is just the group of isotopies of $\mathfrak{J}$ where by an *isotopy* of $\mathfrak{J}$ we mean an isomorphism of $\mathfrak{J}$ into an isotope $\mathfrak{J}^{(c)}$.

We now relate this to $S(\mathfrak{J})$ by indicating first the relation between $S(\mathfrak{J})$ and $S(\mathfrak{J}^{(c)})$. We note first that if σ is an associative specialization of $\mathfrak{J}$ in $\mathfrak{A}$ then σ is an associative specialization of the isotope $\mathfrak{J}^{(c)}$, $c \in \mathfrak{J}$, in $\mathfrak{A}^{(c)}$. If we follow an associative specialization by a homomorphism of associative algebras then we obtain an associative specialization. Since the injection of $\mathfrak{J}$ in $S(\mathfrak{J})$ is an associative specialization the injection of $\mathfrak{J}^{(c)}$ in $S(\mathfrak{J})^{(c)}$ is an associative specialization. Since c_R is an isomorphism of $S(\mathfrak{J})^{(c)}$ into $S(\mathfrak{J})$ it is clear that c_R is an associative specialization of $\mathfrak{J}$ in $S(\mathfrak{J})$. It is not difficult to see that $(S(\mathfrak{J}), c_R)$ is a special universal envelope for $\mathfrak{J}^{(c)}$. This result is due to McCrimmon (unpublished).

Now let $\eta \in \operatorname{Str} \mathfrak{J}$ so η is an isomorphism of $\mathfrak{J}$ into an isotope $\mathfrak{J}^{(c)}$.

Then we have a unique automorphism η_u of $S(\mathfrak{J})$ making

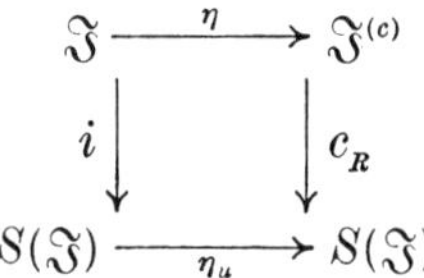

commutative, that is, we have

$$x^{\eta_u} = x^{\eta} c\,, \qquad x \in \mathfrak{J}\,. \tag{5}$$

Applying the main involution π we obtain $x^{\eta_u \pi} = cx^{\eta} = cx^{\eta_u}c^{-1} = cx^{\pi\eta_u}c^{-1}$. Since the elements x generate $S(\mathfrak{J})$ we obtain

$$a^{\eta_u \pi} = ca^{\pi\eta}c^{-1}, \qquad a \in S(\mathfrak{J})\,. \tag{6}$$

Thus the anti-automorphisms $\eta_u \pi$ and $\pi \eta_u$ differ by the inner automorphism $a \to cac^{-1}$ determined by c^{-1}. It is easily seen that if $\mathfrak{J}$ is reflexive then the converse holds: any mapping of the form indicated is in $\mathrm{Str}\,\mathfrak{J}$. This implies the following

THEOREM 1: *Let $(\mathfrak{A}, J)$ be a perfect associative algebra with involution. Then $\mathrm{Str}\,\mathfrak{H}(\mathfrak{A}, J)$ is the set of mappings*

$$x \to x^{\mu} d$$

where μ is an automorphism of $\mathfrak{A}$ and d is an invertible element of $\mathfrak{H}(\mathfrak{A}, J)$ such that μJ and $J\mu$ differ by the inner automorphism determined by d: $a^{\mu J} = d^{-1}a^{J\mu}d$.

We indicate next the infinitesimal or Lie algebra analogue of these results. We define the *structure Lie algebra* $\mathrm{Strl}\,\mathfrak{J}$ of $\mathfrak{J}$ to be the set of Φ-endomorphisms L of $\mathfrak{J}$ so that $1 + \delta L \in \mathrm{Str}\,\mathfrak{J}^{\Delta}$, Δ the algebra of dual numbers over Φ. It is easy to see from this definition that $\mathrm{Strl}\,\mathfrak{J}$ is indeed a Lie algebra of Φ-endomorphisms of $\mathfrak{J}$. The condition that $1 + \delta L \in \mathrm{Str}\,\mathfrak{J}^{\Delta}$ is $U_{a(1+\delta L)} = U_{1(1+\delta L)}(1 - \delta L)U_a(1 + \delta L)$. This follows since $(1 + \delta L)^{-1} = 1 - \delta L$ and if $\eta \in \mathrm{Str}\,\mathfrak{J}^{\Delta}$ then $\eta^* = U_{1^{\eta}}\eta^{-1}$. The condition just given is equivalent to

$$U_{a,aL} = U_{1,1L}U_a + [U_a, L]\,, \qquad a \in \mathfrak{J}\,. \tag{7}$$

It is easily seen that $\mathrm{Strl}\,\mathfrak{J}$ contains the Lie algebra $\mathrm{Der}\,\mathfrak{J}$ of derivations of $\mathfrak{J}$ and in fact $\mathrm{Der}\,\mathfrak{J}$ is the subalgebra of $\mathrm{Strl}\,\mathfrak{J}$ of D

satisfying $1D = 0$. Using the fact that $S(\mathfrak{J}^{\Delta}) = S(\mathfrak{J})^{\Delta}$ one can deduce from Theorem 1 the following result on $\mathrm{Strl}\,\mathfrak{J}$.

THEOREM 2: *Let* $(\mathfrak{A}, J)$ *be a perfect associative algebra with involution. Then* $\mathrm{Strl}\,\mathfrak{H}(\mathfrak{A}, J)$ *is the set of mappings* $D + l_R$ *where* D *is a derivation of* $\mathfrak{A}$ *and* l *is an element of* $\mathfrak{H}(\mathfrak{A}, J)$ *such that* $JDJ - D$ *is the inner derivation* $a \to [a, l]$ *in* $\mathfrak{A}$.

As an illustration of the type of result one can obtain from Theorems 1 and 2 and Martindale's theorem, suppose $\mathfrak{A}$ is an associative algebra containing $n \geqslant 3$ orthogonal idempotents e_i such that $e_i \in \mathfrak{A} e_j \mathfrak{A}$ for all i, j. If every automorphism of $\mathfrak{A}$ is inner then $\mathrm{Str}\,\mathfrak{J}$ is the set of mappings $x \to uxv$ and $x = ux^T v$, u and v invertible, T a fixed anti-automorphism of $\mathfrak{A}$ (assuming $\mathfrak{A}$ has one). If every derivation of $\mathfrak{A}$ is inner then $\mathrm{Strl}\,\mathfrak{J}$ is the set of mappings $x \to ax + xb$, a, b arbitrary in $\mathfrak{A}$.

We shall now indicate more briefly some other notions and results on Jordan algebras which are interesting from the point of view of associative algebras. We mention first the notions of ideals which are natural in this context. If $\mathfrak{J}$ is a Jordan algebra we define an *inner ideal* $\mathfrak{L}$ to be a Φ-submodule such that $bxb \in \mathfrak{L}$ for every $b \in \mathfrak{L}$, $x \in \mathfrak{J}$. An *outer ideal* is a Φ-submodule $\mathfrak{L}$ such that $xbx \in \mathfrak{L}$ for $b \in \mathfrak{L}$, $x \in \mathfrak{J}$. A subset which is both an inner and an outer ideal is called an *ideal*. If $b \in \mathfrak{J}$ then $b\mathfrak{J}b = \mathfrak{J}U_b$ is an inner ideal. This is called the *principal inner ideal determined by b*. (It may not contain b.) Outer ideals are nearly the same thing as ideals. If $\Phi \ni \frac{1}{2}$ then any outer ideal is an ideal and this is the same thing as an ideal in the usual sense in the Jordan structure defined by $a \cdot b = \frac{1}{2}(ab + ba)$. McCrimmon has recently determined the inner ideals for the Jordan algebras determined by Artinian simple associative algebras with involution and a student of mine, J. Katz, has obtained the lattice isomorphisms of the lattice of inner ideals in these cases. For example, if $\mathfrak{A}$ is simple Artinian then the inner ideals of $\mathfrak{A}^{(q)}$ are the intersections $e\mathfrak{A} \cap \mathfrak{A}f$ of a left and a right ideal of $\mathfrak{A}$. If $\mathfrak{A} = \mathfrak{D}_n$, $\mathfrak{D}$ a division algebra with involution j and J is a canonical involution in $\mathfrak{A}$ then the inner ideals of $\mathfrak{H}(\mathfrak{A}, J)$ are principal. $\mathrm{Str}\,\mathfrak{J}$ is always contained in the group of automorphisms of the lattice of inner ideals. However, in the interesting case indicated the latter group is larger and can be determined. This has been done by Katz.

Finally, we should like to mention a direction of research in this area which appears to be natural and in which very little is known at the present time. This concerns the notion of integral domain, division algebras and embedding of the former in the latter. An element $a \in \mathfrak{J}$ is called a *zero divisor* if there exists a $b \neq 0$ in $\mathfrak{J}$ such that

$aba = 0$. If $\mathfrak{J}$ has no zero divisors then $\mathfrak{J}$ is called an *integral domain*. One can formulate the associative condition that $ab = 1 = ba$, $a, b \in \mathfrak{J}$ by the two Jordan conditions: $aba = a$, $ab^2a = 1$, $b^2 = b1b$. Then one defines a Jordan division algebra by the condition that every $a \neq 0$ has an inverse in $\mathfrak{J}$ in the foregoing sense. An analogue of the Ore condition is $\mathfrak{J}U_a \cap \mathfrak{J}U_b \neq 0$ if $a \neq 0$, $b \neq 0$. Can an integral domain satisfying the Ore condition be embedded in a division algebra? Partial results on this have been obtained. One can ask more general questions on quotient rings for rings with zero divisors.

Testo pervenuto il 26 novembre 1970.
Bozze licenziate il **16 dicembre** 1971.

REFERENCES

[1] N. Jacobson, *Structure and Representations of Jordan Algebras*, A.M.S. Colloq. Publ. (1969).

[2] N. Jacobson, *Structure Theory of Quadratic Jordan Algebras*, Tata Institute lecture notes (1970).

[3] K. McCrimmon, *A general theory of Jordan rings*, Proc. Nat. Acad. Sci. U.S.A., 56 (1966), 1072-1079.

[4] W. Martindale III, *Jordan homomorphisms of the ring of symmetric elements of a ring with involution*, Journ. of Algebra, 5 (1967), 232-249.

BULLETIN OF THE
AMERICAN MATHEMATICAL SOCIETY
Volume 80, Number 6, November 1974

ABRAHAM ADRIAN ALBERT
1905–1972

BY NATHAN JACOBSON

Adrian Albert, one of the foremost algebraists of the world and President of the American Mathematical Society from 1965 to 1967, died on June 6, 1972. For almost a year before his death it had become apparent to his friends that his manner had altered from its customary vigor to one which was rather subdued. At first they attributed this to a letdown which might have resulted from Albert's having recently relinquished a very demanding administrative position (Dean of the Division of Physical Sciences at the University of Chicago) that he had held for a number of years. Eventually it became known that he was gravely ill of physical causes that had their origin in diabetes with which he had been afflicted for many years.

Albert was a first generation American and a second generation American mathematician following that of E. H. Moore, Oswald Veblen, L. E. Dickson and G. D. Birkhoff. His mother came to the United States from Kiev and his father came from England.[1] The father had run away from his home in Vilna at the age of fourteen, and on arriving in England, he discarded his family name (which remains unknown) and took in its place the name Albert after the prince consort of Queen Victoria. Albert's father was something of a scholar, with a deep interest in English literature. He taught school for a while in England but after coming to the United States he became a salesman, a shopkeeper, and a manufacturer. Adrian was born when his father was fifty-five and his mother was thirty-five. It was a second marriage for both parents; his father's first wife had died in childbirth, and his mother was a widow with two children when she married his father. Adrian was the middle child of a set of three children which his parents had in common. He grew up in a family that was formally orthodox Jewish but not strongly religious. In common with most immigrant families of the period the family had a strong drive toward assimilation and a determination to make the most of the opportunities offered by a comparatively free society undergoing rapid economic expansion with no limits in sight.

Albert spent all of his school years in the Midwest and all but two of these in Chicago. He attended public schools at Chicago and at Iron

[1] We are indebted to Mrs. Frieda Albert for background material on Professor Albert's family.

Mountain, Michigan, and entered the University of Chicago in 1922 where in rapid succession he earned a B.S. degree in 1926, an M.S. degree in 1927, and a Ph.D. in 1928. His advisor for his master's and his doctoral dissertations was Leonard Eugene Dickson. After his doctorate Albert spent a year at Princeton University as a National Research Council Fellow. He was attracted to Princeton by that great master of associative algebra theory, J. H. M. Wedderburn, who was then a professor at the university. Albert returned to Princeton in 1933, this time as one of the first group of temporary members of the Institute for Advanced Study.

Albert married Frieda Davis in 1927, and they had three children, Alan, Roy, and Nancy, one of whom, Roy, died of diabetes at the age of twenty-three.

Except for two years (1929–1931) as an Instructor at Columbia University and a number of visiting professorships (at Rio de Janeiro, Buenos Aires, University of Southern California, Yale, and the University of California at Los Angeles) all of Albert's academic career was spent at the University of Chicago. In 1960 he was named Eliakim Hastings Moore Distinguished Service Professor, and he served as Chairman of the Department of Mathematics for three years until he became Dean of the Division of Physical Sciences in 1962. He held this position until 1971 when he reached the mandatory retirement age of sixty-five for the deanship.

Of the mathematicians who influenced Albert most directly we should list the following: Dickson, who set the direction for almost all of Albert's research and whose books, *Algebras and their Arithmetics* (1923) and *Algebren und ihre Zahlentheorie* (1927), stimulated the great flowering of associative algebra theory of the 1930's; Wedderburn, whose elegant results and methods were an inspiration to Albert; Hermann Weyl, whose lectures on Lie groups and especially Lie algebras aroused Albert's interest in this subject—an interest which later broadened to encompass the whole range of nonassociative algebras; and above all, Solomon Lefschetz, who introduced Albert to the subject of Riemann matrices during his postdoctoral year (1928–1929) at Princeton.

Mrs. Albert tells the story of this introduction in a charming fashion. Filling in some mathematical details it runs somewhat as follows. Albert had given a lecture on his dissertation at the Princeton mathematics club. In the audience were Dieudonné, J. H. C. Whitehead and Lefschetz, who had worked on the problem of multiplication algebras of Riemann matrices. Lefschetz apparently sensed that here was a brilliant young algebraist whose interests and power made him ideally suited to attack this problem. After Albert's talk he described the problem to him. A lively discussion ensued, mostly in the course of wanderings through the streets of Princeton. This lasted for several hours, well past dinnertime, and Mrs. Albert had become quite concerned before Albert finally returned home, apparently in great excitement over his initiation into a fascinating area of classical mathematics which provided a strong motivation for the study of his chosen field of associative algebras.

Lefschetz was certainly right in his judgment. Albert took to the problem on Riemann matrices with great enthusiasm, and as the structure theory of associative algebras was revealed by Albert, Brauer, Hasse and Emmy Noether, Albert could push forward the theory of multiplication algebras of Riemann matrices until he achieved a complete solution of the central problem (which we shall discuss below). For this achievement Albert was awarded the Cole Prize in algebra in 1939.

This was a memorable year for Albert. Besides the Cole Prize award which he received that year, he was the Colloquium speaker of the Society for 1939. Moreover, he performed a feat, which we believe has never been matched, of having the book, *Structure of Algebras*, the subject of his lectures in print at the same time that the lectures were delivered.

Around 1942 Albert's research interests shifted from associative to nonassociative algebras. He wrote many important papers in this field (which we shall discuss below). In 1965 Albert returned to his first love, structure theory of associative algebras.

Besides his own important contributions to mathematics, Albert was instrumental in a number of ways in improving the status of the profession. He had a good deal to do with the establishment of government research grants for mathematics on more or less an equal footing with those in the other sciences. He was chairman of the Committee to Prepare a Budget for Mathematics for the National Science Foundation, 1950, and chairman of the Committee on a Survey of Training and Research Potential in the Mathematical Sciences, January 1955–June 1957 (which became known as "The Albert Committee"). He demonstrated that pure mathematicians could be useful in applied and directed research by acting as a consultant for a number of government sponsored research agencies. For a number of years he was associated with the Institute for Defense Analysis as a member of its Board of Trustees and for a year as Director of its Princeton group. He directed the research project SCAMP for several summers and organized and directed the project ALP (known as "Albert's little project").

Albert was also a driving force in the creation of the summer research institutes which have become such an important part of the research activities of the Society, supported by the National Science Foundation. He was chairman of the committee which was responsible for the first one of these—on Lie groups and Lie algebras—held at Colby College in Maine in the summer of 1953.

Albert's role as a "statesman" for mathematics included membership on the Board of Trustees of the Institute for Advanced Study, chairman of the Consultative Committee of the Nice Congress, and Vice-President of the International Mathematical Union.

His influence in mathematics extended also through a large number of gifted students. One of the most distinguished of these, Dan Zelinsky, has written a warm appreciation of Albert as a mathematician and as a person **[188]**.

Naturally many important honors came his way. He was elected to the National Academy of Sciences in 1943 and was awarded honorary doctorates from Notre Dame, Yeshiva University, and the University of Illinois. He was elected a corresponding member of the Brazilian Academy of Sciences, honorary member of the Argentine Academy of Sciences, and of the Mexican Mathematical Society. He thoroughly enjoyed these honors, but he derived almost as much pleasure from the honors bestowed on fellow algebraists and on his friends. Most of all he enjoyed seeking out a colleague to whom he could communicate his latest discovery, which excited him greatly.

Most of Albert's important discoveries fall neatly into three categories: I. Associative Algebras, II. Riemann matrices, III. Nonassociative algebras. We proceed to give an indication of these and of some interesting isolated results which we shall mention under IV. Miscellaneous.

I. **Associative algebras.** The Wedderburn structure theorems of 1907 on finite dimensional associative algebras over a field focused attention on the division algebras in this class. In 1906 Dickson had given a construction of a type of algebra called *cyclic* which included division algebras. These contain a maximal subfield $\mathfrak{Z}$ which is cyclic over the base field $\mathfrak{F}$, that is, they are Galois with Galois group $G=\langle s\rangle$, a cyclic group generated by a single element s. Moreover, the algebras are generated by $\mathfrak{Z}$ and an element u for which one has the relations

$$uz = s(z)u, \quad z \in \mathfrak{Z}, \quad u^n = \gamma,$$

where n is the order of G and γ is a nonzero element of $\mathfrak{F}$. The cyclic algebra, denoted as $(\mathfrak{Z}, s, \gamma)$, constructed in this way has dimensionality n^2 over $\mathfrak{F}$. In 1914 Wedderburn **[184]** proved an important sufficient condition for $(\mathfrak{Z}, s, \gamma)$ to be a division algebra. He showed that this is the case if no power of γ, γ^m with $0<m<n$, is a norm $N_{\mathfrak{Z}/\mathfrak{F}}(z)$ of an element $z \in \mathfrak{Z}$. Using this criterion it is easy to construct division algebras of any dimension n^2.

In 1921 Wedderburn published some other important results on division algebras [**185**]. Noting that one may as well consider these as algebras over their centers and so assume that they are *central* in the sense that the center is the base field $\mathfrak{F}$, he showed that the dimensionality over this field is a square, n^2. More generally, if $\mathfrak{A}$ is central simple, by one of Wedderburn's structure theorems, $\mathfrak{A}$ is the algebra $M_r(\mathfrak{D})$ of $r \times r$ matrices with elements in a central division algebra $\mathfrak{D}$. Hence if the dimensionality of $\mathfrak{D}$ over $\mathfrak{F}$ is d^2, then that of $\mathfrak{A}$ over $\mathfrak{F}$ is n^2, where $n=dr$. Then n is called the *degree* of the central simple algebra $\mathfrak{A}$ and d is its *index*. In his 1921 paper, Wedderburn showed also that any maximal subfield $\mathfrak{Z}$ of a central division algebra $\mathfrak{D}$ is a *splitting field*, that is, the algebra $\mathfrak{D}^{\mathfrak{Z}} = \mathfrak{Z} \otimes_{\mathfrak{F}} \mathfrak{D}$ is the matrix algebra $M_d(\mathfrak{Z})$, and he proved that every central division algebra of degree three is cyclic. Wedderburn showed also that Dickson's cyclic algebras were special cases of a more general type of algebra which is now called an abelian crossed product. Here the cyclic field $\mathfrak{Z}$ is replaced by a Galois extension field of the base field with Galois group an abelian group.

Abelian crossed products were rediscovered by Cecioni [**156**], and these were further generalized by Dickson [**158**] and [**159**] to arbitrary crossed products based on any Galois extension field.

Much of Albert's early work was concerned with the study of finite dimensional central simple algebras. His first important result on these was the theorem, proved in his dissertation [**9**], that every central division algebra of degree four (dimension sixteen) is a crossed product. This was the next case to be considered after Wedderburn's theorem that in degree three these algebras are cyclic. Albert improved the result in [**11**] by showing that the degree four central division algebras are crossed products based on abelian extension fields whose Galois groups are direct products of two cyclic groups of order two, and he gave a simpler proof of this result in [**41**]. In both of these papers the algebras of characteristic two were excluded. In a subsequent paper [**53**] he was able to overcome the difficulties of the characteristic two case. Brauer was the first to show that the central division algebras of degree four, unlike those of degree three, need not be cyclic. He constructed an example of such an algebra which was a tensor product of two (generalized) quaternion algebras [**153**]. Subsequently, Albert [**45**] constructed one which is not such a product. This was significant in view of another important theorem, proved by Albert [**39**], stating that a central division algebra $\mathfrak{D}$ of degree four is a tensor product of quaternion algebras if and only if $\mathfrak{D} \otimes_{\mathfrak{F}} \mathfrak{D} \cong M_4(\mathfrak{F})$.

The main goal of the structure theory of algebras of the period 1929–1932 was the determination and classification of finite dimensional division algebras over the field Q of rational numbers, or equivalently, finite dimensional central division algebras over number fields. It was recognized quite early that this problem had two separate aspects: a purely algebraic one concerned with properties of algebras valid for all base fields, and an arithmetic one exploiting the arithmetic of number fields. Albert recognized the importance of the arithmetic method. However, he was handicapped in its use by the fact that he was unaware until rather late of the powerful results of algebraic number theory, notably, class field theory, which had been developed in Germany. He did make use of the arithmetic theory of quadratic forms to achieve definitive results on central division algebras of degree four over number fields and some important early results on the degree 2^n case. For example, he proved that the former are cyclic and are not tensor products of quaternion algebras, and he proved that the only central division algebras over number fields which possess involutions, that is, antiautomorphisms of period two, are the quaternion algebras. This last result was needed for his study of Riemann matrices which we shall discuss below.

Albert's main contributions were on the purely algebraic side. There is a substantial overlap between his results on central simple algebras and those of the German school of algebraists of the period of the early thirties, especially those of Richard Brauer and of Emmy Noether. Albert obtained independently all the algebraic results on splitting fields, extensions of isomorphisms and tensor products which were needed to obtain the fundamental theorems on division algebras over number fields. Of central importance for the algebraic theory is the group of classes of central simple algebras which was introduced by Brauer in 1929 [**153**]. We recall the definition. Two (finite dimensional) central simple algebras $\mathfrak{A}$ and $\mathfrak{B}$ over a field $\mathfrak{F}$ are said to be *similar* ($\sim$) if there exist positive integers m and n such that the matrix algebras $M_m(\mathfrak{A})$ and $M_n(\mathfrak{B})$ are isomorphic. This is an equivalence relation. Denoting the similarity class of $\mathfrak{A}$ as $\{\mathfrak{A}\}$, one defines a product of such classes by $\{\mathfrak{A}\}\{\mathfrak{B}\}=\{\mathfrak{A}\otimes_{\mathfrak{F}}\mathfrak{B}\}$. This gives a commutative group $B(\mathfrak{F})$ called *the Brauer group of the field* $\mathfrak{F}$. The unit of the group is the set of matrix algebras $M_n(\mathfrak{F})$, $n=1, 2, \cdots$, and the inverse of $\{\mathfrak{A}\}$ is $\{\mathfrak{A}^{\mathrm{op}}\}$, where $\mathfrak{A}^{\mathrm{op}}$ is the opposite algebra of $\mathfrak{A}$. In a beautiful paper [**33**] published in 1931, Albert essentially rediscovered the Brauer group. In this paper he proved Brauer's main theorem that $B(\mathfrak{F})$ is a torsion group; more precisely, if $\mathfrak{A}$ has index m, that is, if the degree of the division algebra $\mathfrak{D}$ in $\{\mathfrak{A}\}$ is m, then $\{\mathfrak{A}\}^m=1$. Moreover, if e is the order of $\{\mathfrak{A}\}$ in $B(\mathfrak{F})$, then e and m have the same prime factors. The integer e is called the *exponent* of $\mathfrak{A}$. Albert's proofs are based on Wedderburn's norm condition for cyclic algebras to be division algebras and theorems reducing considerations to the cyclic case; for example, if $\mathfrak{D}$ is a central division algebra of prime degree p, then there exists an extension field $\mathfrak{K}$ of the base field of dimensionality prime to p such that $\mathfrak{D}^{\mathfrak{K}} = \mathfrak{K} \otimes_{\mathfrak{F}} \mathfrak{D}$ is a cyclic division algebra over $\mathfrak{K}$. Another key tool in Albert's method was the following theorem which he called the *index reduction factor theorem*: Let $\mathfrak{D}$ be a central division algebra of degree d and $\mathfrak{K}$ an extension field of the base field $\mathfrak{F}$ with dimensionality r. Then $\mathfrak{D}^{\mathfrak{K}} = M_q(\mathfrak{E})$, where $\mathfrak{E}$ is a central division algebra over $\mathfrak{K}$ and q is a divisor of d and r. Albert's primary interest in the theorem that $\{\mathfrak{A}\}^m = 1$ was its consequence that any central division algebra is a tensor product of division algebras of prime power degrees which are determined up to isomorphism. This reduced most questions on these algebras to the prime power degree case.

The high points of the structure theory of algebras of the 1930's were undoubtedly the theorem that every finite dimensional central division algebra over a number field is cyclic, and the classification of these algebras by a set of numerical invariants. The latter result amounts to the determination of the structure of the Brauer group for a number field. Besides the general theory of central simple algebras we have indicated, the proofs of these fundamental results required the structure theory of central simple algebras over p-adic fields due to Hasse, Hasse's norm theorem ("the Hasse principle"), and the Grünwald existence theorem for certain cyclic extensions of a number field. (Though it was discovered almost thirty years later by S. Wang **[182]** that Grünwald's formulation was incorrect, his error did not affect the theorem on algebras. See also Wang **[183]** and Hasse **[161]**.) The first proof of the cyclic structure of central division algebras over number fields was given by Brauer, Hasse and Noether (**[155]**, 1931). However, it seemed appropriate that Albert should share the honor of this achievement, and at Hasse's suggestion a joint paper (**[42]**, 1932) was published by Albert and Hasse giving another proof of the theorem and the historical background of the problem.

The results which had been obtained up to this point suggested the following two problems: (I) Is every finite dimensional central division algebra a crossed product? (II) Is every one of prime degree cyclic? These are equivalent to the question of existence of a maximal Galois and maximal cyclic subfield, respectively, for these algebras. The results of Wedderburn and Albert imply that the answer to the second question is affirmative for the primes 2 and 3 and for the first for the degrees 2, 3, 4, 6 and 12. Quite recently Amitsur showed that the answer to the first question is negative by showing that for any n divisible by eight or by the square of an odd prime there exists a noncrossed product central division algebra of degree n **[150]**. This leaves intact the second problem, and this is one on which Albert spent a good deal of effort. It is clear from the definition that if $\mathfrak{A}$ is cyclic of degree n, then $\mathfrak{A}$ contains an element u satisfying an irreducible pure equation $x^n - \gamma = 0$, γ in the base field $\mathfrak{F}$. Does the converse hold? Albert showed this is the case if $n=p$, a prime ([**59**], 1934), and is not the case if $n=4$ ([**72**], 1938). For the prime case this reduces the problem (II) to what appears to be a more tractable one: Does every central division algebra $\mathfrak{A}$ of prime degree p contain an element not in $\mathfrak{F}$ whose pth power is in $\mathfrak{F}$? In 1938 Brauer showed that if $\mathfrak{A}$ is of degree 5 there exists a field $\mathfrak{K}$ containing a tower of fields $\mathfrak{F} \subset \mathfrak{K}_1 \subset \mathfrak{K}_2 \subset \mathfrak{K}$ such that the degree $[\mathfrak{K}_1 : \mathfrak{F}] = 2 = [\mathfrak{K}_2 : \mathfrak{K}_1]$ and $[\mathfrak{K} : \mathfrak{K}_2] = 3$, and $\mathfrak{A}^{\mathfrak{K}}$ is cyclic [**154**]. This led Albert to consider the following question: Suppose $\mathfrak{K}$ is a quadratic extension of $\mathfrak{F}$ and $\mathfrak{A}^{\mathfrak{K}}$ is cyclic of prime degree. Then is $\mathfrak{A}$ cyclic? In four papers [**142**], [**144**], [**147**], [**148**] appearing between 1965 and 1970, including his retiring Presidential Address for the Society, Albert considered this problem for algebras of characteristic p and degree p. In spite of many ingenious arguments and partial results, he was unable to completely settle this question.

A beautiful chapter in the structure theory of central simple algebras is the theory of p-algebras which Albert developed in three papers [**66**], [**67**], [**69**] appearing in 1936 and 1937 (cf. also [**181**]). These are the central simple algebras of characteristic p whose division algebras $\mathfrak{D}$ in the Wedderburn theorem ($\simeq M_n(\mathfrak{D})$) have degree a power of p. The main results Albert proved about p-algebras are that any such algebra $\mathfrak{A}$ is cyclically representable, that is, there exists an n such that $M_n(\mathfrak{A})$ is a cyclic algebra, and the exponent of $\mathfrak{A}$ is the minimum of the exponents of purely inseparable splitting fields for the algebra.

A generalization of cyclic algebras in which the cyclic maximal subfield $\mathfrak{Z}$ is replaced by a separable commutative subalgebra on which a cyclic group G acts in such a way that there are no proper subalgebras stabilized by G was considered by Albert in [**75**] following earlier work by Teichmüller in [**180**]. Such *generalized cyclic algebras* arise naturally from cyclic ones when one extends the base field or forms the tensor powers of a cyclic algebra.

Most of the important results on associative algebras which Albert obtained prior to 1939 can be found in an improved form in his AMS Colloquium book, *Structure of Algebras*. This extemely readable and beautifully organized book can still be recommended to a beginning student with a serious interest in structure theory and is an indispensable reference book for certain aspects of the theory, particularly the theory of p-algebras, and of algebras with involution.

II. **Riemann matrices and associative algebras with involution.** The theory of multiplications of Riemann matrices has its origin in algebraic geometry. On a Riemann surface of an algebraic curve of genus p, one chooses p linearly independent integrals of the first kind each with $2p$ periods $\omega_{j\nu}$, $1 \leqq j \leqq p$, $1 \leqq \nu \leqq 2p$. The $p \times 2p$ matrix $\omega = (\omega_{j\nu})$ of complex elements satisfies the *Riemann relations*: there exists a $2p \times 2p$ nonsingular skew symmetric matrix C of rational elements such that $\omega C {}^t\omega = 0$ (${}^t\omega$ the transpose of ω) and $\sqrt{-1}\,\omega C {}^t\bar{\omega}$ is positive definite hermitian. In the theory of so-called singular correspondences on the Riemann surface, one is led to consider the multiplications of ω. These are the $2p \times 2p$ rational matrices A for which there exists a $p \times p$ complex matrix α such that $\alpha\omega = \omega A$. The set of these A's is a finite dimensional algebra of matrices over $\boldsymbol{Q}$, the algebra of multiplications of ω.

Alternatively, the matrices ω and their multiplications arise in the theory of abelian functions, defined to be meromorphic functions of p complex variables having a lattice of periods in $\boldsymbol{C}^p$.

There is another, formally simpler, formulation of Riemann matrices (the foregoing ω) and their multiplications due to Weyl [**186**] which was suggested by geometric considerations. From the purely formal point of view one obtains the passage from the classical formulation to Weyl's by introducing the $2p \times 2p$ matrix

$$W = \begin{pmatrix} \omega \\ \bar{\omega} \end{pmatrix} \quad \text{and} \quad L = \begin{pmatrix} -\sqrt{-1}\, 1_p & 0 \\ 0 & \sqrt{-1}\, 1_p \end{pmatrix}.$$

Put $R = W^{-1}LW$. Then it can be shown that the matrix R has the following properties: (1) R is real, that is $R \in M_{2p}(\boldsymbol{R})$; (2) $R^2 = -1_{2p}$; (3) $S = CR$ is positive definite symmetric. Following Weyl, one calls a matrix $R \in M_n(\boldsymbol{R})$ (here $n = 2p$) a *Riemann matrix* if $R^2 = -1_n$ and there exists a skew symmetric matrix $C \in M_n(\boldsymbol{Q})$ such that $S = CR$ is positive definite symmetric. The matrix C, which is necessarily nonsingular, is called a *principal matrix* of R. The passage from Weyl's R to the classical ω can be reversed. In Weyl's formulation the *multiplications* appear as the matrices $A \in M_n(\boldsymbol{Q})$ commuting with R. The set $\mathfrak{A}$ of these multiplications is a finite dimensional algebra over $\boldsymbol{Q}$ called the *multiplication algebra* of the Riemann matrix R. Weyl observed that for most considerations the condition $R^2 = -1$ $(= -1_n)$ plays no role. Dropping this, one obtains *generalized Riemann matrices*. Subsequently Albert [**2**], [**65**] considered further generalizations (including even a characteristic $p \neq 0$ situation!). For the sake of simplicity we shall stick to the case of Riemann matrices in Weyl's formulation.

The important early work on multiplication algebras is due to Poincaré, Scorza, Lefschetz and Rosati. Poincaré achieved a reduction to so-called *pure* Riemann matrices for which the multiplication algebras are division algebras. Lefschetz considered the situation in which the multiplication algebras are commutative. Rosati observed the important fact that if A is in the multiplication algebra $\mathfrak{A}$ of a Riemann matrix R and C is a principal matrix, then $A^* = C^{-1}\,{}^tAC \in \mathfrak{A}$. The map $A \to A^*$ is an involution (antiautomorphism of period two) in $\mathfrak{A}$. Rosati showed also that if A is symmetric under this involution ($A^* = A$) then its characteristic roots are real, and if $A^* = -A$ then its characteristic roots are pure imaginaries [**176**].

The central problem on multiplication algebras of Riemann matrices is to determine necessary and sufficient conditions that a division algebra over Q be the multiplication algebra of a Riemann matrix. For a proof of sufficiency, one requires a construction of a Riemann matrix whose multiplication algebra is a given algebra $\mathfrak{A}$ satisfying the conditions.

Albert's work on Riemann matrices went hand in hand with the development of the theory of division algebras. It culminated in the complete solution of the principal problem, which he published in three papers appearing in the Annals of Mathematics in 1934 and 1935 ([**51**], [**57**], and [**63**]). To achieve this required the development ab initio of the basic theory of simple algebras with involution. Albert presented improved versions of this theory in [**65**] and in his *Structure of Algebras*. We shall indicate first his results on algebras with involution.

We assume throughout that $\mathfrak{A}$ is finite dimensional simple over a field $\mathfrak{F}$. If $\mathfrak{A}$ has an involution J ($J{:}a \to a^*$ such that $(a+b)^* = a^* + b^*$, $(\alpha a)^* = \alpha a^*$ for $\alpha \in \mathfrak{F}$, $(ab)^* = b^*a^*$), then the center $\mathfrak{C}$ of $\mathfrak{A}$ is stabilized by J and the restriction of J to $\mathfrak{C}$ is either the identity map or an automorphism of period two. Accordingly, the involution is of *first kind* or *second kind*. Albert showed that $\mathfrak{A}$ has an involution if and only if for any $m = 1, 2, \cdots$, the matrix algebra $M_m(\mathfrak{A})$ has an involution having the same effect on the center (which can be identified with the center of the matrix algebra). He showed also that if $\mathfrak{A}$ has an involution J and x is an element of $\mathfrak{A}$ whose minimum polynomial over the center $\mathfrak{C}$ is irreducible and has coefficients that are J-symmetric, then $\mathfrak{A}$ has an involution T leaving x fixed and having the same effect on $\mathfrak{C}$ as J. Assuming $\mathfrak{A}$ is of dimension n^2 over its center $\mathfrak{C}$ and contains a subfield of the form $\mathfrak{X} \otimes_{\mathfrak{F}} \mathfrak{C}$, where $\mathfrak{X}$ is n-dimensional Galois over $\mathfrak{F}$, and $\mathfrak{C}$ is either $\mathfrak{F}$ or a separable quadratic extension of $\mathfrak{F}$, he gave a necessary and sufficient condition in terms of a factor set for $\mathfrak{A}$ to have an involution. This was used to give constructions which in principle yield all simple algebras with involution. Albert also used these results to prove that a central simple algebra has an involution if and only if it has exponent one or two in the Brauer group. One can combine this with one of the results of I to conclude that a central division algebra of degree four has an involution if and only if it is a tensor product of quaternion algebras. It seems unlikely that this is true for degree greater than four but we believe that this remains an open question.

Albert proved that if a division algebra over a number field has an involution of first kind, then the algebra is a quaternion algebra over its center. Moreover, he determined the division algebras over number fields having involutions of second kind. He showed that any such algebra is cyclic $(\mathfrak{Z}, s, \gamma)$ over its center $\mathfrak{C}$, where the cyclic field $\mathfrak{Z}$ over $\mathfrak{C}$ has the form $\mathfrak{Z}_0 \otimes_{\mathfrak{C}_0} \mathfrak{C}$, $\mathfrak{C}_0$ the subfield of symmetric elements of $\mathfrak{C}$, $\mathfrak{Z}_0$ cyclic over $\mathfrak{C}_0$, and $\mathfrak{C}$ separable quadratic over $\mathfrak{C}_0$. Moreover, $\gamma \in \mathfrak{C}$, and if $\bar{\gamma}$ is its conjugate in $\mathfrak{C}$ over $\mathfrak{C}_0$, then $\gamma\bar{\gamma}$ is a norm of an element of $\mathfrak{Z}_0$. Conversely any division algebra having the indicated cyclic structure does have an involution of second kind. In a later paper [**137**], Albert obtained a similar result for division algebras of degree three (over their centers) for arbitrary base fields.

These results, especially those on division algebras with involutions over number fields, provided the machinery for the solution of the problem of multiplication algebras for Riemann matrices. Using Rosati's theorem, one sees that the center $\mathfrak{C}$ of such an algebra is totally real if the (Rosati) involution is of first kind and is a pure imaginary quadratic extension of a totally real field if the involution is of second kind. Besides these conditions on $\mathfrak{C}$, there are supplementary conditions on the quaternion division algebras and the cyclic algebras, which occurred in our description of the division algebras with involution over number fields, that must be fulfilled for these to be multiplication algebras of Riemann matrices. In Albert's proof of the sufficiency of the conditions he had derived, he made use of the Hilbert irreducibility theorem for number fields.

In an exposition [**178**] of the theory of Riemann matrices, C. L. Siegel made some notable improvements on Albert's results. We should mention also that Weyl in [**187**] gave an alternative treatment of the subject based on Brauer factor sets.

III. **Nonassociative algebras.** From about 1942 to 1965, when he returned to the problem of existence of noncyclic associative division algebras of prime degree, most of Albert's research was in the area of nonassociative algebra: structure theory of nonassociative algebras, quasigroups, nonassociative division rings, and nondesarguesian projective planes. In our account of his contributions to this rather broad field of mathematics, we shall be selective, picking out what we consider his most important work—judged from the criterion of general mathematical interest. From this point of view, Albert's discoveries on Jordan algebras are undoubtedly his most important ones in nonassociative algebra, and these are perhaps on a par with his work on associative algebras and Riemann matrices. We shall begin our account with this work, and we shall first sketch the story of Jordan algebras before Albert took them up as a subject of intensive study.

The study of the class of algebras which now bear his name was initiated in 1932 by the physicist, P. Jordan. His declared objective was to achieve a better formalism for quantum mechanics than one based on selfadjoint operators in Hilbert space. Observing that the set of these operators is a vector space over $\boldsymbol{R}$ which is closed under the product $A \cdot B = \frac{1}{2}(AB+BA)$, where AB is the usual associative product, and that this symmetrized product is commutative and satisfies the identity $(A^2 \cdot B) \cdot A = A^2 \cdot (B \cdot A)$, he proposed to consider algebras in which the product composition satisfies these two conditions. He enlisted the help of von Neumann and Wigner in his study, and as a result of their collaboration, there appeared, in 1934, a paper entitled *On an algebraic generalization of the quantum mechanical formalism* which was a gem in the structure theory of algebras. In this paper **[168]**, Jordan, von Neumann and Wigner obtained a complete determination of the finite dimensional (nonassociative) algebras over $\boldsymbol{R}$ satisfying the following conditions: I. Formal reality in the sense of Artin-Schreier, that is, the requirement that the only relations of the form $\sum a_i^2 = 0$ in the algebras are the trivial ones in which every $a_i = 0$. II. Commutativity of the product ($ab = ba$) and the identity $(a^2b)a = a^2(ba)$.

They showed that the algebras satisfying these conditions are direct sums of ideals that are simple algebras, and they determined the simple ones as belonging to one of the following classes:

(1) The vector space over $\boldsymbol{R}$ of $n \times n$ hermitian matrices with entries in $\boldsymbol{R}$, $\boldsymbol{C}$, or Hamilton's quaternion algebra $\boldsymbol{H}$, endowed with the algebra structure in which the product is $a \cdot b = \frac{1}{2}(ab+ba)$ in terms of the usual matrix product ab.

(2) The algebras over $\boldsymbol{R}$ with bases $(1, e_1, e_2, \cdots, e_n)$ and multiplication defined by the table $e_ie_j = \delta_{ij}1$, 1 the unit.

(3) The algebra M_3^8 of 3×3 hermitian matrices with entries in the algebra $\boldsymbol{O}$ of Cayley numbers, endowed with the product $a \cdot b = \frac{1}{2}(ab+ba)$, where, as before, ab is the usual matrix product.

Conversely, the algebras listed satisfy the above conditions.

Now one defines a *Jordan algebra* over a field $\mathfrak{F}$ of characteristic $\neq 2$ as an algebra in which the product ab satisfies

$$ab = ba, \qquad (a^2b)a = a^2(ba).$$

Among these are included the *special* Jordan algebras which are isomorphic to subspaces of an associative algebra closed under the *Jordan product* $a \cdot b = \frac{1}{2}(ab+ba)$ and regarded as algebras relative to this product. Evidently, the algebras in Jordan, von Neumann, Wigner's class (1) are special. It is easy to see, using Clifford algebras, that this is the case also for algebras of their class (2). They conjectured that M_3^8 is not special, and they proposed the proof of this as a problem to Albert, who showed that this is indeed the case: M_3^8 is an *exceptional* (=nonspecial) Jordan algebra **[52]**.

After this brief encounter with the Jordan theory, a number of years elapsed before Albert returned to the subject. In a series of three papers **[94]**, **[97]**, **[105]** appearing in 1946, 1947 and 1950, Albert developed the basic structure theory of finite dimensional Jordan algebras over a field of characteristic not two. Since it is interesting to observe how the subject evolved in Albert's hands, we shall give a brief indication of the contents of each of these papers.

In the first one he considered Jordan algebras defined concretely as Jordan algebras of linear transformations of a finite dimensional vector space, that is, subspaces of End $\mathfrak{V}$ closed under the Jordan product $A \cdot B$. He proved analogues for these algebras of Lie's and Engel's theorems on Lie algebras. Assuming the base field is of characteristic 0 (so trace arguments can be used) he showed that if the algebras contain no nil ideals they are direct sums of simple algebras. Moreover, he determined the simple algebras over an algebraically closed field of characteristic 0. This determination is quite similar to that of Jordan, von Neumann and Wigner's of simple formally real Jordan algebras: Class (2) is unchanged, and the modification required in the definition of class (1) is that $\boldsymbol{R}$, $\boldsymbol{C}$ and $\boldsymbol{H}$ be replaced by the split composition algebras of dimensions 1, 2 and 4 over the given base fields. (These are the algebras which occur in Hurwitz's problem on quadratic forms permitting composition.) Naturally there is no class (3) since the algebras under consideration are special by definition. Actually Albert determined a more general class of so-called reduced algebras over an arbitrary field of characteristic 0. However, the result he obtained in the algebraically closed case is adequate to permit the determination of all the special simple algebras over an arbitrary field using the method of descent. This was done by Kalisch **[169]** and F. D. Jacobson and N. Jacobson **[163]**.

In his second paper, Albert dealt with abstract finite dimensional Jordan algebras over any field of characteristic not two. He showed that nil algebras of this type are nilpotent in the sense that there exists an integer r such that the product of any r elements of the algebra in any association is 0. He proved that if the characteristic is 0 and $\mathfrak{A}$ has no nil ideals $\neq 0$, then $\mathfrak{A}$ is a direct sum of simple algebras, and he determined the simple ones over an algebraically closed field of characteristic 0. Here, one does have exceptional algebras, and the only simple one over an algebraically closed field of characteristic 0 is the analogue of M_3^8 in which the classical Cayley algebra is replaced by a split Cayley algebra. We shall refer to this as the *split exceptional simple Jordan algebra.*

In his third paper, Albert extended these results except for a small gap (which was filled by Jacobson in **[165]**) to the characteristic $p \neq 2$ case.

A particularly interesting class of Jordan algebras is that of the finite dimensional exceptional central simple Jordan algebras. If $\mathfrak{F}$ is the base field and $\bar{\mathfrak{F}}$ is its algebraic closure, then $\mathfrak{A}$ is in this class if and only if the extension algebra $\mathfrak{A}^{\bar{\mathfrak{F}}}$ is the split exceptional simple Jordan algebra previously defined. The theory of these exceptional algebras is intertwined with a number of other "exceptional" phenomena, notably, exceptional Lie groups and exceptional geometries (e.g. Cayley projective planes). Albert made a number of important contributions to this theory, sometimes in collaboration with others. We have already noted that his first paper on Jordan algebras established the exceptional character of M_3^8. In 1959 Albert and Paige, in a joint paper [**129**], proved a much stronger result: M_3^8 is not a homomorphic image of any special Jordan algebra. As a consequence of this and a result of Cohn's [**157**], one can conclude that the free Jordan algebra with three generators is not special.

One can distinguish two types of exceptional simple Jordan algebras: the reduced ones and the division algebras. The first contains idempotents $\neq 0$, 1, and in the second the subalgebras generated by single elements and 1 are associative fields. It can be shown that the reduced ones have the form $\mathfrak{H}(\boldsymbol{O}_3, \gamma)$ the algebra of 3×3 matrices A with entries in some (generalized) Cayley algebra $\boldsymbol{O}$, which are *γ-hermitian* in the sense that $\gamma^{-1}\,{}^t\bar{A}\gamma=A$, where γ is a 3-rowed diagonal matrix with entries in the base field, and $\bar{A}$ is obtained by replacing each Cayley number entry a_{ij} by its conjugate $\bar{a}_{ij}$. The problem of determining conditions for the isomorphism of two such algebras was studied by Albert and Jacobson [**122**]. It was shown in this paper that isomorphism of the Jordan algebras implied isomorphism of the Cayley algebras occurring in their definitions, and they obtained some rather complicated supplementary conditions for isomorphism. These sufficed to give a complete classification of reduced exceptional simple Jordan algebras over number fields. (See also [**179**] and [**175**].)

The first construction of exceptional Jordan division algebras is due to Albert [**126**], [**141**]. He showed also that no such algebras exist over number fields. On the other hand, if $\mathfrak{F}$ is any field over which there exist central associative division algebras of degree three (e.g. a number field), then there exist exceptional Jordan division algebras over the field $\mathfrak{F}(t)$ obtained by adjoining an indeterminate t to $\mathfrak{F}$. Albert used a method of descent in his study of exceptional Jordan division algebras. Subsequently considerably simpler "rational" constructions were given by Tits [**166**, p. 412].

From the abstract point of view, a very natural class of algebras (or rings) is the class satisfying the power associativity condition: subalgebras (or subrings) generated by single elements are associative. This includes Jordan algebras, alternative algebras (defined by the identities

$a^2b=a(ab)$ and $ba^2=(ba)a$), associative algebras, and a number of other interesting types of algebras.

Albert initiated the study of power associative rings (without finiteness conditions) in a paper [**98**] published in 1948. The conditions of power associativity are that for any a one has the power formula

$$a^m a^n = a^{m+n}, \qquad m, n = 1, 2, 3, \cdots,$$

where a^m is defined inductively by $a^1=a$, $a^k=a^{k-1}a$. In [**98**] Albert showed that if the additive group of a nonassociative ring $\mathfrak{A}$ has no torsion, then $\mathfrak{A}$ is power associative if and only if it satisfies the two identities $aa^2=a^3$ and $(a^2)^2=a^4$. These results were obtained by some clever inductive arguments based on linearizations of the assumed identities. These linearizations and commutativity yield also the crucial result that in a commutative power associative rmg the map e_R: $x \to xe$, determined by an idempotent e, satisfies the quadratic equation $(2e_R-1)(e_R-1)e_R=0$. If one assumes that the additive group admits the operator $\frac{1}{2}$, then one obtains the Peirce decomposition relative to

$$e:\mathfrak{A} = \mathfrak{A}_0(e) \oplus \mathfrak{A}_{1/2}(e) \oplus \mathfrak{A}_1(e) \quad \text{where } \mathfrak{A}_i(e) = \{x_i \mid x_i e = ix_i\}.$$

One also has extensions of this to Peirce decompositions relative to orthogonal idempotents. In Albert's hands, these Peirce decompositions became powerful tools for investigating power associative rings. He obtained a number of striking results by this method. We mention two:

Let $\mathfrak{A}$ be a simple commutative power associative ring whose additive group contains no elements of orders 2, 3 or 5 and admits the operator $\frac{1}{2}$. Suppose $\mathfrak{A}$ contains two nonzero orthogonal idempotents e and f such that $e+f$ is not a unit. Then $\mathfrak{A}$ satisfies the Jordan identity $(a^2b)a=a^2(ba)$ [**105**].

Any simple alternative ring containing an idempotent $e\neq 0$, 1 is either associative or a Cayley algebra over its center [**109**].

The ultimate result on simple alternative rings is due to Kleinfeld [**170**], [**171**]. This states that all simple alternative rings are either associative or Cayley algebras. Albert's theorem was used as a step in the first proof of Kleinfeld's theorem.

In [**108**] and [**126**] Albert proved a generalization for power associative rings of Wedderburn's celebrated theorem on the commutativity of finite associative division rings. Call an algebra over a field *strictly power associative* if all the algebras obtained by extending the base field are power associative. Also one defines a (nonassociative) division ring by the property that the left and right multiplications $x\to ax$ and $x \to xa$ are bijective for any $a\neq 0$ in the ring. Then Albert proved that any finite strictly power associative division algebra of characteristic $\neq 2$ is associative and commutative. Albert based his proof on the determination of the simple Jordan algebras over an arbitary field due to F. D. Jacobson and N. Jacobson **[163]** and a result of his own on exceptional Jordan algebras **[126]**. (Later McCrimmon gave alternative proofs which are independent of the structure theory **[173]**, **[174]**.) A number of constructions of nonassociative and noncommutative division rings are due to Albert **[135]**, **[108]**, **[132]**. These yield examples of nondesarguesian projective planes including some finite ones.

Albert had a hand in the discovery of several new classes of simple Lie algebras of prime characteristic (see **[160]**, **[113]** and **[151]**). Recently these results have taken on added luster because of the discovery by Kostrikin and Shafarevitch **[172]** that these Lie algebras can be regarded as characteristic p versions of infinite dimensional Lie algebras which had occurred in Eli Cartan's work on contact transformations.

Albert and his students and followers also studied a number of other classes of nonassociative algebras defined by identities. Until now the results which have been obtained on these appear to be of interest only to specialists in the field. We shall therefore refrain from giving any indication of these results. Albert wrote several papers on general nonassociative theory. In one of these **[84]** he gave a definition of a radical for any finite dimensional nonassociative algebra. Since the theory of the radical is quite interesting and deserves to be better known than it is at present, we take this opportunity to sketch what we believe is an improved version of this theory.

Let $\mathfrak{A}$ be a finite dimensional nonassociative algebra over a field. Then $\mathfrak{A}$ is called *simple* if $\mathfrak{A}^2 \neq 0$ and $\mathfrak{A}$ has no ideals $\neq 0$, $\mathfrak{A}$. $\mathfrak{A}$ is *semisimple* if it is a direct sum of ideals which are simple algebras. Following the pattern of associative ring theory, it is natural to define the *radical* rad $\mathfrak{A}$ to be the intersection of the set of ideals $\mathfrak{B}$ of $\mathfrak{A}$ such that $\mathfrak{A}/\mathfrak{B}$ is simple. This definition implies that if no $\mathfrak{B}$'s, such that $\mathfrak{A}/\mathfrak{B}$ is simple, exist then $\mathfrak{A}=$ rad $\mathfrak{A}$. In any case $\mathfrak{A}/\text{rad}\,\mathfrak{A}$ is semisimple or 0, and rad $\mathfrak{A}$ is contained in every ideal $\mathfrak{D}$ of $\mathfrak{A}$ such that $\mathfrak{A}/\mathfrak{D}$ is semisimple (see, for example, Jacobson, *Structure of Rings*, p. 41). This implies that $A \neq 0$ is semisimple if and only if rad $\mathfrak{A}=0$.

One obtains important information on an algebra $\mathfrak{A}$ in looking at its *multiplication algebra* $M(\mathfrak{A})$. This is the subalgebra of the associative algebra End $\mathfrak{A}$ of linear transformations in $\mathfrak{A}$ generated by 1 and the left and right multiplications (a_L: $x \to ax$, a_R: $x \to xa$) of $\mathfrak{A}$. The centralizer of $M(\mathfrak{A})$ in End $\mathfrak{A}$ is called the *centroid* $C(\mathfrak{A})$ of $\mathfrak{A}$. The study of $M(\mathfrak{A})$ and $C(\mathfrak{A})$ was initiated by Jacobson **[164]** (see also Jacobson, *Lie Algebras*, pp. 290–295). Albert's results on rad $\mathfrak{A}$, as we shall show, amount to a formula for rad $\mathfrak{A}$ in terms of rad $M(\mathfrak{A})$. We shall call $\mathfrak{A}$ *reductive* if $\mathfrak{A}$ is a direct sum of ideals which are simple algebras and the ideal

$\mathfrak{Z}=\{z|\mathfrak{A}z=0=z\mathfrak{A}\}$. The elements of $\mathfrak{Z}$ are called ***absolute zero divisors.*** One can show that $\mathfrak{A}\neq 0$ is reductive if and only if $M(\mathfrak{A})$ is semisimple. Hence $\mathfrak{A}$ is semisimple if and only if $M(\mathfrak{A})$ is semisimple and 0 is the only absolute zero divisor in $\mathfrak{A}$. Now let $\mathfrak{N}$ be the radical of $M(\mathfrak{A})$ and $\mathfrak{R}$ the ideal in $\mathfrak{A}$ such that $\mathfrak{R}/\mathfrak{N}\mathfrak{A}$ is the ideal of absolute zero divisors of $\mathfrak{A}/\mathfrak{N}\mathfrak{A}$. Then $\mathfrak{R}=\operatorname{rad}\mathfrak{A}$. Albert's order of ideas in his paper on the radical is the reverse of what we have indicated; namely, he uses the ideal $\mathfrak{R}$ as his definition of the radical, then proves it has the two basic properties that $\mathfrak{A}/\mathfrak{R}$ is semisimple and $\mathfrak{R}$ is contained in every ideal $\mathfrak{D}$ such that $\mathfrak{A}/\mathfrak{D}$ is semisimple.

For certain important classes of algebras (e.g. associative, alternative, Jordan), rad $\mathfrak{A}$ coincides with the maximal nil ideal. For Lie algebras of characteristic 0, rad $\mathfrak{A}$ is the maximal solvable ideal. On the other hand, Albert has given an example of an algebra in which rad $\mathfrak{A}$ is an associative field. We could not resist recording here a result on the radical which we have known for some time. This is a generalization of a well-known theorem of Hochschild's **[162]** on derivations of associative and Lie algebras.

THEOREM. *If $\mathfrak{A}$ is a finite dimensional nonassociative algebra over a field of characteristic* 0 *then any derivation of $\mathfrak{A}$ stabilizes* rad $\mathfrak{A}$.

This can be proved by using the fact that the Lie algebra Der $\mathfrak{A}$ of derivations is the Lie algebra of algebraic groups of automorphisms of $\mathfrak{A}$. A more direct proof, which is applicable also in some situations in characteristic $p\neq 0$, can be based on Albert's definition of rad $\mathfrak{A}$. We observe first that if D is a derivation in $\mathfrak{A}$, then

$$[D, a_L] \equiv Da_L - a_L D = (Da)_L \quad \text{and} \quad [D, a_R] = (Da)_R.$$

Hence $m\to[D, m]$ is a derivation in $M(\mathfrak{A})$ which we denote as $\tilde{D}$. If $m\in M(\mathfrak{A})$ and $a\in\mathfrak{A}$, then $D(ma)=(\tilde{D}m)a+m(Da)$. It is easily seen that D stabilizes rad $\mathfrak{A}$ if $\tilde{D}$ stabilizes rad $M(\mathfrak{A})$. Our theorem then follows from Hochschild's theorem on associative algebras.

In a paper **[82]** which appeared in 1942, Albert introduced a concept of isotopy for nonassociative algebras. Let $\mathfrak{A}$ and $\mathfrak{B}$ be nonassociative algebras. Then $\mathfrak{A}$ and $\mathfrak{B}$ are called *isotopes* if there exist bijective linear maps P and Q from $\mathfrak{B}$ to $\mathfrak{A}$ and a bijective linear map C from $\mathfrak{A}$ to $\mathfrak{B}$ such that for $x, y\in\mathfrak{B}$ we have

$$xy = C((Px)(Qy)).$$

If $P=Q$ and $C=P^{-1}$, we have $P(xy)=(Px)(Py)$, so P is an isomorphism. Isotopy is an equivalence relation. If $\mathfrak{B}$ and $\mathfrak{A}$ are identical as sets, and $C=1$, we call $\mathfrak{B}$ a *principal isotope* of $\mathfrak{A}$. Define a new multiplication on $\mathfrak{A}$ by $u\circ v=(PCu)(QCv)$. This, along with the given vector space structure, gives a new algebra which is a principal isotope of $\mathfrak{A}$, and since

$$xy = C((Px)(Qy)) = C((PCC^{-1}x)(QCC^{-1}y)) = C(C^{-1}x \circ C^{-1}y),$$

$\mathfrak{B}$ is isomorphic to the principal isotope defined by $\circ$. This reduces the consideration to that of principal isotopes. Albert defined isotopy also for quasigroups [**87**], and he proved a number of interesting results on isotopy of algebras and of quasigroups. While these have not played an important role in structure theory, the concept of isotopy has some importance in nondesarguesian geometry (see [**134**]).

IV. **Miscellaneous.** Albert wrote a number of papers [**50**], [**54**], [**57**], [**58**], [**64**] on the structure of field extensions. He was particularly interested in explicit constructions of cyclic field extensions since these played an important role in his investigations of the structure of division algebras. Albert's results on cyclic extensions are presented in a connected fashion in Chapter IX of his algebra text *Modern Higher Algebra.* There are numerous references to these results in *Structure of Algebras.* For the case of degree p^e and characteristic p, one has an alternative method due to Witt, based on Witt vectors, which provides a better survey of cyclic and abelian extensions (see for example, Jacobson's *Lectures in Abstract Algebra*, vol. III, pp. 124–140). On the other hand, Albert's results on cyclic fields of degree p^e and characteristic $\neq p$ seem not to have been improved upon until now.

Albert was fascinated by the problem of minimum number of generators for algebraic structures. He proved [**92**] that any separable associative algebra is generated by two elements and, with John Thompson [**130**], proved that the projective unimodular group over a finite field is generated by two elements, one of which has order two.

In a joint paper with Muckenhoupt [**120**], he proved that for any field $\mathfrak{F}$, any matrix of trace 0 in $M_n(\mathfrak{F})$ is an additive commutator $[A, B] = AB - BA$. This supplemented an earlier result by Shoda [**177**] for fields of characteristic 0.

In [**77**] Albert proved that a finite dimensional ordered division algebra is necessarily commutative. This does not hold for infinite dimensional algebras, for Hilbert has given an example in the second edition of his *Grundlagen der Geometrie* of a "twisted" power series division ring which is not commutative and which can be ordered. It is interesting to note that Hilbert's first attempt to give such an example in the first edition of *Grundlagen* can be seen to be wrong by invoking Albert's theorem!

Another pretty result of Albert's gives a determination of the finite dimensional absolute valued algebras over $\boldsymbol{R}$. By this we mean a (nonassociative) algebra over $\boldsymbol{R}$ which has a map $a \to |a|$ into $\boldsymbol{R}$ with the usual properties:

(1) $|a| \geqq 0$ and $|a|=0$ if and only if $a=0$;

(2) $|a+b| \leqq |a|+|b|$;

(3) $|\alpha a|=|\alpha|\,|a|$ for $\alpha \in \boldsymbol{R}$;

(4) $|ab|=|a|\,|b|$.

It had been conjectured by Kaplansky that if such an algebra has a unit, then it is alternative, and hence, by a classical result, it is necessarily either $\boldsymbol{R}$, $\boldsymbol{C}$, Hamilton's quaternion algebra $\boldsymbol{H}$, or Cayley's octonions $\boldsymbol{O}$. Moreover, in all cases $|a|=|a\bar{a}|^{1/2}$, where $\bar{a}$ is the usual conjugate. Albert proved this **[96]** and also showed that if all the conditions except the existence of a unit hold, then the algebra is an isotope of $\boldsymbol{R}$, $\boldsymbol{C}$, $\boldsymbol{H}$ or $\boldsymbol{O}$. This result was extended **[101]** to algebraic algebras over $\boldsymbol{R}$ not assumed to be finite dimensional.

Albert's last published paper **[149]**—published posthumously—proves an interesting theorem on quaternion algebras: If $\mathfrak{A}_1$ and $\mathfrak{A}_2$ are two (generalized) quaternion division algebras over a field $\mathfrak{F}$ and $\mathfrak{A}_1 \otimes_{\mathfrak{F}} \mathfrak{A}_2$ is not a division algebra, then $\mathfrak{A}_1$ and $\mathfrak{A}_2$ have a common quadratic subfield.

Our recital of Albert's major achievements gives no indication of his methods or, more broadly speaking, of his mathematical style, which was highly individualistic. Perhaps its most characteristic qualities were the directness of his approach to a problem and his power and stamina to stick with it until he achieved a complete solution. He had a fantastic insight into what might be accomplished by intricate and subtle calculations of a highly original character. At times he could have obtained simpler proofs by using more sophisticated tools (e.g. representation theory), and one can almost always improve upon his arguments. However, this is of secondary importance compared to the first breakthrough which establishes a definitive result. It was in this that Albert really excelled. He regarded himself as a "pure" algebraist and in a sense he was. However, his best work—the solution of the problem of multiplication algebras of Riemann matrices—had its origin in another branch of mathematics. Moreover, he could exploit analytic and number theoretic results when he needed them—as he did in this instance.

Bibliography of A. A. Albert

Books

1. *Modern Higher Algebra*, Univ. of Chicago Press, Chicago, Ill., 1937.

2. *Structure of Algebras*, Amer. Math. Soc. Colloq. Publ., vol. 24, Amer. Math. Soc. Providence, R.I., 1939. MR **1**, 99.

3. *Introduction to Algebraic Theories*, Univ. of Chicago Press, Chicago, Ill., 1941. MR **2**, 241.

4. *College Algebra*, McGraw-Hill, New York, 1946; reprinted by Univ. of Chicago Press, Chicago, Ill., 1963.

5. *Solid Analytic Geometry*, McGraw-Hill, New York, 1949.

6. *Fundamental concepts of higher algebra*, Univ. of Chicago Press, Chicago, Ill., 1958. MR **20** #5190.

7. (with Reuben Sandler), *An Introduction to Finite Projective Planes*, Holt, Rinehart and Winston, New York, 1968. MR **37** #3435.

8. (edited by A. A. Albert), *Studies in Modern Algebra*, Studies in Math., vol. 2, Math. Assoc. of Amer., distributed by Prentice-Hall, Englewood Cliffs, N.J., 1963. MR **26** #3750.

Articles

9. *A determination of all associative algebras in two, three and four units over a non-modular field $\mathfrak{F}$*, M.S. Thesis, University of Chicago, Chicago, Ill., 1927.

10. *Algebras and their radicals, and division algebras*, Ph.D. Dissertation, University of Chicago, Chicago, Ill., 1928.

11. *A determination of all normal division algebras in sixteen units*, Trans. Amer. Math. Soc. **31** (1929), 253–260.

12. *On the rank equation of any normal division algebra*, Bull. Amer. Math. Soc. **35** (1929), 335–338.

13. *The rank function of any simple algebra*, Proc. Nat. Acad. Sci. U.S.A. **15** (1929), 372–376.

14. *On the structure of normal division algebras*, Ann. of Math. **30** (1929), 322–338.

15. *Normal division algebras in $4p^2$ units, p an odd prime*, Ann. of Math. **30** (1929), 583–590.

16. *The structure of an algebra which is a direct product of rational generalized quaternion division algebras*, Ann. of Math. **30** (1929), 621–625.

17. *On the structure of pure Riemann matrices with non-commutative multiplication algebras*, Proc. Nat. Acad. Sci. U.S.A. **16** (1930), 308–312.

18. *On direct products, cyclic division algebras, and pure Riemann matrices*, Proc. Nat. Acad. Sci. U.S.A. **16** (1930), 313–315.

19. *The non-existence of pure Riemann matrices with normal multiplication algebras of order sixteen*, Ann. of Math. **31** (1930), 375–380.

20. *A necessary and sufficient condition for the non-equivalence of any two rational generalized quaternion division algebras*, Bull. Amer. Math. Soc. **36** (1930), 535–540.

21. *Determinations of all normal division algebras in thirty-six units of type R_2*, Amer. J. Math. **52** (1930), 283–292.

22. *A note on an important theorem on normal division algebras*, Bull. Amer. Math. Soc. **36** (1930), 649–650.

23. *New results in the theory of normal division algebras*, Trans. Amer. Math. Soc. **32** (1930), 171–195.

24. *The integers of normal quartic fields*, Ann. of Math. **31** (1930), 381–418.

25. *A determination of the integers of all cubic fields*, Ann. of Math. **31** (1930), 550–566.

26. *A construction of all non-commutative rational division algebras of order eight*, Ann. of Math. **31** (1930), 567–576.

27. *Normal division algebras of order 2^{2m}*, Proc. Nat. Acad. Sci. U.S.A. **17** (1931), 389–392.

28. *The structure of pure Riemann matrices with noncommutative multiplication algebras*, Palermo Rendiconti **55** (1931), 57–115.

29. *On direct products, cyclic division algebras and pure Riemann matrices*, Trans. Amer. Math. Soc. **33** (1931), 219–234; Correction, 999.

30. *On normal division algebras of type R in thirty-six units*, Trans. Amer. Math. Soc. **33** (1931), 235–243.

31. *On the Wedderburn norm condition for cyclic algebras*, Bull. Amer. Math. Soc. **37** (1931), 301–312.

32. *A note on cyclic algebras of order* 16, Bull. Amer. Math. Soc. **37** (1931), 727–730.

33. *On direct products*, Trans. Amer. Math. Soc. **33** (1931), 690–711.

34. *Division algebras over an algebraic field*, Bull. Amer. Math. Soc. **37** (1931), 777–784.

35. *The structure of matrices with any normal division algebra of multiplications*, Ann. of Math. **32** (1931), 131–148.

36. *On the construction of cyclic algebras with a given exponent*, Amer. J. Math. **54** (1932), 1–13.

37. *Algebras of degree* 2^e *and pure Riemann matrices*, Ann. of Math. **33** (1932), 311–318.

38. *A construction of non-cyclic normal division algebras*, Bull. Amer. Math. Soc. **38** (1932), 449–456.

39. *Normal division algebras of degree four over an algebraic field*, Trans. Amer. Math. Soc. **34** (1932), 363–372.

40. *On normal simple algebras*, Trans. Amer. Math. Soc. **34** (1932), 620–625.

41. *A note on division algebras of order sixteen*, Bull. Amer. Math. Soc. **38** (1932), 703–706.

42. (with H. Hasse), *A determination of all normal division algebras over an algebraic number field*, Trans. Amer. Math. Soc. **34** (1932), 722–726.

43. *A note on the equivalence of algebras of degree two*, Bull. Amer. Math. Soc. **39** (1933), 257–258.

44. *On primary normal division algebras of degree eight*, Bull. Amer. Math. Soc. **39** (1933), 265–272.

45. *Non-cyclic algebras of degree and exponent four*, Trans. Amer. Math. Soc. **35** (1933), 112–121.

46. *The integers represented by sets of ternary quadratic forms*, Amer. J. Math. **55** (1933), 274–292.

47. *On universal sets of positive ternary quadratic forms*, Ann. of Math. **34** (1933), 875–878.

48. *A note on the Dickson theorem on universal ternaries*, Bull. Amer. Math. Soc. **39** (1933), 585–588.

49. *Normal division algebras over algebraic number fields not of finite degree*, Bull. Amer. Math. Soc. **39** (1933), 746–749.

50. *Cyclic fields of degree eight*, Trans. Amer. Math. Soc. **35** (1933), 949–964.

51. *On the construction of Riemann matrices*. I, Ann. of Math. **35** (1934), 1–28.

52. *On a certain algebra of quantum mechanics*, Ann. of Math. **35** (1934), 65–73.

53. *Normal division algebras of degree* 4 *over F of characteristic* 2, Amer. J. Math. **56** (1934), 75–86.

54. *Integral domains of rational generalized quaternion algebras*, Bull. Amer. Soc. **40** (1934), 164–176.

55. *On certain imprimitive fields of degree* p^2 *over P of characteristic p*, Ann. of Math. **35** (1934), 211–219.

56. *Normal division algebras over a modular field*, Trans. Amer. Math. Soc. **36** (1934), 388–394.

57. *A solution of the principal problem in the theory of Riemann matrices*, Ann. of Math. **35** (1934), 500–515.

58. *Cyclic fields of degree* p^r *over F of characteristic p, Bull. Amer. Math. Soc.* **40** (1934), 625–631.

59. *On normal Kummer fields over a non-modular field*, Trans. Amer. Math. Soc. **36** (1934), 885–892.

60. *Involutorial simple algebras and real Riemann matrices*, Proc. Nat. Acad. Sci. U.S.A. **20** (1934), 676–681.

61. *The principal matrices of a Riemann matrix*, Bull. Amer. Math. Soc. **40** (1934), 843–846.

62. *A note on the Poincaré theorem on impure Riemann matrices*, Ann. of Math. **36** (1935), 151–156.

63. *On the construction of Riemann matrices*. II, Ann. of Math. **36** (1935), 376–394.

64. *On cyclic fields*, Trans. Amer. Math. Soc. **37** (1936), 454–462.

65. *Involutorial simple algebras and real Riemann matrices*, Ann. of Math. **36** (1935), 886–894.

66. *Normal division algebras of degree p^e over P of characteristic p*, Trans. Amer. Math. Soc. **39** (1936), 183–188.

67. *Simple algebras of degree p^e over a centrum of characteristic p*, Trans. Amer. Math. Soc. **40** (1936), 112–126.

68. *A note on matrices defining total real fields*, Bull. Amer. Math. Soc. **43** (1937), 242–244.

69. *p-algebras over a field generated by one indeterminate*, Bull. Amer. Math. Soc. **43** (1937), 733–736.

70. *Normalized integral bases of algebraic number fields*. I, Ann. of Math. **38** (1937), 923–957.

71. *A quadratic form problem in the calculus of variations*, Bull. Amer. Math. Soc. **44** (1938), 250–253.

72. *Non-cyclic algebras with pure maximal subfields*, Bull. Amer. Math Soc. **44** (1938), 576–579.

73. *Symmetric and alternate matrices in an arbitrary field*. I, Trans. Amer. Math. Soc. **43** (1938), 386–436.

74. *Quadratic null forms over a function field*, Ann. of Math. **39** (1938), 494–505.

75. *On cyclic algebras*, Ann. of Math. **39** (1938), 669–682.

76. *A note on normal division algebras of prime degree*, Bull. Amer. Math. Soc. **44** (1938), 649–652.

77. *On ordered algebras*, Bull. Amer. Math. Soc. **46** (1940), 521–522. MR **1**, 328.

78. *On p-adic fields and rational division algebras*, Ann. of Math. (2) **41** (1940), 674–693. MR **2**, 123.

79. *A rule for computing the inverse of a matrix*, Amer. Math. Monthly **48** (1941), 198–199. MR **2**, 243.

80. *Division algebras over a function field*, Duke Math. J. **8** (1941), 750–762. MR **3**, 265.

81. *Quadratic forms permitting composition*, Ann. of Math. (2) **43** (1942), 161–177. MR **3**, 261.

82. *Non-associative algebras*. I. *Fundamental concepts and isotopy*, Ann. of Math. (2) **43** (1942), 685–707. MR **4**, 186.

83. *Non-associative algebras*. II. *New simple algebras*, Ann. of Math. (2) **43** (1942), 708–723. MR **4**, 186.

84. *The radical of a non-associative algebra*, Bull. Amer. Math. Soc. **48** (1942), 891–897. MR **4**, 130.

85. *An inductive proof of Descartes' rule of signs*, Amer. Math. Monthly **50** (1943), 178–180. MR **4**, 195.

86. *A suggestion for a simplified trigonometry*, Amer. Math. Monthly **50** (1943), 251–253.

87. *Quasigroups*. I, Trans. Amer. Math. Soc. **54** (1943), 507–519. MR **5**, 229.

88. *Algebras derived by non-associative matrix multiplication*, Amer. J. Math. **66** 1944), 30–40.'MR **5**, 227.

89. *The matrices of factor analysis*, Proc. Nat. Acad. Sci. U.S.A. **30** (1944), 90–95 MR **5**, 209.

90. *The minimum rank of a correlation matrix*, Proc. Nat. Acad. Sci. U.S.A. **30** (1944), 144–146. MR **6**, 6.

91. *Quasigroups*. II, Trans. Amer. Math. Soc. **55** (1944), 401–419. MR **6**, 42.

92. *Two element generation of a separable algebra*, Bull. Amer. Math. Soc. **50** (1944), 786–788. MR **6**, 115.

93. *Quasiquaternion algebras*, Ann. of Math. (2) **45** (1944), 623–638. MR **6**, 115.

94. *On Jordan algebras of linear transformations*, Trans. Amer. Math. Soc. **59** (1946), 524–555. MR **8**, 63.

95. *The Wedderburn principal theorem for Jordan algebras*, Ann. of Math. (2) **48** (1947), 1–7. MR **8**, 435.

96. *Absolute valued real algebras*, Ann. of Math. (2) **48** (1947), 495–501; Correction in Bull. Amer. Math. Soc. **55** (1949), 1191. MR **8**, 561; **11**, 76.

97. *A structure theory for Jordan algebras*, Ann of Math (2) **48** (1947), 546–567. MR **9**, 77.

98. *On the power-associativity of rings*, Summa Brasil. Math. **2** (1948), no. 2, 21–33. MR **10**, 97.

99. *Power-associative rings*, Trans. Amer. Math. Soc. **64** (1948), 552–593. MR **10**, 349.

100. *On right alternative algebras*, Ann. of Math. (2) **50** (1949), 318–328. MR **10**, 503.

101. *Absolute-valued algebraic algebras*, Bull. Amer. Math. Soc. **55** (1949), 763–768. MR **11**, 76.

102. *A theory of trace-admissible algebras*, Proc. Nat. Acad. Sci. U.S.A. **35** (1949), 317–322. MR **11**, 6.

103. *Almost alternative algebras*, Portugal. Math. **8** (1949), 23–36. MR **11**, 316.

104. *A note on the exceptional Jordan algebra*, Proc. Nat. Acad. Sci. U.S.A. **36** (1950), 372–374. MR **12**, 5

105. *A theory of power-associative commutative algebras*, Trans. Amer. Math. Soc. **69** (1950), 503–527. MR **12**, 475.

106. *New simple power-associative algebras*, Summa Brasil. Math. **2** (1951), 183–194. MR **14**, 11.

107 *Power-associative algebras*, Proc. Internat. Congress Math. (Cambridge, Mass., 1950), vol. 2, Amer. Math. Soc., Providence, R.I., 1952, pp. 25–32. MR **13**, 527.

108. *On nonassociative division algebras*, Trans. Amer. Math. Soc. **72** (1952), 296–309. MR **13**, 816.

109. *On simple alternative rings*, Canad. J. Math. **4** (1952), 129–135. MR **14**, 11.

110. *On commutative power-associative algebras of degree two*, Trans. Amer. Math. Soc. **74** (1953), 323–343. MR **14**, 614.

111. *Rational normal matrices satisfying the incidence equation*, Proc. Amer. Math. Soc. **4** (1953), 554–559. MR **15**, 94.

112. *The structure of right alternative algebras*, Ann. of Math. (2) **59** (1954), 408–417. MR **15**, 774.

113. (with M. S. Frank), *Simple Lie algebras of characteristic p*, Univ. e Politec. Torino. Rend. Sem. Mat. **14** (1954/55), 117–139. MR **18**, 52.

114. *Leonard Eugene Dickson*, 1874–1954, Bull. Amer. Math. Soc. **61** (1955), 331–345. MR **17**, 2.

115. *On involutorial algebras*, Proc. Nat. Acad. Sci. U.S.A. **41** (1955), 480–482. MR **17**, 9.

116. *On Hermitian operators over the Cayley algebra*, Proc. Nat. Acad. Sci. U.S.A. **41** (1955), 639–640. MR **17**, 123.

117. *A property of special Jordan algebras*, Proc. Nat. Acad. Sci. U.S.A. **42** (1956), 624–625. MR **18**, 375.

118. *The norm form of a rational division algebra*, Proc. Nat. Acad. Sci. U.S.A. **43** (1957), 506–509. MR **19**, 246.

119. *On certain trinomial equations in finite fields*, Ann. of Math. (2) **66** (1957), 170–178. MR **19**, 394.

120. (with B. Muckenhoupt), *On matrices of trace zero*, Michigan Math. J. **4** (1957), 1–3. MR **18**, 786.

121. *On partially stable algebras*, Trans. Amer. Math. Soc. **84** (1957), 430–443. MR **19**, 1156.

122. (with N. Jacobson), *On reduced exceptional simple Jordan algebras*, Ann. of Math. (2) **66** (1957), 400–417. MR **19**, 527.

123. *A property of ordered rings*, Proc. Amer. Math. Soc. **8** (1957), 128–129. MR **18**, 557.

124. (with John Thompson), *Two element generation of the projective unimodular group*, Bull. Amer. Math. Soc. **64** (1958), 92–93. MR **20** #3211.

125. *Addendum to the paper on partially stable algebras*, Trans. Amer. Math. Soc. **87** (1958), 57–62. MR **19**, 1157.

126. *A construction of exceptional Jordan division algebras*, Ann. of Math. (2) **67** (1958), 1–28. MR **19**, 1036.

127. *On the orthogonal equivalence of sets of real symmetric matrices*, J. Math. Mech. **7** (1958), 219–235. MR **19**, 1153.

128. *A solvable exceptional Jordan algebra*, J. Math. Mech. **8** (1959), 331–337. MR **21** #681.

129. (with L. J. Paige), *On a homomorphism property of certain Jordan algebras*, Trans. Amer. Math. Soc. **93** (1959), 20–29. MR **21** #7240.

130. (with John Thompson), *Two-element generation of the projective unimodular group*, Illinois J. Math. **3** (1959), 421–439. MR **21** #5681.

131. *Finite noncommutative division algebras*, Proc. Amer. Math. Soc. **9** (1958), 928–932. MR **21** #1994.

132. *Finite division algebras and finite planes*, Proc. Sympos. Appl. Math., Vol. 10, Amer. Math. Soc., Providence, R.I., 1960, pp. 53–70. MR **22** #6831.

133. *On the collineation groups associated with twisted fields*, Calcutta Math. Soc. Golden Jubilee Commemoration Volume (1958/59), part II, Calcutta Math. Soc., Calcutta, 1963, pp. 485–497. MR **28** #3070.

134. *On the collineation groups of certain non-desarguesian planes*, Portugal. Math. **18** (1959), 207–224. MR **24** #A473.

135. *Generalized twisted fields*, Pacific J. Math. **11** (1961), 1–8. MR **23** #A182.

136. *Isotopy for generalized twisted fields*, An. Acad. Brazil. Ci. **33** (1961), 265–275. MR **25** #3070.

137. *On involutorial associative division algebras*, Scripta Math. **26** (1963), 309–316. MR **31** #3451.

138. *On the nuclei of a simple Jordan algebra*, Proc. Nat. Acad. Sci. U.S.A. **50** (1963), 446–447. MR **27** #3679.

139. *Finite planes for the high school*, The Math. Teacher **55** (1962), 165–169.

140. *A normal form for Riemann matrices*, Canad. J. Math. **17** (1965), 1025–1029. MR **32** #111.

141. *On exceptional Jordan division algebras*, Pacific J. Math. **15** (1965), 377-404. MR **32** #130.

142. *On associative division algebras of prime degree*, Proc. Amer. Math. Soc. **16** (1965), 799-802. MR **31** #3452.

143. *The finite planes of Ostrom*, Bol. Soc. Mat. Mexicana (2) **11** (1966), 1-13. MR **37** #6826.

144. *New results on associative division algebras*, J. Algebra **5** (1967), 110-132. MR **34** #2617.

145. *On some properties of biabelian fields*, An. Acad. Brasil. Ci. **38** (1966), 217-221. MR **34** #7498.

146. *On certain polynomial systems*, Scripta Math. **28** (1967), 15-19. MR **35** #2865.

147. *On associative division algebras* (Retiring Presidential Address), Bull. Amer. Math. Soc. **74** (1968), 438-454. MR **36** #5166.

148. *A note on certain cyclic algebras*, J. Algebra **14** (1970), 70-72. MR **40** #4297.

149. *Tensor products of quaternion algebras*, Proc. Amer. Math. Soc. **35** (1972), 65-66. MR **45** #6855.

References

150. S. A. Amitsur, *On central division algebras*, Israel J. Math. **12** (1972), 408-420. MR **47** #6763.

151. R. Block, *New simple Lie algebras of prime characteristic*, Trans. Amer. Math. Soc. **89** (1958), 421-449. MR **20** #6446.

152. R. Brauer, *Über Systeme hyperkomplexer Zahlen*, Math. Z. **29** (1929), 79-107.

153. ———, *Untersuchungen über die arithmetischen Eigenschaften von Gruppen linearer Substitutionen*. II, Math. Z. **31** (1930), 733-747.

154. ———, *On normal division algebras of index five*, Proc. Nat. Acad. Sci. U.S.A. **24** (1938), 243-246.

155. R. Brauer, H. Hasse and E. Noether, *Beweis eines Hauptsatzes in der Theorie der Algebren*, J. Reine Angew. Math. **167** (1931), 399-404.

156. F. Cecioni, *Sopra un tipo di algebre prive di divisore dello zero*, Rend. Circ. Mat. Palermo **47** (1923), 209-254.

157. P. M. Cohn, *On homomorphic images of special Jordan algebras*, Canad. J. Math. **6** (1954), 253-264. MR **15**, 678.

158. L. E. Dickson, *New division algebras*, Trans. Amer. Math. Soc. **28** (1926), 207-234.

159. ———, *Construction of division algebras*, Trans. Amer. Math. Soc. **32** (1930), 319-334.

160. M. S. Frank, *A new class of simple Lie algebras*, Proc. Nat. Acad. Sci. U.S.A. **40** (1954), 713-719. MR **16**, 562.

161. H. Hasse, *Zum Existenzsatz von Grunwald in der Klassenkörpertheorie*, J. Reine Angew. Math. **188** (1950), 40-64. MR **12**, 677.

162. G. P. Hochschild, *Semi-simple algebras and generalized derivations*, Amer. J. Math. **64** (1942), 677-694. MR **4**, 71.

163. F. D. Jacobson and N. Jacobson, *Classification and representation of semi-simple Jordan algebras*, Trans. Amer. Math. Soc. **65** (1949), 141-169. MR **10**, 588.

164. N. Jacobson, *A note on nonassociative algebras*, Duke Math. J. **3** (1937), 544-548.

165. ———, *A theorem on the structure of Jordan algebras*, Proc. Nat. Acad. Sci. U.S.A. **42** (1956), 140-147. MR **17**, 822.

166. ———, *Structure and representations of Jordan algebras*, Amer. Math. Soc. Colloq. Publ., vol. 39, Amer. Math. Soc., Providence, R.I., 1968. MR **40** #4330.

167. P. Jordan, *Über eine Klasse nichtassoziativer hyperkomplexer Algebren*, Nachr. Ges. Wiss. Göttingen **1932**, 569–575.

168. P. Jordan, J. von Neumann and E. P. Wigner, *On an algebraic generalization of the quantum mechanical formalism*, Ann. of Math. **36** (1934), 29–64.

169. G. K. Kalisch, *On special Jordan algebras*, Trans. Amer. Math. Soc. **61** (1947), 482–494. MR **8**, 561.

170. E. Kleinfeld, *Simple alternative rings*, Ann. of Math. (2) **58** (1953), 544–547. MR **15**, 392.

171. ———, *Alternative nil rings*, Ann. of Math. (2) **66** (1957), 395–399. MR **19**, 383.

172. A. I. Kostrikin and I. R. Šafarevič, *Graded Lie algebras of finite characteristic*, Izv. Akad. Nauk SSSR Ser. Mat. **33** (1969), 251–322=Math. USSR Izv. **3** (1969), 237–304. MR **40** #5680.

173. K. McCrimmon, *Finite power associative division rings*, Proc. Amer. Math. Soc. **17** (1966), 1173–1177. MR **34** #4319.

174. ———, *A note on finite division rings*, Proc. Amer. Math. Soc. **23** (1969), 598–600. MR **40** #7320.

175. M. L. Racine, *A note on quadratic Jordan algebras of degree* 3, Trans. Amer. Math. Soc. **164** (1972), 93–103. MR **46** #3582.

176. C. Rosati, *Sulle matrici di Riemann*, Rend. Circ. Mat. Palermo **53** (1929), 79–134.

177. K. Soda, *Einige Sätze über Matrizen*, Japan J. Math. **13** (1936), 361–365.

178. C. L. Siegel, *Lectures on Riemann matrices*, Tata Institute of Fundamental Research Lectures on Math., no. 28, Tata Institute of Fundamental Research, Bombay, 1963. MR **42** #1861.

179. T. A. Springer, *The classification of reduced exceptional simple Jordan algebras*, Nederl. Akad. Wetench. Proc. Ser. A **63**=Indag. Math. **22** (1960), 414–422. MR **26** #5035.

180. O. Teichmüller, *Multiplikation zyklischer Normalringen*, Deutsch. Math. **1936**, 197–238.

181. ———, *p-Algebren*, Deutsch. Math. **1936**, 362–388.

182. S. Wang, *A counter-example to Grunwald's theorem*, Ann. of Math. (2) **49** (1948), 1008–1009. MR **10**, 231.

183. ———, *On Grunwald's theorem*, Ann. of Math. (2) **51** (1950), 471–484. MR **11**, 489.

184. J. H. M. Wedderburn, *A type of primitive algebra*, Trans. Amer. Math. Soc. **15** (1914), 162–166.

185. ———, *On division algebras*, Trans. Amer. Math. Soc. **22** (1921), 129–135.

186. H. Weyl, *On generalized Riemann matrices*, Ann. of Math. **35** (1934), 714–729.

187. ———, *Generalized Riemann matrices and factor sets*, Ann. of Math. **37** (1936), 709–745.

188. D. Zelinsky, *A. A. Albert*, Amer. Math. Monthly **80** (1973), 661–665. MR **47** #1572.

Department of Mathematics, Yale University, New Haven, Connecticut 06520

PI-ALGEBRAS

N. Jacobson

Yale University

In the first of these lectures we shall outline the history of PI-algebras, that is, algebras satisfying polynomial identities. In the second lecture we shall summarize a theory of central localization of algebras and its application to PI-algebras, which is due to Louis Rowen. This is a part of Rowen's dissertation. His results are based on a recent theorem of Formanek's establishing the existence of central polynomials for the algebra $n \times n$ matrices over a field.

§1. History. The concept of a PI-algebra was introduced by Max Dehn in a paper [1] published in 1922 in Mathematische Annalen. This arose in connection with the following problem in the foundations of geometry: Do there exist configuration theorems in Desarguesian projective geometry which are not implied by Desargues' theorem and which do not imply Pappus' theorem? A configuration theorem states roughly that in a certain construction three points lie on a line or dually three lines meet in a point. It is well-known that Desarguesian geometry can be co-ordinatized by an associative division ring (or skew field) and that Pappus' theorem holds if and only if the division ring is commutative. It can be seen that a configuration theorem in a Desarguesian geometry is equivalent to a rational identity for the co-ordinate division ring, which is not universal, that is, does not hold for every division ring. An example of a universal identity is $(ab)^{-1} = b^{-1}a^{-1}$. A nontrivial example is Hua's identity: If $a \neq 0$, $b \neq 0$ and $ab \neq 1$, then $a^{-1} + (b^{-1} - a)^{-1} \neq 0$ and

$$(a^{-1} + (b^{-1} - a)^{-1})^{-1} = a - aba .$$

We shall not give a precise formulation of the concept of a rational identity. Dehn avoided these and confined his attention to polynomial identities and we shall follow his example. A precise definition of rational identities for rings was given in 1966 by Amitsur in a paper [2] which gave a complete solution of the geometric problem which had been posed by Dehn.

For the present we consider algebras over a field F . To define PI-algebras over F one introduces the free associative algebra $F\{x_1,x_2,...\}$ (freely) generated by a countable set of noncommuting indeterminates. This has a base over F consisting of 1 and the distinct monomials $x_{i_1} \cdots x_{i_r}$, and multiplication is based on the rule that $(x_{i_1} \cdots x_{i_r})(x_{j_1} \cdots x_{j_s}) = x_{i_1} \cdots x_{i_r} x_{j_1} \cdots x_{j_s}$. We consider only algebras with unit 1 over F . If A is such an algebra and $r_1,r_2,...$ is an infinite sequence of elements of A then there exists a unique homomorphism of $F\{x_1,x_2,...\}$ into A sending $x_i \to r_i$, $i \geq 1$. (We always assume homomorphisms map 1 into 1 .) We denote the homomorphism as $f(x_1,...,x_k) \to g(r_1,...,r_k)$. We define <u>an identity</u> for A to be a nonzero element of $F\{x_1,x_2,...\}$ which is mapped into 0 by every homomorphism of $F\{x_1,x_2,...\}$ into A . In other words, it is an element

$f(x_1,\ldots,x_k) \neq 0$ such that $f(r_1,\ldots,r_k) = 0$ for all $r_i \in A$. The simplest example is $x_1x_2 - x_2x_1$. An algebra satisfies this identity if and only if it is commutative. Other examples will be given in a moment.

Dehn initiated the study of PI-algebras, that is, algebras satisfying a polynomial identity in 1922. Aside from introducing the subject, he obtained a few elementary results and he showed that identities of low degree together with the order axioms imply commutativity.

The next important step in the development of PI-algebras was taken by W. Wagner in a paper [3] in Math. Annalen, 1936. In this paper, Wagner showed that any quaternion algebra satisfied the identity

$$(x_1x_2 - x_2x_1)^2x_3 - x_3(x_1x_2 - x_2x_1)^2 , \tag{1}$$

that is, the square of an additive commutator lies in the center of the algebra. Wagner showed also that this identity was satisfied by the algebra $M_2(F)$ of 2×2 matrices. He also gave identities which hold for any $M_n(F)$, the algebra of $n \times n$ matrices over F. Finally, the main result he obtained is that if D is an ordered PI-division algebra, then D is commutative.

After Wagner's paper the theory of PI-algebras was dormant for about ten years. In 1943, Marshall Hall published a paper [4] on a projective planes in which he showed that Wagner's identity characterized quaternion division algebras, that is, if D is a division algebra which satisfies the identity $(x_1x_2 - x_2x_1)^2x_3 - x_3(x_1x_2 - x_2x_1)^2$ then D is a quaternion algebra over its center.

In 1945, the author developed a general structure theory of associative rings without finiteness conditions including a theory of radicals and primitive algebras ([5], [6]). The main application (and, in fact, one of the main motivations for the development of this theory) was to Kurosch's problem which is the ring theoretic analogue of Burnside's problem on groups: Is every algebraic algebra locally finite, in the sense that subalgebras generated by finite subsets are finite dimensional? Using the structure theory, the author proved (in [7]) that the answer is affirmative for semiprimitive (= semisimple in the old terminology) algebras of bounded degree and reduced the general question for algebraic algebras of bounded degree to the case of nil algebras. This was partially settled by Kaplansky and completely by Levitzki in 1946, [8]. As a result, one had the result that any algebraic algebra of bounded degree is locally finite.

During the summer of 1947, the author lectured on structure theory of rings at Chicago. During this visit Kaplansky and the author became aware of Hall's paper. This raised an interesting question: Is every PI-division algebra finite dimensional over its center? Kaplansky showed that the answer is affirmative and, in fact, proved a much more general result which has become a cornerstone of the theory of PI-algebras ([9]).

Kaplansky's theorem. The center of a primitive PI-algebra A is a field and A is finite dimensional over its center C. More exactly, if A/F satisfies an identity of (total) degree d then C is a field and the dimensionality $[A:C] \leq [\frac{d}{2}]^2$.

We recall that an algebra is called primitive if it has a faithful irreducible representation. All simple algebras (with 1) are primitive and hence all division algebras are primitive. Kaplansky's theorem together with the structure theory implied that primitive PI-algebras were of the form $M_n(D)$ the algebra of $n \times n$ matrices over D where D is finite dimensional over its center.

Kaplansky's theorem together with a theorem due to Albert ([10]) that a finite dimensional ordered division algebra is commutative gives immediately Wagner's theorem that ordered PI-division algebras are commutative. Since Albert's theorem can be proved in a few lines we shall indicate it. Suppose D is ordered with center C and $D \supsetneq C$. Choose an element $d \in D$, $\notin C$ and let $f(\lambda)$ be its minimum polynomial over C of degree $m > 1$. By subtracting a multiple of 1 from d one may assume $f(\lambda) = \lambda^m + c_2\lambda^{m-2} + \cdots + c_m$. On the other hand, one has a result of Wedderburn's ([11]) that $f(\lambda) = (\lambda - d_1)(\lambda - d_2)\cdots(\lambda - d_m)$ where $d_1 = d$ and the $d_i \in D$ and the d_i are conjugates, that is $d_i = a_i d a_i^{-1}$. We may assume $d > 0$. Then every $d_i > 0$. The form of $f(\lambda)$ shows that $d_1 + d_2 + \cdots + d_m = 0$ contrary to every $d_i > 0$.

The proof of Kaplansky's theorem is based on our structure theory and on a process of linearization which shows that if an algebra satisfies an identity then it satisfies a multilinear one, that is, one of the form

$$\sum_\pi \alpha_{i_1 \cdots i_r} x_{i_1} \cdots x_{i_r} \tag{2}$$

summed on the permutations $\pi = \begin{pmatrix} 1 & 2 & \cdots & r \\ i_1 & i_2 & \cdots & i_r \end{pmatrix}$. This implies the important fact that the class of PI-algebras is closed under tensor products with commutative algebras. One also has the important and obvious results that subalgebras and homomorphic images of PI-algebras are PI-algebras.

What about the converse of Kaplansky's theorem? More generally what are some conditions that insure that an algebra is PI, and what can be said about the set of identities for a PI-algebra? We have already noted that Wagner had shown that $M_n(F)$ satisfies an identity, in fact, an identity with integer coefficients. It follows from this that if A is an algebra over F which is finite dimensional over a subfield C of its center then A is PI. This follows since the hypothesis implies that A is a subalgebra of $M_n(C)$ where $n = [A:C]$. In 1947, the author showed that any algebraic algebra of bounded degree is PI. This result was published in Kaplansky's paper [9]. Since its proof is quick we shall indicate it. The hypothesis implies that we have an integer n such that for every $a \in A$ one has a relation of the form

$$a^n + \alpha_1 a^{n-1} + \cdots + \alpha_n 1 = 0 \tag{3}$$

where the $\alpha_i \in F$. Let b be any other element of A and take the commutator of the relation (3) with b. This gives

$$[a^n,b] + \alpha_1[a^{n-1},b] + \cdots + \alpha_{n-1}[ab] = 0 .$$

Next take the commutator with $[a,b]$ to obtain

$$[[a^n b][ab]] + \alpha_1[[a^{n-1}b][ab]] + \cdots + \alpha_{n-2}[[a^2 b][ab]] = 0$$

continuing in this way one sees that A satisfies a polynomial identity with integer coefficients of the form

$$f(x_1,x_2) = [\ldots[[[x_1^n x_2],[x_1x_2]],[[x_1^2x_2][x_1x_2]]],\ldots] . \tag{4}$$

It is easy to see that $f \neq 0$.

Another identity which may have been introduced by Kolchin and which has played an important role in the theory is the <u>standard polynomial</u>

$$S_m(x_1,\ldots,x_m) = \sum_{\pi} (sg\pi) x_{i_1} x_{i_2} \cdots x_{i_m} \tag{5}$$

where the summation is taken over all $\Pi = \begin{pmatrix} 1 & 2 & \ldots & m \\ i_1 & i_2 & \ldots & i_m \end{pmatrix}$. This is multilinear and alternating. It follows that if $[A:F] < m$ then S_m is an identity for A . $S_2(x_1,x_2) = x_1x_2 - x_2x_1$ so the identities S_m can be regarded as expressing a weakening of commutativity. This idea can be strengthened by the following result which is quite easy to establish. Define the degree of commutativity to be the infimum of the degrees of the standard identities satisfied by A . (This may be infinite.) Then it is easy to show that if $r(n)$ denotes the degree of commutativity of $M_n(F)$ then $r(n + 1) > r(n) + 1$. The precise formula for $r(n)$ was determined by Amitsur and Levitzki in 1950, [12]. Their result is below.

<u>Theorem of Amitsur and Levitzki</u>. The standard polynomial S_{2n} is an identity for $M_n(F)$.

Since it is easy to see that $M_n(F)$ satisfies no identity of degree $< 2n$, it follows that $r(n) = 2n$. This result permits a sharpening of Kaplansky's theorem: A primitive algebra A satisfies an identity of degree d if and only if A is simple of dimension $\leq [\frac{d}{2}]^2$ over its center.

In 1950, Kaplansky (in [13]) proved another major result on PI-algebras, namely, an affirmative answer to Kurosch's problem for algebraic PI-algebras: Every algebraic PI-algebra is locally finite. This result subsumed the earlier one by Jacobson and Levitzki on algebraic algebras of bounded degree. It also implied an earlier result of Malcev's ([14]) analogous to a theorem of Schur's on groups: Any algebraic algebra which has a faithful representation by finite matrices over some extension field of the base field is locally finite. Kaplansky's proof of his theorem on local finiteness was partly topological. A simply purely algebraic proof of this result was given in 1952, by Amitsur [15]. This proof appears in the author's book, Structure of Rings, Chapter X.

Levitzki and Amitsur have proved a number of important results on nil ideals of PI-algebras. We recall that a nil ideal is an ideal all of whose elements are nilpotent and a nilpotent ideal is one satisfying $I^n = 0$ for some I, that is, there exists an n such that the product of any n elements of I is 0. I is called locally nilpotent if for any finite subset $\{b_1,\dots,b_m\}$ there exists an n such that every product $b_{i_1}b_{i_2}\cdots b_{i_n} = 0$, $1 \le i_j \le m$. The sum of all nil ideals is a nil ideal called the upper nil radical of the ring. The sum of all locally nilpotent ideals is locally nilpotent and this is called the Levitzki nil radical. The sum $N(1)$ of all nilpotent ideals need not be nilpotent and $A/N(1)$ may contain nilpotent ideals. Let $N(2)$ be the ideal containing $N(1)$ such that $N(2)/N(1)$ is the sum of all nilpotent ideals of $A/N(1)$ and repeat this process. For limit ordinals α, do the usual thing of defining $N(\alpha) = \bigcup_{\beta<\alpha} N(\beta)$. The union of the chain of ideals obtained in this way is a nil ideal called the lower nil radical of the ring. It can be proved that this coincides with the intersection of the prime ideals of the ring where, an ideal P is called prime if $B_1B_2 \subset P$ for ideals B_1 and B_2 implies either $B_1 \subset P$ or $B_2 \subset P$. Equivalently, the condition is that if z_1 and z_2 are elements such that $z_1az_2 \in P$ for all $a \in A$ then either $z_1 \in P$ or $z_2 \in P$. A ring A is called prime if the ideal 0 is a prime ideal in A.

The following are the main results of nil radicals of PI-algebras. The upper nil radical is locally nilpotent and satisfies $M^{[d/2]} \subset N(1)$ where $N(1)$ is defined above. This implies that the upper and lower nil radicals coincide and that this is the same as the ideal $N(2)$ defined above. These results were proved by Amitsur in 1951, ([16]), and are based in part on earlier results by Levitzki ([17]). The results are usually stated about subalgebras of algebras without unit. However, it is not difficult to see that the earlier formulations in these terms are equivalent to the ones we have given.

In 1960, Posner, a student of Kaplansky's, determined the structure of prime PI-algebras ([18]). He showed that these can be imbedded as orders in algebras which are simple and finite dimensional over their centers. This result was based on a fundamental theorem due to Goldie ([19]) characterizing prime and semiprime rings satisfying some Noetherian type conditions on one-sided ideals.

The Amitsur-Levitzki theorem that $M_n(F)$ satisfies the identity S_{2n} can be formulated as a theorem on graphs. This seems to have been done in print first by Schutzenberger. However, this had been noted by Kakutani around 1949 or 1950, and he worked for a while on a proof along these lines. A purely graph theoretic proof of the result was given by Swan in two papers ([20]) appearing in 1963 and 1969. (The first of these overlooked a case which was treated in the second.)

In 1958, Kostant wrote a remarkable paper ([21]) relating the Amitsur-Levitzki theorem to two other important results: a theorem of Frobenius on characters of the alternating group and a theorem of Dynkin on cohomology of Lie groups. Also in this paper Kostant proved, using cohomology of Lie groups, a new result on standard identities. If n is even then S_{2n-2} is an identity for $n \times n$ skew symmetric matrices, that is, $S_{2n-2}(s_1,s_2,\ldots,s_{2n-2}) = 0$ for all choices of s_i skew symmetric in $M_n(F)$. This result can also be formulated as a theorem on graphs and a number of attempts to prove it graph theoretically were made by graph theorists during the past ten years or so. Recently three young people more or less simultaneously succeeded in obtaining elementary proofs of Kostant's theorem and showing that it is valid also for odd n. These are Joan Hutchinson, Frank Owens and Louis Rowen. Rowen has apparently obtained the best results, which can be stated as the following theorems ([22]):

Let $M_n^+(F)$ denote the space of $n \times n$ symmetric matrices, $M_n^-(F)$ the space of $n \times n$ skew symmetric matrices. Then one has the following results:

$S_{2n-2}(x_1,\ldots,x_{2n-2})$ vanishes for all specializations of $x_1,\ldots,x_{2n-2}$ in $M_n^-(F)$ for all n.

$S_{2n-2}(x_1,\ldots,x_{2n-2})$ vanishes for all specializations of $x_1,\ldots,x_{2n-3}$ in $M_n^-(F)$ and x_{2n-2} in $M_n^+(F)$ for n odd.

$S_{2n-1}(x_1,\ldots,x_{2n-1})$ vanishes for all specializations of $x_1,\ldots,x_{2n-2}$ in $M_n^-(F)$ and x_{2n-1} in $M_n^+(F)$ for all n .

These results are in a sense best possible if the characteristic is 0 Rowen has shown that one can not increase the number of x_i taken in $M_n^+(F)$ in these statements. He has also obtained complete results for characteristic two and partial ones for characteristic $p \neq 2,0$. Rowen's proofs are largely graph theoretic but make use also of matrix arguments, particularly some results on traces which extend earlier ones due to Kostant.

The results just indicated belong to an area dealing with PI-algebras with involution which was initiated by Herstein. Herstein raised the following question: Suppose A is an algebra with involution and suppose the symmetric elements satisfy an identity, is A necessarily a PI-algebra? Herstein proved this for simple rings ([23]) and Martindale for semiprime rings of characteristic $\neq 2$ ([24]). We recall that a ring is <u>semiprime</u> if it contains no nilpotent ideals $\neq 0$. The general case was settled by Amitsur in 1969, in a paper ([25]) in which the result was formulated in the following improved form. Let $F\{x_1,x_2,\ldots\}$ be the free algebra generated by the x_i as before. Then this has an involution in which $x_i^* = x_{i+1}$ for $i = 1,3,\ldots$. Moreover, this algebra with involution has the universal property that if (A,j) is any associative algebra with involution and $r_1,r_3,\ldots$ are elements of A then there exists a homomorphism of algebras with involution of $F\{x_1,x_2,\ldots\}$ sending $x_i \to r_i$, $i = 2k-1$. (A,j) is called a PI-*algebra with involution* if $F\{x_1,x_2,\ldots\}$ contains a nonzero element $f(x_1,\ldots,x_{2k})$ such that $f(r_1,r_1^*,r_1,r_2^*,\ldots,r_k,r_k^*) = 0$ for all $r_i \in A$. Then Amitsur's theorem is that if (A,j) is a PI-algebra with involution, then A is a PI-algebra.

Two old problems on PI-algebras have been solved quite recently. The first of these is the tensor product problem: Suppose A and B are PI-algebras. Is $A \otimes_F B$ a PI-algebra? This problem is mentioned in the author's book published in 1956. The affirmative answer to this question was published by A. Regev in 1972, in [26]. An earlier result had been proved by Procesi and Small in 1960, in [27]. This showed that $A \otimes B$ is PI if A is PI and B is finite dimensional. Regev's proof is a combinatorial one. We shall indicate the idea.

Let A be a PI-algebra and let I be the ideal in $F\{x_1,x_2,\ldots\}$ of identities of A, that is, the elements which are mapped into 0 by every homomorphism of $F\{x_1,x_2,\ldots\}$ into A. This is a T-ideal, that is, it is stable under every algebra endomorphism of $F\{x_1,x_2,\ldots\}$. For any n let $V_n(x)$ denote the space spanned by the $n!$ monomials $x_{\pi(1)}\cdots x_{\pi(n)}$. π a permutation of $1,2,\ldots,n$ and let $d_n = \dim V_n(x)/(I \cap V_n(X))$ and call the sequence $d_1,d_2,\ldots$ the <u>codimension</u> sequence of the PI-algebra. Then Regev has shown that $d_n \leq (3.4^{d-2})^n$ where d is the minimum degree of identities satisfied by A. If $\{h_n\}$ is the sequence of codimensions for a second PI-algebra B then Regev's estimate implies that for sufficiently large n, $d_n h_n < n!$. If n is chosen so that $d_n h_n < n!$ then it is quite easy to see that $A \otimes B$ satisfies an identity of degree n.

Regev's proof is quite long. This has been simplified in a subsequent paper ([28]) by Klein and Regev. Moreover, in this paper they established in a more complicated fashion a much better bound: $d_n \le (3(d^2 - 7d + 16))^n$.

The second outstanding problem which has been solved recently is that of central polynomials for matrix algebras. We recall that Wagner had shown that $M_2(F)$ and any quaternion algebra satisfies the identity $(x_1x_2 - x_2x_1)^2x_3 - x_3(x_1x_2 - x_2x_1)^2$. On the other hand, $(x_1x_2 - x_2x_1)^2$ is not an identity for $M_2(F)$. A polynomial $f(x_1,\dots,x_n)$ is called <u>central</u> if f is not an identity for A but $[f(x_1,\dots,x_n),x_{n+1}]$ is an identity. Thus $(x_1x_2 - x_2x_1)^2$ is central for $M_2(F)$. In the summer of 1947, Kaplansky and the author discussed the possibility of generalizing this to $M_n(F)$. This was mentioned by Kaplansky in his list of problems on rings which was presented at a conference on Shelter Island, N.Y., in 1955, and was published in the Proceedings of this conference. However, it was settled affirmatively only quite recently by Formanek ([28]) and the result appears to be of central importance in the theory of PI-algebras, as has appeared in the recent work of a number of mathematicians. We shall indicate Rowen's work exploiting the existence of central polynomials in §2.

We conclude this brief sketch of the history of PI-algebras by mentioning its most important application. This is Amitsur's construction of finite dimensional division algebras which are not crossed products, that is, do not have maximal subfields which are Galois over the center ([30]). We recall that one of the crowning achievements of algebra and algebraic number theory of the thirties was the theorem of Albert-Brauer-Hasse-Noether that every finite dimensional central division algebra is cyclic, that is, contains a maximal subfield which is cyclic, and, consequently, has a simple structure which had been defined independently by Dickson and by Wedderburn. It has been known for a long time that this result does not hold for arbitrary base fields and it had been suspected that there exist division algebras finite dimensional over their centers which are not crossed products. Amitsur has proved this in the following way: Let $\xi_{ij}^{(k)}$, $1 \le i,j \le n$, $k = 1,2,\ldots$ be indeterminates and let Ω denote the ring of polynomials with rational coefficients in the countable set of indeterminates $\xi_{ij}^{(k)}$. Form the matrix algebra (over Q) $M_n(\Omega)$ and let B be the subalgebra over Q generated by the "generic" matrices $X^{(k)} = (\xi_{ij}^{(k)})$. Since this is a subalgebra of $M_n(\Omega)$ it is a PI-algebra. It is easy to see that it is a domain. Hence, by Posner theorem, this has a quotient algebra A which is a division algebra finite dimensional over its center. Amitsur has proved that if n is divisible by 8 or by the square of an odd prime then A is not a crossed product.

Appendix. In this appendix we give a slightly improved version of Formanek's theorem and a consequence of it due to Amitsur [3].

Let $F[\eta_1,\ldots,\eta_{n+1}]$ be the polynomial algebra over a field F in the indeterminates $\eta_1,\ldots,\eta_{n+1}$ and let $F\{x,y_1,\ldots,y_n\}$ be the free algebra generated by noncommutative indeterminates $x,y_1,\ldots,y_n$. If $f = \Sigma\, \alpha_{(\nu)}\eta_1^{\nu_1} \cdots \eta_{n+1}^{\nu_{n+1}} \in F[\eta] \equiv F[\eta_1,\ldots,\eta_{n+1}]$ we define

$$(6) \qquad p_f(x,y_1,\ldots,y_n) = \sum_{(\nu)} \alpha_{(\nu)} x^{\nu_1} y_1 x^{\nu_2} y_2 \cdots x^{\nu_n} y_n x^{\nu_{n+1}}$$

which is an element of $F\{y,x\} \equiv F\{y_1,\ldots,y_n,x\}$. Here $(\nu) = (\nu_1,\ldots,\nu_{n+1})$ and the ν_{i_n} are nonnegative integers, and $\alpha_{(\nu)} \in F$. Now specialize $x \to \sum_i \rho_i e_{ii}, y_1 \to e_{i_1k_1}, \ldots, y_n \to e_{i_nk_n}$ in $M_n(F)$ where the e_{ij} are the usual matrix units. Then a simple calculation gives

$$p_f(\Sigma \rho_i e_{ii}, e_{i_1k_1}, \ldots, e_{i_nk_n}) =$$

$$\delta_{k_1i_2}\delta_{k_2i_3}\cdots\delta_{k_{n-1}i_n} f(\rho_{i_1},\ldots,\rho_{i_n},\rho_{k_n})e_{i_1k_n} .$$

Hence this is 0 for all choices of $e_{i_1k_1},\ldots,e_{i_nk_n}$ except the sequences $(e_{i_1i_2}, e_{i_2i_3},\ldots,e_{i_{n-1}i_n}, e_{i_nk_n})$ in which case we get

$$p_f(\Sigma \rho_i e_{ii}, e_{i_1i_2}, e_{i_2i_3},\ldots,e_{i_nk_n}) = f(\rho_{i_1},\ldots,\rho_{i_n},\rho_{k_n})e_{i_1k_n} .$$

We note also that there are only n choices for the subscripts; hence two numbers in the sequence $(i_1,i_2,\ldots,i_n,k_n)$ coincide.

Now suppose f satisfies

(i) $f(\eta_1,\ldots,\eta_{n+1})$ is divisible by every $\eta_i - \eta_j$, $i \neq j$, except $\eta_1 - \eta_{n+1}$.

(ii) $g(\eta_1,\ldots,\eta_n) \equiv f(\eta_1,\ldots,\eta_n,\eta_1)$ is symmetric in $\eta_1,\ldots,\eta_n$.

Then $p_f(\Sigma \rho_i e_{ii}, e_{i_1k_1}, e_{i_2k_2},\ldots,e_{i_nk_n}) = 0$ for all choices of $e_{i_jk_j}$ except those in which $k_j = i_{j+1}$, $k_n = i_1$ and the i_j are distinct. Then

$$p_f(\Sigma \rho_i e_{ii}, e_{i_1i_2}, e_{i_2i_j},\ldots,e_{i_ni_1}) = g(\rho_1,\ldots,\rho_n)e_{i_1i_1} .$$

Hence if we put

$$q_f(x,y_1,\ldots,y_n) = \sum_{i=0}^{n} p_f(x,y_{i+1},\ldots,y_{i+n}) \tag{7}$$

where the subscripts are reduced mod n then

(8) $$q_f(\Sigma\, \rho_i e_i, e_{i_1 k_1}, \ldots, e_{i_n k_n}) = 0$$

for all $(e_{i_1 k_1}, \ldots, e_{i_n k_n})$ except $(e_{i_1 i_2}, e_{i_2 i_3}, \ldots, e_{i_n i_1})$ with distinct i_j in which case

(9) $$q_f(\Sigma\, \rho_i e_{ii}, e_{i_1 i_2}, e_{i_2 i_3}, \ldots, e_{i_n i_1}) = g(\rho_1, \ldots, \rho_n)1 .$$

The simplest choice of a nonzero f satisfying (i) and (ii) is Formanek's.

(10) $$\gamma = \prod_{i=2}^{n} (\eta_1 - \eta_i)(\eta_{n+1} - \eta_i) \prod_{i<j=2}^{n} (\eta_i - \eta_j)^2 .$$

This has integer coefficients and some of these are ±1 . Also in this case

$$\delta(\eta_1, \ldots, \eta_n) = \gamma(\eta_1, \ldots, \eta_n, \eta_1) = \prod_{i<j=1}^{n} (\eta_i - \eta_j)^2$$

which is the discriminant of the polynomial $\prod_1^n (\lambda - \eta_i)$.

Let $g = g(\eta_1, \ldots, \eta_n)$ be a symmetric polynomial with integer coefficients. Then there exists a unique polynomial $h(\eta_1, \ldots, \eta_n) \in Z[\eta_1, \ldots, \eta_n]$ such that

(11) $$g(\eta_1, \ldots, \eta_n) = h(\Sigma\, \eta_i, \sum_{(<)} \eta_i \eta_j, \ldots, \eta_1 \cdots \eta_n) .$$

If K is a commutative ring, the symmetric polynomial g defines a map G of $M_n(K)$ into K by

(12) $$a = (a_{ij}) \to h(\mathrm{tr}\ a, \ldots, \det a)$$

where tr a,...,det a is the sequence of elements of K determined by the coefficients of the characteristic polynomial

(13) $$\det(\lambda 1 - a) = \lambda^n - (\mathrm{tr}\ a)\lambda^{n-1} + \cdots + (-1)^n \det a \ .$$

If $K = F$ is a field and $\rho_1,\ldots,\rho_n$ are the characteristic roots of a in the algebraic closure $\overline{F}$ of F then the map G defined by (11) and (12) is the same as

(14) $$a \to g(\rho_1,\ldots,\rho_n)$$

since $\mathrm{tr}\ a = \Sigma\, \rho_i,\ldots,\det a = \rho_1 \cdots \rho_n$. If F is infinite the maps $a \to \mathrm{tr}\ a,\ldots,a \to \det a$ are polynomial functions on $M_n(F)$. Hence if $g(\eta_1,\ldots,\eta_n) \neq 0$ then the map defined by (12) (or by (14)) is a nonzero polynomial function G on $M_n(F)$. The set of a such that $G(a) \neq 0$ is a Zariski open subset of $M_n(F)$.

All of this applies in particular to $g = \delta$ where $\delta(\eta_1,\ldots,\eta_n) = \prod_{i<j} (\eta_i - \eta_j)$. The matrices a such that $\delta(a) \neq 0$ are those with distinct characteristic roots, hence, those which are similar to diagonal matrices in $M_n(\overline{F})$ with distinct diagonal entries. Also for $g = \delta$ we have

(iii) For any field F there exists an $a \in M_n(F)$ such that $G(a) \neq 0$.

This is clear in the infinite case: any diagonal a with distinct diagonal entries will do. In the finite case there exists irreducible polynomials $h(\lambda)$ of degree n and matrices in $M_n(F)$ having $h(\lambda)$ as characteristic polynomial. Since $h(\lambda)$ is separable this gives a required a .

Theorem. (1). Let $f \in Z[\eta_1,\ldots,\eta_{n+1}]$ satisfy (i), (ii) and (iii). Then $q_f(x,y_1,\ldots,y_n)$ defined by (7) is a central polynomial for $M_n(K)$ for any commutative ring K .

(2). There exist $a \in M_n(K)$ such that $G(a)$ is not nilpotent, and if a is a matrix satisfying this condition, then there exist $b_1,\ldots,b_n \in M_n(K)$ such that $q_f(a,b_1,\ldots,b_n) \neq 0$.

(3). If f is as in (1) and $\ell(\eta_1,\ldots,\eta_n) \in Z(\eta_1,\ldots,\eta_n)$ is symmetric, then

$$q\ell_f(a,b_1,\ldots,b_n) = L(a)q_f(a,b_1,\ldots,b_n) \tag{15}$$

for all $a,b_i \in M_n(K)$, where L is the map defined as in (11) and (12) by $\ell(\eta_1,\ldots,\eta_n)$.

Proof. Suppose first that $K = \overline{F}$ is an algebraically closed field. We show first that $[q_f(x,y_1,\ldots,y_n),z]$, where z is another indeterminate, is an identity in $M_n(\overline{F})$ and (15) holds. Since both of these statements amount to polynomial equations it suffices to prove them for all specializations $x = a$, $y = b_1,\ldots,b_n$, $z = c$ such that $G(a) \neq 0$. Since $g(\eta_1,\ldots,\eta_n)$ is divisible by every $\eta_i - \eta_j$, $i \neq j = 1,\ldots,n$, the condition $G(a) \neq 0$ implies that a has distinct characteristic roots. Hence a is similar to a diagonal matrix, so we may assume a diagonal. In this case the two results hold by (8) and (9) if the b_i are taken to be matrix units. Since $q_f(x,y_1,\ldots,y_n)$ is linear in every y_i the result holds for all choices of b_i in $M_n(\overline{F})$.

Next let $K = F$ be any field. The two results just proved for $M_n(\overline{F})$ hold for $M_n(F)$ by extending F to its algebraic closure $\overline{F}$. Also, by hypothesis, we have an $a \in M_n(F)$ such that $G(a) \neq 0$. Since a is similar to a diagonal matrix, (8) and (9) imply that there exist $b_1,\dots,b_n$ in $M_n(\overline{F})$ such that $q_f(a,b_1,\dots,b_n) \neq 0$. Since $q_f(x,y_1,\dots,y_n)$ is linear in the y_i , it follows that there exist $b_1,\dots,b_n$ also in $M_n(F)$ such that $q_f(a,b_1,\dots,b_n) \neq 0$.

Now let K be arbitrary. The results proved imply that $[q_f(x,y_1,\dots,y_n),z]$ is an identity in $M_n(Z[\zeta_1,\zeta_2,\dots,\zeta_m])$, ζ_i indeterminates. Also we have (15) in this ring. Now choose $a,b_1,\dots,b_n,c$ in $M_n(K)$. Then if m is sufficiently large we have a homomorphism of $M_n(Z[\zeta])$ into $M_n(K)$ sending certain matrices $A,B_1,\dots,B_n,C$ into $a,b_1,\dots,b_n,c$ respectively and $L(A)$ into $L(a)$. Applying this we obtain the two results proved in the first part for $M_n(K)$. Now let P be a maximal ideal in K and let $F = K/P$. The hypothesis implies that there exists an $\bar{a} \in M_n(F)$ such that $G(\bar{a}) \neq 0$. Then if $\bar{a} = a + M_n(P)$, $G(a)$ is not nilpotent. Conversely, let a be a matrix in $M_n(K)$ such that $G(a)$ is not nilpotent. Since the nil radical of K is the intersection of the prime ideals of A , there exists a prime ideal P such that $\overline{G(a)} = G(a) + P \neq 0$. Hence $G(\bar{a}) \neq 0$ for $\bar{a} = a + M_n(P)$. Let $D = K/P$ and let F be a field of fractions of D . Then there exist $\bar{b}_1,\dots,\bar{b}_n \in M_n(F)$ such that $q_f(\bar{a},\bar{b}_1,\dots,\bar{b}_n) \neq 0$. Clearing of fractions we may assume $\bar{b}_1,\dots,\bar{b}_n \in M_n(D)$. Then $q_f(a,b_1,\dots,b_n) \neq 0$.

We now apply part (3) of the foregoing theorem taking f to be Formanek's γ and taking ℓ successively to be the elementary symmetric polynomials $\Sigma\, \eta_i, \sum_{i<j} \eta_i\eta_j, \ldots, \eta_1\eta_2\cdots\eta_n$. Then we obtain the following result using (3) and the Hamilton-Cayley theorem.

<u>Theorem</u> (Amitsur). Let $q_0(x,y_1,\ldots,y_n)$ be Formanek's central polynomial for $M_n(K)$, K a commutative ring. Then there exist central polynomials $q_0(x,y_1,\ldots,y_n),\ldots,q_n(x,y_1,\ldots,y_n)$ such that for any $a,b_i \in M_n(K)$

$$
\begin{aligned}
&q_0(a,b_1,\ldots,b_n)\lambda^n - q_1(a,b_1,\ldots,b_n)\lambda^{n-1} \\
&\quad + \cdots + (-1)^n q_n(a,b_1,\ldots,b_n) \\
&= q_0(a,b_1,\ldots,b_n)f(\lambda)
\end{aligned}
$$

where $f(x)$ is the characteristic polynomial of a . Hence

$$
\begin{aligned}
&q_0(x,y_1,\ldots,y_n)x^n - q_1(x,y_1,\ldots,y_n)x^{n-1} \\
&\quad + \cdots + (-1)^n q_n(x,y_1,\ldots,y_n)
\end{aligned}
$$

is an identity for $M_n(K)$.

§2. <u>Central Localization of PI-Algebras</u>. In this lecture we summarize the principal results of a part of Rowen's dissertation [32]. Some of these have appeared in a brief note [33]; a full account of the theory of central localization for PI-algebras will appear in [34]. The theory of PI-algebras over a commutative ring K was initiated by Amitsur in [35]. One of its most striking results is a theorem of M. Artin [36], as extended by

Procesi in [37], characterizing Azumaya algebras of rank n^2. One of Rowen's applications is to a new proof of the theorem of Artin-Procesi.

As in §1, we deal exclusively with rings and algebras with unit. Let K be a commutative ring (with 1) and let $K\{x_1,x_2,\ldots\}$ be the free algebra over K generated noncommuting indeterminates $x_1,x_2,\ldots$. An element $f(x_1,\ldots,x_n) \in K\{x_1,x_2,\ldots\}$ is called an <u>identity</u> for an algebra A over K if $f(r_1,\ldots,r_n) = 0$ for all $r_i \in A$. Amitsur has called an identity f <u>proper</u> if some coefficient of f does not annihilate A and he has proved the following extension of Kaplansky's theorem: A primitive algebra which satisfies a proper identity of degree d is simple of dimensionality $\leq [\frac{d}{2}]^2$ over its center ([35]). Also Amitsur has proved in [38] that a prime algebra satisfies a proper identity of degree d then it satisfies the standard identity $S_{2[d/2]}$.

Rowen's point of departure is the following definition of a PI-algebra over K: An algebra A over K is a PI-<u>algebra</u> if there exists an $f \in K\{x_1,x_2,\ldots\}$ which is a proper identity for every nonzero homomorphic image of A.

It is easily seen that this is equivalent to the existence of an identity f for A such that the ideal of elements of the form $\Sigma\, \omega_i r_i$, where the ω_i are coefficients of f and the $r_i \in A$, is A. The usual proofs in the field case carry over to show that the results on nil radicals of PI-algebras over a field hold also for PI-algebras over K. Also one has the following result which is essentially due to Amitsur [39]: Any PI-algebra satisfies an identity $S_{2n}(x_1,\ldots,x_{2n})^n$ for suitable n and m. This implies that subalgebras of PI-algebras are PI and also that the tensor product of a PI-algebra with a commutative algebra is PI.

A key result in Rowen's theory is the following Theorem 1.

Theorem 1. Any nonzero ideal in a semiprime PI-algebra has nonzero intersection with the center.

We recall that an algebra is semiprime if it contains no nilpotent ideal $\neq 0$. If A is semiprime PI-algebra it contains no nil ideal $\neq 0$. Hence, by a well-known theorem of Amitsur's, the polynomial algebra $A[\lambda]$, λ an indeterminate, is semiprimitive. This permits a reduction of Theorem 1 to the case of semiprimitive algebras. In this case the proof makes essential use of Formanek's central polynomial for matrix algebras.

An immediate consequence of Theorem 1 is the following corollary.

Corollary. A semiprime PI-algebra is simple (and hence finite dimensional over its center) if and only if its center is a field.

Let A be an arbitrary algebra, C its center, S a submonoid of the multiplicative monoid of C. Thus S is a subset of C closed under multiplication and containing 1. Define an equivalence relation on $A \times S$ by $(r_1,s_1) \sim (r_2,s_2)$ if there exists an $s \in S$ such that $(r_1s_2 - r_2s_1) = 0$. Let rs^{-1} denote the equivalence class of (r,s). Then one can define an algebra structure on the set A_S of these equivalence classes by $r_1s_1^{-1} + r_2s_2^{-1} = (r_1s_2 + r_2s_1)(s_1s_2)^{-1}, (r_1s_1^{-1})(r_2s_2^{-1}) =$

$(r_1r_2)(s_1s_2)^{-1}, \alpha(rs^{-1}) = (\alpha r)s^{-1}, \alpha \in K$. A_S is called the <u>localization of</u> A <u>by</u> S . One has a canonical homomorphism $\nu_S : r \to r1^{-1}$ of A into A_S . The elements $\nu_S(s) = s1^{-1}$, $s \in S$ are invertible in A_S and one has the universal property that if α is any homomorphism of A into an algebra A' such that $\alpha(s)$, $s \in S$, is invertible, then there is a unique homomorphism $\beta : A_S \to A'$ such that $\alpha = \beta\nu_S$. The homomorphism ν_S is a monomorphism if and only if every element $s \in S$ is <u>regular</u> in the sense that $sr = 0$ for $r \in A$ implies $r = 0$. It should be noted also that as additive groups A_S and $C_S \otimes {}_C A$ are naturally equivalent.

One has the following result on identities and centers.

<u>Theorem 2</u>. Any identity for A is one for A_S and A and A_S satisfy the same identities if the elements of S are regular. The center of A_S contains C_S and C_S is the center of A_S if the elements of S are regular.

The set T of elements of C which are regular form a submonoid of the multiplicative monoid of C . The localization A_T is called the <u>algebra of central quotients of</u> A . Rowen has proved, using Theorems 1 and 2, the following sharpening of Posner's theorem (as extended by Amitsur to algebras over K).

<u>Theorem 3</u>. Let A be a prime PI-algebra over K and let A' be its algebra of central quotients. Then A' is simple, hence, by the

Kaplansky-Amitsur Theorem, is finite dimensional over its center. Moreover, the center of A' is the quotient field of the center of A, A and A' satisfy the same identities and A' is the classical left and right algebra of quotients of A.

A part of this result has been obtained independently by other methods by Formanek, Martindale, Procesi and Small.

As in the case of commutative rings, one of the most interesting instances of localization is that relative to the complement $C - P$ of a prime ideal P of C. In this case we write A_P for A_{C-P}.

If $\tilde{B}$ is an ideal in A then $B = \tilde{B} \cap C$ is an ideal in C. On the other hand, if B is an ideal in C then Zorn's lemma provides an ideal $\tilde{B}$ of A maximal relative to $\tilde{B} \cap C \subset B$. If $\tilde{B}$ is a subset of A then one denotes the set of elements $bs^{-1}, b \in \tilde{B}$, in A_S by $\tilde{B}_S$. Then one has the following theorem.

Theorem 4. Let P be a prime ideal in C and let $\tilde{P}$ be an ideal of A maximal relative to $\tilde{P} \cap C \subset P$. Then: (i) $\tilde{P}$ is prime and $\tilde{P}_P$ is maximal in A_P. (ii) Put $\bar{A} = A/\tilde{P}$, $S = C - P$, $\bar{S} = (S + P)/P$. Then there is a canonical monomorphism $\bar{A} \to A_P/\tilde{P}_P$ which induces an isomorphism of $\bar{A}_{\bar{S}}$ onto $A_P/\tilde{P}_P$. Moreover, $\bar{A}_{\bar{S}}$ is canonically isomorphic with the algebra of central quotients of $\bar{A}$. (iii) $\tilde{P}$ is maximal in A if and only if the center of $A/\tilde{P}$ is a field, in which case $A/\tilde{P} \cong A_P/\tilde{P}_P$.

Rowen calls a central polynomial g for an algebra A <u>regular</u> if g is linear in one of its indeterminates. The Formanek polynomial g_n for $M_n(K)$ is regular. If A is a semiprime PI-algebra, A has a certain g_n as central polynomial. Hence any semiprime PI-algebra possesses regular central polynomials. We have the following theorem.

<u>Theorem 5</u>. If A (not assumed to be PI) has a regular central polynomial and has center C which is a field, then A is simple and finite dimensional over C.

It follows that if A has a regular central polynomial and every nonzero element of C is regular then A is prime.

The additive subgroup of C generated by the values taken on by the regular central polynomials of A is called the <u>central kernel</u> I of A ($I = 0$ if A has no regular central polynomials). An ideal $\tilde{B}$ of A (B of C) is called <u>identity faithful</u> if $\tilde{B} \not\supset I$ ($B \not\supset I$). This is equivalent to: There exists a regular central polynomial g for A which is not an identity for $A/\tilde{B}$. The main result on identity faithful ideals is a correspondence between identity faithful prime ideals of A and of C. This is contained in the theorem below.

<u>Theorem 6</u>. If $\tilde{P}$ is an identity faithful prime ideal of A then $P = \tilde{P} \cap A$ is an identity faithful prime ideal of C. On the other hand, if P is an identity faithful prime ideal of C and $\tilde{P}$ is an ideal of A which is maximal relative to $\tilde{P} \cap C \subset P$, then $\tilde{P}$ is prime and

(a) The center of A_P = C_P = central kernel of $A_P = I_P$.

(b) $P = \widetilde{P} \cap C$.

(c) $\widetilde{P}$ contains every ideal $\widetilde{B}$ of A such that $\widetilde{B} \cap C = P$.

Hence $\widetilde{P} \to P = \widetilde{P} \cap C$ is a bijective map of the set of identity faithful prime ideals of A and the set of identity faithful prime ideals of C .

(d) The map $\widetilde{P} \to P$ in (c) is bijective of the set of identity faithful maximal ideals of A on the set of identity faithful maximal ideals of C . Moreover, the inverse map is $P \to \widetilde{P} = AP$.

(e) For any identity faithful prime ideal $\widetilde{P}$ of A and $P = \widetilde{P} \cap A$, $A_P/\widetilde{P}_P$ is isomorphic to the classical algebra of quotients of $A/\widetilde{P}$.

It should be remarked that $\widetilde{P} = AP$ may not hold if P is prime and identity faithful in C . In fact, AP need not be a prime ideal of A . Also it can be shown that if P is a prime ideal of C and $\widetilde{P}$ is maximal relative to $\widetilde{P} \cap C \subset P$ then $\widetilde{P}$ identity faithful implies P identity faithful.

An algebra A is called <u>identity faithful</u> if every proper ideal in A is identity faithful. One has the following theorem.

<u>Theorem 7</u>. (a) A is identity faithful if and only if the central kernel I coincides with the center. (b) If P is an identity faithful prime ideal of C then A_P is identity faithful and has a unique maximal ideal $(PA)_P$. Moreover, $(PA)_P \cap C_P = P_P$. (c) Let $\widetilde{B}$ be an ideal in A and put $\overline{A} = A/\widetilde{B}$, $\overline{C} = (C + \widetilde{B})/\widetilde{B}$, $\overline{I} = (I + \widetilde{B})/\widetilde{B}$. Then if A is identity faithful or if $\overline{A}$ is identity faithful with central kernel $\overline{I}$,

we have $\bar{C}$ is the center of $\bar{A}$.

There are a number of important applications of the theory to semiprime PI-algebras. One of these is the following result which has been obtained independently by S. Steinberg.

<u>Theorem 8</u>. Let A be a semiprime PI-algebra with center C . Suppose C is an order in an algebra, $F_1 \oplus F_2 \oplus \cdots \oplus F_t$, F_i a field. Then A is a left and right order in its algebra of central quotients and this is of the form $S_1 \oplus \cdots \oplus S_t$ where S_i is simple with center F_i and is finite dimensional over F_i .

Let rad A denote the Jacobson radical of A . If A is semiprime but not semiprimitive, so rad A $\neq 0$, then rad A $\cap$ C $\neq 0$ by Theorem 1. Hence, rad C $\neq 0$ and so C is not semiprimitive. On the other hand, we have Theorem 9.

<u>Theorem 9</u>. If A is a semiprimitive PI-algebra then the center of A is semiprimitive.

The proof makes use of Theorem 6 (d) as well as considerations of radicals for rings without units (called "rngs").

We recall that a ring A is called an <u>Azumaya algebra</u> over its center C if A is projective as $A^e = A \otimes_C A^0$ module where A^0 denotes the opposite algebra of A and the module action of A^e on A is defined by $(r_1 \otimes r_2)r = r_1 r r_2$. If A is Azumaya then there is a one-to-one correspondence between the ideals of C and the ideals of A given by $B \to BA = \tilde{B}$ and $\tilde{B} \to \tilde{B} \cap C$. Moreover, $\bar{A} = A/\tilde{B}$ is Azumaya with center $\bar{C} = (C + \tilde{B})/\tilde{B}$. A is called Azumaya of <u>rank</u> n^2 if A is Azumaya and for every prime ideal P of C , A_P is free of rank n^2 over C_P . The theorem of Artin-Procesi gives a characterization of these rings by the two conditions: (i) A satisfies all the Z-identities as Z_n . (ii) S_{2n-2} is not an identity for any nonzero homomorphic image of A . Call a ring satisfying (i) and (ii) an A_n-<u>ring</u>. Rowen shows that the following conditions on a ring A are equivalent: (1) A is an A_n-ring, (2) the Formanek polynomial g_n (of matrix rings $M_n(K)$) is central for all nonzero homomorphic images of A , (3) if $I_n(A)$ denotes the additive subgroup of C generated by the values $g_n(r_1,\ldots),r_i \in A$, then $I_n = C$. Using these conditions and his results on identity faithful algebras, Rowen gives a new proof of the theorem of Artin and Procesi.

REFERENCES

1. Dehn, M., Math. Annalen 85(1922), 184-194.
2. Amitsur, S. A., Journal of Algebra 3(1966), 304-359.
3. Wagner, W., Math. Annalen 113(1937), 528-567.
4. Hall, M., Trans. A.M.S. 54(1943), 229-297.
5. Jacobson, N., Amer. J. Math. 67(1945), 300-320.
6. Jacobson, N., Trans. A.M.S. 57(1945), 228-245.
7. Jacobson, N., Ann. of Math. 46(1945), 695-707.
8. Levitzki, J., Bull. A.M.S. 52(1946), 1033-1035.
9. Kaplansky, I., Bull. A.M.S. 54(1948), 575-580.
10. Albert, A. A., Bull. A.M.S. 46(1940), 521-522.
11. Wedderburn, J. H. M., Trans. A.M.S. 22(1921), 129-135.
12. Amitsur, S. A. and J. Levitzki, Proc. A.M.S. 1(1950), 449-463.
13. Kaplansky, I., Trans. A.M.S. 68(1950), 62-75.
14. Malcev, A., Sbornik 13(55)(1943), 263-286.
15. Anitsur, S. A., Riveon Lematematika 5(1952), 41-48.
16. Amitsur, S. A., Proc. A.M.S. 2(1951), 538-540.
17. Levitzki, J., Proc. A.M.S. 1(1950), 334-341.
18. Posner, E. C., Proc. A.M.S. 11(1960), 180-183.
19. Goldie, A. W., Proc. London Math. Soc. (3)10(1960), 201-220.
20. Swan, R. G., Proc. A.M.S. 14(1963), 367-373, and Proc. A.M.S. 21(1969), 379-380.
21. Kostant, B., J. Math. Mech. 7(1958), 237-264.
22. Rowen, L. H., Trans. A.M.S., to appear.

23. Herstein, I. N., Journal of Algebra 6(1967), 369-375.

24. Martindale, W. S., Journal of Algebra 11(1969), 186-194.

25. Amitsur, S. A., Israel Jour. of Math. 7(1964), 63-68.

26. Regev, A., Israel Jour. of Math. 11(1972), 131-152.

27. Procesi, C. and L. Small, Math. Z. 106(1968), 178-180.

28. Klein, A. A. and A. Regev, Israel Jour. of Math. 12(1972), 421-426.

29. Formanek, E., Journal of Algebra 23(1972), 129-132.

30. Amitsur, S. A., Israel Jour. of Math. 12(1972), 408-422.

31. Amitsur, S. A., Journal of Algebra, to appear.

32. Rowen, L., Yale dissertation, 1973.

33. Rowen, L., Bull. A.M.S. 79(1973), 219-223.

34. Rowen, L., Journal of Algebra, to appear.

35. Amitsur, S. A., C. I. M. E. Conference, II Ciclo, 1965, 1-48.

36. Artin, M., Journal of Algebra 11(1969), 532-563.

37. Procesi, C., Journal of Algebra 22(1972), 307-316.

38. Amitsur, S. A., Proc. London Math. Soc. 17(1967), 470-486.

39. Amitsur, S. A., Israel Jour. of Math. 10(1971), 210-221.

Some Recent Developments in the Theory of PI-algebras

by N. JACOBSON, Yale University

Reprinted from *Universal Algebren Und Theorie Der Radikale*, 1976.

In these lectures we shall give an account of some recent work in the theory of PI-algebras. Our ultimate goal will be to sketch a proof of a recent application of the theory of PI-algebras: the construction of a finite dimensional central division algebra which is not a crossed product. This consists of two parts: a ring theory part and a field theory part. We shall concentrate mainly on the first, which can now be simplified substantially thanks to recent contributions to PI-algebras by FORMANEK, by ROWEN and others.

Unless otherwise stated "algebra" will mean associative algebra with unit over a commutative associative ring K with unit. Let X be a free monoid with a countable set of generators $x_1, x_2, \dots$ and let $K\{X\}$ denote the algebra with base X over K and multiplication determined by that of X. We call this the *free* algebra generated by the x's. It has the universal property that if A is any algebra and $a_1, a_2, \dots \in A$ then there exists a unique homomorphism of $K\{X\}$ into A sending $x_i \to a_i$, $i = 1, 2, 3, \dots$. If we denote an element of $K\{X\}$ as $f(x_1, \dots, x_m)$ its image under the indicated homomorphism is denoted as $f(a_1, \dots, a_m)$. f is called an *identity* for A if $f(a_1, \dots, a_m) = 0$ for all $a_i \in A$.

A polynomial f is called *blended* in x_i if x_i occurs in every monomial occurring in f (with non-zero coefficient). The *height* of a monomial is its degree minus the number of x's occurring in it and the height of a polynomial f is the heighest height of its monomials. A polynomial f is multilinear if it is blended in every x_i and of height 0.

Lemma 1. Any $f \in K\{X\}$ is a sum of blended monomials f_j such that (1) $\deg f_j \leqq \deg f$, $ht\, f_j \leqq ht\, f$ and (2) for any algebra A, f_j is an identity if f is an identity.

Suppose x_i occurs in f and x_j does not. Define $\Delta_{x_j}^{x_i} f = f(\dots, \overset{i}{x_i} + x_j, \dots) - f(\dots, x_i, \dots) - f(\dots, x_j, \dots)$.

Lemma 2. Let f be blended and $\deg_{x_i} f > 1$, $\deg_{x_j} = 0$. Then $\Delta_{x_j}^{x_i} f$ is blended, $\deg \Delta_{x_j}^{x_i} f \leqq \deg f$, $ht\, \Delta_{x_j}^{x_i} f < ht\, f$ and the coefficients in $\Delta_{x_j}^{x_i} f$ form a subset of those of f.

Call an identity *proper* if one of its coefficients α satisfies $\alpha A \neq 0$ (equivalently $\alpha 1 \neq 0$). The foregoing lemmas imply that if A has a proper identity it has a multilinear one of no higher degree. This is one of the steps in the proof of the following basic theorem which was first proved by Kaplansky for K a field and by Amitsur in the general case.

Theorem. Let A be a primitive algebra which has a proper identity of degree d. Then the center C of A is a field and $[A : C] \leq [d/2]^2$.

This theorem has a converse based on the

Theorem of AMITSUR and LEVITZKI. The matrix algebra $M_n(K)$ satisfies the *standard identity* S_{2n} where

$$S_m(x_1, \ldots, x_m) = \sum_{\pi} \operatorname{sg}(\pi)\, x_{\pi(1)}x_{\pi(2)} \cdots x_{\pi(d)}$$

π running over the symmetric group on n elements.

The two results together imply

Theorem of KAPLANSKY—AMITSUR—LEVITZKI. A primitive algebra A satisfies a proper identity if and only if A is simple and finite dimensional over its center. If d is the minimum degree for proper identities for A then $d = 2n$ is even and $[A : C] = n^2$, C the center. Moreover, A satiesfies the standard identity S_d.

In dealing with nil radicals it is preferable to work in the category of algebras without unit. We assume the basic definitions and results on the upper (or KOETHE) nil radical, the lower nil radical. The concept of an identity for an algebra without unit is clear. It is obtained simply by replacing $K\{X\}$ by $K\{X\}'$ which is the ideal in $K\{X\}$ of polynomials with 0 constant term. This has the universal property for algebras without unit that $K\{X\}$ has for algebras with unit.

We now state some results which are well known for algebras over fields. The proof in this case carry over to algebras over a commutative ring K provided we deal with identities thas are *strongly regular* in the sense that they are $\neq 0$ and their nonzero coefficients are units in K. This has the effect that if A satisfies such an identity then so does any homomorphic image (which would not always be the case for proper identities).

Theorem. A nil algebra (without unit) satisfying a strongly regular identity is locally nilpotent.

Theorem. Let A be an algebra (without unit) satisfying a strongly regular identity of degree d. Then any nil subalgebra B of A satisfies $B^{[d/2]} \subset N(0)$ the sum of the nilpotent ideals of A.

Theorem. Same hypotheses. Then the upper and lower nil radicals of A coincide.

A polynomial $f(x_1, \ldots, x_m)$ with 0 constant term is called *central* for an algebra A if $f(a_1, \ldots, a_m) \in C$, the center of A, and there exist a_i such that $f(a_1, \ldots, a_m) \neq 0$. In other words, $[f, x_{m+1}]$ is an identity for A but f is not. In 1936, WAGNER observed that $(x_1x_2 - x_2x_1)^2$ is a central polynomial for $M_2(K)$. In 1955, KAPLANSKY posed the question: do there exist central polynomials for $M_n(K)$ for any n? This remained unanswered until quite recently when it was proved by FORMANEK that this is indeed the case. We shall now indicate FORMANEK's results in a slightly more general form. Let $\mathbf{Z}(\eta_1, \ldots, \eta_{n+1}]$ be the polynomial algebra over $\mathbf{Z}$ in commuting indeterminates η_i and $\mathbf{Z}\{x, y_1, \ldots, y_n\}$ be the free algebra over $\mathbf{Z}$ generated by the noncommuting indeterminates x, y_j. If

$$f = \sum_{(\nu)} \alpha_{(\nu)}\eta_1^{\nu_1}\eta_2^{\nu_2} \cdots \eta_{n+1}^{\nu_{n+1}}, \quad (\nu) = (\nu_1, \ldots, \nu_{n+1})$$

we put

$$p_f = \sum_{(\nu)} \alpha_{(\nu)} x^{\nu_1} y_1 x^{\nu_2} y_2 \cdots x^{\nu_n} y_n x^{\nu_{n+1}} .$$

One obtains $p_f\left(\sum_1^n \varrho_i e_{ii}, e_{i_1 j_1}, e_{i_2 j_2}, \dots, e_{i_n j_n}\right) = \delta_{j_1 i_2} \delta_{j_2 i_3} \cdots \delta_{j_{n-1} i_n} f(\varrho_{i_1}, \dots, \varrho_{i_n}, \varrho_{j_n}) e_{i_1 j_n}$ so this is 0 for all sequences $(e_{i_1 j_1}, e_{i_2 j_2}, \dots, e_{i_n j_n})$ except those of the form $(e_{i_1 i_2}, e_{i_2 i_3}, \dots, e_{i_n j_n})$ in which case we get $f(\varrho_{i_1}, \dots, \varrho_{i_n}, \varrho_{j_n}) e_{i_1 j_n}$. Now suppose f satisfies: (i) f is divisible by every $\eta_i - \eta_j$ except $\eta_1 - \eta_{n+1}$, (ii) $g(\eta_1, \dots, \eta_n) \equiv f(\eta_1, \dots, \eta_n, \eta_1)$ is symmetric. An f satisfying these conditions is FORMANEK's polynomial

$$c = \prod_{i=2}^{n} (\eta_1 - \eta_i)(\eta_{n+1} - \eta_i) \prod_{i<j=2}^{n} (\eta_i - \eta_j)^2 .$$

If K is a field and $a \in M_n(K)$ has the characteristic roots $\varrho_1, \dots, \varrho_n$ in the algebraic closure $\bar{K}$ of K then we define $G(a) = g(\varrho_1, \dots, \varrho_n) \in K$. For FORMANEK's c we have $d(\eta_1, \dots, \eta_n) = c(\eta_1, \dots, \eta_n, \eta_1) = \prod_{i<j=1}^{n} (\eta_i - \eta_j)^2$ and it is easily seen that we have the following condition: (iii) for any field K there exist $a \in M_n(K)$ such that $G(a) \neq 0$.

Theorem (FORMANEK). If $f \in Z[\eta_1, \dots, \eta_{n+1}]$ satisfies the conditions (i), (ii) and (iii) then

$$q_f(x, y_1, \dots, y_n) = \sum_0^{n-1} p_f(x, y_{i+1}, y_{i+2}, \dots, y_{i+n})$$

(where the subscripts are reduced mod n) is a central polynomial for $M_n(K)$ for any commutative ring K.

Let A be a finite dimensional central simple algebra over K, K now a field. If E/K is a splitting field for A we have $A^E = E \otimes_K A \cong M_n(E)$. It follows from this that $q_f(x, y_1, \dots, y_n)$ is a central polynomial for A.

This result will play an important role in the sequel. Another important tool we shall need is commutative localization. Let S be a submonoid of the multiplicative monoid of K. Then we have the ring K_S, the *localization of K at S*, whose elements have the form $s^{-1}k$, $s \in S$, $k \in K$. If A is an algebra over K then we obtain $A_S = A^{K_S} = K_S \otimes_K A$ which can be regarded as an algebra over K or over K_S. If no element of S is a zero divisor in A ($sa \neq 0$ if $a \neq 0$) then $a \to 1^{-1}a$ is an embedding of A in A_S.

Theorem (ROWEN). Any identity f of A is an identity for A_S and the converse holds if every element of S is regular (= not a zero divisor) for A.

Now let A be prime, C the center of A. Then C is a domain and every element $\neq 0$ in C is regular for A. We can regard A as algebra over C and form A_0 the localization of A at $C - \{0\}$. We call this the algebra of *central quotients of A*. This is prime and it can be used to prove

Theorem (AMITSUR). If A is a prime algebra satisfying a proper identity of degree n then A satisfies the standard identity $S_{2[n/2]}$.

This result and the one we noted before on algebras satisfying strongly regular identities implies that if A is prime or more generally is a subdirect product of prime algebras satisfying proper identities then A has no nil ideals $\neq 0$ so the polynomial algebra $A[\lambda]$ is semi-primitive. This and the existence of central polynomials for finite dimensional central simple algebras can be used to prove the following, key

Theorem (ROWEN). Let A be a subdirect product of prime algebras satisfying proper identities such that the degree of these identities is bounded. Then any non-zero ideal I of A has non-zero intersection with the center C of A.

A corollary of this is that if the center of A is a field then A is simple, hence, finite dimensional over its center, by the KAPLANSKY-AMITSUR theorem. This implies the main structure theorem on prime algebras satisfying polynomial identities:

Theorem (POSNER—ROWEN et al.). Let A be a prime algebra satisfying a proper identity. Then:

1. The algebra A_0 is finite dimensional simple over its center which is the field of fractions of the center of A.
2. A is a left and right order in A_0.
3. A and A_0 satisfy the same identities.

There does not seem to be agreement on how one should define PI-algebras over commutative rings. The definition we shall use is one which has been proposed by ROWEN: A is a *PI-algebra* if there exists a polynomial $f(x_1, \dots, x_m) \in K\{X\}$ which is a proper identity for every non-zero homomorphic image of A. If A is a PI-algebra in this sense then so are the homomorphic images of A, subalgebras of A and A^L for any commutative algebra L over K. The first is immediate and the second two follow from

Theorem (AMITSUR). A is a PI-Algebra if and only if A satisfies an identity S_{2n}^m for some n and m.

Let A be a PI-algebra and let $I = I(A)$ be the set of identities of A. I is a *T-ideal* in $K\{X\}$, that is, an ideal which is stabilized by every endomorphism of A. Also I contains S_{2n}^m for some m and n. This suggests the definition: A *universal PI-algebra* is an algebra of the form $U = U_I = K\{X\}/I$ where I is a T-ideal in $K\{X\}$ containing S_{2n}^m for some m and n. This is a PI-algebra.

Theorem (AMITSUR—ROWEN). The JACOBSON radical of a universal PI-algebra is a nilideal.

We now specialize to the case in which $I = I_n$ the T-ideal of identities of $M_n(K)$. We shall study this more closely and indicate how it can be used to settle an old question on finite dimensional algebras: the existence of finite dimensional central division algebras which are not crossed products. From now on *K will be an infinite field.*

Lemma. If f is an identity for a finite dimensional algebra A over K then f is an identity for A^L for any extension field L of K.

We now introduce the polynomial algebra $\Xi = K[\xi_{ij}^{(k)}]$ in a countable set of commutating indeterminates $\xi_{ij}^{(k)}$, $1 \leq i, j \leq n$, $k = 1, 2, 3, \dots$ and the matrix algebra $M_n(\Xi)$. The subalgebra (over K) of $M_n(\Xi)$ generated by the matrices $\xi^{(k)} = (\xi_{ij}^{(k)})$, $k = 1, 2, \dots$ is called the *generic matrix algebra of degree n over K*. This is essential the same thing as the universal PI-algebra $K\{X\}/I_n$. More precisely we have

Theorem. Let $K\{\bar{X}\} = K\{X\}/I_n$ the universal PI-algebra over the infinite field K defined by the T-ideal I_n of identities of $M_n(K)$. If $u \in K\{X\}$ write $\bar{u} = u + I_n$. Then we have an isomorphism of $K\{\bar{X}\}$ onto the generic matrix algebra $K\{\xi\}$ sending $\bar{x}_k \to \xi^{(k)} = (\xi_{ij}^{(k)})$.

Lemma. Let Λ denote the field of fractions of the domain Ξ. Then $\Lambda K\{\xi\} = M_n(\Lambda)$.

Lemma. For any n there exists a division algebra A of dimensionality n^2 over its center F where F is a countable extension field of K.

Using these two lemmas one proves

Theorem. (Amitsur). The universal PI-Algebra $K\{\overline{X}\} = K\{X\}/I_n$ is a domain.

Now let UD (K, n) denote the algebra of central quotients of $K\{X\}$. We shall call this the *universal division algebra of index n* over K. The center is the field of fractions of the center of $K\{\overline{X}\}$. It is easily seen that the latter consists of 0 and the elements $f(\bar{x}_1, \ldots, \bar{x}_m)$ where $f(x_1, \ldots, x_m)$ is a central polynomial for $M_n(K)$. Hence we have

Lemma. The center of $K\{\overline{X}\}$ consists of 0 and the elements $f(\bar{x}_1, \ldots, \bar{x}_m)$ such that $f(x_1, \ldots, x_m)$ is a central polynomial for $M_n(K)$. The center of $UD(K, n)$ is the set of elements $g(\bar{x}_1, \ldots, \bar{x}_m)\, f(\bar{x}_1, \ldots, \bar{x}_m)^{-1}$ where f is a central polynomial and g is another one or 0.

One has the following important homomorphism property of $K\{\overline{X}\}$.

Theorem. Let $K\{\overline{X}\}$ be the universal PI-algebra $K\{X\}/I_n$ and let A be a central simple algebra of n^2 dimensions over an extension field L of K. The given any sequence of elements $a_1, a_2, a_3, \ldots$ of A there exists a unique K-algebra homomorphism of $K\{\overline{X}\}$ into A sending $\bar{x}_i \to a_i$, $i = 1, 2, \ldots$. Moreover, under any K-algebra homomorphism of $K\{\overline{X}\}$ into A the center C of $K\{\overline{X}\}$ is mapped into L and given any finite set $\{f_1, \ldots, f_r\}$ of non-zero elements of $K\{\overline{X}\}$ there exists a K-algebra homomorphism η of $K\{\overline{X}\}$ such that ηf_j is invertible in A for $1 \leqq j \leqq r$.

One can use this to prove

Theorem (Amitsur). If $UD(K, n)$ contains a maximal subfield M which is Galois over its center F and $G = \mathrm{Gal}\, M/F$ then every central division algebra Δ of dimensionality n^2 over an extension field L of K possesses a maximal subfield which is Galois over L and has Galois group isomorphic to G. By using this and some field theory especially valuation theory one can prove

Theorem. If the characteristic is not a divisor of n but n is divisible by the cube of some prime, then $UD(K, n)$ is not a crossed product.

For the case of $K = Q$ one has the stronger result

Theorem (Amitsur). If n is divisible by 8 or by the square of an odd prime then $UD(Q, n)$ is not a crossed product.

References

[1] Amitsur, S. A.: Israel J. Math. **7** (1964), 63–68.
[2] Amitsur, S. A.: Israel J. Math. **10** (1971), 210–221.
[3] Amitsur, S. A., J. Levitzki: Proc. Amer. Math. Soc. **1** (1950), 449–463.
[4] Jacobson, N.: Structure of Rings, AMS Colloquium Publ., Providence, R. I., 1956, 2nd ed. 1963.
[5] Kaplansky, I.: Bull. Amer. Math. Soc. **54** (1948), 575–580.
[6] Posner, E. C.: Proc. Amer. Math. Soc. **11** (1960), 180–183.
[7] Rowen, L.: Yale dissertation, New Haven, 1973.
[8] Wagner, W.: Math. Ann. **113** (1937), 528–567.

PROC. 18th SRI,
CANBERRA 1978, 8-46

16-02
(16A38)

SOME RECENT DEVELOPMENTS IN THE THEORY OF ALGEBRAS WITH POLYNOMIAL IDENTITIES

N. Jacobson

The author gave six lectures at the Summer Research Institute: two survey lectures and four lectures on algebras with polynomial identities. The survey lectures had the titles: *Development of the Concept of a Jordan Structure* and *History of Algebras with Polynomial Identity*. The author decided not to include these lectures in the Proceedings since these would have overlapped substantially with surveys that have appeared or are about to appear, notably, a survey article on Jordan algebras by McCrimmon that will appear shortly in the Bulletin of the American Mathematical Society, a survey lecture "*PI*-algebras" by the author appearing in the Proceedings of a ring theory conference at University of Oklahoma published in 1973 by M. Dekker, and a survey article "Polynomial identities" by S.A. Amitsur in Israel Journal of Mathematics, 19 (1974), 183-199.

The lectures on recent developments in the *PI*-theory have been written up with the following titles:

I Razmyslov's central polynomial,

II The Artin-Procesi Theorem,

III On Shirshov's local finiteness theorems.

Central polynomials for the complete matrix algebra $M_n(K)$, K a commutative ring, have played an important role in recent developments of the *PI*-theory. For example, they can be used to give an improved formulation and proof of the main structure theorem for prime *PI*-algebras. The first construction of central polynomials for

PI-algebras was given in 1972 by Formanek. In Chapter I we present an alternative construction due to Razmyslov that has the advantage that the central polynomial is multilinear and is alternating in n^2 of the arguments. Razmyslov's central polynomial plays an important role in II that is concerned with a theorem connecting *PI*-algebras with Azumaya algebras.

In III we give an account (with some improvements) of the best positive results that have appeared to date on a problem in ring theory, Kurosch's problem, that is analogous to Burnside's problem in group theory.

CHAPTER I

RAZMYSLOV'S CENTRAL POLYNOMIAL

Let A be an algebra (associative with unit) over a commutative ring K and let $K\{x_1, x_2, \ldots\}$ be the free associative algebra generated by $x_1, x_2, \ldots$. An element $f(x_1, \ldots, x_m) \in K\{x_1, x_2, \ldots\}$ is called a *central polynomial* for A if $f(a_1, \ldots, a_m) \in C = C(A)$ the center of A for all $a_i \in A$ and there exist $a_1, \ldots, a_m$ such that $f(a_1, \ldots, a_m) \neq 0$. In other words,

$$[f(x_1, \ldots, x_m), x_{m+1}]$$

is an identity for A but $f(x_1, \ldots, x_m)$ is not. We shall give a construction due to Razmyslov, of a multilinear central polynomial for $M_n(K)$, the algebra of $n \times n$ matrices over K .

We require first some elementary results on $M_n(K)$.

LEMMA 1. *The trace bilinear form* $t(A, B) = \operatorname{tr} AB$ *is non-degenerate on* $M_n(K)$: $t(A, B) = 0$ *for all* B *implies* $A = 0$.

Proof. If $A = (a_{ij})$ and $\{e_{kl} \mid 1 \leq k, l \leq n\}$ are the usual matrix units for $M_n(K)$, then $t(A, e_{kl}) = a_{lk}$. Hence $t(A, e_{kl}) = 0$ for all k, l implies $A = 0$. □

$M_n(K)$ is a free module with base of n^2 elements. Hence any K-endomorphism of $M_n(K)$ determines an $n^2 \times n^2$ matrix relative to a base. The trace of this matrix is independent of the choice of base and is called the *trace of the endomorphism*. In particular, we consider the endomorphism $U \mapsto AUB$ defined by a given pair of matrices A and B. Then we have

LEMMA 2. *The trace of the endomorphism* $U \mapsto AUB$ *of* $M_n(K)$ *is* $(\operatorname{tr} A)(\operatorname{tr} B)$.

This can be verified by direct calculation using the base $\{e_{ij}\}$ of matrix units. We omit this simple calculation.

We note next that if A is any algebra and $A^e = A \otimes_K A^o$ where A^o is the opposite algebra then we have the involution $\sum a_i \otimes b_i \mapsto \sum b_i \otimes a_i$ in A^e. Moreover, A is an A^e-module relative to the action $\left(\sum a_i \otimes b_i\right)x = \sum a_i x b_i$. This action is by K-endomorphisms. For a certain class of algebra, the Azumaya algebras, the representation defined by this action is faithful and the image is $\operatorname{End}_K A$. We can verify this for $A = M_n(K)$:

LEMMA 3. *The map sending* $\sum a_i \otimes b_i$ *into the endomorphism* $x \mapsto \sum a_i x b_i$ *of* $M_n(K)$ *into itself is an isomorphism onto* $\operatorname{End}_K M_n(K)$.

Proof. Since both $M_n(K) \otimes M_n(K)^o$ and $\operatorname{End}_K M_n(K)$ are free K-modules of rank n^4 it suffices to show that the homomorphism is surjective. For this it is enough to show that for any pair $(i, j), (k, l)$ there exists a map of the form $U \mapsto \sum A_i U B_i$ sending e_{ij} into e_{kl} and $e_{i'j'}$ into 0 for every $(i', j') \neq (i, j)$. It is clear that $U \mapsto e_{ki} U e_{jl}$ does this. □

In the situation in which we have an isomorphism of A^e onto $\operatorname{End}_K A$ defined as above we can transfer the involution $\sum a_i \otimes b_i \mapsto \sum b_i \otimes a_i$ to an involution in $\operatorname{End}_K A$. We denote this involution as $l \mapsto l^*$.

Now let $X = K[\xi_{11}, \ldots, \xi_{ij}, \ldots, \xi_{nn}]$ where the ξ_{ij} are indeterminates and consider $M_n(X)$ which is the same thing as $X \otimes_K M_n(K)$. We can regard $M_n(X)$ as $M_n(K)^e$-module obtained by restricting the action from $M_n(X)^e$ to the subalgebra $M_n(K)^e$. Consider the "generic" matrix $X = (\xi_{ij}) \in M_n(X)$ and the cyclic $M_n(K)^e$-module generated by X . This is a free K-module with base $\{\xi_{ij}e_{kl} \mid i, j, k, l = 1, \ldots, n\}$. If $l_1, l_2 \in M_n(K)^e$ and $l_1(X) = l_2(X)$ then $l_1(e_{ij}) = l_2(e_{ij})$ for all i, j and hence $l_1 = l_2$. It follows that we have a well defined K-endomorphism A_X of $M_n(K)^eX$ such that

(1) $$A_X : l(X) \mapsto l^*(X) .$$

We call this the *Razmyslov transposition* in $M_n(K)^eX$. Since

$$\xi_{ij}e_{kl} = e_{ki}Xe_{jl}$$

we have

(2) $$A_X\xi_{ij}e_{kl} = e_{jl}Xe_{ki} = \xi_{lk}e_{ji} .$$

Hence the Razmyslov transposition can also be defined directly as the map

(3) $$\sum a_{ijkl}\xi_{ij}e_{kl} \mapsto \sum a_{ijkl}\xi_{lk}e_{ji}$$

in $M_n(K)^eX$.

LEMMA 4. *If* $H \in M_n(K)$ *then* $(\mathrm{tr}\ X)H \in M_n(K)^eX$ *and*

(4) $$A_X(\mathrm{tr}\ X)H = \mathrm{tr}(XH)1 .$$

Proof. Since $\mathrm{tr}\ X = \sum \xi_{ii}$, $(\mathrm{tr}\ X)H$ is a K-linear combination of the elements $\xi_{ii}e_{kl}$ and so is contained in $M_n(K)^eX$. Write $H = \sum h_{jk}e_{jk}$, $h_{jk} \in K$. Then $(\mathrm{tr}\ X)H = \sum_{i,j,k} h_{jk}\xi_{ii}e_{jk}$. Hence, by (2),

$$A_X(\mathrm{tr}\ X)H = \sum_{i,j,k} h_{jk}\xi_{kj}e_{ii} = (\mathrm{tr}\ HX)1$$
$$= \mathrm{tr}(XH)1 . \qquad \square$$

We now consider the free algebra $K\{x_1, x_2, \ldots; y_1, y_2, \ldots\}$ generated by $x_1, x_2, \ldots; y_1, y_2, \ldots$. Let $h(x_1, \ldots, x_r, y_1, \ldots, y_s)$ be multilinear and alternating in the x's . Let $A = (a_{ij}) \in M_r(K)$ and $x_i' = \sum a_{ij}x_j$. Then direct substitution shows that

(5) $$h(x_1', \ldots, x_r', y_1, \ldots, y_s) = \det(A)h(x_1, \ldots, x_r, y_1, \ldots, y_s) .$$

Let λ be an indeterminate and replace K by $K[\lambda]$, A by $\lambda 1 - A$. Then comparing coefficients of the powers of λ in the relation obtained from (5) we obtain a number of relations like (5). In particular, taking the coefficient of λ^{n-1} we obtain

$$(6)\quad h(x_1', x_2, \ldots, x_r, y_1, \ldots, y_s) + h(x_1, x_2', x_3, \ldots, x_r, y_1, \ldots, y_s) + \ldots + h(x_1, \ldots, x_{r-1}, x_r', y_1, \ldots, y_s) = \operatorname{tr}(A)h(x_1, \ldots, x_r, y_1, \ldots, y_s) .$$

This holds in particular for the *Capelli polynomial*

$$(7)\qquad h(x_1, \ldots, x_{n^2}, y_1, \ldots, y_{n^2-1}) = \sum_{\pi\in\Sigma_{n^2}} (\operatorname{sg} \pi)x_{\pi 1}y_1x_{\pi 2}y_2 \cdots x_{\pi n^2}$$

of degree $2n^2 - 1$. For this we have

LEMMA 5. *If* $(A_1, \ldots, A_{n^2})$ *is a base for* $M_n(K)$ *then we can choose* $(B_1, \ldots, B_{n^2-1})$ *so that*

$$(8)\qquad h(A_1, \ldots, A_{n^2}, B_1, \ldots, B_{n^2-1}) \neq 0 .$$

Proof. Using (5) it suffices to prove the result for the base $\{e_{ij}\}$. It is clear that for any ordering of this base, say, as $A_1 = e_{ij}, \ldots, A_{n^2} = e_{kl}$ there is only one choice of $B_1, \ldots, B_{n^2-1}$ in $\{e_{ij}\}$ so that $A_1B_1A_2B_2 \cdots B_{n^2-1}A_{n^2} \neq 0$. Then the product is e_{il} and $h(A_1, A_2, \ldots, A_{n^2}, B_1, \ldots, B_{n^2-1}) = e_{il} \neq 0$. □

LEMMA 6. *Let* h *be the Capelli polynomial of degree* $2n^2 - 1$ *and let* $A_1, \ldots, A_{n^2}, B_1, \ldots, B_{n^2-1}, C, D \in M_n(K)$. *Then*

$$(9)\quad h(CA_1D, A_2, \ldots, A_{n^2}, B_1, \ldots, B_{n^2-1}) + h(A_1, CA_2D, A_3, \ldots, A_{n^2}, B_1, \ldots, B_{n^2-1}) + \ldots + h(A_1, \ldots, A_{n^2-1}, CA_{n^2}D, B_1, \ldots, B_{n^2-1}) = (\operatorname{tr} C)(\operatorname{tr} D)h(A_1, \ldots, A_{n^2}, B_1, \ldots, B_{n^2-1}) .$$

Proof. Since the relation to be proved is multilinear in all of the variables $A_1, \ldots, A_{n^2}, B_1, \ldots, B_{n^2-1}, C, D$ and since $M_n(K) \cong K \otimes_{\mathbb{Z}} M_n(\mathbb{Z})$, it suffices to prove (9) for $M_n(\mathbb{Z})$. Now (9) will follow for $M_n(\mathbb{Z})$ if we can prove it for $M_n(\mathbb{Q})$. Hence it suffices to take $K = \mathbb{Q}$. Using the Zariski topology on $M_n(\mathbb{Q})$ it suffices to prove (9) on the Zariski open subset on which $(A_1, \ldots, A_{n^2})$ is a base. (This is defined by the non-vanishing of the determinant of the $n^2 \times n^2$ matrix expressing the A_i in terms of the base $\{e_{ij}\}$.) Consider the linear transformation $U \mapsto CUD$ of $M_n(\mathbb{Q})$. We obtain the matrix of this linear transformation by writing

(10) $$CA_iD = \sum_{j=1}^{n^2} p_{ij}A_j \ , \quad 1 \le i \le n^2 \ .$$

By (6), the left hand side of (9) equals $\mathrm{tr}(p)h(A_1, \ldots, A_{n^2}, B_1, \ldots, B_{n^2-1})$ where $(p) = (p_{ij})$. By Lemma 2, $\mathrm{tr}(p) = (\mathrm{tr}\ C)(\mathrm{tr}\ D)$. Hence (9) holds. □

We can now define Razmyslov's central polynomial.

THEOREM 1 (Razmyslov). *Let* h *be the Capelli polynomial of degree* $2n^2 - 1$ *and let* H *be the polynomial obtained from*

(11) $$h(y_{n^2}x_1y_{n^2+1}, x_2, \ldots, x_{n^2}, y_1, \ldots, y_{n^2-1}) + h(x_1, y_{n^2}x_2y_{n^2+1}, \ldots, x_{n^2}, y_1, \ldots, y_{n^2-1}) + \ldots + h(x_1, \ldots, x_{n^2-1}, y_{n^2}x_{n^2}y_{n^2+1}, y_1, \ldots, y_{n^2-1})$$

by "pivoting on y_{n^2} *": replacing every monomial* $wy_{n^2}z$ *occurring in* (11) *by* $zy_{n^2}w$. *Then for any* $A_1, \ldots, A_{n^2}, B_1, \ldots, B_{n^2+1} \in M_n(K)$ *we have the relation*

(12) $$H(A_1, \ldots, A_{n^2}, B_1, \ldots, B_{n^2+1}) = \mathrm{tr}\ B_{n^2+1}\ \mathrm{tr}(B_{n^2}h(A_1, \ldots, A_{n^2}, B_1, \ldots, B_{n^2-1}))1 \ .$$

Moreover, H *is a central polynomial for* $M_n(K)$.

Proof. We consider $M_n(X)$ where $X = K[\xi_{11}, \ldots, \xi_{ij}, \ldots, \xi_{nn}]$ and let $X = (\xi_{ij})$. By (9), we have

(13) $$h(XA_1B_{n^2+1}, A_2, \ldots, A_{n^2}, B_1, \ldots, B_{n^2-1}) + \ldots + h(A_1, \ldots, A_{n^2-1}, XA_{n^2}B_{n^2+1}, B_1, \ldots, B_{n^2-1}) = (\mathrm{tr}\ B_{n^2+1})(\mathrm{tr}\ X)h(A_1, \ldots, A_{n^2}, B_1, \ldots, B_{n^2-1}) \ .$$

Next we apply the Razmyslov transposition A_X to both sides of this equation. By (4) we obtain

$$H(A_1, \ldots, A_{n^2}, B_1, \ldots, B_{n^2-1}, X, B_{n^2+1}) = \mathrm{tr}\ B_{n^2+1}\ \mathrm{tr}(Xh(A_1, \ldots, A_{n^2}, B_1, \ldots, B_{n^2-1}))1 .$$

Specializing $X \mapsto B_{n^2}$ gives (12). It is clear from (12) that $H(A_1, \ldots, A_{n^2}, B_1, \ldots, B_{n^2+1}) \in K1$ for all choices of the A_i, B_j . It remains to show that the A_i, B_j can be chosen so that $H(A_1, \ldots, A_{n^2}, B_1, \ldots, B_{n^2+1}) \ne 0$. To do this we choose B_{n^2+1} so that $\mathrm{tr}\ B_{n^2+1} = 1$, $(A_1, \ldots, A_{n^2})$ to be a base for $M_n(K)$. Then, by Lemma 5, we can choose $B_1, \ldots, B_{n^2-1}$ so that

$h(A_1, \ldots, A_{n^2}, B_1, \ldots, B_{n^2-1}) \neq 0$. By Lemma 1, we can choose B_{n^2} so that $\operatorname{tr} B_{n^2}h(A_1, \ldots, A_{n^2}, B_1, \ldots, B_{n^2-1}) \neq 0$. These choices give a non-zero value for H . □

We shall now extend Theorem 1 to prime algebras satisfying proper identities, where a polynomial $f(x_1, \ldots, x_m)$ is called *proper* for the algebra A if for some coefficient α in $f(x_1, \ldots, x_m)$ we have $\alpha A \neq 0$ or, equivalently, $\alpha 1 \neq 0$, for the unit 1 of A . We define the *PI-degree* of A to be the least integer d for which there exists a proper identity of degree d . If A is a prime algebra satisfying a proper identity and C is the center of A then C is a domain and, by the structure theorem for prime *PI*-algebras, if F is the field of fractions of C then $A \hookrightarrow A_F = F \otimes_C A$ and A_F is finite dimensional central simple over F . Moreover, A and A_F have the same identities and $[A_F : F] = n^2$ (Jacobson [1], p. 57). If $\bar{F}$ is the algebraic closure of F then $A_{\bar{F}} = \bar{F} \otimes_C A \cong \bar{F} \otimes_F A_F \cong M_n(\bar{F})$. By the Amitsur-Levitzki Theorem S_{2n} is an identity for $A_{\bar{F}}$ and since $A \hookrightarrow A_{\bar{F}}$, S_{2n} is an identity for A . On the other hand, if f is a proper identity of degree less than $2n$ for A then A satisfies a proper multilinear identity of degree less than $2n$. Then $M_n(\bar{F}) \cong A_{\bar{F}}$ satifies this identity. This is impossible (Jacobson [1], p. 16). Hence, we see that the *PI*-degree of A is $2n$ if $[A_F : F] = n^2$.

We now prove

THEOREM 2. *If* A *is a prime algebra of PI-degree* $2n$ *and* C *is the center of* A *then* A *and* $M_n(C)$ *have the same identities and the same central polynomials.*

Proof. Since $f(x_1, \ldots, x_m)$ is a central polynomial if and only if it is not an identity but $[f(x_1, \ldots, x_m), x_{m+1}]$ is an identity the second statement follows from the first. Since A_F and A have the same identities and proper identities and $M_n(C)$ and $M_n(F)$ have the same identities and proper identities it suffices to prove that this is the case also for A_F and $M_n(F)$. If $\alpha \in K$ then $\alpha 1 \in C \subset F$. Hence any proper identity for A yields a non-zero identity with coefficients in F for A_F . Hence it suffices to assume $K = F$ in considering A_F and $M_n(F)$. We distinguish two cases:

I F finite and

II F infinite.

In the first case, by Wedderburn's Theorem on finite division rings, $A_F = M_n(F)$. Hence the result holds in this case. Now assume F infinite. We prove first that if

A is a finite dimensional algebra over an infinite field F and E is an extension field of F then A and $A_E = E \otimes_F A$ have the same identities over F. Let $(a_1, \ldots, a_r)$ be a base for A/F, hence for A_E/E. Let $f(x_1, \ldots, x_m) \in F\{x_1, x_2, \ldots\}$ and let $\xi_j^{(i)}$, $1 \le i \le m$, $1 \le j \le r$, be indeterminates. Put $X = F\left[\xi_1^{(1)}, \ldots, \xi_r^{(1)}, \ldots, \xi_r^{(m)}\right]$ and consider the algebra A_X and the "generic" elements $x^{(i)} = \sum_j \xi_j^{(i)} a_j$ of this algebra. Since $(a_1, \ldots, a_r)$ is a base for A_X/X we can write $f(x^{(1)}, \ldots, x^{(m)}) = \sum \varphi_l\left(\xi_1^{(1)}, \ldots, \xi_r^{(m)}\right) a_l$. Then f is an identity for A if and only if $\varphi_l\left(\alpha_1^{(1)}, \ldots, \alpha_r^{(m)}\right) = 0$ for all $\alpha_j^{(i)} \in F$ and all $l = 1, 2, \ldots, m$. Since F is infinite this occurs if and only if every $\varphi_l = 0$. Since $(a_1, \ldots, a_r)$ is a base for A_E/E the same conditions obtain for A_E. Hence A and A_E have the same F-identities. We apply this to A_F, A as before, taking $E = \overline{F}$. This shows that A_F and $M_n(\overline{F})$ have the same F-identities. Similarly $M_n(F)$ and $M_n(\overline{F}) \cong M_n(F)_{\overline{F}}$ have the same F-identities. Hence this holds also for A_F and $M_n(F)$. □

An immediate consequence of this result and Theorem 1 is the following.

COROLLARY. *If* A *is a prime algebra of PI-degree* $2n$ *then the Razmyslov polynomial of degree* $2n^2 + 1$ *is a central polynomial for* A.

References

Nathan Jacobson

[1] *PI-Algebras. An Introduction* (Lecture Notes in Mathematics, 441. Springer-Verlag, Berlin, Heidelberg, New York, 1975).

Ju.P. Razmyslov

[1] "Trace identities of full matrix algebras over a field of characteristic zero", *Math. USSR Izv.* 8 (1974), no. 4., 727-760.

CHAPTER II

THE ARTIN-PROCESI THEOREM

In 1969, Michael Artin [1] established a surprising connection between two important areas of ring theory: Azumaya algebras and algebras with polynomial identity. He gave a characterization by identities of the most interesting class of Azumaya algebras, the Azumaya algebras that have a rank.

There are a number of equivalent definitions of an Azumaya algebra that can be given. The usual one is: An algebra A over a commutative ring K is called an *Azumaya algebra* if A regarded in the natural way as $A^e = A \otimes_K A^o$ module is projective and the map $\alpha \mapsto \alpha 1$ of K into A is an isomorphism onto the center $C(A)$. If the first condition alone holds then the algebra is called *separable*. The second condition states that $K1$ is the center of A and $\alpha 1 = 0$ for $\alpha \in K$ implies $\alpha = 0$. If this condition holds then A is called *central*. Thus central plus separable equals Azumaya.

If K is a field it can be shown that A is separable over K in the above sense if and only if A is classically separable, that is, A is finite dimensional over K and $A_{\overline{K}} = \overline{K} \otimes_K A$ is semi-simple for $\overline{K}$ the algebraic closure of K. Moreover, A is Azumaya over a field K if and only if A is finite dimensional central simple over K. Thus the theory of separable algebras over commutative rings and the theory of Azumaya algebras constitute natural generalizations of the classical theories of finite dimensional separable and central simple algebras.

We need to recall the definition of rank of a finitely generated projective module M over a commutative ring K. Let P be a prime ideal in K and let K_P be the localization of K at P. This is a local ring, that is, it has a unique maximal ideal P_P. Moreover, we can form the localization M_P of M at P, that can be defined as $K_P \otimes_K M$. This is finitely generated projective over K_P and since K_P is local it follows that M_P is a free module of finite rank n_P over K_P. We say that M *has a rank over* K if n_P is constant as P ranges over the set of prime ideals of K. In this case $n = n_P$ is called *the rank* of the finitely generated projective module M over K.

It can be shown that the rank is defined if K contains no idempotents not equal to $0, 1$. More generally, it can be shown that if M is finitely generated projective over K then there are only a finite number of distinct ranks n_P for P a prime ideal in K and if these ranks are $n_1, \ldots, n_s$ then $K = K_1 \oplus \ldots \oplus K_s$ where the K_i are ideals and $M = K_1 M \oplus \ldots \oplus K_s M$ where $K_i M$ has rank n_i over K_i. It can be shown also that if M has a rank n over K and K' is a commutative algebra over K then $M_{K'} = K' \otimes_K M$ has rank n over K'.

It can be shown that if A is an Azumaya algebra over K then A is finitely generated projective over K. Moreover, if K' is a commutative algebra over K then $A_{K'}$ is Azumaya. Let P be a prime ideal of K. Then $A_P = K_P \otimes_K A$ is Azumaya over K_P and its rank over K_P is the same as that of $(K_P/P_P) \otimes_{K_P} A_P$ over K_P/P_P. Since K_P/P_P is a field and $(K_P/P_P) \otimes_{K_P} A_P$ is Azumaya over K_P/P_P its rank is the dimensionality and this is a square. It follows that the rank n_P of A_P over K_P is a square. Hence if A has a rank over K then this is a square. It can be shown also that if A is a ring that is Azumaya over its center C then A has a unique decomposition as direct sum of a finite number of rings A_i that are Azumaya over their centers and have ranks over their centers (Bix [1]).

The theorem of Artin-Procesi that we shall prove characterizes Azumaya algebras that have a rank.

DEFINITION 1. A ring A is called an *A_n-ring* if A (regarded as an algebra over $\mathbb{Z}$) satisfies every identity of $M_n(\mathbb{Z})$ but no non-zero homomorphic image of A satisfies every identity of $M_{n-1}(\mathbb{Z})$.

Then we have the

ARTIN-PROCESI THEOREM. *A ring A is an A_n-ring if and only if it is an Azumaya algebra of rank n^2 over its center.*

This was proved by Artin under the hypothesis that A is an algebra over a field. This restriction was removed by Procesi. After central polynomials became available simpler proofs of the result were given by Amitsur, by Rowen, by Goldie and finally by Schelter. We shall give Schelter's proof which, it should be noted, was confined to proving A_n implies Azumaya for prime rings. While the general case can be reduced to that of prime rings, it is simpler to avoid this reduction and prove directly that A_n implies Azumaya for arbitrary A. We shall do this. The implication that if A is Azumaya of rank n^2 then A is an A_n-ring can be deduced from a deep theorem on splitting rings of Azumaya algebras (see Knus, Ojanguren, p. 104). We shall give a more elementary proof via a series of reductions. This may be of interest to readers who are not experts in the theory of Azumaya algebras.

Following Schelter, we base the proof that A_n implies Azumaya on the following well known characterization of Azumaya algebras: An algebra A over K is Azumaya if and only if A is finitely generated projective and faithful over K and the canonical homomorphism of A^e into $\mathrm{End}_K A$ is an isomorphism. The statement that A is faithful over K means that if $\alpha \in K$ and $\alpha A = 0$ then $\alpha = 0$. A characterization of projective modules that we shall use is in terms of a dual basis (or "projective coordinates"): A module M over a ring R is projective if and only if there exists a set $\{(x_i, f_i)\}$ of pairs (x_i, f_i) where $x_i \in M$, $f_i \in M^* = \mathrm{Hom}_R(M, R)$ such that for any $x \in M$, $f_i(x) = 0$ for all but a finite number of i and $x = \sum f_i(x)x_i$. The set $\{(x_i, f_i)\}$ is called a *dual basis* for M. M is finitely generated projective if and only if it has a finite dual basis.

We shall need the following

LEMMA 1. *Let M be finitely generated projective over a commutative ring K and let $\{(x_i, f_i) \mid 1 \le i \le n\}$ be a dual basis for M. Let E_{ij} denote the endomorphism $x \mapsto f_i(x)x_j$. Then the n^2 endomorphisms E_{ij} span $\mathrm{End}_K M$ over K.*

Proof. Let $L \in \mathrm{End}_K M$ and suppose $Lx_i = \sum_j \lambda_{ij}x_j$, $\lambda_{ij} \in K$. Then

$$Lx = L\left(\sum f_i(x)x_i\right) = \sum \lambda_{ij}f_i(x)x_j$$
$$= \sum \lambda_{ij}E_{ij}x .$$

Hence $L = \sum \lambda_{ij}E_{ij}$.

We shall require also

LEMMA 2. *There exists an element $H(x_1, \ldots, x_{n^2}, y_1, \ldots, y_s)$ in the free associative algebra $\mathbb{Z}\{x_1, x_2, \ldots; y_1, y_2, \ldots\}$ such that:*

(1) H *is multilinear in the* x*'s and the* y*'s ;*

(2) H *is alternating in the* x*'s; and*

(3) H *is central for* $M_n(K)$ *for any commutative ring* K .

This is a consequence of Razmyslov's theorem that was proved in the preceding chapter.

Let H be as in Lemma 2 and put

$$H'(x_1, \ldots, x_{n^2+1}, y_1, \ldots, y_s) = H(x_1, \ldots, x_{n^2}, y_1, \ldots, y_s)x_{n^2+1}$$
$$+ \sum_{j=1}^{n^2} (-1)^{n^2+j+1} H(x_1, \ldots, \hat{x}_j, \ldots, x_{n^2+1}, y_1, \ldots, y_s)x_j \; .$$

It is readily verified that H' is alternating and multilinear in $x_1, \ldots, x_{n^2+1}$. Hence H' is an identity for $M_n(\mathbb{Z})$.

We shall now give the

Proof of the Artin-Procesi Theorem. We begin with the proof of

$$A_n \Rightarrow \textit{Azumaya of rank } n^2 \textit{ over its center.}$$

Let A be a ring satisfying A_n and let $\bar{A}$ be a simple homomorphic image of A . Then $\bar{A}$ satisfies the standard identity S_{2n} and hence, by Kaplansky's theorem, $[\bar{A} : C(\bar{A})] = m^2 \le n^2$. We claim that the polynomial H is not an identity for $\bar{A}$. Otherwise, if $\bar{F}$ is the algebraic closure of $C(\bar{A})$ then H is an identity for $M_n(\bar{F}) \cong \bar{F} \otimes_{C(\bar{A})} \bar{A}$. Since H is central for $M_n(\bar{F})$ this implies that $m < n$. Now any $\mathbb{Z}$-identity for $M_m(\mathbb{Z})$ is an identity for $M_m(\Phi)$ and since Φ is infinite it follows, as in the proof of Theorem 2 of the last chapter, that any $\mathbb{Z}$-identity for $M_m(\mathbb{Z})$ is an identity for $M_m(\bar{F})$ and hence for $\bar{A}$. Thus every identity for $M_{n-1}(\mathbb{Z})$ is an identity for $\bar{A}$ contrary to the hypothesis. Hence H is not an identity for $\bar{A}$. On the other hand, $[H, x_{n^2+1}]$ is an identity for $M_n(\mathbb{Z})$, hence for A , so H is a central polynomial for A .

Now let B be the ideal in A generated by all the values $H(a_1, \ldots, a_{n^2}, b_1, \ldots, b_s)$, $a_i, b_j \in A$. We claim B = A . Otherwise, B is contained in a maximal ideal M of A and $\bar{A} = A/M$ is a simple homomorphic image of A . It is clear that H is an identity for $\bar{A}$ contrary to what we showed before. Thus B = A . Since H is central, B is the set of elements of the form

(2) $$\sum_k d_k H\left(a_1^{(k)}, \ldots, a_{n^2}^{(k)}, b_1^{(k)}, \ldots, b_s^{(k)}\right)$$

where the a's, b's and d's are in A. Since $\mathcal{B} = \mathrm{A}$ we have $a_i^{(k)}$, $b_i^{(k)}$, d_k such that

(3) $$\sum_k d_k H\left(a_1^{(k)}, \ldots, a_{n^2}^{(k)}, b_1^{(k)}, \ldots, b_s^{(k)}\right) = 1 .$$

Hence for any $a \in \mathrm{A}$ we have

(4) $$a = \sum H\left(a_1^{(k)}, \ldots, a_{n^2}^{(k)}, b_1^{(k)}, \ldots, b_s^{(k)}\right) a d_k .$$

Now put

(5) $$f_{kj}(a) = (-1)^{j+n^2} H\left(a_1^{(k)}, \ldots, \widehat{a_j^{(k)}}, \ldots, a_{n^2}^{(k)}, a, b_1^{(k)}, \ldots, b_s^{(k)}\right) .$$

Since H' is an identity for $M_n(\mathbb{Z})$, hence for A, by the definition of H' we have

(6) $$H\left(a_1^{(k)}, \ldots, a_{n^2}^{(k)}, b_1^{(k)}, \ldots, b_s^{(k)}\right) a = \sum_j f_{kj}(a) a_j^{(k)} .$$

Hence, by (4), we have

(7) $$a = \sum_{k,j} f_{kj}(a) a_j^{(k)} d_k .$$

Since f_{kj} is a $C(\mathrm{A})$-linear map of A into $C(\mathrm{A})$ it follows that $\left\{\left(a_j^{(k)} d_k, f_{kj}\right)\right\}$ is a dual base for A as $C = C(\mathrm{A})$-module. Hence A is finitely generated projective over C. Evidently A is C-faithful also.

We show next that the canonical map φ of A^e into $\mathrm{End}_K \mathrm{A}$ is surjective. By Lemma 1, it suffices to show that for any k, j and any b, the map $a \mapsto f_{kj}(a) b$ has the form $a \mapsto \sum e_i a e_i'$ for suitable e_i, e_i' in A. This is clear from the definition of the f_{kj} given in (5).

Next we have to show that φ is injective. Suppose $\varphi\left(\sum e_i \otimes e_i'\right) = 0$, so $\sum_i e_i a e_i' = 0$ for all $a \in \mathrm{A}$. Write $f_{kj} = \varphi\left(\sum_l u_{kjl} \otimes u_{kjl}'\right)$. Then, by (7), we have

$$a = \sum_{k,j,l} u_{kjl} a u_{kjl}' a_j^{(k)} d_k .$$

Hence

$$\begin{aligned}\sum e_i \otimes e_i' &= \sum_{i,j,k,l} u_{kjl} e_i u_{kjl}' a_j^{(k)} d_k \otimes e_i' \\ &= \sum a_j^{(k)} d_k \otimes u_{kjl} e_i u_{kjl}' e_i' \\ &= \sum_{k,j,l} a_j^{(k)} d_k \otimes u_{kjl} \sum_i e_i u_{kjl}' e_i' \\ &= 0\end{aligned}$$

since $\sum_i e_i u_{kjl}' e_i' = 0$. Thus φ is injective and so φ is an isomorphism. Then A is Azumaya by the conditions we noted before.

We show next that A has rank n^2 over C . Let P be a prime ideal of C . We have seen that the rank n_P of $A_P = C_P \otimes_C A$ over C_P is the same as the rank of $(C_P/P_P) \otimes_{C_P} A_P$ over the field C_P/P_P , and this is a square m^2 . Now A satisfies the standard identity S_{2n} . Hence S_{2n} is an identity for A_P and for $(C_P/P_P) \otimes_{C_P} A_P$. Suppose $m < n$. Then it follows from Theorem 2 of the last chapter that $(C_P/P_P) \otimes_{C_P} A_P$ satisfies every identity of $M_m(C_P/P_P)$ and hence every identity of $M_m(\mathbb{Z})$. Then $(C_P/P_P) \otimes_{C_P} A_P$ satisfies every identity of $M_{n-1}(\mathbb{Z})$. On the other hand $(C_P/P_P) \otimes_{C_P} A_P = (C_P/P_P) \otimes_{C_P} (C_P \otimes_C A) \cong (C_P/P_P) \otimes_C A$ and we have the canonical homomorphism $a \mapsto 1 \otimes a$ of A into $(C_P/P_P) \otimes_C A$. This gives a non-zero homomorphic image $\bar{A}$ of A that satisfies every identity of $M_{n-1}(\mathbb{Z})$ contrary to hypothesis. Hence $n_P = n^2$. Since this holds for every prime ideal P of C , A has rank n^2 over C . □

We prove next the converse:

Azumaya of rank n^2 *over the center* $\Rightarrow A_n$.

The second condition in the definition of A_n-ring is easily established for any Azumaya algebra A of rank n^2 . For it is known that we have a bijection $I \mapsto IA$ of the set of ideals of the center C of A with the set of ideals of A and $A/IA \cong (C/I) \otimes_C A$. Hence any non-zero homomorphic image of A is Azumaya of rank n^2 . Now suppose $\bar{A}$ is a non-zero homomorphic image that satisfies every identity of $M_{n-1}(\mathbb{Z})$. Then we may assume $\bar{A}$ is simple. Then $\bar{A}$ is n^2

dimensional over its center and hence $\bar{A}$ does not satisfy $S_{2(n-1)}$. This contradiction proves the result.

It remains to prove that A satisfies every identity of $M_n(\mathbb{Z})$. We shall do this by a series of reductions.

Let I_n denote the set of elements of $\mathbb{Z}\{x_1, x_2, \ldots\}$ that are identities for $M_n(\mathbb{Z})$. This is a T-ideal in $\mathbb{Z}\{x_1, x_2, \ldots\}$, that is, it is stabilized by all the ring endomorphisms of $\mathbb{Z}\{x_1, x_2, \ldots\}$. We shall call a ring A an I_n*-ring* if every element of I_n is an identity for A. We prove first

LEMMA 3. *If* A *is an* I_n*-ring and* A *is an algebra over* K *and* K' *is a commutative algebra over* K *then* $A_{K'} = K' \otimes_K A$ *is an* I_n*-ring.*

Proof. Since $M_n(\mathbb{Z})$ is a free $\mathbb{Z}$-module, if $mf \in I_n$ for $m \in \mathbb{Z}$ then $f \in I_n$. This implies by a Vandermonde determinant argument that if $f \in I_n$ and $f = f_0 + f_1 + \ldots + f_m \in I_n$ where f_j is homogeneous of degree j in one of the x_i then $f_j \in I_n$. Next suppose $f = f(x_1, \ldots, x_m) \in I_n$ is homogeneous of degree m_i in x_i and write

$$f(x_1, \ldots, x_{i-1}, x_i+x_{m+1}, x_{i+1}, \ldots, x_m) = \sum_{j=0}^{m_i} f_{ij}(x_1, \ldots, x_{m+1})$$

where f_{ij} is homogeneous of degree j in x_i (and of degree $m_i - j$ in x_{m+1}). Then every $f_{ij} \in I_m$. These two properties of I_n imply as in Jacobson [2], pp. 27-30, that if A/K is an I_n-ring then so is $A_{K'}$ for any commutative algebra K'/K. □

We need to prove that if A is Azumaya of rank n^2 over its center C then A is an I_n-ring. We first reduce the proof to the case in which C is noetherian by using the following

LEMMA 4. *Let* A *be an Azumaya algebra over* C. *Then there exists an Azumaya algebra* A' *over a noetherian subring* C' *of* C *such that* $A \cong C \otimes_{C'} A'$. *Moreover, if* A *has a rank then so has* A' *and their ranks are equal.*

Proof. The first assertion is proved in Orzech, Small [1]. To prove the second we write $A' = A'_1 \oplus \ldots \oplus A'_s$ where the A'_i are Azumaya over their centers of different ranks. Then $A \simeq A'_C = A'_{1C} \oplus \ldots \oplus A'_{sC}$ where rank A'_{iC} = rank A_i . It follows that $s = 1$ and rank A' = rank A . □

Evidently Lemmas 3 and 4 reduce the proof that any Azumaya algebra of rank n^2 is an I_n-ring to the case in which C is noetherian. The "local-global" principal of commutative ring theory then permits a reduction to the case in which C is local noetherian. If M is the maximal ideal of the local noetherian ring C $\cap M^i = 0$ by Krull's intersection theorem. Hence we can define the completion $\hat{C}$ of C in the M-adic topology and form the algebra $\hat{A} = A_{\hat{C}}$. Since $A \hookrightarrow \hat{A}$ it suffices to prove that $\hat{A}$ is an I_n-ring. We have therefore achieved a reduction to the situation in which C is complete, local and noetherian. We now drop the noetherian condition (which was needed to obtain a completion) and we proceed to prove that if K is a complete local ring and A is Azumaya over K then A is an I_n-ring.

We recall that a ring A is called an SBI-*ring* ("suitable for building idempotents") if given any z in the Jacobson radical, rad A , there exists a $w \in$ rad A such that $w^2 - w = z$ and the centralizer $C_A(w) = C_A(z)$ (Jacobson [1], pp. 53-55). Then we have

LEMMA 5. *If* A *is an Azumaya algebra over a complete local ring* K *then* A *is an* SBI-*ring.*

Proof. If M is the maximal ideal of K then MA is a maximal ideal in A so A/MA is simple. A is a free module of finite rank over K and we have the product topology in A $\left(\cong K^{(n^2)}\right)$ making A a topological ring. The completeness of K implies that if $z_i \in M^i A$ then $\sum z_i$ exists in A . In particular, if $z \in MA$ then z is quasi-regular since $1 - z$ has the inverse $1 + z + z^2 + \dots$. Hence $MA \subset$ rad A and since A/MA is simple MA = rad A . The proof of Proposition 3 on p.54 of Jacobson [1] carries over to prove that A is SBI . □

We recall also the following useful lifting property of SBI-rings.

LEMMA 6. *Let* A *be an* SBI-*ring and let* $\{\bar{e}_{ij} \mid 1 \le i, j \le n\}$ *be a set of matrix units in* $\bar{A} = A/\text{rad } A$ $\left(\sum \bar{e}_{ii} = \bar{1}, \bar{e}_{ij}\bar{e}_{kl} = \delta_{jk}\bar{e}_{il}\right)$. *Then* $\bar{e}_{ij}$ *has a lift* e_{ij} *in* A *such that* $\{e_{ij}\}$ *is a set of matrix units for* A .

This is proved in Jacobson [1], p. 55.

We require also

LEMMA 7. *Let* K *be a complete local ring with maximal ideal* M *and let* $\bar{L} = \bar{K}(\bar{\theta})$ *be a finite dimensional extension field of* $\bar{K} = K/M$. *Then there exists*

a commutative algebra L *over* K *that is a complete local ring having a* K*-rank of* $[\overline{L} : \overline{K}]$ *and such that if* N *is the maximal ideal of* L *then* $L/N \cong \overline{L}$.

Proof. Let $\overline{f}(x) \in \overline{K}[x]$ be the minimum polynomial of $\overline{\theta}$ over $\overline{K}$. Let $f(x)$ be a monic lift of $\overline{f}(x)$ in $K[x]$. Put $L = K[x]/(f(x))$. Then it is easily seen that L satisfies the conditions. □

We can now prove that any Azumaya algebra of rank n^2 over a complete local ring K is an I_n-ring. Let M be the maximal ideal of K so $MA = \text{rad } A$ and $\overline{A} = A/MA$ is central simple over the field $\overline{K} = K/M$. Let $\overline{L}$ be a separable splitting field over $\overline{K}$ of $\overline{A}$ and let L be as in Lemma 7. If we replace A by A_L we may assume $\overline{A} = A/MA = M_n(\overline{K})$. By Lemmas 5 and 6 there exists a set of matrix units $\{e_{ij} \mid 1 \le i, j \le n\}$ in A such that the $\overline{e}_{ij} = e_{ij} + MA$ form a base for $\overline{A}$ over $\overline{K}$. Then the e_{ij} form a base for A over K . Hence $A = M_n(K)$. It is now clear that A is an I_n-ring.

This completes the proof that any Azumaya algebra of rank n^2 over its center is an A_n-ring.

References

M. Artin

[1] "On Azumaya algebras and finite dimensional representations of rings", *J. Algebra* 11 (1969), 532-563.

R. Bix

[1] "Separable Jordan algebras over a commutative ring" (PhD thesis, Yale University, Connecticut, 1977). See also: *J. Algebra* (to appear).

Frank DeMeyer, Edward Ingraham

[1] *Separable Algebras Over Commutative Rings* (Lecture Notes in Mathematics, 181. Springer-Verlag, Berlin, Heidelberg, New York, 1971).

Nathan Jacobson

[1] *Structure of Rings* (American Mathematical Society Colloquium Publications 37. American Mathematical Society, Providence, Rhode Island, 1956).

[2] *Structure and Representations of Jordan Algebras* (American Mathematical Society, Colloquium Publications, 39. American Mathematical Society, Providence, Rhode Island, 1968).

[3] *PI-Algebras. An Introduction* (Lecture Notes in Mathematics, 441. Springer-Verlag, Berlin, Heidelberg, New York, 1975).

Max-Albert Knus, Manuel Ojanguren

[1] *Théorie de la Descente et Algèbres d'Azumaya* (Lecture Notes in Mathematics, 389. Springer-Verlag, Berlin, Heidelberg, New York, 1974).

Morris Orzech, Charles Small

[1] *The Brauer Group of Commutative Rings* (Marcel Dekker, New York, 1975).

Claudio Procesi

[1] "On a theorem of M. Artin", *J. Algebra* 22 (1972), 309-315.

[2] *Rings with Polynomial Identities* (Pure and Applied Mathematics, 17. Marcel Dekker, New York, 1973).

W. Schelter

[1] "On a theorem of Artin and Azumaya algebras", *J. Algebra* (to appear).

CHAPTER III

SHIRSHOV'S THEOREMS ON LOCAL FINITENESS

In this part we shall prove some theorems on local finiteness of *PI*-algebras that were proved by Shirshov in 1957. These results were overlooked by Western algebraists until quite recently and for this reason may qualify for inclusion in this series of lectures. An extension of some of Shirshov's work (dealing with alternative algebras) was published by McCrimmon in 1974. In this paper McCrimmon made use of an earlier exposition of Shirshov's work by Slater. Moreover, McCrimmon gave an account of the associative Shirshov theorems in a lecture at a Summer Institute on Ring Theory at the University of Chicago in 1973. Perhaps this was the first time that associative ring theorists in the West became aware of Shirshov's important results.

Shirshov's results generalized earlier ones by Jacobson, Levitzki and Kaplansky, especially, the local finiteness theorem of Kaplansky for *PI*-algebras that are algebraic over a field. The earlier results had been obtained using structure theory. On the other hand, Shirshov's methods are elementary and combinatorial so they apply equally well to algebras over commutative rings. Shirshov also proved a local finiteness theorem for special Jordan algebras. We shall consider this also in this paper.

To state Shirshov's theorem we need to introduce some definitions. Let A be an algebra over a commutative ring K. An element $a \in A$ will be called *(integral) algebraic* if there exists a monic polynomial $f(\lambda) \in K[\lambda]$ such that $f(a) = 0$. The minimum degree for such polynomials is called the *degree* of a. If this degree is n then the subalgebra $K[a]$ generated by a is spanned over K by $1, a, \ldots, a^{n-1}$. An algebra A is called *locally finite* if the subalgebras generated by finite subsets of A are finitely spanned over K, that is, have finite sets of generators as K-modules.

The polynomial identity condition that we shall require is somewhat stronger than what is needed for some other parts of the theory, notably, for the structure theorem on prime PI-algebras. An element $f(x_1, \ldots, x_m) \in K\{x_1, x_2, \ldots\}$ is called *monic* if one of the monomials of highest (total) degree occurring in f has coefficient 1. We shall require that A satisfies a monic identity. It is not difficult to show, by the usual linearization process, that if A satisfies a monic identity then A satisfies a multilinear monic identity of the same degree and this can be taken to have the form

$$x_1 \ldots x_m - \sum_{\substack{\pi \neq 1 \\ \pi \in \Sigma_m}} \alpha_\pi x_{\pi 1} \ldots x_{\pi d} \tag{1}$$

where $\alpha_\pi \in K$.

We can now state

SHIRSHOV'S THEOREM. *Let* A *be an algebra over a commutative ring* K *satisfying a monic identity of degree* d. *Suppose* A *is generated by a set of elements* $\{a_i\}$ *(not necessarily finite) such that every monomial of degree less than or equal to* d *in these elements is algebraic. Then* A *is locally finite.*

The proof of this theorem is based on some combinatorial results on free monoids. We shall follow an exposition of these results due to McCrimmon which will appear in a forthcoming book of his on alternative algebras.

Let $X_n = \{x_1, x_2, \ldots, x_n\}$ and let $\mathrm{FM}(X_n)$ be the free monoid generated by X_n. Its elements are the word 1 and the words $x_{i_1} \ldots x_{i_r}$, $r \geq 1$. Equality, multiplication and degree of an element are defined as usual.

We introduce a total ordering in $\mathrm{FM}(X_n)$ by specifying that $x_1 < x_2 < \ldots < x_n$ and that $1 > u$ for any $u \neq 1$ and

$$u = x_{i_1} \ldots x_{i_r} > v = x_{j_1} \ldots x_{j_s}$$

if either u is an initial segment of v $(v = uw\ ,\ w \neq 1)$ or

$i_1 = j_1, \ldots, i_k = j_k$ and $i_{k+1} > j_{k+1}$ for $k < \min(r, s)$. We call this the *augumented lexicographic ordering* of $FM(X_n)$.

We now list some properties of this ordering.

I. $u > v \Longleftrightarrow wu > wv$ *for any* $w \in FM(X_n)$.

Proof. Clear. □

II. *If* $u > v$ *and* u *is not an initial segment of* v *then* $uw > vw'$ *for any* w, w'.

Proof. Clear. □

We shall need to consider factorizations of a word w as $w = w_1 w_2 \ldots w_m$ where the w_i are subwords. The integer m is called the *length* of the factorization. A factorization $w_1 w_2 \ldots w_m$ is called *dominant* if

(2) $$w > w_{\pi 1} w_{\pi 2} \ldots w_{\pi m}$$

for every $\pi \neq 1$ in Σ_m.

III. *If* $w_1 \ldots w_m$ *is dominant then* $w_r \ldots w_m$ *is a dominant factorization for every* r, $1 \le r \le m$.

Proof. Clear from I. □

IV. *Let* $w = w_1 \ldots w_m$ *where* $w_i = x_n^{e(i)} v_i$, $e(i) \ge 1$ *and* v_i *begins with some* $x_k \neq x_n$. *Suppose* $w = w_1 \ldots w_m$ *is dominant. Then*

(3) $$e(1) \ge e(2) \ge \ldots \ge e(m) .$$

Proof. If $e(1) < e(2)$ then $w_2 w_1 \ldots w_m$ begins with a higher power of x_n than w, contrary to the hypothesis that $w_1 w_2 \ldots w_m$ is dominant. Thus $e(1) \ge e(2)$. Next, by III, $w_2 \ldots w_m$ is a dominant factorization. Hence $e(2) \ge e(3)$. Continuing in this way we obtain (3). □

V. *The factorization* $w = w_1 w_2 \ldots w_m$ *is dominant if each* $w_i = u_i v_i$ *where* $u_1 > u_2 > \ldots > u_m$ *and no* u_i *is an initial segment of a* u_j *with* $j > i$.

Proof. Let $\pi \neq 1$ be in Σ_m and suppose $\pi(1) = 1, \ldots, \pi(k-1) = k-1$ but $\pi(k) \neq k$. Then $\pi(k) > k$ and $k < m$. By II,

$$w_k \ldots w_m = u_k v_k w_{k+1} \ldots w_m > u_{\pi k} v_{\pi k} w_{\pi(k+1)} \ldots w_{\pi m} .$$

Hence $w_1 \ldots w_m > w_{\pi 1} \ldots w_{\pi m}$ by I. □

Let m and M be two positive integers and let Y be the set of words of the form $x_n^e v$ where $1 \le e \le M$ and v is in the submonoid $\mathrm{FM}(X_{n-1})$ generated by $x_1, \ldots, x_{n-1}$ and $1 \le \deg v \le m$. Since the number of v of degree k in $x_1, \ldots, x_{n-1}$ is $(n-1)^k$ we see that

$$(4) \qquad |Y| = N(n, m, M) = M\big((n-1) + (n-1)^2 + \ldots + (n-1)^m\big) .$$

It is clear that the submonoid generated by Y can be identified with the free monoid $\mathrm{FM}(Y_N)$, $N = N(n, m, M)$. An element of this submonoid has a degree in the y's, called its *Y-degree* (as well as its X-degree). We order the $y_i \in Y = Y_N$ in increasing order given by the ordering $<$ in $\mathrm{FM}(X_n)$ and use this to define the augmented lexicographic order in $\mathrm{FM}(Y_N)$. We denote this order by $\ll$. We have

VI. *The order in* $\mathrm{FM}(Y_N)$ *given by* $\ll$ *coincides with the induced order given by regarding* $\mathrm{FM}(Y_N)$ *as a submonoid of* $\mathrm{FM}(X_n)$.

Proof. It suffices to show that if $u, v \in \mathrm{FM}(Y_N)$ and $u \gg v$ then $u > v$. Suppose $u \gg v$.

Case I. $v = uw$, $w \in \mathrm{FM}(Y_N)$, $w \neq 1$. Then $w \in \mathrm{FM}(X_n)$ and the result is clear.

Case II. $u = wy_iz$, $v = wy_jt$, $i > j$, $w, z, t \in \mathrm{FM}(Y_N)$.

(a) $y_i > y_j$ but y_i is not an initial segment of y_j. Then $u > v$ follows from I and II.

(b) $y_j = y_is$, $s \in \mathrm{FM}(X_{n-1})$, $s \neq 1$. If $z = 1$ then $v = ust$ so $u > v$. If $z \neq 1$ then z begins with x_n. Hence $u = wy_ix_n \ldots$, $v = wy_ist$ so again $u > v$. □

An immediate consequence of this is that if we have a factorization of $u \in \mathrm{FM}(Y_N)$ as $u_1 \ldots u_m$ where the $u_i \in \mathrm{FM}(Y_N)$ then this is dominant in $\mathrm{FM}(Y_N)$ if and only if it is dominant in $\mathrm{FM}(X_n)$.

We are now ready to prove the

FIRST COMBINATORIAL LEMMA. *For given positive integers* n, m, M *there exists a positive integer* $f(n, m, M)$ *such that any word* w *in* $\mathrm{FM}(X_n)$ *with* $\deg w > f(n, m, M)$ *contains a subword of one of the following two forms:*

(i) $w_0 = u^M$, $\deg u \ge 1$;

(ii) w_0 *has a dominant factorization* $w_0 = u_1 u_2 \ldots u_m$ *of length* m .

Proof. We use induction on m . We can start the induction with $f(n, 1, M) = 1$ since any x_i is dominant of length 1 . Also for a given m we can use a sub-induction on n . Here we can start with $f(1, m, M) = M$ since any word in x_1 of degree greater than M contains the subword x_1^M . Assume we have defined $f(n, m-1, M)$ for all n and $f(n-1, m, M)$ to satisfy the conditions for $(n, m-1, M)$ and $(n-1, m, M)$. Put

$$f(n, m, M) = \big(M+f(n-1, m, M)\big)\big(2+f(N, m-1, M)\big) \tag{5}$$

where $N = N\big(n, f(n-1, m, M), M\big)$ as defined in (4).

Let w be a word in $x_1, \ldots, x_n$ of degree greater than $f(n, m, M)$ and write

$$w = \left\{x_n^{e(0)} v_0\right\}\left\{x_n^{e(1)} v_1\right\} \ldots \left\{x_n^{e(r+1)} v_{r+1}\right\} \tag{6}$$

where $v_i \in \mathrm{FM}\big(X_{n-1}\big)$, $e(0) \geq 0$, $e(i) > 0$ if $i > 0$, $\deg v_i > 0$ for $0 \leq i \leq r$. Then

$$w = \left\{x_n^{e(0)} v_0\right\} w' \left\{x_n^{e(r+1)} v_{r+1}\right\} . \tag{7}$$

If some $e(i)$ in (6) is greater than M then we have the subword x_n^M and we are done. Also if $\deg v_i > f(n-1, m, M)$ then the induction on n implies that we have a subword w_0 of v_i satisfying *(i)* or *(ii)* so again we are done. Hence we may assume every $e(i) \leq M$ and every $\deg v_i \leq f(n-1, m, M)$. Then

$$\begin{aligned}\deg w &\leq (r+2)M + (r+2)f(n-1, m, M)\\ &= (r+2)\big(M+f(n-1, m, M)\big) .\end{aligned}$$

On the other hand,

$$\deg w > f(n, m, M) = \big(M+f(n-1, m, M)\big)\big(2+f(N, m-1, M)\big) .$$

Hence

$$f(N, m-1, M) < r . \tag{8}$$

Now consider the set of words $x_n^e v$, $1 \leq e \leq M$, $v \in \mathrm{FM}\big(X_{n-1}\big)$, $1 \leq \deg v \leq f(n-1, m, M)$. The number of these words is $N = N\big(n, f(n-1, m, M), M\big)$ as in (4). The word w' in (7) is $\left\{x_n^{e(1)} v_1\right\} \ldots \left\{x_n^{e(r)} v_r\right\}$ so its Y-degree is $r > f(N, m-1, M)$. Hence by induction on m , w' contains a subword $w_0' = u'^M$, $\deg u' \geq 1$, or a word w_0' with a dominant factorization $u_1' \ldots u_{m-1}'$ in $\mathrm{FM}\big(Y_N\big)$ and hence in $FM(X_n)$ of length $m-1$. In the first case we are done. Hence we assume the second and we proceed to expand $u'_1 \ldots u'_{m-1}$ to a dominant subword $w_0 = u_1 \ldots u_m$ of w.

Taking into account the definition of Y_N we see that $u'_i = x_n^{e'(i)} v'_i$ where $e'(i) \geq 1$ and v'_i begins with an $x_{k_i} \neq x_n$. Then, by IV, we have

(9) $$e'(1) \geq e'(2) \geq \ldots \geq e'(m-1) .$$

Since $e(r+1) \geq 1$ there is an x_n in w after u'_{m-1}. Hence

(10) $$w = pu'_1 \ldots u'_{m-1} q x_n s$$

where q does not involve x_n. We claim that $u'_1 \ldots u'_{m-2}(u'_{m-1}q)$ is also a dominant factorization. Let $\pi \neq 1$ be in Σ_{m-1}. Suppose first that π fixes $m-1$. Then applying π to the factorization $u'_1 \ldots u'_{m-2}(u'_{m-1}q)$ gives $u'_{\pi 1} \ldots u'_{\pi(m-2)}(u'_{m-1}q) < u'_1 \ldots u'_{m-2}(u'_{m-1}q)$ since $u'_1 \ldots u'_{m-1}$ is dominant (using II). Next suppose π moves $m-1$. Then applying π to the factorization $u'_1 \ldots u'_{m-2}(u'_{m-1}q)$ gives the factorization

$$u'_{\pi 1} \ldots (u'_{m-1}q)u'_{\pi j} \ldots < u'_{\pi 1} \ldots u'_{m-1}u'_{\pi j} \ldots q$$

(since $u'_{\pi j}$ begins with x_n and q does not). But

$$u'_{\pi 1} \ldots u'_{m-1}u'_{\pi j} \ldots q < u'_1 \ldots u'_{m-1}q$$

by II and the dominance of the factorization $u'_1 \ldots u'_{m-1}$. Thus $u'_1 \ldots u'_{m-2}(u'_{m-1}q)$ is dominant and so if we replace u'_{m-1} by $u'_{m-1}q$ we may assume $q = 1$ and we have the subword $u'_1 \ldots u'_{m-1}x_n$ in w.

Now $u'_i = x_n^{e'(i)} v'_i$ where $e'(i) \geq 1$, v'_i begins with an $x_{k_i} \neq x_n$ and (9) holds. Then let

(11) $$w_0 = x_n\left(x_n^{e'(1)-1} v'_1 x_n\right)\left(x_n^{e'(2)-1} v'_2 x_n\right) \ldots \left(x_n^{e'(m-1)-1} v'_{m-1} x_n\right) = u_1 u_2 \ldots u_m$$

where $u_1 = x_n$, $u_2 = x_n^{e'(1)-1} v'_1 x_n$, $\ldots$, $u_m = x_m^{e'(m-1)-1} v'_{m-1} x_n$. We claim that $u_1 u_2 \ldots u_m$ is a dominant factorization of w_0. Let $\pi \neq 1$ be in Σ_m. Suppose first that $\pi 1 = 1$. Then

$$u_{\pi 1} \ldots u_{\pi m} = x_n\left(x_n^{e'(\sigma 1)-1} v'_{\sigma 1} x_n\right)\left(x_n^{e'(\sigma 2)-1} v'_{\sigma 2} x_n\right) \ldots \left(x_n^{e'(\sigma(m-1))-1} v'_{\sigma(m-1)} x_n\right)$$

where $\sigma \neq 1$ in Σ_{m-1}. Then

$$u_{\pi 1} \ldots u_{\pi m} = u'_{\sigma 1} \ldots u'_{\sigma(m-1)} x_n < u'_1 \ldots u'_{m-1} x_n = u_1 \ldots u_m$$

by the dominance of the factorization $u'_1 \ldots u'_{m-1}$. Next suppose $\pi 1 \neq 1$. Then applying π gives $x_n^{e'(j)-1} v'_j x_n \ldots < u_1 \ldots u_m$ since $u_1 \ldots u_m$ begins with $x_n^{e'(1)}$ and $e'_1(1) \geq e'(j) > e'(j) - 1$. Thus $w_0 = u_1 \ldots u_m$ is dominant. □

We need to improve the first lemma to the

SECOND COMBINATORIAL LEMMA. *If* n, m, M *are positive integers there exists a positive integer* $g(n, m, M)$ *such that any word* w *in* $\mathrm{FM}(X_n)$ *of degree greater than* $g(n, m, M)$ *contains a subword* w_0 *having one of the following forms:*

(i) $w_0 = u^M$, $1 \leq \deg u \leq m$;

(ii) w_0 *has a dominant factorization of length* m.

The improvement over the first lemma is that we are able to require that $\deg u \leq m$ in case *(i)*. The proof will be based on the following

SUBLEMMA. *Let* l *and* m *be positive integers with* $l > m$ *and let* u *be a word of degree* l. *Then either* $u = v^e$ *for a word* v *and* $e \neq 1$ *a divisor of* l *or* u^{2m} *contains a subword that has a dominant factorization of length* m.

Proof. Write $u = z_1 \ldots z_l$ where the $z_j \in X_n$ and let $\sigma = (12 \ldots l)$. Let H be the subgroup of $\langle \sigma \rangle$ of $\tau = \sigma^k$ such that $z_{\tau 1} \ldots z_{\tau l} = z_1 \ldots z_l$. Then $H = \langle \sigma^d \rangle$ for $1 \leq d \leq l$ and $d | l$, say $l = de$. Then $u = z_1 \ldots z_l = z_{d+1} \ldots z_{d+l}$ (indices reduced mod l). If $d \neq l$ then $e > 1$ and $u = v^e$, $v = z_1 \ldots z_d$. Next suppose $d = l$. Then the words $\sigma^k u \equiv z_{\sigma^k 1} \ldots z_{\sigma^k l}$, $1 \leq k \leq l$, are distinct so we have a permutation π of $1, 2, \ldots, l$ such that

$$\sigma^{\pi 1} u > \sigma^{\pi 2} u > \ldots > \sigma^{\pi l} u .$$

Consider the element u^2. This can be written as

$$u^2 = z_1 \ldots z_l z_1 \ldots z_l = z_1 (\sigma u) z_2 \ldots z_l = z_1 z_2 (\sigma^2 u) z_3 \ldots z_l =$$
$$= \ldots = z_1 \ldots z_k (\sigma^k u) z_{k+1} \ldots z_l = \ldots .$$

Hence for any k, $1 \leq k \leq l$, $u^2 = v_k (\sigma^k u) v'_k$ where $v_k = z_1 \ldots z_k$ and $v'_k = z_{k+1} \ldots z_l$. Since $l > m$,

$$u^{2m} = (u^2)^m = \left\{v_{\pi 1}(\sigma^{\pi 1}u)v'_{\pi 1}\right\} \cdots \left\{v_{\pi m}(\sigma^{\pi m}u)v'_{\pi m}\right\}$$
$$= v_{\pi 1}u_1 \cdots u_m$$

where

$$u_i = (\sigma^{\pi i}u)v'_{\pi i}v_{\pi(i+1)} \quad (1 \le i < m) ,$$

and

$$u_m = (\sigma^{\pi m}u)v'_{\pi m} .$$

Since the $\sigma^{\pi i}u$ all have the same length and $\sigma^{\pi 1}u > \sigma^{\pi 2}u > \ldots$ it follows from V that $u_1 \ldots u_m$ is a dominant factorization and $u_1 \ldots u_m$ is a subword of u^{2m} . □

We now give the

Proof of the Second Combinatorial Lemma. Let $f(n, m, M)$ be as in the first lemma and put $g(n, m, M) = f(n, m, \bar{M})$ where $\bar{M} = \max(2m, M)$. Let w be a word of degree greater than $g(n, m, M) = f(n, m, \bar{M})$. By the first lemma, either w contains a subword $w_0 = u^{\bar{M}}$ with $\deg u \ge 1$ or a word w_0 with a dominant factorization of length m . We are done in the second case and also in the first case if $\deg u \le m$. Hence suppose $\deg u = l > m$. We claim that $u^{\bar{M}}$ either contains a subword u_0^M with $1 \le \deg u_0 \le m$ or a subword with a dominant factorization of length m . We prove this by induction on l . By the sublemma, either $u = v^e$ where $e | l$ and $e \ne 1$ or u^{2m} contains a subword having a dominant factorization of length m . In the first case, $\deg v < l$ and $u^{\bar{M}}$ contains the subword $v^{\bar{M}}$. Then the result follows by the degree induction. In the second case, u^{2m} and hence $u^{\bar{M}}$ contains a subword with a dominant factorization of length m . □

We are now ready to give the

Proof of Shirshov's Theorem. We assume first that $|\{a_i\}| = n < \infty$ and we shall prove that A is finitely generated as a K-module. Consider the free monoid $\mathrm{FM}(X_n)$ and the free algebra $K\{X_n\}$ having $\mathrm{FM}(X_n)$ as base over K with multiplication determined by that in $\mathrm{FM}(X_n)$. Let η be the homomorphism of $K\{X_n\}$ into A such that $x_i \mapsto a_i$, $1 \le i \le n$, and let $I = \ker \eta$. The subset U of monomials in the x's of positive degree less than or equal to d is finite. Hence there exists a positive integer e such that for every $u \in U$ there exists a monic polynomial in $K[\lambda]$ of degree e such that $f(\eta u) = 0$. Let $g(n, d, e)$ be as in the second combinatorial lemma and let V be the set of monomials in the x's of degree less than or equal to $g = g(n, d, e)$. We claim that every monomial in the x's of degree greater than g is congruent modulo I to a linear combination of monomials in V . We use induction on the degree of the monomial and for a given degree induction on the order as defined in $FM(X_n)$. For a given degree r the first monomial (in the ordering) is x_1^r and if $r > g$ then $r > e$ (since the proofs of the combinatorial lemmas show that $g \geq e$ if $d \neq 1$). Now a_1^e is a linear combination with coefficients in K of $1, a_1, \ldots, a_1^{e-1}$. It follows that x_1^r is congruent modulo I to a linear combination of monomials of degree less than r . Then the result follows for x_1^r by the degree induction. Now let w be any monomial in the x's of degree greater than g . By the second combinatorial lemma either w has a factor of the form u^e , $u \in U$, or it has a monomial factor w_0 that has a dominant factorization $u_1 \ldots u_d$ of length d . In the first case the argument we used for x_1^r shows that w is congruent modulo I to a linear combination of elements in the set V . In the second case we use the fact that A satisfies an identity of the form

$$x_1 \ldots x_d - \sum_{\pi \neq 1} \alpha_\pi x_{\pi 1} \ldots x_{\pi d}$$

to conclude that $u_1 \ldots u_d$ is congruent modulo I to a linear combination of monomials of the same degree and lower order. It follows that w is congruent modulo I to a K-linear combination of monomials of the same degree and of lower order. Hence w is congruent to a linear combination of monomials contained in the set V . Since this set is finite it follows that any monomial in $a_1, \ldots, a_n$, and hence any element of A is a K-linear combination of a finite subset of A . Thus A is finitely generated as K-module.

To finish the proof we suppose $\{a_i\}$ is any set of generators satisfying the hypothesis and we let $\{b_1, \ldots, b_r\}$ be a finite subset of A . We have to show that the subalgebra B generated by $b_1, \ldots, b_r$ is finitely generated as K-module. Now the b_j are contained in a subalgebra A' generated by a finite subset of the set $\{a_i\}$, say, $\{a_1, \ldots, a_n\}$. By what we have proved, A' is finitely generated as K-module. Hence the result required will follow from the following:

LEMMA. *Let* A *be an algebra that is finitely generated as* K*-module,* B *a* *subalgebra that is finitely generated as K-algebra. Then B is finitely generated as K-module.*

Proof. Let $\{u_1, \ldots, u_n\}$ be a set of generators of A as K-module and let $\{b_1, \ldots, b_m\}$ be a set of generators of B as K-algebra. We have

$$u_i u_j = \sum \gamma_{ijk} u_k , \quad \gamma_{ijk} \in K ,$$

$$1 = \sum \gamma_i u_i , \quad \gamma_i \in K ,$$

$$b_l = \sum \mu_{li} u_i , \quad \mu_{li} \in K .$$

Let K' be the subring of K generated by the finite set $\{\gamma_{ijk}, \gamma_i, \mu_{li}\}$. Then K' is noetherian and $A' = \sum K' u_i$ is a K'-subalgebra of K containing the K'-subalgebra B' generated by the b_l . Since K' is noetherian B' is finitely generated as K'-module by a subset, say, $v_1, \ldots, v_r$. Then every monomial in the b_l is a K'-linear combination of the v's and hence every element of B is a K-linear combination of the v's . Thus the v's form a set of generators for B as K-module. □

This completes the proof of Shirshov's Theorem. □

Shirshov's Theorem can be carried over to PI-algebras without unit. Moreover, the proof gives a stronger result in the case of nil algebras. To state this we require the concepts of nilpotency and local nilpotency for algebras. An algebra A is called *nilpotent* if there exists an integer s such that $A^s = 0$, which is equivalent to saying that the product of any s elements of A is 0 . A is called *locally nilpotent* if the subalgebras generated by finite subsets are nilpotent. It is an old result of Amitsur's that any nil PI-algebra is locally nilpotent. The following stronger result is due to Shirshov.

THEOREM 1. *Let A be an algebra without unit over a commutative ring K satisfying a monic identity of degree d . Suppose also that A is generated by a subset $\{a_i\}$ such that every monomial of degree less than or equal to d in the a_i is nilpotent. Then A is locally nilpotent.*

Proof. The proof is similar but somewhat simpler than the proof of the general local finiteness theorem. First, let $\{a_i\} = \{a_1, a_2, \ldots, a_n\}$. Then there exists an e such that $b^e = 0$ for every monomial in the a_i of degree less than or equal to d . Let $g = g(n, d, e)$ as in the Second Combinatorial Lemma. Then we claim that any product of $g' = g + 1$ elements of A is 0 , so $A^{g+1} = 0$. It suffices to show this for all products of g' a_i's . As before, let η be the homomorphism of $K\{X_n\}'$ the free associative algebra without unit on the n generators $x_1, x_2, \ldots, x_n$ such that $x_i \mapsto a_i$, $1 \leq i \leq n$, and let $I = \ker \eta$. If u is any monomial of degree less than or equal to d in the x_i then $u^e \in I$. Now consider the set V of monomials in the x's of degree g' . The first of these in the augumented lexicographic ordering is $x_1^{g'}$ and this is in I since $g' \geq e$. Now consider the monomial $z_1 \ldots z_{g'}$, $z_i \in X_n$. By the second combinatorial lemma, $z_1 \ldots z_{g'}$ either contains a subword u^e with $1 \leq \deg u \leq d$, or it contains a subword with a dominant factorization $u_1 \ldots u_d$ of length d . In the first case $u^e \in I$ and hence $z_1 \ldots z_{g'} \in I$. In the second case $u_1 \ldots u_d$ is congruent modulo I to a K-linear combination of monomials of lower order and the same degree. In this case induction on the order implies that $z_1 \ldots z_{g'} \in I$. Hence the product of any g' elements a_i is 0 . The proof for arbitrary sets $\{a_i\}$ is an immediate consequence of the result for finite sets of generators. □

It is a well known result that is easily proved that if A is algebraic of bounded degree then A satisfies a monic identity (Jacobson [2], p. 14). The same is true of algebras without unit and, in particular, of nil algebras. We therefore have the following consequences of the foregoing results.

COROLLARY 1. *If* A *is an algebra over a commutative ring* K *and* A *is algebraic of bounded degree then* A *is locally finite.*

COROLLARY 2. *If* A *is a nil algebra over* K *of bounded degree then* A *is locally nilpotent.*

We shall consider next Shirshov's local finiteness theorem for special Jordan algebras. Again let A be an associative algebra over a commutative ring K . A *special Jordan algebra* J *in* A is a K-submodule of A containing 1 and closed under the binary product aba . Since $a^2 = a1a$, $a^3 = aaa$, $a^{n+2} = aa^na$ it follows that A is closed under the unary compositions $a \mapsto a^n$. It is easily seen also that A is closed under the trilinear product $\{abc\} = abc + cba$ and the bilinear product $a \circ b = \{a1b\} = ab + ba$. A itself is a special Jordan algebra in A . We denote this as A^+ . More interesting examples are obtained from associative algebras with involution (A, j) . Then the subset $H(A, j)$ of j-symmetric elements of A is a Jordan algebra in A . Homomorphisms of special Jordan algebras are defined to be K-module homomorphisms η such that $\eta 1 = 1$ and $\eta(aba) = (\eta a)(\eta b)(\eta a)$.

If J is a special Jordan algebra in A we let $\mathrm{Env}\,J$ denote the (associative) subalgebra of A generated by J. We have the following

PROPOSITION 1. *If* J *is finitely generated as* K*-module then so is* $\mathrm{Env}\,J$. *Conversely, if* $\mathrm{Env}\,J$ *is finitely generated as* K*-module and* J *is finitely generated as Jordan algebra then* J *is finitely generated as* K*-module.*

Proof. Let $\{u_1, \ldots, u_n\}$ be a subset of J such that $J = \sum Ku_i$. Evidently the u_i generate $\mathrm{Env}\,J$ as algebra so every element of $\mathrm{Env}\,J$ is a linear combination of the monomials $u_{i_1} \ldots u_{i_r}$. We claim that every element of A is a linear combination of 1 and the monomials $u_{i_1} \ldots u_{i_r}$ in which the i_j are distinct. Since the number of these is finite this will prove the first statement. It suffices to show that every $u_{j_1} \ldots u_{j_s}$ in which $j_k = j_l$ for some $l > k$ can be expressed as a linear combination of $u_{i_1} \ldots u_{i_r}$ with distinct i_j. We prove this by induction on s and on $l - k$. If $l = k + 1$,

$$u_{j_1} \ldots u_{j_s} = u_{j_1} \ldots u_{j_k}^2 \ldots u_{j_s}$$

and since $u_{j_k}^2 \in J$ we have $u_{j_k}^2 = \sum \alpha_{kl} u_l$. Substituting this gives an expression for $u_{j_1} \ldots u_{j_s}$ as a linear combination of monomials that are products of $s - 1$ u's. Then we can invoke the degree induction. Next let $l - k > 1$. Then we can use the relation $u_{j_k} u_{j_{k+1}} + u_{j_{k+1}} u_{j_k} = u_{j_k} \circ u_{j_{k+1}} = \sum \beta_l u_l$ to replace $u_{j_k} u_{j_{k+1}}$ by $-u_{j_{k+1}} u_{j_k} + \sum \beta_l u_l$. This gives an expression for $u_{j_1} \ldots u_{j_s}$ as a linear combination of monomials to which the induction applies. This proves the first statement.

The second statement is a consequence of the following result: if J' is a special Jordan algebra that is finitely generated as K-module and J is a subalgebra that is finitely generated as Jordan algebra then J is finitely generated as K-module. The proof of this is identical with that of the lemma in the proof of Shirshov's Theorem. The result we require is obtained by taking $J' = \mathrm{Env}\,J$. □

The most natural way of defining identities for special Jordan algebras is to first define free special Jordan algebras. If $X = \{x_1, x_2, \ldots\}$ we can define the *free special Jordan algebra* $\mathrm{FSJ}(X)$ over K to be the subalgebra of $K\{X\}^+$ generated by X. In other words, this is the smallest K-submodule of $K\{X\}$

containing 1 and X and closed under the product aba. It is easily seen that FSJ(X) has the freeness property that any map of X into a special Jordan algebra J can be extended in one and only one way to a homomorphism of FSJ(X) into J. In a similar manner we can define the free special Jordan algebra $\mathrm{FSJ}(X_n) \subset K\{X_n\}$ where $X_n = (x_1, \ldots, x_n)$. The elements of FSJ(X) or $\mathrm{FSJ}(X_n)$ are called *Jordan polynomials* or *Jordan elements* of $K\{X\}$ or $K\{X_n\}$.

If J is a special Jordan algebra in the associative algebra A over K, an element $g(x_1, \ldots, x_m) \in \mathrm{FSJ}(X)$ is called an *identity* for J if g is mapped into 0 by every homomorphism of FSJ(X) into J. If we denote the image of g under the homomorphism such that $x_i \mapsto a_i$, $1 \le i \le m$, by $g(a_1, \ldots, a_m)$ then g is an identity for J if and only if $g(a_1, \ldots, a_m) = 0$ for all $a_i \in J$. g is called *monic* if it is a monic element of $K\{X\}$.

Associative ring theorists have considered a somewhat different concept of identity for the case of the special Jordan algebra of symmetric elements of an associative algebra with involution. They have considered arbitrary elements $f(x_1, \ldots, x_m) \in K\{X\}$ and required that $f(a_1, \ldots, a_m) = 0$ for all $a_i \in J$. This is equivalent to: f if mapped into 0 by every homomorphism of $K\{X\}$ into A such that $x_i \mapsto a_i \in J$, $1 \le i \le m$. We shall call an element of this sort an *associative identity* for J.

It turns out that it does not matter which of these notions we use, for, as we shall show, a special Jordan algebra has a monic identity if and only if it has an associative monic identity. For the proof of this we need to look at the elements of FSJ(X). We recall first that $K\{X\}$ (or $K\{X_n\}$) has a unique involution ρ such that $x_{i_1} \ldots x_{i_r} \mapsto x_{i_r} \ldots x_{i_1}$. This is called the *reversal involution*. Let $H(K\{X\}, \rho)$ denote the subset of $K\{X\}$ of symmetric elements under ρ. This is a special Jordan algebra in $K\{X\}$ containing X. Hence $\mathrm{FSJ}(X) \subset H(K\{X\}, \rho)$. Similarly, $\mathrm{FSJ}(X_n) \subset H(K\{X_n\}, \rho)$. If $n = 2$ we have

PROPOSITION 2. $\mathrm{FSJ}(X_2) = H(K\{X_2\}, \rho)$.

Proof. Write $x = x_1$, $y = x_2$. Any ρ-symmetric element of $K\{x, y\}$ is a linear combination of elements of the following forms:

$$\dots x^{i_2}y^{j_1}x^{i_1}y^{j_1}x^{i_2}\dots$$

$$\dots y^{j_2}x^{i_1}y^{j_1}x^{i_1}y^{j_2}\dots$$

$$x^{i_1}y^{j_1}\dots x^{i_r} + x^{i_r}\dots y^{j_1}x^{i_1}$$

$$y^{j_1}x^{i_1}\dots y^{j_r} + y^{j_r}\dots x^{i_1}y^{j_1}$$

$$x^{i_1}y^{j_1}\dots x^{i_r}y^{j_r} + y^{j_r}x^{i_r}\dots y^{j_1}x^{i_1}\ , \quad i_k, j_k > 0\ .$$

Those in the first two lines are clearly Jordan polynomials and if we can show that the ones in the last line are Jordan polynomials it will follow that those in the third and fourth lines are also Jordan polynomials. For the ones in the last line we prove the result by induction on the *height* r. We have

$$x^{i_1}\left(y^{j_1}x^{i_2}\dots x^{i_r}+x^{i_r}\dots x^{i_2}y^{j_1}\right)y^{j_r} + y^{j_r}\left(y^{j_1}x^{i_2}\dots x^{i_r}+x^{i_r}\dots x^{i_2}y^{j_1}\right)x^{i_1}$$
$$= \left(x^{i_1}y^{j_1}\dots x^{i_r}y^{j_r}+y^{j_r}x^{i_r}\dots y^{j_1}x^{i_1}\right)$$
$$+ \left(x^{i_1+i_r}y^{j_{r-1}}\dots x^{i_2}y^{j_1+j_r}+y^{j_1+j_r}x^{i_2}\dots y^{j_{r-1}}x^{i_1+i_r}\right)\ .$$

The height induction implies that the left hand side and the second parenthesis on the right hand side are Jordan polynomials. It follows that

$$x^{i_1}y^{j_1}\dots x^{i_r}y^{j_r} + y^{j_r}x^{i_r}\dots y^{j_1}x^{i_1}$$

is a Jordan polynomial. □

We can now prove

PROPOSITION 3. *J has a monic identity if and only if it has an associative monic identity.*

Proof. Since a monic identity is an associative monic identity it remains to show that if J has an associative monic identity then it has a monic identity. By linearization we can show that if J has an associative monic identity then it has a multilinear one, say,

$$f = x_1\dots x_n + \sum_{\pi\neq 1}\alpha_\pi x_{\pi 1}\dots x_{\pi n}\ .$$

Now apply the homomorphism of $K\{X\}$ into $K\{x, y\}$ such that $x_i \mapsto xy^i$, $i = 1, 2, \dots$. The image of f under this homomorphism is the monic homogeneous polynomial

$$f(xy, xy^2, \dots, xy^n) = xyxy^2\dots xy^n + \dots\ .$$

Next apply the reversal operator ρ and form

$$g(x, y) = f(xy, \ldots, xy^m)\rho f(xy, xy^2, \ldots) = xyxy^2 \ldots xy^{2n}x \ldots y^2xyx + \ldots .$$

This is monic and symmetric and is an identity for J. By Proposition 2, $g \in \mathrm{FSJ}(x, y)$.

We have seen that any special Jordan algebra is closed under powers. We can therefore define algebraic elements as in the associative case: $a \in J$ is *algebraic* if there exists a monic $f(\lambda) \in K[\lambda]$ such that $f(a) = 0$. The least degree for such polynomials is called the *degree* of a.

To state the local finiteness theorem for special Jordan algebras we require also the concept of *Jordan monomial* of $\mathrm{FSJ}(X)$. We define these inductively by: 1 and the x's are Jordan monomials and if p, q and r are Jordan monomials then so are pqp and $pqr + rqp$. It is clear from the definition that Jordan monomials are homogeneous elements of $K\{X\}$. We note also that if p is a Jordan monomial then so is p^k for any $k \geq 0$ and if p and q are Jordan monomials then so is $p \circ q = pq + qp = p1q + q1p$. Since

$$\left(\sum \alpha_i p_i\right) q \left(\sum \alpha_i p_i\right) = \sum \alpha_i^2 p_i q p_i + \sum_{i<j} \alpha_i \alpha_j \left(p_i q p_j + p_j q p_i\right)$$

it is clear that the set of K-linear combinations of the Jordan monomials is a subalgebra of $\mathrm{FSJ}(X)$ containing X. Hence it coincides with $\mathrm{FSJ}(X)$. Thus any Jordan polynomial is a linear combination of Jordan monomials.

If $a_1, \ldots, a_n$ are elements of a special Jordan algebra and $p(x_1, x_2, \ldots, x_n)$ is a Jordan monomial then $p(a_1, \ldots, a_n)$ will be called a *Jordan monomial in the* a_i.

We are now ready to state an extension of Shirshov's theorem on special Jordan algebras.

THEOREM 2. *Let* J *be a special Jordan algebra in the associative algebra* A *and assume* $\mathrm{Env}\, J = A$. *Suppose* J *satisfies a monic identity and* J *has a set of generators* $\{a_i\}$ *(not necessarily finite) such that every Jordan monomial in the* a_i *is algebraic. Then* A *and* J *are locally finite.*

We remark that the hypothesis $\mathrm{Env}\, J = A$ is not a real restriction since it can be achieved by replacing A by $\mathrm{Env}\, J$.

For the proof of the theorem we shall require the concept of a k-word in $\mathrm{FM}(X_n)$ and another combinational lemma on free monoids. If $1 \leq k \leq n$ we define a *k-word* as either a power x_k^e, $e > 1$, or a word $x_k w x_i$ where $w \in \mathrm{FM}(X_k)$ and $i < k$. The k-words of the first kind are Jordan monomials and any k-word of the second kind is the highest word in a suitable Jordan monomial. More precisely, if

$x_k w x_i$ is a k-word then there exists a Jordan monomial q in FSJ(X) such that

$$q = x_k w x_i + \sum \alpha_{w'} w'$$

where $w' \in \mathrm{FM}(X_k)$ and w' begins with some x_j, $j < k$. Then $w' < x_k w x_i$. To see this we write

$$x_k w x_i = x_k^{e_1} w_1 x_k^{e_2} w_2 \ldots x_k^{e_r} w_r$$

where the $e_i \geq 1$, $w_i \in \mathrm{FM}(X_{k-1})$ and $\deg w_i \geq 1$. Suppose first that $r = 1$ and write $w_1 = x_{i_1} \ldots x_{i_s}$, $1 \leq i_j \leq k-1$, $s \geq 1$. Consider the Jordan monomial

$$q_1 = \left(\ldots\left(\left(x_k^{e_1} \circ x_{i_1}\right) \circ x_{i_2}\right) \ldots\right) \circ x_{i_s} .$$

It is clear that the only term beginning with x_k in q_1 is $w = x_k^{e_1} x_{i_1} \ldots x_{i_s}$. Now suppose we have a Jordan monomial $q_{r-1} \in \mathrm{FM}(X_k)$ such that the only term beginning with x_k in q_{r-1} is $x_k^{e_1} w_1 \ldots x_k^{e_{r-1}} w_{r-1}$ and let $w_r = x_{i_1} \ldots x_{i_s}$. Put

$$q_r = \begin{cases} q_{r-1} x_k^{e_r} x_{i_1} + x_{i_1} x_k^{e_r} q_{r-1} & \text{if } s = 1 , \\ \left(\ldots\left(\left(q_{r-1} x_k^{e_r} x_{i_1} + x_{i_1} x_k^{e_r} q_{r-1}\right) \circ x_{i_2}\right) \ldots\right) \circ x_{i_s} & \text{if } s > 1 . \end{cases}$$

Then it is readily seen that the only term beginning with x_k in q_r is the given k-word $w = x_k^{e_1} w_1 \ldots x_k^{e_r} w_r$.

The combinatorial lemma we require is the

THIRD COMBINATORIAL LEMMA. *Let* m, n *be positive integers,* $M(n)$ *a positive integral valued function of* n. *Then there exists a positive integer* $d(n, m, M(n))$ *such that any word* $w \in \mathrm{FM}(X_n)$ *of degree greater than* $d(n, m, M(n))$ *contains a subword* w_0 *such that for some* k, $1 \leq k \leq m$, *either*

(i) $w_0 = x_k^{M(1)}$; *or*

(ii) $w_0 = u^{M(k)}$ *for a* k*-word* u *such that*

$$1 \le \deg u \le \delta(k) = m\big(M(1)+d(k-1,\ m,\ M(k-1))\big)$$

where $\delta(1) = mM(1)$; *or*

(iii) w_0 *has a dominant factorization* $w_0 = u_1 \ldots u_m$ *of length* m *in which every* u_i *is a* k*-word.*

We delay the proof of this lemma and use it to give the

Proof of Theorem 2. We assume first that $\{a_i\} = \{a_1, \ldots, a_n\}$ and that we have a polynomial

$$f(x_1, \ldots, x_m) = x_1 \ldots x_m - \sum_{\pi \ne 1} \alpha_\pi x_{\pi 1} \ldots x_{\pi m} \tag{12}$$

such that $f(b_1, \ldots, b_m) = 0$ for all $b_j \in J$. Let $M(1)$ be the maximum degree of algebraicity of the a_i and define $M(k)$ inductively as the maximum degree of algebraicity of the Jordan monomials in the a_i of degree (as given in the free algebra) at most $\delta(k) = m\big(M(1) + d(k-1, m, M(k-1))\big)$ where $d(k-1, m, M(k-1))$ is as defined in the Third Combinatorial Lemma. Since the a_i generate J and J generates A, the a_i generate A. We have a homomorphism η of $K\{X_n\}$ into A such that $x_i \mapsto a_i$, $1 \le i \le n$. Let $I = \ker \eta$. To prove that A is finitely generated as K-module it suffices to show that every $w \in \mathrm{FM}(X_n)$ of degree greater than $d = d(n, m, M(n))$ is congruent modulo I to a linear combination of elements of $\mathrm{FM}(X_n)$ of degree less than or equal to d. We use induction on the degree of w and for a given degree induction on the order. We apply the Third Combinatorial Lemma to w and the subword w_0. In case *(i)* we use the fact that $a_k^{M(1)}$ is a linear combination of $1, a_k, \ldots, a_k^{M(1)-1}$ to conclude that w is congruent modulo I to a linear combination of monomials of lower degree. Then the degree induction is applicable. In case *(ii)* we write $u = q + u'$ where q is a Jordan monomial of the same degree as u and u' is a linear combination of elements of $\mathrm{FM}(X_k)$ of the same degree as u and of lower order than u. Then $u^{M(k)} = q^{M(k)} + u''$ where u'' is a linear combination of monomials of the same degree and lower order than $u^{M(k)}$. Since $q^{M(k)}$ is congruent modulo I to a linear combination of $1, q, \ldots, q^{M(k)-1}$, degree induction and order induction give the result in this case. Now suppose we have case *(iii)*: $w_0 = u_1 \ldots u_m$ where the u_i are k-words and the factorization is dominant. We write $u_i = q_i + u_i'$ where q_i is a Jordan monomial and u_i' is a linear combination of monomials of the same degree as u_i and of lower order. Then, by (12) applied to the image of q_i, we have

$$u_1 \cdots u_m = q_1 \cdots q_m + u'' \equiv \sum_{\pi \neq 1} \alpha_\pi q_{\pi 1} \cdots q_{\pi m} + u'' \equiv \sum \alpha_\pi u_{\pi 1} \cdots u_{\pi m} + u''' \pmod{I}$$

where u''' is a linear combination of monomials of lower order than $u_1 \cdots u_m$. Since $u_1 \cdots u_m$ is a dominant factorization the result follows in this case by the order induction.

Now suppose $\{a_i\}$ is arbitrary and let $b_1, \ldots, b_l$ be arbitrary elements of A. Then these elements are contained in a subalgebra of A generated by a finite subset of $\{a_i\}$, say, $\{a_1, \ldots, a_n\}$. By what we have proved, the subalgebra generated by the a_i is finitely generated as K-module. Hence, as we showed before, the subalgebra generated by the b_j is a finitely generated K-module. This shows that A is locally finite. It remains to show that J is locally finite. To see this we now suppose that the $b_j \in J$. Then the subalgebra of A generated by the b_j is finitely generated as K-module. Hence, by Proposition 1, the subalgebra of J generated by the b_j is a finitely generated K-module. □

It remains to give the

Proof of the Third Combinatorial Lemma. For any n, m, M let $g(n, m, M)$ be as in the Second Combinatorial Lemma. Define $d(n, m, M(n))$ inductively by

$$d(1, m, M(1)) = M(1),$$

$$d(n+1, m, M(n+1)) = (g(N, m, M)+1)(M(1)+d(n, m, M(n)))$$

where $M = M(n+1)$ and $N = N(n+1, d(n, m, M(n)), M(1))$ as in (4), so N is the number of words in

$$Y_N = \{x_{n+1}^e v \mid 1 \leq e \leq M(1),\ v \in \mathrm{FM}(X_n),\ 1 \leq \deg v \leq d(n, m, M(n))\}.$$

The lemma will be proved by induction on n. It holds for $n = 1$ since in this case $d(1, m, M(1)) = M(1)$ and any word of degree greater than $d(1, m, M(1))$ contains the subword $x_1^{M(1)}$. Now assume the lemma for n and let $w \in \mathrm{FM}(X_{n+1})$ have degree greater than $d(n+1, m, M(n+1))$. Write

$$w = w_0 x_{n+1}^{e(1)} w_1 x_{n+1}^{e(2)} \cdots x_{n+1}^{e(r)} w_r x_{n+1}^{e(r+1)}$$

where $e(i) > 0$ if $i < r+1$, $w_i \in \mathrm{FM}(X_n)$ and $\deg w_i > 0$ if $i > 0$. If some $e(i) > M(1)$ then w contains the subword $x_{n+1}^{M(1)}$ and we are done. If some w_i has degree greater than $d(n, m, M(n))$ we are done by the induction hypothesis. Thus we may assume every $e(i) \leq M(1)$ and every $\deg w_i \leq d(n, m, M(n))$. Then

$$\deg w \le (r+1)\big(M(1)+d(n,\ m,\ M(n))\big)$$

and since

$$\deg w > d(n+1,\ m,\ M(n+1)) = \big(g(N,\ m,\ M)+1\big)\big(M(1)+d(n,\ m,\ M(n))\big)$$

it follows that

$$r > g(N,\ m,\ M)\ .$$

Hence we may assume that if Y_N is defined as above then w contains a Y_N-subword of Y-degree $r > g(N,\ m,\ M)$. Then the Second Combinatorial Lemma implies that w either contains a subword u^M , $u \in \mathrm{FM}(Y_N)$, where the Y-degree is positive and less than or equal to m or w has a subword that has a dominant factorization in $\mathrm{FM}(Y_N)$ (and hence in $\mathrm{FM}(X_{n+1})$) of length m . In the first case $\deg u \le m\big(M(1) + d(n,\ m,\ M(n))\big) = \delta(n+1)$ and u is an $(m+1)$-word. In the second case we have a dominant factorization of length m into $(m+1)$-words. □

Theorem 2 can be extended also to algebras without unit. If A is an associative algebra without unit, a *special Jordan algebra without unit* J in A is a K-submodule of A such that if $a, b \in J$ then aba and $a^2 \in J$. Homomorphisms between such algebras are defined as K-module maps η such that $\eta(aba) = \eta(a)\eta(b)\eta(a)$ and $\eta(a^2) = \eta(a)^2$. Let $K\{X\}'$ denote the ideal in $K\{X\}$ generated by the x_i . This is the set of elements of $K\{X\}$ with zero constant term. Let $\mathrm{FSJ}(X)'$ denote the smallest K-submodule of $K\{X\}'$ containing X and closed under a^2 and aba . Then it is readily seen that if J is a special Jordan algebra without unit and $x_i \mapsto a_i$, $i = 1, 2, \dots$, is a map of X into J then this has a unique extension to a homomorphism of $\mathrm{FSJ}(X)'$ into J . The various concepts we had before carry over. Moreover, the method used to prove Theorem 1 carries over to prove the following

THEOREM 3. *Let J be a special Jordan algebra without unit contained in an associative algebra A without unit such that $\mathrm{Env}\, J = A$ where $\mathrm{Env}\, J$ is the subalgebra of A generated by J . Assume J satisfies a monic identity and J has a set of generators $\{a_i\}$ such that every Jordan monomial in the a_i is nilpotent. Then A is locally nilpotent.*

The results can be applied also to algebraic and nil special Jordan algebras of bounded degree since it can be shown that these satisfy monic identities.

We shall conclude our discussion of the Jordan case by showing how the results apply to abstract Jordan algebras. For simplicity we restrict our attention to unital algebras. The concept of an (abstract) Jordan algebra arises in attempting to treat special ones in an intrinsic manner. This is highly desirable for many reasons. For one thing, the homomorphic image of a special Jordan algebra need not be special, so this class of algebras is not a variety. The definition of a (unital) Jordan algebra that is due to McCrimmon is as follows. A *Jordan algebra over* K is a triple $(J, U, 1)$ where J is a K-module, $1 \in J$ and U is a map of J into $\mathrm{End}_K J$ such that the following axioms hold:

QJ1. U is quadratic: $U_{\alpha a} = \alpha^2 U_a$ if $\alpha \in K$, $a \in J$.

QJ2. $U_1 = 1$.

QJ3. $U_a U_b U_a = U_{U_a b}$.

QJ4. If $U_{a,b} = U_{a+b} - U_a - U_b$ and $V_{a,b}$ is defined by $V_{a,b}x = U_{a,x}b$ then $U_a V_{b,a} = V_{a,b} U_a$.

QJ5. QJ3 and QJ4 hold "absolutely", that is, for all J_L where L is a commutative (associative) algebra over K.

If A is an associative algebra over K, A defines a Jordan algebra A^+ in which A is the K-module, the unit of A is the 1 in the definition and U_a is defined by $U_a x = axa$. If J and J' are Jordan algebras, a *homomorphism* η of J into J' is defined to be a K-module map such that $\eta(1) = 1'$, $\eta(U_a b) = U_{\eta(a)}\eta(b)$. The class of Jordan algebras over K with this definition of morphism constitutes a category. If A and A$'$ are associative and η is a homomorphism of A into A$'$ then η is a homomorphism of A^+ into A'^+. Then we have a functor from the category of associative algebras to the category of Jordan algebras. This functor has an adjoint. This fact amounts to the following: given any Jordan algebra J, there exists an associative algebra $s(J)$ and a homomorphism σ_u of J into $s(J)^+$ such that if η is any homomorphism of J into A^+ where A is associative then there exists a unique homomorphism ζ of associative algebras such that

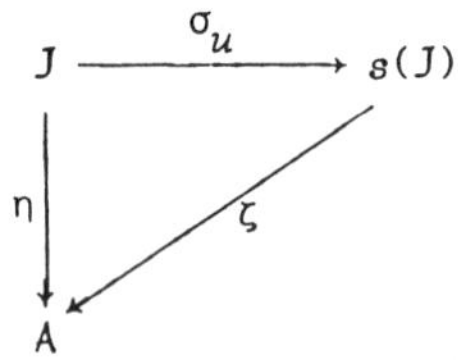

is commutative. $(s(J), \sigma_u)$ or simply $s(J)$ is called the *special universal algebra* of J. Evidently this is analogous to the universal enveloping algebra of a Lie algebra.

We shall obtain a consequence on $s(J)$ of Theorem 2. To formulate this we need to define powers, algebraic elements and monic identities for Jordan algebras. If $a \in J$ we define $a^0 = 1$, $a^1 = a$, $a^{n+2} = U_a a^n$. If $J = A^+$ then a^n defined in J coincides with a^n as defined in A. For arbitrary J we call an element $a \in J$ algebraic if there exists a monic polynomial $f(\lambda) \in K[\lambda]$ such that $f(a) = 0$ with the obvious meaning for $f(a)$. To define identities for Jordan algebras we need the concept of a free Jordan algebra $FJ(X)$ defined by $X = \{x_1, x_2, \ldots\}$. We shall not give a construction for this but will be content to invoke a general theorem on universal algebra that insures the existence of an $FJ(X)$ such that $FJ(X)$ contains X and any map of X into a Jordan algebra J has a unique extension to a homomorphism of $FJ(X)$ into J. In particular we have a unique homomorphism ν of $FJ(X)$ into $FSJ(X)$ such that $\nu x_i = x_i$, $i = 1, 2, 3, \ldots$. We call an element $f \in FJ(X)$ an *identity* for J if $f \mapsto 0$ under every homomorphism of $FJ(X)$ into J. It is known that $FSJ(X)$ has non-zero identities and these will be identities for all special Jordan algebras. We wish to avoid these and so we confine attention to identities f that are *monic* in the sense that νf is a monic element of $FSJ(X)$.

We can now obtain the abstract version of Theorem 2. We shall prove the following somewhat weaker result.

THEOREM 4. *Let* J *be a Jordan algebra such that:*

1. J *satisfies a monic identity, and*
2. J *is algebraic in the sense that every element of* J *is algebraic.*

Then the special universal envelope $s(J)$ *is locally finite.*

Proof. The image $\sigma_u(J)$ is a special Jordan algebra in $s(J)$ and it is well known that $\text{Env}\, \sigma_u(J) = s(J)$. Now we have the hypothesis of Theorem 2 for $\sigma_u(J)$ and $s(J)$. Hence the local finiteness of $s(J)$ follows from Theorem 2. □

References

Nathan Jacobson

[1] *Lectures on quadratic Jordan Algebras* (Tata Institute of Fundamental Research Lectures on Mathematics, 45. Tata Institute of Fundamental Research, Bombay, 1969).

[2] *PI-Algebras. An Introduction* (Lecture Notes in Mathematics, 441. Springer-Verlag, Berlin, Heidelberg, New York, 1975).

Kevin McCrimmon

[1] "Alternative algebras satisfying polynomial identities", *J. Algebra* 24 283-292.

А.И. Ширшов [A.I. Shirshov]

[1] "О некоторых неассоциативных нилькольцах и алгебраических алгебрах" [On some non-associative nil rings and algebraic algebras], *Mat. Sb. (N.S.)* 41 (83) (1957), 381-394.

[2] "О кольцах с тождественными соотношениями" [On rings with identical relations], *Mat. Sb. (N.S.)* 43 (85) (1957), 277-283.

M. Slater

[1] "Local functions for certain alternative algebras", Lecture Notes, University of Bristol, Bristol, 1967.

Department of Mathematics,
Yale University,
New Haven,
Connecticut,
USA.

SEA BULL MATH 5(1981), 27-38

SURVEY OF JORDAN STRUCTURE THEORY*

Nathan Jacobson

In this lecture we shall give a historical survey of the structure theory of Jordan algebras. We shall follow a straight line beginning with the first results obtained in 1934 by Jordan, von-Neumann and Wigner and ending with some remarkable results published during the past year by a young Russian mathematician, E. Zelmanov. Jordan algebras are not as well known as I believe they deserve to be, in view of their applications to many fields, notably, real and complex analysis, symmetric spaces, exceptional geometries, Lie groups and Lie algebras. Time does not permit the indication of any of these, an account of which can be found in an excellent expository article "Jordan Algebras and their Applications" by Kevin McCrimmon in BAMS, v. 84.

We shall begin with an indication of how the subject arose and of two alternative definitions of the concept of an abstract Jordan algebra. We consider the set $HM_n(\mathbb{R})$ of $n \times n$ real symmetric matrices. This is a real vector space containing the unit matrix 1. It is natural to seek to endow it with a richer algebraic structure by introducing other compositions. The simplest one appears to be the bilinear one

$$a \cdot b = \frac{1}{2}(ab + ba) \tag{1}$$

which is preferable to $a \circ b = ab + ba$ since $1 \cdot a = a = a \cdot 1$ whereas $1 \circ a = 2a$. A little experimentation shows that most of the obvious compositions which $HM_n(\mathbb{R})$ permits are expressible in terms of $a \cdot b$. For example, it is closed under a^n, $n \geq 1$, and we have $a^n = a^{n-1} \cdot a$ so a^n is expressible in terms of the $\cdot$ product. Similarly $HM_n(\mathbb{R})$ is closed under aba and this can be expressed in terms of the $\cdot$ product by

---

* Invited address delivered at the 5th Biennial General Meeting and Conference of the Southeast Asian Mathematical Society, Hong Kong, June 1980.

$$(2) \qquad aba = 2(a\cdot b)\cdot a - a^2\cdot b, \quad a^2 = a\cdot a$$

On the other hand, $a\cdot b$ can be expressed in terms of the composition aba, the vector space operations and the unit since we have

$$abc + cba = (a+c)b(a+c) - aba - cbc$$

and $a\cdot b = \frac{1}{2}(a1c + c1a)$.

The first definition of a Jordan algebra was based on the composition $a\cdot b$. The abstract concept was introduced by a German physicist, Pascual Jordan, who first studied the properties of the composition $a\cdot b$ in an associative algebra over $\mathbb{R}$. He discovered a number of properties of this composition and was able to show that all of these were consequences of two: commutativity and the identity $(a^2\cdot b)\cdot a = a^2\cdot(b\cdot a)$, which is easy to verify and which is now called the <u>Jordan identity</u>. This leads to the following definition of a linear Jordan algebra over any commutative ring Φ containing $\frac{1}{2}$.

<u>Definition 1</u>. A linear Jordan algebra over a commutative ring $\Phi \ni \frac{1}{2}$ is a Φ-module equipped with a bilinear product $a\cdot b$ and an element 1 such that the following properties hold

J1 $\qquad a\cdot 1 = a$

J2 $\qquad a\cdot b = b\cdot a$

J3 $\qquad (a^2\cdot b)\cdot a = a^2\cdot(b\cdot a)$.

Standard ways of obtaining such algebras are the following: Let $\mathcal{A}$ be an associative algebra over $\Phi \ni \frac{1}{2}$ and define the product $a\cdot b$ in $\mathcal{A}$ by (1). Evidently this is bilinear and it is easy to check that J1–J3 hold. Thus we have a linear Jordan algebra with composition $a\cdot b$. We denote this as $\mathcal{A}^+$. Evidently this is reminiscent of the process of constructing a Lie algebra $\mathcal{A}^-$ from $\mathcal{A}$ by replacing the given associative product ab by the additive commutator $[a,b] = ab - ba$.

Next let $\mathcal{A}$ have an involution j, that is, a map $a \rightsquigarrow a^*$ that is Φ-bilinear and satisfies $1^* = 1$, $(ab)^* = b^*a^*$, $(a^*)^* = a$. Let $H(\mathcal{A},j)$ denote the set of j-symmetric elements of $\mathcal{A}(h^* = h)$. It is clear that this set is a Φ-submodule of $\mathcal{A}$ containing 1 and closed under $a \cdot b$. In other words, $H(\mathcal{A},j)$ is a subalgebra of $\mathcal{A}^+$. The example we considered first is obtained in this way from the algebra $M_n(\mathbb{R})$ of $n \times n$ matrices over $\mathbb{R}$ and the transposed involution $A \rightsquigarrow {}^tA$. Other examples are $\mathcal{A} = M_n(\mathbb{C})$ where j is the hermitian involution $A \rightsquigarrow {}^t\bar{A}$ (the conjugate transposed of A). Then $H(\mathcal{A},j) = HM_n(\mathbb{C})$ the set of complex hermitian matrices. Similarly we can obtain the Jordan algebra $HM_n(\mathbb{H})$ of $n \times n$ hermitian quaternionic matrices.

We consider next the quadratic point of view for Jordan algebras. Again we begin with an associative algebra $\mathcal{A}$ over Φ but this time we need not restrict Φ in any way. We now focus on the composition aba in addition to the Φ-module structure of $\mathcal{A}$ and the unit 1. The product aba is linear in b and quadratic in a. There are many identities on this product. It turns out that the basic ones on which an abstract theory can be built are the following:

(3) $$a(b(a\,x\,a)b)a = (aba)x(aba)$$

(4) $$ab(a\,x\,a) + (a\,x\,a)ba = a(bax + xab)a.$$

It is preferable to state these in operator form by introducing the Φ-endomorphism

(5) $$U_a : x \rightsquigarrow a\,x\,a$$

Then we have the quadratic map

(6) $$U : a \rightsquigarrow U_a$$

of $\mathcal{A}$ into $\mathrm{End}_\Phi \mathcal{A}$. Next we put

(7) $$U_{a,b} = U_{a+b} - U_a - U_b$$

so $(a,b) \rightsquigarrow U_{a,b}$ is Φ-bilinear. Finally we define $V_{a,b}$ by $V_{a,b}x = U_{a,x}b$. These maps are respectively

$$U_{a,b}: x \rightsquigarrow a\,x\,b + b\,x\,a$$

$$V_{a,b}: x \rightsquigarrow abx + xba.$$

Using U_a, $U_{a,b}$ and $V_{a,b}$ we can write (3) and (4) in the following operator form

(8) $$U_a U_b U_a = U_{U_a b}$$

(9) $$V_{a,b} U_a = U_a V_{b,a}.$$

This leads to the following definition of an abstract quadratic Jordan algebra.

Definition 2. A quadratic Jordan algebra is a triple $(J,U,1)$ where J is a Φ-module over a commutative ring Φ, 1 is an element of J and U is a quadratic map of J into $\mathrm{End}_\Phi J$ such that

QJ1 $U_1 = 1$ (the identity map on J)

QJ2 $U_a U_b U_a = U_{U_a b}$

QJ3 If $U_{a,b} = U_{a+b} - U_a - U_b$ and $V_{a,b}$ is defined by $V_{a,b}x = U_{a,x}b$ then

$$U_a V_{b,a} = V_{a,b} U_a$$

QJ4 QJ2 and QJ3 hold strictly, that is, they hold for any $J_P = P \otimes_\Phi J$ for any commutative algebra P/Φ.

These axioms are due to McCrimmon, who showed also that the concepts of quadratic and linear Jordan algebras are equivalent if $\Phi \ni \frac{1}{2}$. More precisely, if J is a linear Jordan algebra over $\Phi \ni \frac{1}{2}$ then one obtains a quadratic Jordan algebra $(J,U,1)$ by defining

(10) $$U_a = 2L_a^2 - L_{a^2},\quad L_a: x \rightsquigarrow a \cdot x \qquad \text{(cf. (2))}.$$

On the other hand, if $(\mathcal{J},U,1)$ is quadratic over Φ then we obtain a linear Jordan algebra by defining

(11) $$a\cdot b = \frac{1}{2}U_{a,b}1.$$

The two constructions are inverses. Moreover, if we define homomorphism in the quadratic case by: η is linear (= Φ homomorphism) $\eta(1) = 1$, $\eta(U_a b) = U_{\eta(a)}\eta(b)$ then a map is a homomorphism of the linear structure if and only if it is a homomorphism of the quadratic one. It follows that we have a category isomorphism of the category of linear Jordan algebras with the category of Jordan algebras, assuming $\Phi \ni \frac{1}{2}$.

There are many advantages of taking the quadratic point of view. Aside from the removal of the condition $\Phi \ni \frac{1}{2}$ it forces us to focus on the right concepts and it simplifies many proofs. However, it takes a bit of concentration to work in the unfamiliar setting of quadratic maps. For this reason, unless otherwise indicated, we shall assume $\Phi \ni \frac{1}{2}$ and hence our algebras are linear Jordan algebras.

The structure theory of Jordan algebras began with a big bang in the publication in the 1934 Annals of Mathematics of a paper "An Algebraic Generalization of the Quantum Mechanical Formalism" by Jordan, vonNeumann and Wigner. They noted that observables in quantum mechanics were represented by hermitian operators in Hilbert space and the physically meaningful compositions on these were addition, multiplication by real numbers and squaring, from which one deduces the bilinear product $A\cdot B = \frac{1}{2}(A+B)^2 - \frac{1}{2}A^2 - \frac{1}{2}B^2$. As we have noted, the composition satisfies the conditions J2 and J3. Moreover, we have the condition of formal reality: $\Sigma A_i^2 = 0 \Rightarrow$ every $A_i = 0$. They set themselves the problem of studying systems satisfying just these axioms, with the hope of discovering some new systems that might give a better formalism for quantum mechanics. While it appeared that they needed to consider infinite dimensional algebras, in this paper, as a preliminary, they proposed to study finite dimensional ones. Thus they considered finite dimensional Jordan algebras over $\mathbb{R}$ that are formally real ($\Sigma a_i^2 = 0 \Rightarrow$ every $a_i = 0$). They succeeded in determining all of these algebras. First, they showed that any $\mathcal{J}$ satisfying the conditions is a direct sum of simple ones and they determined the simple ones. They showed that the following is the list of these:

I $\quad HM_n(\mathbb{R})$, the Jordan algebra of $n \times n$ real symmetric matrices.

II $\quad HM_n(\mathbb{C})$, the Jordan algebra of $n \times n$ complex hermitian matrices.

III $\quad HM_n(\mathbb{H})$, the Jordan algebra of $n \times n$ quaternionic hermitian matrices.

IV $\quad HM_3(\mathbb{O})$, the algebra of 3×3 hermitian octonion matrices with $A \cdot B = \frac{1}{2}(AB+BA)$.

V $\quad \mathcal{J} = R1 \oplus V$ where V is a real vector space equipped with a positive definite scalar produce $\langle , \rangle$ with multiplication defined by

(12) $\quad (\alpha 1+x)\cdot(\beta 1+y) = (\alpha\beta+\langle x,y\rangle)) + \alpha y + \beta x.$

It is readily seen that (12) defines a Jordan algebra. Also they showed that the system $HM_3(\mathbb{O})$, which they denoted as M_3^8, satisfies the axioms for a Jordan algebra. It is clear that I-III are subalgebras of algebras of the form $\mathcal{A}^+$, $\mathcal{A}$ associative. A Jordan algebra that is isomorphic to a subalgebra of an $\mathcal{A}^+$, $\mathcal{A}$ associative, is called <u>special</u>. It is not difficult to see, using Clifford algebras, that V is also special. Thus I, II, III and V are special. Jordan - von Neumann and Wignor conjectured that IV is not special. They proposed the problem of showing this to Albert, who proved that M_3^8 is <u>exceptional</u>, that is, is not a special Jordan algebra. Since they were looking for algebras that could not be obtained from associative ones the only candidate appeared to be M_3^8. Since the example appeared not to be of interest in physics their quest for an improved framework for quantum mechanics appeared to be a failure from the standpoint of an application to physics.

Nothing more was done about Jordan algebras until 1946 when Albert initiated the study of such algebras over an arbitrary field of characteristic $\neq 2$. In a series of papers that appeared from 1946 to 1950, Albert obtained for finite dimensional Jordan algebras over a field of characteristic $\neq 2$ analogues of the principal theorems of Wedderburn's structure theory for finite dimensional associative algebras. He defined the radical of such an algebra as the maximal nil ideal, that is, the maximal ideal all of whose elements are nilpotent. He showed that the radical is nilpotent in the sense that there exists an integer N such that the product of any N elements of the radical taken in any association is 0. Albert also obtained the structure of the semi-simple algebras, that is, the algebras that have radical 0. He showed that such an algebra is a direct sum of simple ones and he determined the simple ones over an algebraically closed field. The list of these was like that of Jordan-von Neumann and Wigner with the difference that the field $\mathbb{R}$ was replaced by the algebraically closed base field Φ.

There was a small gap in Albert's structure theory. In seeking to fill this I was led to introduce the U operator into Jordan theory, that is, the quadratic map $U_a = 2L_a - L_{a^2}$, and to define inverses in Jordan algebras. Using the quadratic point of view, we call an element a <u>invertible</u> if U_a is invertible in $\mathrm{End}_\Phi J$. It is easy to see that a is invertible if and only if 1 is in the range of U_a. Moreover, if we define $a^{-1} = U_a^{-1}a$ then a^{-1} is invertible and $(a^{-1})^{-1} = a$. If $\mathcal{a}$ is an associative algebra then a is invertible in $\mathcal{a}$ iff it is invertible in $\mathcal{a}^+$, in which case a^{-1} is the same in $\mathcal{a}$ and in $\mathcal{a}^+$.

The elements a and b are invertible if and only if $U_a b$ is invertible. The concept of invertibility permits the definition of a <u>Jordan division algebra</u>: a Jordan algebra in which every non-zero element is invertible.

Another important concept for Jordan algebras that has an associative background is that of isotopy. This arises from the following simple remark: If c is an element of an associative algebra then we can define a new c-multiplication in $\mathcal{A}$ by $a_c b = acb$. This is again associative and, moreover, if c is invertible then the element c^{-1} serves as unit since $c^{-1}{}_c a = a = a_c c^{-1}$. This suggests the following result which we shall state in the quadratic form: Let $(\mathcal{J},U,1)$ be a quadratic Jordan algebra and let c be an invertible element of $\mathcal{J}$. Define $U^{(c)} = UU_c$ or $U_a^{(c)} = U_a U_c$. Then $(\mathcal{J},U^{(c)},c^{-1})$ is a quadratic Jordan algebra. This is called the c-<u>isotope</u> of $\mathcal{J}$. In the linear situation this amounts to replacing the given composition $a \cdot b$ by $a \cdot_c b = U_{a,b} c$. Two algebras are called <u>isotopic</u> if each is isomorphic to an isotope of the other. This concept is broader than isomorphism and is often preferable to isomorphism.

In 1965, following a definition given by Topping for Jordan operator algebras in Hilbert space, I introduced a concept of inner (or quadratic) ideal in abstract Jordan algebras and I used this to develop an analogue of the Wedderburn-Artin structure theorem for associative algebras. If $\mathcal{J}$ is a Jordan algebra then we define an <u>inner ideal</u> B as a Φ-submodule of $\mathcal{J}$ such that $U_b x \in B$ for $b \in B$, $x \in \mathcal{J}$. It turns out that these are analogous to the one-sided ideals of the associative theory. For example, if $\mathcal{A}$ is associative then any left or right ideal of $\mathcal{A}$ is an inner ideal of $\mathcal{A}^+$. More generally, the intersection of a left and a right ideal of $\mathcal{A}$ is an inner ideal of $\mathcal{A}^+$. If a is any element of $\mathcal{J}$ then $U_a \mathcal{J}$ is an inner ideal. We call this the <u>principal inner ideal determined</u> by a. (It need not contain a.) The following is the analogue of the Wedderburn-Artin theorem.

<u>Structure Theorem</u>. Let $\mathcal{J}$ be a Jordan algebra satisfying the following two conditions: (i) $\mathcal{J}$ is <u>non-degenerate</u> in the sense that $U_a = 0 \Rightarrow a = 0$, (ii) $\mathcal{J}$ satisfies the minimum condition for inner ideals. Then $\mathcal{J}$ is a direct sum of a finite number of simple algebras satisfying these conditions. Moreover, the simple ones are isotopic to algebras in the following list:

1. division algebras.

2. Jordan algebras defined by non-degenerate scalar products (as in the Jordan-von-Neumann-Wigner class V).

3. $M_n(\Delta)^+$ where Δ is an associative division algebra.

4. $HM_n(\Delta)$ where Δ is an associative division algebra with involution $j : d \rightsquigarrow d^*$ and $HM_n(\Delta)$ denotes the Jordan algebra of j-hermitian $n \times n$ matrices over Δ.

5. $HM_3(\mathbb{O})$ the Jordan algebra of 3×3 hermitian matrices defined by an octonion algebra over a field P that is a Φ algebra.

It is not difficult to use this to get a similar somewhat more complicated statement on isomorphism in place of isotopy. The condition of non-degeneracy is the substitute for semi-simplicity (or semi-primitivity) in the associative theory. Somewhat closer to the associative case is a condition based on a radical for Jordan algebras that was introduced by McCrimmon. This is the analogue of the (Jacobson) radical of associative algebras. McCrimmon based this on the concepts of quasi-invertibility and proper quasi-invertibility. An element z of a Jordan algebra is called <u>quasi-invertible</u> if $1 - z$ is invertible. For any $c \in \mathcal{J}$ we can define a c-homotope of $\mathcal{J}$ and z is called <u>properly quasi-invertible</u> if z is quasi-invertible in every homotope.

McCrimmon's theory can be improved slightly by taking as a starting point the following definition that is suggested by recent work of Zelmanov.

Definition 3. The Jacobson radical, rad $\mathcal{J}$, is defined by

(13) $$\operatorname{rad} \mathcal{J} = \bigcap_{\substack{B \text{ maximal} \\ \text{inner ideal of } \mathcal{J}}} c(B)$$

where $c(B)$ denotes the largest ideal of $\mathcal{J}$ contained in B.

We have $\operatorname{rad}(\mathcal{J}/\operatorname{rad} \mathcal{J}) = 0$. There are many characterizations of rad $\mathcal{J}$. Here are the most important ones:

(i) rad $\mathcal{J}$ is the largest ideal all of whose elements are quasi-invertible.

(ii) If we call $\mathcal{J}$ primitive if it contains a maximal inner ideal B such that $c(B) = 0$ and semi-primitive if it is a subdirect product of primitive algebras then $\mathcal{J}/\operatorname{rad} \mathcal{J}$ is semi-primitive and rad $\mathcal{J}$ is contained in every ideal K such that $\mathcal{J}/K$ is semi-primitive.

(iii) rad $\mathcal{J}$ is the intersection of all the maximal inner ideals.

(iv) rad $\mathcal{J}$ is the set of properly quasi-invertible elements.

If $\mathcal{J}$ satisfies the minimum condition on principal inner ideals then $\mathcal{J}$ is non-degenerate if and only if it is semi-primitive. Also if this minimum condition holds then rad $\mathcal{J}$ is nilpotent by results of McCrimmon and Zelmanov.

We shall conclude this survey by describing some recent remarkable results on prime Jordan algebras due to Zelmanov. We note first that if B and C are ideals in a Jordan algebra then $U_B C = \{\Sigma U_{b_i} c_i,\ b_i \in B,\ c_i \in C\}$ is an ideal ($B \cdot C$ need not be). $\mathcal{J}$ is called prime if $U_B C \neq 0$ for $B \neq 0$, $C \neq 0$. We remark that primitivity implies primeness. Now call $\mathcal{J}$ an Albert algebra if there exists a field Ω that is a Φ-algebra such that $\mathcal{J}_\Omega = HM_3(\mathbb{O})$ where $\mathbb{O}$ is an octonion algebra over Ω. We can now state

Zelmanov's Theorem on Prime Jordan Algebras. Any prime Jordan algebra containing no non-zero ideals all of whose elements are nilpotent is either a homomorphic image of a special Jordan algebra or is an Albert algebra.

Zelmanov's result was for algebras with $\Phi \ni \frac{1}{2}$. The general case has been proved just recently by McCrimmon who also weakened the first hypothesis of Zelmanov and conjectured that this can be dropped altogether. Zelmanov's proof is based on a new concept on inner ideals that McCrimmon has called the absorber. It is noteworthy also that the proof makes essential use of a variant of the structure theorem for semi-primitive algebras satisfying the minimum condition on inner ideals.

An important consequence of Zelmanov's theorem is the fact that the free Jordan algebra on three or more generators contains non-zero zero-divisors, that is, elements $b \neq 0$, $c \neq 0$ such that $U_b c = 0$.

In another paper Zelmanov has applied his theorem on prime algebras to obtain a complete determination of the Jordan division algebras. The list is as follows.

(i) Δ^+, Δ an associative division algebra.

(ii) $H(\Delta,1)$ the j-symmetric elements of an associative division algebra with involution j.

(iii) Jordan algebras determined by certain symmetric bilinear forms.

(iv) Jordan division algebras that are Albert algebras.

Finally, we should mention that there are structure theories for Jordan systems that are more general than Jordan algebras, notably, Jordan triple systems and Jordan pairs. A very complete structure theory for these systems satisfying minimum conditions has been given by O. Loos in a recent Springer book. It is rumored also that Zelmanov has extended his result on prime algebras to prime Jordan pairs.

References

1. L. Hogben and K. McCrimmon, "Maximal modular inner ideals and the Jacobson radical of a Jordan algebra", to appear.

2. N. Jacobson, Structures and Representations of Jordan Algebras, AMS Colloq. Publ. v. 39, Providence , R.I., 1968.

3. __________, Lectures on Quadratic Jordan Algebras, Tata Institute Notes, Bombay, India, 1969.

4. O. Loos, Jordan Pairs, Springer Notes Series, v. 460, Springer-Verlag, Berlin-Heidelberg-New York, 1975.

5. K. McCrimmon, "A characterization of the radical of a Jordan algebra", Journal of Algebra, v. 18 (1971).

6. __________, "The Zelmanov nilpotence theorem for quadratic Jordan algebras", Journal of Algebra, v. 63 (1980), 76-98.

7. __________, "Zelmanov's prime theorem for quadratic Jordan algebras", to appear.

8. E. Zelmanov, "On prime Jordan algebras" (in Russian), Algebra i Logika, v. 18, 2, (1979), 162-175.

9. __________, "Jordan division algebras", (in Russian), v. 18, 3 (1979), 286-310.

Department of Mathematics
Yale University
New Haven, Connecticut 06520

Introduction

by N. Jacobson

Emmy Noether was one of the most influential mathematicians of this century. The development of abstract algebra, which is one of the most distinctive innovations of twentieth century mathematics is largely due to her – in published papers, in lectures, and in personal influence on her contemporaries. By now her contributions have become so thoroughly absorbed into our mathematical culture that only rarely are they specifically attributed to her. It therefore seems appropriate in this introduction to her collected papers to seek to highlight her principal contributions to the area of mathematics variously known as "abstract", "conceptual" or "modern" algebra.

Abstract algebra can be dated from the publication of two papers by Noether, the first, a joint paper with Schmeidler *Moduln in nichtkommutativen Bereichen, insbesondere aus Differential- und Differenzenausdrücken* (1920, no. 17 in this volume) and *Idealtheorie in Ringbereichen* (1921, no. 19). Of these papers, the subject matter of the first is of somewhat specialized interest and its influence was negligible. Only in retrospect does one observe that it contained a number of important ideas whose rediscovery by others had significant impact on the development of the subject. We shall postpone the discussion of this paper until later and we shall begin with the truly monumental work *Idealtheorie in Ringbereichen.* This belongs to one of the mainstreams of abstract algebra, commutative ring theory, and it may be regarded as the first paper in this vast subject – as distinguished from its precursor: ideal theory in rings of polynomials in several variables with complex number coefficients. In this paper, Noether introduced the class of rings, now called "noetherian", in which every ideal is finitely generated or, equivalently, the ascending chain condition for ideals holds. Noether began her paper with the axioms for a commutative ring. Later in the paper she defined also non-commutative rings. Apparently this is the first time these concepts were formulated as they are now understood. (An earlier definition given by A. Fraenkel in 1914 was marred by the inclusion of some highly artificial ad hoc hypotheses.)

The main results of *Idealtheorie in Ringbereichen* are the now classical decomposition theorems for ideals in noetherian rings. Noether proved that any ideal in such a ring is a finite intersection of primary ideals and she showed that, assuming some natural mild restrictions on the decompositions (e.g. "irredundancy"), then the associated prime ideals and the "isolated" components are uniquely determined. She proved also that if Q

is primary in a noetherian ring with P as its associated prime then $P^e \subset Q$ for some positive integer e.

These results represented a radical departure from what was previously known. In 1905, E. Lasker, in a paper published in Mathematische Annalen, had defined the concepts of prime ideal and primary ideal in the ring of polynomials in several indeterminates with coefficients in $\mathbb{C}$ or $\mathbb{Z}$, and he proved the existence of a decomposition of any ideal as intersection of primary ones in these rings. He showed also that if Q is primary with P as associated prime then $P^e \subset Q$ for some e. The uniqueness theorem in the case of polynomial ideals was proved by F. S. Macaulay in his Cambridge Tract, *The Algebraic Theory of Modular Systems* (1916), p. 42. It is noteworthy that Lasker's "Moduln" and "Ideale" meant the same thing: "ideals". The distinction that he made (p. 39) was that the former applied to $\mathbb{C}[x_1, \ldots, x_n]$ and the latter to $\mathbb{Z}[x_1, \ldots, x_n]$. Lasker's and Macaulay's results were immediate consequences of Noether's by virtue of Hilbert's basis theorem, which can be formulated in post-Noetherian style as: If R is noetherian then so is $R[x_1, \ldots, x_n]$.

Both Lasker and Macaulay based their proofs on elimination theory and the geometry of algebraic sets defined by polynomial equations. On the other hand, Noether's proofs rested entirely on elementary consequences of the ascending chain condition and were (and remain) startling in their simplicity. They were entirely devoid of geometric considerations. She was certainly aware of the geometry in the special case of polynomial rings and under her influence this aspect of the subject was pursued by van den Waerden and others. While the geometry played no role in the general setting in which Noether operated, it reappeared with a vengeance at a later stage in the unfolding of commutative ring theory that began with the introduction of the prime spectrum.

In 1927 Emmy Noether turned her attention to another aspect of ideal theory, the multiplicative theory, that had originated in Dedekind's proof of unique factorization of ideals in the ring of integers of a number field as products of prime ideals. It was known that this result holds also in the ring of integral elements of an algebraic function field of one variable. In *Abstrakter Aufbau der Idealtheorie in algebraischen Zahl- und Funktionenkörpern* (no. 30) Noether gave the first characterization of the class of rings now known as Dedekind rings: the commutative rings in which factorization of ideals ($\neq (0)$, $\neq (1)$) as products of prime ideals holds. She showed that the following conditions were necessary and sufficient for the validity of the prime ideal factorization theorem:

I The ascending chain condition for ideals.

II The descending chain condition modulo every non-zero ideal.

III Existence of a unit element.

IV Non-existence of zero divisors ($\neq 0$).

V Integral closure in the field of fractions.

Emmy Noether's proof that I–V imply the factorization theorem is extremely interesting. The axioms were added one by one in the order listed and the resulting strengthening of the ideal theory was studied. As proved in *Idealtheorie in Ringbereichen,* I implies that every ideal in the commutative ring R is an intersection of primary ideals. The addition of axiom II implies that very non-zero prime ideal is maximal. The uniqueness theorem for isolated components now implies that any non-zero ideal has only one normalized representation as intersection of primary ideals. The addition of axiom III then implies that every non-zero ideal is uniquely a product of primary ideals with distinct associated primes. Axiom IV that R is a domain implies that the powers of a non-zero ideal are distinct. Finally, the addition of the hypothesis of integral closure in the field of fractions implies that the only non-zero primary ideals are the powers of prime ideals. Hence I–V taken together imply that every non-zero ideal has a unique factorization as a product of prime ideals.

Noether then proved the converse: the unique factorization of non-zero proper ideals as products of primes implies I–V. She also considered fractional ideals (of the field of fractions) and showed that these form a group under multiplication. At the beginning of her paper she proved that if R is Dedekind with F as field of fractions and E is a finite dimensional separable extension field of F then the ring O of R-integral elements is also Dedekind. This, of course implied Dedekind's factorization theorem in the number field case. It implied the result also in the case of a function field $F(x, y)$ provided the base field F is $\mathbb{C}$ or, more generally any field of characteristic 0.

There are a number of basic incidental results in *Abstrakter Aufbau der Idealtheorie ....* Here one finds the two Dedekind-Noether isomorphism theorems for modules and for rings, the theorem that if M is a finitely generated module over R then the ascending or descending chain condition for the ideals of R carries over to the same condition for submodules of M, and the equivalence of the existence of a composition series for M with the validity of both chain conditions for submodules.

Noether wrote a number of other important papers on Dedekind rings and orders in algebraic number fields and function fields: no. 31, 37, and 43. Of these the most interesting from the standpoint of novelty of results and especially of methods are perhaps no. 31, *Der Diskriminantensatz für die Ordnungen eines algebraischen Zahl- oder Funktionenkörpers* and the unfinished posthumous paper no. 43, *Idealdifferentiation und Differente,* which had been started in 1927 but was not published until 1950, fifteen years after her death. The first of these is devoted to an extension to a more general class of orders of Dedekind's classical theorem that the rational primes that are ramified in the maximal order of integers of a number field are precisely the ones that divide the discriminant of the order. Noether based the study of the generalization of this theorem on the theory of finite dimensional commutative algebras over a field F. The results she obtained for these can be stated in the following way using current terminology.

14

Call a finite dimensional commutative algebra A over F separable if $A_{\bar{F}} = \bar{F} \otimes_F A$, for $\bar{F}$ the algebraic closure of F, is a direct sum of copies of $\bar{F}$. Then $A_{\bar{F}}$ is separable if and only if A is a direct sum of separable field extensions of F. Moreover, A is separable if and only if the trace bilinear form $T(a, b) = \operatorname{tr} ab$ defined by the regular representation is non-degenerate. This is the case if and only if the discriminant $\det(T(e_i, e_j))$ for a base $(e_1, e_2, \ldots, e_n)$ is not zero.

Now let R be a Dedekind domain with field of fractions F and let E be a finite dimensional separable extension of E. An order O of E is a subring of the ring of R-integral elements of E containing R and a base of E/F. By the results of *Abstrakter Aufbau der Idealtheorie* ... any non-zero ideal of O can be written in one and only one way as a product of primary ideals with distinct associated prime ideals and every non-zero prime ideal is maximal. Let p be a prime ideal $\neq 0$ of R, pO its extension in O. Then p is called ramified in O if one of the primary ideals in the factorization of pO is not a prime or one of the prime factors P of pO in O has the property that O/P is an inseparable extension of R/P (identified with $R + P)/P$). One defines the discriminant (ideal) of O *relative* to R as the ideal in O generated by all the determinants $\det(T(e_i, e_j))$ where $e_1, \ldots, e_n \in O$. This is non-zero since O contains a base for E/F and E is separable over F. The generalization of Dedekind's theorem proved by Noether is

Theorem. A prime p of R is ramified in O if and only if it is a divisor of the discriminant ideal of O relative to R.

Noether proved this result in two stages: the first in which R is a principal ideal domain and the second consisting of a reduction of the general case to this one. The first case was treated by noting that O is a free R-module of rank $[E:F]$ and O/pO is an n-dimensional algebra over R/P. Then p is ramified in O if and only if O/pO is not separable over R/p. The theorem then follows from the discriminant criterion for separability of a commutative algebra. The reduction of the general case to that in which R is a principal ideal domain was made by a localization argument. One of the results required for this is that if R is a Dedekind ring with a single non-zero prime ideal then R is a principal ideal domain.

Noether began her paper *Idealdifferentiation und Differente* by noting the analogy between the main theorem of ramification theory of algebraic number theory with the elementary theorem on muliple roots of a polynomial. The former states that if the highest power of a prime ideal P of a number field that divides the rational prime p is ϱ then the different is divisible by $P^{\varrho-1}$ and by no higher power of P if $\varrho \not\equiv 0 \pmod{p}$. The theorem on multiple roots is that if r is a root of multiplicity ϱ of $f(x)$ then r is a root of $f'(x)$ of multiplicity at least $\varrho - 1$ and exactly $\varrho - 1$ if $\varrho \not\equiv 0$ $(\operatorname{mod} p)$. She then goes on to say that she intends to show that there is more than a formal analogy between the two results, since the different can be defined by a type of "ideal differentiation" that in special cases is related to differentiation of polynomials.

15

The definition of the different and related concepts are based on the concept of tensor products of algebras over commutative rings. (This paper may be the first place in which tensor products are defined in this generality.) Noether assumes that if A and B are algebras over the commutative ring R then the images $A \otimes 1$ and $1 \otimes B$ of A and B respectively have R as intersection. In this case she says that the *direct product* of A and B exists and she denotes it as $A \times B$. She observes that $A \times B$ exists if either A or B is R-free with basis including 1 (assumed to exist for all algebras). If X is a set of generators for A/R and M is the kernel of the homomorphism of the free (non-commutative) algebra $R\{X\}$ onto A matching generators, then we can form the free algebra $B\{X\}$ and mod out the ideal generated by M. The resulting algebra can be identified with $A \times B$ (if it exists).

Now let A be commutative and assume $A \times A$ exists (that is, $A \otimes_R A$ satisfies Noether's condition). Noether defined the difference ideal of A to be the ideal J in $A \times A$ generated by the elements $a \otimes 1 - 1 \otimes a$, $a \in A$ and she calls the ideal $(0\,J) = \mathrm{ann}_{A \otimes A} J$ the *difference quotient* of A. The image of $(0:J)$ under the canonical homomorphism sending $\sum a_i \otimes b_i \mapsto \sum a_i b_i$ (in A) is Noether's *different* of A relative to R. (It is interesting to observe the similarity of these concepts with those that have been used to define separable algebras over commutative rings.)

Noether showed how these ideals can be calculated if A is defined by generators and a relation ideal M as above. In particular, suppose $A = R[x]/(G(x))$ where $G(x)$ is a monic polynomial of degree n. Then A is R-free with base $(1, u, \ldots, u^{n-1})$, $u = x + (G(x))$. Then the different is the principal ideal $(G'(u))$, $G'(x)$ the formal derivative of $G(x)$.

Noether was mainly interested in differents for "orders" defined in the following generality. Let R be an integrally closed domain, F its field of fractions, E a finite dimensional separable extension of F. An order O in E is defined here to be a subring of E which generates E (as field) and which is a finitely generated R-module. To study the different of O relative to R, Noether made a preliminary study of $E \otimes_F E$ and of $E \otimes_F K$ where K is the normal closure of E. The results on these were used to show that the different of O can be defined by complementary bases as in the classical case of number fields. Other important results were obtained in the classical case and it is clear from the sketch that appears at the end of her unfinished paper that she had planned to pursue this aspect of the theory and to complete the theory of the different in the relative case along the lines initiated in this paper.

One of the most important developments in algebra during the late twenties and thirties was the structure theory of finite dimensional algebras and its extension to artinian rings, that is, rings satisfying the descending chain condition for one sided ideals. The original motivation was the goal of classifying finite dimensional algebras. The Wedderburn structure theory of 1908 had in some sense reduced the problem to that of nilpotent algebras and division algebras, and, since the former appeared hopeless, attention focused on division algebras. Outstanding progress on

16

the study of division algebras and the more general class of simple algebras was made during 1927–1935. This culminated in the proof by Brauer, Hasse and Noether in *Beweis eines Hauptsatzes in der Theorie der Algebren* (no. 38) of the theorem that all finite dimensional central division algebras over an algebraic number field are cyclic, that is, can be obtained by a canonical construction from a cyclic field that had been given by Dickson in 1906. (See also a joint paper by Albert and Hasse in the Transactions of the American Mathematical Society, vol. 34, 1932).

The structure theory of simple algebras was pursued on both sides of the Atlantic – in the United States by Albert and in Germany by Artin, Brauer, Hasse and Noether. The activity in the theory of algebras in Germany was stimulated by the publication of Dickson's book *Algebras and their Arithmetics* (1923) and its German version, *Algebren und ihre Zahlentheorie.* Emmy Noether's contributions in this area were, first, a number of fundamental concrete results: the characterization of finite dimensional splitting fields, the theory of centralizers and automorphisms in central simple algebras, the theory of crossed products and her share of the proof of the cyclicity of division algebras over number fields. Of equal importance with these specific achievements were Noether's contributions in unifying the field and providing the proper framework for future research. She bridged the gap between the structure theory as developed by the American school and the representation theory due to Schur and Brauer. Moreover, she formulated the representation theory in module terms instead of the more concrete matrix point of view that had been used exclusively before.

The two most comprehensive papers on algebras by Noether are *Hyperkomplexe Größen und Darstellungstheorie* (no. 34, 1929) and *Nichtkommutative Algebren* (no. 40, 1933). The first of these is a write-up done jointly with van der Waerden of lectures that had been given by Noether in Göttingen, in the winter semester of 1927–1928. In a paper *Zur Theorie der hyperkomplexen Zahlen* published in v. 5 (1928) of the Hamburg Abhandlungen, Artin extended the Wedderburn structure theory of algebras to rings satisfying both chain conditions for right ideals.

The first chapter of *Hyperkomplexe Größen und Darstellungstheorie* gave a review of the concepts of groups with operators (due to Krull and O. Schmidt), modules, vector spaces and algebras over fields, and the fundamental results on homomorphisms. The second chapter gave a new derivation of Artin's extension of the Wedderburn theory. The emphasis here is on direct sum decompositions into one sided ideals, the relation between these and direct decompositions into ideals (two-sided) and decomposition of the center. All of this is fundamental for representation theory, which was the main objective of the paper.

In this paper Noether gave the first formulation of representation theory in module terms. She defined a *representation of degree n* of a ring R into a ring K with unit to be a homomorphism of R into the ring $M_n(K)$ of $n \times n$ matrices with entries in K. She defined a *representation module* of R with respect to K to be an $(R\text{-}K)$-bimodule that is K-free with a base of n-elements. A representation module together with a choice of base of n-elements defines a representation of degree n of R into K. A change of base to another containing n elements yields an equivalent (or similar) representation. Conversely, any representation of degree n of R into K defines a representation module. The standard concepts of reducibility, decomposability and complete reducibility have their counterparts for representation modules. Of particular interest are representations and representation modules for algebras over a commutative ring P. For these it is required that K is an algebra over P, the homomorphism of R into $M_n(K)$ is an algebra homomorphism of R so we have the condition $\varrho(a\,m) = a(\varrho\,m) = (\varrho\,a)m$, $a \in R$, $m \in M$, $\varrho \in P$.

The basic theorems on representation theory of semi-simple artinian rings were given here for the first time. Complete reducibility was proved (assuming $1 \mapsto 1$) and it was shown that any irreducible module is isomorphic to a minimal left ideal of the semi-simple artinian ring R and that the number of isomorphism classes of irreducible modules is the number of simple components R_i in the decomposition $R = R_1 \otimes \ldots \otimes R_s$, R_i simple. She showed also that for arbitrary artinian R, the irreducible modules are isomorphic to minimal left ideals of R/N, N the maximal nilpotent ideal of R.

After developing these general results, Noether specialized the theory to finite dimensional algebras over a field. She showed that the results implied Burnside's theorem and the generalized Burnside theorem on irreducible and completely reducible representations of algebras over an algebraically closed field. She studied the behavior of an algebra A and of a representation of A on extension of the base field F to an algebraically closed field Ω and she showed that if A is semi-simple then A_Ω remains semi-simple if and only if the centers of the simple components are separable fields. She defined the discriminant of an algebra as the determinant of the matrix $(T(a_i\,a_j))$ where T is the trace defined by the regular representation and she used this to prove the semi-simplicity of the group algebra FG of a finite group G over a field F of characteristic not dividing the order of G.

The avowed purpose of *Nichtkommutative Algebren*, as stated in the introduction to the paper, was to develop a Galois theory of non-commutative rings that is analogous to the Galois theory of fields. The paper contains the elements of such a theory for finite dimensional simple algebras and groups of inner automorphisms. As such it initiated the Galois theroy of non-commutative rings that has been cultivated by a number of algebraists beginning around 1940 to the present time (see, for example, Jacobson, *Structure of Rings*, Chapters VI and VII and Hsia Tominaga and Takasi Nagahara, *Galois Theory of Simple Rings*). *Nichtkommutative Algebren* goes considerably beyond what is now regarded as properly belonging to Galois theory. For example, it includes all the key results on simple subalgebras of finite dimensional central simple algebras. The main results of this paper were first presented in lectures at Göttingen in 1928 and 1929. It should be noted also that many of the results on finite dimensional algebras were obtained independently by Albert.

The principal tool employed by Noether in this paper is that of representations and anti-representations of a ring R by matrices with entries in a second ring S, especially in the case in wich the latter is a division ring. As Noether pointed out, the concepts of representation and anti-representation, defined to be homomorphisms and anti-homomorphisms of R into a ring $M_n(S)$ of $n \times n$ matrices with entries in S, are equivalent to those of bimodule M with respect to R and S such that M is S-free of rank n. Noether noted that these are equivalent to modules for a "product ring" of R and S (or R^{op} and S). It is apparent from earlier work that she had in mind a tensor product. At any rate, in the case of primitive interest for this paper, that in which R is finite dimensional over a field F and S is an algebra over F, Noether defined the tensor product (which we shall denote in the usual manner by $R \otimes_F S$) by using a basis for R over F.

Noether showed that if R is finite dimensional over F and A is a central division algebra over F (not necessarily finite dimensional) then there is a $1-1$ correspondence between the ideals of R and the ideals of $R \otimes_F A$, and the center of R can be identified with the center of $R \otimes A$. In particular, if R is simple then so is $R \otimes A$ and if R is central so is $R \otimes A$. Moreover, for any finite dimensional R and central division algebra A, $R \otimes A$ is artinian. Accordingly, the representation theory of artinian algebras is applicable. Using this she showed that if R_1 and R_2 are finite dimensional simple subalgebras of $M_n(A)$, A a central division algebra, then any isomorphism of R_1 onto R_2 can be extended to an inner automorphism of of $M_n(A)$ (Theorem of Skolem-Noether). She showed also that if R is a simple subalgebra of $M_n(A)$ then the centralizer S of R in $M_n(A)$ is simple artinian whose associated division algebra B in the Wedderburn-Artin theorem is anti-isomorphic to the division algebra associated with $R \otimes_F A$. If A is also finite dimensional then R is the centralizer of S in $M_n(A)$ and the product of the dimensionalities of R and S is the dimensionality of $M_n(A)$. The results on centralizers in the finite dimensional case had been proved earlier by Brauer. These results yield also the main theorem on finite dimensional splitting fields, that had been proved first by Brauer, assuming the base field is perfect. We recall the definition: An extension field E of the base field F is a splitting field if $A_E = E \otimes_F A \cong M_n(E)$. The Brauer-Noether theorem (stated without proof in no. 32) gives the following characterization: A finite dimensional extension field E of F is a splitting field for a central division algebra A if and only if the dimensionality $[E:F]$ is a multiple mf of the degree m of $A([A:F] = m^2)$ and E is a subfield of $M_f(A)$.

The Galois theory developed by Noether is concerned with a finite dimensional simple algebra R and certain subgroups of the inner automorphism group of R that she called "einfach abgeschlossen" and which we shall refer to more briefly as "closed". If F is the center of R and H^* is a subgroup of the group of units of R containing F^* (the multiplicative group of F) then we have an exact sequence $1 \to F^* \to H^* \to I_{H^*} \to 1$

where I_{H^*} is the group of inner automorphisms $i_h: x \mapsto h\, x\, h^{-1}$, $h \in H^*$. H^* is uniquely determined by I_{H^*} and I_{H^*} is called *closed* if the subspace FH^* (which is a subalgebra) is a simple subalgebra of R. The theorem on centralizers of simple subalgebras of R and the fact that the F-subspace spanned by the units of any finite dimensional simple algebra S is S implies the fundamental Galois correspondence between simple subalgebras of R and closed groups of inner automorphisms.

Noether also solved the problem of determining the structure of R_Ω where R is finite dimensional simple over F with center Z and Ω is an extension field of F. The structure is determined by that of the commutative algebra Z_Ω.

The theory developed in *Nichtkommutative Algebren* leads naturally to Noether's crossed products and factor sets. However, by the time she got around to writing up her results on these, most of which dated from 1928 and 1929, a presentation of her crossed product theory had already appeared as Chapter II of Hasse's paper *Theory of cyclic algebras over an algebraic number field* published in vol. 34 (1932) of the Transactions of the American Mathematical Society. Moreover, another presentation was in the offing: Deuring's in *Algebren* (*Ergebnisse*, 1935). Perhaps for this reason Noether did not include the crossed product theory in *Nichtkommutative Algebren.* While Noether has noted in a footnote that Deuring's treatment was close to her own, the editor has felt that it is of interest to reproduce here the lecture notes of Noether's Summer Semester Course of 1929 at Göttingen that were prepared by Deuring and that contained the first account of the crossed product theory. Moreover, these notes contain some additional bonuses, notably, Noether's development of Brauer factor sets, which is different from any other version that we have seen.

Some background material is needed to put Noether's crossed products into perspective. First, the definition: An associative algebra A over a field F is a *crossed product* if A contains a subfield E that is Galois of degree n with Galois group G and n elements u_S, $S \in G$, such that every element of A can be written in the form $\sum_{S \in G} a_S u_S$, $a_S \in E$ and we have the following relations

$$u_S\, a = a^S u_S\,, \quad u_S\, u_T = k_{S,T}\, u_{ST}$$

where $a \in E$ and $k_{S,T} \neq 0$ in E. This definition is due to Dickson in a paper *New division algebras* in Transactions AMS vol. 28 (1926). Dickson, however, failed to give the conditions on the "factor set" $k = \{k_{S,T} \mid S, T \in G\}$ that are equivalent to associativity of A, namely,

$$k_{S,T}\, k_{ST,V} = k_T^S\, k_{S,TV}\,, \quad S,\ T,\ V \in G\,.$$

Considerably earlier (1906) Dickson had introduced the special case of cyclic algebras for which G is a cyclic group. In this case the u_S can be replaced by the powers of a single element u and the defining relations are

$$u\, a = a^S u\,, \quad u^n = \gamma \neq 0 \in F$$

20

where $G = \langle S \rangle$. Dickson was interested primarily in constructing division algebras that are crossed products but are not cyclic. He did not succeed in doing this. Such algebras were first constructed by Brauer.

Noether's interest was not in constructing new examples of division algebras but in using crossed products as a tool for studying the Brauer group of similarity classes of central simple algebras. We recall that for a fixed field F we call two (finite dimensional) central simple algebras A and B over F *similar* ($A \sim B$) if there exist integers m and n such that the $n \times n$ matrix algebra $M_n(A) \cong M_m(B)$. The similarity classes $\{A\}$, $\{B\}, \ldots$ form an abelian group if one defines multiplication by $\{A\}\{B\} = \{A \otimes B\}$. This is the *Brauer group* $Br(F)$ of the field F. Its unit is the similarity class consisting of the matrix algebras $M_n(F)$, $n = 1, 2, \ldots$ and the inverse is $\{A\}^{-1} = \{A^{\mathrm{op}}\}$. If E is an extension field of F we have the homomorphism of $Br(F) \to Br(E)$ defined by $\{A\} \mapsto \{A_E\}$. The kernel is a subgroup denoted as $Br(E/F)$. It is clear from the definitions that $\{A\} \in Br(E/F)$ if and only if E is a splitting field for A.

Noether used crossed products to give what has become the classical determination of $Br(E/F)$ for E a finite dimensional Galois (= separable and normal) extension field of F. She showed that this group is isomorphic to the factor group of the group of factor sets with respect to the subgroup consisting of the factor sets that have the form $k_{S,T} = l_S\, l_T^S\, l_{ST}^{-1}$, $l_S \neq 0$ in E, $S \in G$. In the language of homological algebra (which was not available to her) the result is the isomorphism of $Br(E/F)$ with the second cohomology group $H^2(G, E^*)$ where E^* is the multiplicative group of E. In the special case in which G is cyclic we have the isomorphism $Br(E/F) \cong F^*/NE^*$ where NE^* is the group of norms $N_{E/F}(a)$, $a \in E^*$.

One of the high points of the theory of algebras is the theorem proved by Brauer, Hasse and Noether in no. 38 that any central division algebra over a number field is cyclic. There are two elements in the proof of this theorem: 1) the "Hasse principle" that if A is central simple over a number field Ω and $A_P \sim 1$ (that is, is a matrix algebra over Ω_P) for every valuation P of Ω then $A \sim 1$, and 2) the Grünwald existence theorem in its weakest form that asserts the existence of a cyclic extension Z/Ω whose degree at P for every P is a multiple of the index of A_P. The first result is established by Brauer, Hasse and Noether by purely algebraic reduction to Hasse's norm theorem for cyclic extensions of prime degree. The second result follows from Hasse's theory of simple algebras over p-adic fields. One must conclude that the major share of this achievement should be attributed to Hasse.

After this description of Noether's contributions to two main areas of algebra, commutative ideal theory and the theory of algebras, we shall mention briefly, without attempting at completeness, some other important papers of hers.

11. (1918) *Gleichungen mit vorgeschriebener Gruppe.* The problem considered in this paper is that of obtaining a parametrization of the set of polynomials having a given Galois group Γ. Such a parametrization is given by a polynomial $f(x) = x^n + F_1(\lambda_1, \ldots, \lambda_r)^{n-1} + \ldots + F_n(\lambda_1, \ldots, \lambda_r)$

where the F_i are rational expressions in indeterminates λ_i with coefficients in an infinite field Ω. It is required that if $g(x) = x^n + a_1 x^{n-1} + \ldots + a_n \in \Omega$ has the Galois group Γ and $(a_1, \ldots, a_n)$ is contained in a Zariski open subset of $\Omega^{(n)}$ then $g(x)$ can be obtained from $f(x)$ by a suitable specialization of the λ's to elements of Ω. Noether obtained such a parametrization whenever there is an affirmative answer for the group Γ to the so-called "Noether problem" which is first mentioned in her paper no. 5 *Rationale Funktionenkörper* where it is attributed to E. Fisher. The problem is the following: Suppose Γ is subgroup of the symmetric group S_n and let Γ act as group of automorphisms of $L = \Omega(\lambda_1, \ldots, \lambda_n)$, λ_i indeterminates, so that $\sigma \lambda_i = \lambda_{\sigma_i}$ for $\sigma \in \Gamma$. Let L^Γ be the subfield of fixed elements under this group of automorphisms. Is L^Γ purely transcendental over Ω? Then $L^\Gamma = \Omega(\mu_1, \ldots, \mu_n)$. If this is the case then

$$f(x) = \Pi(x - \lambda_i) = x^n + a_1(\mu_1, \ldots, \mu_n)\, x^{n-1} + \ldots + a_n(\mu_1, \ldots, \mu_n)$$

gives a parametrization of the type required. An improvement of Noether's result is given in a paper by W. Kuyk, *On a theorem of E. Noether,* in Nederl. Akad. Wetenich. Proc. Ser. A. 67, 1964, 32–39. The Fisher-Noether problem is a very interesting one independent of its application to Galois theory. Noether showed that the answer is affirmative for $n = 2$, 3 and 4. On the other hand, it was first shown by R. Swan in *Invariant Rational Functions and a Problem of Steenrod,* Inventiones Mathematicae, vol. 7, 1969, 148–158, that the problem has a negative answer for certain cyclic groups (e.g. the cyclic group of order 47).

17. (1920) (with W. Schmeidler). *Moduln in nichtkommutativen Bereichen, insbesondere aus Differential- und Differenzenausdrücken.*

The authors consider in this paper domains J that include and generalize rings of linear partial differential operators with function coefficients as well as rings of difference operators. The coefficients are taken in any field K and the elements of J are assumed to be expressible in one and only one way as polynomials of the form $\Sigma\, a_{i_1 \ldots i_r} \xi_1^{i_1} \ldots \xi_r^{i_r}$ where the a's are in K. The most important case is that in which the ξ_i commute and satisfy commutation relations of the form

$$\xi_i\, a = a_i\, \xi_i + b_i \qquad (*)$$

for any $a \in K$ where the a_i, $b_i \in K$. If $r = 1$ we have the skew polynomial domains that were studied by Ore in 1933 (who allowed K to be a skew field). The paper is concerned with left ideals (called "Moduln") and especially with the representation of such an ideal as intersection of relatively prime left ideals. A left ideal I determines the cyclic left module J/I and associated with the type of representation indicated we have a direct decomposition of J/I as sum of submodules. The concept of (module) isomorphism between J/I_1 and J/I_2 is defined. The concept of similarity ("von gleicher Art") that generalizes a known concept for linear differential operators is given here and it is shown that this holds if and only if J/I_1 and J/I_2 are isomorphic. The authors show also that if (*) and $\xi_i \xi_j = \xi_j \xi_i$ hold then J is left noetherian. They investigate the situation in which J/I is completely reducible and study the question of uniqueness up to morphism and absolute uniqueness of the irreducible components.

It is clear from this sketch that this paper belongs to modern algebra and comprises a number of "firsts" in this domain.

12. *Invarianten beliebiger Differentialausdrücke.*

13. *Invariante Variationsprobleme.*

These papers constitute Noether's contributions to physics. Professor Feza Gursey has given the following account of their significance.

"They key to the relation of symmetry laws to conservation laws in physics is Emmy Noether's celebrated theorem which states that a dynamical system described by an action invariant under a Lie group with n parameters admits n invariants (conserved quantities) that remain constant in time during the evolution of the system. The theorem is very general, being applicable to both discrete and continuous, classical and quantum mechanical systems, although originally it was derived in the classical case. Field theories like hydrodynamics, Maxwell's Electrodynamics and Einstein's theory of gravitation can all be reformulated as continuous dynamical systems derivable from an action principle. Hence, Noether's Theorem also tells us how to construct explicitly the conserved quantities (integral invariants) in classical or quantized field theories if we know the symmetry group of the differential equations of the field. For instance, in the examples given above the associated Lie groups are respectively the Galilean group, the Poincaré group and the group of diffeomorphisms for hydrodynamics, electrodynamics and Einstein's gravitation.

"Noether's theorem associates each element of the Lie algebra (generator of a one parameter transformation of the group) with a corresponding conserved quantity.

"A famous example is the association of invariance under time translation with energy conservation, the generator of an infinitesimal time translation being the Hamiltonian. Similarly conservation of momentum and angular momentum follow respectively from invariance under spatial translation and rotation. Before Noether's Theorem the principle of conservation of energy was shrouded in mystery, leading to the obscure physical systems of Mach and Ostwald. Noether's simple and profound mathematical formulation did much to demystify physics.

"Noether also showed that in the case of field theories each element of the Lie algebra is associated with a conserved current $j_\mu^{(\alpha)}$ where α is a Lie algebra label and $\mu = 0, 1, 2, 3$ a space-time vector label, so that

$$\alpha_\mu j^{(\alpha)\mu} = \partial_\varepsilon j_o^{(\alpha)} + \vec{V} \cdot \vec{j}^{(\alpha)} = 0\,.$$

"This equation leads to the constancy in time of the corresponding integral invariant, called a charge in modern physics

$$Q^{(\alpha)} = \int j_o^{(\alpha)}\, d^3x\,, \qquad \frac{dQ^{(\alpha)}}{dt} = 0\,.$$

"In Quantum field theory these charges are operators which satisfy the original Lie algebra even if the action is not invariant under the group and the charges are time dependent. This is the basis of M. Gell-Mann's current algebra, of paramount importance in Particle physics*. The explicit expression for the divergence of the current in the general case was derived by Noether and rediscovered half a century later by Gell-Mann. Here is the outline of Noether's variational method recast in modern notations.

"The field equations are derived as the Euler-Lagrange equations from the action principle

$$\delta \int L(\varphi_a, \partial_\mu \varphi_a)\, d^4x = 0\,, \qquad \left(\partial_\mu = \frac{\partial}{\partial x_\mu}\,,\ d^4x = dt\, dx_1\, dx_2\, dx_3\right)$$

and they read

$$\partial_\mu \frac{\partial L}{\partial(\partial_\mu \varphi_a)} - \frac{\partial L}{\partial \varphi_a} = 0\,.$$

"Here the fields $\varphi_a(x)$, $(a = 1, \ldots, N)$ belong to a N-dimensional representation of the group G which can be connected with spacetime or internal symmetries. Under the action of the group associated with the infinitesimal parameter $\Lambda(x)$ at the point x, the change in the fields is given by

$$\delta\varphi_a = \Lambda(x)\, G_a[\varphi_b(x)]\,.$$

Then, the current connected with this transformation is

$$J_\mu = -\frac{\partial L}{\partial(\partial_\mu \Lambda)} = -\sum_a \frac{\partial L}{\partial(\partial_\mu \varphi_a)} G_a\,.$$

It is called the Noether current and its divergence takes the form

$$\partial_\mu J^\mu = -\frac{\partial L}{\partial \Lambda}$$

so that, if the Lagrangian is invariant under the group the current is conserved.

"It is interesting that Emmy Noether was led to her great discovery after being motivated by a physics problem, namely, the invariance properties of the action in General Relativity. She had then become interested in Hilbert's discovery of the Lagrangian formulation of Einstein's theory. There followed two papers, both written in 1918 in Göttingen. The first treats the invariance properties and the associated conserved quantities of systems of differential equations, while the second starts from the action formulation of dynamical systems and field theories and employs variational methods. The second paper also contains the expression for the current divergences and the construction of conserved charges (integral invariants) in the exact symmetry case.

* For a review of current algebra, see S. L. Adler and R. F. Dashen, Current Algebras (Benjamin, N.Y. 1968).

24

"Since all the laws of fundamental physics can be expressed in terms of quantum fields which are associated with symmetry groups at each point and satisfy differential equations derived from an action principle, the conservation laws of physics and the algebra of time dependent charges can be all constructed using Noether's methods. The only additional conserved quantities not connected with the Lie algebra are topological invariants that are related to the global properties of the fields. These have also become important in the last few years. With this exception, Noether's work is of paramount importance to physics and the interpretation of fundamental laws in terms of group theory."

29. (1926) *Der Endlichkeitssatz der Invarianten endlicher linearer Gruppen der Charakteristik p.* This paper is devoted to the proof of the following theorem. Let G be a finite group and let G act on the polynomial ring $P[x_1, \ldots, x_n]$ (x_i indeterminates) by automorphisms over P that stabilize the vector subspace $\sum Px_i$. Let Inv G be the set of fixed points under this action. Then Inv G is a finitely generated algebra over P. The case in which the characteristic is 0 had been proved by Hilbert. Moreover, a simple proof for this case had been given by Noether in *Der Endlichkeitssatz der Invarianten endlicher Gruppen* (no. 7). The proof of the general case given here is based on the following two finiteness results that are of independent interest.

1. Any subfield over P of the field $P(x_1, \ldots, x_n)$, is finitely generated as a field over P, that is, it has the form $P(y_1, \ldots, y_m)$. For P of characteristic 0, Noether had proved earlier in *Körper und Systeme rationaler Funktionen* (no. 6) that if the transcendence degree of a subfield K/P of $P(x_1, \ldots, x_n)$ is ϱ then K has the form $P(y_1, \ldots, y_{\varrho+1})$.

2. Let $\mathfrak{J}$ be a subalgebra of $P(x_1, \ldots, x_n)$. Then $\mathfrak{J}$ is finitely generated over P if and only if $\mathfrak{J}$ contains a finitely generated subalgebra $\mathfrak{K}$ such that $\mathfrak{J}$ is integral over $\mathfrak{K}$. In this case any finite set of generators for $\mathfrak{K}/P$ can be supplemented by a finite set of generators for $\mathfrak{J}$ as $\mathfrak{K}$-module to obtain a finite set of generators for $\mathfrak{J}/P$.

In the proof of 2, Noether makes use of the *Normalization lemma* (without displaying or naming it): If $\mathfrak{K}$ is a finitely generated domain over a field then $\mathfrak{K}$ has a transcendency basis $y_1, \ldots, y_t$ such that $\mathfrak{K}$ is integral over $P[y_1, \ldots, y_t]$. This result in its homogeneous form was proved by Hilbert in *Über die vollen Invariantensysteme,* Mathematische Annalen, vol. 42, 1893, p. 1316 or *Gesammelte Abhandlungen,* vol. 2, p. 290. Noether refers to this paper in *Algebraische und Differentialinvarianten* (no. 23), in which the finiteness criterion 2 is stated for the first time.

41. (1943) *Der Hauptgeschlechtssatz für relativ-galoissche Zahlkörper.* An important idea initiated by Emmy Noether was that of the applicability of non-commutative algebra, especially, the theory of central simple algebras, to commutative algebra. This was the theme of her address to the 1932 International Congress of Mathematicians at Zürich (no. 39). In *Nichtkommutative Algebren* she applied non-commutative methods to derive the Galois theory of fields. In her Zürich address she emphasized the applicability of these methods to arithmetic. Under her influence,

Hasse in *Die Struktur der R. Brauerschen Algebrenklassengruppe über einem algebraischen Zahlkörper,* Mathematische Annalen, vol. 107, 1932, 731–760 and Chevalley in *Sur la theorie des symboles de restes normiques,* Journal für Mathematik, vol. 169, 1933, 140–157, applied these methods to obtain some of the main results on global and local class field theory.

In the present paper (no. 41) Noether gave an extension of a known result of class field theory on cyclic extensions of number fields to arbitrary Galois extensions of such fields. The principal tool for formulating and proving the extension was the theory of crossed products and Hasse's principle mentioned before. The result is somewhat too complicated to state here. Before giving the main result, Noether formulates the generalization of Hilbert's "Satz 90" that she calls the "Hauptgeschlechtssatz im Minimalen" which is the cohomology theorem that if E/F is finite dimensional Galois with Galois group G then the cohomology group $H^1(G, E^*)$ is trivial. She proves this by applying the Noether-Skolem theorem to the crossed product of E with G and trivial factor set. This proof, while not as elementary as the standard proof due to Artin, has the advantage of applicability in a number of other important situations. We note also that in the proof of the main theorem, Noether used a connecting homomorphism argument. This is perhaps the first place in which the connecting homomorphism appears.

Reprinted from
Emmy Noether: Collected Papers
Springer-Verlag, 1983.

26

Brauer Factor Sets, Noether Factor Sets, and Crossed Products*

NATHAN JACOBSON†

The role of Noether's crossed products and factor sets in the study of the Brauer group Br(F) of a field F is well known. In particular, it is central in the determination of the Brauer group of a number field and in the proof of the Albert–Brauer–Hasse–Noether theorem that central division algebras over number fields are cyclic ([2], [5], [8], [9]). The central algebraic result of Noether's theory is the isomorphism of the subgroup Br(E/F) of Br(F) consisting of the algebra classes having a finite dimensional Galois extension field E/F as a splitting field with the cohomology group $H^2(G, E^*)$ where $G = \text{Gal } E/F$. This leads to an isomorphism (given later) of the full Brauer group Br(F) with a cohomology group of the Galois group of the separable algebraic closure of F.

On the other hand, the main algebraic results on Br(F) (e.g. the fact that this group is a torsion group) were first derived by Brauer using Brauer factor sets. It is interesting to note also that in what may have been Noether's first presentation of crossed products, namely, that given in her lectures in Göttingen in the summer semester of 1929, the crossed product theory was preceded by an account of Brauer's factor sets. The notes for these lectures were prepared by Deuring and will appear in [14]. Brauer factor sets were used by Weyl in his study of Riemann matrices and they are also implicit in Albert's original treatment of the subject ([1], [15]). A lesser known aspect of Brauer's theory is a general construction of central simple algebras from a separable extension field K/F and a Brauer factor set with non-zero values in the normal closure E/F of K ([3]). This construction was used by Brauer to show that if $\mathcal{A}$ is a central division algebra of degree 5 over F then there exists an extension field E/F obtained by successive adjunctions of three square roots and a cube root such that $\mathcal{A}_E$ is a cyclic algebra ([4]).

* This work was partially supported by the National Science Foundation Grant MCS79-05018.

† Department of Mathematics, Yale University, P.O. Box 2155, Yale Station, New Haven, CT 06520, U.S.A.

In this paper we give a new derivation and extension of the theory of Brauer factor sets. Our starting point is a special type of generation of any finite dimensional central simple algebra $\mathscr{A}$ over F of degree n by a commutative separable subalgebra $K = F[u]$ of dimension n and another element v. ($\mathscr{A}$ contains plenty of such subalgebras.) We show that there exists an element v such that $\mathscr{A} = KvK$. If E/F is a splitting field for K in the sense that E is a minimal extension field of F such that $K_E = Ee_1 \oplus Ee_2 \oplus \ldots \oplus Ee_n$ then E is a splitting field for $\mathscr{A}$. Hence we have an imbedding of $\mathscr{A}$ in the matrix algebra $M_n(E)$ so that $M_n(E) = E\mathscr{A}$ and $u = \mathrm{diag}\{r_1, r_2, \ldots, r_n\}$ where the r_i are distinct elements of E. Then v is a matrix (v_{ij}) all of whose entries are $\neq 0$ and $c = \{c_{ijk}\}$ where $c_{ijk} = v_{ij}v_{jk}v_{ik}^{-1}$ is a Brauer factor set. This leads to the general construction of central simple algebras due to Brauer.

Formulated in a more abstract fashion, we can define a Brauer factor set c in the following way. Let K be a finite dimensional commutative separable algebra over F, E/F its splitting field (unique up to isomorphism) and let M be the set of homomorphisms of K/F into E/F. Then $|M| = [K:F]$. A Brauer factor set is a map of $M \times M \times M$ into E^* that is homogeneous in the sense that for any $\sigma \in \mathrm{Gal}\, E/F$, $c(\sigma\alpha, \sigma\beta, \sigma\gamma) = \sigma c(\alpha, \beta, \gamma)$, $\alpha, \beta, \gamma \in M$ and

$$c(\alpha, \beta, \gamma)c(\alpha, \gamma, \delta) = c(\alpha, \beta, \delta)c(\beta, \gamma, \delta).$$

Let $\mathscr{B}(K, c)$ denote the set of homogeneous maps l of $M \times M$ into E. These constitute a vector space over F and we can define a product in $\mathscr{B}(K, c)$ by

$$ll'(\alpha, \beta) = \sum_{\gamma \in M} l(\alpha, \gamma)c(\alpha, \gamma, \beta)l'(\gamma, \beta).$$

Then $\mathscr{B}(K, c)$ is central simple and every finite dimensional central simple algebra over F can be obtained in this way.

The Brauer factor sets form an abelian group under multiplication of images in E^*. This contains the subgroup of elements of the form

$$(\alpha, \beta, \gamma) \rightsquigarrow l(\alpha, \beta)l(\beta, \gamma)l(\alpha, \gamma)^{-1}$$

where l is a homogeneous map of $M \times M$ into E^*. We can form the factor group which we denote as $H^2(K/F)$. On the other hand, the subset $\mathrm{Br}(K/F)$ of the Brauer group $\mathrm{Br}(F)$ of the classes of central simple algebras over F that contain an $\mathscr{A}$ of degree n containing K is a subgroup of $\mathrm{Br}(F)$ and $\mathrm{Br}(K/F) \cong H^2(K/F)$. If K is a field then $\mathrm{Br}(K/F)$ is the kernel of the homomorphism of $\mathrm{Br}(F)$ into $\mathrm{Br}(K)$ defined by $\{\mathscr{A}\} \rightsquigarrow \{\mathscr{A}_K\}$.

The Noether theory can be obtained by specializing to the case in which $K = E$. In this case if $\mathscr{A}$ is a central simple algebra of degree n containing the Galois extension E/F of dimension n then $\mathscr{A}$ is a crossed product (E, G, k) of E with its Galois group G and its Noether factor set k. On the other hand, we know also that $\mathscr{A} = \mathscr{B}(E, c)$ for a Brauer factor set c. It is easy to establish the relations between the two constructions and between the Brauer factor sets and the Noether factor sets. In this way we obtain Noether's results, in particular, the basic isomorphism of $\mathrm{Br}(E/F)$ with the second cohomology group $H^2(G, E^*)$.

In the last section we give an application to the theory of central simple algebras with involution. If the Brauer factor set c is symmetric in the sense that $c(\alpha, \beta, \gamma) =$

$c(\gamma, \beta, \alpha)$ for $\alpha, \beta, \gamma \in M$ then $l \rightsquigarrow {}'l$ where ${}'l(\alpha, \beta) = l(\beta, \alpha)$ is an involution of orthogonal type in $\mathscr{B}(K, c)$. Moreover, we can show that any central simple algebra with an involution of orthogonal type is isomorphic as algebra with involution to a pair $(\mathscr{B}(K, c), t)$.

1. Reduced Characteristic Polynomial, Trace and Norm

The concept of the reduced characteristic (or generic or minimum) polynomial of a finite dimensional associative algebra has been defined by the author in [10]. An extension of the theory to strictly power associative algebras is given in [12], where many properties of the reduced characteristic polynomial are derived. In this section we shall derive some properties and applications that will be required.

Let $\mathscr{A}$ be a finite dimensional associative algebra over a field F, $(u_1, u_2, \ldots, u_n)$ a base for A/F. If $a \in \mathscr{A}$ we denote the minimum polynomial of a in $\mathscr{A}$ by $\mu_a(\lambda)$. We recall some well-known results on $\mu_a(\lambda)$ in the special case in which $\mathscr{A} = M_m(F)$ (so $n = m^2$). In this case we have the characteristic polynomial $\chi_a(\lambda) = \det(\lambda 1 - a)$ and the Hamilton–Cayley theorem that $\chi_a(a) = 0$. It follows that $\mu_a(\lambda) | \chi_a(\lambda)$. We recall also that we can diagonalize the matrix $\lambda 1 - a$ in $M_m(F[\lambda])$, that is, we can find invertible matrices $P(\lambda)$ and $Q(\lambda)$ in $M_m(F[\lambda])$ such that

$$P(\lambda)(\lambda 1 - a)Q(\lambda) = \operatorname{diag}\{d_1(\lambda), d_2(\lambda), \ldots, d_m(\lambda)\}, \tag{1.1}$$

where the $d_i(\lambda)$ are monic polynomials and $d_i(\lambda) | d_j(\lambda)$ if $i \le j$. Then $\chi_a(\lambda) = \prod_1^m d_i(\lambda)$ and we have the sharpening of the Hamilton–Cayley theorem due to Frobenius: $d_m(\lambda) = \mu_a(\lambda)$. Evidently this implies that $\mu_a(\lambda)$ and $\chi_a(\lambda)$ have the same irreducible factors in $F[\lambda]$ and the same roots in the algebraic closure $\bar{F}$ of F.

We consider $\mathscr{A}$ again with the base $(u_1, u_2, \ldots, u_n)$. We introduce n indeterminates $\xi_1, \xi_2, \ldots, \xi_n$ and the field $F(\xi) \equiv F(\xi_1, \xi_2, \ldots, \xi_n)$ of rational expressions in the ξ_i with coefficients in F. Consider the algebra $\mathscr{A}_{F(\xi)}$ obtained by extending the base field of $\mathscr{A}$ to $F(\xi)$. Put $x = \sum \xi_i \mu_i \in \tilde{\mathscr{A}} = \mathscr{A}_{F(\xi)}$. We call this a *generic element* of $\mathscr{A}$. We now denote the minimum polynomial of x by $m_x(\lambda)$ (rather than $\mu_x(\lambda)$) and write this as

$$m_x(\lambda) = \lambda^m - \tau_1(\xi)\lambda^{m-1} + \cdots + (-1)^m \tau_m(\xi), \tag{1.2}$$

where $\tau_i(\xi) \in F(\xi)$. We now have

1.3 Lemma. *$m_x(\lambda) \in F[\lambda, \xi] \equiv F[\lambda, \xi_1, \ldots, \xi_n]$. More precisely, $\tau_i(\xi)$ is a homogeneous polynomial of degree i in the ξ's.*

Proof. Let ρ be a faithful representation of $\mathscr{A}$ by matrices. For example, we can take ρ to be the regular representation. Now ρ has a unique extension to a faithful representation $\tilde{\rho}$ of $\tilde{\mathscr{A}}$ and $\tilde{\rho}(x) = \sum \xi_i \rho(u_i)$ so the entries of this matrix are homogeneous linear expressions in the ξ's. Hence the characteristic polynomial $\chi_{\tilde{\rho}(x)}(\lambda)$ has the form

$$\lambda^N - t_1(\xi)\lambda^{N-1} + \cdots + (-1)^N t_N(\xi), \tag{1.4}$$

where $t_i(\xi)$ is a homogeneous polynomial of degree i in the ξ's and N is the degree of the representation. Since $\tilde{\rho}$ is a monomorphism, $m_x(\lambda)$ is the minimum polynomial of $\tilde{\rho}(x)$ and hence $m_x(\lambda) | \chi_{\tilde{\rho}(x)}(\lambda)$. It follows from Gauss' lemma that $m_x(\lambda) \in F[\lambda, \xi]$. Moreover, since $\chi_{\tilde{\rho}(x)}(\lambda)$ is homogeneous of degree N in λ and the ξ's, $m_x(\lambda)$ is homogeneous of degree m in λ and the ξ's. It follows that $\tau_i(\xi)$ in (1.2) is homogeneous of degree i in the ξ's. □

We now write $x^j = \sum_{i=1}^n \mu_{ji}(\xi)u_i$, $j = 0, 1, 2, \ldots$ where it is clear that $\mu_{ji}(\xi)$ is a homogeneous polynomial of degree j in the ξ's. The equation $m_x(x) = 0$ is equivalent to n equations

$$\sum_{j=0}^{m} (-1)^j \tau_j(\xi)\mu_{m-j,i}(\xi) = 0, \tag{1.5}$$

where $1 \leq i \leq n$, $\tau_0 = 1$. Now let $a = \sum \alpha_i u_i$, $\alpha_i \in F$ and put

$$m_a(\lambda) = \lambda^m - \tau_1(a)\lambda^{m-1} + \cdots + (-1)^m \tau_m(a), \tag{1.6}$$

where $\tau_i(a) = \tau_i(\alpha) \in F$. Then we have the relations $\sum (-1)^j \tau_j(\alpha)\mu_{m-j,i}(\alpha) = 0$ obtained by specializing $\xi_k \rightsquigarrow \alpha_k$ in (1.5). These imply that

$$m_a(a) = 0. \tag{1.7}$$

Now $m_a(\lambda)$ is independent of the choice of the base $(u_1, u_2, \ldots, u_n)$. For, if $(v_1, v_2, \ldots, v_n)$ is a second base and $v_i = \sum \beta_{ij}u_j$, $(\beta_{ij}) \in \mathrm{GL}_n(F)$ then $\sum \xi_i v_i = \sum \xi'_j u_j$ where $\xi'_j = \sum \xi_i \beta_{ij}$ and $m_{\sum \xi_i v_i}(\lambda)$ is obtained by replacing ξ_j by ξ'_j, $1 \leq j \leq n$ in $m_x(\lambda)$. Then if $a = \sum \alpha_i v_i$, $a = \sum \alpha'_j u_j$ where $\alpha'_j = \sum \alpha_i \beta_{ij}$. It follows that $m_a(\lambda)$ is unaltered in passing from the base $(u_1, \ldots, u_n)$ to the base $(v_1, \ldots, v_n)$. We shall call $m_a(\lambda)$ the *reduced characteristic polynomial* of the element a and

$$t(a) = \tau_1(a), \qquad n(a) = \tau_m(a) \tag{1.8}$$

the *reduced trace* and *reduced norm* respectively of a. The integer m is called the *degree of the algebra* $\mathscr{A}$.

Let E be an extension field of F and consider the algebra $\mathscr{A}_E$. We can regard $\mathscr{A}$ as contained in $\mathscr{A}_E$. Then the base $(u_1, \ldots, u_n)$ of $\mathscr{A}$ is also a base for $\mathscr{A}_E$. It follows that the reduced trace and reduced norm functions on $\mathscr{A}_E$ are extensions of these functions on $\mathscr{A}$. This remark implies that in developing the properties of these functions there is no loss in generality in assuming the base field is infinite or even algebraically closed. The advantage of dealing with infinite base fields is that we can use the Zariski topology of the vector space $\mathscr{A}/F$.

We define the *degree* of $a \in \mathscr{A}$ to be the degree of the minimum polynomial $\mu_a(\lambda)$ of a.

1.9 Proposition. *If F is infinite then the elements of $\mathscr{A}$ whose degrees are the degree of $\mathscr{A}$ constitute a non-vacuous Zariski open subset of $\mathscr{A}$.*

Proof. As before, we write

$$x^j = \sum_{i=1}^{n} \mu_{ji}(\xi)u_i, \qquad j \geq 0 \tag{1.10}$$

for the generic element x. Since the degree of the minimum polynomial $m_x(\lambda)$ of x in $\tilde{\mathscr{A}} = \mathscr{A}_{F(\xi)}$ is m, the elements x^k, $0 \le k \le m-1$ are linearly independent in $\tilde{\mathscr{A}}$. Hence there exists a non-zero m-rowed minor in the $m \times n$ matrix $(\mu_{ji}(\xi))$. Let these non-zero m-rowed minors be $D_1(\xi), \ldots, D_q(\xi)$. Then it is clear that the set of elements $a = \sum \alpha_i u_i$ such that $\deg a = m$ is the union of the sets defined by $D_j(\alpha) \ne 0$. Evidently this is a non-vacuous open subset of $\mathscr{A}$. □

We again consider the faithful representation ρ of $\mathscr{A}$. We have the three polynomials $\mu_a(\lambda)$, $m_a(\lambda)$ and the characteristic polynomial $\chi_{\rho(a)}(\lambda)$ of the matrix $\rho(a)$. Since $m_a(a) = 0$, $\mu_a(\lambda) | m_a(\lambda)$. We also have $m_a(\lambda) | \chi_{\rho(a)}(\lambda)$ since $m_x(\lambda) | \chi_{\tilde{\rho}(x)}(\lambda)$. Also $\mu_a(\lambda)$ is the minimum polynomial of the matrix $\rho(a)$ so $\mu_a(\lambda)$ and $\chi_{\rho(a)}(\lambda)$ have the same irreducible factors. Hence $\mu_a(\lambda)$ and $m_a(\lambda)$ have the same irreducible factors and the same roots in $\bar{F}$.

We recall that $\mathscr{A}$ is *separable* if $\mathscr{A}_{\bar{F}} = M_{n_1}(\bar{F}) \oplus M_{n_2}(\bar{F}) \oplus \cdots \oplus M_{n_s}(\bar{F})$. It follows from the Hamilton–Cayley theorem that the degree of $M_n(F)$ is m and if $a \in M_m(F)$ then $m_a(\lambda) = \chi_a(\lambda)$. It follows easily that the degree of

$$\mathscr{A}_{\bar{F}} = M_{n_1}(\bar{F}) \oplus \cdots \oplus M_{n_s}(\bar{F})$$

and hence of $\mathscr{A}$ is $m = \sum_1^s n_i$ ([12], p. 228).

We shall call an element a of an algebra *separable* if $F[a]$ is a separable algebra. This is equivalent to: the minimum polynomial $\mu_a(\lambda)$ has distinct roots. We can now prove

1.11 Theorem. *Let $\mathscr{A}$ be separable of degree m over an infinite field F. Then the subset of $\mathscr{A}$ of elements a that are separable of degree m is non-vacuous Zariski open.*

Proof. Let x be a generic element of $\mathscr{A}$, $\delta(x)$ the discriminant of $m_x(\lambda)$. We claim that $\delta(x) \ne 0$. To see this we note that since $\mathscr{A}_{\bar{F}} = M_{n_1}(\bar{F}) \oplus \cdots \oplus M_{n_s}(\bar{F})$ there exist $m = \sum_1^s n_i$ non-zero orthogonal idempotents a_i in $\mathscr{A}_{\bar{F}}$ such that $\sum e_i = 1$. Let $\alpha_1, \ldots, \alpha_m$ be distinct in $\bar{F}$ and put $a = \sum \alpha_i e_i$. Then $\mu_a(\lambda) = \prod (\lambda - \alpha_i)$ is of degree m. Hence $m_a(\lambda) = \mu_a(\lambda)$ has distinct roots. Then $\delta(a) \ne 0$ and hence $\delta(x) \ne 0$. We have seen in Proposition 1.9 that the set of elements of $\mathscr{A}$ of degree m is non-vacuous open. The subset of these that are separable is defined by $\delta(a) \ne 0$. This is non-vacuous Zariski open. □

Theorem 1.11 gives a very natural proof of a classical result of Koethe's:

1.12 Theorem. *Any finite dimensional central simple algebra $\mathscr{A}$ has a finite dimensional separable splitting field.*

Proof. Since $\mathscr{A} = M_r(\mathscr{D})$ where $\mathscr{D}$ is a central division it suffices to prove the theorem for $\mathscr{A} = \mathscr{D}$ a division algebra. In this case if F is finite $\mathscr{D} = F$ by Wedderburn's theorem on the commutativity of finite division rings. Hence we may assume F infinite. In this case Theorem 1.11 shows that there exists a separable subfield K/F of $\mathscr{D}$ of degree $m = \deg \mathscr{D}$. Such a subfield is a splitting field (BA II, p. 224). □

2. Brauer Factor Sets

We recall that a finite dimensional associative algebra $\mathscr{A}/F$ is called a *Frobenius algebra* if there exists a non-degenerate bilinear form on $\mathscr{A}$ that is associative in the sense that

$$f(ab, c) = f(a, bc). \tag{2.1}$$

If $\mathscr{A} = \mathscr{A}_1 \oplus \mathscr{A}_2$ then $\mathscr{A}$ is Frobenius if and only if $\mathscr{A}_i$, $i = 1, 2$, is Frobenius. The tensor product of Frobenius algebras is Frobenius and any algebra $F[u]$ with a single generator is Frobenius (see Curtis and Reiner [7] or Jacobson [11]). If $\mathscr{A}$ is separable then $\mathscr{A}$ is Frobenius since the *reduced trace bilinear form* $t(a, b) = t(ab)$ is non-degenerate and associative. The following theorem has been proved by the author in [11]:

2.2 Theorem. *Let $\mathscr{A}$ be a central simple algebra of degree n, K a commutative Frobenius subalgebra of dimension n over F. Then there exists a $v \in \mathscr{A}$ such that $\mathscr{A} = KvK$.*

In particular this holds if K is a separable subalgebra with $[K:F] = n$. If F is infinite, by Theorem 1.11, there exists a $K = F[u]$ with u separable of degree n. The same can be seen readily also if F is finite (so $\mathscr{A} = M_n(F)$). We now assume that $K = F[u]$ is separable of degree n and we choose a v so that $\mathscr{A} = KvK$. Let $f(\lambda)$ be the minimum polynomial of u and E a splitting field over F of $f(\lambda)$ so $E = F(r_1, r_2, \ldots, r_n)$ where the r_i are distinct and $f(\lambda) = \prod (\lambda - r_i)$ in $E[\lambda]$. Consider the algebra $K_E = E[u] \cong E[\lambda]/(f(\lambda))$. In this algebra we have n non-zero orthogonal idempotents

$$e_i = \frac{(u - r_1)\cdots(u - r_{i-1})(u - r_{i+1})\cdots(u - r_n)}{(r_i - r_1)\cdots(r_i - r_{i-1})(r_i - r_{i+1})\cdots(r_i - r_n)} \tag{2.3}$$

such that $\sum e_i = 1$. Hence $K_E = \bigoplus_1^n Eu_i$. Now $\mathscr{A}_E$ is central simple of degree n over E and this contains K_E and hence the n orthogonal idempotents e_i. It follows that $\mathscr{A}_E = M_n(E)$ so E is a splitting field for $\mathscr{A}$. Thus we may regard $\mathscr{A}$ as an F-subalgebra of $M_n(E)$ such that $E\mathscr{A} = M_n(E)$. Also we may suppose that

$$u = \operatorname{diag}\{r_1, r_2, \ldots, r_n\}. \tag{2.4}$$

If $v = (v_{ij})$ then $u^k v u^l = (r_i^k r_j^l v_{ij})$ and since $\mathscr{A} = KvK$ the elements $u^k v u^l$, $0 \leq k, l \leq n - 1$, form a base for $\mathscr{A}/F$. Hence every element of $\mathscr{A}$ is a matrix

$$L = (l_{ij} v_{ij}), \tag{2.5}$$

where

$$l_{ij} = \sum_{k,l=1}^{n} a_{kl} r_i^{k-1} r_j^{l-1} \tag{2.6}$$

and the $a_{kl} \in F$ and are uniquely determined. Since $E\mathscr{A} = M_n(E)$ it is clear that every $v_{ij} \neq 0$.

Let $G = \operatorname{gal} E/F$. If $\sigma \in G$, $\sigma r_i = r_{i'}$ and σ is determined by the permutation

$i \rightsquigarrow i'$ of $\{1, 2, \ldots, n\}$. We denote this permutation by σ also, so we have $\sigma r_i = r_{\sigma i}$. If l_{ij}, $1 \le i, j \le n$, is defined by (2.6) then the l_{ij} satisfy

$$\sigma l_{ij} = l_{\sigma i, \sigma j}, \qquad \sigma \in G. \tag{2.7}$$

These conditions, which we shall call the *conjugacy conditions* on $l = (l_{ij})$, are also sufficient that the l_{ij} have the form (2.6); for we have

2.8 Lemma. *Let $l = (l_{ij})$ be a matrix of elements $l_{ij} \in E$ satisfying the conjugacy conditions* (2.7). *Then there exist $a_{kl} \in F$ such that* (2.6) *holds for all i, j.*

Proof. Let V be the Vandermonde matrix

$$V = \begin{bmatrix} 1 & r_1 & r_1^2 & \cdots & r_1^{n-1} \\ 1 & r_2 & r_2^2 & \cdots & r_2^{n-1} \\ \vdots & \vdots & \vdots & \cdots & \vdots \\ 1 & r_n & r_n^2 & \cdots & r_n^{n-1} \end{bmatrix}. \tag{2.9}$$

Then V is invertible. Hence there exists a unique matrix $a = (a_{ij}) \in M_n(E)$ such that

$$Va({}^tV) = l. \tag{2.10}$$

This matrix relation is equivalent to the equations (2.6). Applying $\sigma \in G$ to these equations we obtain

$$l_{\sigma i, \sigma j} = \sum_{k,l} (\sigma a_{kl}) r_{\sigma i}^{k-1} r_{\sigma j}^{l-1},$$

or $l_{ij} = \sum_{k,l} (\sigma a_{kl}) r_i^{k-1} r_j^{l-1}$. By the uniqueness of a we have $\sigma a_{kl} = a_{kl}$ for every $\sigma \in G$. Hence $a_{kl} \in F$. $\square$

(The foregoing proof is due to Walter Feit.)

We now put $L = (l_{ij} v_{ij})$, $L' = (l'_{ij} v_{ij})$ where the l_{ij} and l'_{ij} satisfy the conjugacy conditions, so $L, L' \in \mathscr{A}$. Then $LL' = L'' = (l''_{ij} v_{ij})$ where

$$l''_{ij} = \sum l_{ik} c_{ikj} l'_{kj}, \tag{2.11}$$

$$c_{ikj} = v_{ik} v_{kj} v_{ij}^{-1}. \tag{2.12}$$

2.13 Lemma. *The c_{ijk} satisfy*

$$\sigma c_{ijk} = c_{\sigma i, \sigma j, \sigma k}, \tag{2.14}$$

$$c_{ijk} c_{ikl} = c_{ijl} c_{jkl}. \tag{2.15}$$

Proof. Apply σ to (2.11) to obtain $l''_{\sigma i, \sigma j} = \sum_k l_{\sigma i, \sigma k} (\sigma c_{ikj}) l'_{\sigma k, \sigma j}$. On the other hand, $l''_{\sigma i, \sigma j} = \sum_k l_{\sigma i, \sigma k} c_{\sigma i, \sigma j, \sigma k} l'_{\sigma k, \sigma j}$. Hence we have

$$\sum_k l_{\sigma i, \sigma k} (\sigma c_{ikj} - c_{\sigma i, \sigma k, \sigma j}) l'_{\sigma k, \sigma j} = 0,$$

or

$$\sum_k l_{ik} d_{ikj} l'_{ij} = 0, \qquad d_{ikj} = \sigma c_{\sigma^{-1} i, \sigma^{-1} k, \sigma^{-1} j} - c_{ikj} \tag{2.16}$$

and these relations hold for all l_{ik}, l'_{jk} satisfying the conjugacy conditions. We can write also

(2.17)
$$\sum_k l_{ik} e_{ikj}(v_{kj} l'_{kj}) = 0, \qquad e_{ikj} = d_{ikj} v_{kj}^{-1}.$$

Now $M_n(E) = E\mathscr{A}$. Hence taking a suitable E-linear combination of the matrices in $\mathscr{A}$ we obtain a matrix whose jth column is $(0, \ldots, 0, 1, 0, \ldots, 0)$ where the 1 is in any chosen position. Using this linear combination of the relations (2.17) we obtain $l_{ik} e_{ikj} = 0$ for all k. Then $e_{ikj} = 0$ and $d_{ikj} = 0$ which is (2.14). Now (2.15) follows by direct verification using the definition (2.12). □

We now define a *Brauer factor set* c to be an indexed set of elements $c_{ijk} \in E^*$, $1 \leq i, j, k \leq n$, such that

(i)
$$\sigma c_{ijk} = c_{\sigma i, \sigma j, \sigma k},$$

(ii)
$$c_{ijk} c_{ikl} = c_{ijl} c_{jkl}.$$

The foregoing lemma states that the c_{ijk} defined by (2.12) from the element $v = (v_{ij})$ constitute a Brauer factor set. We shall call (i) the conjugacy conditions on the c_{ijk}. We note that these imply that

(iii)
$$c_{ijk} \in F(r_i, r_j, r_k).$$

For, if $\sigma \in \mathrm{Gal}\, E/F(r_i, r_j, r_k)$ then $\sigma i = i, \sigma j = j, \sigma k = k$ and hence, by (i), $\sigma c_{ijk} = c_{ijk}$. Since this holds for every $\sigma \in \mathrm{Gal}\ E/F(r_i, r_j, r_k)$, $c_{ijk} \in F(r_i, r_j, r_k)$ by the Galois correspondence. If we put $i = j = k$ and $j = k = l$ successively in (ii) we obtain

(iv)
$$c_{iij} = c_{iii} = c_{jii}.$$

We have seen that if we define $c_{ijk} = v_{ij} v_{jk} v_{ik}^{-1}$ then $c = \{c_{ijk}\}$ is a Brauer factor set. Here $v = (v_{ij})$ was any element of $\mathscr{A}$ such that $\mathscr{A} = KvK$. We now observe that c is independent of the imbedding of $\mathscr{A}$ in $M_n(E)$ provided that

$$u = \mathrm{diag}\{r_1, r_2, \ldots, r_n\}$$

in the imbedding. For, if we have a second imbedding with this property then it follows from the Skolem–Noether theorem and the fact that the only matrices that commute with u are diagonal matrices that in the second imbedding we have $v = (d_i v_{ij} d_j^{-1})$ where $d_i \in E^*$. Then

$$(d_i v_{ik} d_k^{-1})(d_k v_{kj} d_j^{-1})(d_i v_{ij} d_j^{-1})^{-1} = v_{ik} v_{kj} v_{ij}^{-1} = c_{ijk}.$$

We shall now normalize v so that the corresponding factor set c is *reduced* in the sense that every $c_{iii} = 1$. By (iv) this implies $c_{iij} = 1 = c_{jii}$ for all i, j. We remark that if $f(\lambda)$ is irreducible or, equivalently, K is a field then c is reduced if $c_{111} = 1$. For, in this case the permutation group of the r_i determined by G is transitive. Then $c_{111} = 1$ implies $c_{iii} = 1$. We now note that, by (2.12), $c_{iii} = v_{ii}$ so $\sigma v_{ii} = v_{\sigma i, \sigma i}$, $\sigma \in G$. Hence if we put $l_{ii} = v_{ii}^{-2}$, $l_{ij} = 0$ if $i \neq j$ then the conjugacy conditions hold for the l_{ij} so

(2.18)
$$y = \mathrm{diag}\{v_{11}^{-1}, v_{22}^{-1}, \ldots, v_{nn}^{-1}\} \in \mathscr{A}.$$

Since y commutes with u, $y \in F[u]$ and we can replace v by yv. This normalization permits us to assume $v_{ii} = 1$ and hence $c_{iii} = 1$, that is, c is reduced.

We can now prove

2.19 Theorem. *Let $K = F[u]$ be finite dimensional separable, $f(\lambda)$ the minimum polynomial of u over F and let $E = F(r_1, \ldots, r_n)$ be the splitting field of $f(\lambda)$ over F where $f(\lambda) = \prod(\lambda - r_i)$ in $E[\lambda]$. Suppose $c = \{c_{ijk}\}$ is a reduced Brauer factor set with values in E^* and let $\mathscr{B}(K, c)$ denote the subset of $M_n(E)$ of matrices $l = (l_{ij})$ such that $\sigma l_{ij} = l_{\sigma i, \sigma j}, \sigma \in G = \mathrm{Gal}\, E/F$. Then $\mathscr{B}(K, c)$ is an F-subspace of $M_n(E)$ and if we define a product $l_c l'$ for $l = (l_{ij})$, $l' = (l'_{ij}) \in \mathscr{B}(K, c)$ as $l'' = (l''_{ij})$ where*

$$l''_{ij} = \sum_k l_{ik} c_{ikj} l'_{kj}, \tag{2.20}$$

then $\mathscr{B}(K, c)$ becomes a central simple associative algebra of degree n over F containing a subalgebra isomorphic to K. Moreover, the map

$$l = (l_{ij}) \rightsquigarrow L = (c_{ij1} l_{ij}) \tag{2.21}$$

is an isomorphism of $\mathscr{B}(K, c)$ with an F-subalgebra $\mathscr{A}$ of $M_n(E)$.

Conversely, every central simple algebra of degree n containing K as subalgebra can be obtained by this construction.

Proof. It is clear that $\mathscr{B} = \mathscr{B}(K, c)$ is an F-subspace of $M_n(E)$ and if l''_{ij} is defined by (2.20) then

$$\begin{aligned} \sigma l''_{ij} &= \sum_k (\sigma l_{ik})(\sigma c_{ikj})(\sigma l'_{kj}) \\ &= \sum_k l_{\sigma i, \sigma k} c_{\sigma i, \sigma k, \sigma j} l'_{\sigma k, \sigma j} \\ &= \sum_k l_{\sigma i, k} c_{\sigma i, k, \sigma j} l'_{k, \sigma j} \\ &= l''_{\sigma i, \sigma j}. \end{aligned}$$

Hence $\mathscr{B}$ is closed under the "c-multiplication." Consider the map defined by (2.21). Evidently this is F-linear and injective. The (i, j)-entry of the matrix product $(c_{ij1} l_{ij})(c_{ij1} l'_{ij})$ is

$$\begin{aligned} \sum_k c_{ik1} c_{kj1} l_{ik} l'_{kj} &= \sum_k c_{ij1} c_{ikj} l_{ik} l'_{kj} \qquad \text{(by (ii))} \\ &= c_{ij1} l''_{ij}. \end{aligned}$$

Hence the map is a homomorphism for multiplication. The image $\mathscr{A} = \{L\}$ of $\mathscr{B}$ is an F-subspace of $M_n(E)$ closed under multiplication. Observe next that since $c_{ii1} = 1$, any diagonal matrix satisfying the conjugacy conditions is fixed under (2.21). Then $1 \in \mathscr{A}$ and is the unit of $\mathscr{A}$ and of $\mathscr{B}$. Hence $\mathscr{A}$ is an F-subalgebra of $M_n(E)$.

We note next that $r = \mathrm{diag}\{r_1, r_2, \ldots, r_n\} \in \mathscr{A}$ and $F[r]$ is a subalgebra of $\mathscr{A}$ isomorphic to K. Next let $l_{ij} = 1$ for all i, j and let s be the corresponding matrix $(c_{ij1} l_{ij}) = (c_{ij1})$. Note that every entry of s is $\neq 0$. Now every matrix unit $e_{ii} \in E[r]$, and since $e_{ii}se_{jj}$ is a non-zero multiple of e_{ij} it is clear that $E\mathscr{A} = M_n(E)$. Since $M_n(E)$ is simple $\mathscr{A}$ contains no nilpotent ideals $\neq 0$ and $\mathscr{A}$ is not a direct sum of more than one non-zero ideal. Hence by the Wedderburn structure theory, $\mathscr{A}$ is simple. Any element of the center of $\mathscr{A}$ is in the center of $M_n(E)$ and so is a scalar matrix. Such an element has pre-image under (2.21) that is a diagonal matrix $\operatorname{diag}\{l_1, \ldots, l_n\}$. The conditions $\sigma l_i = l_{\sigma i}$ and $\operatorname{diag}\{l_1, \ldots, l_n\}$ is a scalar matrix imply that this element is in $F1$. Hence $\mathscr{A}$ is central simple. Then $M_n(E) \cong E \otimes_F \mathscr{A}$ (BA II, Theorem 4.7) and consequently $\mathscr{A}$ has degree n. The isomorphism of $\mathscr{B}(K, c)$ with $\mathscr{A}$ now implies that $\mathscr{B}(K, c)$ is central simple of degree n and $\mathscr{B}(K, c)$ contains a subalgebra isomorphic to K.

Conversely, assume $\mathscr{A}$ is central simple of degree n containing $K = F[u]$. We have seen that $\mathscr{A} = KvK$ and we can identify $\mathscr{A}$ with the F-subalgebra of matrices $(v_{ij}l_{ij})$ where $v = (v_{ij})$ has all its entries $\neq 0$ and (l_{ij}) satisfies the conjugacy conditions. Moreover, if we define $c_{ikj} = v_{ik}v_{kj}v_{ij}^{-1}$ then $c = \{c_{ijk}\}$ is a Brauer factor set. By normalizing v we may assume c is reduced. Now we have $(v_{ij}l_{ij})(v_{ij}l'_{ij}) = (v_{ij}l''_{ij})$ where l''_{ij} is given by (2.20). Hence the map $(l_{ij}) \rightsquigarrow (v_{ij}l_{ij})$ is an isomorphism of $(\mathscr{B}, c)$ onto $\mathscr{A}$. □

We shall now determine the elements $w \in \mathscr{A}$ such that $\mathscr{A} = KwK$. We claim that these are just the elements $w = (l_{ij}v_{ij})$ such that every $l_{ij} \neq 0$. We have seen that $E[u] = D = \sum Ee_{ii}$. It is clear that $DwD = M_n(E)$ for a matrix $w = (w_{ij})$ if and only if every $w_{ij} \neq 0$. On the other hand, $D = EK$ and hence $DwD = EKwEK = EKwK$. Now if $w \in \mathscr{A}$ then $KwK \subset \mathscr{A}$ and hence $EKwK = E \otimes_F KwK$. Hence $\mathscr{A} = KwK$ for $w \in \mathscr{A}$ if and only if $w = (l_{ij}v_{ij})$ with every $l_{ij} \neq 0$.

We have associated with an element $v \in \mathscr{A}$ such that $\mathscr{A} = KwK$ a factor set $c = \{c_{ijk}\}$ where $c_{ijk} = v_{ij}v_{jk}v_{ik}^{-1}$ for $v = (v_{ij})$. If $w = (l_{ij}v_{ij})$ where the l_{ij} satisfy the conjugacy conditions and every $l_{ij} \neq 0$ then the Brauer factor set determined by w is $c' = \{c'_{ijk}\}$ where

$$c'_{ijk} = l_{ij}l_{jk}l_{ik}^{-1}c_{ijk}. \tag{2.22}$$

Two Brauer factor sets related in this way by l_{ij} satisfying the conjugacy conditions are called *associates*. These constitute an equivalence class. We denote the equivalence class of the Brauer factor sets all of which $c_{ijk} = 1$ by 1 and the relation of associativeness by $\sim$.

3. Condition for Split Algebra. The Tensor Product Theorem

We retain the notations of the last section. We prove first

3.1 Theorem. *$\mathscr{B}(K, c) \sim 1$ in the Brauer group* Br(F) (*that is,* $\mathscr{B}(K, c) \cong M_n(F)$) *if and only if $c \sim 1$.*

Proof. Suppose $c \sim 1$. Then we may assume every $c_{ijk} = 1$. Hence the subalgebra $\mathscr{A}$ of $M_n(E)$ isomorphic to $\mathscr{B}(K, c)$ contains the matrix v all of whose entries are 1. This matrix has rank 1 and hence the left ideal $M_n(E)v$ of $M_n(E)$ is minimal and so is n dimensional over E. It follows that $[\mathscr{A}v:F] = n$. Then $\mathscr{A}$ has a representation by $n \times n$ matrices over F determined by the module $\mathscr{A}v$. It follows that $\mathscr{A} \cong M_n(F)$. Conversely, suppose $\mathscr{B}(K, c) \cong M_n(E)$. Then $\mathscr{B}(K, c) \cong \mathscr{B}(K, 1)$. We have shown in [11] that if $\mathscr{A}$ is central simple of degree n and K_1 and K_2 are isomorphic commutative Frobenius subalgebras of $\mathscr{A}$ with $[K_i:F] = n$ then any isomorphism of K_1 onto K_2 can be extended to an inner automorphism of $\mathscr{A}$. Hence if $\mathscr{B}(K, c) \cong \mathscr{B}(K, 1)$ then we may assume that we have an isomorphism between these algebras that is the identity map on K. Let $\mathscr{A}_1$ and $\mathscr{A}_2$ be the subalgebras of $M_n(E)$ isomorphic to $\mathscr{B}(K, 1)$ and $\mathscr{B}(K, c)$, respectively, so that $\mathscr{A}_i$ contains the matrix $u = \operatorname{diag}\{r_1, r_2, \ldots, r_n\}$, $\mathscr{A}_1$ contains the matrix v_1 all of whose entries are 1 and $\mathscr{A}_2$ contains $v_2 = (v_{ij})$ so that $c_{ijk} = v_{ik}v_{kj}v_{ij}^{-1}$. Then $\mathscr{A}_1 = F[u]v_1F[u]$, $\mathscr{A}_2 = F[u]v_2F[u]$ and we have an isomorphism η of $\mathscr{A}_2$ onto $\mathscr{A}_1$ that is the identity on u. Then $w_1 = \eta(v_2)$ satisfies $\mathscr{A}_1 = F[u]w_1F[u]$ and we have seen that the Brauer factor set determined by w_1 is c. Since that determined by v_1 is 1 we have $c \sim 1$. $\square$

We consider next the tensor product of two central simple algebras $\mathscr{A}_i, i = 1, 2$, of degree n containing $K = F[u]$ as subalgebra. Let v_i be an element of $\mathscr{A}_i$ such that $\mathscr{A}_i = Kv_iK$ and let $v_i = (v_{jk}^{(i)})$ in an imbedding of $\mathscr{A}_i$ in $M_n(E)$. The algebra $\mathscr{A}_1 \otimes_F \mathscr{A}_2$ contains $K \otimes_F K$. We have the exact sequence of algebra homomorphisms

$$K \otimes_F K \xrightarrow{\nu} K \to 0, \tag{3.2}$$

where $\nu: \sum a_i \otimes b_i \rightsquigarrow \sum a_ib_i$. Since $K \otimes_F K$ is semi-simple we have

$$K \otimes_F K = (K \otimes_F K)e \oplus (K \otimes_F K)(1 - e), \tag{3.3}$$

where e is an idempotent and $(K \otimes_F K)(1 - e) = \ker \nu$. Then $(K \otimes K)e \cong (K \otimes K)/\ker \nu \cong K$. Moreover, since $a \otimes 1 - 1 \otimes a \in \ker \nu$ for $a \in K$, we have

$$(a \otimes 1)e = (1 \otimes a)e, \qquad a \in K. \tag{3.4}$$

We now consider the algebra

$$\mathscr{A} = e(\mathscr{A}_1 \otimes_F \mathscr{A}_2)e. \tag{3.5}$$

Since $\mathscr{A}_1$ and $\mathscr{A}_2$ are central simple so is $\mathscr{A}_1 \otimes_F \mathscr{A}_2$ and hence so is $\mathscr{A}$. Moreover, $\mathscr{A}$ and $\mathscr{A}_1 \otimes_F \mathscr{A}_2$ determine the same element of the Brauer group Br(F) and $\mathscr{A}$ contains $e(K \otimes_F K)e \cong K$ which we can identify with K. Then we have

3.6 Theorem. *$\mathscr{A}$ is of degree n containing K, and a Brauer factor set associated with $\mathscr{A}$ is $c^{(1)}c^{(2)} = \{c_{jkl}^{(1)}c_{jkl}^{(2)}\}$ where $c^{(i)} = \{c_{jkl}^{(i)}\}$ is a Brauer factor set associated with $\mathscr{A}_i$.*

Proof. If $a^{(1)} = (\alpha_{ij}^{(1)})$, $a^{(2)} = (\alpha_{ij}^{(2)}) \in M_n(E)$ we define

$$a^{(1)} \otimes a^{(2)} = \begin{bmatrix} \alpha_{11}^{(1)}a^{(2)} & \alpha_{12}^{(1)}a^{(2)} & \cdots & \alpha_{1n}^{(1)}a^{(2)} \\ \alpha_{21}^{(1)}a^{(2)} & \alpha_{22}^{(1)}a^{(2)} & \cdots & \alpha_{2n}^{(1)}a^{(n)} \\ \vdots & \vdots & \cdots & \vdots \\ \alpha_{n1}^{(1)}a^{(2)} & \alpha_{n2}^{(1)}a^{(2)} & \cdots & \alpha_{nn}^{(1)}a^{(n)} \end{bmatrix} \tag{3.7}$$

and we use this tensor multiplication of matrices to obtain an imbedding of $\mathscr{A}_1 \otimes_F \mathscr{A}_2$ in $M_{n^2}(F)$. Since $u = \operatorname{diag}\{r_1, r_2, \ldots, r_n\}$ in $M_n(E)$ it is clear that the matrix for any $a \in K \otimes_F K$ in $M_{n^2}(E)$ is a diagonal matrix. Hence the matrix for e is diagonal with entries 0 and 1. Also we have

$$u \otimes 1 = \begin{bmatrix} r_1 1_n & & & 0 \\ & r_2 1_n & & \\ 0 & & \ddots & \\ & & & r_n 1_n \end{bmatrix}, \tag{3.8}$$

$$1 \otimes u = \begin{bmatrix} u & & & 0 \\ & u & & \\ 0 & & \ddots & \\ & & & u \end{bmatrix}, \tag{3.9}$$

Hence the condition (3.4) for $a = u$ implies that all the diagonal entries of e are 0 with the exception of those in the positions $1, n + 2, 2n + 3, \ldots, n^2$. This implies that $eM_{n^2}(E)e$ has degree $\leq n$ and hence the degree of $\mathscr{A} = e(\mathscr{A}_1 \otimes \mathscr{A}_2)e$ is $\leq n$. On the other hand, this degree is $\geq n$ since $\mathscr{A} \supset K$. Hence $\mathscr{A}$ has degree n and the diagonal entries of e in the positions $1, n + 2, \ldots, n^2$ are 1 and the remaining ones are 0. Then matrix $e(v_1 \otimes v_2)e$ in $M_{n^2}(E)$ has non-zero entries only in the

$$((k - 1)n + k, (l - 1)n + l)$$

positions $1 \leq k, l \leq n$, and the entry in this position is $v_{kl}^{(1)}v_{kl}^{(2)}$.

By performing a similarity transformation by a permutation matrix and cutting down to a diagonal block we obtain an imbedding of $\mathscr{A}$ in $M_n(E)$ in which $u = \operatorname{diag}\{r_1, \ldots, r_n\}$ and $v = e(v_1 \otimes v_2)e = (v_{kl}^{(1)}v_{kl}^{(2)})$. Since all the entries of v are $\neq 0$, $M_n(E) = (\sum Ee_{ii})v(\sum Ee_{ii})$ and hence $\mathscr{A} = KvK$. It follows that we can use v to determine a Brauer factor set for $\mathscr{A}$. Evidently this set is $c^{(1)}c^{(2)}$. □

4. The Brauer Group Br(*K*/*F*)

From now on we assume the base field F is infinite. As before, let K be a finite dimensional commutative separable algebra over F. If $\bar{F}$ is the algebraic closure of F then $K_{\bar{F}} = \bar{F}e_1 \oplus \cdots \oplus \bar{F}e_n$ where the e_i are orthogonal idempotents and $n = [K:F]$. It follows that the degree of $K = \deg K_{\bar{F}} = n$. Hence, by Theorem 1.11, $K = F[u]$ where u is separable with minimum polynomial $f(\lambda)$ of degree n. We shall say that an extension field E/F *splits* K if $K_E = Ee_1 \oplus \cdots \oplus Ee_n$ where the e_i are orthogonal idempotents and we call E a *splitting field* for K if E splits K and no proper subfield of E splits K. It is readily seen that E is a splitting field for K/F if and only if E is a splitting field in the usual sense for the polynomial $f(\lambda)$. Hence any two splitting fields E/F and E'/F of K/F are isomorphic.

Now let E/F be a splitting field for K/F where $K = F[u]$ and $f(\lambda)$ is the minimum polynomial of u. Then E is a splitting field of $f(\lambda)$. For each root r of $f(\lambda)$ we have a homomorphism α of K/F into E/F such that $u \rightsquigarrow r$. In this way we obtain $n = [K:F]$ homomorphisms of K/F into E/F such that $\alpha_i u = r_i$ where $f(\lambda) = \prod (\lambda - r_i)$ in $<[\lambda]$. Moreover, this gives all the homomorphisms of K/F into E/F.

Thus

$$(4.1) \qquad M = \{\alpha_1, \alpha_2, \ldots, \alpha_n\}$$

is the set of homomorphisms of K/F into E/F. If $\sigma \in G = \mathrm{Gal}\, E/F$ then $\sigma\alpha_i \in M$. In fact, we have $\sigma\alpha_i u = \sigma r_i = r_{\sigma i}$ so $\sigma\alpha_i = \alpha_{\sigma i}$.

Now let $c = \{c_{ijk}\}$ be a Brauer factor set. We can regard this as a map c of $M \times M \times M$ into E^* such that

$$(4.2) \qquad c\colon (\alpha_i, \alpha_j, \alpha_k) \rightsquigarrow c_{ijk}.$$

Accordingly, we write $c(\alpha_i, \alpha_j, \alpha_k)$ for c_{ijk}. Then the defining conditions on the c_{ijk} are first that

$$(4.3) \qquad \sigma c(\alpha_i, \alpha_j, \alpha_k) = \sigma c_{ijk} = c_{\sigma i, \sigma j, \sigma k} = c(\sigma\alpha_i, \sigma\alpha_j, \sigma\alpha_k)$$

or, independently of the indexing,

$$(4.3') \qquad \sigma c(\alpha, \beta, \gamma) = c(\sigma\alpha, \sigma\beta, \sigma\gamma), \qquad \alpha, \beta, \gamma \in M.$$

We shall now call these conditions *homogeneity* and, more generally, if $g\colon \overbrace{M \times \cdots \times M}^{r} \to E$ or E^* then E is *homogeneous* if

$$(4.4) \qquad g(\sigma\alpha, \sigma\beta, \ldots, \sigma\varepsilon) = \sigma g(\alpha, \beta, \ldots, \varepsilon)$$

for $\alpha, \beta, \ldots, \varepsilon \in M$. In addition to this condition on c we have

$$(4.5) \qquad c(\alpha, \beta, \gamma)c(\alpha, \gamma, \delta) = c(\alpha, \beta, \delta)c(\beta, \gamma, \delta)$$

for $\alpha, \beta, \gamma, \delta \in M$. c is reduced if $c(\alpha, \alpha, \alpha) = 1$ for all $\alpha \in M$. This implies that $c(\beta, \alpha, \alpha) = 1 = c(\alpha, \alpha, \beta)$ for all α, β. If K is a field then c is reduced if $c(\alpha, \alpha, \alpha) = 1$ for a single $\alpha \in M$.

Similarly, a matrix $l = (l_{ij}) \in M_n(E)$ can be regarded as a map $(\alpha_i, \alpha_j) \rightsquigarrow l_{ij}$. The usual matrix product of l and l' can then be defined by $ll'(\alpha, \beta) = \sum_{\gamma \in M} l(\alpha, \gamma)l'(\gamma, \beta)$. Homogeneity of l as map of $M \times M \to E$ is equivalent to the conjugacy conditions $l_{\sigma i, \sigma j} = \sigma l_{ij}$.

We can now re-state Theorem 2.19 in the following way:

4.6 Theorem. *Let K/F be a finite dimensional separable commutative algebra, E/F a splitting field for K/F, c a reduced Brauer factor set with values in E^*. Let $\mathscr{B}(K, c)$ denote the F-space of homogeneous maps of $M \times M$ into E and define a product in $\mathscr{B}(K, c)$ by*

$$(4.7) \qquad ll'(\alpha, \beta) = \sum_{\gamma \in M} l(\alpha, \gamma)c(\alpha, \gamma, \beta)l'(\gamma, \beta)$$

for $l, l' \in \mathscr{B}(K, c)$. Then $\mathscr{B}(K, c)$ becomes a central simple algebra of degree $n = [K\colon F]$ containing a subalgebra isomorphic to K. Moreover, for any fixed $\gamma \in M$ the map $l \rightsquigarrow L$ where

$$(4.8) \qquad L(\alpha, \beta) = c(\alpha, \beta, \gamma)l(\alpha, \beta)$$

is an isomorphism of $\mathscr{B}(K, c)$ with an F-subalgebra $\mathscr{A}$ of the matrix algebra of maps of $M \times M$ into E.

Conversely, any central simple algebra of degree n containing K as a subalgebra can be obtained in this way.

We shall call $\mathscr{B}(K, c)$ the *Brauer algebra determined by the Brauer factor set c*. The condition that K is a commutative separable subalgebra of dimension equal to the degree is equivalent to two other conditions given in

4.9 Theorem. *Let $\mathscr{A}$ be central simple of degree n over F, K/F a commutative separable subalgebra of $\mathscr{A}$. Then the following conditions on K are equivalent*:

(i) $[K:F] = n$;
(ii) *K is a maximal commutative separable subalgebra of* $\mathscr{A}$;
(iii) $\mathscr{A}^K = K$ *for the centralizer* $\mathscr{A}^K$ *of K in* $\mathscr{A}$.

Proof. (i) ⇒ (ii). Suppose L is a commutative separable subalgebra of $\mathscr{A}$ containing K. Then $L = F[v]$ and the degree of the minimum polynomial of $v \leq \deg \mathscr{A} = [K:F]$. Hence $[L:F] \leq [K:F]$ so $L = K$.

(ii) ⇒ (iii). Let K be a maximal commutative subalgebra of $\mathscr{A}$. Then $K \subset \mathscr{A}^K$. Now $\mathscr{A}^K$ is separable. For, if $\bar{F}$ is the algebraic closure of F then $\mathscr{A}_{\bar{F}} = M_n(\bar{F})$, $K_{\bar{F}} = \bar{F}e_i \oplus \cdots \oplus \bar{F}e_m$ where the e_i are non-zero orthogonal idempotents such that $\sum e_i = 1$. Then $(\mathscr{A}^K)_{\bar{F}} \cong \mathscr{A}_{\bar{F}}^{K_{\bar{F}}} = M_n(\bar{F})^{\Sigma \bar{F}e_i}$. It is clear that the last algebra is a direct sum of algebras $M_{n_i}(\bar{F})$. Hence $\mathscr{A}^K$ is separable. Then the center of $\mathscr{A}^K$ is separable and since it contains K it coincides with K by the maximality of K. Now $\mathscr{A}^K = \mathscr{A}_1 \oplus \cdots \oplus \mathscr{A}_s$ where $\mathscr{A}_i$ is separable with separable center K_i and $K = K_1 + \cdots + K_s$. Suppose for some i, $\mathscr{A}_i \supsetneqq K_i$. If $\mathscr{A}_i$ is not a division algebra then $\mathscr{A}_i$ contains $m \geq 2$ non-zero orthogonal idempotents f_j such that $\sum f_j = 1_i$ the unit of $\mathscr{A}_i$. Then $K_1 + \cdots + K_{i-1} + \sum K_i f_j + K_{i+1} + \cdots + K_s$ is a commutative separable subalgebra of $\mathscr{A}$ properly containing K contrary to the maximality of K. The same conclusion holds if $\mathscr{A}_i$ is a division algebra since in this case $\mathscr{A}_i$ contains a separable subfield properly containing K_i. These contradictions show that $\mathscr{A}_i = K_i$ for every i and hence $\mathscr{A}^K = K$.

(iii) ⇒ (i) Suppose $\mathscr{A}^K = K$. Then $\mathscr{A}_{\bar{F}}^{K_{\bar{F}}} = K_{\bar{F}}$ for $\bar{F}$ the algebraic closure of F and hence $M_n(\bar{F})^{\Sigma_1^m \bar{F}e_i} = \sum \bar{F}e_i$ where the e_i are non-zero orthogonal idempotents such that $\sum e_i = 1$ and $m = [K:F]$. It follows that $m = n$ and $[K:F] = n$. □

The Brauer factor sets (regarded as maps of $M \times M \times M$ into E^*) form a group under multiplication of images in E^*. This contains the subgroup of factor sets such that

$$c(\alpha, \beta, \gamma) = l(\alpha, \beta)l(\beta, \gamma)l(\alpha, \gamma)^{-1}, \tag{4.10}$$

where $l: M \times M \to E^*$ is homogeneous. We can form the factor group which we shall denote as $H^2(K/F)$. If c is a factor set then l defined by $l(\alpha, \alpha) = c(\alpha, \alpha, \alpha)^{-1}$, $l(\alpha, \beta) = 1$ if $\alpha \neq \beta$ is homogeneous and $c(\alpha, \beta, \gamma)l(\alpha, \beta)l(\beta, \gamma)l(\alpha, \gamma)^{-1}$ is reduced. It follows that $H^2(K, F)$ is the factor group of the group of reduced Brauer factor sets with respect to its subgroup of reduced Brauer factor sets of the form (4.10).

We now let $\mathrm{Br}(K/F)$ denote the subset of the Brauer group $\mathrm{Br}(F)$ consisting of the classes $\{\mathscr{B}(K, c)\}$. If K is a field then we have the homomorphism of Brauer groups defined by $\{A\} \rightsquigarrow \{A_K\}$. The kernel is the subgroup of classes $\{A\}$ having K as splitting field. By the Brauer–Noether theorem on splitting fields, a central simple algebra has K as splitting field if and only if it is similar to an $\mathscr{A}$ containing K such as subfield such that $\mathscr{A}^K = K$ (BA II, p. 221). Thus it is clear that if K is a field then $\mathrm{Br}(K/F)$ is the kernel of the homomorphism of $\mathrm{Br}(F)$ into $\mathrm{Br}(K)$. We shall now prove for arbitrary commutative separable K the following;

4.11 Theorem. *$\mathrm{Br}(K/F)$ is a subgroup of $\mathrm{Br}(F)$ isomorphic to $H^2(K/F)$.*

Proof. We have the surjective map $c \rightsquigarrow \{\mathscr{B}(K, c)\}$ of the group of reduced Brauer factor sets with values in E^* onto $\mathrm{Br}(K/F)$. Now let $c^{(1)}$ and $c^{(2)}$ be reduced Brauer factor sets. Then it follows from Theorem 3.6 that $\mathscr{B}(K, c^{(1)}) \otimes_F \mathscr{B}(K, c^{(2)}) \sim \mathscr{B}(K, c^{(1)}c^{(2)})$. This implies that $\mathrm{Br}(K/F)$ is a subgroup of $\mathrm{Br}(F)$ and $c \rightsquigarrow \{\mathscr{B}(K, c)\}$ is a homomorphism of the group of reduced Brauer factor sets onto $\mathrm{Br}(K/F)$. By Theorem 3.1 the kernel of this homomorphism is the group of reduced $c \sim 1$. Hence $\mathrm{Br}(K/F) \cong H^2(K/F)$. □

We also have the following generalization of the theorem of Speiser–Noether that $H^1(G, E^*) = 1$ for G the Galois group of E/F.

4.12 Theorem. *Let K be a finite dimensional commutative separable algebra over F, E/F a splitting field for K/F, $M = \{\alpha\}$ the set of homomorphisms of K/F into E/F. Let $(\alpha, \beta) \rightsquigarrow b(\alpha, \beta)$ be a homogeneous map of $M \times M$ into E^* such that*

$$b(\alpha, \beta)b(\beta, \gamma) = b(\alpha, \gamma) \tag{4.13}$$

for $\alpha, \beta, \gamma \in M$. Then there exists an invertible $a \in K$ such that

$$b(\alpha, \beta) = (\alpha a)(\beta a)^{-1}. \tag{4.14}$$

Proof. Consider the algebra $\mathscr{B}(K, 1)$ which is the F-space of homogeneous maps of $M \times M$ into E with multiplication defined by $ll'(\alpha, \beta) = \sum_{\gamma \in M} l(\alpha, \gamma)l'(\gamma, \beta)$. For $a \in K$ we define a homogeneous map a' of $M \times M$ into E by $a'(\alpha, \alpha) = \alpha a$, $a'(\alpha, \beta) = 0$ if $\alpha \neq \beta$. Then $a \rightsquigarrow a'$ is a homomorphism of K into $\mathscr{B}(K, 1)$. This is a monomorphism since $K \otimes_F E = Ee_1 \oplus \cdots \oplus Ee_n$ where the e_i are orthogonal idempotents and for any $a \in K$, $a = \sum (\alpha_i a)e_i$ where $\alpha_i a \in E$. Then $\alpha_i \in M$ and if $\alpha_i a = 0$ for all i, $a = 0$. Thus we can identify K with its image in $\mathscr{B}(K, 1)$ and write a for a'. Then $\mathscr{B}(K, 1)^K = K$ by Theorem 4.9. We now consider the map $\eta: l \rightsquigarrow l'$ where $l'(\alpha, \beta) = l(\alpha, \beta)b(\alpha, \beta)$ for $l \in \mathscr{B}(K, c)$. The condition (4.13) implies that η is an automorphism of $\mathscr{B}(K, 1)$. Moreover, (4.13) gives $b(\alpha, \alpha)^2 = b(\alpha, \alpha)$ so $b(\alpha, \alpha) = 1$. Hence $\eta a = a$ for $a \in K$. It follows from the Skolem–Noether theorem and $\mathscr{B}(K, 1)^K = K$ that there exists an invertible $a \in K$ such that $\eta = I_a$ the inner automorphism $x \rightsquigarrow axa^{-1}$. Now let v be defined by $v(\alpha, \beta) = 0$ for all α, β. Then v is homogeneous and $v' = \eta v$ satisfies $v'(\alpha, \beta) = b(\alpha, \beta)$. Since $(ava^{-1})(\alpha, \beta) = (\alpha a)(\beta a)^{-1}$ we have $b(\alpha, \beta) = (\alpha a)(\beta a)^{-1}$ for $\alpha, \beta \in M$. □

5. Crossed Products

We shall now specialize to the case $E = K$, $M = G = \text{Gal } E/F$ in the foregoing considerations. In this case one has the crossed product representation, due to Emmy Noether, of an algebra $\mathscr{A}$ containing E and having degree $n = [E:F]$. Let $\sigma \in G$ then σ can be extended to an inner automorphism I_{u_σ} of $\mathscr{A}$. By Theorem 4.9, $\mathscr{A}^K = K$. Hence the element u_σ is determined up to a multiplier in E^*. Moreover, since $I_{u_\sigma}I_{u_\tau}$ and $I_{u_\sigma u_\tau}$ for $\sigma, \tau \in G$ have the same restriction $\sigma\tau$ to E we have $u_\sigma u_\tau = k_{\sigma,\tau}u_{\sigma\tau}$, $k_{\sigma,\tau} \in E^*$. Also $u_\sigma a u_\sigma^{-1} = \sigma a$, $a \in E$. Thus we have

$$u_\sigma a = (\sigma u)u_\sigma, \qquad u_\sigma u_\tau = k_{\sigma,\tau}u_{\sigma\tau} \tag{5.1}$$

for $a \in E$, $\sigma, \tau \in G$. The associativity $(u_\sigma u_\tau)u_\rho = u_\sigma(u_\tau u_\rho)$ gives the relations

$$k_{\sigma,\tau}k_{\sigma\tau,\rho} = k_{\sigma,\tau\rho}(\sigma k_{\tau,\rho}), \qquad \sigma, \tau, \rho \in G. \tag{5.2}$$

It is clear from (5.1) that the E-subspace $\sum_{\sigma \in G} Eu_\sigma$ is a subalgebra. On the other hand, it is easily seen by a Dedekind independence argument that the u_σ are linearly independent over E. Hence $[\sum Eu_\sigma : E] = |G| = n$ and hence $[\sum Eu_\sigma : F] = n^2 = [A:F]$. Thus

$$\mathscr{A} = \sum Eu_\sigma. \tag{5.3}$$

We now consider the converse in which we begin with the Galois extension field E/F and the Galois group G. Then a map k of $G \times G$ into E^*: $(\sigma, \tau) \rightsquigarrow k_{\sigma,\tau}$ is called a *Noether factor set* if (5.2) holds. We form the (left) vector space over E with base $\{u_\sigma | \sigma \in G\}$ and we define a product in $\mathscr{A} = \sum Eu_\sigma$ by

$$(\textstyle\sum a_\sigma u_\sigma)(\sum b_\tau u_\tau) = \sum\limits_{\sigma,\tau} k_{\sigma,\tau}a_\sigma(\sigma b_\tau)u_{\sigma\tau}. \tag{5.4}$$

Then (5.2) implies that this is associative. Moreover, if we put $\sigma = \tau = 1$ and $\tau = \rho = 1$ successively in (5.2) we obtain

$$k_{1,\rho} = k_{1,1}, \qquad k_{\sigma,1} = \sigma k_{1,1}, \tag{5.5}$$

which imply that $1 = k_{11}^{-1}u_1$ is the unit of $\mathscr{A}$. Moreover, $\mathscr{A}$ is a vector space over $F \subset E$ and we have

$$\alpha(xy) = (\alpha x)y = x(\alpha y) \tag{5.6}$$

for $x, y \in \mathscr{A}$, $\alpha \in F$. Thus $\mathscr{A}$ is an algebra over F (associative with 1). This is called the *crossed product of E with G and Noether factor set k* and is denoted as $\mathscr{A} = (E, G, k)$. The result we proved above can now be stated as

5.7 Theorem. *If $\mathscr{A}$ is a central simple algebra containing E and the degree of $\mathscr{A}$ is $n = [E:F]$ then $\mathscr{A}$ is a crossed product (E, G, k).*

It is quite easy to prove directly the converse that any crossed product is central simple over F of degree $n = [E:F]$. We shall obtain this result by establishing the connection between crossed products and Brauer algebras. We note first that if we replace u_1 by 1 we may assume that the Noether factor set is normalized in the sense that $k_{1,\sigma} = 1 = k_{\sigma,1}$, $\sigma \in G$. Then we have

5.8 Theorem. *If k is a normalized Noether factor set then c defined by*

$$c(\rho, \sigma, \tau) = \rho k_{\rho^{-1}\sigma, \sigma^{-1}\tau} \tag{5.9}$$

is a reduced Brauer factor set and $(\dot{E}, G, k) \cong \mathscr{B}(E, c)$.

Conversely, if c is a reduced Brauer factor set then

$$k_{\sigma,\tau} = c(1, \sigma, \sigma\tau) \tag{5.10}$$

defines a normalized Noether factor set and $\mathscr{B}(E, c) \cong (E, G, k)$.

Proof. For $a \in \mathscr{A} = (E, G, k)$ we write

$$u_\sigma a = \sum_\tau a(\sigma, \tau) u_\tau, \tag{5.11}$$

where $a(\sigma, \tau) \in E$. Let $\mu(a)$ denote the matrix $(a(\sigma, \tau))$ (regarded as a map of $G \times G$ into E). Then $a \rightsquigarrow \mu(a)$ is a homomorphism of $\mathscr{A}$ into $M_n(E)$ since

$$u_\sigma ab = \textstyle\sum_\tau a(\sigma, \tau) u_\tau b = \sum_{\tau, \rho} a(\sigma, \tau) b(\tau, \rho) u_\rho$$

so $\mu(ab) = \mu(a)\mu(b)$. For $a \in E$, $\mu(a)$ is a diagonal matrix with σa in the (σ, σ)-position. Hence $E\mu(E) = \sum Ee_{ii}$. Also, by (5.1), if $v = \sum_{\sigma \in G} u_\sigma$ then $\mu(v) = (v(\sigma, \tau))$ where

$$v(\sigma, \tau) = k_{\sigma, \sigma^{-1}\tau} \neq 0. \tag{5.12}$$

It follows as in the proof of Theorem 2.19 that $\mu(\mathscr{A})$ is central simple of degree n. Since $[\mathscr{A}:F] = n^2$, μ is a monomorphism and $\mathscr{A} = KvK$. Then the Brauer factor set determined by v is c where

$$c(\rho, \sigma, \tau) = v(\rho, \sigma) v(\sigma, \tau) v(\rho, \tau)^{-1} = k_{\rho, \rho^{-1}} k_{\sigma, \sigma^{-1}\tau} k^{-1}_{\rho, \rho^{-1}\tau} = \rho k_{\rho^{-1}\sigma, \sigma^{-1}\tau}$$

by the factor set condition (5.2). We have $c(\rho, \rho, \rho) = \rho k_{1,1} = 1$ so c is reduced. It is clear that $\mathscr{A} \cong \mathscr{B}(K, c)$.

Our result shows also that if K is a Noether factor set then c defined by (5.9) is a Brauer factor set. Direct verification shows that if c is a Brauer factor set then K defined by (5.10) is a Noether factor set and that the maps $k \rightsquigarrow c$ and $c \rightsquigarrow k$ are inverses. Now let $\mathscr{B}(E, c)$ be the Brauer algebra defined by c and let k be the corresponding Noether factor set. Consider the crossed product (E, G, k). Then since c is given by (5.9) the result we proved shows that $\mathscr{B}(E, c) \cong (E, G, k)$. □

The Noether factor sets form a group under multiplication which contains the subgroup of factor sets of the form

$$(\sigma, \tau) \rightsquigarrow l_\sigma (\sigma l_\tau) l_{\sigma\tau}^{-1}, \tag{5.13}$$

where $\sigma \rightsquigarrow l_\sigma$ is any map of G into E^*. The factor group is the cohomology group $H^2(G, E^*)$. As usual, we write $k \sim 1$ if k is in the subgroup defined by (5.13) and $k \sim k'$ if k and k' differ by an element of this subgroup.

We now observe that the map $k \rightsquigarrow c$ is an isomorphism of the group of Noether factor sets onto the group of Brauer factor sets. If l is any map of G into E^* then $l(\sigma, \tau) = \sigma l(\sigma^{-1} \tau)$ is a homogeneous map of $G \times G$ into E^*. It follows that if

$k \rightsquigarrow c$ in our homomorphism then $k \sim 1$ if and only if $c \sim 1$. Hence we have an induced isomorphism of $H^2(G, E^*)$ onto $H^2(E, F)$. This isomorphism together with Theorem 5.8 permit us to carry over the results of Section 4 to Noether factor sets and crossed products. We obtain in this way

5.14 Theorem. (i) *The crossed product* $(E, G, k) \sim 1$ *if and only if* $k \sim 1$.
(ii) $(E, G, k_1) \otimes_F (E, G, k_2) \sim (E, G, k_1 k_2)$.
(iii) *Let* $\{k\}$ *denote the element of* $H^2(G, E^*)$ *determined by* k. *Then* $\{k\} \rightsquigarrow \{(E, G, k)\}$ *is an isomorphism of* $H^2(G, E)$ *onto* $\mathrm{Br}(E/F)$. □

6. Central Simple Algebras with Involution of Orthogonal Type

In this section we assume char $F \neq 2$ (as well as F is infinite). Let $\mathscr{A}$ be a finite dimensional central simple algebra with an involution j (= anti-automorphism j such that $j^2 = 1_{\mathscr{A}}$). If $\bar{F}$ is the algebraic closure of F then j has a unique extension to an involution $\tilde{j}$ in $\tilde{\mathscr{A}} = \mathscr{A}_{\bar{F}} \cong M_n(\bar{F})$ and we can identify $\tilde{\mathscr{A}}$ with $M_n(\bar{F})$ and $\tilde{j}$ with either the transpose map $a \rightsquigarrow {}^t a$ or the map $a \rightsquigarrow s({}^t a)s^{-1}$ where

$$s = \mathrm{diag}\left\{\begin{pmatrix} 0 & 1 \\ -1 & 0 \end{pmatrix}, \begin{pmatrix} 0 & 1 \\ -1 & 0 \end{pmatrix}, \ldots, \begin{pmatrix} 0 & 1 \\ -1 & 0 \end{pmatrix}\right\}. \tag{6.1}$$

In the first case j is said to be of *orthogonal type* and in the second of *symplectic type*. These can be distinguished by the dimensionality of the space $\mathscr{H}(\mathscr{A}, j)$ of symmetric elements which is $n(n+1)/2$ in the orthogonal case and $n(n-1)/2$ (with n even) in the symplectic case. We can also distinguish the two cases by the fact that in the symplectic case if the degree of $\mathscr{A}$ is $2n$ then the degree of every element of $\mathscr{H}(\mathscr{A}, j)$ does not exceed n ([12], p. 231). On the other hand, we have

6.1 Theorem. *Let* $\mathscr{A}$ *be a central simple algebra of degree* n *over* F *with involution* j *of orthogonal type. Then* $\mathscr{H}(\mathscr{A}, j)$ *contains separable elements of degree* n *and for any such element* u *there exist* $v \in \mathscr{H}(\mathscr{A}, j)$ *such that* $\mathscr{A} = KvK$, $K = F[u]$.

Proof. As in Theorem 1.11, the subset of $\mathscr{H} = \mathscr{H}(\mathscr{A}, j)$ of separable elements of degree n is Zariski open in $\mathscr{H}$. Hence to show that it is not vacuous it suffices to show that the corresponding subset of $M_n(\bar{F})$ is not vacuous. This is the set of symmetric matrices having minimum polynomials with n distinct roots. Now $\sum r_i e_{ii}$ with distinct r_i is such a matrix. Now let u be any separable element of degree n in $\mathscr{H}$. If $v \in \mathscr{H}$ then $\mathscr{A} = F[u]vF[u]$ if and only if the elements $u^i v u^j$, $0 \leq i, j \leq n-1$ are linearly independent. It is readily seen that the set of these v's is Zariski open in $\mathscr{H}$. Hence to prove that there exist such v's it suffices to show that the corresponding set in $M_n(\bar{F})$ is not vacuous. Now if $u \in M_n(\bar{F})$ is separable of degree n then there exists an orthogonal matrix s such that

$$sus^{-1} = \mathrm{diag}\{r_1, r_2, \ldots, r_n\}$$

where the r_i are distinct. Then we may take $v = \sum_{i,j} e_{ij}$, which is symmetric, to obtain n^2 linearly independent elements $u^i v u^j$. □

6.2 Remark. The argument we have used in the foregoing proof can be used to prove Theorem 2.2 for the case in which K is separable and F is infinite.

Now let K be a commutative separable algebra and let $\mathscr{B}(K, c)$ be a Brauer algebra defined by a reduced factor set c that is symmetric in the sense that

$$c(\alpha, \beta, \gamma) = c(\gamma, \beta, \alpha) \tag{6.3}$$

for $\alpha, \beta, \gamma \in M$. Then the map $l \rightsquigarrow {}^t l$ where

$${}^t l(\alpha, \beta) = l(\beta, \alpha) \tag{6.4}$$

is an involution of orthogonal type in $\mathscr{B}(K, c)$. For, we have $ll' = l''$ where

$$l''(\alpha, \beta) = \sum_{\gamma \in M} l(\alpha, \gamma)c(\alpha, \gamma, \beta)l'(\gamma, \beta)$$

and

$$\begin{aligned}({}^t l')({}^t l)(\alpha, \beta) &= \sum_{\gamma \in M} l'(\gamma, \alpha)c(\alpha, \gamma, \beta)l(\beta, \gamma) \\ &= \sum l(\beta, \gamma)c(\beta, \gamma, \alpha)l'(\alpha, \gamma) = l''(\beta, \alpha) = {}^t l''(\alpha, \beta).\end{aligned}$$

Hence $({}^t l')({}^t l) = {}^t l''$ and t is an involution. The subalgebra of $\mathscr{B}(K, c)$ consisting of the elements l such that $l(\alpha, \beta) = 0$ if $\alpha \neq \beta$ is isomorphic to K and is contained in $\mathscr{H}(\mathscr{B}(K, c), t)$. Since K contains an element of degree n it follows that t is of orthogonal type. We shall call t the *transpose involution* in $\mathscr{B}(K, c)$.

Two algebras with involution $(\mathscr{A}_1, j_1)$ and $(\mathscr{A}_2, j_2)$ are *isomorphic* if there exists an isomorphism n of $\mathscr{A}_1$ onto $\mathscr{A}_2$ such that $\eta j_1 = j_2 \eta$. With this definition we have

6.4 Theorem. *Let $\mathscr{A}$ be a central simple algebra with an involution j of orthogonal type. Then there exists a Brauer algebra $\mathscr{B}(K, c)$ with c symmetric such that $(\mathscr{A}, j)$ is isomorphic to $(\mathscr{B}(K, c), t)$.*

Proof. Let u and v be as in Theorem 6.1. Then every element of $\mathscr{A}$ can be written in one and only one way in the form $\sum_{k,l=1}^{n} a_{kl}u^{k-1}vu^{l-1}$. Since $u, v \in \mathscr{H}(\mathscr{A}, j)$, the involution j is

$$\begin{aligned}\sum_{k,l=1}^{n} a_{kl}u^{k-1}vu^{l-1} &\rightsquigarrow \sum a_{kl}u^{l-1}vu^{k-1} \\ &= \sum_{k,l} a_{lk}u^{k-1}vu^{l-1}.\end{aligned} \tag{6.5}$$

Now let E/F be a splitting field of $K = F[u]$ so $E = F(r_1, r_2, \ldots, r_n)$ where $f(\lambda) = \prod (\lambda - r_i)$ is the minimum polynomial of u over F. As in Section 2, we can take an imbedding of $\mathscr{A}$ in $M_n(E)$ so that $u = \operatorname{diag}\{r_1, r_2, \ldots, r_n\}$ and $v = (v_{ij})$ where every $v_{ij} \neq 0$. Then $\sum a_{kl}u^{k-1}vu^{l-1} = (l_{ij}v_{ij})$ where $l_{ij} = \sum a_{kl}r_i^{k-1}r_j^{l-1}$. Under our imbedding the involution j given by (6.5) becomes

$$(l_{ij}v_{ij}) = \sum a_{kl}u^{k-1}vu^{l-1} \rightsquigarrow \sum a_{lk}u^{k-1}vu^{l-1} = (l_{ji}v_{ij}). \tag{6.6}$$

Now we have the algebra $\mathscr{B}(K, c)$ which we regard, as in Section 2, as the set of matrices (l_{ij}) where the $l_{ij} \in E$ and $\sigma l_{ij} = l_{\sigma i, \sigma j}$ for $\sigma \in G = \text{Gal } E/F$. The multiplication in $\mathscr{B}(K, c)$ is $ll' = l''$ where $l = (l_{ij}), l' = (l'_{ij}), l'' = (l''_{ij})$, and $l''_{ij} = \sum_k l_{ik} c_{ikj} l'_{kj}$, $c_{ikj} = v_{ik} v_{kj} v_{ij}^{-1}$. Since $(l_{ij}) \rightsquigarrow (l_{ji})$ is an involution we have $\sum_k l'_{ki}(c_{ikj} - c_{jki}) l_{jk} = 0$ or

$$\sum_k l_{ik}(c_{ikj} - c_{jki}) l'_{kj} = 0 \tag{6.7}$$

for all l_{ik} and l'_{kj} satisfying the conjugacy conditions. As in the proof of (2.14) we can conclude that $c_{ikj} = c_{jki}$ for all i, k, j. If we now pass to the definition of $\mathscr{B}(K, c)$ as the algebra of homogeneous maps of $M \times M$ into E we see that this algebra has a symmetric Brauer factor set and the isomorphism $(l_{ij} v_{ij}) \rightsquigarrow$ is an isomorphism of $(\mathscr{A}, j)$ onto $(\mathscr{B}(K, c), t)$. □

References

[1] A. A. Albert. The structure of matrices with any normal division algebra of multiplications. *Ann. Math.*, **32** (1931), 131–148.

[2] A. A. Albert and H. Hasse. A determination of all normal division algebras over an algebraic number field. *Trans. Amer. Math. Soc.*, **34** (1932), 722–726.

[3] R. Brauer. Untersuchungen über die arithmetischen Eigenschaften von Gruppen linearer Substitutionen. *Math. Z.*, **8** (1928), 677–696.

[4] ———. On normal division algebras of degree 5. *Proc. Nat. Acad. Sci.*, **24** (1938), 243–246.

[5] R. Brauer, H. Hasse and E. Noether. Beweis eines Hauptaatzes in der Theorie der Algebren. *J. Math.*, **167** (1931), 399–404.

[6] R. Brauer and E. Noether. Uber minimale Zerfallungskorper irreduzibler Darstcllungen. *Sitzungsber. Preuss. Akad. Wiss.*, **32** (1927), 221–228.

[7] C. W. Curtis and I. Reiner. *Representation Theory of Finite Groups and Associative Algebras.* Interscience Publishers: New York, London.

[8] H. Hasse. Theory of cyclic algebras over an algebraic number field, *Trans. Amer. Math. Soc.*, **34** (1932), 171–214. Also Additional additional note to the authors: "Theory of cyclic algebras over an algebraic number field." *ibid*, 727–730.

[9] ———. Die Struktur der R. Brauerschen Algebren-Klassengresppen. *Math. Ann.*, **107** (1933), 731–760.

[10] N. Jacobson. *The Theory of Rings.* American Mathematical Society Surveys, 1943.

[11] N. Jacobson. Generation of separable and central simple algebras. *J. Math.*, **36** (1957), 217–227.

[12] ———. *Structure and Representations of Jordan Algebras.* Ann. Math. Soc. Colloquium Publ., 1968.

[13] ———. *Basic Algebra* II. W. H. Freeman and Company: San Francisco, 1980.

[14] E. Noether. *Gesamelte Abhandlungen.* Springer-Verlag: Berlin, Heidelberg, New York, 1982.

[15] H. Weyl. Generalized Riemann matrices and factor sets. *Ann. Math.*, **37** (1936), 709–745.

Reprinted from
Emmy Noether in Bryn Mawr
Springer-Verlag, 1983.

Permissions